WITHDRAWN BY THE
UNIVERSITY OF MICHIGAN

Multivariable Analysis: An Introduction

Multivariable Analysis: An Introduction

Alvan R. Feinstein

....Yale University Press   New Haven and London

Copyright © 1996 by Yale University.
All rights reserved.
This book may not be reproduced, in whole or in part,
including illustrations, in any form (beyond that copying
permitted by Sections 107 and 108 of the U.S. Copyright
Law and except by reviewers for the public press),
without written permission from the publishers.

Designed by Sally Harris/Summer Hill Books.
Set in Meridien and Gill Sans types by D&T/Bailey Typesetting, Inc.,
Nashville, Tennessee.
Printed in the United States of America by BookCrafters, Inc.,
Chelsea, Michigan.

Library of Congress Cataloging-in-Publication Data

Feinstein, Alvan R., 1925–
   Multivariable analysis : an introduction / Alvan R. Feinstein.
     p. cm.
   Includes bibliographical references and index.
   ISBN 0-300-06299-0
   1. Multivariable analysis.   I. Title.
QA278.F45 1996
519.5'35'02461—dc20

A catalogue record for this book is available
from the British Library.

The paper in this book meets the guidelines for permanence
and durability of the Committee on Production Guidelines
for Book Longevity of the Council on Library Resources.

10 9 8 7 6 5 4 3 2 1

*To the members
of the Robert Wood Johnson Clinical Scholars Program,
a constant pleasure and joyful stimulus
in the challenge to search, learn, create, and teach*

Contents

Preface   ix

Part I   Outline of Goals and Methods

1   Introduction   1
2   Goals of univariate and bivariate analysis   5
3   Goals of multivariable analysis   34
4   Taxonomy of multivariable methods   59

Part II   Preparing for Multivariable Analysis

5   Classification and coding decisions   85
6   Appraisal of simple displays   105
7   Simple linear regression: the algebraic prototype   128
8   Stratified rates: the cluster prototype   164
9   Bivariate evaluations of illustrative data set   188

Part III   Basic Strategies of Targeted Algebraic Methods

10   Complexity of additional independent variables   207
11   Multivariable explorations and reductions   227
12   Evaluations and illustrations of multiple linear regression   264

Part IV   Regression for Nondimensional Targets

13   Multiple logistic regression   297
14   Evaluations and illustrations of logistic regression   331
15   Proportional hazards analysis (Cox regression)   370
16   Evaluations and illustrations of Cox regression   398
17   Discriminant function analysis   431
18   Illustrations of discriminant function analysis   458

Part V   Targeted-Cluster Methods

19   Basic strategies of cluster operations   477
20   Multicategorical stratification   507
21   Recursive partitioning   529

22   Additional discussion and conclusions   559

Master List of References   585

Index   605

Preface

Many workers in the field of health care—medicine, nursing, public health, dentistry, pharmacy, administration—entered it because they liked people, liked science, and hated mathematics. A remarkable irony of their subsequent professional lives is the constant onslaught of statistics.

Although instruction in statistics may have been received (or evaded) during professional education, the recipients usually hoped to be untroubled afterward during their working careers. Having become an essential part of medical research, however, statistical communications constantly intrude today in ordinary clinical practice or other professional activities, at meetings and conferences, and particularly in the "literature" published in professional journals. For investigators who prepare the published papers, the intrusion is more direct and threatening when statistical analyses are demanded by granting agencies, manuscript reviewers, and journal editors.

For many years the readers of literature and doers of research could survive the statistical assault by knowing a few relatively simple things about means, medians, correlation coefficients, and the t or chi-square "tests of significance". During the past decade, however, things have become much more complicated. The statistical activities now regularly include the fearsome complexity of multivariable analyses. In such general medical publications as *The New England Journal of Medicine* and *The Lancet*, results of a multivariable analytic method were reported, on average, more than once in each weekly issue during 1989, and the annual incidence has been increasing.[1]

The multivariable analytic methods are particularly intimidating not only because they are unfamiliar (having been omitted during previous instruction), but also because they seem intricate, difficult to explain, and even more difficult to understand. Nevertheless, the multivariable results are constantly used for important substantive decisions. The decisions are expressed in such statements as "The value of treatment A was confirmed by multiple linear regression", "The possible confounding variables were adjusted with proportional hazards analysis", "The apparent effect of sex was refuted with multiple logistic regression", "B is a much more important risk factor than C", and "You live 2 years longer for every 3-unit drop in substance X".

Not understanding what was done during the mathematical manipulations, and not knowing how to interpret the results, most readers accept these statements with acts of faith. The hope is that the journal's editors and statistical reviewers knew enough to check things out properly and to verify that all the "significant" conclusions were indeed significant. Nevertheless, despite the journal's approbation, the mathematical conclusions may sometimes be misleading or occasionally wrong, because important biologic distinctions, clinical goals, or even simple algebraic principles were violated during the automated "number crunching". The deceptions or errors may often be difficult to discern, however, because the concomitant mathematical mysteries may be so overwhelming that an otherwise knowledgeable reader fails to look beneath the numbers, apply appropriate common sense, and note the critical defects.

For researchers, the multivariable methods produce an additional challenge. Many years ago, the methods were seldom used because they required extensive, laborious calculations. Today, however, most of the calculations can be done quickly and easily with a digital computer. The programs that do the work are readily available in automated "packages"; and many of the packages can be promptly applied through a personal computer at the researcher's desk. With this convenience, many researchers can operate their own programs and are then confronted with deciding what program to choose and how to interpret the results. Even if consultative help is sought from a professional data analyst, the researcher may still want to see, and preferably to understand, what appears in the printout.

This book is aimed at improving the quality of this aspect of intellectual professional life for readers and researchers. They may work in the field of health care, from which most of the illustrations are derived, but the ideas and principles can also be helpful for psychologists, sociologists, and other people who are regularly confronted with multivariable challenges. The text is intended to demystify the mathematical methods so that you can understand things well enough to think about what has happened, rather than be awed or flustered by inscrutable statistical ideas and expressions.

The text is concerned mainly with four multivariable methods that commonly appear[1] in health-care literature. They are linear regression, logistic regression, the proportional hazards procedure often called Cox regression, and discriminant function analysis. These methods are often used because they seem suitable for the research data and because they can readily be carried out with the increasingly wide availability of statistical software and personal computers. The main problems for most readers (and medical users) of the methods are that the basic analytic strategies are seldom understood, the basic assumptions behind the strategies are seldom tested, and the limitations of the methods are not well known. One main goal of this text is to clarify the basic strategies of these four methods, to indicate how their assumptions can be checked, and to highlight their strengths and weaknesses.

The book is also intended, however, to emphasize and call attention to some unconventional new methods that can be used to confirm, augment, and sometimes replace what is produced by the customary mathematical approaches. The new methods rely on categorical strategies that are fundamentally different from the algebraic models used in the conventional procedures. Some of the new methods (such as "recursive partitioning") are becoming well known and have even been automated with statistical software. Other new methods (such as "conjunctive consolidation") are relatively unfamiliar and require judgmental operation.

The new methods have intellectual and scientific advantages that will appeal to many readers and investigators.

The dual objectives of clarifying the old and introducing the new are thus analogous to offering a road map for the jungle of multivariable statistics, while simultaneously demonstrating and improving some little-known but highly valuable paths. A third objective, which is literary rather than scientific or statistical, is to write in a manner appropriate for readers who are not statisticians. These are the readers whose interests and needs I know best. The illustrations come from the general field of health care, which I know well, and can easily use to find or construct suitable examples. Nevertheless, the basic discussions (as well as the examples) should be easily comprehensible to interested readers with other backgrounds.

Earlier drafts of the manuscript have been used as background text for a set of seminars on multivariable analysis for postresidency physicians. The text might be applied in more formal teaching, but is not aimed at statisticians, and may not be mathematical enough to be attractive for advanced courses in statistics. The intended audience consists of persons with a "substantive" background in a field of science, but without a major background in statistics. Statisticians may find the text enlightening, however, as an indication of what their "clients" really want and are (or should be) thinking about.

This is the text that I would have avidly welcomed years ago in searching for authors who recognized that readers in a substantive field, such as the health sciences, know a lot about data, but relatively little about mathematics. A book on multivariable statistics will never become fascinating light reading, but the individual doses here should not be too hard to swallow; the explanations should be clearer than what is available elsewhere; and the consequences—if you want to understand what is and might be done with the analyses—should help you get through the journals (or your data) with an increased sense of clarity and security, if not necessarily mastery.

I have tried to explain the basic strategies and principles of the analytic methods while avoiding sophisticated mathematical details or other intricacies of the matrix algebra and "black box" computations. The methods cannot be suitably discussed (or understood) without appropriate mathematical symbols and equations, but my goal has been to introduce them sparingly, explain them clearly, and generally use nothing more elaborate than high school algebra. Nevertheless, any discussion of mathematical principles will regularly require you to stop, think, ruminate, and digest, rather than proceed rapidly from one sentence or paragraph to the next. Several previous readers of the text have said that it became much easier to follow and understand when they began reading rather than racing.

Except for a few contrived examples, all illustrations are taken from real-world research and computer printouts. The text will succeed if it gives you a reasonable account of what procedures are available, what principles underlie their operation, when they are used, what emerges in the computer printout, what to do with it thereafter, and how to evaluate what is reported in the published literature.

The book may seem quite long for something that is "an introduction." If you want only a general outline of multivariable analysis, the easiest part of the text is in the first four chapters and in the last one, which offers a summary and some simple suggestions about what to beware of and when to be skeptical. Those chapters have relatively few mathematical details and should be relatively easy to read, absorb, and use.

If you actually get yourself involved in data processing, Chapters 5 through 9 indicate the main preparation for what to do and how to check it. As a reader rather than maker of literature, however, you may want to examine those chapters to find out what the investigators might (or should) have checked *before* all the multivariable activities began. Chapters 7 and 8 are particularly important because they discuss the conventional bivariate analyses—simple regression, correlation coefficients, two-way tables, and "tests of significance"—that constantly appear in ordinary medical literature and that serve as prototypes for the more complex multivariable procedures.

The bulk of the text (Chapters 10–18) is devoted to the "big four" algebraic methods that are most commonly used in multivariable regression (and "survival") analysis. In Chapters 10–12, the simple linear regression procedure is expanded to its multivariable counterpart. The complexities and new ideas involve the "covariance" effects of the additional variables (Chapter 10), and the special methods used (Chapter 11) to explore combinations of variables and to eliminate those that seem unimportant. Chapter 12 then provides many illustrations of the pragmatic use of multiple linear regression.

The next six chapters (13–18) contain descriptions and illustrations of three additional regression procedures: multiple logistic regression, Cox (proportional hazards) regression, and discriminant function analysis. The logistic and Cox procedures have become particularly common[1] in current health-care literature. Although much less popular now than formerly, discriminant analysis is discussed because it regularly appears in older literature and is sometimes seen today.

Chapters 19–21 (in Part V of the text) describe the unconventional new methods of multivariable analysis. Their titles and strategies have such unfamiliar names as *multicategorical stratification, recursive partitioning,* and *conjunctive consolidation.* Unlike the conventional procedures, these new methods do not rely on fitting the data with the "models" of algebraic configurations. Instead, the new methods arrange things into clusters of categories that directly show the results for a selected target variable. In addition to being directly displayed, the results of the unconventional methods are relatively easy to understand; and they can also be used as a "gold standard" for checking what emerges from the conventional algebraic approaches.

The last chapter contains reports of empirical comparisons among the methods, some summary remarks, and a set of guidelines, caveats, and precautions for both investigators and readers of the literature.

My next set of comments here is offered to explain (and perhaps justify) why a book on the formidable subject of multivariable statistics is being written by a physician. I have had many multivariable adventures as researcher, research consultant, reviewer, editor, and reader. The adventures began more than twenty years ago when I was trying to develop better systems of prognostic prediction, using clinical manifestations that had usually been ignored in most formal statistical analyses. Because the automated mathematical methods had not yet become readily available, I had to think about what I was doing. Being generally unfamiliar at that time with the "established" techniques—such as multiple linear regression—or uncomfortable with the little that I knew about them, I developed a new "judgmental" multicategorical procedure that was originally called *prognostic stratification*,[2] and later retitled *conjunctive consolidation*.[3] With the aid of a splendid computer programmer, I also developed an automated multicategorical procedure that was called *polarized stratification*.[4]

None of these procedures used the conventional mathematical models that were becoming increasingly popular as they became increasingly available in commercial computer packages. Trying to learn more about these models, however, I became increasingly frustrated. As a reader of published literature, I was often puzzled by the confident acceptance of methods that often seemed somewhat "shady", but I felt too ignorant to raise questions. As a reviewer and editor, I could raise questions and could learn various things from the *ad hoc* responses, but my education was sporadic and incomplete. The mathematical methods were not well explained in the descriptive manuals for the associated commercial computer programs; and private conversations with knowledgeable statisticians were seldom successful. We did not often speak the same language; they may not have understood my questions; or I may not have understood their answers.

The publications in professional statistical literature were even more unhelpful. Regardless of how well the authors may try to communicate, statistical editors seem to have a tradition of insisting on prose written so concisely that even professional statisticians may have a hard time trying to get successfully from one sentence to the next. As an interested amateur with a reasonably good educational background (I had obtained an M.S. degree in mathematics before going to medical school), I usually became lost in the early paragraphs of most papers amid the flurry of unexplained symbols, arcane abbreviations, and sophisticated matrix concepts that may have been well known to the statistical authors but not to me.

Textbooks on what was called *multivariate* analysis were also unhelpful because (for reasons noted in Chapter 1) they excluded all of the regression methods that were most useful in health-care research and most common in health-care publications. Statistical books on regression analysis, particularly the 1966 text by Draper and Smith[5] and the 1973 text by Kerlinger and Pedhazur,[6] were somewhat instructive, but not fully satisfactory. They usually told me much more than I wanted to know about mathematical details and not enough about the basic ideas and challenges. Besides, the texts were confined almost exclusively to multiple linear regression, and did not discuss the logistic and proportional hazards methods that were beginning to appear regularly in health-care literature.

In 1978 David G. Kleinbaum and Lawrence L. Kupper[7] published a splendid book called "Applied Regression Analysis and Other Multivariable Methods". (A revised second edition,[8] with Keith E. Muller as additional author, appeared in 1988.) Emanating from a school of public health, the Kleinbaum-Kupper text was aimed at an "application-oriented reader who has a limited mathematical background", and was developed from class notes for a course "designed for nonstatistics majors in the health and related fields". The text had many examples from the health sciences; it discussed discriminant function analysis as well as multiple linear regression; and, although not always free of excessive mathematical emphasis, it was generally much easier to follow than anything I had previously read. With Muller as an additional author, the second edition includes a section on logistic regression and maximum likelihood methods, although the discriminant-function discussion remains confined to an analysis of two groups, and nothing is said about the "survival analysis" procedure called Cox regression. I have learned a lot from the Kleinbaum-Kupper-Muller text; I respect it greatly; and I regularly use it somewhat like a reference encyclopedia.

In 1990, several years after I began writing the current manuscript, Stanton Glantz and Bryan K. Slinker published[9] a "Primer of Applied Regression and Analysis of Variance". The text is aimed at "people interested in the health and life sciences"; it describes multiple linear and logistic regression but not Cox regression or discriminant function analysis; it contains many real-world and contrived illustrations, including "data" from a survey of Martians; and it also has an abundance of technical details. (For anyone who believes that the current "introduction" is too long, the Glantz-Slinker "primer" has 568 pages of text, and an additional 181 pages of Appendixes.)

These two excellent texts were written by biostatisticians with wide experience in teaching graduate students and consulting with clients. Neither text, however, was written in the language or from the viewpoint of a non-statistical client who wants to understand the basic strategies via supplemental reading, without taking a long, intensive, special course. Neither text emphasizes the client's goals rather than the goals of the mathematical methods; both texts seem to regard the fitting of algebraic models rather than identification of important variables as the main purpose of the analyses; and neither text mentions the existence of the new multicategorical procedures that have been developed as an attractive alternative to all the mathematical modeling.

After my original adventures with the new multicategorical procedures, I continued to like them, but did little to help their development. The algebraic-model methods had such popular and widespread usage that they obviously seemed "the way to go". While occupied with other professional tasks, I continued making sporadic efforts to learn more about the big four algebraic methods.

My efforts were transformed in 1987, when Prof. Burton H. Singer came to Yale as head of the Division of Biometry. I soon discovered that I did not have to hide my surreptitious blasphemy in liking the multicategorical procedures. Singer, a mathematical statistician who has extensively collaborated with investigators in the biomedical and social sciences, was an outright heretic. He not only liked the decision-rule cluster procedures; he also used them in his own research.[10] We began a pleasant collaboration trying to improve the conjunctive-consolidation method, while engaging in mutual education about the pros and cons of the algebraic models.

The collaboration led to the plan for this book. Its purposes would be to inform investigators and readers about the main goals and problems, not just methodologic details, and to help non-statistical clients realize that optimal analytic strategies require a mutual exchange of substantive and mathematical ideas. While clarifying the algebraic methods, the text would also be aimed at inspiring life-science investigators and statisticians to use and improve the new categorical methods.

When I began writing the text, I had hoped that Singer would collaborate, read and correct the drafts, and join as co-author. As often happens to creative academicians, however, he was enticed to a higher administrative job; and the time available for writing textbooks vanished. Accordingly, although I am deeply grateful for his major contributions to my multivariable education and am willing to credit him with almost anything that is good in the text, I absolve him of any blame for its malefactions. He has read almost none of it and should not be held responsible for it.

The text is not written in the "pablum for kiddies" approach that is sometimes used to simplify statistical discussions aimed at practitioners in the health sciences. I assume that the reader has been confronted and puzzled by some of the products or printout of multivariable procedures, wants to understand the mathematical principles that are being applied to manipulate the data, recognizes that those principles cannot be absorbed without reasonable amounts of thoughtful effort, and is willing to make those efforts. The basic mathematical ideas here are no more difficult than the basic principles of molecular biology or computerized imaging, which also cannot be suitably "digested" and well understood without appropriate time and thought. If you make that effort, you should emerge with a reasonably comfortable feeling about when to be uncomfortable with what the printouts and published literature are trying to "sell" you.

Before concluding, I want to thank many other people and institutions for their help. The text was evoked by the challenge of teaching multivariable analysis to fellows in the Robert Wood Johnson Clinical Scholars Program at Yale. I am grateful to the Johnson Foundation for their long-term, continuing support of the Program; and I am particularly grateful to the fellows, who have been exposed to the drafts and have made constructive comments that improved the text. John P. Concato, a former fellow and now a faculty colleague, has made particularly valuable contributions from his experience in using the text for teaching. Two biostatisticians at other institutions have been gracious, kind, and patient in answering some of my questions about procedures. They are Frank E. Harrell, Jr. at Duke University (for logistic regression and Cox regression) and Peter Lachenbruch, formerly at the University of California at Los Angeles and now with the Food and Drug Administration (for discriminant function analysis). Peter Peduzzi, a biostatistical colleague at Yale and at the West Haven Cooperative Studies Coordinating Center, has also been particularly helpful in offering constructive suggestions for the chapters on logistic and Cox regression.

The time required for doing the work was supported by the Johnson Foundation and the Veterans Administration, and by grants from agencies associated with the tobacco industry. In all instances, the support was given unconditionally, and I thank the donors for their help.

I have two last sets of particularly important acknowledgements and expressions of gratitude. First, I want to thank Mary N. Lyman, Elizabeth S. Vassiliou, Maria D. Stahl, Joan B. Small, and Matthew H. Lewis, who successfully accomplished (at various times) the difficult task of typing a manuscript whose mathematical symbols required particularly ingenious maneuvers with a word processor. And finally, but perhaps most importantly, the entire work would have been impossible without the dedicated efforts of Carolyn K. Wells, who offered many useful suggestions, who helped many features of the textual organization, and who excellently managed the crucial role of making the computer programs do their job. To master the complex instructions and operations of a large array of diverse technical procedures is a heroic task, which she carried out splendidly.

Alvan R. Feinstein, M.D., M.S.
Sterling Professor of Medicine and Epidemiology,
Yale University School of Medicine,
New Haven, 1995

References

Throughout the text the references are numbered sequentially as they appear in each chapter. At the end of the chapter, the numbers are accompanied by the name of the first author, year of publication, and (when needed) additional details of identification. At the end of the text, the full citations for all references appear in alphabetical order of first authors, together with indication of chapters where the references were used.

For the *Preface*, the citations are as follows: 1. Concato, 1993; 2. Feinstein, 1972; 3. Feinstein & Wells, 1990; 4. Feinstein, 1973; 5. Draper, 1966; 6. Kerlinger, 1973; Kleinbaum, 1978; 8. Kleinbaum, 1988; 9. Glantz, 1990; 10. Levy, 1985.

Part I Outline of Goals and Methods

The four chapters of Part I offer a background for the concepts and strategies of multivariable analysis. After a brief introduction in Chapter 1, the goals and activities of univariate and bivariate analysis are outlined in Chapter 2. They are the foundation from which the multivariable procedures are expanded. Their additional goals are discussed in Chapter 3; and the most commonly used analytic methods are classified and briefly illustrated in Chapter 4.

1 Introduction

Without consciously thinking about it, most people do multivariable analysis every day. *Variable* is the name given to a particular characteristic—such as age, race, sex, or occupation—that will vary with different persons or occasions. *Multivariable* is the term used when three or more variables are considered simultaneously, rather than one or two at a time. In deciding what to wear when going outdoors, you think about such multiple variables as the time of day, temperature, and the possibility of rain or snow, as well as where you are going and what clothes you have available. Among these multiple variables, some may receive a stronger emphasis or "weight" than others. Regardless of weather, your chosen attire may depend on such destinations as the beach, the opera, or the football stadium. If especially eager not to get wet, you may take a raincoat *and* an umbrella.

For the decisions just discussed, the analysis was informal, using no special strategies, patterns of logic, or documentary data. You thought about the situation, applied your own common sense, and decided what to do. In many other circumstances, however, you (or a researcher) may be confronted with a large collection of data, for many different variables in a group of people. When challenged to interpret and draw conclusions from the formal data, you cannot immediately apply the common sense that usually works so well for simple personal decisions.

To get the data suitably arranged and organized, you may first want to do *univariate* analyses, examining the pattern of information for each individual variable. For example, in data for a group of patients with a particular cancer, the individual variables might include each person's age, sex, cell type, anatomic extensiveness, types of symptoms, severity of co-morbidity, and duration of survival. In those analyses, you might examine such univariate summary statistics as the mean and standard deviation for age, or the proportions of patients who are in stages **I, II,** or **III** for anatomic spread.

The next step would be *bivariate analyses,* examining relationships between pairs of any two variables. During this procedure, you might note the association (or correlation) between age and cell type, or between anatomic spread and types of symptoms. If the goal is to predict survival, you might do a bivariate check of the mean survival durations or the 5-year survival rates associated with each of the other individual variables: age, sex, cell type, anatomic spread, and so on.

After completing these inspections, you might then turn to the multivariable analysis, examining such interrelationships as age, sex, and cell type, or noting the simultaneous effect of all the other six variables on duration of survival. During this process, you will want to preserve and apply common sense, but the job of analyzing a large amount of data will often require some form of quantitative assistance, usually from a computer.

Using the computer, you can arrange the data in diverse ways and see what happens with each arrangement. The multivariable arrangements can be made informally according to "clinical sensibility", which is a combination of ordinary common sense plus the knowledge acquired from appropriate instruction and experience. The arrangements can also be constructed with formal mathematical methods that are deliberately intended for multivariable analysis.

The informal judgmental methods are usually easy to understand, but not easy to do. There is no single, universally accepted clinical approach; and the results achieved with the same set of data may be inconsistent when different analysts apply different ideas and judgments about what is important and how to arrange it. On the other hand, the formal mathematical methods are easy to use, but difficult to understand. Having been automated and made ubiquitously available in computer packages, the mathematical techniques are now constantly applied for multivariable analysis, and the results constantly appear in ordinary biomedical literature.

The next 21 chapters of this text are aimed at demystifying, explaining, and encouraging the suitable use of both the formal and informal methods, particularly those that are applied for analyzing multivariable effects on a single target variable, such as *survival*. The judgmental and the automated procedures each have their own mixture of virtues and demerits, achievements and problems. The good and not-good components of each are important background for a reader who sees reports of what emerged from the methods, for a researcher who chooses the methods and interprets their output, and for a reviewer (or editor) who evaluates what seems worthy of publication.

Before the formal text begins, we can try to demystify the nomenclature itself. The word *multivariable* appears in the title of this book, but the words *univariate* and *bivariate* were used earlier rather than *univariable* and *bivariable*. Why not refer to *multivariate* rather than *multivariable* analysis? Are there distinctions in the two words? This question can be answered yes or no, according to how strict you want to be. The distinctions arise because multiple variables can be related to one another in at least three different ways, as shown in Fig. 1-1. On the left side of the figure, a set of five variables—labeled as X_1, X_2, X_3, X_4, and X_5—have a *many-to-one* relationship with a single "target" variable, Y_1. This situation would occur in patients with cancer if *survival* (as a specific external variable, Y_1) is related simultaneously to such multiple "independent" variables as *age* (X_1), *anatomic extensiveness* (X_2), *cell type* (X_3), *weight loss* (X_4), and *co-morbidity* (X_5).

In the middle section of Fig. 1-1, the five "independent" variables $X_1, \ldots, X_5$ have a *many-to-many* relationship. They are related simultaneously to three external "dependent" variables, which might be $Y_1 = $ *survival,* $Y_2 = $ *functional status,* and $Y_3 = $ *costs of care*. On the right side of Fig. 1–1, the five variables $X_1, \ldots, X_5$ are neither "independent" nor "dependent". They have a *many-internal* relationship, being interrelated to one another, but not to an external variable.

Any of the three types of simultaneous relationship shown in Fig. 1-1 can be called *multivariable*, but mathematical purists do not use *multivariate* for the many-

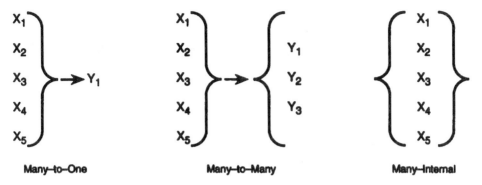

Fig. 1.1 Three main types of relationships among multiple variables. On the left, the five variables X_1, X_2, X_3, X_4, X_5 are related to a single external variable, Y1. In the middle, they are related to several external variables, Y_1, Y_2, Y_3. On the right, the five variables are related internally to one another.

to-one relationship on the left, where only a single variable is the external target. In the strict definition, *multivariate* is reserved for the middle and right situations, where multiple variables are examined in either a many-to-many or many-internal relationship. In medical literature, however, most multivariable analyses involve many-to-one relationships, with the individual target being such entities as survival, development of a disease, occurrence of a "success" or "failure", or a change in blood pressure or serum cholesterol. The many-internal types of relationship seldom appear in medical reports, but are often examined in psychology and the social sciences, using such procedures as *factor analysis* and *principal component analysis*. The formal analysis of many-to-many relationships is a rare phenomenon in either medical or psychosocial research.

Since this book is concerned almost exclusively with many-to-one relationships, a strict definition of *multivariate* would keep the word from being applied to most of the topics to be discussed. Rather than lose both a good word and its direct linguistic analogy to familiar terms such as *univariate* and *bivariate*, *multivariate* will often be used in its broader sense, for any analysis that involves multiple variables, regardless of where and how they appear. Most people already use (or abuse) *multivariate* in this way, without regard to the special mathematical requirement. The occasional transgression here can be forgiven if fastidious statistical colleagues recognize that the evolution of language often alters words to reflect popular usage rather than professional restrictions.

To give the text a logical sequence, while giving readers easy access to pertinent individual sections when desired, the book is divided into five parts. Part I offers a general background, outline, and classification for the pertinent concepts and strategies. Part II discusses what is done to prepare and check multivariable information *before* an analysis begins. Parts III and IV are devoted to the conventional, well-known procedures that use the big four mathematical methods of multiple linear regression, multiple logistic regression, proportional hazards analysis (often called Cox regression), and discriminant function analysis. Part V describes the uncommon but valuable (and currently underdeveloped) methods that can be used for easy-to-understand forms of "stratified" analysis. The last chapter contains summary comments as well as guidelines, caveats, and precautions for both users of the multivariable methods and readers of the published results.

2 Goals of Univariate and Bivariate Analysis

Outline . . . .

2.1 Goals of univariate analysis
    2.1.1. Types of data and scales
        2.1.1.1. Dimensional scales
        2.1.1.2. Nominal scales
        2.1.1.3. Ordinal scales
        2.1.1.4. Binary scales
        2.1.1.5. Quasi-dimensional scales
    2.1.2. Summary indexes
    2.1.3. Additional procedures

2.2. Orientation and patterns of bivariate data
    2.2.1. Pairs of rating scales
    2.2.2. Orientation of variables
    2.2.3. Patterns of data and models
        2.2.3.1. Patterns of algebraic models
        2.2.3.2. Conformity of model and data
        2.2.3.3. Monovalent patterns
        2.2.3.4. Monotonic or non-monotonic patterns
        2.2.3.5. Diffuse monovalent patterns
        2.2.3.6. Polyvalent patterns
    2.2.4. Analytic importance of patterns
    2.2.5. Role of algebraic models
    2.2.6. Zonal partitions
    2.2.7. Strategies for clusters or scores

2.3. Goals of bivariate analysis
    2.3.1. Trends
    2.3.2. Impacts
        2.3.2.1. Measuring impact
        2.3.2.2. Types of associations
        2.3.2.1. Regressions
        2.3.2.2. Contrasts
        2.3.2.3. Correlations
        2.3.2.4. Concordances
    2.3.3. Estimates
        2.3.3.1. Substitutions
        2.3.3.2. Predictive estimates
        2.3.3.3. Estimates vs. impacts
    2.3.4. Compressions
    2.3.5. Configurations
        2.3.5.1. Advantages of algebraic configuration
        2.3.5.2. Candidate formats
            2.3.5.2.1. Straight lines
            2.3.5.2.2. Exponential curves
            2.3.5.2.3. "Powers" of X

        2.3.5.2.4. Transcendental expressions
        2.3.5.2.5. Polynomials
        2.3.5.2.6. Step functions and splines
      2.3.5.3. Problems in choosing formats
  2.4. Correspondence of goals and analytic procedures
    2.4.1. Zonal partitions and clusters
    2.4.2. Algebraic models and scores
      2.4.2.1. Problems in curvilinear trend
      2.4.2.2. Advantages of straight-line models
      2.4.2.3. Disadvantages of straight-line models

. . . .

The relatively simple methods of univariate and bivariate analysis are worth reviewing carefully because they are the source from which multivariable procedures are expanded. The simple methods are discussed in this chapter, but the emphasis is on goals, rather than operational procedures. To analyze data effectively, the main questions to be answered are *not* "What do you do?" and "How do you do it?", but rather "What do you want?" and "Do you really get it?". Statistical packages and computer programs offer a series of attractive analytic methods, but the methods are often aimed at inherently mathematical rather than substantive (or scientific) goals. If the two sets of goals happen to agree, everything is splendid. If they disagree, however, an investigator (or reader) who neglects the substantive goals may not recognize the disparate aims and possible deceptions of the mathematical results.

The extensive discussion of bivariate methods in this text is particularly important because the basic goals of bivariate and multivariable analyses are essentially similar. If you clearly understand the fundamental tactics (and limitations) of the relatively simple bivariate methods, the task of comprehension will be eased when analogous procedures (and problems) later appear in the multivariable activities.

2.1. Goals of univariate analysis

Univariate analysis is so simple that it is usually not regarded as an "analysis". The analytic goal is to summarize the data collected for a single variable, such as *age, occupation, stage of disease,* or *survival time.*

2.1.1. Types of data and scales

A univariate collection of data consists of the values of a single variable observed in each member of a group. Each value is chosen from the variable's scale, which contains the expressions available for describing each member. The scale for *age in years* can be **1, 2, 3, 4, . . . , 99, 100, . . .** ; and the collection of observed values in a particular group of people might be **24, 57, 31, 8, 62, ...** The scale for *severity of pain* might be **none, mild, moderate, severe;** and the individual values collected for a group could be **mild, none, none, severe, mild, ...**

The four main types of scales for variables are named according to their different degrees of precision in ranking.

2.1.1.1. Dimensional scales

As the conventional form of scientific measurement, dimensional scales have precise quantitative rankings, with a measurably equal interval between any two adjacent ranked values, such as . . . , 7, 8, 9, . . . , 50, 51, 52, . . . for *years of age,* or . . . , 1.8, 1.9, 2.0, 2.1, . . . for *serum bilirubin* (in mg/dl).

Dimensional scales have diverse alternative names. They are often called *continuous* by mathematicians, and *interval* or *ratio* by psychosocial scientists. (A ratio scale, such as *height* or *age,* has an absolute zero value. An interval scale, such as *Fahrenheit temperature,* does not.) The dimensional results are usually regarded as quantitative measurements; and non-dimensional scales are often called *categorical.*

2.1.1.2 Nominal scales

At the other extreme of ranking, nominal scales have categories that cannot be ranked. Examples of nominally scaled variables are *religion* (having such alphabetically arranged categories as **Buddhist, Catholic, Hindu, Jewish, Moslem, Protestant, . . .**) and *occupation, birthplace, color of eyes,* and *principal diagnosis.*

The categories of nominal scales may be given coding digits such as 1, 2, 3, 4, . . . for electronic data processing, but the digits are arbitrary designations, somewhat like telephone numbers. They have no quantitative meaning, and cannot be subjected to any mathematical procedures such as ranking, addition, multiplication, etc. The main mathematical operation for nominal categories is to count their frequency in a group.

2.1.1.3. Ordinal scales

Ordinal scales contain graded categories such as **none, mild, moderate, severe** for *severity of pain,* or **Stage I, Stage II, Stage III** for *anatomic extensiveness of cancer.* The categories can be ranked, but the intervals between categories do not have measurably equal magnitude. Such scales are commonly used in modern medicine, having originated[1] many years ago with ratings such as **0, trace, 1+, 2+, 3+, 4+** for the intensity of positive results in certain blood and urine tests.

For electronic data processing, ordinal scales are commonly given coding digits such as **0, 1, 2, 3** for *severity of pain* or **1, 2, 3** for *stage of cancer.* Because the arbitrary intervals between categories are unequal, the values of data in these scales should not be mathematically added or multiplied. Nevertheless, despite the impropriety, such mathematical operations are commonly applied, as noted later.

2.1.1.4. Binary scales

Binary scales have two complementary categories for existence of a particular attribute, such as **absent/present, no/yes, dead/alive, or failure/success.** They are commonly coded as **0/1** for data analysis.

In older literature, binary scales are sometimes regarded as nominal, particularly when used for categories such as **male/female,** or **Caucasian/Non-Caucasian.** Nevertheless, the 0/1 coding allows binary scales to be summarized in a way that resembles dimensional scales, because the mean of a set of 0/1 data is identical to the corresponding binary proportion. Thus, in a group of 50 people, if 19 dead persons are coded 0 and 31 alive persons are coded 1, the mean is 0.62[={(19 x 0) + (31 x 1)}/50], which is the same value as the binary proportion of alive people, 31/50 = .62.

2.1.1.5. Quasi-dimensional scales

A potentially fifth type of scale can be called quasi-dimensional. It is produced either as a sum of ordinal ratings or by marks on a visual analog line.

The well-known Apgar score,[2] used to rate the status of a newborn baby, contains the sum of 0, 1, or 2 ordinal ratings for five variables. If an arbitrary set of scores is assigned to right, wrong, or unanswered questions in a verifying or licensure examination, each score is itself as ordinal rating. The sum of the scores, however, may have apparently dimensional values such as 194 or 328.

With a visual analog scale, the respondent makes a mark on a demarcated line, usually 100 mm. long to indicate subjective perception of a magnitude that otherwise cannot be quantified. The following visual analog scale might be used for marking a patient's severity of pain:

None　　　　　　　　　　　　　　　　　　　　　　　　　　　Worst
　　　　　　　　　　　　　　　　　　　　　　　　　　　　　　pain
　　　　　　　　　　　　　　　　　　　　　　　　　　　　　　ever

After the mark is placed, its measured distance becomes the "dimensional" rating.

Neither a sum of ordinal ratings nor a visual analog scale has the exact equi-interval quantitative characteristics of dimensional data. Nevertheless, despite occasional complaints,[3] both types of quasi-dimensional scales are regularly managed mathematically as though they were dimensional. A particularly common example occurs in studies of analgesic agents when ordinal ratings for intensity of pain at different time points are used to construct total scores for pain relief.[4]

Since the quasi-dimensional data can be regarded either as ordinal or dimensional, they do not require special consideration as a "fifth" type of scale.

2.1.2. Summary indexes

The values that form each univariate collection of data produce a spectrum or "distribution". For most univariate collections, the main "analysis" usually consists of summarizing each distribution with appropriate statistical indexes for the central tendency and spread of the data. The choice of these indexes depends on the associated scale of expression.

For variables expressed in measured dimensions—such as *age* or *survival duration*—the index of central tendency is usually a median or a mean. The index of spread is usually a range, standard deviation, or inner percentile range.

For binary data, scaled in only two categories such as **live/dead** or **male/female,** a central tendency can be listed as the proportion of one of the two pertinent categories, such as 32% **alive** or 57% **men.** If the data are expressed in more than two categories—for such variables as *occupation* or *stage of disease*—the frequencies of values in each category are counted; and the summary consists of listing the proportions (or relative frequencies) either for each category or for the category with the largest proportion.

Although the median is generally a better central index for descriptive summaries of dimensional data,[5] the mean is almost always used when statistical analyses employ the mathematical models discussed later. The reason is that most of the current mathematical strategies and "tests of statistical significance" were developed for means, not medians. In those analyses, means and standard deviations, although not wholly appropriate mathematically, are also regularly used to summarize data for binary and ordinal variables. Nominal variables, however, cannot be ranked for any mathematical calculations, and are regularly decomposed into "dummy" binary variables. (The strategies used to code data for analysis and to avoid some of the mathematical problems of nondimensional variables are considered in Chapters 5 and 6.)

2.1.3. Additional analyses

Sometimes a set of univariate data may receive additional statistical analyses to test whether the distribution is Gaussian or whether the central index is located at a specified value. In most instances, however, the main decision in the univariate "analysis" is to check that the selected summary indexes are satisfactory for the pattern formed by the distribution of data. For example, the values of a mean and standard deviation may be unsuitable for data having an eccentric distribution, made non-Gaussian by outliers or by a nonsymmetrical "skew" shape. Data spread diffusely among many ordinal or nominal categories may not be suitably summarized if a "central" proportion is cited for only a single category.

The analyses and "transformations" used to detect and manage these problems are discussed in Chapter 6.

2.2. Orientation and patterns of bivariate data

When two variables are examined simultaneously, the presence of the second variable immediately creates new analytic possibilities. They arise from the component scales of the variables themselves, from the orientation of the two variables, and from the pattern formed by the bivariate distribution of data.

2.2.1. Pairs of rating scales

A collection of bivariate data contains pairs of values for two variables, X and Y, observed in each member of a group. The two variables might be *age* and *occupation,* or *stage of disease* and *survival time.*

Since each variable can be expressed in one of the four main types of scales, the bivariate possibilities consist of sixteen pairs of component scales, shown as follows:

Rating scale of first variable	Rating scale of second variable			
	Dimensional	Ordinal	Binary	Nominal
Dimensional	•			
Ordinal		•		
Binary			•	
Nominal				•

For the four pairings in the main diagonal of this table (marked with dots), the two variables have similar types of scales. The remaining twelve pairings contain six symmetrically opposite pairs, such as the binary-ordinal and ordinal-binary pairs, or the nominal-dimensional and dimensional-nominal pairs. Despite the apparent symmetry, however, the analytic process for two different variables may differ if the variables are oriented as X vs. Y, or as Y vs. X, or neither.

2.2.2. Orientation of variables

With a *dependent* orientation, the dependent or "target" variable, Y, is believed to be affected by X, the independent variable. (The two variables can also be arranged so that X appears to depend on Y.) Thus, *occupation* might be affected by *age*, or *survival time* might depend on *stage of disease*.

In a *nondependent* or *interdependent* orientation, X and Y coexist in the pairs of data points, but one variable does not necessarily affect the other. Thus, we can examine the bivariate relationship of *age* or *sex* in a group of data without believing that age has a biologic effect on sex, or vice versa. In growing children, we expect *weight* to depend on *age*, but the two variables might not be biologically dependent in adults. In patients with renal disease, we would expect hemoglobin level to be affected by, i.e., to depend on, the level of creatinine or blood urea nitrogen. Hemoglobin and hematocrit, however, would be nondependent or interdependent: the two variables might have a close relationship with one another, but one would not necessarily affect the other. A similar nondependent relationship might exist for the two serum aminotransferases (or transaminases) prefixed as the aspartate (AST or SGOT) and alanine (ALT or SGPT).

In Chapter 1, the multiple variables shown in Fig. 1-1 were oriented in a dependent manner for the *many-to-one* and *many-to-many* external relationships. The orientation was nondependent or interdependent for the *many-internal* relationship.

The orientation of the variables determine the analytic arrangements. They will differ according to the type of dependent variable, if any, that has been chosen as a target; and the arrangements used for nondependent relationships may not be applicable for dependent relationships.

The terms *dependent* and *nondependent* (or *interdependent*), however, are merely mathematical parlance. They describe the way in which the data will be regarded and analyzed mathematically, but the terms do not necessarily denote the actual biologic relationship of the variables. We could mathematically analyze a child's age as depending on weight, or blood urea nitrogen as depending on hematocrit, even though the mathematical arrangement might be biologically silly or wrong.

In scientific parlance, the dependent variable is believed to be a specific target affected by the independent variable. Consequently, the terms *targeted* and *untargeted* can be used instead of *dependent* and *nondependent* to indicate scientific knowledge of the variables' biologic meaning and relationship. This type of knowledge was used earlier to decide that a child's weight depended on age, or that hemoglobin and hematocrit were interdependent.

2.2.3. Patterns of data and models

When plotted as points on a graph, the data for each pair of X, Y values form a distinctive visual pattern. In the usual arrangement (called *Y vs X*), X appears as the independent variable on the horizontal axis, and Y is the dependent variable on the vertical axis. For nondependent variables, the choice of axes is arbitrary; they can be oriented either as Y vs. X or as X vs. Y.

If each variable has a dimensional scale, such as *weight* or *age*, the pairs of points can easily be located on a conventional graph; and the collection of points will have a shape that shows the pattern of their distribution. Although usually presented as frequency counts in tables, nondimensional categorical data can also be displayed as points in graphs. The counted frequency of each bivariate category will appear as clusters of points at appropriate locations on the graph.

2.2.3.1. Patterns of algebraic models

Most statistical analyses of bivariate data are based on the idea of using a mathematical model to fit the points on the graph. The model is usually expressed algebraically as a relatively simple line, having a straight or curved shape. ("Mathematical model" can also be used as a label for diverse structures that include disconnected lines, composite collections of different lines, and algorithmic clusters. To avoid this broad scope of possibilities, the term *algebraic model* will be used here for the simpler lines that are basic staples of statistical analysis.)

After the analyst chooses the basic shape of the model, the algebraic calculations arrange for the line to receive a best possible fit to the pattern of points. With only a small number of points, the line may have a close-to-perfect fit. In the realities of data analysis, however, hundreds of points may be involved, and they may not have the shape of a relatively simple line. Furthermore, even when relatively few points are involved, their pattern may not closely conform to the shape of the mathematical line.

2.2.3.2. Conformity of model and data

The disparity between the shape of the points and the shape of the line is particularly important, because the algebraic characteristics of the line (as discussed later) are used for crucial analytic decisions about the relationship of the variables. For those decisions, a statistical index is calculated for the "goodness of fit" of the line, and subsequent evaluations often depend on the average "goodness".

Although the basic strategy is effective and the "average" result is accurate, no attention is given to the fit's *conformity*—an idea that refers to similarities in shape, rather than closeness of fit. Is the pattern of points similar to the shape of the line?

Can the collection of points be properly fit with *any* line? If the points are too numerous or their pattern too erratic to be well fit with a simple line, the customary "linear" model may be inappropriate. If the points do not suitably conform to the model, conclusions about a linear trend may be erroneous or deceptive.

Despite this potential problem, the "line-ability" of a set of points seldom receives formal analytic attention. To determine whether the shape can easily be fitted with a relatively simple straight or curved line, graphical patterns of points can be classified into four basic types. They can be monovalent or polyvalent, and the monovalent patterns can be monotonic, non-monotonic, or diffuse.

2.2.3.3. Monovalent patterns

In a *monovalent* pattern, each value of X is associated with essentially a single value of Y, as shown in Fig. 2-1. Two or more closely similar values of Y at a single value of X can also be regarded as a monovalent point. For example, at several locations in Fig. 2-2, more than one value of Y occurs at an individual value of X, but the individual sets of Y values are similar enough for the pattern to be essentially monovalent.

Fig. 2-1 A monovalent pattern of data.

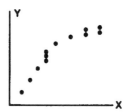

Fig. 2-2 Essentially monovalent pattern, with several close values of Y at three locations of X.

Any monovalent pattern is "line-able", i.e., it can be almost perfectly fit by a straight or suitably curved line. The line can be relatively simple, as in Figs. 2-1, 2-2, 2-3, or 2-4, but can also be somewhat complex, as in Fig. 2-5, or highly complex, as in Fig. 2-6.

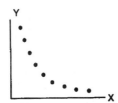

Fig. 2-3 Monovalent curved pattern.

Fig. 2-4 Monovalent, non-monotonic pattern.

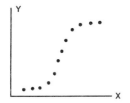
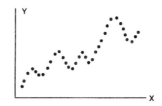

Fig. 2-5 Monovalent monotonic pattern, with two zones of "flattening".

Fig. 2-6 Monovalent non-monotonic pattern with several reversals.

2.2.3.4. Monotonic or non-monotonic patterns

In a *monotonic* monovalent pattern, the values of Y consistently rise or fall in the same direction with increasing values of X. In the monotonic patterns of Figs. 2-1 and 2-2, Y increased as X increased; in Fig. 2-3, Y decreased.

A monotonic trend may sometimes seem to flatten out, as shown in the far right of Fig. 2-2 or at the two ends of Fig. 2-5. As long as the trend continues in the same basic direction, however, without a distinct reversal, zones of flattening do not affect the monotonicity of the pattern.

A monovalent pattern is *non-monotonic* if the trend has distinct zones of reversal. A single major reversal is shown in the simple pattern of Fig. 2-4, and multiple reversals occur in the complex pattern of Fig. 2-6.

2.2.3.5. Diffuse monovalent patterns

Fig. 2-7 shows a pattern that commonly occurs in medical research. The pattern is monovalent, since only a single value of Y appears at each value of X. The pattern, however, is too diffuse to immediately suggest the shape of a simple line. Nevertheless, a non-monotonic line can be perfectly fitted to the data if it directly connects adjacent pairs of points, as in Fig. 2-8. The interconnecting straight lines can then be "smoothed" to produce the curve shown in Fig. 2-9.

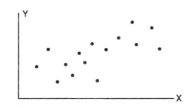

Fig. 2-7 Diffuse monovalent pattern.

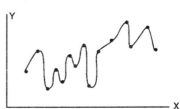

Fig. 2-8 Straight lines connecting points of Fig. 2-7.

Fig. 2-9 Curved smoothing of line connecting points in Figs. 2-7 and 2-8.

2.2.3.6. Polyvalent patterns

In a polyvalent pattern, many different values of Y occur at the same individual points or zones of X. In Fig. 2-10, where X is an ordinal variable, each of its graded categories forms a zone containing diverse values of Y. In Fig. 2-11, X and Y are each dimensional variables, but the data are spread in a manner that suggests a "cloud" or "swarm" of points.

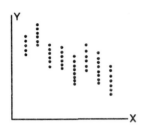 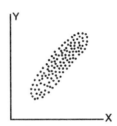 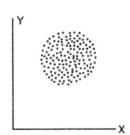

Fig. 2-10 Polyvalent patterns of dimensional points of Y for ordinal values of X.   **Fig. 2-11** Polyvalent "swarm" of points for two variables that are each dimensional.   **Fig. 2-12** Another polyvalent swarm of points.

A polyvalent pattern cannot be excellently fitted with a reasonably simple line. Certain patterns, however, may seem to have distinctive linear trends. Thus, the trend in Fig. 2-11 seems to go generally upward monotonically whereas Fig. 2-12 looks like "scattered buckshot", and does not show a trend.

Fig. 2-13 looks unusual, but the polyvalent "cannonball" pattern is a common occurrence in medical research, appearing whenever a graph is plotted for data that are *not* dimensional. In this instance, each point of data represents a single occurrence at the appropriate location of bivariate categories for X and Y. The ordinal values of X have been coded as 1 = low, 2 = medium, and 3 = high. The binary

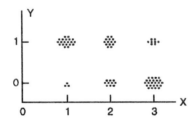

Fig. 2-13 "Cannonball" clusters showing many points at same locations on graph.

values of Y have been coded as 0 = failure and 1 = success. The cannonball effect appears when multiple points must be shown individually at the same location.

If a deliberate effort were not made to show these data in a graph, the clusters of points in Fig. 2-13 could be displayed more conventionally in the arrangement of Table 2-1.

Table 2-1 Tabular arrangement of data in Fig. 2-13

Values of X	Frequency counts for Y	
	Failure	Success
Low	3	20
Medium	10	14
High	24	8

2.2.4. Analytic importance of conformity in patterns

A perfect or really close fit cannot be achieved by *any* reasonably simple line for polyvalent patterns, but is possible for monovalent data. If the monovalent data are non-monotonic or diffuse, however, the line may have to be a highly complex curve to get a really good fit.

Monotonic monovalent patterns are analytically important because they can often be fit perfectly or quite well by a *straight line,* as in Fig. 2-1. Even if the monotonic pattern has curves, it can still receive a reasonably good rectilinear fit. Thus, the straight lines drawn in Figs. 2-14 and 2-15 will produce relatively good

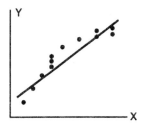
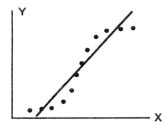

Fig. 2-14 Straight line through points of Fig. 2-2.   **Fig. 2-15** Straight line through points of Fig. 2-5.

fits for the corresponding data patterns shown in Figs. 2-2 and 2-5. For non-monotonic patterns, a straight line can sometimes give a good indication of trend, as in Fig. 2-16, which shows the rectilinear fit for the data of Fig. 2-6.

Nevertheless, a straight line will seldom (if ever) have a really excellent fit for substantially non-monotonic patterns, which almost always require curves. Because of the inability to show reversals or flattenings, a straight line for non-

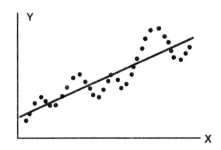

Fig. 2-16 Straight line through points of Fig. 2-6.

monotonic patterns will often produce highly distorted or deceptive results. For example, Fig. 2-17 shows *two* straight lines that produce a relatively good fit for the reversal pattern in Fig. 2-4. If we insisted on a *single* straight line, however, the best line through those points, as shown in Fig. 2-18, would offer a false portrait of the true relationship between the X and Y variables.

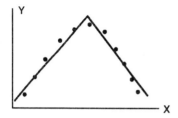 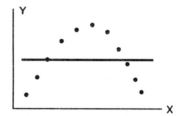

Fig. 2-17 Two straight lines through points of Fig. 2-4.

Fig. 2-18 Single straight line through points of Fig. 2-4.

A straight line can sometimes be deceptive even for monotonic data. Thus, the straight line in Fig. 2-14 correctly indicates the general upward monotonic pattern of the data, but fails to show that the pattern has two distinctive gradients: a sharp rise at the beginning and a flattening off at the end. A single straight line for the data in Fig. 2-3 would produce an analogous problem: a correct indication of the general downward pattern, but failure to show the different gradients. In Fig. 2-15, the straight line misses the possibly important flattening at the two ends of the pattern; and in Fig. 2-16, the straight line shows none of the possibly important oscillations.

2.2.5. Role of algebraic models

An *algebraic model* is a mathematical expression, such as $Y = a + bX$, that fits a line to the collection of data. The most common algebraic models are straight lines ($Y = a + bX$), curved lines with a quadratic term ($Y = a + bX + cX^2$), exponential lines ($Y = e^{a + bX}$), and logarithmic expressions ($\ln Y = a + bX$; or $Y = a + b \ln X$). Although a single model is preferred, the collection of data can sometimes be divided into zones that are fitted with piecemeal connections, such as "step functions" or "splines". The single model (or the piecemeal connector) can have one, two, or many terms, according to the different formats used for expressing X. In the expression $Y = a + bX$, the model has two terms and one format of X. In $Y = a + bX + c\,e^{X-4} + d \log X + f \sin X$, the model has five terms and four formats of X.

For only two variables, X and Y, the algebraic model will form a continuous line (except for values that cannot be determined, such as square roots or logarithms for negative values of X). For multiple variables, the model forms a "surface" that will be discussed later.

The fitting of data with the lines (or surfaces) of algebraic models has been a fundamental strategy in both bivariate and multivariable analysis. For most analyses, the information shown in Table 2-1 would *not* be depicted as points on a graph, and no attempt would be made to fit a line to the data. The polyvalent graph was shown in Fig. 2-13, however, because such cannonball data are commonly fitted (as noted lated) with algebraic "linear" models in multivariable analysis.

The implied graphical portrait should always be visualized whenever an algebraic model is fitted to the data. Regardless of the types of variables, the pattern of points will determine whether the algebraic result can achieve the desired goal in suitably fitting the data. What you see indicates whether what you get is really what you want.

2.2.6. Zonal partitions

The foregoing illustrations have shown that any set of bivariate data can be displayed as a table or as a graph. In a table, each pair of bivariate categories forms a cell, containing the corresponding number of counted frequencies. In a graph, each pair of bivariate values forms a location, showing the corresponding number of points. Although a tabular arrangement seems more appropriate and "natural" for data expressed in categories of nominal, binary, or ordinal variables, such data can nevertheless be shown in graphs such as Fig. 2-13.

Conversely, a graph seems more appropriate for data expressed in dimensional variables, such as the diffuse pattern of Fig. 2-7. Nevertheless, the graph of Fig. 2-7 can readily be converted into a table if the X and Y variables are partitioned into ordinal or binary zones. For example, the values of X can be divided at their tertile locations to form the three zones shown in Fig. 2-19.

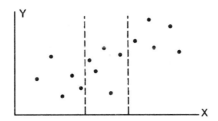

Fig. 2-19 Zonal partition for abscissa (X-variable) in Fig. 2-7.

The Y variable can then be divided at the median (or some other appropriately selected location) to form the two vertical zones shown in Fig. 2-20.

Fig. 2-20 Zonal partition added for ordinate (Y-variable) in Fig. 2-19.

The six bivariate zones of Fig. 2-20 can then be converted to the following table of frequency counts:

	Zones of Y	
Zones of X	Low	High
Lowest third	4	1
Middle third	3	2
Upper third	0	5

2.2.7. Strategies for clusters or scores

Just as any pair of two variables can be displayed with either a table or a graph, the subsequent analysis can be conducted with either of two strategies. In one strategy, the tabular data are arranged in conjoined categories or zonal partitions that can be called *clusters*. In the other strategy, the graphical points are fitted with an algebraic expression. When values of X are entered, the algebraic expression produces an estimate for Y that can be called a *score*.

For example, suppose the data of Fig. 2-7 were fitted with a straight-line algebraic model, $Y = a + bX$. When each of the various values of X is inserted into this model, the subsequent calculation would produce a score as the estimated value of Y at that location of X. When the same data are demarcated into zonal partitions, as in Fig. 2-20, the values of X and Y are converted into six main categories. To estimate values of Y from the categorial clustered values of X, the corresponding results would be as follows:

Zones of X	Proportion of high values of Y
Lowest	1/5 (20%)
Middle	2/5 (40%)
Highest	5/5 (100%)

Since two different strategies—clusters or scores—are available for displaying and analyzing bivariate data, the choice of an approach will depend on the goals of the analysis.

2.3. Goals of bivariate analysis

Bivariate analyses can be done to identify trends, quantify impacts, achieve estimations, produce taxonomic compressions, or discern configurations. Each of these goals will be affected by the orientation and pattern of data.

2.3.1. Trends

The pattern shown by the two variables indicates the trend as one set of values moves for corresponding movement in the other set. Does Y tend to rise or fall as X rises or falls? Does Y remain relatively stationary while X increases? Does the relationship have more complex trends of rises and falls, with different directions in different zones of the data?

The discernment of these trends is a prime goal of bivariate analysis, but the analyst often wants to quantify the trends rather than merely note their directions.

For most purposes, therefore, the concept of "trend" usually implies the concept of "impact", discussed in the next section.

2.3.2. Impacts

Magnitudes can always be determined for the directions of a trend, but if Y is a dependent variable, the effect shown in the trend is called the *impact* or *influence* of X on Y. As a name for this effect, the word *influence* is better than *impact*, but can no longer be used because it has been reserved for a different purpose in statistical analysis. Instead of referring to the influence of one variable on another, statistical analysts now talk about the influence of "observations", i.e., different members of the group. The influence is shown as the effect of eliminating individual persons or "outlier" points from the set of data. To avoid confusion between influential variables and influential people, the term *impact* can be used for the effect of variables.

2.3.2.1. Measuring impact

The impact of X on Y is noted from the gradient of change in Y with changes in X. The larger the gradient, the greater the impact. The gradient can be determined as a single average value for the overall relationship between the two variables, or for changes from one zone to the next. In Figs. 2-1 and 2-14, the slope of the straight line would indicate the gradient. For the categorical data in Fig. 2-13 and Table 2-1, the gradient is best shown when the results are expressed in the targeted arrangement of Table 2-2. The overall gradient would be determined by subtracting the results of Y for the highest and lowest values of X. Thus, the overall gradient for proportion of failures in Y would be 62% (= 75% − 13%). Since this overall gradient involves two changes, the "crude" average gradient would be 62%/2 = 31%. The changes can be determined directly as intermediate gradients of 29% (= 42% − 13%) from the low to medium value of X, and 33% (= 75% − 42%) from the medium to high value.

Table 2-2 Targeted arrangement for data of Fig. 2-13 and Table 2-1

Values of X	Proportion of Failures in Y	
Low	3/23	(13%)
Medium	10/24	(42%)
High	24/32	(75%)
Total	37/79	(47%)

Measuring these gradients is a fundamental scientific goal in bivariate (and multivariate) analysis. The idea is to determine how much Y rises, falls, or is essentially unchanged with changes in X. This is the idea when "risks" are expressed in phrases such as, "You live 2 years less for every 5-point rise in substance W." The gradients can be found with the two tactics just illustrated: by noting either the slopes of algebraic lines or the incremental results between clustered zones. The mechanisms used for the measurements are further discussed in Chapters 7 and 8.

2.3.2.2. Types of associations

The name *association* is commonly used for the mathematical relationships of two (or more) variables, but trends in the gradients are commonly cited as the *regressions, contrasts, correlations,* or *concordances* discussed in the next four subsections.

2.3.2.2.1. Regressions

With the assumption that variable Y depends on X, an algebraic model will try to fit the pattern of data with a straight or curved line, called the *regression* of Y on X. Although any of the regression models can be called *linear,* the term *rectilinear* will be reserved here for the straight-line model, $Y = a + bX$. In such a line, the gradient is the slope of the regression coefficient, b.

2.3.2.2.2. Contrasts

Suppose the X variable is binary, representing membership in one of two groups, such as treatment A and treatment B. The association with the Y variable can then be expressed by contrasting the univariate results for Y in each group. Thus, if Y represents *blood pressure* or *success* in each member, we can compare the mean blood pressures or the proportions of success in Group A and Group B.

This type of contrast, usually symbolized as $\bar{Y}_A$ vs. $\bar{Y}_B$, or p_A vs. p_B, is probably the most common analytic activity in ordinary statistics, but would seldom be regarded as a "bivariate analysis". Nevertheless, two variables are involved; and the increment in the corresponding results of Y would indicate the "trend" produced as the X variable "moves" from Group A to Group B.

If the X variable represents a set of ordinal categories, the trend is the gradient formed by contrasting results in adjacent categories. This approach was used earlier for the intermediate contrasts of 29% and 33% in the three categories of X in Fig. 2-13. In Fig. 2-10, a mean or median could be the central index for the dimensional values of Y at each ordinal location of X. The ordinal trend would then be shown in successive contrasts of adjacent pairs of central indexes. A similar process could have been used in the three zones of Fig. 2-19 to discern trend from the means or medians of the Y values, rather than from the binary proportions demarcated in Fig. 2-20.

2.3.2.2.3. Correlations

If the variables are nondependent, a trend can still be noted and quantified, but it cannot be oriented. We can say that one variable rises as the other falls, but the two variables do not have a dependent relationship. The mutual effect is a *correlation,* not a regression, between the two variables. As discussed later, however, the correlation coefficient for the two variables, symbolized as r, is a unit-free "standardized" regression coefficient. Since a bivariate correlation is analogous to, and calculated as, a mutual regression, the construction of *correlation coefficients* will be considered later, in the discussion of regression analysis.

2.3.2.2.4. Concordances

In certain circumstances, such as studies of observer variability or quality control in laboratory measurements, the two variables are assessed for exact agreement rather than trend. A correlation analysis is inadequate for these assessments. The two variables could have nonexistent agreement but perfect correlation if one is always twice the value of the other, or 10 units higher. For this reason, concordance is assessed with special indexes of proportional agreement, kappa coefficients, intraclass correlation coefficients, or other mathematical mechanisms that differ from those used for ordinary correlations.[6-8]

2.3.3. Estimates

For a third main analytic goal, estimation, variable X is used to produce either a substitute replacement or a reasonably predictive "guess" for the value of variable Y. If close enough to be highly accurate, the result obtained from X can actually substitute for Y. If not always accurate, but expected to improve the identification of Y, the analytic result is a predictive estimate.

2.3.3.1. Substitutions

Bivariate analyses commonly lead to substitute replacements in measurements of laboratory data. For example, many chemical substances are measured in a system that converts the magnitude of their concentration into a photometric voltage. Fig. 2-21 shows the basic "calibration" obtained as the voltage values of X for known concentrations of Y. After a line is fitted to the points, the concentration of "unknown" specimens is indicated by the value of Y at the measured voltage of X. Fig. 2-22 shows another calibration pattern that could be applied in a similar

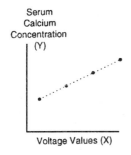

Fig. 2-21 Relationship of serum calcium concentration and voltage values in a chemical measurement system. The four points are the known "calibration" values of calcium specimens. The dotted line represents the "relationship" used to estimate future values of Y from observed values of X in unknown specimens.

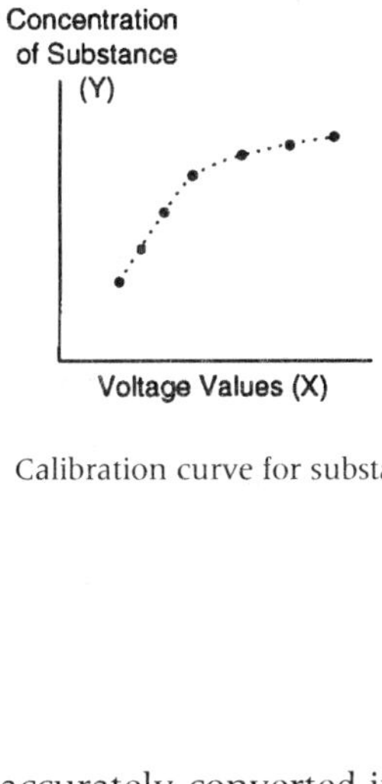

Fig. 2-22 Calibration curve for substance Y.

manner, with observed values of X that could be accurately converted into corresponding values of Y.

If the analytic goal is to get a substitute replacement, the most important criterion is high accuracy. Because the analyst wants to fit the data almost perfectly, an algebraic line is almost always preferred over a categorical cluster. As long as the line offers the quasi-perfect fit of a highly accurate transformation from X to Y, the analyst may not care about any other analytic goals, such as finding the effect of one variable on the other. The purpose of substitution is to get accurate fit, not to determine impact.

An analogous approach is used when economists fit a "time series" polynomial curve, as $a + bt + ct^2 + dt^3 + ft^4 + \ldots$, to the temporal movements of gross national product or the price of a particular stock. This type of pattern would be portrayed in Fig. 2-6 if Y represented the price of a stock on each of a series of dates shown by X. If the algebraic equation produces accurate substitutions (and successful predictions of future stock prices), the impact of the time variable need not be specifically quantified, and the exact configuration of the curve is relatively unimportant.

2.3.3.2. Predictive estimates

In most analytic circumstances, X and Y refer to different entities that may have a distinct but not extremely close relationship. Although an improvement over what might be obtained from Y alone, the bivariate estimates made from X will often be far from perfect.

Despite the imperfections, estimates of this type are a common goal in bivariate and particularly in multivariate analysis. We might try to estimate a person's life expectancy from his current age, the length of hospitalization anticipated for the severity of current illness, or the hepatic diagnosis that best corresponds to the level of serum bilirubin. The estimated results are often called *predictions,* and the independent variables are often called *predictor variables.* This parlance is not linguistically correct, however, if the "prediction" is aimed at identifying a diagnosis or other concurrent phenomenon rather than forecasting (or prognosticating) a future outcome. Nevertheless, to avoid using two terms—*predictive estimates* and *diagnostic* (or other) *estimates*—the estimate is regularly called a *prediction*.

For example, a "predictive" estimate occurs when the result of a diagnostic marker test is used to indicate that a particular disease is present or absent. In these diagnostic predictions, the biologic and mathematical orientations of the analysis are usually reversed. Biologically, the disease is the independent variable that affects the result of the diagnostic marker test; but mathematically, the result of the test, such as *serum bilirubin,* is used as the X variable in estimating Y for existence of the target diagnosis, such as hepatic cirrhosis. (This same reversal occurred in Section 2.3.3.1. Because the voltage depends on the chemical concentration, the latter variable should biologically be regarded as X. Nevertheless, in Fig. 2-21, voltage was the X variable because it becomes used for the subsequent estimates of concentration.)

Although imperfect, predictions made from variable X can substantially improve what might otherwise occur using results only for variable Y. For example, in Table 2-2, the results for only variable Y would show that 37 patients had failed and 42 were successful. From the overall failure rate of 37/79 *(47%),* the only possible prediction for each individual patient would be **success.** The prediction would be correct in 42 (= 79 − 37) patients and wrong in 37. With bivariate values available for both Y and X, however, Table 2-2 shows a dramatic gradient that demonstrates the predictive value of X from its substantial effect on Y. For **low** values of X, the prediction of **success** in Y would be correct in 20 of the 23 members. For **medium** values of X, the **success** prediction would be correct in 14 of the 24 instances. The prediction of **failure** in Y would be correct in 24 of the 32 members of the **high** X group. The total number of accurate predictions would rise from 42 to 58 (= 20 + 14 + 24), and the errors would drop from 37 to 21. The percentage reduction in number of errors—expressed as (37 − 21)/37 = 43%—would denote the proportionate improvement in the predictive rate of errors.

The improvement here occurs because of a distinctive trend in the relationship between Y and X. If X has little or no effect on Y, a knowledge of X should hardly alter the estimate of Y. If X has an apparently distinct impact, however, the bivariate analysis can distinctively improve the estimates.

2.3.3.3. Estimates vs. impacts

Because the accuracy of an estimate from X depends on how well the data for Y are fitted by the analytic arrangement, large impacts do not necessarily imply better estimates, and vice versa. Highly accurate estimates can sometimes be obtained with data showing a small impact, and relatively poor estimates can come from data having a large impact. The reason for this apparent contradiction is that impact refers to change, whereas estimation refers to individual points. Furthermore, with the straight-line models that are often used to fit the data, the slope of the line may not give an accurate account of the true trend. A particularly striking example of this problem was shown in Fig. 2-18, and other examples will be discussed in Chapter 7.

Another important distinction between estimates and impacts is shown in the algebraic and categorical arrangements of data. With an algebraic model, such as Y = a + bX, the impact is shown immediately as the slope, b, but estimates of Y must be calculated when appropriate values of X are entered into the equation. On the other hand, with a categorical arrangement such as Table 2-2, the estimate is shown immediately as the value of Y for each category of X, but the impacts must be determined from subtraction (or other calculations) of the Y values in adjacent zones. Thus, a rectilinear algebraic model shows impact immediately, but the estimate must be calculated; a categorical cluster shows the estimate immediately, but impact must be calculated.

2.3.4. Compressions

The word *compression* is used here for a taxonomic classification that forms a new set of categories or rating scale by descriptively joining two (or more) variables. The values of each constituent variable are no longer evident when the compressed combinations form the single new composite variable. Examples of bivariate compressions are the temperature-humidity index and wind-chill index in daily life, or the anion gap, free thyroxine index, or Quetelet index in medical activities. The Apgar score is a compression of five variables.

The compressions are not well described by the more general term, *classification*, which can refer either to a taxonomic construction (as in the process here) or to the "diagnostic" assignment made from a set of established taxonomic categories. For example, several different taxonomies—containing such categories as *fiction, biography,* and *surgery*—have been developed for classifying books in a library. The decisions about what categories to establish were an act of taxonomic classification. A different type of classification consists of deciding which available category to assign for cataloguing a particular book. With the latter type of "diagnostic" decision, we might determine how to categorize a fictionalized biography of a surgeon.

Analogously, in medical work, a large array of categories has been established as a taxonomy containing the authorized names of human diseases. The categories, which are regularly altered as new knowledge becomes available, are listed in such catalogs as the International Classification of Disease.[9] In pragmatic clinical activities, after observing a patient's various manifestations, a clinician makes a diagnosis by choosing one (or more) of these categories of disease.

Although taxonomic compressions and diagnostic decisions both involve classification of variables, the activities are quite different mathematically. In bivariate procedures, a taxonomic compression combines two variables to form a separate new composite variable, whereas a diagnostic assignment uses one variable to estimate the value of another.

With univariate data, a taxonomic process is used to demarcate "cutpoints" that classify a single variable such as *age* into categorical zones. The zones might be **young** for ≤ 40 years old, **middle** for 40–69, and **old** for ≥ 70. Taxonomic demarcations also produce "ranges of normal", expressed in zones such as **too low, normal,** and **too high** for single variables such as level of *serum calcium.*

In a taxonomic compression of two variables, the new composite variable can often be formed as a simple matter of sensible judgment, without any elaborate or formal mathematical strategy. An example is the wind chill index, in Fig. 2-23, where the values of X and Y are combined in the clusters of a grid-like pattern. The interior cells show *the wind chill index* as values of the new variable formed by the conjunction of each pair of values for the two main variables, *temperature* and *wind,* shown in the margins. Compressions are more commonly formed when the values of X and Y are mathematically converted into scores, using such arithmetic as subtraction (as in the anion gap), multiplication (as in the free thyroxine index), or division (as in the Quetelet index).

The bivariate graphic pattern may sometimes demarcate distinctive groups that are not apparent from either variable alone. For example, in Fig. 2-24, a univariate inspection of all the X values alone or Y values alone would not show anything remarkable beyond the spread of data for each variable. When X and Y are examined simultaneously, however, the pattern in Fig. 2-24 shows two distinct groups that could be taxonomically categorized as A and B, according to the appropriate demarcation of bivariate zones.

In all the examples here, the compressions were formed by combining the constituent variables without regard to a specific external variable. Other compressions may be developed as discussed later, however, during a targeted multivariable analysis that produces prognostic scores, risk groups, or staging systems.

2.3.5. Configurations

Configuration refers to the straight or curved linear shape that is formed either by the graphical pattern of points or by the selected algebraic model for the rela-

WIND CHILL INDEX

Temp. F	WIND (MPH)								
	0	5	10	15	20	25	30	35	40
35	35	33	21	16	12	7	5	3	1
30	30	27	16	11	3	0	-2	-4	-4
25	25	21	9	1	-4	-7	-11	-13	-15
20	20	16	2	-6	-9	-15	-18	-20	-22
15	15	12	-2	-11	-17	-22	-26	-27	-29
10	10	7	-9	-18	-24	-29	-33	-35	-36
5	5	1	-15	-25	-32	-37	-41	-43	-45
0	0	-6	-22	-33	-40	-45	-49	-52	-54
-5	-5	-11	-27	-40	-46	-52	-56	-60	-62
-10	-10	-15	-31	-45	-52	-58	-63	-67	-69
-15	-15	-20	-38	-51	-60	-67	-70	-72	-76
-20	-20	-26	-45	-60	-68	-75	-78	-83	-87
-25	-25	-31	-52	-65	-76	-83	-87	-90	-94
-30	-30	-35	-58	-70	-81	-89	-94	-98	-101
-35	-35	-41	-64	-78	-88	-96	-101	-105	-107
-40	-40	-47	-70	-85	-96	-104	-109	-113	-116
-45	-45	-54	-77	-90	-103	-112	-117	-123	-128

Fig. 2-23 Construction of the wind chill index as combinations of values for temperature and wind. (Courtesy of WTNH, Channel 8, New Haven, Conn.)

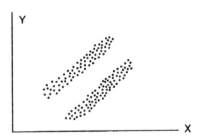

Fig. 2-24 Two groups of points distinguishable with bivariate but not univariate characterizations.

tionship of Y vs. X. This shape sometimes offers valuable knowledge for its own sake. For example, in Fig. 2-21, the analyst would be reassured about the accuracy of the measurement process if X and Y have a straight-line relationship. If they have a curvilinear pattern, as in Fig. 2-22, the reassurance would be reduced because the measurement process seems to become relatively insensitive at high values of X, where major changes produce relatively little effect in the values of Y.

In Fig. 2-3, if X represents time, the L-shaped exponential curve could depict the physical phenomenon of radioactive "decay", or the biologic phenomenon of diminishing survival proportions in a longitudinally followed cohort. In Fig. 2-4, the curve could represent the parabolic shape of muscle response to tension in Starling's physiologic "law of the heart".

In most bivariate situations, however, configuration is not analyzed for its own sake. It is used only to achieve the two other goals of estimation and/or impact. For substitute estimates in Figs. 2-1 through 2-6, the line must have a proper configuration to let Y be accurately identified from the value of X. For determining impact, the rectilinear or curvilinear shape is simple enough in five of those lines to indicate the effect of X in different zones. With the straight line of Fig. 2-1, changes in X produce a constant rate of change in Y; and the changing effect of X is also readily evident in different zones of the curves for Figs. 2-2, 2-3, 2-4, and 2-5. In the complex curve of Fig. 2-6, however, the influence of the X-variable (which might be time) is too diverse to be easily characterized either from the graph or from an algebraic formulation.

2.3.5.1. Advantages of algebraic configuration

The algebraic expression of configuration is appealing because it can denote three things: shape, estimation, and impact. *Shape* is shown by the format of the expression. It is a straight line if X is cited as $a + bX$. It is a curve if X is expressed in any other format, such as X^2, e^X, or sin X; and different formats produce different curves. *Estimation* occurs when the observed values of X are inserted into the algebraic expression to produce the predicted values of Y. *Impact* is noted from the "weight" or magnitude of the coefficients attached to the expressions for X. In the straight line $Y = 3.0 - 1.5X$, the value of Y constantly drops 1.5 units for every unitary increase in X. If the line is $Y = 2.0 + 4.1X$, the corresponding value of Y constantly rises 4.1 units. In an expression such as $Y = e^{a-bX}$, the impact of X becomes easy to interpret with the transformation, $\ln Y = a - bX$, which shows that the natural logarithm of Y declines b units for each unitary rise in X.

2.3.5.2. Candidate formats

A great many algebraic expressions are available as candidate formats to cite the configuration of the Y vs. X relationship.

2.3.5.2.1. Straight lines

The expression of a straight line is

$$Y = a + bX.$$

In this expression, *b* is the *regression coefficient*, showing the slope of the line, which denotes the impact of X on Y. The value of *a* is called the *intercept*, which is the value of Y when $X = 0$.

2.3.5.2.2. Exponential curves

A simple exponential curve is expressed as

$$Y = e^{a+bX}.$$

With this expression, the rectilinear format of $a + bX$ is retained when the algebra is transformed as

$$\ln Y = a + bX.$$

The simple exponential curve will have an ascending "J" shape or a descending "L" shape, according to the positive or negative sign of the coefficient for *b*. The curve in Fig. 2-3 has a descending exponential shape.

A more complex exponential expression is

$$Y = \frac{e^{a+bX}}{1 + e^{a+bX}}.$$

This expression (as discussed later) is called the *logistic transformation*. A rectilinear format for X is retained when the expression is algebraically transformed into the *logit* of Y, which is

$$\ln\left(\frac{Y}{1-Y}\right) = a + bX.$$

The curve for Y has a sigmoidal, or S, shape that was shown in the configuration of Fig. 2-5.

2.3.5.2.3. 'Powers' of X

X can also be expressed in "powers", which can be positive, such as X^2, X^3, X^4, or negative, such as X^{-2}, X^{-3}, ... The powers can also be fractions or decimals, such as $X^{1/2}$ or $X^{-.75}$.

The expression X^{-1} is usually cited as the reciprocal, $1/X$, and forms a curve called a hyperbola, which also has an L-shaped pattern. Expressions such as $X^{1/2}$ or $X^{1/3}$ are usually written as the square root ($\sqrt{X}$) or cube root ($\sqrt[3]{X}$). The expression of X^2 is often called *quadratic*; X^3 is called *cubic*, and X^4 is called *quartic*.

2.3.5.2.4. Transcendental expressions

The oscillating curves of trigonometry—such as sines and cosines—are expressed in the "transcendental" format of sin X, cos X, etc.

2.3.5.2.5. Polynomials

The algebraic expression is called a polynomical when it contains more than one format of X. The most common polynomials have positive powers of X, such as

$$a + bX + cX^2,$$

$$a + bX + cX^2 + dX^3,$$

and so on. The "time series" in Fig. 2-6 could be expressed with a polynomial arrangement. A polynomial, however, can also be a mixture of different formats, such as

$$a + bX + e^{cX} + e^{-dX^{-3}} + f \sin X^4 + g\sqrt{X}.$$

2.3.5.2.6. Step functions and splines

When all else fails, the data can be fitted with piecemeal expressions for different parts of the total pattern. In Fig. 2-17, the data received piecemeal fits with two straight lines; and a third line could have been introduced to improve fit for the flattening at the top of the curve.

The piecemeal expressions can be horizontal lines called *step functions*, or algebraic formulations called *splines*. You may previously have seen displays of step functions in the "staircase" pattern used for Kaplan-Meier survival curves in medical publications. A spline has a specific algebraic pattern that often contains terms in X^2, X^3, or X^4. Splines have been advocated[10,11] to improve the fit of multivariate algebraic formats by producing connecting links for zones where the main model is unsatisfactory.

In exchange for the mathematical advantage of improving fit, piecemeal expressions have the major scientific disadvantage of increasing complexity. When a piecemeal expression is inserted, the pattern of data is no longer described by the single algebraic model that allows a simple decision about configuration and impact.

2.3.5.3. Problems in choosing formats

Despite their intellectual (and esthetic) attraction, algebraic formats create a major problem in data analysis. The format is easy to choose if the investigator has sound substantive reasons for anticipating that the pattern will be a straight line, parabola, exponential decay, or other type of curve. For example, the relationship of height vs. age might need a format that shows a monotonic rise in early life, a leveling off after adulthood, and a possible slight decline in the very late years. In most instances, however, no substantive anticipations exist. The configuration either emerges from the pattern of data or is chosen arbitrarily for convenience in the data analysis.

Unless the investigator previously knows what the format should be, the diverse candidates offer an enormous (actually infinite) series of choices for fitting an algebraic line to a collection of data for X and Y. If the data are monovalent, a perfectly fitting line can always be drawn with a set of piecemeal expressions that connect consecutive points. Beyond the perfect fit, however, the connected piecemeal lines will make little or no contribution to the interpretive analysis of the data. In particular, the piecemeal expressions will not indicate the general configuration of the X–Y relationship, and will not show the general impact of X on Y.

Nevertheless, for scientific goals, we almost always want to assess the impact of X, even when the data are polyvalent and not well fitted by any line. Consequently, to determine the impact of X, a suitable single algebraic expression is needed for the line. Since each expression offers a format for the pattern of data, the challenge is to choose a suitable format. This is not a simple task.

If the pattern shows an obvious straight line, as in Fig. 2-1, the decision is easy. If the pattern is a curve, however, the decision is much more difficult, because the same graphical pattern can be fit with many different curves. Thus, although the pattern in Fig. 2-3 looks like the classical shape of a descending exponential curve, the pattern might also be part of a larger parabola, hyperbola, or transcendental expression. Even a pattern that looks like an obvious straight line might really be the relatively rectilinear segment of a much larger curve whose remaining extent is not shown on the graph.

If the pattern is a polyvalent swarm of points, the choice of an algebraic format is even trickier. For example, the swarm of points in Fig. 2-11 can be fitted with the straight line shown in Fig. 2-25, with the oscillating line shown in Fig. 2-26, with the sigmoidal curve in Fig. 2-27, or with the spiral pattern of Fig. 2-28. Although one of these four lines might mathematically have a slight advantage in offering the best fit, each of the four lines would be best justified not by any mathematical calculations, but by the topic described in the data. Accordingly, none of the four lines can immediately be regarded as most appropriate.

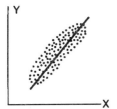 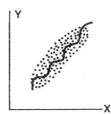

Fig. 2-25 Straight line fit to a swarm of points.   **Fig. 2-26** Oscillating line to fit points of Fig. 2-25.

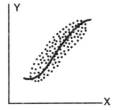 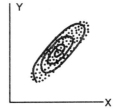

Fig. 2-27 Sigmoidal line to fit points of Fig. 2-25.   **Fig. 2-28** Spiral line to fit points of Fig. 2-25.

In all of the foregoing challenges, we wanted to choose an algebraic format, but were uncertain about which one to pick. It is an arbitrary choice; and the format that seems to offer the best fit to the data may not necessarily be best for denoting the true scientific impact and configuration of the X-Y relationship.

2.4. Correspondence of goals and analytic procedure

The reason for extensively discussing analytic goals is that they are crucial decisions, but seldom receive adequate attention. The analyst often "plugs" the data into a convenient computer program, receives the results, and offers apparently appropriate interpretations—without establishing exactly what was wanted and whether it was, in fact, obtained. Nevertheless, the goals must be determined, in advance, to choose a suitable analytic procedure. The procedure can use either the clusters of zonal partitions or the scores of algebraic formats; and if the latter strategy is chosen, a particular format must also be selected.

In bivariate analysis, where the simple pattern of the data is readily visualized, the choice of a procedure is relatively easy. For compressions, a zonal cluster or algebraic score will be chosen according to the substantive content and format of the data. For configurations and accurate substitutions, an algebraic format can be readily fitted if the data show a suitable monovalent pattern. If the pattern is not monovalent, the algebraic format may still fit relatively well—but the configuration may reflect an arbitrary mathematical rather than truly biologic relationship.

The decisions become more difficult if the goals are to quantify trends and make predictions. These goals can each be approached with either a cluster or score strategy, and each strategy has its advantages and disadvantages.

2.4.1. Zonal partitions and clusters

The clusters of a zonal partition can always be used to show trends in the data, no matter what pattern appears in points on the graph. The choice and demarcation of zones is entirely arbitrary, and will depend on the extensiveness and pattern of the data. The Y variable is often divided into two parts (customarily at the overall median value) because binary proportions are particularly easy to interpret; but means or medians could be used to summarize the univariate values of Y in each zone of X. The X variable is usually divided into at least three zones to allow discernment of a trend (rather than a mere contrast of two groups) in the summary values for Y. If no obvious cutpoints are suggested by the data, X can be partitioned in a "natural" manner at its tertiles, quartiles, quintiles, etc.

The zonal-partition strategy offers a simple and highly effective method to discern trends in the data, and is particularly useful for distinguishing zones in which different gradients occur because X does *not* have a consistently similar (straight-line) type of effect on Y. The zonal method is especially applicable for nondimensional data that are expressed in "zones" of nominal, binary, and ordinal categories. To be cited with an algebraic model, such data become clusters or cannonballs of points at appropriate locations on a graph. The visual arrangement forms such odd-looking graphs as Fig. 2-13; and the algebraic model regularly produces a poor-fitting line through the polyvalent points. Thus, for categories of nondimensional data, the zonal method of analysis may often seem more "natural" than the artifice of an algebraic model.

The zonal-partition strategy would not be applied, of course, in circumstances where the goal is to use X for getting an *accurate* substitution for Y. In such situations, however, the analyst is concerned with fit, not with trend (or shape). If the data are adequately monovalent, an algebraic format can always be found to provide an excellent fit.

The main apparent flaw of the zonal-partition strategy is that it lacks the "elegance" of a mathematical model for configuration of the relationship: we discern trends without identifying shapes. The zonal-partition strategy may also involve arbitrary judgments in choosing boundaries for the zones, summary indexes for results in each zone, and methods for appraising the summary indexes.

An additional problem occurs when more than two variables are under analysis. When the partitioned variables are evaluated simultaneously, an excessively large number of multivariate zones can be formed. For example, if four independent variables are under analysis, and if each variable is partitioned in only three categories, the results of the dependent target variable must be appraised in 81 (= 3 × 3 × 3 × 3) tetravariate zones. To avoid these unwieldy appraisals, a special additional strategy is needed to condense the multiple zones into a cogent smaller number of clustered categories. This strategy, which is also arbitrary and devoid of mathematical elegance, leads to the decision-rule clusters that will be discussed later as an alternative to algebraic methods of targeted multivariable analysis.

2.4.2. Algebraic models and scores

Although the problem of arbitrarily choosing a suitable curve can be avoided by abandoning algebraic formats and using zonal partitions, the zonal approach will involve other types of arbitrary decisions, and will lack the "intellectual" attraction of a mathematical configuration for the data. As noted in the next subsection, however, the specified configuration may make things harder rather than easier if the goal is to identify trend and impact, rather than shape.

2.4.2.1. Problems in curvilinear trend

The effect of X on Y can easily be discerned from the rectilinear expression of $a + bX$. This expression can be applied in several arrangements that denote Y as a straight line, $\ln Y$ as an exponential curve, or $\ln [Y/(1 - Y)]$ as a logistic transformation. The regression coefficient b indicates the corresponding change in the expression for Y with each corresponding unitary change in X.

For curvilinear models that involve powers of X in polynomial arrangements (or in splines), however, a series of regression coefficients will be attached to the different formats of X. The impact of X may then be difficult to discern amid the different coefficients and formats. For example, consider the terms in the expression

$$a + bX - cX^2 - dX^3 + fX^4.$$

The effect of the bX term is obvious, but the others are not. The terms in X^2 and X^4 will rise with increasingly positive *and negative* values of X; the term in X^3 will enlarge in positive or negative directions with corresponding changes in X; and each of the effects of X^2, X^3, and X^4 will be altered by the magnitude and direction of the corresponding coefficients $-c$, $-d$, and $+f$.

The polynomial expression may thus provide a perfect fit for the set of data to which it is modeled, but the general trend of Y in response to X may be difficult or impossible to identify amid the plethora of different coefficients and formats for X. This problem, which can occur whenever X is cited in a polynomial model, is a major impediment to the analytic goal of evaluating trend. The polynomial curves can be excellent for suggesting configuration and achieving a replacement fit, but not for indicating the impact of X on Y. Besides, the formats and coefficients of the X constituents may be highly arbitrary—inserted by an automated data-processing mechanism that is mathematically aimed only at good fit, not at the true effect of X on Y, or the true configuration of the biologic relationship. Thus, the selected polynomial curve may not necessarily be the best or correct biologic curve, despite its apparent excellence in fitting the data.

2.4.2.2. Advantages of straight-line models

All of these arbitrary or difficult problems in zonal partitions and polynomial curves can be avoided with yet another arbitrary decision: choosing a straight-line algebraic model. It may not fit perfectly, and it may not offer the best configuration for the data; but it does provide a mathematical format that is both "standardized" (because of the uniqueness of the straight line) and easy to interpret for identifying the general effect of X on Y.

This advantage is responsible for the popularity of rectilinear models. They first became valuable when applied in systems of chemical measurement where a straight line could easily and properly achieve the goals of getting an accurate fit and a suitable configuration and impact. The popularity was increased by the ability of straight-line models to produce a reasonably good fit for many patterns of data while offering a simple, useful index of general trend. Thus, as noted earlier (in Section 2.2.4), a straight line offers a reasonably good fit in Figs. 2-14 and 2-15, while also correctly denoting the general trend. Straight lines would also produce reasonable fits for the polyvalent data in Figs. 2-10 and 2-11, and would also correctly denote the general trends.

A straight-line model, $a + bX$, also has some appealing mathematical advantages. In addition to being easy to interpret, the best values for the intercept a and regression coefficient b are easy to calculate (with the "least-squares" principle discussed later). The main virtue, however, is that the use of a unique, "standard" straight-line model eliminates the problem of deciding what to choose among the many curvilinear candidates.

2.4.2.3. Disadvantages of straight-line models

The main disadvantages of straight-line models have already been noted: distortion and imprecision. If the data are dramatically non-monotonic, with the striking reversal shown in Fig. 2-4, a rectilinear model will be highly deceptive, producing the unsatisfactory fit and erroneous trend shown in Fig. 2-18. If the data are non-monotonic without striking reversals, as in Fig. 2-6, the straight line will distort the oscillating pattern and fail to show zones of reversed gradient in the trend. If the data are monotonic with zones of varying gradient, as in Figs. 2-2, 2-3, and 2-5, the straight line will distort these variations and produce an imprecise single value for the different trends.

Furthermore, as noted later (in Chapter 7), the accomplishments of a straight-line model are usually expressed with statistical indexes for fit and for impact. When examined concomitantly, these two indexes can sometimes produce misleading contradictions. The impact of X on Y may be indicated correctly despite a poor index for fit, or erroneously despite an excellent index for fit.

For all these reasons, the advantages of rectilinear models can be threatened by their possible deceptions. Although statistical indexes are calculated to indicate the fit and impact of each model, there is no statistical index of deception. Consequently, the possibility of deception must be checked with separate judgments.

This judgmental process involves the principles of "regression diagnostics," which will be discussed later.

Chapter References

1. Feinstein, 1987; 2. Apgar, 1953; 3. Altman, 1991, p. 180; 4. Schachtel, 1984; 5. Tukey, 1977; 6. Landis, 1977; 7. Fleiss, 1981, Chapter 13; 8. Kramer, 1981; 9. ICD-10. International statistical classification of diseases and related health problems, 1992; 10. Harrell, 1988; 11. Yandell, 1993.

3 Goals of Multivariable Analysis

Outline ....

3.1 Similarities to bivariate goals
   3.1.1. Trends
   3.1.2. Estimates
      3.1.2.1. Substitutions
      3.1.2.2. Diagnostic estimates
      3.1.2.3. Predictive estimates
   3.1.3. Compressions
3.2. Special multivariable goals
   3.2.1. Screening variables
   3.2.2. Reducing variables
   3.2.3. Modifying impacts
      3.2.3.1. Persistent distinctiveness
      3.2.3.2. Collinear ineffectiveness
      3.2.3.3. Interactive alterations
         3.2.3.3.1. Exaggerations
         3.2.3.3.2. Reversals
         3.2.3.3.3. Interpretation of 'interactions'
   3.2.4. Quantifying impact
      3.2.4.1. Standard magnitudes of gradients
      3.2.4.2. Constancy of gradient
      3.2.4.3. 'Double gradient' effects
   3.2.5. Organization of estimates
      3.2.5.1. Clustered categorical stages
      3.2.5.2. Complex scores
      3.2.5.3. Simple scores
      3.2.5.4. Accuracy of estimates
   3.2.6. 'Adjustments' of baseline disparity
3.3. Other differences from bivariate analysis
   3.3.1. Categorization
   3.3.2. Configuration
   3.3.3. Multiple interdependent correlations
3.4. Role in medical research
   3.4.1. Interventional maneuvers, baselines, and outcomes
   3.4.2. Formation of groups
   3.4.3. Effects of maneuvers
      3.4.3.1. Distinctive subgroup effects
      3.4.3.2. Deception from disparate effects in subgroups
      3.4.3.3. Adjustment for susceptibility bias
      3.4.3.4. Mechanisms of adjustment
         3.4.3.4.1. Stratification
         3.4.3.4.2. Standardization
         3.4.3.4.3. Algebraic coefficients

  3.4.3.5. Improved efficiency of design
  3.4.3.6. Precision in future applications
3.5. Problems in 'confounding' variables
 3.5.1. Absence of true confounding
 3.5.2. Location in causal pathway
3.6. Omission of important variables

....

The basic goals of bivariate analysis are retained when the analysis extends to three or more variables, but the additional variables lead to new aims and challenges. The similar and new goals are the focus of this chapter.

3.1. Similarities to bivariate goals

Like bivariate analysis, multivariable analysis can be aimed at the basic goals of identifying trends, making estimates, and producing compressions.

3.1.1. Trends

Identifying or confirming trends in the effects of independent variables is probably the main goal of most multivariable analyses that appear in medical and public health literature today. These are the activities that produce quantitative expressions for "risk factors," "relative risks," and statements such as, "You live 2 years longer for every 5-point drop in substance X".

Before any of the multivariable work begins, the investigator usually analyzes the simple bivariate relationships between a target (or dependent) variable, such as survival time, and each of a set of independent variables, such as age, occupation, and stage of disease. If some of the independent variables are shown to have a significant effect on the target, the next step is to check whether the effect is maintained when all the other variables are examined simultaneously. The goal is not just to identify the trend, but to confirm it.

For example, suppose the 5-year survival rate is 88/200 (44%) for a group of patients with a particular cancer. The additional data shown in Table 3-1 indicate the relationships of survival to each of four independent variables.

Table 3-1 5-year survival rates for each of four independent variables in 200 patients with a cancer

Variable	Categories	5-year survival rate	
Cell type	Differentiated:	73/140	(52%)
	Undifferentiated:	15/60	(25%)
Anatomic stage	Localized:	69/90	(77%)
	Spread:	19/110	(17%)
Educational Background	No higher than high school:	59/135	(44%)
	Beyond high school:	29/65	(45%)
Age	Young:	52/100	(52%)
	Old:	36/100	(36%)
Total		88/200	(44%)

The bivariate results of Table 3-1 indicate that *educational background* seems to have no effect: the survival rates are essentially the same whether patients did or did not complete high school. Of the other three variables, *anatomic stage* has the greatest effect, with a gradient of 60% (= 77% − 17%) between the two categories. The other two variables, *cell type* and *age*, seem to have impressive but somewhat smaller effects, with gradients of 27% (= 52% − 25%) and 16% (= 52% − 36%), respectively.

For *cell type, anatomic stage,* and *age,* the individual bivariate results suggest that each variable has a distinctive impact on survival. The multivariable question, however, is whether the effects persist when the variables are examined simultaneously. In the concomitant presence of the other variables, does each variable continue to have a distinctive impact? If so, are its direction and magnitude consistent with what would be anticipated from the bivariate results?

When these questions are answered with a multivariable analysis of trends, the two main goals are to confirm and to quantify the effects noted in the simpler bivariate analyses. (Because the effect is examined separately for each independent variable, the simpler analyses are sometimes called *univariate.* Since the independent variable's effects are concomitantly checked in the dependent variable, however, *bivariate* is a more accurate term for the process.) Multivariable *confirmation* is aimed at showing that the simple bivariate effects continue to have the same impact and direction. Multivariable *quantification* is aimed at determining their magnitude. These two goals and other distinctions of multivariable trends are further discussed in Section 3.2.

3.1.2. Estimates

The estimates made with multivariable analysis can be intended to substitute, diagnose, or predict.

3.1.2.1. Substitutions

A dependent variable Y will almost always be more closely estimated from several independent variables, rather than one. The improvement, however, is seldom good enough to achieve the goal of highly accurate substitution. In most pragmatic research, if an almost perfect fit cannot be obtained with one or at most two independent variables, multiple variables will rarely (if ever) bring things close to perfection. Multivariable analysis is therefore commonly aimed at improving the general accuracy of an estimate, but the fit of the multivariable arrangement will seldom be close enough to let it become a substituted replacement.

3.1.2.2. Diagnostic estimates

The results of *diagnostic marker tests* for bivariate estimates of diagnosis are usually expressed with statistical indexes such as *sensitivity* and *specificity.*[1] Multivariable analyses have been increasingly applied,[2] however, in the hope that more variables will provide better accuracy. The strategy can use either an algebraic model or an algorithmic decision rule that works with categorical clusters.

The diagnostic target can be the binary presence or absence of a particular disease, or the selection of a single disease from a series of possible candidates. The algebraic methods most commonly used for these purposes are *multiple logistic regression* and *discriminant function analysis*. The different clustering techniques are often subsumed under the title of *recursive partitioning*. Both sets of strategies are outlined in Chapter 4, and discussed thereafter in greater detail.

3.1.2.3. Predictive estimates

Although diagnostic estimates can be regarded as predictions, a truly predictive estimate will forecast a target that occurred *after* the "baseline state" delineated by the independent variables. As a prime purpose of many multivariable analyses in medical research, predictive estimation usually involves risk factors for occurrence of a future disease, or prognostic factors for the outcome of an established disease. The identified factors are then used for predictions in individual persons, or to form predictively "homogeneous" groups in whom etiologic or therapeutic agents can be evaluated.

The approaches will be discussed in Section 3.2.5.

3.1.3. Compressions

For trends and estimates, some of the independent variables may be eliminated because they have little or no effect on the target variable. The variables that remain will then be a reduced or "compressed" group of what was originally available. For many compressions, however, the multiple variables are analyzed, without a target variable, as though they have a nondependent many-internal relationship. The taxonomic compressions are done either judgmentally or with formal mathematical methods, and the results are expressed either as categorical clusters or as algebraic scores.

A well-known example of judgmental compression for multivariate data is the TNM staging system for cancer,[3] which is expressed as clusters of multivariable categories. Another example, the Apgar score,[4] is a sum of ratings for five variables. Both the TNM stages and Apgar score were originally constructed without any formal mathematical procedures.

The formal procedures for multivariable taxonomic compressions, as outlined in Chapter 4, have such names as *factor analysis* and *cluster analysis*. They use mathematical principles to find and combine the most "important" members of a non-dependent set of multiple categories or variables.

3.2. Special multivariable goals

The special goals of multivariable analysis arise from additional challenges produced by the extra variables. The array of independent variables can be "screened" in search of those with important impacts. The number of variables can be reduced by removing the ones that seem unimportant. The trends noted in the apparently important variables can be checked, confirmed, and quantified. The results can then be organized in diverse ways to simplify the subsequent estimations. These special goals are discussed in the next five sections.

3.2.1. Screening variables

Before the multivariable analysis begins, the bivariate analyses will already have produced hints or clues about which variables seem important. Sometimes, however, the data are immediately placed into an automated multivariable procedure, with the hope of quickly screening the available variables to determine their statistical importance. The statistical tactic often works well, but if a data analyst has no idea of which variables are important, someone else should probably be doing or guiding the analysis. Nevertheless, if no other knowledge is available, and even if it is, the multivariable screening process can be helpful in corroborating previous ideas or suggesting new possibilities.

3.2.2. Reducing variables

Since a multiplicity of variables (and categories) will impair the ability to understand what is happening, an important goal of multivariable analysis is to reduce the number of variables (or categories). The ones that seem unimportant can be removed immediately after the screening process, or after more careful consideration of their substantive as well as statistical attributes.

The fewer the variables (and categories) to be considered, the easier will be the job of analyzing them and interpreting the results. The methods used for the reduction procedure are discussed later in the text.

3.2.3. Modifying impacts

A particularly common role of multivariable procedures is to modify (or confirm) the impacts found in the simpler bivariate analyses. In the context of multiple other variables, each independent variable may retain its distinctive individual impact, but sometimes the apparent impact is altered, showing effects different from what was previously expected.

The phenomena that make the multivariable effects disagree with the expectations are often called *collinearities* and *interactions*. In a collinearity, the variable turns out to be deceptively effective, having exerted its apparent impact mainly through a close or "collinear" relationship with some other variable that is truly effective. In the interaction problem, the direction or magnitude of a variable's expected effect is altered by the concomitant effects of other variables. The three types of phenomena—persistent distinctiveness, collinear ineffectiveness, and interactive alterations—are illustrated in the next three subsections.

3.2.3.1. Persistent distinctiveness

Table 3-2 shows the simultaneous multivariable effects of *anatomic stage* and *cell type* in the data of Table 3-1. Note the difference between Table 3-2 and the usual 2 × 2 (or "fourfold") table that often appears in medical literature. Table 3-2 is a 2 × 2 × 2 trivariate table. It has two categories in the row variable (*cell type*) and two categories in the column variable (*anatomic stage*). The third variable is shown in the interior cells of the table, where the two categories of alive or dead appear as proportions denoting survival rates at 5 years.

Table 3-2 Simultaneous effect of cell type and anatomic extensiveness of cancer on 5-year survival rates

	Anatomic Stage		
Cell type	Localized	Spread	Total
Differentiated	55/60	18/80	73/140
	(92%)	(22%)	(52%)
Undifferentiated	14/30	1/30	15/60
	(47%)	(3%)	(25%)
Total	69/90	19/110	88/200
	(77%)	(17%)	(44%)

The marginal totals of Table 3-2 show the bivariate effects previously seen in Table 3-1: survival was distinctly altered as *anatomic stage* went from localized to spread, and as *cell type* went from differentiated to undifferentiated. The trivariate cells of Table 3-2 show that the bivariate effects of each independent variable persist in the presence of the other. Anatomic stage had a distinctive effect on survival within each of the two categories of cell type, and cell type had a distinctive effect within each of the two categories of anatomic stage. The interior cells of the table thus confirm the directional effect expected from the marginal totals. Survival was reduced within each row from left to right, and within each column from top to bottom.

3.2.3.2. Collinear ineffectiveness

Since *age* also seemed to affect survival in Table 3-1, the distinctiveness of the effect can be checked within the context of other variables. When *age* and *anatomic stage* are examined simultaneously in Table 3-3, however, we find a surprise. Although a prognostic gradient is maintained for anatomic stage, the gradient for age vanishes in the cells. The two age groups had identical survival rates, 77%, within the anatomically localized group, and essentially similar survival rates (15% and 19%) within the anatomically spread group.

Table 3-3 Simultaneous effect of age and anatomic extensiveness of cancer on 5-year survival rates

	Anatomic Stage		
Age	Localized	Spread	Total
Young	46/60	6/40	52/100
	(77%)	(15%)	(52%)
Old	23/30	13/70	36/100
	(77%)	(19%)	(36%)
Total	69/90	19/110	88/200
	(77%)	(17%)	(44%)

The multivariable results thus show that age had no real prognostic impact in this group of patients. The source of the apparent effect was an unbalanced distribution that produced a correlation between anatomic stage and age. The denominators of each cell in Table 3-3 show a correlation that can be outlined as $\begin{Bmatrix} 60 & 40 \\ 30 & 70 \end{Bmatrix}$.

The young patients tended to have localized cancers, which occurred in 60% (= 60/100) of this age group, whereas 70% (= 70/100) of the old patients had tumors that were spread. Having proportionately more localized cancers, the young group could readily be expected to have higher survival rates even if age itself had no distinct effect on prognosis.

The phenomenon just noted occurs when a third variable (such as survival) is affected by two other variables (age and anatomic stage) that are distributed in a collinear, relatively close relationship. In this instance, the co-relationship made an ineffectual variable seem to have an impact that is really produced by some other effective variable.

The crucial contribution of the multivariable analysis was to reveal this distinction, which would not have been detected with bivariate analysis alone.

3.2.3.3. Interactive alterations

The directional consistency illustrated in Table 3-2 may be altered even though each variable continues to have a distinctive effect. Sometimes the two independent variables act synergistically to increase the magnitude of the expected directional effect. At other times the two variables act antagonistically to reverse the expected direction.

These unanticipated conjunctive effects of two (or more) variables are what statisticians call "interactions," but medical people tend to think of interactions as active events, such as "drug-drug interactions" or doctor-patient relationships, rather than relatively "passive" statistical phenomena. To avoid ambiguity, the term *conjunctive effect* can be used instead of *interaction*.

An interaction or conjunctive effect occurs when two variables simultaneously produce results that alter the directional magnitude of the effect anticipated from either variable alone. The alteration can be a synergistic exaggeration in the same direction, or an antagonistic reversal in the opposite direction. The phenomena are illustrated in the next two subsections.

3.2.3.3.1. Exaggerations

Consider a hypothetical disease, omphalosis, for which the 5-year survival rate is 213/310 (69%) for the entire group under study. The bivariate rates for three categories of *age* are as follows: Young: 96/120 (80%); Middle: 79/108 (73%); and Old: 38/80 (48%). The corresponding rates for three categories of weight are: Thin: 87/108 (79%); Medium: 68/95 (72%); and Fat: 58/105 (55%).

These bivariate results would make us expect survival to go down as age increases and as weight increases. In Table 3-4, when the two variables are examined simultaneously, the survival rates show that each variable has a distinctive effect within the context of the other, and that the effect goes in the anticipated direction. The bottom right-hand cell of the table, however, is surprising. Instead of the modest decrement in survival that would be expected from the marginal totals and from the successive adjacent cells of the table, the old fat people have a striking drop, to 0%. The conjunction of old age and fat weight thus produced an *exaggeration* at one of the corners of the table where each variable was at the extreme end of its scale.

Table 3-4 "Terminal exaggeration" for effect of weight and age on 5-year survival rates in 308 patients with "omphalosis"

Age	Weight			Total
	Thin	Medium	Fat	
Young	36/40 *(90%)*	30/37 *(81%)*	30/43 *(70%)*	96/120 *(80%)*
Middle	28/35 *(80%)*	23/32 *(72%)*	28/41 *(68%)*	79/108 *(73%)*
Old	23/33 *(70%)*	15/26 *(58%)*	0/21 *(0%)*	38/80 *(48%)*
Total	87/108 *(81%)*	68/95 *(72%)*	58/105 *(55%)*	213/308 *(69%)*

The exaggeration need not occur at a "terminal" cell. It can also appear as an unexpected internal "peak" or "bump" among the interior cells.

3.2.3.3.2. Reversals

In a more striking type of conjunction, the anticipated effect is reversed. Suppose the overall survival rate for omphalosis was 199/320 (62%) and the individual rates for *age* were young, 89/110 (81%); middle, 56/100 (56%)' and old, 54/110 (49%). Suppose the individual rates for weight were thin, 89/106 (84%); medium 55/99 (56%); and fat, 55/115 (48%). From these bivariate results we would expect each variable to have the same progressive effect in lowering survival, with the effect stronger between the first and middle category for each variable than between the middle and last.

When the two variables are examined simultaneously in Table 3-5, however, the results in the central cell are contrary to expectations. The people of middle age and medium weight had substantially lower survival than their neighbors in the adjacent cells. The prognostic gradients reversed their direction and formed a dip or "valley" in the interior middle cell.

3.2.3.3.3. Interpretation of 'interactions'

The "interactions" shown in Tables 3-4 and 3-5 can be exaggerations or reversals of the expected results, and can occur in terminal or interior cells of the table. Regard-

Table 3-5 "Internal valley" of reversal for effect of age and weight on 5-year survival rates of patients with "omphalosis"

Age	Weight			Total
	Thin	Medium	Fat	
Young	36/40 *(90%)*	32/40 *(80%)*	21/30 *(70%)*	89/110 *(81%)*
Middle	33/39 *(85%)*	6/30 *(20%)*	17/31 *(55%)*	56/100 *(56%)*
Old	20/27 *(74%)*	17/29 *(59%)*	17/54 *(31%)*	54/110 *(49%)*
Total	89/106 *(84%)*	55/99 *(56%)*	55/115 *(48%)*	199/320 *(62%)*

less of direction and location, the interactions are difficult to interpret. They may sometimes be statistical artifacts, but they may also have a distinct biologic meaning.

For example, in Table 3-4, it is possible that fat old people with omphalosis have a distinct biologic disadvantage in survival. Another explanation, however, is that prognostically unfavorable fat old people have been selectively referred to the institution where these statistics were collected. In Table 3-5, similar possibilities of biology or referral patterns may explain why middle-aged people of medium weight should have a particularly low survival.

Regardless of what the biologic or statistical explanations may be, the main point is that something odd is happening in both tables, that it needs further attention, and that it would not have been recognized without the multivariable analyses. In particular, the analyses contradict the conclusion that the variables consistently exert the effects anticipated from the bivariate marginal totals. In certain zones of the data, the effects may be substantially greater or smaller than what is otherwise expected.

3.2.4. Quantifying impact

As an independent variable goes through its range of values, its effect is quantified from the gradient of change in the dependent variable. The gradient can be shown by the slope of a regression line, or by changes in the summary values from one category to the next. If the group sizes are large enough for the gradients to be numerically stable, their magnitudes will directly reflect the statistical impact or importance of the independent variable.

Although a goal of the analysis is to quantify these impacts, the quantitative decisions often require that gradients be expressed in a standard manner, that the constancy be checked for gradients in algebraic models, and that the impact of additional variables be suitably evaluated or adjusted.

3.2.4.1. Standard magnitudes of gradients

In many situations, gradients are reported directly in the actual units of observation, with statements such as "Systolic blood pressure rises two points for each year of life after age 70." For more general comparisons, however, the quantitative magnitude can be expressed in a standard manner that is free of the units of measurements and/or coding. The standardized result allows impacts to be computed for diverse studies and diverse systems of coding.

For example, the slope of the same set of data will differ if *age* is expressed in years or months, *weight* in kilograms or pounds, and *severity of illness* in ordinal ratings of **1, 2, 3,** or **1, 2, 3, 4, 5**. To avoid the inconsistency produced by arbitrary units of measurement or coding, standardized expressions can be obtained by converting the independent and dependent ranked variables into "standardized Z scores", which can be calculated regardless of the measurement units or coding system used for each variable. The procedure is described in Section 7.5.4.

For categorical data that have been coded as **0/1**, or as **1, 2, 3, . . .** , a standardized expression may not always be necessary, but is sometimes used to allow consistency when gradients are cited for different variables within the same study or across studies.

3.2.4.2. Constancy of gradient

The use of a straight-line model, such as Y = a + bX, is based on the assumption that the gradient in Y is constant throughout the entire range of variable X. As shown in many examples of Chapter 2, however, this assumption may often be incorrect. Consequently, to avoid misleading conclusions from a rectilinear algebraic model, the pattern of data should be checked for constancy of gradient.

No specific statistical test has been developed for this purpose, but the examination is easily accomplished by checking the gradient in adjacent categorical zones of the data. If expressed in dimensions, rather than categories, the independent variable can be partitioned (as illustrated earlier in Fig. 2-20) into tertiles, quartiles, quintiles, or other zones that can be used to discern a trend. If the intercategorical gradients are reasonably equal, the overall gradient is essentially constant. If not, the inequalities demonstrate that the data do not have a rectilinear pattern, and that the slope of a straight line will offer a deceptive conclusion about the pattern.

If an independent variable has only two categories, the single gradient between them is always "constant." With three or more categories, however, the gradient will have a chance to vary. A nonconstant gradient was shown in the marginal totals for categories of *age* in Table 3-4. The overall gradient from the first to last category is 29% (= 80% − 51%), but the intercategorical gradients are unequal. They are 9% (= 80% − 71%) from the first to second category, and 20% (= 71% − 51%) from the second to the last. If an algebraic "linear" model shows that the average gradient is about 14.5% for these data, the real difference in gradients will be misrepresented.

A check for conformity between the data and the analytic model is an important part of the "regression diagnostics" to be discussed later. If the actual pattern of the data does not conform to the constant gradient indicated by the algebraic model, any quantitative conclusions about a variable's role in risk, prognosis, or other impacts may be correct on average, but wrong in many particulars. Unless the model was checked for its conformity to the data, a single quantitative statement may be misleading because the gradient may be much higher or lower in different zones of X.

3.2.4.3. 'Double gradient' effects

With algebraic methods, the concomitant effects of the multiple variables on one another can be suitably analyzed and adjusted (as noted later) before each variable's average multivariable impact appears as its "partial regression coefficient." With categorical analyses, either directly or as "gold standard" checks for the algebraic results, the approach begins by examining the effects of conjoined pairs of variables. For example, Table 3-1 contains four independent variables, of which one (educational background) seems ineffectual. The remaining three independent variables would be examined in 3 [= (3 × 2)/2] tables formed by conjoined pairs of the variables. One such pair was the conjunction of *anatomic stage* and *cell type* in Table 3-2; another pair was the conjunction of *age* and *anatomic stage* in Table 3-3.

When the trivariate data in these pairs of conjoined variables were examined earlier (Section 3.2.3) for decisions about distinctiveness, collinearity, and interactions, the double-gradient phenomenon in Table 3-2 was prominent evidence that each member of a pair of independent variables has a distinctive effect. In the rows of Table 3-2, anatomic stage creates a gradient within each category of cell type; and in the columns, cell type creates a gradient within each category of anatomic stage. Furthermore, the gradients produced by anatomic stage within each row are larger than the gradients produced within each column by cell type. The gradients in the two rows, respectively, are 70% (= 92% − 22%) and 44% (= 47% − 3%). The gradients in the two columns, respectively, are 45% (= 92% − 47%) and 19% (= 22% − 3%).

The double-gradient phenomenon indicates that both variables have distinctive individual impacts in the anticipated directions. In addition, the differential gradients within rows and columns indicate that anatomic stage has a greater prognostic effect than cell type.

In Table 3-3, the gradient for anatomic stage persisted within each age category, but the gradient for age vanished within the categories of anatomic stage. The absence of the double gradient helped indicate that age was an ineffectual variable, deriving its apparent effect from the collinear relationship with anatomic stage. With a mathematical analytic model for age and anatomic stage, as discussed later, the collinearity of age would be revealed either by a very small coefficient for its slope, or by its elimination during a "sequential" or "stepped" analysis.

Simultaneous gradients in conjoined categories for two variables must be examined for certain "judgmental" decisions in multivariable analysis, but are not displayed in the customary computer printout for most automated procedures. The examination can be separately arranged as a valuable way of "getting close" to the data and seeing what they really show.

3.2.5. Organization of estimates

After determining which independent variables are most important, the investigator will often use the results for diverse estimates. For example, the treatment in a clinical trial may be assigned with separate randomization schedules for patients previously stratified into groups with an estimated good, fair, or poor prognosis. The multivariable prognostic estimates can be organized and expressed as clustered categorical stages, as complex scores, or as simple scores.

3.2.5.1. Clustered categorical stages

The TNM system of staging for cancer[3] is a well-known example of clustered categorical stages. Each stage is itself a ranked ordinal category, with such labels as **I, II, III, IV,** . . . . Each ordinal category, however, is a composite cluster of categories derived from multiple variables, such as *size of primary tumor, fixation of pri-*

mary tumor, metastasis in regional lymph nodes, metastasis to distant sites, etc. The prognostic estimate depends on the survival rate (or other target variable) associated with each stage. Thus, a patient in **Stage II** might be estimated as having a 30% chance of 5-year survival.

3.2.5.2. Complex scores

For algebraic analyses, each of the multiple variables included in the model receives a "weighting" coefficient that multiplies its values. The sum of the products becomes a complex score, having such forms as 0.31 (Hematocrit) − 0.02 (Age) − .44 (Anatomic Stage) + .78 (Sex). The result of these complex scores for each person can then be used to estimate the probability of survival (or some other target event).

3.2.5.3. Simple scores

To avoid the intricate arithmetic of complex scores and the difficulty of remembering which categories are located in different stages, both types of expression are often converted into simple scores. They are usually constructed from easy-to-remember rating values or "points" (such as **0, 1, 2, etc.**) that are assigned to categories of the component variables, and then added. For this process, which will be discussed later in greater detail, dimensional variables are usually converted into demarcated ordinal (or binary) categories, and the weighting coefficients supplied by the algebraic analysis may guide the assignment of the rating points for each category.

The classic example of a simple scoring system is the *Apgar score,* constructed as a sum of three possible ratings (**0, 1, 2**) for each of five variables.[4] Two of the variables (heart rate and respiratory rate) are measured dimensionally, but are partitioned into three ordinal zones for the Apgar ratings. The other three variables (color, reflexes, and tone) are expressed directly in ordinal categories.

3.2.5.4. Accuracy of estimates

The accuracy of an estimate is almost always improved when effective independent variables (or categories) are identified during a multivariable analysis. For example, with only the total results available for the 200 patients in Table 3-1, a prediction of **death** would be correct in 112 (56%). With only the bivariate results for *cell type,* we would predict **alive** for the 140 patients with differentiated cell types, and **dead** for the 60 others. The prediction would be correct in 73 members of the first group and in 45 members of the second. The overall predictive accuracy in the group of 200 patients would rise to 59% (= 118/200). With only the bivariate results for *anatomic stage,* we would predict **alive** for the localized group, and **dead** for the spread group; and the predictions would be correct in 69 and 91 (= 110 − 19) respective members of the two groups. The overall accuracy of prediction would rise to 160/200 = 80%.

With the multivariable combinations in Table 3-2, the four groups in the interior cells would allow some highly accurate predictions. In the 60 patients with localized cancer and differentiated cell type, the prediction of **alive** would be correct in 53 or 92%. In the 30 patients with spread cancer and undifferentiated cell type, the prediction of **dead** would be correct in 29, or 97%. For the other two groups, whose survival rates are below 50%, the prediction of **dead** would be correct in 16 of the 30 people with localized cancer and undifferentiated cell type, and in 62 of the 80 people with spread cancer and differentiated cell type. The total number of correct predictions would be 55 + 29 + 16 + 62 = 162, so that the overall accuracy of prediction would be 162/200 = 81%.

Although the overall predictive accuracy would be raised only slightly (to 81%) from the 80% accuracy previously achieved with anatomic stage alone, the multivariate analysis has the advantage of identifying two specific categorical zones in which predictive accuracy would be higher than 90% for the corresponding groups of patients. The individual predictions for such patients could be made with more confidence about accuracy than for any of the groups produced in previous arrangements of the data.

3.2.6. 'Adjustments' of baseline disparity

In many medical analyses, the investigator wants to compare the effects of such maneuvers as therapeutic interventions or etiologic risk factors that can alter each person's baseline state. The comparisons will be unfair if the compared groups, before the maneuvers are imposed, have disparate prognostic susceptibility for the outcome event. For example, long-term survival rates for maneuver A and maneuver B would receive a biased comparison if members of group A are substantially older than members of group B. To achieve a fair comparison of maneuvers, the variables that produce these prognostic disparities in the baseline state can be 'adjusted' with several multivariable methods that are further discussed in Section 3.4.3.3.

3.3. Other differences from bivariate analysis

Multivariable analysis also differs from simple bivariate analysis in three other distinctions that involve the frequent focus on categories rather than variables, the general unimportance of configurations, and the occasional measurement of special interdependent correlations.

3.3.1. Categorization

In all the examples thus far in this chapter, the discussion was concerned with variables such as *anatomic stage, cell type, age,* and *weight.* The results were displayed, however, in tables rather than graphs, and the variables were expressed in categories, such as **localized, differentiated, old,** or **thin.**

One main reason for this type of display was intellectual convenience. The basic ideas are easier to show and discuss with simple tables of categories than with dimensional graphs of variables. A more important reason, however, is that identifying distinctive categories is often a more important analytic goal than revealing distinctive variables. For example, the results of a statistical analysis might be expressed as "Survival is related to weight;" "Survival decreases as weight increases;" or "Fat people are particularly likely to die." All three of these expressions are correct and proper, but the second is more informative than the first, because it indicates direction. The third statement, which demarcates the category of fat from among the different zones of *weight,* is particularly informative for scientific clinical decisions.

In Table 3-2, the results might be described with the comment that *cell type* and *anatomic stage* are distinctive individual variables in affecting survival. A more cogent description, however, is to state that almost everyone survives if the cancer is both **localized** *and* **differentiated,** but dies if the cancer is neither. In Tables 3-4 and 3-5, the data could all be described as showing an interaction between age and weight, but the statements would be more informative if they noted not only the general survival trends for each variable, but also the particular conjoint category in which the conjunctive alteration occurred.

For all these reasons, a frequent goal of multivariable analysis is not just to appraise variables, but to identify cogent categories (or zones) within a single variable, and cogent conjunctions of categories formed from two or more variables. Medical people tend to think about categories, such as **fat old men,** rather than weighted combinations of rated values for variables such as *weight, age,* and *sex.* Unlike the general term, *variable,* categories can indicate the zones where important things are happening, and will also identify the particular groups who are affected. Consequently, many acts of multivariable analysis are really aimed at multicategorical analysis, even if the analytic principles and language do not specifically acknowledge the categories.

3.3.2. Configuration

Although sometimes important in bivariate analysis, configuration itself is rarely a major goal in multivariable analysis. The patterns of points in multivariable data are difficult to visualize; the shape of the multivariate points or the associated mathematical model seldom has a specific biologic meaning (like the bivariate Starling's curve of cardiac function); and the mathematical configuration almost never fits the points well enough to be used as an accurate substitute.

The problem of visualizing a multivariable configuration is usually formidable. For two variables, things are easy. Their relationship is easily portrayed with the graphic plots shown in Chapter 2. For three variables, the relationship can be shown with a plane or other surface, but is much more difficult to draw on two-dimensional paper. Even if artistic ingenuity shows the points well on paper (or sometimes in three-dimensional structures), the fitted surfaces are seldom easy to envision or construct. If more than three variables are involved, the configuration becomes a "hyperplane" or "multidimensional surface" that is physically impossible to display. The basic mathematical idea is that each variable is "orthogonal," i.e., located on an axis that is perpendicular to the axes for all the other variables. This type of orthogonal arrangement can be physically constructed and displayed for the axes of three perpendicular variables, but is impossible for four or more.

When an algebraic model is applied to multivariable data, a configuration is arbitrarily imposed by the model rather than directly determined from the data. The arbitrary configuration may later be altered to produce a better "fit," but the alteration is almost always guided by mathematical strategies, not by biologic perceptions or implications.

Furthermore, if the analyst's goal is to get a perfect fit for the data, a multivariate estimate is almost never successful. If the "best" two or three variables do not produce a perfect fit, adding more variables will seldom help.

For all these reasons, the actual configuration of the relationship is almost never an important goal in multivariable analyses. The data analyst may want to compress variables, distinguish trends, or make estimates, but is seldom interested in the complex spatial shape of multivariate data. The only real attention to configuration occurs *after* an algebraic analysis, when regression diagnostics are used to check that the arbitrarily imposed algebraic shape has not distorted the effects of the independent variables. In the clustered-categorical analyses discussed later, no configurations are imposed or needed.

3.3.3. Multiple interdependent correlations

In all of the foregoing discussion of trends, effects, and categories, a set of independent variables was being evaluated for impact on a particular target variable. The process was oriented in a dependent many-to-one relationship. Before the analytic process begins, pairs of individual independent variables may often be examined for the trends shown in their *bivariate* correlations with one another.

In most multivariable analyses, however, no attempt is made to get a single expression for the trend of simultaneous correlation among *all* of the available variables. When a single "multivariable correlation coefficient" is calculated, it reflects the simultaneous effect of all the independent variables on the selected target variable, not the simultaneous interdependent correlation of all the available variables.

For the taxonomic compressions discussed in Section 3.1.3, however, there is no dependent variable. The collection of variables may be examined for their interdependent trends, but the process is intended to produce compressed combinations, not to identify *overall* trends. After the combinations are developed, however, certain trends may be checked for the constituent variables in each combination.

In an additional type of untargeted multivariable analysis that seldom occurs in medical research, the goal is to show overall trend among multiple interdependent variables. Such analyses are often done in psychometric activities to evaluate the intercorrelation or relative "homogeneity" among a set of multiple items in a survey questionnaire or educational test of competence. The multi-intercorrelation result is often expressed with statistical indexes such as *Cronbach's alpha* or the *Kuder-Richardson coefficient*.[5,6] This type of multivariate analysis is discussed elsewhere, and is beyond the scope of the current text.

To illustrate the distinctions, however, suppose we have data for five variables: age, weight, anatomic stage, cell type, and survival time. Conventional multivariable analyses would produce a multiple correlation coefficient for the simultaneous regression impact of the first four variables on survival. In taxonomic compressions, the first four variables might be expressed as two new variables or "factors." The first factor might contain a mixture of age, weight, and cell type; and the second factor might include a mixture of cell type and anatomic stage. In a multiple interdependent correlation, the simultaneous "internal" association among all five variables (or perhaps just the first four) would be cited with the Cronbach alpha coefficient.

3.4. Role in medical research

In using multivariable analysis, medical investigators can have any of the general or special goals that have just been described. When applied to real-world data, however, the multivariable goals are often given different names adapted to the specific function served by the analysis. Thus, an investigator compresses variables to develop a new rating scale (such as the Apgar score), examines trends to eliminate unimportant variables, such as *age* in the data of Table 3-3, and makes estimates to prepare a scoring or staging system for prognosis, such as the predictions achieved from Table 3-2.

A particularly valuable role of multivariable analysis is in appraising the results of interventional maneuvers,[7] such as etiologic agents that may cause disease or therapeutic agents used to treat disease. Because these maneuvers are imposed on people whose underlying baseline conditions can substantially affect the subsequent outcomes, multivariable analyses help clarify the separate effects that baseline conditions and maneuvers may have on the outcomes. The analytic aims may be to identify therapeutic distinctions in pertinent clinical subgroups, to adjust (or standardize) for possible imbalances or biases in the data, and to increase efficiency in the research plan.

The rest of this chapter illustrates the way in which the general aims of multivariable analysis are used for the pragmatic goals of evaluating interventional maneuvers.

3.4.1. Interventional maneuvers, baselines, and outcomes

In any analysis of a cause-effect relationship for interventional maneuvers, three sets of variables are involved. One set describes the interventional maneuvers (or agents) themselves. Therapeutic agents include pharmaceutical, surgical, or other forms of remedial treatment, as well as prophylactic therapy that raises antibody levels or lowers lipids, blood pressure, or blood sugar. Etiological agents include smoking, diet, occupation, chemical exposures, or other risk factors.

A second set of variables refers to the baseline condition of the people receiving the maneuvers. The baseline conditions can be described with variables that are *demographic* (age, sex, ethnic background), *genetic* (familial longevity or hereditary diseases), *clinical* (patterns of symptoms, severity of illness), or *paraclinical* (laboratory and imaging tests, electrocardiograms).

The third set of variables refers to the particular outcomes or consequences that are examined as the presumed effects of the interventional maneuvers. These target variables can be changes in blood pressure, duration of survival, development of a particular disease, relief of symptoms, or occurrence of an event regarded as "success."

(Psychosocial descriptions of mood, behavioral states or traits, and social relationships can appear in any or all three roles as baseline conditions, interventional maneuvers, or outcomes. For example, an investigator might ask whether a change in social environment [intervention] favorably affects the associated mood [outcome] of someone with chronic anger and hostility [baseline state].)

In an unfortunate custom, the results of interventional maneuvers are often reported merely as a summary of data for the outcomes and the maneuvers, ignoring crucial distinctions in the baseline conditions. Thus, post-therapeutic success rates are often cited merely as 80% for treatment A vs. 50% for treatment B; the occurrence rates of a particular disease might be listed only as .071 in persons exposed to agent C and .032 in persons not so exposed; mortality rates might be compared for hospital D vs. hospital E, or for region F vs. region G. To be properly evaluated, however, the maneuvers must be compared in groups of recipients who had suitably similar baseline conditions. Afterward, because the therapeutic or etiologic agents may have different effects in groups with different baseline conditions, the distinctions must be appropriately cited or adjusted when the analytic results are applied in the future.

Whenever the baseline conditions become considered in addition to the maneuvers and outcomes, the analysis immediately becomes multivariate, since at least three sets of variables are involved.

3.4.2. Formation of groups

The variables under analysis can be expressed in dimensional measurements, such as *amount of ingested alcohol, dose of medication,* and *blood pressure,* or in categorical groups, such as **ex-smoker, modified radical mastectomy,** and **young men.** For many medical decisions, and for the illustrations in this chapter, the data are expressed in categorical groups.

A group of people can be formed statistically as a clustered category or a demarcated score. The clustered category can be simple, such as *men,* or a composite combination of individual categories, such as *thin young men,* or *students who attend Arizona, Yale, or Iowa State Universities.* The term *cluster* usually refers to categorical attributes that have been joined via the "unions" and "intersections" of Boolean logic. In the two examples just cited, the first composite cluster, "thin young men," contains a group of people having the Boolean-logic intersection of categories for thin *and* young *and* male. The second composite cluster was formed when the union of categories for three universities—*Arizona, Yale,* and/or *Iowa State*—was intersected with the category of *student.*

Groups are formed by a demarcated score when age is divided into zones such as ≤ **40, 41–70,** and ≥ **71,** or when these (or analogous) zones are called **young, middle,** and **old.** The sum of ratings that produces the Apgar score, which ranges from **0** to **10,** could be partitioned to form groups of babies who have the individual scores of . . . **3, 4, 5, 6,** . . . , or scores in the zones of **9–10, 6–8, 4–5,** and ≤ **3.**

These two mechanisms for forming groups—from categorical clusters or scores—also appear as fundamental distinctions in methods of multivariable analysis. All of the analytic strategies produce either clusters or scores. The multivariate pattern that classifies a cancer's anatomic stage as **I, II, III,** or **IV** (in the TNM system) is formed with categorical clusters. The multivariate pattern that expresses cardiac prognosis in the Norris index[8] as $3.9\ X_1 + 2.8\ X_2 + 10\ X_3 + 1.5\ X_4 + 3.3\ X_5 + 0.4\ X_6$ is an additive score, analogous to the Apgar score.

3.4.3. Effects of maneuvers

In the illustrations that follow, multivariable analyses are used to detect distinctive subgroup effects, to avoid deception from disparate effects in subgroups, to reveal distortions caused by susceptibility bias, to improve efficiency of research design, and to allow precision in future applications. The cited examples all refer to effects of therapy as the interventional maneuver, but counterpart examples could easily be given for etiologic agents.

3.4.3.1. Distinctive subgroup effects

Clinicians constantly evaluate the outcome of treatment not just for a particular disease, such as *diabetes mellitus, coronary disease, hypertension, cancer,* or *meningitis,* but for distinctive categorical subgroups of patients designated with such titles as *brittle diabetes mellitus, unstable angina pectoris, refractory hypertension, Stage III cancer,* or *comatose meningitis.* The analysis becomes multivariate when the outcomes are evaluated simultaneously for treatments within subgroups having different baseline conditions.

To illustrate the procedure, suppose that patients with a particular disease can be divided into two subgroups, having *good clinical condition* or *poor clinical condition* before treatment. Now consider the success results for two treatments, A and B, that are used for this disease. Treatment A had a success rate of 38% in 100 patients, and treatment B was successful in 62% of 100 patients. Before concluding that treatment B is definitely better than A, we examine the results in the two clinical subgroups shown in Table 3-6.

Table 3-6 Success rates for two treatments

Clinical Condition	Treatment A	Treatment B	Total
Good	30/50 (60%)	40/50 (80%)	70/100 (70%)
Poor	8/50 (16%)	22/50 (44%)	30/100 (30%)
Total	38/100 (38%)	62/100 (62%)	100/200 (100%)

The data for the subgroups confirm the distinctions found in the totals: treatment B was better in both clinical subgroups. (The confirmatory result is statistically analogous to what was previously noted in Table 3-2.)

On the other hand, for the same marginal totals of Table 3-6, suppose the cells of the subgroups had the data shown in Table 3-7. These subgroup results are very different from before. Table 3-7 shows a conjunctive effect (or "interaction") for treatment A in the poor clinical subgroup, where the results are surprisingly better than those of treatment B. Therefore, despite the overall superiority of treatment B, treatment A might be preferred for patients in poor clinical condition.

Table 3-7 Success rates for two treatments with conjunctive subgroup effects

Clinical Condition	Treatment A	Treatment B	Total
Good	16/40 (40%)	54/60 (90%)	70/100 (70%)
Poor	22/60 (37%)	8/40 (20%)	30/100 (30%)
Total	38/100 (38%)	62/100 (62%)	100/200 (100%)

3.4.3.2. Deception from disparate effects in subgroups

If two treatments seem to have similar total effects, a multivariable examination of clinical subgroups can be particularly valuable. For example, at the end of a randomized clinical trial, suppose treatments C and D each achieved success rates of 44% in 100 patients. Since the 95% confidence interval for this zero increment would show that neither treatment is more than 10% better than the other, the investigators might conclude that the two treatments are similar. Among individual clinical subgroups, however, this conclusion might be dramatically wrong.

Table 3-8 Success rates for two treatments with disparate subgroup effects

Clinical Condition	Treatment C	Treatment D	Total
Good	38/40 (95%)	26/40 (65%)	64/80 (80%)
Poor	6/60 (10%)	18/60 (30%)	24/120 (20%)
Total	44/100 (44%)	44/100 (44%)	88/200 (44%)

As shown in the cells of Table 3-8, the total results for the two treatments were similar only because they acted in opposite directions, having strikingly disparate effects in the subgroups. Treatment D was much better than C for the patients in poor condition, but treatment C was much better than D for patients in good condition. The distinctive but opposite effects in the subgroups were canceled in the total results.

Collinear relationships or conjunctive interactions cannot be held responsible for this "deception." The treatments were equally distributed among the subgroups; and the total results for each treatment do not evoke any expectations about a direction for the differences. The two treatments had similar total results because of opposite effects in the component clinical subgroups. (Table 3-8 will be discussed later in the text when we consider mathematical models for determining whether an interaction occurred between treatments and clinical conditions.

Strictly speaking, however, the deception is not really due to an interaction, because the pattern of results for conditions within treatments is consistent with what might be anticipated from the marginal totals: each treatment had better results in the *good* group than in the *poor* group. The marginal totals would also suggest similar effects for treatments within conditions, but the results in the cells show a reversal of therapeutic effects, rather than an interaction in the *good* and *poor* groups.)

These three different kinds of effects—superiority in all subgroups, a conjunctive action in a subgroup, and deceptive similarity due to opposing effects in subgroups—would not have been distinguished without the multivariable analyses.

3.4.3.3. Adjustment for susceptibility bias

In Tables 3-6 and 3-8, the clinical subgroups of patients were equally distributed in each of the compared treatments before treatment was given. This type of baseline equality commonly occurs if the treatments were assigned with randomization. If the treatments were not randomly assigned, however, the compared groups may have substantial baseline disparities, produced by the clinical (or other) judgments with which the treatments were selected. Treatment A may be chosen mainly for the prognostically good patients, and treatment B may be given mainly to the poor group. The subsequent comparison of treatment is then distorted by the baseline disparity that is called *susceptibility bias*.[7]

For example, suppose the 5-year survival rates in a *nonrandomized* comparison are 52% (= 52/100) for treatment E and 36% (= 36/100) for treatment F. Before concluding that treatment E is superior to F, we examine their results in clinical subgroups with good and poor prognosis. In the cells of Table 3-9, the data show that the two treatments had similar results for patients in the same prognostic subgroups. The apparent superiority of treatment E arose from a baseline prognostic disparity: the good prognostic group was the source for 52 of the 100 patients receiving treatment E, but only for 28 of the 100 patients in treatment F. The imbalance produced a baseline susceptibility bias that led to the misleading results afterward. (The problem is statistically analogous to the collinearity noted earlier in Table 3-3.)

Table 3-9 5-year survival rates for two treatments assigned with susceptibility bias

Prognostic Condition	Treatment E	Treatment F	Total
Good	42/52 (81%)	22/28 (79%)	64/80 (80%)
Poor	10/48 (21%)	14/72 (19%)	24/120 (20%)
Total	(52/100 (52%)	36/100 (36%)	88/200 (44%)

Susceptibility bias is particularly common in nonrandomized comparisons of treatment, but can also arise, due to the "luck of the draw," in randomized trials.

The need to seek and adjust for this (and other types of) bias is another important reason for using suitable forms of multivariate analysis in the comparative evaluation of therapeutic or etiologic maneuvers. Thus, Table 3-9 shows how susceptibility bias can be both detected and adjusted by comparing therapeutic results in patients with similar prognostic susceptibility.

3.4.3.4. Mechanisms of adjustment

Several mechanisms can be used for the multivariate adjustment process. They include stratification, standardization, and algebraic coefficients.

3.4.3.4.1. Stratification

In Table 3-9, the baseline conditions are divided or "stratified" into good and poor prognostic groups. The therapeutic results can then be directly examined and compared in each of the pertinent groups. In this instance, the groups were identified as the clustered categorical stages (see Section 3.2.5.1) that were labeled prognostically good and poor.

In other instances, the groups can be formed from persons having the same score or zones of scores (see Sections 3.2.5.2 and 3.2.5.3). Regardless of how the multivariate groups are formed, however, the maneuvers are compared within members of each group.

3.4.3.4.2. Standardization

Although stratification offers precise results for the maneuvers' effects in each pertinent group, the investigator may prefer to compare a single "adjusted" value for each maneuver, rather than examining results in several stratified groups. This adjustment can produced by an epidemiologic process called *standardization* or *standardized adjustment*. Details of the process, which can be done with "direct" or "indirect" methods, are presented elsewhere.[7, 9-11] In a simple example here, a "direct standardization" for the data in Table 3-9 can be achieved from the sum of products obtained when the survival rate in each prognostic stratum is multiplied by the proportion of that stratum in the total cohort. The direct-adjusted survival rates here would indicate that the two treatments have essentially similar results, which are (42/52) (80/200) + (10/48) (120/200) = 44.8% for treatment E, and (22/28) (80/200) + (14/72) (120/200) = 43.1% for treatment F. Standardization is thus a mechanism that converts each set of stratified results into a single adjusted result for each treatment.

3.4.3.4.3. Algebraic coefficients

In the multivariable algebraic methods that are discussed in Parts III and IV of the text, an adjustment is achieved by including the maneuvers as a separate independent "dummy" variable along with the other independent variables that characterize the baseline condition. The coefficient that emerges for the dummy variable will indicate the effect of the treatment (or other maneuver), adjusted for the simultaneous effect of all the baseline "covariates."

The mechanism of this adjustment is discussed later in the text. Although statistically popular and frequently used today, the algebraic adjustment has two scientific disadvantages: (1) the single "overall" coefficient for the maneuvers may obscure their different or opposite effects in individual clinical subgroups; and (2) data for the baseline state and the maneuvers are analyzed as though they were "concurrent" variables, although the maneuver is always imposed afterward on a baseline state that exists beforehand. The problems produced by these disadvantages will be discussed in Part IV.

3.4.3.5. Improved efficiency of design

A problem quite different from susceptibility bias can arise from unrecognized inefficiency in planning a comparison of therapy.

Suppose a new treatment, G, has been compared against placebo in a randomized trial of 240 patients with a particular illness. The total results show that treatment G was better, as expected. Its success rate was 60/120 *(50%)*, compared with 48/120 *(40%)* for placebo. When the results are statistically tested, however, the investigators are chagrined to find they did *not* achieve "statistical significance" ($\chi^2 = 2.42$; $P > .05$), despite the relatively large number of patients admitted to the trial. Although a larger sample size might have yielded statistical significance, the investigators are particularly distressed at not getting a conclusive result after all the time, effort, and expense that went into the research.

On the other hand, if the patients were classified at baseline according to mild, moderate, or severe clinical states of the treated illness, the multivariable results would show the interesting data of Table 3-10. Patients with mild illness almost all had successful outcomes, and patients with severe illness were almost all unsuccessful, regardless of treatment. The main therapeutic differences occurred in the group with moderate illness, where the success rates were 42/70 *(60%)* for treatment G and 28/70 *(40%)* for placebo. In this clinical subgroup, treatment G was distinctively better than placebo, and the difference is also "statistically significant" ($\chi^2 = 5.6$; $P < .025$).

Table 3-10 Success rates according to severity of illness and treatment in a trial with inefficient design

Severity of Clinical Illness	Treatment		Total
	G	Placebo	
Mild	17/18 *(94%)*	19/20 *(95%)*	36/38 *(95%)*
Moderate	42/70 *(60%)*	28/70 *(40%)*	70/140 *(50%)*
Severe	1/32 *(3%)*	1/30 *(3%)*	2/62 *(3%)*
Total	60/120 *(50%)*	48/120 *(40%)*	108/240 *(45%)*

The multivariable analysis thus shows that the trial was not efficiently designed. Patients with the extremes of either mild or severe illness should have been excluded, since they had uniformly successful or unsuccessful results, regardless of treatment. If the trial had been confined to patients with moderate illness, the value of treatment G could have been demonstrated with both statistical significance and with a substantially smaller number of patients.

3.4.3.6. Precision in future applications

Although all the foregoing analyses were used to evaluate effects of the compared treatments, the results are also important for future clinical application. To choose appropriate treatment, a clinician would want to know results for the patient's pertinent subgroup, not just the overall results for the total treatment.

The multivariate analyses are valuable, therefore, not only for appraising past results, but also for adding precision in future application of the findings.

3.5. Problems in 'confounding' variables

In section 3.4.3.3, the susceptibility denoted by good or poor prognosis was a confounding variable, being strongly correlated with both the outcome event and the selected maneuver. If undetected, the susceptibility bias would lead to a false conclusion about the effects of the compared therapeutic (or etiologic) agents. The effects of variables suspected of having analogous confounding roles are frequently detected and adjusted by multivariable analysis.

3.5.1. Absence of true confounding

The analytic procedures are particularly useful if the chosen baseline variables indeed have a confounding effect. If they do not actually have a distinctive impact on data for both the maneuver and the outcome event, however, their multivariate adjustment may be a mathematically impressive but scientifically futile activity. For example, if patients with cancer are in a similar stage of anatomic extensiveness and clinical severity, their *prognosis* is usually unaffected by such demographic variables as race, occupation, and educational status. Nevertheless, different treatments are sometimes evaluated with a multivariable analysis that makes provision for all the ineffectual demographic variables that are called "confounders," while omitting the truly confounding role of clinical severity. To be more than a theoretical confounder, the variable must first be shown to exert suitable confounding effects.

3.5.2. Location in causal pathway

Another problem in confounding variables is their location in the causal pathway between maneuvers and outcome events. Anatomic stage, clinical severity, demographic features, and other variables that occur *before* treatment can readily be checked for their susceptibility bias in affecting both outcome and choices of treatment. Variables that reflect events occurring *after* the main treatment (or other "causal" agent) is initiated, however, many represent phenomena that are not suitably adjusted merely by analyzing the variables themselves. For example, when a patient with cancer receives chemotherapy after surgery, the chemotherapy may have been given deliberately as part of a routine plan to follow the surgery immediately with "adjuvant" treatment. Alternatively, however, the chemotherapy may have been chosen by ad hoc clinical decision when the patient was later found to have widespread metastases. A multivariate adjustment that considers the chemotherapy alone, without regard to the possibly altered clinical condition that led to the chemotherapy, will also yield spurious results that do not really cope with the main source of bias.

3.6. Omission of important variables

In Tables 3-6 through 3-10, the analyses relied on the assumption that patients could readily be classified into clinically important prognostic categories, such as good and poor. In clinical reality, however, the most important predictive attributes may be unknown or unspecified.

Sometimes the key data are missing because the crucial characteristics are unknown. For example, while identifying certain persons who are more susceptible to cancer than other people, many risk factors may have very low predictive discrimination. If a particular cancer occurs in .005 (five per thousand) of people who have a designated risk factor, and in .001 of people who lack that factor, the risk ratio of 5 (= .005/.001) is impressive. Nevertheless, the risk factor is ineffectual for individual predictions. The cancer will not occur in 995 of 1000 people with that factor. Thus, despite the statistical value of certain risk factors in assessing possible predispositions to cancer, the really effective predictors may be unknown.

In other circumstances, the important prognostic attributes may be known or suspected, but may not be specifically identified and coded for analysis. For example, observant clinicians know that survival in cancer is substantially reduced for patients who have lost a great deal of weight, or who have become constantly fatigued, or who have major difficulties in daily function. Nevertheless, this "soft" information is often excluded from the analytic variables, which are usually confined to such "hard" data as anatomic stage, cell type, and demographic features.[12,13]

The omission or lack of crucial predictive information is an important, frequent problem, no matter what multivariable method is used. The results that emerge from the analysis may seem impressive but may be highly misleading if crucial variables are missing. Warnings about this problem are constantly offered in mathematical discussions[14-19] of multivariable methods, but the warnings are often neglected. Impressed by the apparent mathematical accomplishments in achieving diverse quantitative desiderata, the data analysts may forget that the achievements are often meager. Although "statistically significant," the results may be substantively unimportant and possibly deceptive, because the really powerful variables have not been identified or included in the analysis.

Chapter References

1. Yerushalmy, 1947; 2. Wasson, 1985; 3. American Joint Committee on Cancer, 1988; 4. Apgar, 1953; 5. Nunally, 1978; 6. Wright, 1992; 7. Feinstein, 1985, (*Clinical epidemiology*);

8. Norris, 1969; 9. Chan, 1988; 10. Armitage, 1971; 11. Kahn, 1989; 12. Feinstein, 1987, pp. 93, 169–70; 13. Feinstein, 1985, (Arch. Intern. Med.); 14. Gordon, 1974; 15. Breslow, 1980 and 1987; 16. Schlesselman, 1982; 17. Glantz, 1990; 18. Kleinbaum, 1988; 19. Kelsey, 1986.

4 Taxonomy of Multivariable Methods

Outline . . . .

4.1. Orientation of arrangement
    4.1.1. Targeted (dependent) orientation
        4.1.1.1. Multiple target variables
        4.1.1.2. Single target variable
    4.1.2. Nontargeted (nondependent) orientation

4.2. Patterns of combination
    4.2.1. Formation of mathematical scores
    4.2.2. Formation of categorical clusters
    4.2.3. Strategies of reduction
        4.2.3.1. Algebraic structures for scores
        4.2.3.2. Decision rules for clusters
    4.2.4. Other analytic goals
        4.2.4.1. Configuration
        4.2.4.2. Compression
        4.2.4.3. Estimation
        4.2.4.4. Trends

4.3. Constraints of orientation and data
    4.3.1. Constraints of orientation
    4.3.2. Constraints of target variable
        4.3.2.1. Categorical methods
        4.3.2.2. Algebraic methods
    4.3.3. Constraints of independent variables

4.4. Basic taxonomy of multivariable analysis

4.5. Focus of subsequent text

4.6. Illustration of decision rules for clustered categories
    4.6.1. Example of a nontargeted decision rule
    4.6.2. Construction of a targeted decision rule
        4.6.2.1. Statistical guidelines
        4.6.2.2. Biological guidelines
            4.6.2.2.1. Rules of coherence for two ordinal variables
            4.6.2.2.2. Data set for rectal cancer
        4.6.2.3. Bivariate arrangements
        4.6.2.4. Multivariable results
        4.6.2.5. Reduction of cells

4.7. Illustration of algebraic models for mathematical scores
    4.7.1. Bivariate results
        4.7.1.1. Linear regression
        4.7.1.2. Logistic regression
        4.7.1.3. Comparison of bivariate results
    4.7.2. Multivariate results
        4.7.2.1. Linear regression
        4.7.2.2. Logistic regression

4.7.3. Comparison of multivariate results
4.8. Choice of targeted analytic patterns

....

Having considered the *why* and *what* of multivariable analysis, we can turn to the *how*. Because of the diverse goals, data, and orientations, so many multivariable procedures have been developed that just a list of their names can be frightening. With a few distinctions that have already been cited, however, the complexity can be simplified, and the methods can be classified into four main groups that are discussed and illustrated in this chapter.

Much of the apparent complexity arises not from major differences in each method, but from different options for playing the same basic theme. The existence of diverse multivariable options should not be surprising if you recall the plethora of choices available in elementary statistics as a simple index to summarize the "average" value or "central tendency" for a single variable such as age. The three best known indexes are the *mean,* the *median,* and the *mode,* but you may also have met a *geometric mean* or a *log mean.* If you have been widely adventurous, you may have met a *harmonic mean,* a *tri-mean,* a *trimmed mean,* or a *Winsorized mean.* If you wanted to avoid all those means, you could demarcate age into **young** and **old** categories, which would be summarized as the binary *proportion* for either category. After realizing that the "average" value of a single variable can be expressed in at least ten ways, you may be less dismayed by the diversity of multivariable methods.

The diversity can be organized into manageable segments by classifying the methods according to three basic distinctions that arise from the orientation, pattern of combinations, and types of data being analyzed. These three distinctions are briefly discussed and illustrated here, before being used as the main axes of classification for multivariable methods.

4.1. Orientation of arrangement

Multivariable analyses can be oriented according to the same *dependent* and *nondependent* (or *targeted* and *nontargeted*) directions that were discussed for bivariate analysis.

4.1.1. Targeted (dependent) orientation

In a targeted or dependent orientation, one or more variables is the external focus of attention; and the independent variables are examined for their relationship to this target. In Fig. 1-1, the *many-to-one* relationship was aimed at a single target variable, and the *many-to-many* relationship was aimed at several.

4.1.1.1. Multiple target variables

Because the results are difficult to interpret for multiple target variables, many-to-many relationships are hardly ever analyzed in medical (or most other forms of scientific) research.

For example, suppose three target variables—such as survival, costs of care, and patient satisfaction—are outcome events to be examined as simultaneous targets of a predictive analysis for such independent variables as age, sex, anatomic stage, cell type, and symptom severity. Most investigators would not want the three target variables to be combined according to arbitrary decisions made by an automated process. Instead, the investigators would usually analyze the separate effect of the independent variables on each outcome variable—survival, costs, and satisfaction—one at a time. If someone insisted on combining the three outcome variables, they would probably be first compressed judgmentally into a single composite variable. In such an arrangement, the composite combination of survival, costs, and satisfaction would form a single *outcome,* which might then be either given a set of suitably selected point scores or rated as **excellent, good, fair,** or **poor.** The composite single variable would then become the target of a *many-to-one* analysis.

Multivariable algebraic methods called *canonical analysis* and *multivariate analysis of variance* are available for examining many-to-many relationships, but the methods are used so rarely in medical research that they will receive no further consideration until a brief mention in Section 22.1.4. All of the targeted multivariable relationships under discussion will be the dependent *many-to-one* arrangements.

4.1.1.2. Single target variable

A many-to-one dependent orientation was used throughout Chapter 3, where each target was the presence or absence of a particular outcome event, such as **survival at 3 years** or **success** after treatment. The independent variables, such as *cell type, anatomic extensiveness, age, weight, clinical condition, prognostic status,* or *therapy,* were examined for their relationship to this target, with the results expressed as binary proportions for success or survival.

Although the mathematical target is called a *dependent variable,* the term *independent* can be a confusing label for the other variables, because they may sometimes be closely associated or have special conjunctive interrelationships with one another. Nevertheless, the other variables are called *independent* because they are explored for their relationship to the dependent (target) variable.

In scientific parlance, the independent variables are often called *predictor* variables, even though the "prediction" may aim at a concomitant phenomenon, such as the current diagnosis of disease, rather than a future forecast, such as survival. In both scientific and mathematical parlances, the independent variables can also be called *explanatory.* Scientifically, they help identify phenomena that may "explain" what happens in the target variable; statistically, they help reduce or "explain" certain mathematical phenomena, such as error rates or group variance, in the results.

4.1.2. Nontargeted (nondependent) orientation

In a nontargeted or nondependent orientation, as shown earlier in the right-hand part of Fig. 1-1, a target variable is not specifically chosen or identified. The many-internal relationships are examined among two or more variables, without being specifically "aimed" at an external target.

In the absence of an external target, the orientation is *nondependent*, but the goal is often a search for *interdependence* in the way that all the available variables, or selected subsets, relate to one another. A high degree of interdependence can suggest that members of the selected group of variables are reasonably "homogeneous" in measuring a particular entity, called a "construct," which is identified by their combination. The nondependent multivariable procedure can produce the type of compression noted in Sections 2.3.3 and 3.1.3, or the measurement of homogeneity noted in Section 3.3.3.

At other times, however, the goal is to find and combine variables that are not specifically homogeneous or interdependent, and that will each make a distinctive contribution to their combined effect. This approach was used by Virginia Apgar[1] when she created the multivariate score that commemorates her. The Apgar score for the construct called *condition of a newborn baby* was *not* specifically designed to predict a target event, such as an immediate need for ventilator therapy or the baby's future development at one year of age. Many other multivariate scores in medicine have been developed in a nontargeted manner to grade the severity of clinical illnesses.[2] Despite the original nontargeted construction, however, the scores have often been subsequently used to predict a target such as survival, costs, or other outcome events.

Probably the best-known nontargeted multivariate procedure is called *factor analysis*, which is used to compress the original variables. After various assessments of interdependence, the available variables are aggregated into combinations that form new variables, called *factors*. The total number of factors is chosen to be smaller than the original number of variables, while preserving and perhaps illuminating the most useful information contained in the original variables.

Thus, the three variables X_1, X_2, and X_3 might be converted into two factors, V_1 and V_2, expressed with appropriate coefficients as

$$V_1 = a_0 + a_1 X_1 + a_2 X_2 + a_3 X_3, \text{ and}$$
$$V_2 = b_0 + b_1 X_1 + b_2 X_2 + b_3 X_3.$$

The Apgar score offers an informal, nonmathematical illustration of this type of factor analysis. Apgar began by considering the many different variables that might describe a newborn baby. Using clinical judgment, she then eliminated all but five of them: heart rate, respiratory rate, color, muscular tone, and reflex responses. She gave each retained variable a rating of **0, 1,** or **2;** and the sum of the five ratings forms the Apgar score. The process resembles a factor analysis that yielded a *single* factor,

$$V = a_0 + a_1 X_1 + a_2 X_2 + \ldots + a_5 X_5.$$

In the single Apgarian factor, $a_0 = \mathbf{0}$; each value of $a_1, a_2, \ldots, a_5$ is **1**; and each value of $X_1, \ldots, X_5$ can range from **0** to **2.**

4.2. Patterns of combination

Regardless of orientation, multiple variables can be arranged according to one of two main patterns of combination that have already been mentioned: mathematical scores or categorical clusters.

4.2.1. Formation of mathematical scores

The Apgar rating is an example of a mathematical algebraic score. In the customary scoring pattern, each variable (marked as X_1, X_2, X_3, . . .) has its own mathematical format of original or transformed citation, such as X_1, X_2^2, $\sqrt{X_3}$, etc. (In the Apgar example, the variables for *heart rate* and *respiratory rate* were demarcated into three ordinal categories, coded as **0, 1, 2,** but no special transformations were used.) Each expressed variable has a weighting coefficient, such as the a_1, a_2, a_3, . . . of the foregoing example. The weighted expressions of the variables are then mathematically combined, usually by addition, as in the example $a_1X_1 + a_2X_2 + a_3X_3 + \ldots$. The score is formed when the mathematical operations of multiplication and addition are carried out after the actual values of each variable for each person—as **0, 1, 2** or **93.7,–.09**, etc.—are inserted into the total expression.

4.2.2. Formation of categorical clusters

The TNM staging system for cancer[3] is an example of categorical clusters. Each variable is initially expressed in a categorical scale, and the categories are then combined in a specified manner.

In the current TNM convention, the T variable (for primary tumor) can be cited in a scale containing six ordinal categories designated as **T_0, T_1, T_2, T_3, T_4,** or **T_5**. The N variable (for regional lymph nodes) is cited as **N_0, N_1,** or **N_2**. The M variable (for distant metastases) is cited as **M_0, M_1,** or **M_2**. When the categorical values for these three variables are conjoined in tandem expressions such as **$T_3N_1M_0$** or **$T_1N_0M_2$**, a total of 54 (= 6 × 3 × 3) conjoined categories or "cells" can be formed. In this instance, each cell will be trivariate, containing conjoined categories from each of the three main T, N, and M variables.

In a more general instance, with k variables, each original cell would have conjoined categories from k components. The total number of possible cells will be the product of the number of categories for each of the component variables. For example, since each of the five Apgar variables can be cited in three categorical ratings of **0, 1, 2,** a total of 3 × 3 × 3 × 3 × 3 = 243 cells would be available if we wanted to examine multivariate Apgar categories, rather than add ratings to form an Apgar score.

The large potential array of multicategorical cells can be reduced to a more manageable and useful collection by applying a *decision-rule* strategy. For the TNM categories (as illustrated later), the rule combines the cells into three main clusters, called **Stages I, II,** and **III**. Unlike mathematical scores, categorical clusters do not have specific weighting coefficients for each variable. Instead, the relative effects of the variables are considered in diverse ways during the analysis, and the relative weights of the clusters are shown by their rank in the ordinal collection of clustered categories. Thus, in the TNM system, Stage I is regarded as a "better" prognostic condition than Stage II, which is better than Stage III.

4.2.3. Strategies of reduction

When multiple variables are available, an additional analytical goal, as noted in Section 3.2.2, is to reduce and simplify the multivariate complexity.

One type of reduction is aimed at eliminating variables by excluding those that are "unimportant," and then forming the final scores or clusters from the "important" variables that are retained. Thus, from the many variables that can characterize the condition of a newborn baby, only five were retained for inclusion in the Apgar score. From the many attributes that can characterize a patient with cancer, only three main variables are retained in TNM staging.

A second type of reduction is aimed at condensing or consolidating the retained multivariate components, which consist of individual variables or categories. They are reduced when combined into new expressions that form the "final" scores or clusters. In the Apgar procedure, the values for each variable become multidimensional components that are added to produce a single score. In the TNM procedure, the multiple categories are condensed into a single composite variable whose scale has three ordinal clusters, called "stages."

The judgmental reductions that led to the Apgar score or TNM staging were done without a standard "formal" strategy. Each multivariable arrangement was reduced according to the scientific "common sense," "intuition," or "judgment" derived from observational experience and substantive knowledge of the phenomena under consideration. When a formal strategy is used, however, a standard plan is chosen and applied. A major difference between the score and cluster procedure is in the formal analytic strategy used for these plans.

4.2.3.1. Algebraic structures for scores

In the score procedures, an algebraic structure or "model" is chosen as an appropriate mathematical configuration for the data. This algebraic structure, which can have a multivariable "linear" or other format, is used as the underlying basis for both reductions, intended to eliminate unimportant variables and to combine those that are retained. Consequently, the configuration of the selected algebraic format dominates the subsequent analyses: variables are retained or eliminated according to the way they affect the fit of the chosen configuration; and the retained variables are eventually combined in the form of the selected algebraic expression.

4.2.3.2. Decision rules for clusters

The cluster procedures, on the other hand, have no underlying mathematical configuration or format. The two sets of reductions are accomplished with a variety of mechanisms that differ for each multivariable procedure. In some instances (as illustrated by TNM staging), apparently unimportant variables are specifically identified and eliminated before the categories of the retained variables are aggregated. In other instances, the search is aimed directly at finding important clusters of categories; and the unimportant variables become excluded afterward if they did not participate in any of the important clusters. Regardless of which approach is used, the cluster procedures rely on a sequence of different strategies, rather than a single basic mathematical configuration. The sequential operation of each cluster strategy is cited as an *algorithm* or *decision rule.*

4.2.4. Other analytic goals

The other main goals of multivariable analysis are achieved before, after, or during the reduction processes.

4.2.4.1. Configuration

As noted previously, configuration is not an important goal of multivariable analysis. In the cluster procedures, no configuration is used or sought. In the score procedures, a configuration is essential, but is chosen *before* the analysis begins. The selected configuration may receive various alterations during or after the analysis, but the changes are aimed at an improved mathematical fit for the variables, not at better substantive understanding of their true relationships. The pattern of combinations that produce multivariable scores does not *reveal* configuration; the pattern is adjusted to fit a configuration that is either chosen before the analysis begins or modified afterward.

4.2.4.2. Compression

In nondependent orientations, compressions are produced by the combinations formed from the retained variables or categories. With a categorical procedure, the combination is always a compression of the original collection of multivariate categories. With a score procedure, the compressions lead to reductions, but the original variables may sometimes all be retained. The "reduction" is produced either as the score or as the smaller number of "factors" formed as new variables from weighted combinations of the original variables. For example, suppose the plans to form an Apgar score began with the five variables that it now includes. The five variables would all be retained in the final product, but would be reduced to a single new composite variable or factor, which is the Apgar score itself.

4.2.4.3. Estimation

If the reduction process has been aimed at estimating a target variable, the estimate for a particular person occurs when the corresponding values of that person's variables are either entered into the weighted combination of variables, or located in the clustered aggregations of categories. For example, when a person's values for independent variables are inserted into the algebraic combination, a score of .30 might emerge as the predicted probability for the target event of survival at one year. If so, the person's chance of surviving one year would be estimated at 30%. With categorical-cluster methods, if the values of a person's variables indicate membership in Stage II of a TNM staging system, the person's chance of surviving one year will be estimated at 42% if .42 is the proportion of one-year survivors associated with that stage.

4.2.4.4. Trends

In a dependent orientation, the trend of the individual variables and categories is discerned from their impact on the target variable. During the reduction processes, the score and cluster procedures use different strategies to discern these trends.

In the mathematical-score procedures, the main decisions are made according to the way that individual variables fit the algebraic format. Their trends are demonstrated, as a by-product of the fitting procedure, by the coefficients that produce the best fit. In the clustered-category procedures, categories are eliminated or retained according to their impact, rather than fit. (It so happens that the more effective categories also produce better estimates, but the key decisions usually depend on assessing trend rather than fit.)

This difference in operational strategy becomes important later when regression diagnostics are used to determine whether an algebraic model has produced deceptive results. Because the categorical-cluster strategy is aimed directly at trend rather than fit, categorical results can be used as a "gold standard" for demonstrating the trends of variables within different zones. Thus, if an algebraic straight line were fitted to the data in Fig. 2-7, the effect of X on Y would be demonstrated by the regression coefficient, b, in the equation $Y = a + bX$. If we wanted to determine whether the trend is constant throughout the full scope of the X variable, however, we would examine its impact on Y in the three categorical zones of X shown in Fig. 2-19, or in some other zonal partition of Fig. 2-7.

4.3. Constraints of orientation and data

The discussion in Chapter 2 demonstrated that any set of categorical tabular data can be displayed as points on a graph, and that any set of graphical points can be converted to categories in a table. Because of this interchangeability, almost any form of multivariable analysis can be applied to almost any collection of multivariable data.

4.3.1. Constraints of orientation

A prime constraint in choosing a multivariable procedure is the orientation of the analysis. If a specific target variable has been identified, the analyst will want to use a targeted, dependent method. The nondependent approach, which does not distinguish a target variable from among the independent variables, is seldom used because most medical analyses are aimed at a target.

4.3.2. Constraints of target variable

Among the targeted methods, the type of data in the target variable will not constrain the choice of a method, but will often make certain methods preferable to others.

4.3.2.1. Categorical methods

If the target variable is a binary event, the effect of each categorical cluster is readily evident in the corresponding binary proportion for occurrence of the target event. These effects on the binary targets were easily seen in the previous tables in Chapter 3.

The impact of clusters is more difficult to discern if the target variable is dimensional. With a dimensional *duration of survival* (rather than a binary event, such as *alive at 5 years*) the survival in each clustered category could be summarized as a mean, median, a mean of the logarithm of survival durations, or in some other appropriate expression. The choice among these summary indexes may lead to problems because the gradients among different categories will not always be the same for different summary expressions. On the other hand, the survival summary for a binary event is easily and consistently expressed as a "rate," showing the proportion of people who are alive (or dead) at the stipulated time.

The choice of an appropriate summary expression becomes particularly difficult if the target variable is expressed in more than two categories. Such arbitrary ordinal ratings as **0, 1, 2, 3, 4** for *severity of illness* can be summarized as though they were dimensional values, or they can be split into two zones and managed as a binary variable. Both approaches seem unattractive because the dimensional strategy violates the customary mathematical demand for measurably equal intervals between categories, and the binary strategy tends to "lose information" by ignoring some of the ordinal distinctions.

For variables with nominal categories—such as **hepatitis, cirrhosis, obstructive jaundice,** and **other** for *diagnosis of liver disease*—the categories cannot be readily summarized with a single expression. In the easiest approach—which cites the proportion of the single, most frequent category—information about the proportionate occurrence of all the other categories is "lost".

4.3.2.2. Algebraic methods

Algebraic analyses can also be constrained by a formal set of mathematical requirements. For example, in ordinary *linear regression*, the target value estimated by Y is expressed with a dimensional scale. If Y is not a dimensional variable, linear regression is mathematically inappropriate. Nevertheless, as shown in Section 4.7, linear regression is sometimes used for ordinal or binary targets (and generally gets reasonably satisfactory results) despite the impropriety.

If the target variable, Y, is a binary event, a particularly appropriate algebraic method is *logistic regression*, which aims the independent variables at the logistic transformation of the probability estimated for Y. In this transformation, the probability for Y is converted to an odds, which is $Y/(1 - Y)$. The mathematical model is then aimed at the natural logarithm of the odds, i.e., $\ln[Y/(1 - Y)]$. This process, illustrated in Section 4.7, can be used with appropriate adaptations when Y is an ordinal rather than binary variable.

If the target variable contains unranked nominal categories, such as a set of possible diseases to be diagnosed, neither linear nor logistic regression is directly pertinent. *Discriminant function analysis* was devised to deal with this type of challenge.

In a different type of analysis, the Y variable represents the duration of time elapsed for each person before a "failure event," such as death. If the failure event has not yet occurred at the end of the observation period, the person is designated as "censored" at that time. In this situation, the target variable is really bivariate: it indicates a timed duration of observation, and also shows each patient's binary state as **censored** or **failed** at the "exit" time. Data of this type are regularly summarized with a "survival curve," showing a group's declining survival proportions over time.

To determine the way that the survival curve is affected by multiple independent variables—such as age, clinical stage, anatomic extensiveness, etc.—we cannot use linear or logistic regression. Linear regression would require that each person have a known duration of survival, which will be unknown for all the censored people. Logistic regression could be applied if the target were converted to the static occurrence (or nonoccurrence) of the failure event at specified time intervals, such as 1-year, 3-years, etc. The length of the chosen time interval would be arbitrary, however, and the logistic procedure would not reflect the "dynamic" attributes of the survival curve. To analyze the effects of independent variables on a curve of survival (or death), the most commonly used multivariable procedure is the proportional hazards analytic method proposed by Cox[4] and often called *Cox regression*.

These four methods of regression—linear, logistic, discriminant function, and Cox's proportional hazards—will be the main focus of the targeted algebraic models discussed in this book.

4.3.3. Constraints of independent variables

A great deal has been written about mathematical requirements for the *independent* variables entered into algebraic multivariable analyses. One conventional requirement for the mathematical operations is that the variables be expressed dimensionally, in scales for which adjacent items have measurably equal intervals.

This requirement, however, is constantly ignored in actual usage. Nominal variables are regularly "decomposed" into constituent categories that are cited as "dummy binary variables." Ordinal and binary variables regularly have their **0, 1, 2, 3, . . .** or **0, 1** coding digits entered into analysis as though the data were dimensional. A special coding[5] (discussed in Section 5.4.4) can be used to avoid this impropriety for ordinal variables, but the coding eliminates the possibility of getting a *single* coefficient for the trend in that variable.

Chapter 5 will contain an extensive discussion of the mechanisms used to prepare multivariable data for analysis. The main point to be noted now is that the choice of analytic methods is seldom affected by the type of data contained in the *independent* variables. They can always be suitably arranged or transformed to fit whatever analytic method is chosen.

4.4. Basic taxonomy of multivariable analysis

With orientation and pattern of combination as the two main axes of classification, the diverse methods cited in the foregoing discussion (and a few not yet mentioned) can be arranged into a simple taxonomy of multivariable analytic procedures. The procedures can be classified in four groups according to whether they are targeted or nontargeted, and whether the pattern of combination produces algebraic-model scores or decision-rule clusters.

Table 4-1 A simple taxonomy of methods of multivariable analysis*

Orientation	Analytic Pattern	
	Algebraic Models	Decision-rule Clusters
Targeted	Multiple linear regression Multiple logistic regression Proportional hazard (Cox) analysis Discriminant function analysis	Recursive partitioning Polarized stratification Sequential sequestration Conjunctive consolidation
Non-targeted	Factor analysis Principal component analysis	Cluster analysis Numerical taxonomy

*Several other methods, as noted in the text, have been omitted to keep this table relatively simple.

For the classification in Table 4-1, the upper left corner contains the big four targeted algebraic methods that commonly occur in medical literature. The upper right corner contains the decision-rule targeted-cluster procedures that will receive extensive discussion in Part V of the text, although the conjunctive consolidation procedure will be briefly illustrated in Section 4.6.2.4. The lower part of Table 4-1 shows the nontargeted procedures. On the left are factor analysis and its mathematical "sibling," principal component analysis; on the right are two cluster procedures—cluster analysis and numerical taxonomy—that have often been used for biologic classifications. For medical classifications, factor analysis[6-12] and principal component analysis[13,14] have been applied more often than the cluster procedures,[15-18] but none of these methods has produced particularly valuable results, and the strategies are seldom used.

All of the multivariable procedures are given names that reflect differences in the operating strategy used for arranging variables and/or categories. Because Table 4-1 is intended to be relatively simple, several other multivariable procedures have either been omitted or reserved for brief discussion later in the text. For example, the log-linear analysis of contingency tables will be outlined in Chapter 19. Several omitted multivariable methods, such as analysis of covariance and path analysis, can be regarded as variations or extensions of procedures listed in Table 4-1. Multivariate analysis of variance has been omitted for the same reasons cited earlier for its counterpart, canonical analysis.

The big four algebraic methods, which are a prime focus of this text, are all aimed at an individual target variable. The main distinctions in these four methods arise from the type of expression (discussed in Section 4.3.2.2) used for the target variable. The distinctions are summarized now in Table 4-2.

Multiple linear regression is intended for a *dimensional* target variable, such as blood pressure. The method has sometimes been applied, however, to ordinal or binary targets. *Multiple logistic regression* is intended for a *binary* target, such as **alive** or **dead,** but the method can also be used for ordinal targets. *Cox regression* is best intended for a *moving binary target,* such as the sequential proportions of survival in

Table 4-2 Types of target variables and corresponding algebraic methods

Type of intended target variable	Algebraic method	Also sometimes used for target variables that are:
Dimensional	Multiple linear regression	Ordinal, binary
Binary	Multiple logistic regression	Ordinal
"Moving" binary (e.g., survival curve)	Proportional hazard function analysis (Cox regression)	—
Nominal	Discriminant function analysis	Binary

a survival curve. Although applicable to other situations, the Cox method is seldom used for other purposes. The *discriminant function* procedure is best intended for a *nominal* target variable, such as diagnostic category. The method has most often been applied, however, to binary target variables.

All four of these procedures employ linear algebraic models, discussed later, that take the form $G = b_0 + b_1X_1 + b_2X_2 + b_3X_3 + \ldots + b_pX_p$. The expression for G contains p (for "parameter") independent variables designated as $X_1, X_2, \ldots, X_p$. In linear regression, the model expressed as G is used to make the estimate as $\hat{Y}_i = G$. (The "hat" symbol, "^", is placed over the Y to indicate that it is an estimated rather than observed value. Thus, Y_i is an observed value for the i-th person; $\hat{Y}_i$ is the corresponding estimated value.)

In logistic regression, with the model expressed as G, the estimate is made as $\ln[\hat{Y}_i/(1 - \hat{Y}_i)] = G$. In Cox regression, the estimate of each person's survival is $\hat{Y}_{i,t} = S(t)^{e^G}$. In this expression, which uses G as the multivariable model, $S(t)$ is the proportion of survivors at any time, t, in the summary survival curve for the entire group; e is the "natural" scientific constant (2.718 . . .); and e^G is the power to which S(t) is raised for estimating survival of person i at time t. In discriminant function analysis, each of the target categories, designated as C_h, is estimated with a separate linear multivariable model, G_h. The estimated value, $\hat{Y}_h = G_h$, is determined for each category; and the chosen estimate for a particular person is the category that has the highest of the $\hat{Y}_h$ values.

These expressions are listed here merely to indicate the "flavor" of the models for readers who do not go beyond this chapter in the text. The construction and interpretation of the models are discussed in Chapters 10–18.

4.5. Focus of subsequent text

To avoid becoming too long and unwieldy, the text hereafter will be confined to the targeted methods of multivariable analysis. They commonly appear in medical literature, where the analyses are almost always aimed at a particular target variable or event. Nontargeted algebraic methods, such as factor analysis and principal component analysis, have been particularly popular in the psychosocial sciences, where cohort studies and clinical trials are seldom done to provide an outcome as the target variable. The nontargeted decision-rule methods of cluster analysis and numerical taxonomy have been used mainly in botany and invertebrate biology by investigators trying to develop taxonomic classifications for the myriads of new data made available by modern biochemistry. Readers who want to know more about the nontargeted methods can find appropriate discussions in many textbooks, of which only a few are referenced here.[19-22]

The rest of this chapter contains a "preview of coming attractions" to illustrate the operation of decisional-cluster and algebraic-score procedures in targeted analysis.

4.6. Illustration of decision rules for clustered categories

Although Part V of the text is devoted to targeted-cluster analysis, two advance illustrations are supplied here to demonstrate the decision-rule methods. They are shown for both nontargeted and targeted approaches.

4.6.1. Example of a nontargeted decision rule

The TNM stages discussed earlier were originally formed with a nontargeted decision rule developed from "clinical judgment." Although the stages were initially prepared without direct relationship to an external target event, such as survival, the prognostic impact of the individual characteristics was doubtlessly an "implicit" consideration during the judgments.

The first main set of judgments involved decisions about the allocation, rather than aggregation, of individual variables. For example, a patient with lung cancer can be characterized according to the **presence** or **absence** of a series of binary variables denoting metastasis to liver, metastasis to brain, metastasis to abdominal lymph nodes, metastasis to extrathoracic bone, etc. In an allocation decision, this array of variables became components of a binary "union" variable, called *distant metastasis*. The union of distant metastasis would be **present,** if a patient had any one (or more) of the component metastatic characteristics.

An analogous type of allocation decision produced the binary union variable called *regional metastasis*. It would be cited as **present** if a patient had any of the component characteristics such as metastasis to hilar lymph nodes, metastasis to mediastinal lymph nodes, direct invasion of chest wall, etc. Finally, another allocation decision led to criteria for the binary variable, *local tumor*. This variable would be marked present if a primary site of the cancer could be identified, regardless of the associated size, mobility, cell type, or other characteristics of the cancer at that site.

The second set of judgments involved decisions about aggregating the three main binary variables that denoted distant metastasis, regional metastasis, and local tumor. The Venn diagram in Fig. 4-1 shows the seven distinctive sectors (or anatomic subgroups) formed as Boolean categories when each of the three main

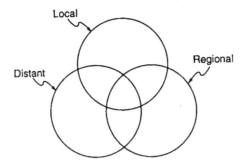

Fig. 4-1 Subgroups of cancer. Venn diagram showing seven subgroups formed by presence or absence of three binary variables in patients with cancer. (An eighth subgroup, outside the circles, occurs when none of the three variables is present.)

binary variables is present or absent. The decision rule combined these seven trivariate sectors into three clusters, according to the following plan:

The distant metastatic category, Stage III, is the "worst" neoplastic characteristic, regardless of the status of regional metastasis or local tumor. In patients without distant metastasis, the regional metastatic category, Stage II, is the next worst group, regardless of the status of the local tumor. The remaining patients form the best group, Stage I, having local tumor only, without evidence of regional or distant metastasis. The Venn diagram in Fig. 4-2 shows the consolidations that converted

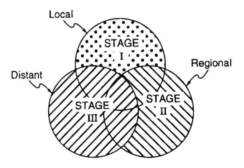

Fig. 4-2 Stages of cancer. Illustration of decision rule that combines the seven subgroups of Fig. 4-1 into three clustered categories, labelled as Stages I, II, and III.

the seven trivariate categories into the clusters of a single composite variable. The clusters became ordinal stages for the anatomic extensiveness of a cancer.

The contents of these three stages are summarized as follows:

Stage I: Localized cancer *and* no evidence of regional spread *and* no evidence of distant metastasis.
Stage II: Regional spread *and* no evidence of distant metastasis.
Stage III: Distant metastasis.

4.6.2. Construction of a targeted decision rule

In a targeted decision rule (which is usually a series of rules, rather than one), the decisions are usually guided by statistical effects noted in the target variable, but biologic judgments can also be applied.

4.6.2.1. Statistical guidelines

The statistical guidelines for targeted clusters will be extensively discussed in Chapters 19–21. The guidelines involve decisions about the magnitude of gradients in an effective variable, the modicum size of a numerically stable group (or "sample"), and the mechanism for consolidating groups (or cells) that are too small in either size or gradient. For example, two adjacent cells in a table, or adjacent categories in a variable, might be consolidated if they are "statistically homogeneous," i.e., having essentially similar values for occurrence of the target variable. Two independent variables, when conjoined in the types of tables shown in Chapter 3, might each be regarded as having an important impact if a suitable gradient in the target is produced by each variable within the categories of the other variable. This double-gradient phenomenon was well shown in Tables 3-2, 3-4, and 3-6.

4.6.2.2. Biologic guidelines

Statistical guidelines are the only rules used in the automated targeted-cluster procedures that have such names as *recursive partitioning* and *polarized stratification.* In several other procedures, however, certain biologic guidelines may also be applied to insist that the consolidated cells be biologically as well as statistically homogeneous. Although statistical homogeneity requires similar results for the target variable, biological homogeneity requires that the combined categories have a "plausible coherence." For example, **metastasis to liver** and **metastasis to brain** can be plausibly combined as **distant metastasis** in patients with lung cancer, but **hemoptysis** and **ingrown toenail** would not be a biologically coherent combination. Both biologic and statistical guidelines are used in two multivariable targeted-clustering procedures, discussed in Chapter 20, that are called *sequential sequestration* and *conjunctive consolidation.*[23] The conjunctive consolidation method is used for the illustration later in this section.

Because the concept will always vary according to the situation under analysis, a rigid guideline cannot be established for biologic homogeneity or coherence. For example, suppose *age* and *sex* have been conjoined to produce four bivariate categories: **boy, girl, man,** and **woman.** We would coherently combine **boy** and **man** as **male,** and **girl** and **woman** as **female,** but we could just as coherently combine **boy** and **girl** as **child,** and **man** and **woman** as **adult.** Either set of combinations would be biologically homogeneous, according to the goal of the analysis.

Biological coherence is easier to delineate for variables that are cited in ordinal categories such as **mild, moderate, severe.** With a *single* variable, we could plausibly combine the two adjacent categories of **mild-moderate** or **moderate-severe,** but not the extreme categories of **mild** and **severe.** If bivariate categories are being examined for two ordinal variables, as in several tables of Chapter 3, each conjoint category has a mixture of ratings for each variable, such as **low-low, high-high, low-medium, high-low,** and so on. If the bivariate categories are combined, a **low-low** cell would not be joined with a **high-high** cell, but a combination of a **high-low** cell and a **low-high** cell might be plausible. Such a combination would represent a symmetrically opposite mixture of the highest value in one variable and the lowest in the other.

4.6.2.2.1. Rules of coherence for two ordinal variables

The concepts of coherence for adjacent and symmetrical cells in a conjunction of two ordinal variables could be stipulated as follows: Suppose the two ordinal variables form a table whose categories are cited in r rows and c columns. Each cell in the table will have a row and column location that can be cited, with the row listed first, in numbers such as **(2,3)** or **(4,1)**. For two variables that each have five ordinal categories, the cellular locations in the table would be identified as follows:

Categories of row variable	Categories of Column Variable				
	1	2	3	4	5
1	(1,1)	(1,2)	(1,3)	(1,4)	(1,5)
2	(2,1)	(2,2)	(2,3)	(2,4)	(2,5)
3	(3,1)	(3,2)	(3,3)	(3,4)	(3,5)
4	(4,1)	(4,2)	(4,3)	(4,4)	(4,5)
5	(5,1)	(5,2)	(5,3)	(5,4)	(5,5)

Two adjacent cells in row 2 could be **2,3** and **2,4;** and two adjacent cells in column 3 might be **4,3** and **5,3**. Two cells would be symmetrically opposite if they have diagonally transposed locations such as **2,3** and **3,2;** or **4,1** and **1,4**. Cells in locations such as **1,3** and **3,1** might also be regarded as diagonally symmetrical for a cell in the **2,2** location.

With these labels to show the ordinal characteristics of the bivariate cells, we can set up the rule that two cells are biologically coherent if they have adjacent positions in the same row or the same column. After two coherent cells are combined, they form a single cell for evaluating adjacency to other cells. Thus, if the (3,2) and (3,3) cells are combined to form a (3, 2-3) cell, the adjacent cells of that row would be in the (3,1) and (3,4) positions. Two cells might also be regarded as biologically coherent if they have diagonally transposed locations or a similar sum for the numbers that indicate their cellular locations. Thus, the (5,2), (4,3), (3,4), and (2,5) cells can be regarded as biologically coherent because they contain a combination of similar but opposite "strengths" in the ordinal ranks of the two constituent variables.

The cited statistical and biologic principles could lead to the following decision rule for consolidating the cells of conjoint categories in a table showing outcome events for two ordinal variables:

Rule 1. Combine two adjacent cells in the same row or column if they have no substantial increment in the rates of the target event.

Rule 2. Combine two cells in symmetrically opposite rows and columns if they have no substantial increment in the rates of the target event.

Rule 3. Otherwise, preserve the cells as separate categories.

The guidelines for statistical and biologic decisions are used in the illustration that follows.

4.6.2.2.2. Data set for rectal cancer

Table 4-3 shows the trivariate data for 243 patients who received surgery for rectal cancer.[24] One independent variable, *anatomic stage*, is cited in the ordinal categories of 1, 2, 3. The other independent variable, *symptom stage*, is also coded in three ordinal categories of 1, 2, 3. The dependent or target binary variable, *5-year survival*, is coded as 0 for dead and 1 for alive. The fourth column of the table shows the number of patients in each of the 18 (= 3 × 3 × 2) trivariate cellular categories.

Table 4-3 Anatomic stage, symptom stage and 5-year survival in 243 patients after surgery for cancer of the rectum

Anatomic stage	Coding for: Symptom stage	5-year survival	Number of patients in each category
1	1	0	20
		1	36
	2	0	42
		1	37
	3	0	2
		1	0
2	1	0	9
		1	6
	2	0	29
		1	6
	3	0	0
		1	0
3	1	0	17
		1	1
	2	0	26
		1	4
	3	0	8
		1	0
		Total	243

This same information could also be shown in the alternative format of Table 4-4, which allows an easier demonstration of subtotals for some of the cells.

Table 4-4 Alternative arrangement of data for table 4-3

Symptom stage	Anatomic stage 1 Survival 0	1	2 Survival 0	1	3 Survival 0	1	Subtotal 0	1	Total
1	20	36	9	6	17	1	46	43	89
2	42	37	29	6	26	4	97	97	144
3	2	0	0	0	8	0	10	0	10
Subtotal	64	73	38	12	51	5	153	90	—
Total	137		50		56		—		243

Nevertheless, neither Table 4-3 nor Table 4-4 offers a clear arrangement of trends in the data.

4.6.2.3. Bivariate arrangements

The bivariate trend for anatomic stage alone can be clearly shown if the data of Table 4-4 are arranged as follows:

Anatomic stage	5-year survival rate	
1	53%	(73/137)
2	24%	(12/50)
3	9%	(5/56)
Total	37%	(90/243)

The counterpart bivariate trend for symptom stage is:

Symptom stage	5-year survival rate	
1	48%	(43/89)
2	33%	(47/144)
3	0%	(0/10)
Total	37%	(90/243)

The striking gradients in the survival results show that each independent variable seems to have a strong impact on the target variable.

4.6.2.4. Multivariable results

For multivariable analysis, the results are best displayed as shown in the 3 × 3 × 2 arrangement of Table 4-5. The marginal totals of columns and rows in Table 4-5 show respectively the bivariate results previously found for anatomic stage and symptom stage. The main new information of Table 4-5 is in the cells. The double gradient phenomenon is clearly demonstrated within the first two categories of

Table 4-5 5-year survival rates for conjoined categories of anatomic and symptom stages (Stratified arrangement of data in Tables 4-3 and 4-4)

Symptom stage	Anatomic Stage			Total
	1	2	3	
1	36/56 (64%)	6/15 (40%)	1/18 (6%)	43/89 (48%)
2	37/79 (47%)	6/35 (17%)	4/30 (13%)	47/144 (33%)
3	0/2 (0%)	—	0/8 (0%)	0/10 (0%)
Total	73/137 (53%)	12/50 (24%)	5/56 (9%)	90/243 (37%)

each variable. Within symptom stage 3, no gradient is possible, since no one survived. Within anatomic stage 3, a substantial gradient would not be expected, since the total survival in the stage is only 9%, and some of the cells have relatively few members. Accordingly, Table 4-5 confirms that each independent variable has a distinct double gradient impact on the target variable.

4.6.2.5. Reduction of cells

For the reduction process, we want a single composite result that combines the two predictor variables. The reduction can be attained with the following principles of statistical and biological coherence:

1. The highest survival rate (64%) occurs in the **1,1** cell that represents the best stage in both variables. This cell can be maintained as a separate category, since no other cell contains a similarly high rate.

2. The next highest survival rates (47% and 40%) are reasonably close statistically, and occur in cells that are in the symmetrically opposite locations of **1,2** and **2,1**. Therefore, these two cells can be combined.

3. The remaining cells show low survival rates, ranging from 0% to 17%; and some of the cells have relatively few members. Rather than demarcating another separate subgroup (such as the **2,2** cell with 17% survival), we can simply cluster the remaining cells together.

The single new composite variable will contain three ordinal stages that can be designated **A, B,** and **C,** formed in the following pattern of conjunctive consolidation:

Symptom Stage	Anatomic Stage		
	1	2	3
1	A	B	C
2	B	C	C
3	C	C	C

After the conjunctive consolidation, the survival results for the new composite variable will be:

Composite Symptom-Anatomic stage	5-year survival rate	
A	64%	(36/56)
B	46%	(43/94)
C	12%	(11/93)
Total	37%	(90/243)

4.7. Illustration of algebraic models for mathematical scores

The operation of algebraic methods will be discussed in Parts III and IV of the text, but the results of two mathematical procedures—linear regression and logistic regression—are shown in advance here to illustrate what might happen if they were applied to the data of Table 4-3 (or 4-4).

4.7.1. Bivariate results

The linear regression and logistic regression methods are first applied here in a bivariate manner, with only one independent variable related to the outcome.

4.7.1.1. Linear regression

In simple linear regression, the target variable, Y, would be expressed as **0** or **1** for 5-year survival. The categorical ratings of **1, 2, 3** would be expressed as variable

X for either anatomic stage or symptom stage. (These categorical variables are not strictly appropriate for the underlying mathematical principles, but the algebraic model is "robust" enough for the illustration.) The mathematical format for each expression would be $Y = a + bX$.

When *anatomic stage* is used as X, the algebraic model that emerges for these data is $\hat{Y} = .75 - .23X$. When *symptom stage* is used as X, the algebraic model is $\hat{Y} = .68 - .19X$. The estimates of $\hat{Y}$ for anatomic stages 1, 2, and 3 will be .52, .29, and .06, respectively. The corresponding estimates of $\hat{Y}$ for symptom stages 1, 2, and 3 will be .49, .30, and .11, respectively.

4.7.1.2. Logistic regression

In logistic regression, which is particularly appropriate for a **0/1** target variable, the algebraic format is $\ln[\hat{Y}/(1 - \hat{Y})] = a + bX$. The estimated value of $\hat{Y}$ is derived from the algebra as $\hat{Y} = 1/[1 + e^{-(a+bX)}]$. (The algebra and operational strategy of logistic regression will be discussed in Chapter 13.)

When anatomic stage is used as X, the algebraic model is calculated to be $1.37 - 1.24X$. To illustrate the use of the formula, suppose $X = 2$. The value of $a + bX$ will be $1.37 - (1.24 \times 2) = -1.11$. The value of $e^{-(a+bX)}$ will be $e^{1.11} = 3.034$, and $\hat{Y}$ will be $1/(1 + 3.034) = .25$. The respective estimates of $\hat{Y}$ for anatomic stages 1, 2, and 3 will be .53, .25, and .09.

When symptom stage is used as X, the corresponding result of the logistic algebraic model is $0.84 - 0.83X$. For symptom stage 2, the estimate will be $\hat{Y} = 1/[1 + e^{-.84+1.66}] = .31$. The respective estimates of $\hat{Y}$ for symptom stages 1, 2, and 3 will be .50, .31, and .16.

4.7.1.3. Comparison of bivariate results

Table 4-6 compares the three sets of bivariate results obtained with clustered categories, linear regression, and logistic regression.

Table 4-6 Comparison of bivariate results when data of Table 4-4 are analyzed with categorical clusters, linear regression, and logistic regression

		Survival rate shown in categorical cluster	Survival rate estimated by:	
			Linear regression	Logistic regression
Anatomic stage	1	.53	.52	.53
	2	.24	.29	.25
	3	.09	.06	.09
Symptom stage	1	.48	.49	.50
	2	.33	.30	.31
	3	.00	.11	.16

For anatomic stage, the results of the "gold standard" clusters are estimated quite closely with both regression methods, and are almost identical with logistic regression. For symptom stage, the estimates are reasonably close, but not as close as for anatomic stage; and logistic regression did not perform as well as linear regression.

[An obvious reason for defects in the algebraic performance is that the two independent variables do not have a constant gradient. The two intercategorical gradients for *anatomic stage* are, respectively, 29% (= 53% − 24%) and 15% (= 24% − 9%). The corresponding results for *symptom stage* are 15% (= 48% − 33%) and 33% (= 33% − 0%). Since the gradients are less disparate for *anatomic stage*, it will get a somewhat better fit with a rectilinear model. Because the effect of anatomic stage seems to decline with increasing values, we might get an improved fit by entering $\sqrt{X}$ rather than X into the algebraic model. Conversely, since the effect of symptom stage increases with larger values, it might be better fitted with X^2 rather than X. When these transformations are used, the fit of the algebraic models improves slightly, but the results are biologically strange. What biological meaning can be attached to the square root or square of a coding digit arbitrarily assigned to an ordinal stage?]

4.7.2. Multivariate results

To illustrate the multivariable analytic process, the *anatomic stage* variable is designated as X_1 and *symptom stage* is X_2. The value of 5-year survival is estimated from a rectilinear combination expressed as $b_0 + b_1X_1 + b_2X_2$.

4.7.2.1. Linear regression

The multivariable linear regression model (calculated by methods discussed later) produces $\hat{Y} = .96 - .21X_1 - .14X_2$ as the estimate for these data. Thus, when $X_1 = 2$ and $X_2 = 3$, the estimate is $.96 - (.21)(2) - (.14)(3) = .12$. Other estimates are shown in Table 4-7.

Table 4-7 Comparison of trivariate results in different methods of analysis for data of Table 4-5

Identification code		Survival rate shown in categorical cluster	Survival Rate estimated by:	
Anatomic stage	Symptom stage		Multiple linear regression	Multiple logistic regression
1	1	.64	.61	.65
1	2	.47	.47	.46
2	1	.40	.40	.35
1	3	.00	.33	.28
2	2	.17	.26	.20
3	1	.06	.19	.14
2	3	—	.12	.11
3	2	.13	.05	.07
3	3	.00	−.09	.03

4.7.2.2. Logistic regression

Multivariable logistic regression would produce $2.57 - 1.21X_1 - .76X_2$ as the model for these data. When $X_1 = 2$ and $X_2 = 3$, the combination is $2.57 - (1.21)(2) - (.76)(3) = -2.13$. The estimate is derived as $\hat{Y} = 1/(1 + e^{2.13}) = .11$. Other estimates are in Table 4-7.

4.7.3. Comparison of multivariable results

In Table 4-7, which compares the multivariable algebraic results against the "gold standard" of the categorical clusters, the algebraic methods perform less impressively than in the previous bivariate analyses.

The multiple linear regression estimates are excellent for the top three categories (with coordinates of **1,1; 1,2;** and **2,1**) but are much less successful thereafter. The estimates in the **3,1** and **1,3** categories are reversals of the actual results in the clusters; and an impossible negative estimate ($-.09$) is produced for the **3,3** category.

The multiple logistic estimates are also quite good for the top three categories, and (although still not very good) are somewhat better thereafter than the estimates produced by linear regression. The logistic estimates did not yield any impossible negative values, but both types of regression produced estimates for the **2,3** category, although it contained no members in the categorical cluster.

Since the purpose of these analyses is simply to illustrate the procedures, the comparative results will not be further discussed here. The main point is merely to show what happens when the same set of data is analyzed in different ways.

(For readers who may be curious about the role of interactions, the algebraic models—as discussed later—could contain X_1X_2 as an additional interaction term. The algebraic arrangement would then be $b_0 + b_1X_1 + b_2X_2 + b_3X_1X_2$. When this model is used in multiple linear regression for these data, the values of $b_0, \ldots, b_3$ are 1.25, $-.38$, $-.31$, and .10. The interaction coefficient, b_3, however, is not "statistically significant" [P = .09]. With multiple logistic regression, the corresponding values for $b_0, \ldots, b_3$ are 3.19, -1.64, -1.14, and .26, but neither the b_2 nor b_3 coefficients are "statistically significant." Their respective P values are .07 and .50. Because the coefficients were not "significant," specific estimates have not been shown for the interaction models.)

4.8. Choice of targeted analytic patterns

Since the data can be transformed in many ways, very few patterns of multivariable analysis are uniquely associated with a particular type of data. In most circumstances, the total collection of information will determine the orientation of the analysis. It will be targeted if one (or more) of the variables is identified as dependent. Since the dependent orientation will usually be obvious, the main decision will be to select either an algebraic-method or a decision-rule cluster pattern of analysis.

Algebraic methods are currently much more popular than decision-rule approaches to multivariable analysis. The main reasons are that the algebraic methods are automated, readily available in various computer packages, and operated without any apparent subjective judgments. The computations create no difficulty, and the results can be provided to anyone who wants them. If the main goal is to fit a "line" to the data, the algebraic methods are directly aimed at the goal. If the main goal is to identify important variables, the results of the algebraic methods seem relatively easy to interpret because each of the "important" variables is concisely identified in the computer printout, and each variable has a coefficient that seems to "weight" its relative impact.

In contrast, the decision-rule procedures are arbitrary; they are not good for fitting lines to data; and they may not be automated and readily available for use in all computer systems. The categories from different variables may sometimes be combined in a way that shows the effect of the combinations, but not of the individual variables. A particular problem is that the final results may need much more space for citation than those of the algebraic models. For example, consider the data of Table 4-3. The best cluster display of this information occupied the many cells of Table 4-5; and the best reduction occupied the three tabular lines needed for the results of composite stages **A, B,** and **C** in Section 4.6.2.5 By contrast, the multiple linear regression in Section 4.7.2.1 summarized everything succinctly in a single expression as $\hat{Y} = .96 - .21X_1 - .14X_2$.

On the other hand, the decision-rule procedures have some substantial scientific advantages:

1. They allow the results to be inspected directly and understood almost immediately. Thus, the categories in Table 4-5 show the exact data for patients with different anatomic and symptomatic stages. To note this same effect, the equation $\hat{Y} = .96 - .21X_1 - .14X_2$ must receive a detailed further evaluation.

2. A categorical table requires no assumptions about the data. It shows exactly what happened. An algebraic approach, however, depends on fitting the data with a particular mathematical pattern. In the instance just cited, the pattern happened to fit the data moderately well, but it may not always do so.

3. As shown in Section 4.6.2.5, a decision-rule procedure involves a set of arbitrary judgments, but the judgments are all "up front" and readily evident. The algebraic methods also involve arbitrary judgments—about choice of methods, patterns, and various operational decisions—but the judgments are usually obscured by the mathematical formulas or hidden during the computerized "number crunching."

4. The decision-rule procedures routinely search for the kinds of conjunctive effects (or interactions) explored throughout the tables of Chapter 3. The algebraic methods can also do these searches, but the interactions are not sought routinely; they must be anticipated and entered into the mathematical models.

5. The decision-rule procedures yield results that are particularly compatible with clinical and biologic thought processes, which seldom involve mathematical combinations of weighted variables.

After the analytic preparation of multivariable data is discussed in Part II of the text, the big four algebraic methods appear in Parts III and IV. The decision-rule procedures, which are being constantly improved and are becoming more popular, will occupy Part V.

Chapter References

1. Apgar, 1953; 2. Feinstein, 1987, p. 175; 3. American Joint Committee on Cancer, 1988; 4. Cox, 1972; 5. Walter, 1987; 6. Stitt, 1977; 7. Mason, 1988; 8. Coste, 1991; 9. Ries, 1991; 10. Bailey, 1992; 11. Lauer, 1994; 12. Peters, 1993; 13. Cowie, 1985; 14. Henderson, 1990; 15. Winkel, 1970; 16. Schlundt, 1991; 17. Furukawa, 1992; 18. Persico, 1993; 19. Sokal, 1963; 20. Everitt, 1993; 21. Jackson, 1991; 22. Yates, 1987; 23. Feinstein, 1990; 24. Feinstein, 1975.

Part II Preparing for Multivariable Analysis

Before a multivariable analysis begins, the data and the analyst should be suitably prepared. To be processed, the raw information must be transferred into stored codes. The coded information should be verified for accuracy and then checked for potential problems in univariate distributions. In the next step, the bivariate relationships are evaluated according to two basic statistical methods that also serve as prototypes for the subsequent multivariable analyses. The five chapters of Part II are devoted to these preparatory activities, which are illustrated with a set of data that will be used thereafter for examples throughout the rest of this book.

Chapter 5 describes the process that converts raw data into suitably classified and coded variables. The conversion process seems so elementary that it is usually ignored in most discussions of statistical analysis. Nevertheless, like many other elementary activities in "basic" science, what is done (or not done) during the conversion is so important that it can affect everything that happens afterward.

In Chapter 6, the univariate distributions of the coded data are checked for missing information and "eccentric" variables. The defects can either be easily fixed at this early phase of the analysis, or they can offer useful warnings about problems to beware of later on.

Chapter 7 describes simple linear regression for two variables, as well as statistical distinctions in the evaluation of "significance." As a prototype from which all the more complex regression activities are derived, simple linear regression is the basic strategy for fitting an algebraic model and appraising the co-related trend of variables.

Chapter 8, devoted to stratified rates, describes another prototype method that later becomes expanded for the analysis of targeted multivariable clusters. The method is illustrated now for its application to appraise the structure and bivariate effects of categorical variables.

The two prototype methods of algebraic-model and stratified-rate analysis are then used, in Chapter 9, to evaluate bivariate relationships in the illustrative data set. After these preparations are completed, the multivariable activities begin in Part III.

5 Classification and Coding Decisions

Outline ....
5.1. Illustrative multivariable data set
    5.1.1. Target variable
    5.1.2. Independent variables
5.2. Classification of data
    5.2.1. Missing or ambiguous data
    5.2.2. Consolidation of categories
5.3. Construction of Boolean unions
    5.3.1. Timing of construction
    5.3.2. Binary-scale unions
    5.3.3. Ordinal-scale unions
5.4. Principles of data coding
    5.4.1. Dimensional variables
    5.4.2. Binary variables
    5.4.3. Nominal variables
        5.4.3.1. 'Dummy' binary variables
        5.4.3.2. Other coding patterns
    5.4.4. Ordinal variables
    5.4.5. Choice of ordinal codings
    5.4.6. 'Dummy' coding for zones of dimensional variables
    5.4.7. Coding the target variable
    5.4.8. Intervariable coding
5.5. Importance of classification and coding decisions
    5.5.1. Choice of important variables
    5.5.2 Effectiveness of taxonomy
    5.5.3 Interpretation of coefficients
5.6. Analytic procedures and 'packaged programs'
    5.6.1. Choice of packaged programs
    5.6.2. Beginning the operations
    5.6.3. Choice of necessary and 'optional' options
    5.6.4. Inspecting the printout
        5.6.4.1. Uninvited options
        5.6.4.2. Unsatisfactory presentations and labels
        5.6.4.3. Inadvertent errors
        5.6.4.4. General comments
5.7. Verification of data entry
5.8. Preliminary statistical appraisals

....

For a multivariable analysis to make sense, the investigator, analyst, and reader of the results must understand what has been collected in the data and what is sought in the analysis. If these ideas are not clear, the analysis itself cannot be properly arranged or interpreted. At the most elemental level of decisions, the analyst will not know the most satisfactory ways to choose and classify the raw data, to code the variables, and to determine how well the coded results can be analyzed.

To offer concrete illustrations, a specific set of multivariable data will be used repetitively throughout this book. In this chapter and in Chapters 6 and 9, the data set will help demonstrate the process by which variables are chosen, classified, coded, and checked. In these and subsequent chapters, the results are shown with various tables, graphs, or photographed excerpts of computer printout that emerged when the illustrative data were analyzed with different procedures.

The discussion in this chapter is concerned first with the contents of the illustrative data set, and then with the general strategies used to classify and code multivariable information.

5.1. Illustrative multivariable data set

The illustrative multivariable data set was obtained during research[1] aimed at developing an improved new prognostic staging system for patients with primary cancer of the lung. In an inception cohort[2] of 1266 patients, *zero time* for each person was defined as either the date of the first antineoplastic therapeutic intervention or the date of the decision not to give such treatment. Each patient was classified for a baseline status, which was the corresponding medical condition at zero time, and for an outcome, which was the survival noted in follow-up observations for at least eight years thereafter.

To get a data set large enough for multivariable analyses, but small enough to be fully displayed in the examples, 200 patients were chosen as a random sample from the original group of 1266, after exclusion of 238 patients who did not have complete data for each of the independent variables under analysis. Those 200 patients are the subjects of the illustrations and analyses here. The diskette attached on the inside back cover contains the identification number for each of those patients and the corresponding coding digits for the variables described in this chapter.

The description of the data has two useful roles. First, it will acquaint you with basic information that will appear repeatedly in subsequent examples of multivariable procedures. Second, and more importantly, the elementary work in classifying and coding multiple variables involves many subtle judgments, not apparent in any of the subsequent statistics, that affect the understanding and interpretation of the statistical results.

5.1.1. Target variable

In most illustrative analyses, the target variable was survival after zero time. Each of the 200 patients with lung cancer was known to be either dead at a specific duration after zero time, or alive at a time that exceeded 99.9 months.

In some analyses (such as linear regression), the target variable was expressed dimensionally as *duration of survival* in months, but was transformed, for reasons cited in Section 6.3.3, to its natural logarithm. The mean survival duration will be somewhat lower than its correct value in the total group, because 7 of the 200 patients were coded as 99.9 months for survival times that exceeded those three digits. In certain analyses (such as Cox regression), the untransformed survival durations were analyzed from zero time until a closing date 5 years later.

In other analyses (such as logistic regression), the target variable was a binary event: alive or dead at 6 months after zero time. The 6-month interval was chosen because it was close to the median survival of the cohort, thereby letting the 6-month survival rate be almost 50%. (Because the variance of a proportion, p, is equal to $(p)(1 - p)$, the variance available for subsequent analytic partitioning is at a maximum when p is .50, or 50%.)

[With the three different uses of the target variable, survival was coded in three ways in the data set: SURVIVE shows the original survival time in months after zero time; LNSURV is the natural logarithm of SURVIVE; and SURV6 contains no/yes coding of **0/1** to denote status as dead or alive at 6 months after zero time.]

In one particular analysis (used to illustrate discriminant functions), the target variable was expressed in more than two nominal categories. For this purpose, the selected target was *cell type*, cited in four nominal histologic categories of the cancer. (Their coding is described in Section 5.1.4.1.) With this change in the target variable, the discriminant analysis would be aimed at "predicting" cell type, whereas all the other analyses were aimed at predicting survival. In one illustration, however, multiple linear regression and discriminant function analysis were compared—for reasons discussed later—in predicting the binary event of survival at 6 months.

5.1.2. Independent variables

Of eight independent variables available as "predictors," three were demographic attributes: *age, sex,* and *customary amount of cigarette smoking*.

Three clinical variables—*symptom stage, percentage of weight lost,* and *progression interval*—came from the patient's history. The contents and categories of *symptom stage* are described in Section 5.3.3. *Percentage of weight lost* (or "percent weight loss") was calculated as the amount of nondeliberately lost weight at zero time, divided by the patient's customary weight. The *progression interval* was the duration of the cancer's manifestations before zero time. This interval was taken to be the longer of two candidate possibilities: the duration of symptoms associated with the cancer; or the length of time elapsed since the first unequivocally positive chest X-ray film.

Three variables came from special technologic tests. One of these "paraclinical" variables was a laboratory measurement for *hematocrit* (or *hemoglobin*). Another paraclinical variable, *TNM stage*, was morphologic. It indicated the extensiveness of the cancer's anatomic spread at zero time, as noted from X-rays, bronchoscopy or endoscopy, biopsy, cytology, and sometimes physical examination. Yet another morphologic attribute, the microscopic report for *cell type* of cancer, was sometimes used as an additional independent variable and, in one instance, as a dependent variable.

5.2. Classification of data

To be analyzed, each item of information must be expressed as a value in a variable. The *classification* of data refers to a taxonomy that contains the scale of possible values for each variable chosen for analysis.

In the illustrative data set, four of the selected variables—*age, percent weight loss, progression interval,* and *survival duration*—were classified in the same dimensional scale in which they had been initially recorded. (*Percent weight loss* was coded as **0** if the patient had lost no weight or had gained weight.) *Sex* was also classified in its original binary scale of **male/female.**

Except for these five variables, data for all of the other original variables were classified, before entering the analysis, with various modifications that are discussed in the next few sections.

5.2.1. Missing or ambiguous data

One set of modifications was done to avoid having missing or ambiguous data for two variables, *hematocrit* and *progression interval.*

Although used as an index of red blood status, the value of hematocrit was sometimes not available. In these situations, the hemoglobin value, multiplied by 3, was used as a surrogate. Consequently, the *hematocrit* variable in the analyses contained either a direct value of hematocrit or a value of (3 × hemoglobin).

The duration of *progression interval* was ambiguous if the patient's antecedent lung cancer symptoms were cited by different historians as having different lengths of time. To eliminate these ambiguities, we used principles discussed elsewhere[3] to code ambiguous durations of symptoms as the longest of the candidate intervals.

A different problem in *progression interval* was tricky to manage. Patients who were asymptomatic often had long survival; but if these patients were omitted from analysis because they had an "unknown" duration of symptoms, the analytic results would be distorted. A definite duration could be obtained for everyone by ignoring symptoms and using the pre-zero-time duration of positive X-ray films. With this tactic, however, some patients would be fallaciously coded as having very short durations because their only previous film was taken just before the date of diagnosis and zero-time treatment. The solution to this problem was to use the median duration of symptoms in all other patients as an arbitrary "imputed" value for symptoms of unknown duration. The progression interval for asymptomatic patients was then classified as the longer of either the radiographic duration or the imputed duration of symptoms. (The imputation strategy is discussed in Section 6.3.1.3.3.)

5.2.2. Consolidation of categories

A common analytic tactic is to consolidate the categories of a variable originally expressed in an array of nominal groups or ordinal grades.

In the lung cancer data set, the histologic cell types had originally been reported in a plethora of possible nominal categories. They had been consolidated, for certain previous analyses,[4] into nine categories. For the current activities, these nine categories were condensed, using criteria described elsewhere,[5] into four. Thus, the many groups of *cell type* were nominally categorized as **well-differentiated; small cell; anaplastic; and cytology only.** (The last category was used for patients with a positive pap smear for sputum or other site, but no histologic evidence of the cancer.)

The information for *cigarette smoking* was deliberately classified according to the patient's *customary* amount of smoking before any illness-induced changes. The number of packs per day was originally reported dimensionally. Since these dimensions were usually approximations, *customary cigarette smoking* was cited in the ordinal categories of **0** = None; **1** = less than 1/2 pack per day; **2** = at least 1/2 pack per day but less than one; **3** = at least one pack per day but less than 2; and **4** = two or more packs per day. (The duration of smoking had not been reported routinely enough to allow a citation for "pack-years.")

5.3. Construction of Boolean unions

The construction of Boolean unions is a particularly powerful tool in data analysis. In Boolean logic, a *union* is a combination of one or more binary attributes. The union is regarded as **present** whenever any one of the attributes is **present.** For two attributes, A and B, the Boolean union is verbally described as "A and/or B," and symbolized as A ∪ B. In contrast, the Boolean logical intersection of the two attributes is described as "both A and B," and symbolized as A ∩ B. With overbars (such as $\bar{A}$) to indicate absence of the attribute, the logical expressions can identify many complex distinctions. For example, the symbol (A ∩ B) ∪ (C ∩ $\bar{D}$) would represent the union achieved either when A and B are present together, or when C is present in the absence of D, or when both of the two main conditions exist.

As a more concrete example, the category *pulmonic symptom* in lung cancer is defined[1] as a union in which the patient has a bronchial symptom, a parenchymal symptom, a parietal symptom, or any combination of those symptoms. Each of those symptom categories is itself also defined as a union. Thus, the *brochial symptom* category contains patients who have hemoptysis, a new cough, a distinct change in an old cough pattern, subjective wheezing, or any combination of one or more of those four manifestations. *Parenchymal* and *parietal* symptoms are unions with analogous clinical definitions.[1]

A Boolean logical strategy was used to construct the categories of the staging system described in Section 4.6.1. In one of the decisions, a union called *distant metastasis* was defined as present if any of several appropriate binary variables was present. Another union, called *regional metastasis,* was defined in an analogous manner. These two unions, and a binary category, called *local tumor,* were then demarcated with appropriate Boolean unions and intersections to form the three ordinal categories (Stages I, II, and III) of the staging system. The ordinalizing clusters were shown in Fig. 4-2.

The analytic power of Boolean unions is that they allow a substantive knowledge of biologic phenomena to be effectively applied in data analysis. If a set of clinical data were originally cited with such binary variables as *dyspnea on exertion, peripheral edema, hepatomegaly,* and *distended neck veins,* an analyst unfamiliar with clinical biology might preserve each variable individually and analyze it separately. A knowledgeable clinical analyst, however, could put the four variables together (with suitable diagnostic criteria) into a single union called *congestive heart failure.* An analogous union, which could be defined without any complex diagnostic criteria, was the *distant metastasis* category in Section 4.6.1.

A Boolean union can have two important analytic distinctions.[6,7] Qualitatively, the union creates a substantive category that cannot be attained with any form of statistical or computer analysis. Although mathematical strategies can easily be automated to look for logical intersections (which may then be called "interactions"), no exclusively mathematical strategy can readily be established to form a union that is sure to be biologically "sensible." Because such unions as *pulmonic symptoms, congestive heart failure,* or *distant metastasis* are qualitative biologic concepts, they must be defined and demarcated biologically. A computer or statistical procedure can be instructed to look for these unions after they have been previously identified, but their initial discernment must come from someone who recognizes the pertinent substantive biology. An understanding of these potential unions is a prerequisite for scientific plausibility or "biologic coherence" when categories from two or more variables are combined, as discussed in Section 4.6.2.2 and in Chapters 19 and 20.

The second important analytic distinction of unions is quantitative. Instead of small numbers being diffused among a series of individual variables that may not yield "statistically significant" results, the composite category of the union contains a larger frequency count and can therefore be quantitatively more effective in the total analysis.

5.3.1. Timing of construction

The binary variables that form constituents of a Boolean union can be entered into the analysis as separate independent variables, and the unions can be constructed afterward as a retrospective analytic procedure. The retrospective approach, however, might involve analyzing a multitude of independent variables. For example, the clinical and morphologic manifestations of lung cancer—entities such as hemoptysis, hepatomegaly, or a positive bone biopsy—had been listed as **present** or **absent** in binary expressions that originally occupied seventy-eight variables.[1] A suitably large statistical program and computer could manage all those variables, but the analysis and interpretation could become awkward and the megavariable procedure could itself[8] lead to unreliable results.

In most situations, therefore, and in the illustrative data set here, the most cogent unions are formed beforehand, and then entered into analysis as individual variables.

5.3.2. Binary-scale unions

Some unions, such as those just described for *pulmonic symptoms* and for *bronchial symptoms*, are expressed in a binary scale as **present** or **absent**.

5.3.3. Ordinal-scale unions

Many other unions, however, are organized as combinations of categories that form a cluster variable having an ordinal scale.

An ordinal-scale union in the illustrative data set was *TNM stage*. It was demarcated according to conventional criteria[9] that outline the cancer's anatomic spread as follows: **Stage I:** localized cancer with no evidence of spread; **Stage II:** involvement of adjacent regional, hilar, or mediastinal lymph nodes; **Stage IIIA:** "isothoracic" involvement elsewhere in the same side of the chest; **Stage IIIB:** "contrathoracic" involvement of the opposite side of the chest; **Stage IV:** distant involvement beyond the chest. In the "union" that defined each category, the patient was required to have evidence of the category together with no evidence of any higher category. In other words, the stage of a particular patient is assigned in a reverse sequence, starting with IV and ending with I, according to the first encountered stage whose criteria are satisfied.

A second ordinal-scale union, *symptom stage*, was a variable developed during previous research.[1,10] The classification depended on symptoms attributable or possibly attributable to the lung cancer. Symptoms that were unequivocally attributable to a co-morbid disease were excluded from these classifications. Many of the eligible symptoms were themselves unions; and the symptom stage variable had originally been cited in six ordinal categories: *distant*, for symptoms at a distant metastatic site (such as bone pain in lumbar spine); *regional*, for symptoms emanating from a regional metastasis (such as thoracic bone pain); *mediastinal* for symptoms of pertinent mediastinal metastases (such as hoarseness due to a paralyzed vocal cord, or the clinical syndrome of superior vena cava obstruction); *systemic* for complaints of significant weight loss or manifestations of paraneoplastic "hormonal" effects, such as hypertrophic osteoarthropathy; and *pulmonic* for a previously discussed union that included hemoptysis, dyspnea, manifestations of pneumonia, and other symptoms that could be ascribed to the cancer at its primary site. The final category was an *asymptomatic* group, containing patients with none of the foregoing manifestations.

For the analysis here, *symptom stage* was consolidated into four ordinal categories according to the highest of the cited manifestations. The four categories were **1:** asymptomatic; **2:** pulmonic or systemic symptoms (or both) but no higher-order symptoms; **3:** mediastinal or regional (or both) but no distant symptoms; and **4:** distant symptoms.

5.4. Principles of data coding

Because all numerical entries for an independent variable can be analyzed as though they are measured dimensions, the assignment of suitable coding digits is an important step in preparing multivariable data for computer processing. The discussion that follows will describe general principles of the coding strategy, as well as the particular choices made for the illustrative data set. The main concern now is with the actual coding digits given to the data entered for analysis. (Chapter 6 will describe some of the transformations used for "analytic coding" when the data were analyzed.)

Most of the main coding decisions are *intravariable:* they assign coding digits to categories in the scale of an individual variable. A few of the decisions discussed later, however, are *intervariable,* used to maintain consistency when coding digits are assigned among different variables.

5.4.1. Dimensional variables

Dimensional variables are preferably coded directly in their original dimensional expressions. They can be later transformed, if desired, to "improve" the statistical distribution of the variable, or to demarcate it into categories.

In the illustrative data set, dimensional codings were used for each patient's *survival duration* (in months), *age* (in years), *percent weight loss, progression interval* (in months), and *hematocrit* (in volume percent). Unknown values were left blank, rather than given an arbitrary designation, such as "99," because an additional set of instructions would have been needed to keep the "99" from being analyzed as a dimensional value.

5.4.2. Binary variables

A binary variable that refers to the existence of an entity can be simply coded **0** for *absent* and **1** for *present*. Aside from being easy to remember, this convention allows the mean value of the variable to show the relative frequency (or proportionate "rate") of occurrence of the entity.

Some data analysts will code binary variables as −1 and +1. The main purpose of this coding is to facilitate the algebraic analysis of interactions. Thus, if X_1 and X_2 are two binary variables each coded as **0** and **1**, the X_1X_2 interaction will be coded as **1** only when both entities are present; the code will be **0** if either one or both are absent. If the respective codes for each variable are −1 and +1, however, the X_1X_2 interaction will be coded as +1 either when both entities are **present** or when both are **absent.** If either one is absent, but not both, the interaction will be coded as −1. The appraisal of this coding convention depends on whether you think of "interaction" as an event, i.e., both things occur, or as a "symmetrical" relationship, with both variables going up or down together.

The **0/1** convention seems generally easier to understand, remember, and interpret. Besides, interactions for bivariate categories are probably best examined in a categorical rather than algebraic analysis.

The only binary independent variable in the lung cancer data set was *sex.* It was coded **0** for *female* and **1** for *male*. The alternative coding (**0** for *male* and **1** for *female*) would have been equally acceptable. The choice between **0/1** and **1/0** (or any other binary) codes will have an important effect, however, in analyses in which an associated regression coefficient must be interpreted to discern the impact of the variable.

5.4.3. Nominal variables

The categories of nominal variables can be directly coded with numerical digits such as **1, 2, 3, 4, ...** These digits cannot be used for algebraic analysis, however, because nominal categories do not have a quantitative ranking. To be used as independent algebraic variables, therefore, the categories receive the transformations discussed in the next subsection. In a discriminant function analysis, when nominal categories become the dependent target, they are managed separately; and their original coding digits serve only as identifications.

5.4.3.1. 'Dummy' binary variables

A customary coding tactic for nominal variables is to convert each category into a "dummy" binary variable. For example, the four main categories of *cell type* (well-differentiated, small cell, anaplastic, and cytology only) could each be expressed as a separate binary variable that is either **present** or **absent.**

If a particular person can have more than one set of nominal attributes (for such variables as *occupation* or *diagnosis*), each categorical attribute must be cited separately as a binary variable. In the *cell type* variable here, however, and in many other nominal variables, the categories are mutually exclusive, i.e., a patient can have only one of them. For m mutually exclusive nominal categories, only $m - 1$ binary variables are needed for the coding, since a person cited as **0** in all the others must have the one that remains. Thus, if three binary variables with **0/1** ratings are used for *well-differentiated, small cell,* and *anaplastic* cell types, a code of **0** in all three binary variables would denote patients with *cytology only.*

Despite the economy in coding, this $m - 1$ tactic has a distinct disadvantage. The omitted category, chosen arbitrarily, does not have a direct opportunity to demonstrate its analytic importance or to receive its own weighting coefficient. To prevent this problem, an equal-opportunity data analyst might want to construct a dummy binary variable for each of the m nominal categories.

The latter tactic is both satisfactory and desirable in categorical cluster analyses, because each category is retained for the analysis. In algebraic-score analyses, however, statistical relationships among all of the candidate variables are often appraised simultaneously before the formal calculations begin. The appraisal may show that **1** values for patients in the mth category-variable have a perfect negative correlation with the collection of **0** values in the other $m - 1$ transformed variables. For example, for the four categories of cell type, patients coded as **1** for *cytology only* will be perfectly correlated with the combined group of patients coded **0** for *well-differentiated, small cell,* or *anaplastic.* Patients coded **1** for *well-differentiated* will be correlated with **0** values in the other three categories; and so on.

Because perfect correlations cause major problems (discussed later) during the multivariate computations, a computerized algebraic program will eliminate one of the *m* binary variables. In the arbitrary choice made by different computer programs, the eliminated binary variable may be the one with the fewest members, or the one that happened to have the last number in the identifying coded digits. To avoid an arbitrary deletion, the data analyst can make an advance choice of which dummy binary variable to eliminate.

Because of the difficulty of interpreting results for the dummy category, *cell type* was seldom used as a concomitant independent variable in the illustrative algebraic analyses. As noted later, the effect of cell type could be examined in a different way, by doing a special regression procedure that analyzed each cell type separately within their combined results as a group.

5.4.3.2. Other coding patterns

The **0/1** binary transformations just described for m − 1 variables are sometimes called "reference-cell codings," because the mth original category, which is coded as **0** in each of the m − 1 variables, becomes a "reference cell" or "control," for all the others. In computer jargon, this pattern is also sometimes called a "partial" coding system.

Two other systems of coding for nominal variables both employ m − 1 dummy variables, but the coding patterns are different. In a second system, called "effect" or "marginal" coding, the m − 1 binary variables are cited with **0** or **1,** as in reference coding. The last (i.e., the mth) original category, however, is coded as −**1** instead of **0** in each of the new m − 1 binary variables.

In a third system, called "orthogonal coding," each of the m − 1 binary variables is given a set of calculated coding digits that will produce what is called "orthogonal comparability" for the variables. The relative merits of the different coding systems receive a particularly lucid discussion in the book by Kerlinger and Pedhazur.[11]

The effect and the orthogonal coding systems have some theoretical advantages for certain mathematical purposes that seldom occur in pragmatic algebraic analyses. For customary procedures, the 0-1 reference-cell coding system is usually recommended;[12] and it was used here.

5.4.4. Ordinal variables

Because each category is retained as a category, the coding of ordinal variables creates no problems in categorical-cluster analysis. In algebraic analyses, however, the coding creates two tricky decisions. The first is whether to use the arbitrary values of ordinal coding digits as though they were dimensions. If so, the second decision is to choose the arbitrary values of those digits.

Arguing that dimensional status should not be given to arbitrary ordinal ratings, some data analysts prefer to transform the ordinal data while preserving a ranked connotation. In the transformation, a special coding scheme[13] is used for a series of dichotomous splits that transform the m ordinal categories into m - 1 "ranked" new binary variables. A value of 1 in each new binary variable represents the presence of a particular ordinal category as well as any category that has a higher ranking in the ordinal scale. A **0** is entered in the new variable for any of the lower-ranked categories.

The pattern of transformed ranked binary coding is illustrated below for an original ordinal variable that contains four categories (**1, 2, 3, 4**). They are converted to three new binary dummy variables, V_1, V_2, and V_3, that are each coded as **0/1**:

Code in original ordinal variable	Code in three new binary dummy variables		
	V_1	V_2	V_3
1	0	0	0
2	1	0	0
3	1	1	0
4	1	1	1

Thus V_1, the first new binary variable, would be coded as **1** for patients with an original ordinal rating of **2, 3,** or **4,** and coded as **0** for the original rating of **1**. The codes for V_2, the second binary variable, would be **1** for patients with an original rating of **3** or **4;** and **0** for the other two original ratings. V_3 would be coded as **1** for the original rating of **4;** and **0** for all others. Ratings of **1** in the original variable would be represented by a code of **0** in all three new variables.

This type of transformation has the mathematical advantage of preserving an ordered arrangement while avoiding arbitrary quantitative values for the ordinal ratings. The transformation has a substantial disadvantage, however, when the algebraic results are interpreted: the overall effect or impact of the original ordinal variable is not indicated by a single coefficient. The coefficients of the transformed binary variables will denote the effect of each binary partition when the original ordinal variable is decomposed, but will not allow the variable's overall effect to be compared directly with other variables that remained intact. Therefore, if a prime goal of the analysis is to determine the impact of the ordinal variable, its binary ranked decomposition is not a desirable transformation.

When the individual ordinal categories are retained, the decision about codes depends on whether the ordinal variable is null-based or conditional. A null-based coding—such as **0, 1, 2, 3,** . . . —is used for an entity, such as *magnitude of pain,* that can be either absent or present in different degrees of severity. A conditional coding—such as **1, 2, 3, 4,** . . . —is used for a condition, such as stage of cancer, that is always present in some degree for each person under analysis. In the illustrative data set here, *TNM stage* and *symptom stage* were coded as conditional ordinal variables with respective values of **1, 2,** . . . , **5** and **1, 2, 3, 4.** *Cigarette smoking* received a null-based ordinal coding of **0, 1,** . . . , **4.**

5.4.5. Choice of ordinal codings

For algebraic analyses of ordinal variables, reasonable arguments can be offered both for and against the policy of maintaining the original codes rather than using ranked binary transformations. The most common policy in the past has been to preserve the original codes, probably because the results of the transformed binary variables are not easy to understand and interpret. In a comparison of results when different multivariable methods were applied to the same sets of data, Walter et al.[14] used the new ranked binary transformations to preserve mathematical "propriety" in the comparisons. In the algebraic analyses here, the customary old approach was chosen despite its mathematical imperfection, because it is easy to understand and particularly because it lets the regression coefficients indicate the effect of the *entire* ordinal variable, rather than its binary decompositions.

5.4.6. 'Dummy' coding for zones of dimensional variables

In certain algebraic analyses in which multivariable regression is used mainly for activities such as "age-adjustment" or "age-standardization," a dimensional variable such as age can be converted to "standard" ordinal zones, such as . . . , 50–59, 60–69, etc. Each of these zones is then entered in binary form as a separate dummy variable. Thus, **age 50–59** would be one variable that is present or absent; **age 60–69** would be a second such variable; and so on. With m ordinal zones, $m - 1$ dummy variables can be prepared in a manner analogous to the reference cell coding system.

The results will not show the impact of *age* with a single coefficient, but the other (non-age) variables attain coefficients that are presumably "adjusted" for each age group. With this technique, the multiple regression procedure can produce the coefficients that were formerly attained with a direct age-standardization process.

In the algebraic analyses here, *age* and the other dimensional variables were retained in their dimensional expressions. For categorical-cluster analyses, however, and for certain regression diagnostics, the dimensional variables were divided into ordinal zones that will be discussed later.

5.4.7. Coding the target variable

For multiple linear regression, survival time was entered into analysis as the natural logarithm of the dimensional values. This transformation has two virtues for this set of data. It helps make the distribution of survival times more Gaussian (by reducing the effect of a few very high outlier values) and it ensures positive values for all estimates of survival time. For all other analyses, survival was coded as **0/1** for the patient's dead or alive status as various intervals after zero time.

5.4.8. Intervariable coding

In an algebraic analysis, the scheme of coding for individual variables should be maintained consistently across the variables. For example, if **0/1** rather than **1/2** or **−1/+1** is chosen for coding binary variables, it should be used for all such variables. If nominal variables are routinely converted into $m - 1$ reference-cell binary categories, all nominal variables should be coded in that manner. Ordinal variables should be coded in concordant directions, so that **1, . . . , 5** represents a consistent increase (or decrease) in the magnitude of the presumed effect for each variable.

This type of consistency helps preserve accuracy when the data are coded, and is also helpful when the results are interpreted. For example, unless the codings are consistent, ordinal variables that have similar directions in their effects may get opposite signs in their regression coefficients; another problem is the reversal of directions of gradients, making categorical arrays difficult to interpret.

5.5. Importance of classification and coding decisions

The pre-analytic classification and coding decisions have crucial effects on what can be done with the results. These decisions determine whether important variables will be omitted, whether the included variables will receive a useful taxonomy, and whether the coefficients in algebraic analyses can be interpreted effectively.

5.5.1. Choice of important variables

The classification decisions are important in both categorical-cluster and algebraic-score analyses. For example, although prognostic importance has now been unequivocally demonstrated[1,15] for such critical variables as *symptom stage, symptom severity,* and *severity of co-morbid diseases,* they have regularly been omitted from multivariable analyses of cancer therapy. The main reason for the omission has been the previous absence of a suitable set of taxonomic categories for the "judgmental" information. In the illustrative analyses here, *symptom stage* was included, but *symptom severity* and *severity of co-morbid diseases* were omitted to avoid a surfeit of variables.

5.5.2. Effectiveness of taxonomy

Some of the most important variables in medical research are expressed[16] in ordinal grades of severity for such features as symptoms, illness, co-morbidity, morphologic manifestations, behavioral patterns, or diverse sociopersonal factors. Consequently, the results of an analysis can be substantially affected by arbitrary taxonomic decisions that arrange the ordinal variables in different ways and in different numbers of categories. Whether the taxonomic choices are optimal or inadequate will often require a judgmental evaluation, using substantive knowledge of what each variable represents in the biologic context of each analysis. The merits or demerits of different taxonomic approaches can sometimes be documented, as shown later in Part V of the text, if suitable data are available for the taxonomic categories.

An excellent example of taxonomic problems recently occurred when a multivariable analysis of long-term mortality after prostatic surgery led to a highly publicized claim[17] that the commonly used transurethral prostatectomy was more lethal than an open (transabdominal) operation. In the original analysis, the information was obtained from the discharge summaries entered in "claims data;" and co-morbidity was classified in only two categories. In a subsequent study,[18] however, the co-morbidity information was obtained directly from medical records and was classified in three or more ordinal categories of severity. In the new analyses—

which used improvements in data, taxonomy, and adjustment for severity of co-morbidity—the two prostatic operations were found to have similar effects on long-term mortality.

In patients with lung cancer, the TNM staging-system variable had been classified for many years in three ordinal categories.[19] In more recent years, however, the number of categories was expanded[9] to the five used here in the illustrative data set. The ordinal variable called *symptom stage* was originally expressed in nine categories, but was consolidated, after suitable evaluations in research elsewhere,[1] to the four used in the illustrative analyses.

The main point of these remarks is to emphasize that ordinal variables are often critically important but are constructed arbitrarily. The statistical impact of each construction can be analytically quantified, but the true impact of each variable cannot be properly evaluated without going beneath the statistical results to appraise the original construction itself.

5.5.3. Interpretation of coefficients

In categorical-cluster analyses, the impact of different categories or variables is shown *directly* by the gradient in the target variable. In algebraic analyses, the impact of each variable is shown *indirectly* by the calculated regression coefficient. The magnitude and direction of this coefficient will depend on the taxonomy used for coding the variable. For example, the coefficient for *sex* will have different meanings if the men and women are coded as **0/1, 1/0, 1/2, 2/1,** or in some other digits. In the multivariable procedures of logistic regression and Cox regression, coefficients such as b are raised to the e^b power and then regarded as "relative risks" or "relative hazards." The values of e^b will be more than doubled if the same binary variable is coded as **−1/+1** rather than **0/1**.

For these reasons, an evaluation of regression coefficients requires careful attention to the original coding digits used for the variables. (Unfortunately, coding digits are often not listed[20] in published reports, thus suggesting that the authors, reviewers, or editors were probably unaware that the omission might have made the results difficult or impossible for readers to interpret.)

5.6. Analytic procedures and 'packaged programs'

After the coding systems are developed, the individual items of raw data will be coded and transferred to suitable analytic media, such as magnetized disks. The immediate next step is to check that these transfers in typing (or "key punching") were carried out accurately for each item of data. After this verification, the next step is to check certain features of the univariate and bivariate *distributions*, before the actual multivariable analysis begins. Because both of these steps (as well as the multivariable procedures themselves) are usually carried out with "packaged" computer programs, the systems of packages warrant separate discussion.

The popularity of multivariable analyses has led to many computer-operated data processing systems or packages that contain each multivariable procedure as a program in the system. The packaged systems of programs are usually known by such acronyms as BMDP, GENSTAT, GLIM, OSIRIS, SAS, and SPSS. Some of the systems are prepared for operation on a personal computer; others (designed for particularly huge data sets) require a mainframe computer.

The mainframe computations described in this text were done in batch mode on the IBM 3090 computer at the Yale University Computer Center. The data were entered through a personal IBM PC/AT, which serves as a remote terminal connection to the mainframe computer. The personal computer (PC) operations were executed in batch mode on an IBM/PS2 model 386 machine with 6 megabytes of RAM, under MS/DOS version 4.1.

5.6.1. Choice of packaged programs

The packaged programs that illustrate this text were taken from the SAS and BMDP systems, which are commonly employed in medical research throughout North America. (Both the BMDP and SAS acronyms have now become so well known and so well developed commercially that the components of the acronyms are not identified in the most recent publications[21,22] issued either by the "SAS Institute" or by "BMDP Statistical Software." BMDP stands for *BioMeDical* programs, P series; and SAS stands for *Statistical Analysis System*.)

Each packaged system contains a set of programs identified with distinctive symbols, abbreviations, or acronyms. For example, a program that produces a univariate display of data is designated as PROC UNIVARIATE in the SAS system, and as BMDP2D in the BMDP system. The packages contain procedures for essentially the same set of programs, which are usually managed in relatively similar ways by the "black box" computations within each system. The two external ends of the programs, however, will differ. The systems have different "input" mechanisms for their instructional commands, and different "output" arrangements for their printout.

The proprietors of the BMDP and SAS systems periodically issue new "versions" that may slightly or substantially revise the format and operations of the programs that appear in previous versions. In addition, when prepared for use on a relatively small personal computer, multivariable regression programs may be somewhat changed from the corresponding operation on a large mainframe computer.

The latest available versions of each package were used for the text here, but in a few instances (noted later), an earlier version of a program offered some distinct advantages. The illustrative displays were generally chosen from printouts produced by the mainframe computer because they are "prettier." The printer attached to the mainframe computer uses a horizontal page, which provides additional space for clear and compact displays, avoiding the "wrap-around" continuations that are frequently needed for the narrower vertical page of a PC printer. The sources of each program have been carefully identified in the text, but the printed results may not always coincide with what emerges in the versions and computers available elsewhere. SAS Version 6.07 (and occasionally 5.18) was used for the mainframe, and version 6.04 for the PC. The BMDP version 1990 (IBM/OS) was used for the mainframe, and BMDP/386 Version 1991 for the PC.

The selection of the BMDP and SAS systems for this text does not constitute either an endorsement or a recommendation of either system. They were used because they are well known and were conveniently available. Like all computer packages, the BMDP and SAS systems have advantages and disadvantages (some of which will be noted later).

5.6.2 Beginning the operations

Statistical packages are not as easy to operate as other programs, such as word processing, that can be used with a personal computer. To apply any packaged statistical program in any of the systems, the user must first become familiar with a complex set of instructions and operational commands. The details of those instructions, stated in the pertinent manuals for each system, are beyond the scope of this text, which assumes that the reader (or an associated colleague) is able to make the programs work.

5.6.3. Choice of necessary and 'optional' options

To increase versatility, each program has many "options." Some of them are indeed optional: if not solicited, they are not used. Other options are not really options. Constructed as an inherent feature of the program, they will appear even if uninvited. Yet other options are necessary requirements for the program to operate. If the user does not specify a desired value (or decision) for the option, the program will use its own "default" value. For example, in the sequential stepping procedures for algebraic analyses, certain boundary values are needed for entrance or deletion of variables. If no boundary option has been specified by the user, the program operates with its own default choice. Because the default boundaries are chosen differently in different systems, a user who omits the optional specifications should always determine whether the default values are satisfactory for the desired analytic goals. (The different results that can be produced by different default values are discussed in Section 11.9.)

The components of any program's operation and printout can therefore be divided into the inherent, necessary, and extra options. The inherent options, over which the data analyst has no control, represent the basic operations and display arrangements prepared by the computer programmer(s). The necessary options are the instructions required for the program to make certain decisions. If no specifications are offered for these options, the program will use its own default selection. The extra options are used to solicit additional embellishments that will appear only if specifically requested.

Most of the activities here were done in a "no frills" manner, with each program's own default decisions and without the adornment of the extra options. In various instances noted in the text, however, the default values were replaced by other specifications; and some or all of the extra options were sometimes solicited.

5.6.4. Inspecting the printout

The many options that give the computer programs their versatility are also responsible for many problems in the printout of results. The problems occur so often and can sometimes be so confusing that some users refer to the inspection process as "surviving the printout."

The problems in printout may arise because of uninvited inherent options, unsatisfactory presentations and labels, or inadvertent errors.

5.6.4.1. Uninvited options

The persons who prepare each program would like to make it as versatile as possible. Accordingly, the results often include many of the enormous numbers of analytic strategies that have been proposed by creative statisticians. Because certain inherent options may appear even if unsolicited, the consequences are reminiscent of what sometimes happens in medicine today in an automated chemical analysis of blood. Wanting to know only a patient's blood sugar and creatinine levels, the clinician also receives many unsolicited measurements, such as serum sodium, calcium, and bilirubin. The multiple additional chemical measurements can often be dismissed as merely unnecessary and useless, but sometimes they create difficulties if an unexpected abnormality occurs in an unsolicited test, or if an unfamiliar (but possibly important) new test is listed in the group.

The problem is easy to manage if the user understands the unsolicited results appearing in the printout. Each uninvited result can be ignored, or possibly given attention if it seems to offer something worthwhile. If unfamiliar with the option and its result, however, the user will look for help in the associated instruction manual. Unfortunately, the published instructions are usually intended to describe how to operate the programs and identify the results, not to explain the strategy of the procedures or the interpretation of the output. Consequently, when results of certain unfamiliar options appear in the printout without being clearly described in the corresponding manual, users may be baffled by unknown things that were not solicited, expected, or adequately explained. For example, a result for *boundary on condition* appears uninvitedly but repeatedly in the SAS printout for stepwise multiple linear regression. Because this phrase is not clearly defined or explained in the associated instruction manual, a user would have to trace and check the idea through a published reference.

5.6.4.2. Unsatisfactory presentations and labels

A different set of problems arises because the person(s) who wrote the packaged programs, being intimately familiar with the contents, may forget that the familiarity is not universally shared. Sections of the printout may be presented in a nonlogical sequence or in arrangements that are difficult to comprehend; or the labels may be ambiguous in location or in name. For example, the term *error variance* is disconcerting for readers who may begin worrying that something has gone wrong. They may start to think about scientific sources of error until realizing that the statistical term refers not to "error," but to a group's *residual variance* around the estimates made by an algebraic model. Several other problems will be noted later during illustrations of individual procedures.

5.6.4.3. Inadvertent errors

From time to time, despite the apparent omnipotence of computers, the program may make an error, usually because it was given a wrong formula or strategy for the calculations. For example, in reviewing a new packaged program for data in sequential clinical trials, Farrington[23] noted several errors in the determination of the median and 90th percentile values. Because the BMDP and SAS systems have been widely used for many years, almost all the errors have been discovered and corrected. Nevertheless, an occasional defect may still survive. For example, in the current BMDP program for all possible multiple linear regressions, as discussed in Section 11.7.3, a "best" regression is chosen according to an index proposed by Mallows, but the program does not make its choice according to Mallows criteria for decisions. In a previous version of one of the programs, an index called "adjusted R^2" was not correctly calculated, but the error has been repaired in the most recent version.

When you first use a particular program in a particular system, it is a good idea to check the computed results (whenever possible) with a hand calculator. If things come out as they should, you need not worry about undiscovered inadvertent errors.

5.6.4.4. General comments

None of the foregoing remarks is intended to detract from the hard work and admirable efforts that have gone into making the packaged programs so readily accessible to both sophisticated and unsophisticated users, and that produce a generally excellent job. The main purpose of the cautionary remarks is to warn users about problems that arise because the programs were generally planned for experienced professionals, not for amateurs.

This text is intended to help both amateur and professional users understand and interpret the results of the analyses, particularly in published literature where the printouts themselves are not presented. An amateur user who is directly confronted by the printouts, however, will often require help from a sophisticated professional; and even the sophisticates will not be immune from occasional uncertainty or confusion.

5.7. Verification of data entry

Before any univariate distributions or other analytic results are examined, the data analyst's first scientific obligation is to check the accuracy of data entry when information is transferred into the magnetized disk (or other storage medium). Many slips can occur during the typing process that converts the written data of forms or sheets into coded electronic storage.

In laboratory work, this type of checking involves the concept of replication. Each specimen is divided into aliquots, and each of two or three aliquots is measured separately. The idea of replication is also used for checking the accuracy of coded conversions in data entry. In the early days of electronic computation, the data were first keypunched into special cards and then "verified" when the cards were keypunched a second time in a separate machine called a verifier. For particularly fastidious work, the process was repeated again to produce double verification; and arrangements were made to have different persons do each job of keypunching.

With modern forms of magnetic media, this process is carried out differently, and the data are entered directly onto a magnetic disk or tape. The data entry should be done at least twice, preferably by different persons. Verification is accomplished with computer programs that compare the two "copies" of the data, character by character, and then print out a listing of discrepancies. In an excellent account of problems and operations in the verification process, Blumenstein[24] has offered a vigorous argument in favor of double-keying, rather than a single-keying process that is checked by visual instead of automated "audit."

Unfortunately, a concern with accuracy of the data-entry process is not always apparent in many research protocols; and published reports of computerized analyses almost never indicate the existence or even awareness of a need to verify the electronically coded information. In fact, with certain modern techniques, the original data may be entered directly into electronic storage, without a formal written record. The opportunity for verification by replication is thus destroyed.

The absence of adequate attention to the basic scientific quality and accuracy of data is not a unique phenomenon, but the surprising feature of the modern scene is that scientists who are extraordinarily careful about the quality of laboratory procedures may give no attention to the need for analogous care about the quality of electronic coding in computerized data processing. In an era of automation, scientists may have forgotten that the basic computerized data are not themselves the product of automation. The data are entered by human beings, who have all the potential for human errors of omission or commission during the entry process.

Newspapers and other media regularly report some of the outstanding blunders that occur in computer work when this process is not adequately supervised or checked. In medical research projects, the errors may not always have dramatic public impact, but the effects can be devastating for individual projects, particularly in those with relatively small sample sizes. A major error in a single value of a critical variable can make unimportant results seem important, or vice versa.

The magnitude and effects of these errors have seldom been investigated or reported. One recent study[25] was done after a hospital purchased a new information technology system to enable "decentralization," in which the responsibility for coding diagnoses and procedures was transferred from clerks to the medical staff. When the "completeness and accuracy of clinical coding" were checked before and after the system was installed, the doctors were found to be much more accurate than the clerks, although the initial "reluctance and resentment" of the medical staff required special efforts for "motivating . . . change."

For reasons of both economy and quality, the investigators in the Cardiac Arrhythmia Suppression Trial[26] decided to use a distributed system, rather than collecting all data for entry at a single "central" station. Within that system, however, a randomized crossover experiment compared single-entry (by one person) vs. double-entry (by two persons or twice by one person) methods of "keying" data from written forms to the microcomputer codings. Seeking a desirable error rate below 10 per 10,000 fields, the investigators found that the best participating clinics had single-entry error rates of 17 and double-entry error rates of 8 per 10,000 fields. (These rates were as high as 250 or 292 for some of the individual operators, and as low as 0 for others.) The investigators concluded that the added quality of the double-entry process outweighed the extra time required. The error rates of single-key entry were determined without comparison in three studies,[27–29] and were found to be substantially lowered by a double-entry method in another study.[30]

I know of only one investigation[31] in which the consequences of errors in data coding were directly evaluated. When the computerized data were compared with the original raw information for a subset of cases in a large cooperative enterprise,[32] so many errors were found that the reviewers doubted the scientific credibility of several prominent conclusions in the original research.

5.8. Preliminary statistical appraisals

After the encoded data have been verified, certain preliminary statistical appraisals are done before the multivariable analysis begins. The purpose of these appraisals is to "improve" certain univariate distributions, and to help guide the judgmental decisions made during algebraic and categorical analyses. The appraisals are described in the next four chapters.

Chapter References

1. Feinstein and Wells, 1990; 2. Feinstein, 1985 (*Clinical epidemiology*); 3. Feinstein, 1969; 4. Yesner, 1973; 5. Feinstein, 1974; 6. Feinstein, 1967; 7. Feinstein, 1993; 8. Concato, 1993 (Clin. Res.); 9. Mountain, 1987; 10. Feinstein, 1966; 11. Kerlinger, 1973; 12. Kleinbaum, 1988; 13. Walter, 1987; 14. Walter, 1990; 15. Feinstein, 1985 (Arch. Intern. Med.); 16. Feinstein, 1987; 17. Roos, 1989; 18. Concato, 1992; 19. American Joint Committee on Cancer, 1979; 20. Concato, 1993 (Ann. Intern. Med.); 21. SAS/STAT User's Guide, 1990; 22. Dixon, 1990; 23. Farrington, 1993; 24. Blumenstein, 1993; 25. Yeoh, 1993; 26. Reynolds-Haertle, 1992; 27. Crombie, 1986; 28. Birkett, 1988; 29. Prud'homme, 1989; 30. Neaton, 1990; 31. Wiseman, 1984; 32. Heinonen, 1977.

6 Appraisal of Simple Displays

Outline . . . .

6.1 Individual observations
6.2 Univariate distributions
    6.2.1. Gross summaries
    6.2.2. Individual distributions
        6.2.2.1. Categorical variables
        6.2.2.2. Dimensional independent variables
        6.2.2.3. Special 'normal' plots
6.3 Checks for univariate problems
    6.3.1. Missing data
        6.3.1.1. Patterns of missing data
        6.3.1.2. Hazards of missing data
            6.3.1.2.1. Reduced sample size
            6.3.1.2.2. Biased sample
            6.3.1.2.3. Statistical deception
        6.3.1.3. Management of problems
            6.3.1.3.1. Isolation
            6.3.1.3.2. Transfer
            6.3.1.3.3. Imputation
            6.3.1.3.4. Segregation
    6.3.2. Maldistributions
        6.3.2.1. Grossly unbalanced categorical distributions
        6.3.2.2. Eccentric dimensional distributions
        6.3.2.3. Detection of eccentricities
        6.3.2.4. Eccentricities in illustrative data set
        6.3.2.5. Hazards and remedies for eccentricity
            6.3.2.5.1. Distortion of summary indexes
            6.3.2.5.2. Transformation of variables
            6.3.2.5.3. Problems in tests of 'significance'
            6.3.2.5.4. Hazards of transformations
            6.3.2.5.5. 'Robustness' of data and tests
    6.3.3. Appraisal of target variable
6.4. Examination of bivariate displays
    6.4.1. Purpose of examination
    6.4.2. Methods of evaluation

. . . .

The relatively simple displays to be checked before the start of a multivariable analysis consist of individual observations, univariate distributions, and certain bivariate relationships. They are examined to find defects that can be repaired, and clues and evidence that can be used when the multivariate procedures are applied.

106 Preparing for Multivariable Analysis

This chapter discusses the displays of observations and the appraisal of problems in univariate distributions. Chapters 7-9 describe bivariate evaluations and other immediate preparations for multivariable analysis.

6.1. Individual observations

The "multivariate matrix" of Fig. 6-1 shows coded observations for the first 25 of 200 persons in the illustrative data set.

```
              S     S          L           A  M  S  T  S  P     H     P     W  S  A  C
              U     U          N           G  A  M  N  X  C     C     R     E  M  N  Y
              R     R          S           E  L  K  M  S  T     T     O     L  A  A  T
              V     V          U              E  A  S     W           G     L  L  P  O
              6     I          R              M  T  A     T           I        L     L
                    V          V              T  A  G     L           N
                    E                         A  G  E     O
  CASE                                        G     E     S
   NO.                                        E
  ----  -  -----  ----------  --  -  -  -  -  ----  ----  ----  -  -  -  -
    1   1   12.8   2.549450   45  1  3  1  2   0.0  46.0  54.1  1  0  0  0
    2   1   99.9   4.604170   62  1  2  1  1   0.0  42.0   4.6  0  0  1  0
    3   1   94.0   4.543289   58  1  3  1  1   0.0  42.0   4.6  1  0  0  0
    4   1   14.4   2.667230   54  1  3  1  1   3.0  51.0   9.0  0  0  1  0
    5   1   17.8   2.879200   58  1  3  1  2   5.7  34.5   7.3  0  0  1  0
    6   0    5.8   1.757859   67  1  3  1  2   0.0  42.0   7.3  1  0  0  0
    7   1   99.9   4.604170   53  1  2  1  2   0.0  43.5   7.7  1  0  0  0
    8   0    5.1   1.629239   55  0  4  1  4   7.9  42.0   1.2  0  0  1  0
    9   1   38.5   3.650660   64  1  3  1  1   0.0  47.0  24.7  1  0  0  0
   10   1   82.3   4.410370   54  1  4  1  2   0.0  44.0  36.5  1  0  0  0
   11   1   99.9   4.604170   44  1  3  2  2   0.0  54.0   1.3  1  0  0  0
   12   1   66.4   4.195700   70  0  0  1  2   3.8  39.3  99.0  1  0  0  0
   13   1   22.1   3.095579   52  1  4  1  1   6.4  37.8   5.8  1  0  0  0
   14   0    5.2   1.648660   47  1  3  2  2   8.8  33.0   2.3  0  0  1  0
   15   1   23.2   3.144150   60  1  4  1  2   8.1  42.0   7.7  0  0  1  0
   16   1   13.5   2.602690   56  1  3  1  2   4.9  45.0   1.8  1  0  0  0
   17   0    5.3   1.667709   56  1  4  2  2   0.0  45.6  58.0  1  0  0  0
   18   1   99.9   4.604170   56  1  3  2  2   0.0  45.0   5.3  1  0  0  0
   19   1   29.6   3.387770   63  1  2  3  2  19.4  40.5   3.5  1  0  0  0
   20   1   15.8   2.760010   61  1  4  1  2  11.1  34.5  54.1  1  0  0  0
   21   0    2.9   1.064710   64  1  3  1  4  10.1  33.3   2.6  1  0  0  0
   22   0    3.8   1.334999   67  1  1  1  4  24.7  33.0  24.5  1  0  0  0
   23   0    0.0  -3.506559   65  1  3  1  2   7.5  36.3  18.2  1  0  0  0
   24   1   15.2   2.721299   63  1  3  1  2  18.6  35.0  20.1  1  0  0  0
   25   1   98.4   4.589040   59  1  3  1  2   2.2  41.7   3.1  0  0  1  0
```

Fig. 6-1 Matrix of observed multivariable data for 25 persons. (Produced as option in BMDP1D program for simple data description.)

In the BMDP1D program that produced Fig. 6-1, the patients are numbered ("CASE NO.") in rows 1 through 25. The columns show fifteen variables, of which the first three represent different formats for survival time. Survival at six months (SURV6) is shown with **1** for alive and **0** for dead; survival duration in months is shown as SURVIVE; and the natural logarithm of the duration is marked LNSURV. In case number 23, who died within one day (.03 months) of zero time, SURVIVE is reported as 0.0, while LNSURV, the natural logarithm of 0.03, is cited as −3.506559. (Note that the six or more decimal places often shown in the computer printouts will usually be "rounded" in the text here to two or three places.)

The remaining twelve variables sequentially show age (in years), sex (marked **1** or **0** for MALE), customary amount of cigarette smoking (marked SMKAMT), TNM stage, symptom stage (SXSTAGE), percent weight lost (PCTWTLOS), hematocrit (HCT), and progression interval (PROGIN). The last four columns show the four binary-transformed dummy variables for cell type as *well-differentiated, small cell, anaplastic,* or *cytology only.*

This data set had been thoroughly checked in various ways before being chosen for illustration here. If such checks have not yet occurred, the individual observations should first be inspected for errors in coding, of which the most obvious are unauthorized or out-of-range codes. For example, SURV6, MALE, and the four cell types (WELL, SMALL, ANAP, and CYTOL) are all binary variables, coded as **0** or **1**. TNMSTAGE and SXSTAGE are ordinal variables, coded respectively as **1, . . . , 5** and **1, . . . , 4**. Anything other than the authorized codes in these columns would be an error. The five dimensional variables—SURVIVE, LNSURV, PCTWTLOS, AGE, and PROGIN—can be checked for values that exceed the expected range. In this instance, SURVIVE can extend from 0.03 to 99.9 months and LNSURV, from -3.51 to 4.60. The AGE of the 200 persons ranged from 32 to 85. PCTWTLOS extended from 0 to 33.6. Any values outside these ranges would be erroneous.

In a second set of checks, the multivariable matrix of observations would be inspected for patterns of omission that are discussed in Section 6.3.1.1.

6.2. Univariate distributions

The easiest way to check for coding errors is to examine the univariate distributions for the frequency of each value of data in the set for each variable. Out-of-authorized-range values can be promptly detected in printouts showing the extent of dimensional values or categorical coding digits.

Categorical variables are usually displayed in a table showing the individual values of data, their associated frequency, and their relative frequency as a proportion of the total distribution. For dimensional variables, the display may be expanded to include certain special summary statistics and graphical portraits. In addition to conventional means, medians, and standard deviations, the display can show "stem-leaf plots" that correspond to histograms, and "box plots" that indicate selected percentiles and other special demarcations of the data. The univariate display may also offer a diversity of additional statistical indexes and graphs, discussed later, that often show much more than the analyst usually wants to know.

6.2.1. Gross summaries

The single table in Fig. 6-2 shows gross univariate summaries for the fifteen variables listed in Fig. 6-1. The summaries contain each variable's total frequency, mean, standard deviation, standard error of mean, coefficient of variation, and the largest and smallest values with their corresponding standardized Z scores. In any variable X, the Z score for individual items, marked with subscripts as i, is obtained as $Z_i = (X_i - \bar{X})/s$, where $\bar{X}$ is the mean and s is the standard deviation of the variable. For example, case number 4 in Fig. 6-1 has a hematocrit value of 51.0. From the mean of 41.089 and standard deviation of 5.1792 shown for hematocrit in Fig. 6-2, the corresponding Z score would be $Z_4 = (51.0 - 41.089)/5.1792 = 1.91$.

VARIABLE NO	NAME	TOTAL FREQUENCY	MEAN	STANDARD DEVIATION	ST. ERR OF MEAN	COEFF. OF VARIATION	SMALLEST VALUE	Z-SCORE	LARGEST VALUE	Z-SCORE
1	SURV6	200	.47000	.50035	.03538	1.0646	0.0000	-0.94	1.0000	1.06
2	SURVIVE	200	14.312	23.609	1.6694	1.6496	0.0000	-0.61	99.900	3.63
3	LNSURV	200	1.4829	1.7725	.12534	1.1953	-3.5066	-2.81	4.6042	1.76
4	AGE	200	61.084	9.4814	.67043	.15522	32.000	-3.07	85.000	2.52
5	MALE	200	.91000	.28690	.02029	.31527	0.0000	-3.17	1.0000	0.31
6	SMKAMT	200	3.0050	.91056	.06439	.30302	0.0000	-3.30	4.0000	1.09
7	TNMSTAGE	200	3.2150	1.6316	.11537	.50749	1.0000	-1.36	5.0000	1.09
8	SXSTAGE	200	2.6150	1.0642	.07525	.40695	1.0000	-1.52	4.0000	1.30
9	PCTWTLOS	200	8.2670	8.0069	.56617	.96855	0.0000	-1.03	33.600	3.16
10	HCT	200	41.089	5.1792	.36623	.12605	24.000	-3.30	56.000	2.88
11	PROGIN	200	14.414	18.248	1.2903	1.2660	.70000	-0.75	99.900	4.64
12	WELL	200	.41500	.49395	.03493	1.1903	0.0000	-0.84	1.0000	1.18
13	SMALL	200	.12000	.32577	.02304	2.7148	0.0000	-0.37	1.0000	2.70
14	ANAP	200	.37000	.48401	.03422	1.3081	0.0000	-0.76	1.0000	1.30
15	CYTOL	200	.09500	.29395	.02079	3.0943	0.0000	-0.32	1.0000	3.08

Fig. 6-2 Summaries of univariate distribution for fifteen variables in illustrative data set. (Produced by BMDP1D simple data description program.)

The printout from the corresponding SAS program, PROC MEANS, is not shown here. It displays each variable's total frequency, mean, standard deviation, sum of values, and minimum and maximum values, regardless of the variable's dimensional or arbitrary numerical codings.

The summaries for each variable in Fig. 6-2 have the following useful roles:

1. The smallest and largest values can be checked for errors occurring as out-of-range codes.

2. The results for total frequency will immediately indicate variables with missing values. In this instance, since no data were missing, each variable had 200 values in Fig. 6-2.

3. The maximum and minimum Z-score values can act as clues to variables that have maldistributed data, as discussed in Section 6.3.2. The maldistributions are more easily detected, however, when each variable's distribution is later examined individually in greater detail.

4. The standard deviations for the target and independent variables will later be useful, as discussed in Chapter 10, for standardizing any unstandardized partial regression coefficients in multivariable analysis).

6.2.2. Individual distributions

For nondimensional (i.e., categorical) variables, the univariate distributions are best shown as frequencies and relative frequencies. For dimensional data, the univariate display becomes much more extensive.

6.2.2.1. Categorical variables

Fig. 6-3 shows a tabulation for each of the eight categorical variables in the lung cancer data set. For these displays, a binary dummy variable was used for each of the four nominal categories of *cell type*.

UNIVARIATE STATISTICS

TNM STAGE

TNMSTAGE	Frequency	Percent	Cumulative Frequency	Cumulative Percent
I	51	25.5	51	25.5
II	27	13.5	78	39.0
IIIA	18	9.0	96	48.0
IIIB	36	18.0	132	66.0
IV	68	34.0	200	100.0

SYMPTOM STAGE

SXSTAGE	Frequency	Percent	Cumulative Frequency	Cumulative Percent
ASX	28	14.0	28	14.0
PULM/SYST	82	41.0	110	55.0
REG/MEDIAST	29	14.5	139	69.5
DISTANT	61	30.5	200	100.0

MALE GENDER

MALE	Frequency	Percent	Cumulative Frequency	Cumulative Percent
ABSENT	18	9.0	18	9.0
PRESENT	182	91.0	200	100.0

RANK CUSTOMARY PPD CIGS

SMKAMT	Frequency	Percent	Cumulative Frequency	Cumulative Percent
NONE	9	4.5	9	4.5
< 1/2	4	2.0	13	6.5
1/2 TO <1	16	8.0	29	14.5
1 TO <2	119	59.5	148	74.0
2 OR >2	52	26.0	200	100.0

WELL DIFF HISTOLOGY

WELL	Frequency	Percent	Cumulative Frequency	Cumulative Percent
ABSENT	117	58.5	117	58.5
PRESENT	83	41.5	200	100.0

SMALL CELL HISTOLOGY

SMALL	Frequency	Percent	Cumulative Frequency	Cumulative Percent
ABSENT	176	88.0	176	88.0
PRESENT	24	12.0	200	100.0

ANAPLASTIC HISTOLOGY

ANAP	Frequency	Percent	Cumulative Frequency	Cumulative Percent
ABSENT	126	63.0	126	63.0
PRESENT	74	37.0	200	100.0

CYTOLOGIC PROOF ONLY

CYTOL	Frequency	Percent	Cumulative Frequency	Cumulative Percent
ABSENT	181	90.5	181	90.5
PRESENT	19	9.5	200	100.0

Fig. 6-3 Univariate frequencies for data of eight categorical variables. (SAS PROC FREQ program.)

6.2.2.2. Dimensional independent variables

Fig. 6-4 shows the univariate distribution of the dimensional independent variable *Hematocrit*. The upper left-hand section of the display, marked "moments", shows the mean at 41.09, with standard deviation of 5.18. Most of the other information in this section can be ignored. It offers statistical indexes for decisions about whether the distribution is Gaussian and whether the mean differs from 0.

The upper middle section of 6-4 shows numerical values for various quantiles, including the median and quartiles (marked Q3 and Q1) that form the horizontal lines of the box plot discussed shortly. [Quantiles can be determined in five ways in the SAS program. The "(Def = 5)" statement here indicates that the program used its default option, which is the most conventional method of calculation.] The upper right of Fig. 6-4 shows the five lowest and highest values in the distribution. The Frequency Table occupying the entire lower part of Fig. 6-4 displays the individual counts, and the percentages for individual and cumulative frequencies of each item (or "value") in the distribution.

The stem-leaf plot and the box plot (appearing just below the upper displays) were originally devised by John Tukey[1] and are now often shown in computer portraits of univariate distributions. The stem-leaf plot is a type of horizontal histogram, created by using the leading digits of the data as "stems", instead of demarcating the intervals used in conventional vertical histograms. The terminal digits of data for each stem became the adjacent "leaves", with their bulk showing relative frequencies. The shape of the stem-leaf plot will immediately indicate the general configuration of the distribution.

The box (sometimes called "box-and-whiskers") plot shows some special summary indexes and outlier values. The upper and lower boundaries of the "box" are formed by lines at the upper and lower quartiles of the data. These lines in the box plot of the SAS displays are placed at essentially the same level that the values occupy in the stem-leaf plot. The median is shown with a horizontal line across the interior of the box, and the mean is indicated with an appropriate sign, here a "+", which is usually inside the box, and which happens, in this instance, to coincide with the median. The "whiskers" are lines extending vertically above and below the box to certain demarcated boundaries that vary among different computer programs. In the SAS box plot here, the whiskers extend above and below the box for a distance of no more than 1.5 interquartile ranges on each side. The outlier values, beyond the whiskers, are shown with special symbols. In the SAS program, they are marked with 0 if within 3 interquartile ranges and with * if more extreme.

As discussed in Section 6.3.2.2., the stem-leaf plot and the box plot are probably the most useful part of the display in Fig. 6-4, because they demonstrate the shape and other summary characteristics that allow prompt decisions about the eccentricity of the distribution. (The "normal probability plot" is discussed in Section 6.2.2.3.)

Variable=HCT   HEMATOCRIT

                Moments

N   200   Sum Wgts   200
Mean   41.09   Sum   8218
Std Dev   5.179307   Variance   26.82523
Skewness   0.029179   Kurtosis   0.270536
USS   343015.8   CSS   5338.22
CV   12.60479   Std Mean   0.366232
T:Mean=0   112.1965   Pr>|T| 0.0001
Num ^= 0   200   Num > 0   200
M(Sign)   100   Pr>=|M| 0.0001
Sgn Rank   10050   Pr>=|S| 0.0001
W:Normal   0.989689   Pr<W   0.9308

           Quantiles(Def=5)   Extremes

100% Max   56   99%   53.5   Lowest   Obs   Highest   Obs
75%   Q3   44   95%   50   24(193)   52(146)
50%   Med   41   90%   48   28(65)   53(78)
25%   Q1   38   10%   34.1   30(64)   53(185)
 0%   Min   24   5%   33   31(31)   54(11)
                      1%   29   31(50)   56(115)
Range   32
Q3-Q1   6
Mode   42

Stem Leaf   #   Boxplot
 56 0   1
 54 0   1
 52 000   3
 50 00000000   8
 48 000000000   9
 46 000000000001   12
 44 000000000000000000000000006   27   +-----+
 42 0000000000000000000000090000000000   34 | |
 40 000000000000000000000500000047   27   *--+--*
 38 000000000000000040000000000003399   29 | |
 36 0000000330000000000558   19   +-----+
 34 0000002550000000000   19
 32 070000003   9
 30 000   3
 28 0   1
 26
 24 0   1
    ----+----+----+----+----+----+----+----

Normal Probability Plot

57+   *
                                               ++
                                          ****+++
                                       ******++
                                     ******+
                                   *****+
                                 ****+
41+   ******
                          ******
                       ******
                    ******+
                 *****++
              ****+++
           +++++
25+**+
   +----+----+----+----+----+----+----+----+----+----+
  -2   -1   0   +1   +2

Frequency Table

                Percents   Percents   Percents
Value Count   Cell   Cum   Value Count   Cell   Cum   Value Count   Cell   Cum
 24   1   0.5   0.5   36   4   2.0   17.0   41   9   4.5   51.5
 28   1   0.5   1.0   36.3   2   1.0   18.0   41.4   1   0.5   52.0
 30   1   0.5   1.5   37   10   5.0   23.0   41.7   1   0.5   52.5
 31   2   1.0   2.5   37.5   2   1.0   24.0   42   18   9.0   61.5
 32   1   0.5   3.0   37.8   1   0.5   24.5   42.9   1   0.5   62.0
 32.7   1   0.5   3.5   38   10   5.0   29.5   43   12   6.0   68.0
 33   6   3.0   6.5   38.4   1   0.5   30.0   43.5   3   1.5   69.5
 33.3   1   0.5   7.0   39   14   7.0   37.0   44   12   6.0   75.5
 34   6   3.0   10.0   39.3   2   1.0   38.0   45   14   7.0   82.5
 34.2   1   0.5   10.5   39.9   2   1.0   39.0   45.6   1   0.5   83.0
 34.5   2   1.0   11.5   40   15   7.5   46.5   46   8   4.0   87.0
 35   7   3.5   15.0   40.5   1   0.5   47.0   47   3   1.5   88.5

Value Count   Cell   Cum
 47.1   1   0.5   89.0
 48   5   2.5   91.5
 48.9   1   0.5   92.0
 49   3   1.5   93.5
 50   6   3.0   96.5
 51   2   1.0   97.5
 52   1   0.5   98.0
 53   2   1.0   99.0
 54   1   0.5   99.5
 56   1   0.5 100.0

Fig. 6-4 Univariate display for Hematocrit. (SAS PROC UNIVARIATE program. PLOT option produces the stem-leaf, box, and normal probability plots. FREQ option produces the frequency table.)

112 Preparing for Multivariable Analysis

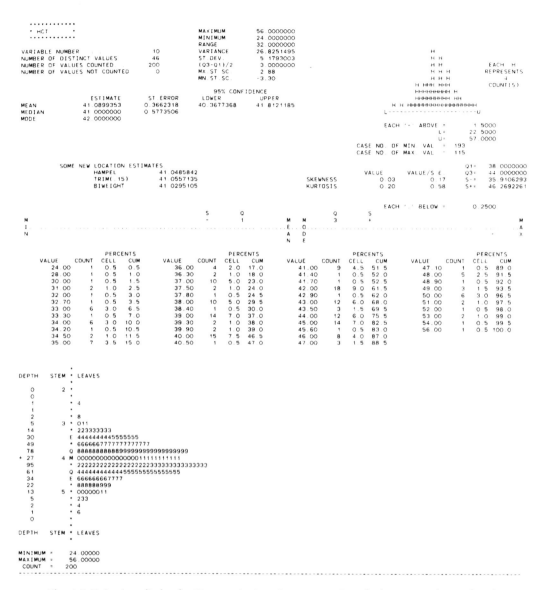

Fig. 6-5 Univariate display for Hematocrit. (BMDP2D program. Stem-leaf output and "new location estimates" are options.)

Fig. 6-5 shows the BMDP univariate display for this same variable (hematocrit). The BMDP and SAS programs show relatively similar information, but each has features missing in the other. The BMDP program provides a standard error for the median, 95% confidence interval for the mean, a vertical histogram (in the upper right-hand corner), and three "new location estimates" that are not offered in the SAS display. The BMDP program, however, does not show a conventional box plot. Instead, just above the individual values and their frequency counts, the BMDP printout has a single dashed line extending horizontally across the entire page from "MIN" to "MAX". The relative locations are cited on that line for standard deviations, lower and upper quartiles, mean, and mode, but the actual values at those locations are not easily identified. The simple, effective communication of a conventional Tukey-style box plot is lost.

Because the SAS displays have conventional box plots and are somewhat more visually compact, they have been used for the univariate displays of the other dimensional variables here. The displays appear and are examined for eccentricity in Section 6.3.2.4.

6.2.2.3. Special 'normal' plots

Although not routinely useful, special "normal" graphs are available for analysts who want to check specifically for Gaussian distributions. In the SAS program that produced Fig. 6-4, the "normal probability plot" on the right shows what the ranks of the observed values of the variable (hematocrit) would look like if they had a Gaussian distribution.

For this graph, the location of quantiles in the observed distribution is plotted against the quantile locations expected for the ranks of the observations in a normal Gaussian distribution. The asterisks show the observed values. The "+" signs show a reference straight line drawn for the expected Gaussian ranks. Observed data from a Gaussian distribution should coincide with this line.

A more intensive evaluation of Gaussian normality is available in the BMPD5D program, which produces three types of normal probability plots (not shown here) for each variable. One plot is similar to the SAS normal probability graph just discussed, but the axes are reversed, with observed values on the abscissa and "expected" values on the ordinate. The BMDP5D printout uses * marks for the actual values and / lines for those that would occur with a Gaussian distribution. A BMDP, "half-normal plot" is used mainly for ignoring the sign of the residual values when they are examined *after* a multivariable analysis. The graph is similar to the normal plot except that (according to the BMDP manual) the "expected values are computed by using only the positive half of the normal distribution".

The third "normal" plot, which is perhaps the most useful, is called *detrended normal probability*. Before the plot is printed, the linear trend is removed by using a counterpart of "standardized residual values". Each observation is converted into a standardized Z score, which is subtracted from the corresponding score expected at that quantile of the Gaussian distribution. The vertical scale of the graph shows these deviations, which should form a horizontal line through zero if everything were perfectly Gaussian.

Fortunately, unlike the unnecessary diagnostic tests in medical examinations, these additional displays of deviations from Gaussian distributions involve almost no increase in cost except computer time and paper.

6.3. Checks for univariate problems

The "physical examination" of univariate distributions is aimed at finding problems in missing data and maldistributions.

6.3.1. Missing data

The collection of data for each variable of an individual person is usually called an *observation*. In algebraic methods of analysis, the entire observation is eliminated if data are missing for either the dependent Y variable or any one of the independent variables. The elimination creates no problem if many other observations are available for analysis. With multiple variables, however, the elimination of observations can produce diverse problems, according to the way the missing data are scattered among different variables. The different patterns of missing data lead to different hazards, requiring different methods of management.

6.3.1.1. Patterns of missing data

The most common pattern of omission is a "univariate concentration", in which many items of data are missing for a single variable. For example, in a multivariable analysis of the relationship of survival to age, sex, weight, hematocrit, white blood count, serum cholesterol, and CD-4 cell count, information may be available for the first six variables in a group of 150 patients, but the CD-4 cell count may be unknown in 40.

In another important pattern of omission, called "multivariate porosity", small amounts of missing data are diffusely scattered across different variables for different persons. The porosity pattern differs from what would occur if data were missing for each of several variables in the *same* persons. For example, suppose 9 items of data are missing for each of the seven variables in the foregoing group of 150 patients. If the missing items occur in the same 9 persons, the observations will be complete for 141 persons. If the missing data do not overlap, however, 9 items might be missing in *age* for one set of patients, in *sex* for another set, in *weight* for yet another set, and so on. The effect of 9 nonoverlapping losses for 7 variables can thus eliminate 63 observations from analysis.

6.3.1.2. Hazards of missing data

The main hazards of missing data are reduced sample size, bias, and statistical deceptions.

6.3.1.2.1. Reduced sample size

Because the entire observation is omitted from algebraic analyses if *any* of the variables has an unknown value, the "effective" sample size will be reduced to the persons who have known values for *all* of the variables. In the foregoing group of 150 people, the multivariable analysis would be confined, in the first instance, to the 100 persons (= 150 − 40) with complete observations, and in the second instance, to 87 (=150 − 63). All the other persons will be omitted, together with their data for all the other variables.

The reduced sample size is easy to detect if the missing data occur in univariate concentrations, where the absences can be noted promptly in the frequency counts of univariate displays for each variable. Multivariate porosity cannot be discerned from univariate displays, however, and requires an inspection of the matrix for the total array of data. The simplest way to search for multivariate porosity is to note which variables have missing data (from a summary such as Fig. 6-2) and then to examine the multivariable pattern of omissions in a complete matrix of observations (such as Fig. 6-1).

Data analysts may sometimes become so occupied with interpreting the multivariable results for goodness of fit, effective variables, etc., that the appropriate information is not checked, and the disparity is overlooked between the sample size in the submitted data and what was actually used in the analysis. Unless the missing-data hazard is recognized, an investigator may be deluded into thinking that data were analyzed for the full group of 150 people, when in fact the actual analyzed group contained a much smaller number. To avoid transmitting the delusion to readers, the report of an algebraic multivariable analysis should always state the effective sample size, not just the number of people whose data were entered into analysis.

The illustrative data set here contains no problems in multivariate porosity because the random sample of 200 was chosen only from those patients with complete data for all variables.

6.3.1.2.2. Biased sample

Both of the other two hazards of missing data arise from the reduced sample size. One of these hazards is a biased sample. The residual group of people who are actually analyzed may be a biased subset of the entire group under study, particularly if the data are missing for a reason that is in some way related to the outcome event. For example, the CD-4 cell count may have been ordered only for patients who are most seriously ill.

6.3.1.2.3. Statistical deception

An additional hazard of the reduced sample size is statistical deception. Important variables that might have achieved "statistical significance" in the full group may be dismissed as unimportant if their P values become too high in the smaller set of data.

6.3.1.3. Management of problems

Prevention is the best way to deal with the problems of missing data. During the research itself, vigorous efforts should be made to obtain and suitably record all the necessary information. In many situations, however, the problems are inevitable, and arrangements must be made for remedial rather than prophylactic management. The remedial methods consist of isolation, transfer, or imputation for data, and segregation for variables.

6.3.1.3.1. Isolation

Unknown values cannot be processed in algebraic analysis, but in categorical analyses, the **unknown** category can often be isolated, given its own identity as a category, and analyzed appropriately along with all the other categories. The tactic can work well for nominal and binary variables, where the categories need not be ranked. For the ranked arrangements of ordinal and dimensional variables, however, an appropriate rank cannot be established for the **unknown** category. In these circumstances, the missing data will have to be managed in some other way.

The problem can sometimes be resolved if an ordinal or dimensional variable is transformed to a binary scale. For example, *customary amount of cigarette smoking* may be **unknown** either because no data are available or because the amount is unstated for someone known to be a smoker. If the variable is changed so that it indicates binary existence rather than ranked amount of smoking, a smoker whose amount is unknown can be coded as yes for *customary cigarette smoking*, but a person whose existence of smoking is unknown must remain as **unknown.**

6.3.1.3.2. Transfer

Data from a suitable additional variable can sometimes be appropriately transferred, if the other variable offers a reasonable surrogate. For example, in the illustrative data set here, the value of 3 × hemoglobin was transferred and used instead of hematocrit when the value of hematocrit was unknown. This tactic offers a pleasant solution to the problem if a suitable additional variable is available, but the opportunities are uncommon. Serum ferritin and transferrin are another example of two variables having a close relationship[2] that could allow a surrogate transfer if necessary.

6.3.1.3.3. Imputation

A statistical method of dealing with missing univariate data or multivariate porosity is to "fill the holes" by using a process called *imputation*. In this process, the available values of data for each variable are used to assign or "impute" values for the data that are missing. From the known information for the variable, a suitably chosen central value (such as the mean or median) is assigned to each of the unknown items. For example, if we did not know someone's age, it might be imputed as the mean of 62.3 or the median of 59 found in the rest of the patients. In the illustrative data set, as described in Section 5.2.1, the median duration of symptoms in the rest of the cohort was imputed for patients who were asymptomatic.

The imputation procedure can be tricky; and its optimal use has received diverse statistical recommendations.[3-7] The tactic has obvious advantages in preventing the loss of observations, and obvious hazards in emphasizing central values, particularly if the missing data were missing because they represented extreme rather than central values in the distribution.

6.3.1.3.4. Segregation

After all the foregoing repairs, the remaining set of "corrected" data may still contain some missing items. The problem is then managed by doing segregated analyses.

In one form of segregation, a variable that contains a large amount of missing data is removed from the analysis. Thus, in the earlier cited example, if *CD-4 cell count* were missing for 40 patients, this variable could be removed. The analysis would then be done with all remaining variables for the total group of 150 people.

If worried about the effects of the omitted *CD-4 cell count* variable, we could do a separate segregated analysis on the 110 patients for whom all data are available. If the first and second sets of results are essentially similar, everything is splendid. If not, a solution will usually have to emerge not from the statistical analyses, but from appraisals of the clinical biology of what is being studied in the research itself.

In simple linear regression for two variables, X and Y, none of the cited problems are particularly important. The analyst can readily note what is missing, and can analyze the available data according to the principle of "what you see is what you get". If a great many variables are being analyzed, however, a diffuse scatter of missing data in each variable can produce extensive multivariate porosity. This problem cannot be readily solved by removing individual variables; and the removal of individual observations may eliminate so many people that the analytic results become untrustworthy. The general solution would be to see if the results agree in segregated analyses done for the set of complete observations, and in segregated analyses when variables with missing data are individually added one at a time.

6.3.2. Maldistributions

The decision that a variable is maldistributed will vary with the type of analytic method. During an algebraic analysis, each variable is essentially replaced with a pair of summary values (such as the mean and standard deviation) that can greatly misrepresent eccentric distributions. Furthermore, tests of "statistical significance" in algebraic analyses may rely on distributional assumptions that can be grossly violated by the shape of the data. In categorical analyses, these two problems are relatively unimportant because *algebraic* summary statistics are usually avoided, and the tests of "significance" may not rely on Gaussian distributions. With categorical analyses, therefore, the main problem is a grossly unbalanced distribution that will have little or no discriminatory power.

6.3.2.1. Grossly unbalanced categorical distributions

A categorical distribution is grossly unbalanced if most of its values occur in only one or two of several possible categories. The remaining categories may then be ineffective because of sparse members. The problem is particularly evident in distributions of binary data, but can usually be ignored unless one of the binary categories has too few members. For example, if 147 men and 3 women are the distribution for *sex*, the variable will not have enough women for the results to make useful discriminations. With sparse categories, such grossly unbalanced variables are usually omitted from the analysis.

If the imbalance occurs in nominal data, the problem of sparsity can be managed by combining the categories with relatively few members into an **other** or **miscellaneous** group. For example, suppose the frequency distribution for principal *diagnosis* shows **cancer of breast,** 2; **coronary heart disease,** 37; **liver disease,** 4; **pneumonia,** 46; and **miscellaneous,** 8. In this situation, the **miscellaneous** frequency could be expanded to 14 if the category incorporates the cases of breast cancer and liver disease.

For an unbalanced sparsity in ordinal data, adjacent categories can be consolidated. Suppose *severity of illness* is distributed as **none,** 42; **mild,** 53; **moderate,** 6; and **severe,** 4. For these data, the **moderate** and **severe** categories might be combined to include 10 people. The previous digital codes, if cited as **0, 1, 2, 3,** would be changed to **0, 1, 2.**

In the categorical independent variables of the illustrative data set shown in Fig. 6-3, several distributions are not well balanced, but sparse categories (with frequency values below 10) occur only in *amount of cigarette smoking* (RANK CUSTOMARY PPD CIGS,"). To deal with this problem, the "none" and "<1/2" categories of cigarette smoking could have been consolidated, but they were left intact here at least until their bivariate performance could be examined.

6.3.2.2. Eccentric dimensional distributions

A distribution of dimensional data is eccentric if it distinctively departs from the Gaussian (or "normal") pattern that is assumed or required for many statistical operations and inferences. In the customary Gaussian pattern, the curve of frequencies or relative frequencies is convex, symmetrical around the center, and tails off at the two ends. It looks like the profile of a bell or an old-fashioned admiral's cocked hat.

Eccentricities occur most commonly when the entire distribution is asymmetrically skewed, or when a few drastic outlying values appear at one or the other end (sometimes at both ends) of the distribution.

6.3.2.3. Detection of eccentricities

Eccentric distributions are easily detected from stem-leaf and box-plot displays. In a box plot, asymmetrical (or skew) distributions will be manifested by a wide separation between median and mean values and/or by substantial asymmetry in the respective distances inside the box from median to upper and lower quartiles. Outliers will be indicated by their excessive distances from the extreme ends of the distribution to the corresponding upper and lower quartiles.

The computer programs for univariate distributions may also show other measures of eccentricity, such as indexes of skewness and kurtosis, tests of normality, and the Gaussian probability plots discussed earlier. The extra information is interesting but seldom offers much beyond the things that can be readily discerned from the box plot.

6.3.2.4. Eccentricities in illustrative data set

The univariate distributions of each independent dimensional variable in the illustrative data set are appraised in this section.

For the box plot of *hematocrit,* as shown in Fig. 6-4, the median value of 41 is essentially the same as the mean, and the two halves of the box (at boundary levels of 44 and 38) are not substantially unequal—consistent with a basically Gaussian shape. The two 0 symbols, above and below the box beyond the whiskers, show the two highest and lowest values that might be outliers. They do not seem substantially unbalanced.

Appraisal of Simple Displays   119

The upper half of Fig. 6-6 shows the univariate distribution for PERCENT WEIGHT LOSS. Because 60 people were coded as having values of <2% for weight loss, the

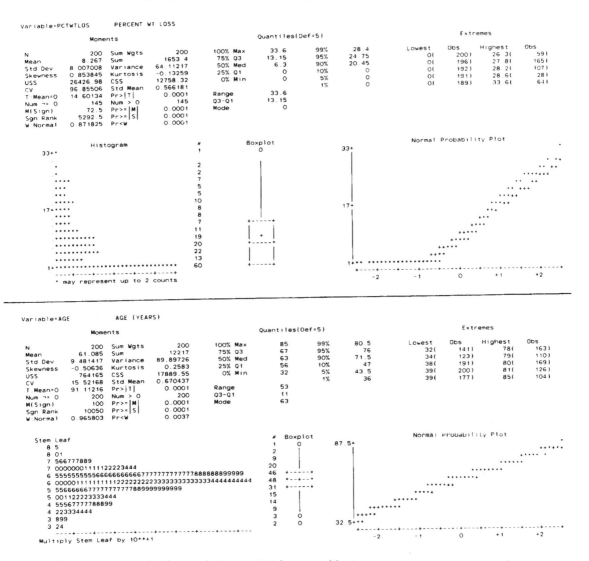

Fig. 6-6 Univariate distributions for Percent Weight Loss and for Age. (SAS PROC UNIVARIATE program.)

variable has an obviously skew distribution, which is shown best in the horizontal histogram on the left side. (When 48 or more observations occur in a single stem interval, the SAS program uses asterisks to produce a horizontal histogram rather than a stem-leaf plot.) Somewhat surprisingly, despite the skew, the upper and lower quartiles are at relatively similar distances from the median of 6.3, and the mean value (8.2) is not too far displaced. The normal probability plot for this variable shows the difficulty of trying to interpret a visual display that contains intermingled * and + signs. The non-Gaussian character of this plot is shown most obviously by the long horizontal tail of * marks on the left.

The univariate distribution for AGE is shown in the lower half of Fig. 6-6. The spacing of the box plot does not clearly show the asymmetry noted from the actual upper and lower quartile values. They are, respectively, 4 years above and 7 years below the median, which was at 63 years, compared with the mean of 61.1. The stem-leaf plot, however, shows that there asymmetry is not strikingly eccentric.

Fig. 6-7 shows PROGRESSION INTERVAL, which also had an eccentric distribution because of the many patients for whom the value was not more than 2.5 months. About half the group (97 members) had progression intervals less than 6 months. The box plot here does not look strikingly asymmetrical because the axis has been compressed to allow full coverage of the wide range of values, extending from 0.7 to 99. The asymmetry is better shown, however, in the printed values of Q3, median, and Q1, which occur, respectively, at 18.3, 6.3, and 3.2, with the mean at 14.4 months. The normal probability plot in this instance shows a relatively clear separation of the + and * values for the distinctly non-Gaussian distribution.

6.3.2.5. Hazards and remedies for eccentricity

The remedies for eccentricity depend on how serious its consequences seem to be.

6.3.2.5.1. Distortion of summary indexes

When each variable is statistically summarized by the mean and the variance (or standard deviation) calculated around the mean, the algebraic strategy may produce misleading or distorted results if the variable is not suitably represented by those indexes.

6.3.2.5.2. Transformation of variables

The main precaution against the threat of a distorted dimensional summary is to transform the variable into a suitable alternative expression. A particularly common arrangement, which frequently fits the pattern of medical data, uses logarithmic transformations. The logarithms are often designated as *ln* rather than *log* to indicate their calculation with the "natural" constant e (= 2.718 . . .) rather than with 10 as the base. As discussed earlier, a logarithmic transformation can be applied to the X variable (e.g., $Y = a + b \ln X$) or to the Y variable (e.g., $\ln Y = a + bX$).

Logarithmic transformations are particularly useful for demonstrating relationships in monotonic curving patterns, such as in Figs. 2-3 and 2-5. The logarithmic formats will improve fit, but are especially valuable for avoiding the deceptive relationships that can occur despite apparently well-fitting straight lines.

A logarithmic transformation is also valuable for distributions that form convexly shaped asymmetrical curves, and for reducing the effects of outliers. A different precaution, however, is to regard the outlying values as unknown, particularly if they seem peculiar enough to be errors rather than extreme values. Sometimes an outlier obviously contains wrong data, such as 250 years old for *age;* but 250 kg. for *weight* cannot necessarily be dismissed as an error in the data.

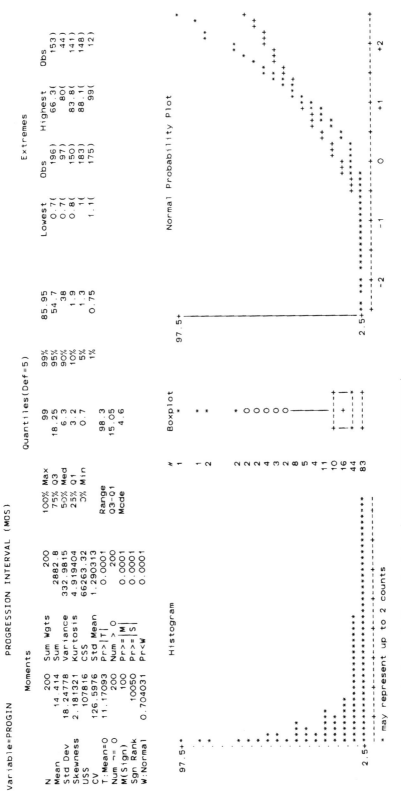

Fig. 6-7 Univariate distribution for Progression Interval. (SAS PROC UNIVARIATE program.)

6.3.2.5.3. Problems in tests of 'significance'

In the statistical analytic activities, Gaussian distributions are assumed for many inferential tests of "statistical significance", as well as for certain descriptive operations. For example, discriminant function analysis makes the assumption that not only is each variable Gaussian, but also that their simultaneous multivariate distribution is Gaussian. (For two variables, a bivariate Gaussian distribution would look like a Gaussian hill.) Because most medical variables are non-Gaussian, the violation of these assumptions can be distressing, particularly if the validity of crucial decisions becomes suspected.

Logarithmic transformations of data have the additional virtue, beyond those cited in the previous section, of helping reduce this distress. The data of many non-Gaussian medical variables often become much more Gaussian when expressed in logarithms rather than in the original measured values. Logarithmic transformations are often applied, therefore, to add pragmatic "legitimacy" to the mathematical theory that lies behind the analytic procedures.

6.3.2.5.4. Hazards of transformations

Although transformations have the advantage[8] of improving the fit of the model and its mathematical propriety, the main disadvantages occur when the results are interpreted. In the expression $Y = bX$, the value of Y rises b units for each unitary change in X. If $Y = b \ln X$, however, Y rises b units for each unitary change in $\ln X$. The result is easy to interpret if X is the only independent variable, but becomes difficult when impact is compared for multiple variables, some of which are expressed as $X_1, X_2, \ldots$ and others as $\ln X_3, \ln X_4, \ldots$. If the dependent variable is transformed to produce $\ln Y = bX$, the actual expression is $Y = e^{bX}$. Because of the exponentiation, the impact of X will be substantially different in this format from what is expected with the untransformed $Y = bX$. For example, if $\ln Y = 3X$, so that $Y = e^{3X}$, Y will be 1 when $X = 0$. If X rises by one unit from 0 to +1, Y will rise to $e^3 = 20.1$ units. If X rises from +1 to +2, the new value of Y will be $e^6 = 403.4$ units, so that the change is "multiplicative" [i.e., $(20.1) \times (20.1)$] rather than "additive".

6.3.2.5.5. 'Robustness' of data and tests

Medical literature regularly contains a striking disparity between what is actually done (often by sophisticated biostatisticians) and what is stated as precautions or warnings in published discussions of statistical principles. The warnings are commonly ignored.

A reason for the neglect is that data analysts are usually interested in bivariate (or multivariable) relationships, rather than univariate distributions. Thus, the analysts may wait to see what happens in those relationships before making adjustments, transformations, or other alterations in the individual variables. Although individual eccentric variables may not be adequately represented by means and standard deviations, the inadequacies may not affect relationships among variables. Besides, binary variables, dummy binary variables, and ordinal variables are constantly expressed algebraically with the impropriety of means and standard deviations. For all these reasons, the independent variables in the illustrative data set here were not changed until the bivariate analyses could be examined.

Another reason for ignoring the theoretical warnings, however, is that eccentricity is often a "paper tiger" in inferential tests of significance. Most of the tests are quite "robust" and will work reasonably well even when their basic assumptions and strategies have been grossly violated by the data. Besides, in situations where the algebraic methods are being used merely to "screen" the variables and where the "serious" analyses will be done by other procedures, the serious problems can be managed later.

The main value of discovering eccentricities, therefore, is to raise flags of warning about potential problems. If the problems seem really troublesome, suitable precautions can be taken before the analysis begins. Otherwise, the analyst may want to wait and explore again afterward to see which problems have really occurred.

6.3.3. Appraisal of target variable

As the focus of attention for all other variables, the target variable warrants a careful separate appraisal. Fig. 6-8 shows a univariate summary of the original values for *survival* (in months). In the frequency table in the lower half of the figure, the last line in the far right columns shows the seven survival-time values, mentioned earlier, that exceeded 99.9 months. In the horizontal histogram (replacing a stem-leaf plot), the data obviously do not have a Gaussian distribution. The box plot confirms the non-Gaussian features by showing an asymmetrical array of high outliers. In addition, the median (at 5.3 mos.) is far from the mean (14.3 mos.), with an upper-quartile-to-median zone (from 13.3 to 5.3) that is more than twice the median-to-lower-quartile zone (from 5.3 to 1.7). The latter distinctions are not displayed well in the excessively concise box plot, and must be noted directly from the quartile values. The normal probability plot also shows major disparities from a Gaussian shape.

Because the principles of multiple linear regression require a Gaussian target variable, survival time was transformed to get a better distribution. Fig. 6-9 shows the univariate results for *ln survival time*, which has a negative value whenever survival was less than 1 month. The stem-leaf plot immediately indicates that the transformed distribution is relatively Gaussian. In addition, the box plot shows a median (1.67) that is reasonably close to the mean (1.48), roughly equal interquartile zones (from 2.6 to 1.7 and from 1.7 to 0.5), and only 1 outlier location. The normal probability plot also looks much more Gaussian than before.

Regardless of these Gaussian desiderata, however, another important reason for logarithmically transforming the original survival times was to ensure positive results in any algebraic estimates. Thus, if the algebraic model is $\hat{Y}_i = a + bX_i$, the values of the estimated $\hat{Y}_i$ may sometimes be negative—an impossible phenomenon for survival times. If the model is $\ln \hat{Y}_i = a + bX_i$, however, negative values of the logarithm will still yield a positive result when transformed back to $\hat{Y}_i$.

124 Preparing for Multivariable Analysis

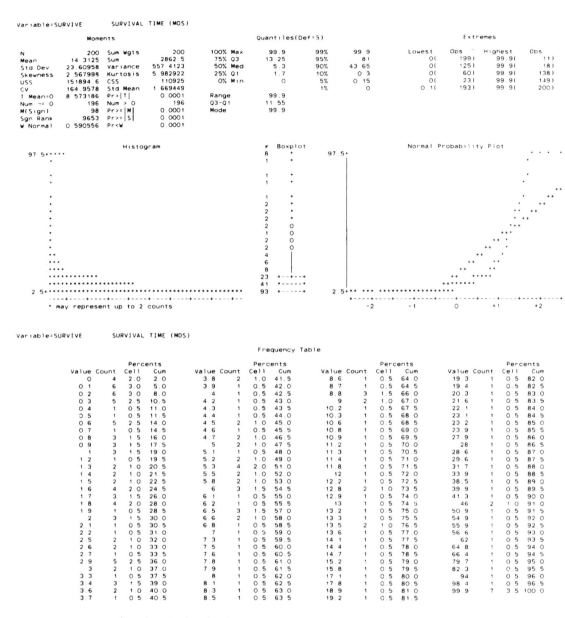

Fig. 6-8 Univariate distribution for survival time. (SAS PROC UNIVARIATE program.)

6.4. Examination of bivariate displays

Although multivariable relationships are difficult to visualize and analyze, bivariate relationships are relatively easy. Whether the multivariable analysis is done with categorical or algebraic methods, the bivariate relationships should always receive a preliminary screening examination.

Appraisal of Simple Displays 125

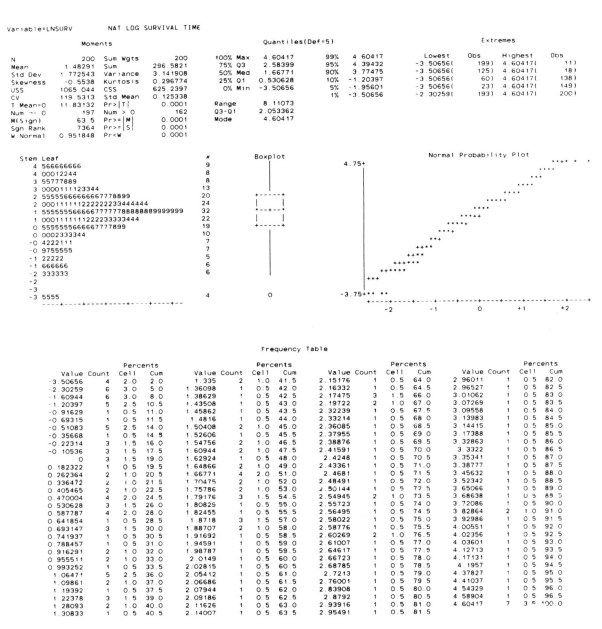

Fig. 6-9 Univariate distribution for natural logarithm of survival time. (SAS PROC UNIVARIATE program.)

As symbols for the discussion, X_j will represent any one of a total of k independent multiple variables, where j takes on the values of 1, 2, ..., k. The symbol $\{X_j\}$ will represent the set of independent variables. The k independent variables contained in $\{X_j\}$ will each have a targeted bivariate relationship with the target variable, Y. They will be examined as Y vs. X_1, Y vs. X_2, Y vs. X_3, ..., Y vs. X_k.

Among the k independent variables, there will also be (k) (k - 1)/2 pairs of bivariate relationships which will be examined as X_1 vs. X_2, X_1 vs. X_3, X_2 vs. X_3, X_1 vs. X_4, and so on. For example, with 8 independent variables, there will be 8 targeted relationships and 28 (= 8 × 7/2) intervariable relationships to be inspected.

6.4.1. Purpose of examination

The screening of bivariate displays has two purposes. First, problems in collinearity can be suggested for pairs of independent variables that are highly correlated. Second, and more importantly, the bivariate examinations help promptly identify the independent variables that may have important effects on the target variable. If none of the independent variables shows any interesting effects on the target, the total analysis may sometimes be terminated at that point. An independent variable that has no impact when examined for itself alone is seldom likely to "develop" effects when placed in a multivariable context.

Sometimes, however, the multivariable analysis can be aimed at revealing important effects that were obscured by counterbalancing or inefficient designs. Table 3-8 showed an otherwise hidden counterbalancing, when two treatments seemed similar because they had opposite effects in the different clinical subgroups; and Table 3-10 revealed the inefficient design with which the similar results in extreme prognostic groups for two treatments mitigated the effect of their different impacts in the middle prognostic group.

In most instances, however, the bivariate analyses will show distinctive targeted effects for one or several independent variables. The multivariable analyses will then be done mainly to confirm these effects or to disclose those that are "false positive" results. Occasionally—as in the examples of counterbalancing and inefficient design—the analyses may demonstrate an important "false negative" effect that was not previously apparent.

6.4.2. Methods of evaluation

The bivariate screenings are best done by examining two simple statistical procedures—correlation coefficients and stratified rates—that are extensively discussed in the next two chapters. These procedures are necessary background for the bivariate evaluations, but are particularly important because the simple analytic methods are prototypes for everything that happens later in multivariable analysis. The correlation coefficients emerge from simple linear regression, presented in Chapter 7, which uses a basic strategy employed thereafter in all the multivariable algebraic methods in Parts III and IV of the text. The stratified rates, presented in Chapter 8, involve statistical structures and evaluations that later appear in all the multivariable targeted cluster methods in Part V.

After these two types of relatively simple analysis are discussed, the bivariate displays of the illustrative data set will be presented and evaluated in Chapter 9.

Chapter References

1. Tukey, 1977; 2. Guyatt, 1992; 3. Afifi, 1966, 1967, and 1969; 4. Little, 1987; 5. Boswick, 1988; 6. Vach, 1991; 7. Schenker, 1993; 8. Grambsch, 1991.

7 Simple Linear Regression: The Algebraic Prototype

Outline . . . .

7.1. Basic data and symbols
   7.1.1. Basic fit
   7.1.2. Index of individual discrepancy
   7.1.3. Index of total discrepancy
7.2. Formation of a model
7.3. Operations of the model
   7.3.1. Index of individual discrepancy
   7.3.2. Index of total discrepancy
   7.3.3. Partition of deviation
   7.3.4. Partition of variance
7.4. Finding the coefficients
   7.4.1. Illustration of calculations
   7.4.2. Least-squares principle
7.5. Indexes of accomplishment
   7.5.1. r^2 and 'explained variance'
      7.5.1.1. Nomenclature for r^2
      7.5.1.2. Range of r^2
   7.5.2. F ratio of mean variances
      7.5.2.1. Mean of a group variance
      7.5.2.2. Concept of degrees of freedom
      7.5.2.3. Expressing the ratio
      7.5.2.4. Alternative expression and appraisal of F
   7.5.3. r as a correlation coefficient
   7.5.4. Standardized regression coefficient and impact of X
7.6. Stability ('statistical significance') of the indexes
   7.6.1. t test for r^2
   7.6.2. F test for F
   7.6.3. t test for b
   7.6.4. Confidence intervals for small values
7.7. Quantitative vs. stochastic 'significance'
7.8. Interpretation of quantitative accomplishments
   7.8.1. Fit of the model
   7.8.2. Effect of the independent variable(s)
   7.8.3. Stochastic vs. quantitative accomplishments
7.9. Selection of independent variables
7.10. Conformity of the model
   7.10.1. Reasons for checking conformity
   7.10.2. 'Regression diagnostics'
   7.10.3. Analytic goals and algebraic structures
   7.10.4. Inspecting the pattern of data
   7.10.5. Problems in indexes of accomplishment
      7.10.5.1. Distortions in good fits

      7.10.5.2. Distortions in poor fits
      7.10.5.3. Concomitant values for r^2 and b
      7.10.5.4. Stochastic deceptions
7.11. Preventing problems beforehand
  7.11.1. Problems in univariate distributions
  7.11.2. Transformations for 'sensibility'
  7.11.3. Problems of additive and multiplicative effects
  7.11.4. Examination of bivariate correlations
7.12. Special tests
  7.12.1. Pattern of residuals
    7.12.1.1. Univariate patterns
    7.12.1.2. Bivariate patterns
      7.12.1.2.1. 'Bivariate' checks for multiple variables
      7.12.1.2.2. 'Terminal' sources of distortion
      7.12.1.2.3. Connotations of pattern
      7.12.1.2.4. Standardized residuals
    7.12.1.3. Summary of diagnostic contributions by residual patterns
  7.12.2. Examination of ordinal zones
    7.12.2.1. Regressions for each zone
    7.12.2.2. Dummy binary variables
    7.12.2.3. Ordinal variables
    7.12.2.4. Interpretation of results
    7.12.2.5. Mechanism of calculations
  7.12.3. Check of quadratic regression
7.13. Management of problems
7.14. Subsequent procedures

- -

Each algebraic method of analysis for two or more variables begins by getting a "basic fit" for data of the target variable alone, without using any independent variables. At the end of the analysis, a selected mathematical model has accomplished the two main tasks of improving basic fit and indicating the relative impact of the chosen independent variable(s). The beginning and end of the analysis are connected by mechanisms for deciding which independent variables to include, and what coefficients to assign to each variable. These mechanisms create the complexity of multivariable algebraic procedures, but the basic operational strategies are similar to what happens in ordinary, simple linear regression for two variables. The simple procedure is therefore thoroughly reviewed in this chapter as a prototype for everything that happens later.

    Another prime reason for discussing simple linear regression now, however, is that it is essential background for doing the bivariate appraisals, conducted in Chapter 9, that should precede any multivariable activities. The correlation coefficients and other statistical indexes developed for simple regression are invaluable for screening the pair-wise bivariate relationships of the multiple variables, and for determining what to anticipate, avoid, or worry about when the multivariable procedures are chosen and applied.

7.1. Basic data and symbols

In simple linear regression, the basic data available for N persons have the following arrangement for a target variable, Y, and a single independent variable, X:

	Values of	
Person	X	Y
1	X_1	Y_1
2	X_2	Y_2
.	.	.
.	.	.
.	.	.
i	X_i	Y_i
.	.	.
.	.	.
.	.	.
N	X_N	Y_N

Note that the i subscripts here represent the individual persons identified as 1, 2, 3, . . . , i, . . . , N. The subscript j, which was introduced in Section 6.4, refers to individual independent variables: X_1, X_2, . . . , X_j, . . . , X_k. With strictly accurate symbols, double subscripts would be used to denote X_{ij} or $X_{i,j}$ as the value of variable j for person i. Thus, X_{38} would be the value of the eighth variable for the third person and $X_{43,17}$ would the value of the 17th variable for the 43rd person. The complex double subscripts can generally be avoided, since the context will usually be clear without them. Thus, since only a single X variable is discussed throughout this chapter, the single subscripts all indicate the identity of i for individual persons. Since Y is always a single target variable, its subscripts always represent i for persons.

In linear regression, the X and Y variables are usually each expressed in dimensional data, such as age, survival time, systolic blood pressure, serum cholesterol level, or hematocrit. For a simple illustration here, columns 1–3 of Table 7-1 show basic data for the relationship between X = blood urea nitrogen (BUN) and Y = hematocrit.

Table 7-1 Basic data for X_i and Y_i; and deviations for estimated values of $\hat{Y}_i$ in 6 persons

(1)	(2)	(3)	(4)	(5)	(6)	(7)
Person	X_i = Blood urea nitrogen	Y_i = Hematocrit	$Y_i - \bar{Y}$	$\hat{Y}_i$	$Y_i - \hat{Y}_i$	$\hat{Y}_i - \bar{Y}$
1	80	29	−2.33	28.06	0.94	−3.27
2	32	40	8.67	37.37	2.63	6.04
3	138	18	−13.33	16.81	1.19	−14.52
4	36	31	−0.33	36.60	−5.60	5.27
5	70	27	−4.33	30.00	−3.00	−1.33
6	23	43	11.67	39.12	3.88	7.79

7.1.1. Basic fit

Before any independent variables are introduced, the data for Y alone can be "fit" with their own "model." It is usually the mean, $\bar{Y} = \Sigma Y_i/N$. For the six values of Y in Table 7-1, the mean is $\bar{Y} = (29 + 40 + 18 + 31 + 27 + 43)/6 = 188/6 = 31.33$. If data were available only for Y_i, the results could be displayed as shown in Fig. 7-1.

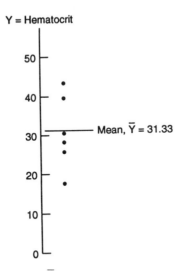

Fig. 7.1 Values of Y_i and $\bar{Y}$ for Hematocrit in Table 7-1.

7.1.2. Index of individual discrepancy

Whenever a model offers estimates for the actual values of Y_i, the discrepancy in fit can be noted between each estimate and the corresponding actual value. For dimensional data, with Y as the basic model, this discrepancy is the amount of the *deviation*, $Y_i - \bar{Y}$. In Table 7-1, with $\bar{Y} = 31.33$, the deviations are shown in column 4.

7.1.3. Index of total discrepancy

The index of total discrepancy for the entire group is obtained for dimensional data by squaring each deviation (to get rid of negative values) and then taking the sum of the squared deviations. It can be called the *group variance* or *basic group variance*, symbolized as

$$S_{yy} = \Sigma(Y_i - \bar{Y})^2. \qquad [7.1]$$

The "yy" subscript indicates that the sum of squared deviations comes from the products of $(Y_i - \bar{Y})(Y_i - \bar{Y})$. For the sum of each $(X_i - \bar{X})(X_i - \bar{X})$, the symbol is S_{xx}. Later on, when codeviation products of X and Y form *covariances*, the symbol is $S_{xy} = \Sigma(X_i - \bar{X})(Y_i - \bar{Y})$.

A more rapid computational formula for an electronic hand calculator is an algebraic conversion that restates expression [7.1] as

$$S_{yy} = \Sigma Y_i^2 - N\bar{Y}^2. \qquad [7.2]$$

Because formulas [7.1] or [7.2] could lead to rounding errors if $\bar{Y}$ is not extended beyond several decimal places, a better way to calculate S_{yy} is with another algebraic conversion,

$$S_{yy} = \Sigma Y_i^2 - [(\Sigma Y_i)^2/N], \qquad [7.3]$$

which allows no possible rounding until the division by N.

If formula [7.1] is applied to the data in column 4 of Table 7-1, we get S_{yy} = $(-2.33)^2 + (8.67)^2 + \ldots + (11.67)^2 = 413.3334$. For formula [7.2] or [7.3], we first calculate $\Sigma Y_i^2 = 29^2 + 40^2 + \ldots + 43^2 = 6304$. Using $\bar{Y} = 31.33$ and substituting in formula [7.2], we would get $S_{yy} = 6304 - (6)(31.33)^2 = 414.5866$. Using $\Sigma Y_i = 188$ and substituting in formula [7.3], the result is $S_{yy} = 6304 - [(188)^2/6] = 413.3333$. Of the three forms of calculation, the last is the most accurate.

Many names have been given to S_{yy}. They include *deviance, total variance, group variance, basic group variance, sum of squares, total sum of squares,* and *corrected sum of squares*. For simplicity, S_{yy} can be called *basic group variance*, because it represents the basic total fit when each Y_i is estimated by the mean, $\bar{Y}$. The subsequent regression activities are intended to improve this fit.

7.2. Formation of a model

The values of the dependent target variable, Y_i, can also be estimated from the "score" produced when each value of X_i is inserted into a particular algebraic format called a *model*. The algebraic estimates, as noted later, will almost always produce a better fit than what is achieved with the basic summary value, $\bar{Y}$, alone.

If several independent variables are available, they can all be candidates for inclusion in the model. In simple regression, however, the only independent variable is X. Fig. 7-2 shows a graph of the values for Y_i and X_i in Table 7-1. The main

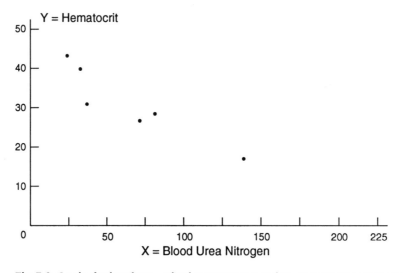

Fig. 7-2 Graph of values for X = Blood Urea Nitrogen and Y = Hematocrit in Table 7-1.

decision in relating these two variables is to choose the particular format of the algebraic model. For reasons noted earlier, Y is usually assumed to have a straight line relationship, expressed as Y = a + bX. This rectilinear shape seems quite suitable for the display in Fig. 7-2. In the algebraic model, the value of *b* is called the *regression coefficient*, and *a* is called the *intercept*, showing the value of Y when X is 0. The best values for *a* and *b* are determined by the strategy described in Section 7.3.

After *a* and *b* are found, each X_i can be used to estimate the corresponding value of Y_i as

$$\hat{Y}_i = a + bX_i.$$

For the data under discussion, appropriate calculations (shown in Section 7.4.1) produce $a = 43.58$ and $b = -.194$. The line, expressed as $\hat{Y}_i = 43.58 - .194X_i$, is shown as the solid "first" line in Fig. 7-3. (The "second" line is discussed in Section 7.5.3.)

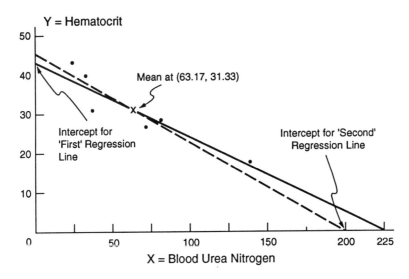

Fig. 7-3 Graph showing the two regression lines for data of Table 7-1 and Fig. 7-2. The solid "first" line is the regression of Y on X. The dashed "second" line is the regression of X on Y.

7.3. Operations of the model

The model's accomplishments in improving the basic fit of the data are determined from the indexes of discrepancy and total discrepancy for the estimated values of Y_i.

7.3.1. Index of individual discrepancy

With the model estimate from $\hat{Y}_i$, each discrepancy in fit can be calculated as $Y_i - \hat{Y}_i$. It is called the *residual deviation* between the actual and estimated values. The values estimated as $\hat{Y}_i = 43.58 - .194X_i$ are shown in column 5 of Table 7-1; and the residual deviations for $Y_i - \hat{Y}_i$ are in column 6. For example, when $X_i = 80$,

$\hat{Y}_i = 43.58 - (.194)(80) = 43.58 - 15.52 = 28.06$; and $Y_i - \hat{Y}_i = 29 - 28.06 = 0.94$.

7.3.2. Index of total discrepancy

To get a non-zero total index of discrepancy,* the $Y_i - \hat{Y}_i$ deviations are squared and added to form

$$S_R = \Sigma(Y_i - \hat{Y}_i)^2. \qquad [7.4]$$

Squaring and adding the values in column 6 of Table 7-1 produces $S_R = (0.94)^2 + (2.63)^2 + \ldots + (3.88)^2 = 64.631$.

S_R is called the *residual deviance* or *residual group variance*. The R subscript indicates that the square deviations are determined as residuals from the estimated values of $\hat{Y}_i$. S_R is also sometimes called *error variance*, because the deviations are regarded as "errors" in the estimates. (Later on, when $\hat{Y}_i$ is estimated from more than one independent variable, the symbol for S_R will be altered to denote the number of variables.)

7.3.3. Partition of deviation

Formation of the algebraic model allows each Y_i to be associated with three values. They are

Y_i: the actual observed value;
$\overline{Y}$: the mean of the Y_i values; and
$\hat{Y}_i$: the value estimated by the model from each observed value of X_i.

With these three values, the basic discrepancy in fit is divided into two components, expressed as an algebraic identity:

$$Y_i - \overline{Y} = (Y_i - \hat{Y}_i) + (\hat{Y}_i - \overline{Y}).$$

The arrangement is shown in Fig. 7-4, where the basic deviation, $Y_i - \overline{Y}$, becomes the sum of two deviations: $Y_i - \hat{Y}_i$ and $\hat{Y}_i - \overline{Y}$.

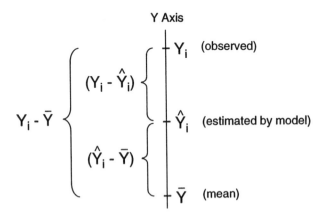

Fig. 7-4 Division of $Y_i - \overline{Y}$ into $Y_i - \hat{Y}_i$ and $\hat{Y}_i - \overline{Y}$, by the algebraic model, $\hat{Y}_i = a + bX_i$.

---

*Since $\hat{Y}_i = a + bX_i$, the value of $\Sigma\hat{Y}_i$ will be $\Sigma a + \Sigma bX_i = Na + Nb\overline{X}$. As shown in Section 7.4, $a = \overline{Y} - b\overline{X}$. Therefore, $Na = N\overline{Y} - Nb\overline{X}$, and $\Sigma\hat{Y}_i = N\overline{Y} - Nb\overline{X} + Nb\overline{X} = N\overline{Y}$. Consequently, if the deviations are not squared, $\Sigma(Y_i - \hat{Y}_i) = \Sigma Y_i - \Sigma\hat{Y}_i = N\overline{Y} - N\overline{Y} = 0$.

The additional deviation that has just been introduced, $\hat{Y}_i - \overline{Y}$, is called the *model deviation*. It expresses the difference between the mean $\overline{Y}$ and each value estimated by the algebraic model as $\hat{Y}_i$. The values of these model deviations are shown in column 7 of Table 7-1.

7.3.4. Partition of variance

When $Y_i - \overline{Y}$ is divided into two components, a form of group variance can be calculated for the original deviation and for each component. The value of the original $S_{yy} = \Sigma(Y_i - \overline{Y})^2$ is the *basic group variance*. The sum of the squared residual deviations, $S_R = \Sigma(Y_i - \hat{Y}_i)^2$, is the *residual group variance*. The *model group variance* is

$$S_M = \Sigma(\hat{Y}_i - \overline{Y})^2,$$

which is the sum of the squared values for deviations of the estimated model from the mean.

A glorious attribute of partitioning group variance is that the basic *deviation* and the basic group *variance* are divided in exactly the same way. Thus, the partition for deviation is

$$Y_i - \overline{Y} = (Y_i - \hat{Y}_i) + (\hat{Y}_i - \overline{Y});$$

and the partition for basic group variance is

$$\Sigma(Y_i - \overline{Y})^2 = \Sigma(Y_i - \hat{Y}_i)^2 + \Sigma(\hat{Y}_i - \overline{Y})^2.$$

In symbols, it is

$$S_{yy} = S_R + S_M, \qquad [7.5]$$

and in words,

Basic group variance	=	Residual group variance	+	Model group variance

When basic group variance is partitioned, the value of S_R shown in formula [7.5] will always be smaller than S_{yy}, except in the unlikely case where $S_M = 0$. Thus, the residual group variance around $\hat{Y}_i$ will always be smaller than the basic group variance around $\overline{Y}$, indicating that the algebraic model has improved the fit between the actual and the estimated values of Y_i. For the example cited here, we have already noted that $S_{yy} = 413.3333$ and that S_R is a smaller value, 64.631. By squaring and adding the data in column 7 of Table 7-1, we would get $S_M = \Sigma(\hat{Y}_i - \overline{Y})^2 = (-3.27)^2 + (6.04)^2 + \ldots + (7.79)^2 = 348.2308$. This same value could have been obtained more quickly by using formula [7.5] to state $S_M = S_{yy} - S_R$, and then substituting $413.3333 - 64.631 = 348.702$. (The small difference in the two values of S_M is due to rounding.)

As a general principle, a well-fitting algebraic model should have small values in the residual deviations of $Y_i - \hat{Y}_i$. With those small values for the residual deviations, S_R should be much less than S_{yy}. In fact, with a perfect fit, S_R would be 0, and S_M would equal S_{yy}.

7.4. Finding the coefficients

The best-fitting coefficients for a and b in the equation, $\hat{Y}_i = a + bX_i$, will produce the smallest possible results for the residual variance of $S_R = \Sigma(Y_i - \hat{Y}_i)^2$. When $a + bX_i$ is substituted for $\hat{Y}_i$, this formula becomes

$$S_R = \Sigma(Y_i - a - bX_i)^2.$$

To find the minimizing values of a and b, S_R is differentiated with respect to a and b, according to some calculus shown in the footnote.* When set equal to 0, the two simultaneous equations are solved for a and b. When everything is finished, it turns out that

$$b = \frac{S_{xy}}{S_{xx}} \qquad [7.6]$$

and

$$a = \overline{Y} - b\overline{X}. \qquad [7.7]$$

The calculations for a and b often make use of algebraic identities (analogous to the computation for S_{yy}) that show

$$S_{xx} = \Sigma X_i^2 - [(\Sigma X_i)^2/N]$$

and

$$S_{xy} = \Sigma X_i Y_i - [(\Sigma X_i \Sigma Y_i)/N].$$

7.4.1. Illustration of calculations

In the data of Table 7-1, $\Sigma X_i^2 = 80^2 + 32^2 + \ldots + 23^2 = 33193$; $\Sigma X_i = 379$; and $(\Sigma X_i)^2/N = 379^2/6 = 23940.167$. Therefore $S_{xx} = 33193 - 23940.167 = 9252.833$. The calculation for $\Sigma X_i Y_i$ is $(80 \times 29) + (32 \times 40) + \ldots + (23 \times 43) = 10079$. Since $\Sigma Y_i = 188$, the value of $(\Sigma X_i \Sigma Y_i)/N = (379 \times 188)/6 = 11875.333$. Therefore $S_{xy} = 10079 - 11875.333 = -1796.333$.

With formula [7.6], we get $b = -1796.333/9252.833 = -0.194$. With formula [7.7], we use $\overline{Y} = 31.33$ and $\overline{X} = \Sigma X_i/N = 379/6 = 63.167$ to get

$$a = 31.33 - (-.194)(63.167) = 31.33 + 12.25 = 43.58.$$

These calculations show how a and b were previously obtained in Section 7.2 for the equation

$$\hat{Y}_i = 43.58 - .194X_i.$$

7.4.2. Least-squares principle

The name *least-squares principle* is given to the strategy of finding a and b by getting a minimum value for the sum of squares that constitutes S_R. This principle

---

*$\delta S_R/\delta a = 2\Sigma(Y_i - a - bX_i)(-1)$ is set equal to 0. The sum of the components then becomes $N\overline{Y} - Na - bN\overline{X} = 0$; and so $a = \overline{Y} - b\overline{X}$. Substituting this value of a and differentiating with respect to b, we get $\delta S_R/\delta b = 2\Sigma[Y_i - \overline{Y} - b(X_i - \overline{X})](-X_i)] = 2\Sigma[-(Y_i - \overline{Y})X_i + b(X_i - \overline{X})(X_i)] = 2\Sigma[b(\Sigma X_i^2 - N\overline{X}^2) - (\Sigma X_i Y_i - N\overline{X}\overline{Y})$. When this set is equal to 0, we get $b = (\Sigma X_i Y_i - N\overline{X}\overline{Y})/(\Sigma X_i^2 - N\overline{X}^2) = S_{xy}/S_{xx}$.

is readily used in simple linear regression because the dimensional deviations of $Y_i - \bar{Y}$ and $Y_i - \hat{Y}_i$ can be directly subjected to all the mathematical processes used in the calculations. Later on, in Part IV of the text, when Y_i is *not* expressed as a dimension, other methods and principles will be needed for citing the discrepancies between Y_i and $\hat{Y}_i$, and for calculating the best-fitting coefficients.

An important but often overlooked feature of the least-squares strategy is the differential calculus used to find the best-fitting coefficients. The differentiation works with the assumption that X is being held constant at a point that need not be identified because, in the rectilinear model, the same value of *b* pertains for all values of X. These assumptions will later become the source of a major problem in linear nonconformity, which occurs when the pattern of the actual data does not conform to the straight-line shape assumed in the algebraic model. In the latter circumstance, the slope of the data of Y may differ substantially in different locations of X, so that the calculated value of a "point-less" *b* is correct only as an average, but not in many individual zones of X. The problem, which was mentioned and illustrated in Sections 2.2.4 and 2.4.2.3, will be discussed later as part of regression diagnostics.

7.5. Indexes of accomplishment

After *a* and *b* have been determined, the basic computations are finished. The regression line, $\hat{Y}_i = a + bX_i$, has *a* as its intercept when $X_i = 0$. The slope of the line is the regression coefficient, *b*, which indicates that $\hat{Y}_i$ will be changed by *b* units for every unitary change in X_i. We can also draw the line by determining any two points through which it passes. The easiest approach is to note that the line always passes through the location of the bivariate mean ($\bar{X}, \bar{Y}$) and through the intercept $(0, a)$. We can thus draw the line here, as shown in Fig. 7-3, through the two points: (63.17, 31.33) and (0, 43.58).

The next step is to determine the model's accomplishment in fitting the data and in denoting the impact of variable X. The statistical indexes used to express these two accomplishments are discussed in the rest of this section.

7.5.1. r^2 and 'explained variance'

An obvious way to express the model's accomplishment in fit is to determine the proportionate reduction in the basic group variance. It is

$$r^2 = \frac{\text{Basic group variance} - \text{Residual group variance}}{\text{Basic group variance}} = \frac{S_{yy} - S_R}{S_{yy}} \quad [7.8]$$

Formula [7.8] can be calculated in an alternative way if we substitute $\hat{Y}_i = a + bX_i$ into $S_R = \Sigma(Y_i - \hat{Y}_i)^2$, which then becomes $S_R = \Sigma[(Y_i - \bar{Y}) - b(X_i - \bar{X})]^2$. When suitably expanded and rearranged, this expression becomes

$$S_R = S_{yy} - 2bS_{xy} + b^2 S_{xx}. \quad [7.9]$$

When S_{xy}/S_{xx} is substituted for *b* in formula [7.9], S_R becomes $S_{yy} - (S_{xy}^2/S_{xx})$; and $S_{yy} - S_R$ will become S_{xy}^2/S_{xx} in formula [7.8]. Division by S_{yy}

will then produce

$$r^2 = \frac{S_{xy}^2}{S_{xx}S_{yy}} \quad [7.10]$$

To find the value of r^2 for the data of Table 7-1, we already know (from Section 7.1.2) that $S_{yy} = 413.33$. In Section 7.4, we found that $S_{xx} = 9252.833$ and $S_{xy} = -1796.33$. All we need do now is square S_{xy} and then substitute appropriately in formula [7.10]. The result is

$$r^2 = \frac{(-1796.33)^2}{(9252.833)(413.33)} = .844$$

7.5.1.1. Nomenclature for r^2

In simple regression, r^2 is often called the *coefficient of determination*. In multivariable analysis, a counterpart expression, cited as R^2, is called the (squared) *coefficient of multiple correlation*.

The value of r^2 (or R^2) shows the proportion of the original basic group variance that has been statistically reduced by imposition of the algebraic regression model. Although often called *explained variance*, the explanation for the proportionate reduction is purely statistical. It shows what happens when X and Y are mathematically associated in the algebraic model, but it does not necessarily indicate that the variables are biologically or substantively related. They might be as substantively disparate as the annual volume of tea exported from China and the sales of videocassette recorders in New York.

7.5.1.2. Range of r^2

The value of r^2 ranges from 0 to 1. When the algebraic model fits perfectly, S_R will equal 0 and r^2 will be $S_{yy}/S_{yy} = 1$. When the model shows no relationship between X and Y, the regression line will be parallel to the X axis and will have a slope of $b = 0$. S_R will equal S_{yy}, and so r^2 will be 0.

Accordingly, when r^2 is close to 1, the model has an excellent fit. For the data of Table 7-1, the relatively high value of $r^2 = .844$ indicates the very good fit that would be expected for the reasonably linear pattern of the points shown in Fig. 7-3. When r^2 is close to 0, however, we cannot conclude that the model has a poor fit. All we can say is that the two variables, X and Y, do not have an effective straight-line relationship. The model may or may not fit well. For example, in Fig. 7-5, where the cloud of points looks like scattered buckshot, a straight-line model will fit the data poorly and r^2 will be close to 0. By contrast, in Fig. 7-6, the points form a linear array and a straight line will fit the data quite well. The data in Fig. 7-6 will also be fit quite well, however, by a horizontal straight line placed at the level of the mean, $\overline{Y}$. In this situation, the straight-line model of $\hat{Y}_i = a + bX_i$ will have a value of b that is close to zero, with the value of a (when calculated as $\overline{Y} - b\overline{X}$) being close to $\overline{Y}$. Thus, the algebraic estimation will be essentially $\hat{Y}_i \simeq \overline{Y}$. The basic value of Syy will be small, and the basic variance will not be greatly reduced when the algebraic model is imposed. Nevertheless, because the relatively small value of Sxy is squared and then divided by the even smaller value of Syy, the value of r^2 may sometimes be relatively high, although the value of b, calculated as Sxy/Sxx, may be quite close to 0. Consequently, r^2 may sometimes correctly indicate a good fit for the line, even though X has essentially no impact on Y.

These distinctions in low and high values of r^2 will be considered again later, when their further interpretation is discussed in Section 7.10.5.

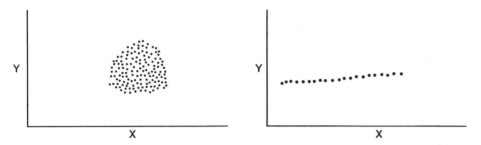

Fig. 7-5 Pattern of data in which the regression coefficient, b, will be close to 0, indicating that Y is essentially unaffected by X, but r^2 will be close to zero, indicating poor fit for a straight line.

Fig. 7-6 Pattern of data in which the regression coefficient, b, will be close to 0, but r^2 will be relatively high, indicating a good fit..

7.5.2. F ratio of mean variances

A different index for evaluating a model's accomplishment is almost never used in simple regression, but is discussed here in preparation for its frequent application in multiple-variable regression. The index is the F ratio of the mean variances for S_M and S_R.

7.5.2.1. Mean of a group variance

Determining the mean of a *group* variance involves one of the most "counter-intuitive" principles of statistics. Having been calculated as a sum of N squared deviations, each of the group variances S_M and S_R has an average, or "mean," value. When these means are calculated for statistical purposes, however, the group variances are *not* divided by N. Instead, they are divided by the corresponding "degrees of freedom" in each group.

7.5.2.2. Concept of degrees of freedom

The concept of degrees of freedom is not easy to understand, because it is an idea of inferential rather than descriptive statistics. The concept rests on statistical ideas about using the results of a sample to make inferential estimates for the parameters of the "parent population."

The algebraic model $Y = a + bX$ is a direct *descriptive* method for estimating values of Y from values of X in the observed data. When the results are tested for "statistical significance," however, the estimate is regarded as inferential rather than descriptive. In that inferential thought, the observed set of data (such as Table 7-1) is assumed to be a sample, containing N members of data, randomly selected from a larger parent population. In that parent population, the Y values have a mean, which is the parameter μ_Y. We do not know its value, but for the inferential activities, we assume that μ_Y has the value of $\overline{Y}$ found in our sample. The parent population also has a parametric regression equation, $Y = \alpha + \beta X$. We do not know the values of α and β, but we infer that they are estimated, respectively, by the values of *a* and *b* calculated from the sample.

The ideas about degrees of freedom refer to the variations that can occur when these estimates are made during the sampling process. Any sample of N members from the parent population can have N different values (or degrees of freedom) for the individual results of Y_i. Because we have assumed (for the statistical inference) that the sample's value of $\overline{Y}$ is the *correct* value of the parametric mean, μ_Y, any other sample of N members should theoretically also have $\overline{Y}$ as the mean. Consequently, the first $N - 1$ members in any sample can roam through any range of possible values for Y_i. The last member is constrained, however, to make $\Sigma Y_i/n = \overline{Y}$ for the sample. The assumption about μ_Y thus gives $\overline{Y}$ its one degree of freedom. With this constraint, the values for $Y_i - \overline{Y}$ have $N - 1$ degrees of freedom: the N degrees in Y_i minus the one in $\overline{Y}$.

Analogously, the values of *a* and *b* each have a single degree of freedom when they estimate the respective parameters of α and β in the expression $Y = \alpha + \beta X$. Consequently, when $\hat{Y}_i$ is estimated as $a + bX_i$, the values of $\hat{Y}_i$ have 2 degrees of freedom. When we determine values of $Y_i - \hat{Y}_i$, there are $N - 2$ degrees of freedom: the N in Y_i minus the two in $\hat{Y}_i$.

Finally, the terms $\hat{Y}_i - \overline{Y}$ will have the two degrees of freedom contained in the *a* and *b* of Y_i, minus the one degree of freedom in $\overline{Y}$. Thus, the degrees of freedom for $\hat{Y}_i - \overline{Y}$ in simple regression are $2 - 1 = 1$.

7.5.2.3. Expressing the ratio

With these ideas about degrees of freedom, we can return to getting the mean values for the group variances, S_M and S_R. For purposes of statistical inference, a group variance is divided by its degrees of freedom to get the mean value. Because S_M arises from $\Sigma(\hat{Y}_i - \overline{Y})^2$ and thus has 1 degree of freedom, the mean value of S_M will be $S_M/1 = S_M$. Because S_R arises from $\Sigma(Y_i - \hat{Y}_i)^2$ and has $N - 2$ degrees of freedom, the mean of S_R will be $S_R/(N - 2)$.

Accordingly, the ratio of the mean variances in simple linear regression will be

$$F = \frac{S_M/1}{S_R/(N-2)} = (N-2)\frac{S_M}{S_R} \qquad [7.11]$$

For the data under analysis in Table 7-1, $F = (6-2)(348.2308)/64.631 = 21.55$.

In multivariable analysis, if $\hat{Y}_i$ contains variables for p parameters, rather than the two parameters estimated in simple regression, the degrees of freedom for S_M will be $p - 1$, rather than $2 - 1$. For p parameters, the degrees of freedom for S_R will be $N - p$. Thus, in multiple regression, the corresponding expression would be

$$F = \frac{S_M/(p-1)}{S_R/(N-p)}. \qquad [7.12]$$

Note that the degrees of freedom for S_M and S_R always add up to he degrees of freedom for S_{yy}. In either simple or multiple regression, S_{yy} has $N - 1$ degrees of freedom because it is determined from $Y_i - \overline{Y}$. In simple regression the sum of degrees of freedom for S_M and S_R is $1 + (N - 2) = N - 1$. In multiple regression, the corresponding sum is $(p - 1) + (N - p) = N - 1$.

7.5.2.4. Alternative expression and appraisal of F

The formula for F can readily be converted to an expression containing only r^2 (and N) if we substitute $S_M = S_{yy} - S_R$ and $S_{yy} - S_R = r^2 S_{yy}$ in formula [7.11]. The result becomes

$$F = (N - 2)\frac{r^2}{1 - r^2}. \qquad [7.13]$$

With this interchangeability, either F or r^2 could be used as the index of accomplishment for the model's goodness of fit. The value of r^2 is preferred, however, because it is "standard," varying from 0 to 1, so that its relative magnitude can immediately be interpreted. The value of F, on the other hand, will range from 0 to infinity. As shown in formula [7.13], when $r^2 = 0$, $F = 0$; and when r^2 approaches 1, F approaches infinity. A zone of values that range from 0 to 1 is easier to interpret in a standard manner than a range from 0 to infinity.

Because the unlimited range of the index is undesirable for expressing goodness of a model's fit, F is seldom needed or used in simple regression; but it has a highly important role, as noted later, in various operational decisions of multivariable analysis.

7.5.3. r as a correlation coefficient

In elementary statistics, r (as the square root of r^2) is regularly called the *correlation coefficient*. It indicates the interdependent association between X and Y, without the idea that one variable specifically depends on the other.

This interdependency is an important distinction that acknowledges the possibility of establishing an alternative regression line for the two variables. If we had arbitrarily decided to reverse the orientation, making X the *target* variable and Y the *independent* variable, the same strategy and tactics could have been used to develop an equation in which X is regressed on Y, instead of vice versa. The alternative equation would have different coefficients, with a' as its intercept and b' as its slope or regression coefficient. The expression would be

$$\hat{X}_i = a' + b'Y_i. \qquad [7.14]$$

With a reversal of independent and dependent variables, the coefficients for this equation would be calculated as

$$a' = \overline{X} - b'\overline{Y}$$

and

$$b' = \frac{S_{xy}}{S_{yy}}.$$

For the data in Table 7-1, $b' = \frac{-1796.33}{413.33} = -4.346$, and so $a' = 63.167 - (-4.346)(31.33) = 63.167 + 136.160 = 199.327$. Thus, the alternative regression equation would be

$$\hat{X}_i = 199.327 - 4.346 Y_i.$$

The alternative line is the dashed "second" line shown in Fig. 7-3. Although both lines pass through the mean value of $\overline{X}, \overline{Y}$, they differ in their intercept values when $X = 0$ and when $Y = 0$. The two lines here seem reasonably similar because the data have a strongly rectilinear pattern. With other patterns, the two lines can be strikingly different.

The existence of two possible regression equations for the same set of data may seem disconcerting, but is entirely reasonable for circumstances in which we are interested in correlation, not regression, and in which we do not particularly care whether Y depends on X, or X depends on Y. The statistical expression for correlation must account for the two different slopes of the two possible regression equations. In the first instance, $b = S_{xy}/S_{xx}$ is the slope for the dependency of Y on X; and in the second instance, $b' = S_{xy}/S_{yy}$ for the dependency of X on Y. To get a suitable expression for the interdependency of X and Y, the calculation of r uses an ingenious approach. It multiplies the two regression coefficients and then takes their square root. Thus, the *correlation coefficient* is a geometric mean of the two slopes, calculated as

$$r = \sqrt{bb'} = \sqrt{\frac{S_{xy}}{S_{xx}} \times \frac{S_{xy}}{S_{yy}}} = \sqrt{\frac{S_{xy}^2}{S_{xx}S_{yy}}} = \frac{S_{xy}}{\sqrt{S_{xx}S_{yy}}}. \qquad [7.15]$$

Since S_{xy}^2 will always be positive, care must be taken with the sign of its square root. Ordinarily, any square root can be positive or negative, but S_{xy} (except when zero) will always have a definite positive or negative sign. Consequently, the best calculational formula for r is

$$r = \frac{S_{xy}}{\sqrt{S_{xx}S_{yy}}}. \qquad [7.16]$$

If the value of r^2 is already know, the value of its square root, r, receives the sign of S_{xy}.

For the data in Table 7-1, we know that r^2 is .844. Its square root is $\pm .917$. Since S_{xy} is -1796.33, the value of r will be negative, i.e., $-.917$. The negative value here reflects the close but inverse relationship of X and Y. As one variable rises, the other falls. Alternatively, with appropriate substitution into formula [7.16], r could have been calculated as

$$r = \frac{-1796.33}{\sqrt{(9252.833)(413.33)}} = -.919.$$

(The small difference in the two calculations is due to rounding.)

As the square root of r², the value of r can range from −1 to +1. Absolute values close to 1, with either sign, will represent a close correlation; and values close to 0 will represent little or no correlation. In this instance the high *r* value of −.92 indicates the strong inverse correlation expected from the graph of Fig. 7-3.

An important thing to note about the calculations is that r² is almost always *smaller* than r, because for numbers between 0 and 1, the squared value is smaller than the original. Thus, if r = .4, r² = .16 and if r = .2, r² is .04. This distinction should be remembered when correlation coefficients are interpreted. An r coefficient as high as .3 reflects a proportionate "explained variance" of only r² = .09.

7.5.4. Standardized regression coefficient and impact of X

If the only goal were to fit an algebraic model to the data, the expression for r² (or F) would indicate the effectiveness of the fit. In many, and perhaps most, of the regression analyses used in medical research, however, fitting a model is merely a process used en route to the main goal, which is to determine the impact of the independent variable(s) on the target variable. For this purpose, we are interested not in r² or F, but in *b*. As the regression coefficient in the equation $\hat{Y}_i = a + bX_i$, *b* shows the slope of the line, indicating the amount by which $\hat{Y}_i$ changes for each unit of change in X_i. Thus, if $Y_i = 43.58 - .194X_i$, an increase of one unit in X_i will be accompanied by a fall of .194 units in $\hat{Y}_i$.

Except for one major problem, the value of *b* will immediately indicate the impact of X on Y. The problem is that the value of *b* is unstandardized; its magnitude will vary according to the arbitrary magnitude of the units in which X and Y have been measured. If X = blood urea nitrogen, the value of *b* will change if BUN is expressed in millimoles rather than milligrams. If X = age and Y = weight, the value of *b* will differ if age is expressed in days, months, or years and if weight is expressed in kilograms or pounds.

To avoid the problem of varying units of measurement, the regression coefficient, b, is often standardized before its impact is evaluated. The standardization can be done in two ways. In one way, the regression line is calculated with Y and X each expressed in their standardized Z-score form, as discussed in Section 6.2.1. When simple regression is carried out for the standardized values of Z_y and Z_x, the standardized regression coefficient turns out to be *r*, the correlation coefficient. In this format,

$$\left(\frac{\hat{Y}_i - \overline{Y}}{s_y} \right) = r \left(\frac{X_i - \overline{X}}{s_x} \right). \quad [7.17]$$

An alternative simpler approach for finding the standardized regression coefficient, however, is to multiply the regression coefficient, *b*, by s_x/s_y. The conversion is

$$\text{Standardized regression coefficient} = b_s = b \left(\frac{s_x}{s_y} \right). \quad [7.18]$$

Formula [7.18] can easily be demonstrated if we express the ordinary regression equation as $Y_i - \bar{Y} = b(X_i - \bar{X})$. Dividing both sides by $s_y s_x$, we get

$$\frac{Y_i - \bar{Y}}{s_y s_x} = \frac{b(X_i - \bar{X})}{s_y s_x}.$$

With terms rearranged, this expression becomes

$$\frac{Y_i - \bar{Y}}{s_y} = \left(\frac{bs_x}{s_y}\right)\left(\frac{X_i - \bar{X}}{s_x}\right).$$

Thus, if Y_i and X_i are expressed in their standardized (or Z-score) formats, the standardized regression coefficient is bs_x/s_y.

Expression [7.18] may at first seem strange because you intuitively might expect b to become standardized with division rather than multiplication by s_x. The s_x/s_y factor, however, accounts for the fact that b is a ratio of S_{xy}/S_{xx}, not just an expression in X alone.

Formula [7.18] offers an easy way to convert any regression coefficient to a standardized regression coefficient. The formula is also applicable to the coefficients found in multiple regression, but $b_s = r$ can be used only in simple regression. In multiple regression, the value of R^2 (or R) applies only to the entire algebraic model, not to any individual variables. Each of the multiple regression coefficients must therefore be standardized separately to find the value that corresponds to b_s. The interpretation of standardized b_s values will be discussed in Section 7.8.

For the data in Table 7-1, the standardized value of b could be noted immediately as $b_s = r = .917$. If we wanted to use formula [7.18], we would first need to get $s_x = \sqrt{S_{xx}/(N-1)} = \sqrt{9252.833/5} = 43.018$ and $s_y = \sqrt{S_{yy}/(N-1)} = \sqrt{413.33/5} = 9.092$. The value for b_s would then be $(.194)(43.018)/9.092 = .917$.

7.6. Stability ('statistical significance') of the indexes

Before any conclusions are drawn about r^2, F, b, or b_s, the indexes should be checked for their stability. They will be unstable if the set of data is too small or has excessive variability. The customary method of checking stability is with the P value obtained in a test of "statistical significance." An index is usually regarded as stable if its P value is below a designated boundary level called α, which is usually set at .05.

7.6.1. test for r^2

The value of r^2 is tested for "statistical significance" with a conventional t test, calculated as

$$t = \frac{r\sqrt{N-2}}{\sqrt{1-r^2}}. \qquad [7.19]$$

In simple regression, the P value for this value of t is interpreted with $N - 2$ degrees of freedom. If N is large, e.g., above 40, the result of formula [7.19] can be interpreted as a conventional Gaussian Z test.

For the data in Table 7-1, $t = -.919\sqrt{6-2}/\sqrt{1-.844} = -.919\sqrt{4}/\sqrt{.156} = -4.65$. At 4 degrees of freedom, a t of -4.60 has a P value of .01. Therefore the observed value of t is "statistically significant" at $P < .01$, and we can assume that r has a stable value.

7.6.2. F test for F

Unless you have previously studied the analysis of variance, you will not have met the F test before. As a ratio of two mean variances, F has a statistical sampling distribution (analogous to t) that allows a direct interpretation for P values, but two separate entries are needed for the degrees of freedom (d.f.) in S_M and S_R. In simple regression, d.f. for S_M is 1 and d.f. for S_R is $N - 2$. In an appropriate set of tables, F would be interpreted with 1 and $N - 2$ degrees of freedom.

In the previous calculation in Section 7.5.2.3 for the data of Table 7-1, we found that $F = (4)(348.2308)/64.631 = 21.55$. At 1 and 4 degrees of freedom, $P = .01$ when $F = 21.20$. Therefore, the P value for F here is $< .01$.

In simple regression, a much simpler approach for finding F is to look closely at formulas [7.13] and [17.19], and note that F is simply the squared value of t. Therefore, in simple regression, a separate test of "statistical significance" is not needed for *F*. The t test for *r* will yield the same result as an F test on *F*. In multiple regression, however, F, requires and receives separate evaluation.

7.6.3. t test for b

To test the stability of b, a t test can be done on the ratio formed when b is divided by its standard error. The formula is

$$t = \frac{b}{\text{standard error of b}}. \qquad [7.20]$$

It can be shown that the standard error of b in simple regression is $\sqrt{S_R/[S_{xx}(N-2)]}$. Since $b = S_{xy}/S_{xx}$, and since $S_R = S_{yy}(1-r)^2$, formula [7.20] becomes $t = [S_{xy}/\sqrt{S_{xx}S_{yy}}]/\sqrt{(1-r^2)/(N-2)}$, which becomes

$$t = \frac{r\sqrt{N-2}}{\sqrt{1-r^2}},$$

which is identical to formula [7.19].

These distinctions give a unique utility to the t test for r in simple linear regression. The result of this single test of "statistical significance" will show the stability of all three main indexes of accomplishment: r^2, F, and b. In multiple regression, however, each index will require separate tests.

7.6.4. Confidence intervals for small values

Although now often recommended as replacements for P values in statistical inference about stochastic significance, confidence intervals are seldom used in ordinary regression analysis, and seldom appear in computer printouts from conventional package programs.

Perhaps the main value of confidence intervals in linear regression is to offer assurance that a small, "nonsignificant" value is indeed stable. The interval is calculated around the observed value of b as

$$b \pm Z_\alpha \text{ (standard error of b)},$$

where Z_α is a selected Gaussian coefficient that corresponds to the value of $1 - \alpha$ chosen for the confidence interval. The value of Z_α is 1.96 for a 95% confidence interval, and 1.675 for a 90% interval. If calculated for r, the interval is

$$r \pm Z_\alpha \text{ (standard error of r)}.$$

The standard error of r is seldom shown in computer printouts. For most practical purposes, its value is $\sqrt{(1 - r^2)/(N - 2)}$. (The confidence interval procedure will be illustrated in Chapter 9.)

7.7. Quantitative vs. stochastic 'significance'

The interpretation of results in any statistical analysis is always confused by the term "statistical significance." Referring to two different ideas, it has become a major source of ambiguity and misconception in modern science. One idea is descriptive: it refers to the quantitative magnitude of an observed distinction,[1,2] such as the difference in two proportions, or a correlation or regression coefficient for two variables. The other idea is inferential: it refers to a probability value (or confidence interval) that reflects the possible role of numerical chance in producing the observed distinction.

To illustrate the two ideas, consider the performance of two otherwise similar baseball players whose batting averages are .400 and .250. The difference in averages seems quantitatively impressive or "significant" because .400 is substantially larger than .250. If the numerical basis for those averages is 2/5 vs. 1/4, however, the difference will be numerically unstable. With each player having had so few times at bat, the observed quantitative distinction could be due to chance alone, even if the two players have identical batting abilities.

On the other hand, suppose we compare two players whose lifetime batting averages are .280 and .275. For practical purposes, these two averages are similar: the increment of .005 seems quantitatively trivial. Nevertheless, if the lifetime performances show 22075/78840 as the source of the first player's batting average, and 21684/78843 as the source of the second's, the distinction will be "statistically significant." A suitable calculation will show that X^2 (for chi-square) = 4.86 and P <.05.

These two examples indicate the two different forms of "statistical significance." One of them depends on the descriptive magnitude of an observed substantive distinction. Descriptively, the difference of .400 vs. .250 was impressive and the difference of .280 vs. .275 was relatively trivial, regardless of the underlying numbers from which the batting averages were calculated. The other component of "significance" is a purely mathematical decision. Inferentially, we decided that 2/5 vs. 1/4 was unimpressive because the basic numbers seemed unstable, but that 22075/78840 vs. 21684/78843 was "impressive" or stable because it had a P value of < .05.

The inferential decision refers to the stability of the underlying numbers. Although often expressed with calculations of probability values or confidence intervals, stability can be appraised in other ways.[3] Thus, in the first example, we could sense intuitively, i.e., without any formal calculations, that the results for 2/5 vs. 1/4 were "chancy." The two players would have identical batting averages if the player whose record is now 1/4 gets a hit in his next time at bat. In the second example, we could also sense intuitively that neither player's average would be substantially affected by whatever happened in one or a few more times at bat. The role of P values, confidence intervals, and other mathematical calculations is to provide specific guidelines when the intuitive decisions about stability are not so obvious or easy.

The current concept of "statistical significance" was an editorial intervention intended to keep investigators from making wrong intuitive decisions when impressive quantitative distinctions—such as .250 vs. .400 —come from relatively small groups (or "sample sizes"). Certain boundaries (such as P < .05) were established as guidelines for the acceptable levels of probability that could help avoid the errors.

Like the interventions used in clinical therapy, the statistical intervention has had benefits and risks. The beneficial effect has been an invaluable contribution to the "editorial" appraisal of scientific data. The adverse effect has been the frequent neglect of quantitative distinctions when "statistical significance" is decided only on the basis of mathematical issues in probability. Because sample size has a dominant role in the probabilistic calculations, a tiny distinction—as small and unimportant as the .005 difference in batting averages—can become "statistically significant" if the sample size is large enough.[1]

To separate and emphasize these two different components of "significance," the term "statistical significance" will generally be avoided in the rest of this book. *Quantitative significance* will be used for descriptive distinctions, which depend on such things as the amount of an increment in proportions for two groups, the magnitude of a regression coefficient, or other expressions that describe the quantity of an observed distinction.[2] *Stochastic significance* will be used for probabilistic appraisals, which are expressed in P values or confidence intervals, and which emerge from statistical calculations for such entities as t, Z, F, or chi-square. (The word *stochastic*, which is derived from the Greek στοχος, a target and στοχαστικος, proceeding by guesswork, originally referred to accuracy in forecasting. In modern statistical usage,[4] the term has been applied for diverse forms of random variation, and in ordinary English[5] it is defined as "pertaining to, or arising from chance; involving probability.")

Because multivariable analysis is seldom done unless the sample sizes are relatively large, the two types of "significance" must be carefully distinguished. Although special boundaries can be applied for P or other statistical demarcations to ensure that the analytic procedure operates with numerically stable, stochastically significant data, the stochastic boundaries are relatively unimportant after the operation is finished and the data are shown to be numerically stable. The interpretation of the final results depends on the quantitative rather than merely stochastic distinctions of what has emerged.

Most of the statistical indexes to be discussed in this text offer a quantitative assessment of magnitude for either fit or trend. The assessments will be expressed in such descriptive indexes as proportionate reduction in variance or error, and slope of a gradient. Each assessment will be checked for its numerical stability, using tests of probabilistic variability. The latter tests produce the customary P values (or confidence intervals) that are conventionally given the ambiguous title "statistically significant," which will almost always here be called "stochastically significant." The interpretation of each descriptive index will then be accompanied by criteria for magnitudes that can be regarded as "substantial" or "quantitatively significant."

7.8 Interpretation of quantitative accomplishments

If the values of r^2, F, and b are statistically stable, i.e., stochastically significant, the indexes will denote the fit of the model and the apparent effect of X on Y. If the values are not stochastically significant, no conclusions can be drawn. The data may be too sparse; the model may fit too poorly; or X and Y may not be significantly related.

Although specific boundaries have been drawn, usually at $\alpha = .05$, for probabilistic conclusions about stochastic significance, no analogous boundaries have been established for descriptive decisions about good fit or effective variables.

7.8.1. Fit of the model

An algebraic model is usually regarded as having relatively poor fit if it fails to reduce at least 10% of the original basic group variance. Thus, when $r^2 < 0.1$, the model has not done a good job. The closer the value of r^2 to 1, the better is the fit.

For descriptive purposes, F is usually converted to a value of r^2, and is interpreted as r^2. For stochastic decisions, F is converted to a P value and interpreted accordingly.

7.8.2. Effect of the independent variable(s)

Because the standardized value of b is r (the correlation coefficient), the analytic effect of X in simple linear regression can be interpreted by examining either b_s or r. If r^2 must be $< .1$ for the model to fit at least modestly well, then r should exceed $\sqrt{.1}$, which is .32. In general, values of .30–35 in simple regression can be used as lower boundaries to regard r (or b_s) as having an effective impact.[2,6]

In multiple regression, the standardized coefficients of the independent variables seldom reach as high as .3. Accordingly, the effectiveness of the different independent variables is usually appraised not by examining the isolated individual values of coefficients, but by comparing the relative magnitudes of standardized coefficients, or by using other methods to be discussed later.

7.8.3. Stochastic vs. quantitative accomplishments

Although the values of r^2 and b_s indicate quantitative accomplishments, their stochastic interpretation will be profoundly affected by the sample size of the group under study.

For the associated P value to be ≤ .05, the stochastic boundary for t will vary with different group sizes, dropping downward from 2.31 to 2.10 to 2.05 as N goes, respectively, from 10 to 20 to 30. The boundary reaches 2.00 when N reaches 60. Consequently, values of $t \geq 2$ are usually stochastically significant.

Because t is calculated as $(r/\sqrt{1-r^2})(\sqrt{N-2})$, and because $\sqrt{N-2}$ appears in the numerator, a large enough value of N can bring the value of t across the "stochastic border" no matter how small r may be. As group sizes enlarge, t can become ≥ 2 and $P \leq .05$ for relatively trivial values of r. For example, if $N \geq 67$, a stochastically significant $P < .05$ can be obtained for $r^2 = .06$, indicating that only 6% of the variance has been explained. With an almost wholly ineffective value of r^2 as tiny as .02, t will exceed 2 if $N = 10{,}004$. On the other hand, a moderately impressive value of $r^2 = .5$, will not get t higher than 2 if $N = 10$.

This distinction should make you leery of any published results that report P values without indicating the actual magnitude of r or r^2. If quantitative decisions about fit and impact are made only according to the P values of stochastic significance, the conclusions can be erroneous or deceptive. With a huge sample size, a model with a poor fit may be acclaimed as good, and a variable with a trivial effect may called significant. Conversely, if the sample size is too small, an excellent fit or a highly effective variable may be erroneously dismissed as nonsignificant. For this reason, the descriptive indexes of accomplishment should always be carefully examined (and published), rather than the P values alone.

7.9. Selection of independent variables

In simple linear regression, only a single independent variable, X, is used in the analytic process. When other independent variables are available, however, their individual effects on the target variable may be obscured or distorted by their special interrelationships with one another. A prime role of multivariable analysis is to examine the variables *simultaneously,* rather than individually, and to reveal these relationships. During the multivariable analytic examinations, some of the independent variables may be eliminated as ineffectual. The algebraic model is then constructed with a smaller subset of the original variables.

The process of selecting these individual variables is a special multivariable activity, discussed in Chapter 11, that does not occur in simple linear regression.

7.10. Conformity of the model

At this point in the activities, everything seems to be finished. The algebraic model has done its work. The accomplishments have been revealed and evaluated for a single independent variable, X, or for the best collection of independent variables, if multiple variables were available. Before any final conclusions are drawn, however, a crucial part of the analysis still remains: the algebraic model must be checked for its linear conformity.

7.10.1. Reasons for checking conformity

A dishwasher can be used for laundry, and a hammer can bang in a screw. Nevertheless, despite misuse of the equipment, the results may be quite acceptable if the laundry gets clean and the screw holds.

Like any other well-designed tool, an algebraic model will do the best job it can, even if given an inappropriate task. As the model carries out its own goal in fitting itself to the data, it does not know and does not care about nonconforming patterns. It goes ahead anyhow and produces the best fit it can get. This best fit may often satisfy the analyst's goal, even though the fit seems poor and the model unsuitable, but at other times, a highly suitable model yielding an apparently excellent fit may be unsatisfactory.

As noted earlier, most of the multivariable analyses that appear in medical literature have the scientific goal of identifying cogent variables and categories, not fitting models. The analyst wants to determine which variables, and often which categorical zones within the variables, have the most effective or important impacts. This disparity creates an intriguing irony in data processing today, as the analytic goal of finding *impact* is regularly pursued with algebraic models that are aimed at the goal of *estimation*. When the achievement in estimation is expressed with an index of fit (e.g., r^2), the index of impact (e.g., magnitude of regression coefficient) emerges merely as a secondary by-product.

The published literature seldom mentions any evaluation of conformity between the single "slope" that is algebraically identified for each ranked variable, and the trends that actually occur in different zones of the variables. Although most algebraic methods have a robustness that makes them flexible enough to bend with different violations of the fundamental assumptions, the investigators seldom check (or report the results of checks) for violations that may make a method break rather than bend.

This omission may not cause problems if the multivariate analysis is used mainly to confirm impact, and if the expected confirmation occurs. Thus, when simple *bivariate* analyses are done to examine each independent variable's impact on the target variable, the apparently effective independent variables (and perhaps some of the others) may then be analyzed simultaneously to confirm that they retain their impact in the multivariable context. If all the confirmations occur as expected, the algebraic results may not be further examined for their conformity with the data. If certain apparently important variables are not confirmed, however, the conformity of the model should be checked before final conclusions are drawn. As discussed later, checking for conformity is a crucial requirement when the impact of a variable is quantified, rather than merely confirmed.

With the customary rectilinear models, however, the problems of inappropriate configuration receive no attention in either type of statistical index. Occupied with interpreting these indexes, the analyst may not recognize configurational problems that distort the impact of the independent variables. For example, in the expression $Y = a + bX$, the coefficient b implies that Y constantly changes b units for each unitary change of X, *throughout the entire range of X*. Suppose, however, that the pattern of the data have a sigmoidal shape, as in Fig. 2-5. A straight line will fit these data well, as shown in Fig 2-15, so that r^2 will be relatively high. The value of *b*, however, although correct on average, will be wrong for all zones of the values of X.

No matter how well or poorly an algebraic model seems to fit the data, the rectilinear implication will be wrong if the data have a nonconstant gradient that changes significantly in different zones. To avoid this problem, an analyst whose main goal is to identify trends or impacts cannot rely alone on statistical indexes calculated with the goal of getting good fit. Because conformity cannot be evaluated from a statistical index of fit or from ordinary regression or correlation coefficients, additional methods must be used to determine whether the constant impact implied by a rectilinear regression coefficient for X conforms to its actual impact in different zones of the data.

7.10.2. 'Regression diagnostics'

The evaluation of defects in an algebraic model is called *regression diagnostics*. As a recently fertile field in statistical research, regression diagnostics has been discussed in three textbooks[7-9] and in many individual publications.

In accord with the mathematical focus on fit of the model rather than impact of the variables, however, the regression diagnostics are aimed almost exclusively at improving general fit of the models, not at evaluating conformity. The "diagnostic" procedures are used to (1) find and perhaps remove influential outlier observations, (2) identify and perhaps eliminate variables that have been inadequately measured, or (3) augment or alter the format of the algebraic models. All three of these approaches will be discussed later when they are applied to multivariable analyses in Chapter 12.

In the discussion now, however, the customary scope of regression diagnostics is expanded to focus on the assessment of conformity. The aim is to check whether the algebraic model's goal in getting a best fit has coincided with the analyst's goal in getting an undistorted fit.

7.10.3. Analytic goals and algebraic structures

In certain biochemical, physiologic, economic, or other analyses in which the investigator wants to get a really close, almost perfect fit for the data, the format of the model is unimportant. Any or all of the possible curvilinear, polynomial, or other formats can be explored; and the best model will be chosen according to best fit, rather than best format, because the goal is to fit, not to relate or explain. For relating and explaining, however, a complex curvilinear format is an impediment, because the impact of the variables is difficult to discern from the multiple terms and formats. Although straight-line models have the advantage of offering a simple, straightforward indication of impact for each variable, their main disadvantage is that the simple indication may be misleading of the data do not conform to the straight-line format.

If aiming at a closely fitting line, the analyst can first determine whether *any* line can give an almost perfect fit. When the pattern of points is relatively monovalent, with essentially one value of Y appearing at each value of X, a closely fitting algebraic format can be found if the mathematical possibilities are vigorously explored. On the other hand, if the pattern of points is relatively polyvalent, a really close linear fit is impossible. For example, the swarm of points in Fig. 2-12 could not be perfectly fit with any linear portrait, but the straight line in Fig. 2-25 would give a reasonably good idea of the general relationship between X and Y.

The first step in regression diagnostics, therefore, is to determine the goal of the analysis. If the purpose is to get a particularly close fit for the data, rectilinear conformity is relatively unimportant. The fit cannot be extremely close for polyvalent swarms of data, but to make the fit as good as possible, the analyst may happily use complex curvilinear formats in which the impact of individual variables is difficult or impossible to determine. If the purpose, however, is to discern individual impact and predictive relationships of variables (or categories of variables), a straight-line model will be preferred and its linear conformity will be an important issue, no matter how well the model seems to fit.

The diagnostic process for the "ailments" of a regression analysis resembles what is done for clinical diagnoses. Judgmental decisions are made from the primary information that comes from the routine "physical" examination and "lab" tests. Special procedures are then ordered afterward when pertinent. In regression diagnostics, the counterpart of a physical examination involves inspecting patterns of the data; the routine laboratory tests are the ordinary indexes of accomplishment; the special procedures contain the extra "diagnostic workup."

7.10.4. Inspecting the pattern of data

In the next few sections, the discussion refers mainly to the diagnostic process for simple bivariate relationships.

To examine monovalent patterns and analyze relationships among variables, a single picture is often better than a horde of algebraic formulas. No matter what else is done, a graphical "portrait" of bivariate data should always be inspected. For simple regression between two variables, Y and X, a two-dimensional graph of points is easy to construct. It should always be examined, and should preferably also be displayed when the results are reported.

If the graph is produced in computer printouts, however, the pattern of points may be obscured, as noted later, when letters (such as A, B, C, etc.) or numbers (such as 1, 2, 3, etc.) are used to indicate multiple points at a single location. The immediate communication offered by visual density is lost when the multiple points are replaced by letters or numbers that have approximately equal size and optical intensity. If desired, the visual density produced by a single bunching of multiple points can readily be shown in a drawing made by hand. The points are spread about slightly, as illustrated earlier in Fig. 2-13, to form a contiguous cannonball pack surrounding their common location.

By inspecting a well-displayed graph, the analyst can immediately determine whether the data show a spreading swarm or a monovalent array, and can usually get an excellent idea of what kind of relationship is manifested in the swarm or the array. The inspection is important for deciding whether the selected format of the algebraic model (straight line, quadratic curve, etc.) is likely to produce deceptions when the accomplishments of the model are evaluated solely from statistical indexes such as r^2, P values, etc.

7.10.5. Problems in indexes of accomplishment

For the customary linear regression model, the statistical indexes of accomplishment are r^2 for the fit of the model and b (or b_s) for the impact of X on Y. Whether the line appears to fit well or poorly, however, these indexes cannot indicate several important distortions.

7.10.5.1. Distortions in good fits

The data shown in Fig. 7-7 have an exponential shape that could represent the decay of a radioactive substance over time. The mathematical configuration of such

Fig. 7-7 Straight line fitted to data having an 'exponential-decay' shape.

a curve is $\ln Y = -cX$, or $Y = e^{-cX}$, where c is a constant. The same type of pattern can appear in survival curves for a group of people. If these data are fit, however, with a straight-line (rather than exponential) regression model, the result shown in Fig. 7-7 would not be grossly misleading. The line would fit the data reasonably well and the value of r^2 would be reasonably high. Nevertheless, an important distinction of the X-Y relationship would be distorted. In particular, the analyst would not know that the curve for Y levels off and becomes almost horizontal as the value of X increases. The striking effect at low values of X would be falsely reduced by the straight line, and the minimum effect at high values of X would be falsely magnified.

A similar type of problem occurred in Fig. 2-15, where a well-fitting straight line distorted the zonal distinctions that create the sigmoidal shape (strong effects in the middle of the data and weak at the two extremes). The single regression coefficient for the straight line would be correct on average, but would not accurately show what is happening in the zones.

A different demonstration of good fit but distorted effect would occur for Fig. 7-6. The pattern of data shows that X seems to have almost no effect on Y. Its values hardly change as X goes through its full range of expression. In fact, the data for Y would be well fit merely by a line drawn through the mean, Y. Nevertheless, the linear regression of Y on X would produce an excellent straight line that has an almost perfect fit. An evaluator who examines only the high value of r^2 for this line might fallaciously conclude that variable X has proportionately reduced almost all the variance in Y, and has therefore had a major impact on Y.

7.10.5.2. Distortions in poor fits

A straight line will usually produce a relatively good fit if the bivariate data have a monotonically increasing or decreasing pattern, which can bend, curve, or sometimes be flat without any major reversals in direction. Monotonic patterns were present in Figs. 2-1, 2-2, 2-3 and 2-5.

If the pattern has a major reversal in direction, however, a straight line cannot capture the reversed movement. The pattern may be well fit by a curve, but a straight line will conform poorly. This problem was dramatically illustrated by the strong parabolic relationship in Fig. 2-4. Although any single straight line is unsuitable for this set of data, an algebraic model that is asked to do a straight-line regression will work valiantly and produce the result shown in Fig. 2-18, where the r^2 value is close to 0, indicating poor fit. The regression coefficient b will also be close to 0, suggesting that variable X has little or no relationship to Y—a completely erroneous conclusion for this parabolic pattern.

7.10.5.3. Concomitant values of r^2 and b

When values of r^2 are particularly high or low, the concomitant values of b will not help indicate whether deceptive results have occurred.

For example, if r^2 and b are both low, we could not distinguish between two types of poor fit: a reversal pattern that could be well approximated with a curved line (as in Fig. 2-18) and a nonlinear swarm of points, shown in Fig. 2-12, that cannot be closely fit with *any* line. The straight lines produced by linear regression models for data in Figs. 2-12 and 2-18 will have similar values for r^2 and b, thus failing to distinguish the poor fit in Fig. 2-12 from the misfit in Fig. 2-18.

If r^2 is low and b is high, we could not differentiate among a less-than-well-fitted linear swarm, as in Fig. 2-11, a misfitted but striking exponential or sigmoidal pattern, as in Figs. 2-3 and 2-5, or a set of clustered categories with a strong gradient, as in Fig. 2-13.

If r^2 is high but b is low, as in the pattern of Fig. 7-6, we might draw the false conclusion that X has had a major impact on Y because so much of the basic variance has been proportionately reduced.

If both r^2 and b are high, the model may fit well and the overall effect of X may be correctly indicated, but the changing gradient in different zones may be obscured. Examples of this problem were discussed in Section 7.10.5.1

These problems can be reduced by using polynomial expressions and piecewise functions that may produce an excellent fit and particularly high r^2. The diffusion of multiple coefficients for formats of X, however, will make the exact impact of X itself difficult or impossible to discern.

7.10.5.4. Stochastic deceptions

A different type of deception can occur, particularly in multivariable analysis, when decisions are made to ignore or remove an independent variable that has an unstable regression coefficient. The decisions usually rest on assessments of stochastic significance, which are greatly influenced by sample size. With a small overall sample size, or with maldistributed variables, a variable that has an important effect may be dismissed as unimportant because the associated P value failed to get past the required stochastic boundary. Thus, at the $\alpha = .05$ level of stochastic significance, an unimportant variable with a "good" distribution might be retained because its P value is .049; an important variable with a "poor" distribution might be eliminated because its P value is .051.

This problem is an inevitable consequence of the rigid demarcations of boundaries for decisions about stochastic significance in an automated analysis.

7.11. Preventing problems beforehand

Since preventing problems is usually an easier task than repairing them, certain prophylactic procedures can always be applied before the analysis begins. If not done previously, the prophylaxis can be carried out afterward, when the data are displayed; and the analysis can then be repeated.

7.11.1. Problems in univariate distributions

As discussed in Chapter 6, univariate distributions can be checked for problems of missing data and maldistributions. The problems of missing data are managed by various forms of isolation, transfer, or imputation. The problems of maldistributions are usually managed, if possible, by various types of transformation for the eccentric variables.

7.11.2. Transformations for 'sensibility'

Certain transformations may be intended, however, not to improve maldistributions, but to help produce apparently "sensible" results in the analysis. For example, as noted near the end of Section 6.3.3, the target variable, *duration of survival*, in the illustrative data set here was transformed to a logarithm, so that the estimated value would always be positive. As noted later, a special logarithmic arrangement called a *logistic transformation* is done so that $\hat{Y}$ will always range between **1** and **0** for estimating the presence or absence of a binary event.

7.11.3. Problems of additive and multiplicative effects

Although transformations for sensibility will ensure that the analytic estimate always lies in a selected range of values, the subsequent interpretation of results becomes complicated. For the simple expression $Y = a - bX$, the changes were additive. Y has the value *a* when X = 0, and the value is additively decremented by

b units for each added unit of X. If $\ln Y = a - bX$, however, the actual value for Y becomes $Y = e^{a-bX} = (e^a)(e^{-bX})$. Since e^a will be constant, the previous value of Y will be multiplied by a factor of (e^{-b}) units for each unit of change in X.

The different results of the multiplicative changes can be shown as follows for $a = 2$ and $b = 3$:

Value of X	Value of $Y = a - bX$	Value of $Y = e^{a-bX}$
0	2	7.38
1	−1	0.37
2	−4	0.018
3	−7	0.00009

The successive values of Y were obtained by adding −3 to each previous value for $Y = a - bX$, and by multiplying by $e^{-3} = .0498$ for each previous value of $Y = e^{a-bX}$. In the multiplicative model, the actual *increment* for each unitary change in X will no longer be constant, because it depends on the pervious value of Y. With some algebra that you will be spared here, it can be shown that as X_i advances to $X_i + 1$, the incremental change will be $(1 - e^{-b})Y_i$, which in this example will be $.9502\ Y_i$.

The distinction between additive and multiplicative effects will become particularly important later on when logarithmic models are used in multivariable analysis.

7.11.4. Examination of bivariate correlations

In simple regression, the two variables X and Y form the only bivariate relationship to be considered. It can readily be depicted by a graphical display that can immediately reveal problems for which solutions might be obtained by transformations in the X or Y variable, by changing from a straight line to some other algebraic model, or by using a categorical instead of algebraic approach.

In multivariable analysis, each of the available independent variables can be checked for its bivariate relationship with the target variable. Because the regression surface cannot be easily shown when more than one independent variable is under analysis, the analyst (or a suitably programmed computer) can easily plot the simple regression lines for the target variable, Y, against each of the independent variables. These bivariate regression lines can then be checked to see how well Y is fitted or affected by each of the candidate independent variables. The process will be illustrated in Chapter 9.

7.12. Special tests

After all the relatively simple "clinical" examinations just discussed, special "laboratory" tests can also be applied. They involve checking the pattern of "residuals" and doing additional ordinalized analyses for conformity of gradients. The tests are described here for ordinary simple regression, but they are particularly pertinent later for multiple regression.

7.12.1. Pattern of residuals

After the regression line has been fitted, each observed Y_i value has an estimated value, $\hat{Y}_i$, and a residual deviation, $Y_i - \hat{Y}_i$. These residual deviations can be examined for their univariate distribution or for their bivariate relationship to the corresponding values of X_i.

7.12.1.1. Univariate patterns

The univariate pattern of the residual deviations, $Y_i - \hat{Y}_i$, can be shown in either a histogram or box plot (which the computer will readily prepare, if asked). Because the algebraic linear-regression model makes the mathematical assumption that the "residuals" have a Gaussian distribution, with a mean of 0, the histogram or box plot can be checked to conform this assumption. If the residual distribution is substantially skewed or shows other patterns that grossly depart from a Gaussian shape, the result may indicate a basic flaw in the model.

7.12.1.2. Bivariate patterns

If plotted against X, the residual values of $Y_i - \hat{Y}_i$ should vary randomly around 0, with no particular pattern evident as X goes from low to high values. A pattern occurs when distinctive sequences of residual values for $Y_i - \hat{Y}_i$ are all positive or all negative within a particular zone of the X values, instead of randomly fluctuating around 0. The pattern can be seen in a specific plot of residuals, where the deviations will fluctuate around the horizontal line where $Y_i - \hat{Y}_i = 0$.

The plot of $Y_i - \hat{Y}_i$ versus X is really unnecessary in simple regression, where the pattern is readily seen in the scatter of Y_i and X points around the ordinary regression line. The plot of the residuals is particularly useful, however, when shown for each of the X_j variables in multivariable regression.

7.12.1.2.1. 'Bivariate' checks for multiple variables

When the residual values of $Y_i - \hat{Y}_i$ are plotted separately against each of the X_j independent variables in a multiple regression model, the results should resemble what occurs in simple regression. For each X_j variable, the estimated values of $\hat{Y}_i$ and the actual values of Y_i should have a reasonably good fit; and the residual values of $Y_i - \hat{Y}_i$ should show random fluctuations around residual values of 0. If striking patterns are evident in the residuals or in a misfit of Y_i vs. $\hat{Y}_i$, something is probably wrong with the format used for the corresponding independent X_j variable. These evaluations will be further discussed in Chapter 12.

In simple regression, however, patterns can easily be found in the original plot of $\hat{Y}_i$ vs. X, as shown in the next two figures.

7.12.1.2.2. 'Terminal' sources of distortion

The two ends of an independent variable can commonly act as sources of "terminal" distortion by exaggerating, truncating, or reversing the expected effect.

In Fig. 7-8, the terminals are exaggerated, causing the straight line to show a reduced impact for X at each end. A pattern of truncated terminals was shown in Fig. 2-15, where the straight line distorts the impact of X throughout its entire course. The impact is too low in the center of the data, and too high at the ends. In Fig. 7-9, the effect of X is reversed at the upper end of the data, and the impact shown by the straight line is too high at each terminal.

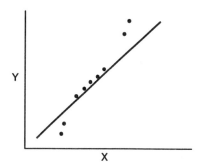 

Fig. 7-8 Pattern of exaggerated terminals. The straight line makes the impact of X too small at each end of the data.

Fig. 7-9 Pattern of "contradictory terminals" with reversal at upper terminal. The straight line makes the impact of X too high at each terminal. The best fit may be a reversal curve, rather than a straight line.

7.12.1.2.3. Connotations of pattern

The discovery of "patterned residuals" can help distinguish poor fits from misfits. For example, if a low r^2 suggests that the model fits poorly, random variation in residuals will indicate that X has no particular relationship to Y. If the residuals show distinctive patterns, however, X is probably related to Y but has been used in the wrong format.

This type of inspection can separate the poor fit that would occur in Fig. 2-12 from the misfit of Fig. 2-18, although both fits have low values of r^2. In Fig. 2-12, as X goes from low to high values, the $Y_i - \hat{Y}_i$ residuals would show random variations around 0. With similar increases for X in Fig. 2-18, however, the residual values will show a pattern of being all negative, then positive, then negative. A similar but less striking positive-negative-positive-negative pattern would be shown for the residuals in Fig. 2-15. In Fig. 12-2, the pattern of residuals would be positive-negative-positive.

When the occurrence of patterned residuals suggests that X has been used in a wrong or deceptive format, a closer fit can be achieved by changing the algebraic format. Alternatively, a categorical analysis can be used either for confirmation or for the main evaluation. The choice of what to do will depend on the analyst's goals and on what is found in either the residual patterns or the basic graph of Y vs. X.

Two sets of patterned residuals can sometimes look similar despite striking differences in the fit of the algebraic model. For example, the residuals would show a negative-positive-negative pattern for both the misfit in Fig. 2-18 and the reasonably good fit in Fig. 2-14. For this reason, residual patterns are valuable for suggesting curved relationships, but are not intended to denote goodness of fit.

7.12.1.2.4. Standardized residuals

For evaluating goodness of fit, the $Y_i - \hat{Y}_i$ residuals can be converted to standardized values when divided by their standard deviation, which is $\sqrt{S_R/(N-2)}$. The magnitude of the standardized residuals can then be used for gross comparisons of goodness of fit for regression lines in different sets of data. If the inspections are also intended to convey ideas about relationships in the fit, or to compare the closeness of the fit, the X values should also be standardized.

7.12.1.3. Summary of diagnostic contributions by residual patterns

Because configurational distortions can occur regardless of the values achieved for r^2 and b, the special "diagnostic" contributions of the patterns of residuals are summarized in Table 7-2.

Table 7-2 *"Diagnoses" from statistical indexes and residual patterns*

Value of r^2	Value of b	Potential problem	Do residuals have a distinctive pattern?	Diagnostic conclusion
High	High	Distorted shape and impact (e.g., exponential or sigmoidal curve)	No	Satisfactory model with good fit
			Yes	Distorted model: change format or analytic procedure
High	Low	Distorted impact	No	Line fits properly, but X has little impact
			Yes	Bivariate relationship may be a reversal curve
Low	High	Unsuitable "line-ability" (No problems in shape or impact)	No	Linear swarm of points
			Yes	Configuration of swarm may be distorted. Try different format or analytic procedure
Low	Low	Unsuitable format (Distortion depends on whether data form a circular swarm or reversal curve)	No	Data form a circular swarm. No distortion of shape or impact
			Yes	Data may form a reversal curve, distorting both shape and impact

7.12.2. Examination of ordinalized zones

A useful way to check for conformity of a single rectilinear coefficient is to divide the dimensional X variable into a set of ordinalized quantile zones. At least three tertile zones are needed to search for trend; five quintile zones will usually suffice; and some analysts prefer four quartile zones, to compare results in the lowest and highest fourths of the data.

7.12.2.1. Regressions for each zone

In one ordinalized approach, Y is regressed separately for the data in each zone of the X variable. If the data have a truly rectilinear pattern, each of the regression lines will have similar values for the intercept a and the slope b.

The approach is simple and direct, but it involves many pairwise comparisons of a and b values, and criteria for decisions about when a pair is substantially different.

7.12.2.2. Dummy binary variables

A simpler approach, which uses only a single additional regression, converts the ordinalized zones into dummy binary variables that are given a reference-cell coding. For dimensional variables, the ordinal partition should produce interior zones of equal widths, but the two extreme zones of X can be unbounded.

For example, the hematocrit variable in the illustrative data set could be divided into five ordinal zones coded as follows:

Ordinalized hematocrit category	Reference-cell coding Dummy variables			
	X_1	X_2	X_3	X_4
< 33	0	0	0	0
33–37	1	0	0	0
38–42	0	1	0	0
43–47	0	0	1	0
≥ 48	0	0	0	1

With this arrangement, the data for Variable X would be fitted with a multivariable regression model expressed with the four dummy variables as

$$Y_i = b_0 + b_1X_1 + b_2X_2 + b_3X_3 + b_4X_4.$$

In the lowest zone of the data, where $X_1 = X_2 = X_3 = X_4 = 0$, the average value of Y_i will be indicated by b_0. In the second zone, the average value of Y_i will be shown by $b_0 + b_1\overline{X}_1$ where $\overline{X}_1$ is the mean of the values in that zone. The value of b_1 will indicate the average slope of Y for the section of the regression line where X moves from the lowest to the next lowest zone of the data. In the middle zone of the data, the average value of Y_i will be shown by $b_0 + b_2\overline{X}_2$. If the mean values of X are equidistant between the second and third zones, the value for b_2 should be twice that of b_1. In the next-to-highest zone of the data, the average value of Y_i will be shown by $b_0 + b_3\overline{X}_3$. With equal distances between the mean X values in the four zones, b_4 should be four times the value of b_1.

If demarcations produce equal distances between the mean values of X in each of the five zones, an alternative "increasing-rank" coding could be established as follows:

Ordinalized hematocrit category	Increasing-Rank Coding Alternative dummy variables			
	X_1	X_2	X_3	X_4
< 33	0	0	0	0
33–37	1	0	0	0
38–42	0	2	0	0
43–47	0	0	3	0
≥ 48	0	0	0	4

With this tactic, the coefficients for b_1, b_2, b_3, and b_4 should be approximately equal.

When the illustrative data set was analyzed for hematocrit cited in the foregoing reference-cell codes, the equation that emerged was ln survival = .51 + .64X_1 + 1.06X_2 + 1.28X_3 + .90X_4. With the increasing-rank coding for hematocrit, the equation was ln survival = .51 + 0.64X_1 + .53X_2 + .43X_3 + .22X_4. Since none of the dummy variables had a stochastically significant coefficient in either equation, the results are difficult to interpret. Nevertheless, the pattern of coefficients is consistent in both equations; and the results do *not* suggest a rectilinear relationship. They are compatible with a monotonic rise of ln survival in the first four zones, but not in the fifth.

7.12.2.3. Ordinal variables

If a variable is originally expressed in ordinal categories, such as *TNM stage* and *symptom stage* in the illustrative data set, the ordinal "zones" have already been identified, and can be immediately examined with any of the three coding systems just proposed for dimensional variables.

For example, when the first reference-cell coding (with each dummy variable coded as **0** or **1**) is applied to the four categories of symptom stage in the illustrative data set, with Y_i = ln survival, the linear regression equation becomes $\hat{Y}_i$ = 2.21 − .049X_1 − 1.25X_2 − 1.72X_3. When a similar coding is used for the five categories of TNM stage, the result is $\hat{Y}_i$ = 2.43 − .130X_1 − 1.03X_2 − 1.32X_3 − 1.77X_4. (In both of these equations, all coefficients and intercepts except b_1 are stochastically significant.) The results show a monotonic progression, as expected, for magnitude of the b_i coefficients, but they do not have the relative magnitudes that would be expected in a rectilinear relationship. These results confirm what will be noted in Chapter 8, where the tabulated results show that intercategorical gradients are unequal, and hence not rectilinear.

When symptom stage and TNM stage are given the increasing-rank system of coding (with dummy variables successively coded as 1, 2, 3, etc.), the linear regression equation for symptom stage is $\hat{Y}_i$ = 2.21 − .49X_1 − .64X_2 − .572X_3, with stochastic significance (at $P < .05$) in the coefficients for b_0, b_2, and b_3. The corresponding result for TNM stage is $\hat{Y}_i$ = 2.431 − .130X_1 − .514X_2 − .440X_3 − .441X_4, with stochastic significance in the coefficients for b_0, b_2, b_3, and b_4. The disparities in the appropriate coefficients for the successive dummy variables again confirm that the relationships are not strictly linear.

7.12.2.4. Interpretation of results

If the special dummy-variable coefficients have inconsistent values, the impact of the X variable will not be suitably represented with a single rectilinear coefficient. This discovery may not be a problem if the research goal is merely to demonstrate that X has an important impact on Y and to get a general idea of the "average" impact. If the goal is to give an accurate quantitative indication of the impact, however, the linear nonconformity shows that a single average value will be deceptive. Thus, a statement such as, "You live 5 years longer (or shorter) for every 2-unit rise (or fall) in substance X" is correct if X has a rectilinear effect on Y, but potentially misleading or wrong if it does not.

7.12.2.5. Mechanism of calculations

If unfamiliar with the mechanisms (discussed later) for determining coefficients in multivariable analysis, you may suspect intuitively that a coding system for ordinalized zones will give erroneous results. For example, consider an X variable having three ordinal categories for which the survival rates in the Y variable are as follows:

$$a_1: 10/100\ (10\%)$$
$$a_2: 20/100\ (20\%)$$
$$a_3: 30/100\ (30\%)$$

Intuitively, you might have the following expectation: When results in zone a_1 are compared against those in the other two zones, the expression is 10/100 (*10%*) vs. 50/200 (*25%*). When results in zone a_3 are compared against those in the other two, the expression is 30/100 (*30%*) vs. 30/200 (*15%*). When the middle zone, a_2, is compared against the two extremes, however, the comparison is 20/100 (*20%*) vs. 40/200 (*20%*) —and so the a_2 zone would seem to lose its distinctiveness because of the cancelled effects.

This intuitive expectation, however, does not correspond to the way things happen when a single variable is converted into binary dummy variables that receive an algebraic "multivariable" analysis. With a multivariable expression such as $\hat{Y}_i = b_0 + b_1X_1 + b_2X_2 + b_3X_3$, the coefficients are *not* calculated by comparing results for an intermediate ordinal zone against results for all other zones. Instead, the effect of any zone (as a dummy variable) is determined, as noted in Chapter 12, with all other zones (as other dummy binary variables) being held constant. Thus, if X_2 is the dummy variable coded as **1** for the intermediate zone, all other variables are coded as **0**; and the effect is determined from the expression $\hat{Y}_i = b_0 + b_2X_2$. Because the extreme variables on either side do not participate at that point in the model's "action," the cancelling effects feared by intuition do not occur.

7.12.3. Check of quadratic regression

A relatively simple algebraic method of checking for rectilinear conformity between X and Y is to examine the regression obtained with the "quadratic" model

$$Y = a + bX + cX^2.$$

As noted by Grambsch and O'Brien,[10] this model may not be "sensitive to every possible form of non-linearity, [but] it may be expected to be reasonably sensitive to a broad class of non-linear relationships that commonly occur." The relationship between X and Y is nonlinear if the *c* coefficient for the X^2 quadratic term is stochastically significant at an appropriately chosen level of α.

7.13. Management of problems

If the residuals show a distinctive pattern, if the ordinalized zones do not confirm the linear pattern, or if the quadratic term has a "significant" coefficient, the true relationship between X and Y has been distorted, even if the rectilinear model seems to fit well.

The problem can be managed in three ways. The first approach is to ignore it. If the goal is merely to demonstrate and get an average general quantitative summary of impact, the results are acceptable if a distinctive impact has been shown. If an impact is expected but is not shown, however, the algebraic model may be incorrect.

In the latter situation, or for certain other goals—such as getting a really close fit—the second approach is to change the format of X, expressing it as X^2, $\sqrt{X}$, log X, or some other transformation. (Alternatively, the format can be changed for Y, or for both Y and X.) The change in format is particularly useful if the main goal is to get a close-fitting equation to the data. On the other hand, if the main goal is to find distinctive variables and categories, the transformations may improve fits but may sometimes impede interpretations.

Consequently, a third way to manage the problem is to express the impact of X in zones, rather than with a single coefficient. If this problem occurs for many of the independent variables, the entire analysis may be best transferred from an algebraic method to one of the stratified-cluster procedures discussed in Chapters 19–21. With this change, the algebraic method will have done its job as a screening procedure for the data, but the more definitive results will come from what is found in the direct categorical clusterings and relationships.

7.14. Subsequent procedures

If the regression diagnostics indicate that the model is "unhealthy," a satisfactory treatment may or may not be available. If the model fits poorly and the residuals show no specific pattern, nothing can be done about the model because the independent variable X is unrelated to Y. If something is to be related to Y, some other independent variable must be sought.

If a poor-fitting model, however, shows distinctive patterns in the residuals or nonconforming linear zones, the fit can often be improved by changing the format of X. Whatever the new format may be, X becomes transformed into a different variable. An entirely new regression analysis must be done, and the subsequent process of evaluation resumes at a "new" beginning.

Chapter References

1. Feinstein, 1985; 2. Burnand, 1990; 3. Feinstein, 1990; 4. Kendall, 1971; 5. Guralnick, 1972; 6. Fleiss, 1981; 7. Belsley, 1980; 8. Cook, 1982; 9. Fox, 1991; 10. Grambsch, 1991.

8 Stratified Rates: The Cluster Prototype

Outline . . . .
8.1. Structure of targeted clusters
8.2. Assessing fit for target variable
   8.2.1. Distinction in nomenclature
   8.2.2. Expressions of discrepancy
      8.2.2.1. Deviations and variance
      8.2.2.2. Errors
         8.2.2.2.1. Example of estimation process
         8.2.2.2.2. 'Spanning the meridian'
      8.2.2.3. Congruences
   8.2.3. Appraisal of stochastic stability
   8.2.4. Criteria for quantitative significance
      8.2.4.1. Reductions in variance and error rate
      8.2.4.2. Conflicting results
8.3. Assessing trend in target variable
   8.3.1. Gradient in targeted clusters
      8.3.1.1. Adjacent incremental changes
      8.3.1.2. Intrasystem and intercategory gradients
   8.3.2. Guidelines and criteria for categorical gradients
      8.3.2.1. Minimum intercategorical gradients
      8.3.2.2. Minimum intrasystem gradients
   8.3.3. Algebraic model of slope
   8.3.4. Standardized slope
   8.3.5. "Quantitative significance" in published literature
   8.3.6. Tests of stochastic significance
      8.3.6.1. Intercategory contrasts
      8.3.6.2. Chi-square test for linear trend
   8.3.7. Simultaneous quantitative and stochastic appraisals
8.4. Demarcation of strata
   8.4.1. Mininum (modicum) size of strata
   8.4.2. Criteria for collapsibility of strata
   8.4.3. Distribution of strata
      8.4.3.1. Number of strata
      8.4.3.2. 'Shape' of distribution
      8.4.3.3. Indexes of 'diversity'
   8.4.4. Selection of dimensional boundaries
      8.4.4.1. 'Esthetic' boundaries
      8.4.4.2. Clinical boundaries
      8.4.4.3. Quantiles
      8.4.4.4. Theoretical statistics
      8.4.4.5. Correlated choices
8.5. Validation of results
   8.5.1. Concept of validation
   8.5.2. External validation

8.5.3. Resubstitution
8.5.4. Cross-validation
   8.5.4.1. Jackknife method
   8.5.4.2. Split-group method
   8.5.4.3. V-fold validation
8.5.5. Bootstrapping

. . . .

When the target variable is expressed in binary or ordinal categories, the algebraic model of simple linear regression is not mathematically appropriate. The dependent categories are usually (and more properly) fitted with alternative algebraic formats, such as the logistic, proportional hazards, and discriminant function methods that will be described in Part IV of the text.

When only two variables are under analysis, however, a categorical target regularly receives a clustered rather than an algebraic form of analysis. The simple form of targeted categorical analysis is discussed now, not only because it is applied in Chapter 9 for bivariate evaluations of the illustrative data set, but also because the strategy is a prototype for the multivariable targeted cluster analyses that arrive in Part V.

8.1. Structure of targeted clusters

The term *cluster* refers to a category that is simple or compound. A simple category contains a single component, such as **old,** from only one variable, such as *age*. A compound category contains components from two or more categories. **Tall old men** is a compound category with components from *height, age,* and *sex*. Compound categories can have a complex structure, such as the composite category delineated as: **tall old men** and/or **young women,** but not anyone who is **fat** or who is a **molecular biologist.**

The term *cluster* is used to avoid the confusion that might arise if the simple word *category* were used for a compound entity that contains many categories. Nevertheless, every cluster, whether simple or compound, is a category.

Chapter 19 contains further discussion of the Boolean logic and other mechanisms used to form clusters of compound categories. The main point to be noted now is that an array of related clusters, containing contributions from two or more variables, often forms the categories of a separate composite variable. The composite categories often have an ordinal ranking and are designated with numbers (such as **I, II, III,** . . .) or letters (such as **A, B, C, D,** . . .). The array of ranked composite categories is often called a *staging system,* and the individual composite categories (or clusters) are identified as **Stage I, Stage II, Stage III** . . . .

In the illustrative data set here, the variables called *TNM stage* and *symptom stage* contain clustered categories that form an ordinal array. Regardless of whether the components are simple or complex, the clusters of a categorical variable form a set of binary, ordinal, or nominal categories, and can be analyzed accordingly.

Thus, the relationship of *symptom stage* vs. *TNM stage* would be examined in a simple bivariate analysis, despite the many variables that were used to form the categories of each stage. Part V of the text is concerned with constructing and analyzing clusters formed when two or more categories are combined from composite variables. The rest of this chapter is devoted to bivariate analyses of relationships for two categorical variables.

For simple bivariate analysis, the target variable is usually expressed in binary categories, such as **alive/dead** or **success/failure.** If the target is an ordinal or nominal variable, the categories can be compressed into a binary split, such as **less than moderate/at least moderate,** or **cardiovascular disease/noncardiovascular diagnoses.** With the binary arrangement, the target event is usually summarized for the relative frequency, proportion, or "rate" of its occurrence.

The single independent variable in such analyses is also expressed in categories that can be binary or ordinal. To receive this structure, dimensional variables can be demarcated into ordinal zones, and nominal variables can be dichotomized by compression into two categories.

The table that shows frequency counts (or "contingency counts") for the bivariate results is called an $r \times 2$ table, having r rows for the r categories of the independent variable, and 2 columns for the two states of the binary dependent variable. The simplest arrangement is the common 2×2 or "fourfold" table that occurs when the independent variable also has two categories, such as **treated/untreated, exposed/nonexposed** or **Treatment A/Treatment B.** When the partitioned categories in the independent variable are associated with the corresponding binary proportions (or rates) of the target variable, the table is called a *stratification,* and the results are called *stratified rates.*

8.2. Assessing fit for target variable

As in algebraic analyses, the stratified rates can be used for making estimates and determining trends. Also as in algebraic analyses, the goal of estimation is assessed from the fit achieved for the target variable.

8.2.1. Distinction in nomenclature

In statistical discussions, the phrases "predictive fit" or "predictive accuracy" are often used for the agreement between a set of estimated values and the actual observed results. These phrases have three misleading implications in time, direction, and challenge. In time, the word *predictive* implies a forecast of the future, but the target being estimated may occasionally be a diagnosis or other attribute that exists concurrently. In direction, the idea of something "prospective" is untrue if the estimating model or system has been fitted retroactively to the collected data. Finally, the idea of *prediction* is best tested when the results of the retroactive fit are challenged for their effectiveness in a new set of data, not in the same data used to develop the model or system.

Consequently, the word *predictive* will generally be avoided in the discussion here. In previous writings,[1] the word *congruent* was used for the agreement between a system/model and the data from which it was developed. In this text, however, the word *congruence* has been reserved for the type of agreement that is converted to a "likelihood", as discussed shortly. Accordingly, the term *concordant fit* will be used here for what is elsewhere often called "predictive accuracy" or "predictive fit".

8.2.2. Expressions of discrepancy

As noted earlier, the term *discrepancy* refers to the disagreement between each observed value of the target, Y_i, and the corresponding estimated value, $\hat{Y}_i$. For binary target variables, discrepancies can be expressed in indexes of deviation, error, or congruence.

8.2.2.1. Deviation and variance

The common, familiar types of discrepancy are the *deviations*, discussed in Chapter 7, that are measured as incremental differences, $Y_i - \hat{Y}_i$, between the observed and estimated amounts of a dimensional variable. The total of the squared deviations in a group is the group variance, expressed as $S_{yy} = \Sigma(Y_i - \hat{Y}_i)^2$.

The basic estimate for the values of the target variable, $\{Y_i\}$, is the mean, $\bar{Y}$. For a binary target coded as **0/1** in a group of N numbers, there will be T values of **1** and N − T values of **0**. The mean can be cited either as the binary proportion P = T/N, or as its compliment Q = 1 − P = (N − T)/N.

The calculation of proportional reduction of variance in a binary target variable uses the same tactics as in dimensional data, but applies some new symbols and simplified formulas. The counterpart of S_{yy} as basic group variance for the Y variable is NPQ.*

Although a linear algebraic model can be used (as discussed later), binary targets are often fitted with categorical clusters. With h as the identifying subscript, each cluster will have n_h members, for whom p_h is the corresponding rate of the target event in the cluster, and q_h is the complementary rate. The group variance in each fitted categorical cluster will be $n_h p_h q_h$, and the sum of group variances across the clusters will be nhpnqh, and the sum of group variances across the clusters will be $\Sigma n_h p_h q_h$.

For example, consider the 6-month survival rate of 100/200 *(50%)* for the total group "staged" with variable X_1 on the left of Table 8-1. The basic group variance, before any independent variables are added, is NPQ = (200) (100/200) (100/200) = 50. When the data were divided according to the three categories of variable X_1, the group variance became $n_1 p_1 q_1 + n_2 p_2 q_2 + n_3 p_3 q_3$ = [70(56/70)(14/70)] =[65 (33/65)(32/65)] + [65(11/65)(54/65)] = 36.585.

---

*To derive this formula, recall that T members of the group have $Y_i = 1$, and N − T members have $Y_i = 0$. Using the formula $\Sigma(Y_i - \bar{Y})^2$, the basic group variance will be $T(1 - P)^2 + (N - T)(0 - P)^2$. Substituting NP = T, NQ = N − T, and Q = 1 − P, the basic variance becomes $NPQ^2 + NQP^2 = NPQ (P + Q)$ = NPQ.

Table 8-1 Three staging systems for a group with total success rate of 50% = 100/200*

	Ordinal variable used for staging	
X_1	X_2	X_3
a_1: 56/70 *(80%)*	b_1: 45/80 *(56%)*	c_1: 35/70 *(50%)*
a_2: 33/65 *(51%)*	b_2: 33/60 *(55%)*	c_2: 34/70 *(49%)*
a_3: 11/65 *(17%)*	b_3: 22/60 *(37%)*	c_3: 31/60 *(52%)*

*The individual numerators in the three arrangements always sum to 100, which is the total numerator for the entire group; and the individual denominators always sum to the total 200.

The proportionate reduction in variance for a binary target is symbolized as ϕ^2, and is sometimes called the *variance reduction score*. Its expression is

$$\phi^2 = \frac{NPQ - \Sigma n_h p_h q_h}{NPQ}. \qquad [8.1]$$

For the data here, $\phi^2 = (50 - 36.585)/50 = .2683$.

The calculations can be simplified by recognizing that $p_h = t_h/n_h$ and $q_h = (n_h - t_h)/n_h$. Each value of $n_h p_h q_h$ then becomes $(t_h)(n_h - t_h)/n_h$, and NPQ becomes $T(N - T)/N$. The values of N, T, n_h, and t_h can therefore be used directly from the available data without any problems of rounding in calculation of p_h or q_h. These simplifications produce identical results for Variable X_1 in Table 8-1, since the value of NPQ becomes $(100)(100)/200 = 50.00$, and the corresponding values for the categories are $[(56)(14)/70] + [(33)(32)/65] + [(11)(54)/65] = 36.585$.

For Variable X_2 in Table 8-1, $\Sigma n_h p_h q_h$ can be calculated as $[(45)(35)/80] + [(33)(27)/60] + [(22)(38)/60] = 48.47$, and the corresponding value for Variable X_3 is 49.969. The respective values of ϕ^2 are .0306 for X_2 and .00062 for X_3.

If all the algebra is suitably worked out, an even simpler formula can be used to provide the value of the chi-square statistic, X^2, which is then converted to ϕ^2 as

$$\phi^2 = X^2/N. \qquad [8.2]$$

The simple formula is

$$X^2 = [\Sigma t_h^2/n_h - T^2/N][N^2/T(N - T)] \qquad [8.3]$$

Thus, for Variable X_1 in Table 8-1, $X^2 = [(56^2/70) + (33^2/65) + (11^2/65) - (100^2/200)][200^2/(100)(100)] = 53.6615$; and $\phi^2 = 53.6615/200 = .2683$.

8.2.2.2. Errors

Each estimate of a binary (or other) category can be converted to a specific category, which will be either right or wrong. For example, if a person is in a category with a 90% success rate, we would estimate success for that person. The estimate will be correct if the person was actually successful, and wrong otherwise. The number of right and wrong estimates for a group can then be counted to form an index such as a total rate (or number) of errors. The total rate (or number) of errors is then compared for estimates made without and with the independent variable(s).

If the error rate of estimates from variable Y alone is E_y and the error rate with estimates from independent variable X is E_x, the proportionate reduction in error is an index called

$$\text{lambda} = \frac{E_y - E_x}{E_y}. \qquad [8.4]$$

To correspond to the proportionate reduction in variance, the lambda index is cited as a reduction in error rather than an increase in accuracy.

8.2.2.2.1. Example of estimation process

The general strategy for making estimates and determining "errors" is as follows: Suppose the binary rate of success for any category is t_h/n_h. If the rate exceeds 50%, we would estimate **success** for everyone in that category; and the estimate will be correct in t_h persons and wrong in $n_h - t_h$ persons. If t_h/n_h is <50%, the estimate of **failure** for everyone in the group would be right in $n_h - t_h$ persons and wrong in t_h. If the success rate is exactly 50%, the accuracy will be the same regardless of which estimate is used. In the original group of N persons, the estimate will be correct in T persons if T/N > .50 or in N − T persons if T/N <.50. (A precise formula cannot be written for the decisions because each selected estimate will depend on whether t_h/n_h or T/N exceeds the meridian value of .50.)

An example of the calculations for lambda was presented in Section 2.3.3.2. Another example can come from Table 8-1, where the overall 6-month survival rate for the 200 patients was 50%. With only this information from the Y variable, an estimate of **dead** (or **alive**) for everyone would have 50% accuracy, with a total of 100 errors in the 200 estimates. After the table is partitioned according to the categories of Variable X_1, the estimate of **alive** would have 14 errors in category a_1 (where survival is 80%), and 32 errors in category a_2 (where survival is 51%). The estimate of **dead** would have 11 errors in category a_3 (with survival 17%). The total number of errors would be 14 + 32 + 11 = 57, and the total error rate would be proportionately reduced to a lambda value of (100 − 57)/100 = 43%.

The total errors for two other sets of estimates in Table 8-1 would be 84 (= 35 + 27 + 22) for Variable X_2 and 98 (= 35 + 34 + 29) for Variable X_3. The respective values of lambda would be 16% and 2%.

8.2.2.2.2. 'Spanning the meridian'

If the original data have a success rate that is close to the polar values of either 0% or 100%, stratification with the X variable can sometimes produce a valuable separation of subgroups without improving the "batting average" cited in the lambda index of proportionate error reduction. For improvement of lambda, the results in at least one of the strata must go across the "meridian" value of 50%. A stratification that does not succeed in "spanning the meridian", with results on both sides of 50%, will not improve the total score for error reduction.

For example, suppose a group of 320 persons with a success rate of 72/320 (23%) is stratified into four categories with the following results:

a_1:	36/80	*(45%)*
a_2:	24/80	*(30%)*
a_3:	12/80	*(15%)*
a_4:	0/80	*(0%)*
Total	72/320	*(23%)*

This stratification will get .16 = [(55.8 − 46.8)/55.8] for its variance reduction score, but the rate of errors will be unchanged and the value of lambda will be 0. To demonstrate the latter point, note that the estimate of **failure** from the unstratified results for everyone would have been wrong in 72 cases. Because none of the stratified results crosses the meridian value of 50%, the estimate would also be **failure** in each of the strata. This estimate would be respectively wrong for 36 persons in the first stratum, for 24 in the second, for 12 in the third, and for 0 in the fourth. The sum of 36 + 24 + 12 + 0 is 72; and so the original number of errors would be unchanged.

A poor improvement in the lambda score for error reduction, despite good or excellent results for other indexes of accomplishment, can regularly be expected if the original success rate is very low or high, i.e., near 0% or 100%. In this situation, excellent accuracy can be achieved simply by estimating **failure** or **success** for everyone, without any further analyses. When stratified according to an additional variable, the results may be unable to improve the already high rate of accuracy.

Conversely, when the original group has a target rate that is near the meridian value of 50%, the lambda score for error reduction may be higher than the ϕ^2 score for variance reduction, as in the values of 43% vs. 27% for variable X_1 in Table 8-1.

The poor values of lambda have often led to doubts about the value of "risk stratification" for individual persons. Thus, if a general population has a 7% rate of occurrence for disease D, the investigators may be pleased to find a stratification of "risk factors" for which the rates of occurrence are 12% in the **high** group, 6% in the **medium** group, and 1% in the **low** group. Despite the statistical achievement, however, the estimations for individual persons will be unchanged; and everyone's risk is still relatively small. (On the other hand, the stratification may be particularly useful for evaluating results of a clinical trial or for deciding which groups of patients have risks high enough to warrant interventions.)

8.2.2.3. Congruences

The estimate for categorical targets is often a value of probability, such as the common meteorological prediction: "80% chance of rain today". Unlike measured dimensions, probabilities are not "properly" subtracted to form deviations; and unlike specific statements, such as yes or no, the probability estimates cannot be unequivocally called right or wrong.

The probabilistic discrepancy, however, can be expressed with a special index, which can be called *congruence*. Its complex mathematical formulation can be simplified as $1 - |Y_i - \hat{Y}_i|$. For example, if the weatherman predicts an 80% chance of rain, $\hat{Y}_i = .80$. If it rains today, the observed $Y_i = 1$, and so congruence is $1 - |1 -$

.80| = .80. If it does not rain, $Y_i = 0$, and the congruence is $1 - |0 - .80| = .20$. The value of congruence is 1 for a perfect estimate, and 0 for the worst estimate.

To get an index of total discrepancy for a group, individual congruences are multiplied to form an entity called *likelihood*. Indexes of congruence and likelihood are unfamiliar because they are almost never used in elementary statistics. These indexes will receive no further discussion here until they appear in Chapters 13–14, where they become a basic strategy in the multivariable procedure of logistic regression.

8.2.3. Appraisal of stochastic stability

Before making decisions about the magnitude of indexes of fit, most analysts will check for stability with a stochastic test of significance.

For a binary target, the value of ϕ^2 is checked with a chi-square test, using the formula $X^2 = N\phi^2$, where N is the total sample size. If the binary target variable has r rows of categories, this X^2 result is converted to a P value at $r - 1$ degrees of freedom. For variable X_1 in Table 8-1, $X^2 = (200)(.268) = 53.6$ At 2 d.f., P is <.000001. The value of X^2 is 6 (=200 x .03) for variable X_2 in Table 8-1, and 0.12 (=200 x .0006) for variable X_3. At 2 d.f., the corresponding P value is just below .05 for variable X_2 but is not significant for X_3. (If the sample sizes are too small, the chi-square test can be replaced by a Fisher Exact Probability test, which yields an "exact" P value.)

With any of the cited stochastic tests, the P values are regarded as significant, i.e., the variance reduction score is stable, if below a boundary called α, which is usually set at $\le .05$. Although a commonly used boundary for stochastic significance, the α level of $\le .05$ need not be rigidly maintained. For example, if the total group is relatively small, the demand for $P \le .05$ may be too draconian to permit anything to emerge as significant, and so the α boundary may be raised to .1 or even higher. Conversely, with large sample sizes and a multiplicity of stochastic comparisons, an argument can be offered that the α levels should be lowered to stricter boundaries, according to a suitable adjustment process. Thus, with the commonly used Bonferroni adjustment, if stochastic significance is tested on k occasions (or "comparisons"), the decisive boundary for α is lowered to α/k. For 8 tests, the α level for each test might be lowered from .05 to .00625.

The P value will usually exceed .05 in situations where the variance reduction score or other descriptive index is small and quantitatively nonsignificant. Nevertheless, stochastic support can be obtained for the conclusion of nonsignificance if an appropriate confidence interval around the "small" result does not exceed a quantitatively significant value.

The stability of lambda in a stratified categorical array is seldom appraised with a special stochastic procedure. If stochastic appraisal is desired, the results often receive the same chi-square or Fisher test used for variance reduction. On the other hand, if the categories have an ordinal ranking (as in the three variables of Table 8-1), the statistical achievement is often evaluated for trend rather than for fit, with a stochastic procedure called the chi-square test for linear trend, discussed later in Section 8.3.6.2.

8.2.4. Criteria for quantitative significance

The choice of a boundary for quantitative significance in estimation is particularly tricky. The decision can involve[2] mathematical appraisals of a magnitude, clinical judgment about the biologic importance of the entity, and epidemiologic projections about the size of the general population that might be affected.

8.2.4.1. Reductions in variance and error rate

In decisions about quantitative magnitude, reasonably good agreement[2,3] exists that an independent variable is unimpressive if it fails to "explain" or proportionately reduce at least 10% of the basic group variance. Consequently, values of ϕ^2 are usually regarded as quantitatively nonsignificant if <0.1. Their quantitative significance increases as they get closer to 1. In Table 8-1, ϕ^2 for the X_1 variable has the quantitatively significant value of .268, but is unimpressive for variables X_2 and X_3.

Like the proportionate reduction in variance, the value of lambda can be regarded as quantitatively significant if it equals or exceeds 10%.

8.2.4.2. Conflicting results

Occasional conflicts can occur in the results of proportionate reduction in group variance vs. concordant accuracy for fit.

Sometimes the reductions are similar for error rate and for group variance. Consider a total group whose success rate is 200/400 = 50%. Suppose an independent variable, having three ordinal categories, stratifies the success rates into **small** = 25/25 *(100%);* **medium** = 175/350 *(50%);* and **large** = 0/25 *(0%)*. With this stratification, the original error rate of .50 is reduced to 175/400 = .4375, so that lambda is (.50 − .4375)/.50 = .125. The original group variance, which is (200)(200/400) = 100, is reduced to (25)(0/25) + (175)(175/350) + (0)(25/25) = 87.5, so that ϕ^2 is also .125 [= (100 − 87.5)/100]. On the other hand, as shown in Section 8.2.2.2.2, the group variance may sometimes be "significantly" reduced without *any* reduction in error rate. Conversely, as shown by variable X_1 in Table 8-1, the error rate reduction may be quantitatively higher than the score for variance reduction.

These conflicts will be discussed in greater detail later. The main point to be noted now is simply that goodness of fit can be measured in more than one way, and that the different measurements may not yield similar conclusions.

8.3. Assessing trend in target variable

The trend in the target variable can be assessed either directly from the results for the clusters, or indirectly with the aid of a special algebraic model.

8.3.1. Gradient in targeted clusters

When the independent variable, X, is stratified in targeted binary categories, the impact of X is shown by the gradient of the Y values in successive strata (or zones) formed by the categories of X. The evaluation of these gradients involves a relatively unfamiliar set of statistical principles, because gradients between and among *categories* are neither noted nor appraised in algebraic analyses.

8.3.1.1. Adjacent incremental changes

The most direct and simple measurement of gradient in a stratified analysis is the incremental change in means, binary proportions, or other summary values of Y as X advances from one category to the next. With only two categories of X, only a single increment is available; but with three or more categories, two or more successive increments can be calculated and checked.

If the X variable really has a constant impact, the successive increments should be reasonably equal. If not, a rectilinear model does not suitably fit the trend of the data, and conclusions based on such a model may be misleading. For example, in Table 8-1, the gradients are monotonic for variables X_1 and X_2 but not for X_3. The successive categorical increments are reasonably close (at 29% and 34%) for variable X_1, but a distinctly unequal 1% and 18% for variable X_2. Neither variable's changes in impact would be accurately expressed by the simple rectilinear coefficient of b in a model such as Y = a + bX.

8.3.1.2. Intrasystem and intercategory gradients

Regardless of whether the final product contains simple or composite categories, the array of clustered categories forms a *system*. This term is analogous to the word *scale* for the categories available to express a single variable, but *system* connotes the idea that the categories may be composite, rather than simple. The term is regularly used in scientific discourse when a scale of composite categories such as **Stage I, Stage II, Stage III,** . . . is called a *staging system*.

The total quantitative gradient in a system of categories can be divided in two ways. The *intrasystem gradient* is the increment in target-variable results between the highest and lowest ranked categories in the system. For this ranking, binary or ordinal categories are always listed according to their status in the *independent* variable. Suppose the survival rates for four categories in an ordinal array are 70%, **none;** 85%, **mild;** 25%, **moderate;** and 35%, **severe.** The intrasystem gradient from the highest to the lowest *ranked* categories is 70% − 35%, even though a larger gradient could be calculated from results in the intermediate categories.

On the other hand, because nominal categories cannot be ranked intrinsically, they may be arranged according to their results in the *target* variable. For example, if survival rates according to *color of eyes* are cited alphabetically as 75% for **blue,** 90% for **brown,** and 55% for **other,** the intrasystem gradient is 35% (=90% − 55%) when the nominal array is rearranged as **brown, blue, other.**

Intercategory gradients for an ordinal variable are formed by the incremental results between each pair of adjacent ranked categories. The sum of the intercategory gradients in any system will equal the intrasystem gradient. For an independent variable with only two binary categories, only one gradient can be determined: the intrasystem and intercategory gradients are identical. A variable with m ranked categories has $m - 1$ intercategorical gradients. In the foregoing variable for *color of eyes*, the intercategorical gradients are, respectively, 15% (= 90% − 75%) and 20% (= 75% − 55%).

To illustrate how the principles are used, consider the three ordinal staging systems marked for simple or composite variables X_1, X_2, and X_3 in Table 8-1, where the overall success rate is 50% (= 100/200) for the target event.

For variable X_1, the intrasystem gradient is large (63% = 80% − 17%), and the intercategorical gradients are also reasonably large. They are 29% (= 80% − 51%) and 34% (= 51% − 17%). In variable X_2, the gradient is moderately large (19% = 56% − 37%) for the system, but is very small (1% = 56% − 55%) between categories b_1 and b_2. Variable X_3 also shows very small intercategorical gradients, and the intrasystem gradient is −2% (= 50% − 52%).

We might therefore conclude that variable X_1 has a strong effect, that variable X_3 has no effect, and that variable X_2 has a distinct but inconsistent effect. In fact, because categories b_1 and b_2 in X_2 have similar effects, we might want to collapse them into a single category whose result would be 78/140 = 56%.

8.3.2. Guidelines and criteria for categorical gradients

All of the comments in the foregoing example required judgments about what is *strong, large, small,* and *collapsible*. Some additional judgments would also be needed about the sizes of subgroups whose results will be considered statistically *stable*. For example, suppose the group of 200 people in Table 8-1, when partitioned into the ordinal categories of Variable X_4, had the following results:

d_1: 2/2 (100%)
d_2: 48/196 (50%)
d_3: 0/2 (0%)

The intrasystem and intercategorical gradients for this variable have the largest magnitudes that are possible for three categories. The intrasystem gradient is 100% and the intercategorical gradients are each 50%. Nevertheless, the small numbers in the d_1 and d_3 strata make the results so unstable that we might be reluctant to draw any conclusions about them. The "modicum" size for a stable stratum is discussed in Section 8.4.1.

Decisions about the magnitude of gradients are difficult because they involve purely quantitative evaluations, for which the usual stochastic tests of significance, being based on probabilities, offer no real guidelines. The tests can tell us whether a particular difference is stochastically significant, but not whether it is biologically impressive or clinically important. Tests of stochastic significance can be immensely helpful in evaluating the numerical *stability* of the observed results, but the probabilistic evaluations will not indicate which gradients are important, when to collapse categories, or whether predictions can be confidently made from the observed values.

For example, suppose the success rates in two adjacent ordinal (or binary) categories are e_1: 3/3 (100%) and e_2: 10/50 (20%). The gradient is obviously large and the difference is stochastically significant. (The Fisher exact test gives a two-tailed P value of 0.012.) Nevertheless, we might be unwilling to make a prediction of certain success for a future person in category e_1. We might be reluctant, however, to collapse the two categories and ignore the distinctions noted in category e_1, despite its small group size.

Because the operations cannot proceed without guidelines for decisions, a set of general criteria can be offered to illustrate the pertinent reasoning. The actual boundaries in the criteria, however, may often need ad hoc adjustments for the particular situations considered in each analysis.

8.3.2.1. Minimum intercategorical gradients

If success rates are being considered for a particular event, a gradient of about 10% might be required for concluding that two adjacent categories have different effects. A smaller increment might be acceptable if the observed rates and intrasystem gradient have small values. For example, if the target event is development of a particular disease, we might be impressed with an increment as small as 0.3% if the compared attack rates in two categories are .001 and .004, forming a risk ratio of 4.

8.3.2.2. Minimum intrasystem gradients

If a system contains m categories, we would expect the overall gradient to be at least $(m-1)$ times the magnitude of the minimum intercategory gradient. Thus, if the minimal acceptable intercategory gradient is 10% in a three-category system, the intrasystem gradient should be no less than $2 \times 10\% = 20\%$. The average intrasystem gradient can be calculated as the overall gradient divided by $m - 1$.

8.3.3. Algebraic model of slope

Although the direct average intrasystem gradient offers a good "crude" index of the overall gradient, the direct average may be inappropriately calculated if the categorical strata have different group sizes.

A mathematically appealing alternative method of getting an average gradient is to fit the ordinal categories with a straight-line algebraic model. Since the binary values of Y will be coded as **0** or **1**, the ranked categories of X can be assigned an arbitrary ranked set of coded values. The coded values or "weights", symbolized as w_h for each category, h, are given equal intercategory increments, such as **0, 1, 2, 3, 4,** . . . for four or more categories. The weighted codes might also be balanced around 0 as **−1, 0,** and **+1** for three categories. With the assigned coded values for each category of X, a regression equation, such as $Y = a + bX$, can be calculated for the straight line that best fits the X and Y values. The slope of this line, or the standardized slope, can then be interpreted in exactly the same way as the b and b_s coefficients calculated for dimensional values of X.

If t_h/n_h is the rate of the target event in each category, with $T = \Sigma t_h$ and $N = \Sigma n_h$, and with w_h as the coded weight for each category, the values needed to calculate the algebraic slope as $b = S_{xy}/S_{xx}$ can be worked out to be

$$S_{xy} = (N\Sigma t_h w_h - T\Sigma n_h w_h)/N, \qquad [8.5]$$

and

$$S_{xx} = [N\Sigma n_h w_h^2 - (\Sigma n_h w_h)^2]/N. \qquad [8.6]$$

In the results for variable X_1 in Table 8-1, we can assign the coded weights of -1, 0, and $+1$ for the three consecutive categories. The calculation then becomes $S_{xy} = [200(-56 + 0 + 11) - 100(-70 + 0 + 65)]/200 = -42.5$; and $S_{xx} = [200 (70 + 0 + 65) - (-70 + 0 + 65)^2]/200 = 134.75$. The slope will be $-42.5/134.75 = -.315$. This value is almost identical to the average intrasystem gradient obtained as 31.6%—a two-step drop of $(56/70) - (11/65) = 63.1\%$ in the tabular data. For variable X_2 in Table 8-1, the same coded weights lead to $S_{xy} = [200(-45 + 22) -100(-80 + 60)]/200 = -13$ and $S_{xx} = [200 (80 + 60) - (-20)^2]/200 = 138$. The slope is $-13/138 = -.094$. This result is also close to what is found with the tabular data. They show a drop from 45/80 to 22/60, which is 19.6% across two steps, producing an average value of 9.8%.

8.3.4. Standardized slope

For the two-variable regression that was just done in Section 8.2.3, the standardized slope, b_s, can be calculated as $b(s_x/s_y)$. Since $s_x/s_y = \sqrt{S_{xx}/S_{yy}}$, and since S_{xx} has already been determined, we can promptly get the standardized value after recalling from Section 8.2.2.1. that

$$S_{yy} = NPQ = T(N - T)/N. \qquad [8.7]$$

The conversion factor corresponding to s_x/s_y is therefore

$$\sqrt{[N\Sigma n_h w_h^2 - (\Sigma n_h w_h)^2]/[T(N - T)]} \qquad [8.8]$$

For the X_1 variable in Table 8-1, this factor will be $\sqrt{[200(135) - (5)^2]/[(100)(100)]} = 1.6424$. The standardized slope will be $(-.315)(1.6424) = -.5174$. The corresponding value of r^2 will be .2677. (In this instance, the extra decimal points are needed to distinguish the close values of r^2 and ϕ^2.)

8.3.5. 'Quantitative significance' in published literature

In a recent assessment[2] of quantitative decisions published in general medical journals, most researchers were found to use 0.3 as the minimum boundary for an "impressive" correlation or standardized regression coefficient. For comparing the gradient (or contrast) between two categories, the commonly chosen boundaries were ≥ 1.2 for the ratio of a larger to a smaller mean and 2.2 for the odds ratio of two compared proportions. The investigators used odds ratio rather than incremental differences for the latter comparisons, because the contrasted proportions often had the low values found in "public health" research (<.1) rather than the larger values (>.1) usually found in "clinical" research. With the small public health values (such as .003 for incidence or prevalence of a disease), incremental differences would always be much less than the 10% suggested for quantitative significance in Section 8.3.2.1.

8.3.6. Tests of stochastic significance

In stratified cluster analyses of a binary target variable, stochastic tests can be done for the intercategory and intrasystem gradients.

8.3.6.1. Intercategory contrasts

The chi-square or Fisher procedure can be used to test stochastic stability of the contrasts in rates that form the intercategory gradients or trends between any two *adjacent* categories in a single variable or in a variable formed as a composite cluster.

8.3.6.2. Chi-square test for linear trend

For the *intrasystem* gradient calculated as an algebraic slope in Section 8.3.3, the binary proportions of Y in the ordinal categories can be stochastically checked with a special procedure called the *chi-square test for linear trend*.[4] For this purpose, the total X^2 score in the entire system, calculated as $X^2 = N\phi^2$, is divided into two components, so that

$$X^2 = X_L^2 + X_R^2. \qquad [8.9]$$

The value of X_L^2 represents the effect of imposing a "linear model" on the categories, and is converted to a P value at 1 degree of freedom. The value of X_R^2, which represents the residual effect after the linear trend is removed, is interpreted at $r - 2$ degrees for freedom for an independent variable or system that has r rows of categories. (The partitioning is analogous to the division of S_{yy} into $S_R + S_M$. For this arrangement, the original X^2 reflects the value of ϕ^2, and X_L^2 reflects the linear effect denoted by r^2.)

A stochastically significant result for X_L^2 indicates stability of the intrasystem gradient, and will often justify retaining the proposed array of categories in situations where the *inter*categorical results are quantitatively impressive, but *not* stochastically significant. If X_R^2 is also stochastically significant for *nonlinear* variance, the gradient has significantly nonlinear components.

A valuable algebraic feature that will be cited here (but not proved) is that

$$X_L^2 = Nr^2 \qquad [8.10]$$

where r^2 is calculated as discussed in Section 8.3.4. Thus, for the data of variable X_1 in Table 8-1, ϕ^2 was previously calculated as .2683 in Section 8.2.2.1. and r^2, in Section 8.3.4, was .2677. With $X^2 = N\phi^2$ and $X_L^2 = Nr^2$, and with $N = 200$, the respective results will be $X^2 = 53.66$ and $X_L^2 = 53.54$, so that $X_R^2 = .12$. Both X^2 and X_L^2 have very small P values, at levels below .000001. The tiny value for X_R^2 indicates that the linear algebraic model here fits the data particularly well. The fit is not as good for variable X_2 in Table 8-1, where ϕ^2 was .0306 and the slope $b = -.094$. The conversion factor (with formula [8.6]) will be $\sqrt{27600/[(100)(100)]} = 1.6613$, so that $r = (-.094)(1.6613) = .1562$, $r^2 = .0244$ and $Nr^2 = X_L^2 = 4.877$. Since $X^2 = (200)(.0306) = 6.12$, the value of $X_R^2 = 1.24$. The P value is just below .05 for X^2 (with 2 d.f.), and is almost below .025 for X_L^2 (with 1 d.f.). For variable X_3 in Table 8-1, none of the diverse chi-square values is stochastically significant.

8.3.7. Simultaneous quantitative and stochastic appraisals

The quantitative and stochastic results for gradients should always be evaluated simultaneously, so that suitable conclusions can be drawn in the contradictory situation where one result is significant but the other is not. In one type of contradiction, which is particularly common in large data sets, the large group sizes bring stochastic significance to gradients that are too small to be quantitatively impressive.

On the other hand, an important quantitative value for gradient may come from small numbers that are too unstable to be stochastically significant. In these circumstances, the data analyst must decide whether to dismiss the result or to take it seriously. The decision will require the type of judgmental appraisal discussed briefly earlier and considered later in greater detail.

8.4. Demarcation of strata

The demarcation of strata for targeted clusters is a challenging problem that does not occur in algebraic methods of analysis.

With algebraic methods, the individual variables may be transformed or given diverse kinds of coding digits, but thereafter the variables (if not yet in categorical form) remain intact. With targeted clusters, however, the independent variables are converted into categories; and categories are often collapsed within the same variable or joined from more than one variable to form composite single categories. Variables that were expressed in dimensional scales are often divided into ordinal zones or binary splits; and variables expressed in ordinal or nominal scales can also be divided into binary splits. Since the single or composite categories that are ultimately formed in relation to the target variable are called *strata*, the general process can be called *demarcation of strata*.

The rest of this section is concerned with that process, which involves decisions about minimum size of the strata, criteria for consolidation, shape of the distribution, and choice of boundaries.

8.4.1. Minimum (modicum) size of strata

In the analysis of stratified clusters, a special statistical criterion is the *modicum*. It is the smallest size of a group (or stratum) that will be accepted as large enough to use in future decisions. The issue here is clinical and scientific, rather than statistical; it involves subjective confidence in a decision, rather than the numerical magnitudes offered by statistical "confidence intervals".

For example, suppose a group with a total success rate of 50% (= 4/8) is partitioned into two strata, A and B, having the respective success rates of 100% (= 4/4) and 0% (= 0/4). The incremental difference of 100% in the two strata is quantitatively impressive: it is the highest possible difference in two proportions. The distinction is also stochastically significant at P = .029 (by two-tailed Fisher exact test).

Nevertheless, because each stratum has so few members, you might be reluctant to predict success for all future persons in Group A and failure for Group B. If groups with only four patients produced this reluctance, how large a group would you want for true scientific confidence in a judgment made from the results? Would you be willing to make unequivocal predictions of success or failure if results of 100% (or 10%) were found in groups containing 6 members, 10, 15, 20, or more than 20?

Different people will give different answers to this question, because of variations in personal levels of comfort. In some clinical situations, a group size of 1 was convincing on the first occasion that insulin promptly relieved diabetic acidosis or penicillin cured bacterial endocarditis. Such situations are rare, however, and for most purposes, investigators usually want results from a case series, not from a single case. How large should that series be?

Although the answer that satisfies one person may not satisfy another and will vary for different situations, most persons in general circumstances would accept a modicum of 20 as the minimum size of a subgroup. If the *total* group size is small, however, the modicum boundary for a subgroup might be reduced to 10 or even fewer.

The minimum value chosen for the modicum in stratified-cluster analyses is an additional decision boundary, beyond those set for quantitative significance in either gradients or proportionate reduction of variance, and for stochastically significant P values (or confidence intervals). As a special feature of categorical analyses, the modicum indicates when a stratum (or cell) has too few members. It may then be collapsed (or "consolidated") into a suitably adjacent stratum (or cell).

(In algebraic methods of multivariate analysis, as discussed later, a counterpart of the *modicum* is the ratio called *events per independent variable*.)

8.4.2. Criteria for collapsibility of strata

Because reduction of categories is often an important analytic goal, and because reductions can be achieved by collapsing suitably adjacent categories, the criteria for collapsibility are important. The criteria that might be used for *biologic* decisions about collapsing categories were discussed in Section 4.6.2.2. The comments that follow are concerned exclusively with statistical decisions.

Two adjacent categories (or strata) should be *retained* if they satisfy all of three criteria: (1) a sufficiently large intercategory gradient, (2) which is stochastically significant, (3) with each stratum exceeding the modicum size. Conversely, the strata should be collapsed if none of the three criteria is satisfied.

The decision becomes tricky if one or two, but not all three criteria are fulfilled. A small group is probably the most vulnerable feature that allows collapsibility. Despite a strong quantitative gradient and the associated stochastic significance, the retention of a stratum with only 3 persons would often be difficult to justify, particularly if the other strata in the system have many more members. In some situations, however, the investigator might want to relax the modicum criterion and retain a stratum that seems particularly impressive. For example, if the modicum is set at 20, a stratum with only 18 members would ordinarily not be retained. Nevertheless, if the success rates are $17/18 = 94\%$ in that stratum and $10/50 = 20\%$ in the adjacent stratum, the investigator would probably want to lower the modicum boundary to 18.

Of the three criteria, the absence of stochastic significance might be ignored, and the strata retained, if they have an impressive gradient and adequate sizes. The presence of stochastic significance is probably the least valuable single criterion. It can be achieved with an unimpressive or tiny gradient if the group sizes are large enough.

8.4.3. Distribution of strata

Because the same set of data can usually be partitioned into different arrangements of strata, an optimal system (whenever possible) should fulfill all of the stipulated criteria for modicum size of strata, quantitative gradients, tests of stochastic significance, and results that span the meridian. After all these desiderata are achieved, however, the denominators will have size n_h in each final stratum, and a pattern of distribution. What would we like the pattern to be?

Should it be *centripetal*, resembling a Gaussian distribution, with most of the persons located in the central or middle strata? Should it be *centrifugal*, or U-shaped, with most people located at the extreme ends of strata in the system? Should it have a *uniform* distribution, with roughly equal numbers in each stratum? Furthermore, how many strata should there be: 3, 7, 12? None of these questions has a uniquely correct or optimal answer.

8.4.3.1. Number of strata

A perfect stratification would have two binary categories, of equal size, with an abundance of members in each stratum. The target rates would have a nadir value of 0% in one stratum and a zenith value of 100% in the other. The variance in the system would be completely reduced, and the concordant accuracy would be perfect. In the world of pragmatic reality, alas, this ideal is almost never attained.

The number of strata will generally be limited by the minimum value chosen for the intercategory gradients. If the maximum instrasystem gradient is 100%, and if values of 10–15% are chosen as minimums for the intercategory gradients, there can be no more than 7–9 strata. In general, the intercategory gradients will seldom be perfectly equal, and their scope can usually be covered well by five strata. Cochran[5] has pointed out that five categories (but not two) are sufficient to demarcate variables in "controlling for confounding". Besides, if a "staging" system contains more than five strata, the individual strata will often be biologically difficult to distinguish and remember. Accordingly, most systems should preferably contain no more than five final strata. (This concept may also be responsible for the common use of quintiles when dimensional variables are partitioned into ordinal zones.)

8.4.3.2. 'Shape' of distribution

If the target results span the meridian and have nearly polar levels (i.e., 0% and 100%) in the extreme strata, a centrifugal or U shape is highly desirable. It will concentrate the subgroups into the nearly polar strata, where the estimates will be particularly accurate.

If this goal cannot be achieved, a relatively *uniform* shape is probably next most desirable, since it will improve stability for the results in each stratum, and allow higher scores in tests of stochastic significance, as noted earlier. (Uniformity of shape is probably another reason for the popularity of quintile partitions.)

In the recursive partitioning method described later in Chapters 19 and 21, the automated algorithm can make its splits according to several policies: maximum reduction in variance, maximum progress toward finding a polar stratum, or avoidance of certain "costs" or errors. The polarizing approach (if attainable) is particularly likely to produce a U-shaped distribution. The customary variance-reduction approach will often lead toward a uniform distribution.

A Gaussian-like centripetal shape is probably the least desirable pattern for good concordance of fit. The subgroups will tend to be concentrated in the middle-most strata where, if the results span the meridian, the outcome rates will be close to 50%, thus producing relatively poor estimates. Whatever the desired goal may be, however, pragmatic reality will often lead to centripetal shapes. The analytic procedure is often able to "chip away" and demarcate only relatively small subgroups at the two extreme ends of the spectrum, leaving a larger bulk in the middle.

On the other hand, a centripetal distribution is not lamentable and may be highly desirable if the stratification is used for classifying patients admitted to randomized trials. The extreme ends of the strata, representing small numbers of patients with very severe or very mild illness, may sometimes be excluded (as noted in Section 3.4.3.4) to increase the efficiency of the trial. The larger number of eligible patients in the middle strata may then have success rates close to 50%, with a high "variance" available to be affected by the impact of treatment.

In general, however, the data analyst can demand a modicum for stratum size, a minimum boundary for intercategorical gradients, a maximum number of strata, and appropriate tests of stochastic significance. What happens afterward usually depends on what is in the data, rather than what the investigator would like to have.

8.4.3.3. Indexes of 'diversity'

If a uniform distribution is desired, the achievement can be measured with an "index of diversity", proposed by Shannon,[6] which gives its highest score when the categories are equally distributed. Shannon's index is

$$H = -\sum_{i=1}^{c} p_i \log p_i, \qquad [8.10]$$

where c is the number of categories and p_i is the proportion of the total group size N, in each category i. If each category contains n_i members, $p_i = n_i/N$.

For example, the total group stratified in Table 8-1 contains 200 people. After the partition with variable X_1, the diversity index for the denominators becomes $-[(70/200)\log(70/200)] + [(65/200)\log(65/200)] + [(65/200)\log(65/200)] = -[-.477] = .477$. For the partition with variable X_2, the diversity index is .473. For variable X_3, the index is .476. Thus, the three diversity scores are close, but highest for variable X_1.

For the total of c categories, the maximum possible value of Shannon's index is

$$H_{max} = \log c. \qquad [8.11]$$

An alternative expression, therefore, is to cite the achieved diversity as a proportion of the maximum possible diversity. The result will be

$$J = \frac{H}{H_{max}}. \quad [8.11]$$

For these categories, H_{max} is log 3 = .477 and so the value of J for the X_1 system here is .477/.477 = 1. The other two values are also quite close to 1.

An alternative way to calculate Shannon's H is to note that if each categorical frequency is n_i, the value of p_i is n_i/N in each cell and $N = \Sigma n_i$. The value of Σp_i log p_i will be $\Sigma(n_i/N)[\log n_i - \log N]$, which becomes $(1/N)[\Sigma n_i \log n_i - N \log N]$. Converting to the negative expression produces

$$H = (N \log N - \Sigma n_i \log n_i)/N \quad [8.13]$$

This formula leads to more rapid calculations than formula [8.10]. For example, with 200 people divided into groups of 70, 65, and 65, formula [8.13] produces H = (200 log 200 − 70 log 70 − 65 log 65 − 65 log 65)/200 = .477, which is the same result obtained with formula [8.10].

8.4.4. Selection of dimensional boundaries

For binary, ordinal, or nominal variables, the strata are originally created by the categories in which the variables are expressed. The strata can be collapsed or (in the case of nominal categories) rearranged according to some of the principles that have just been discussed. A quite different form of stratification, however, takes place when dimensional variables are demarcated into zones that form binary or ordinal categories. The boundaries for these zones can be chosen with 'esthetic', clinical, quantile, or theoretical-statistical boundaries, and in either a univariate or correlated bivariate manner.

8.4.4.1. 'Esthetic' boundaries

"Esthetic" boundaries demarcate equal intervals that span 3, 5, 10, 20 units or other fixed-size zones of the dimensional variable. A classical example is the demarcation of *age* in decades such as **0–9, 10–19, 20–29, 30–39,** . . . or in zones such as **< 20, 20–34, 35–49, 50–64, 65–79,** and $\geq$ **80**. The extreme zones of the strata can have open-ended boundaries, as in the example just cited, but the interior categories retain equi-interval lengths.

8.4.4.2. Clinical boundaries

Boundaries can also be chosen according to clinical or pathophysiologic principles that demarcate such zones as **too low, normal,** or **too high** for variables such as blood pressure, white blood count, or serum calcium. An example of such boundaries might be **< 60, 60–134,** and $\geq$ **135** mm Hg. for *systolic blood pressure*, or **< 3500, 3500–10,000,** and **> 10,000** for *white blood count*.

When dimensional variables were divided into categories for the illustrative data set here, clinical principles were used to classify *hematocrit* values as anemic if **< 37**, normal if $\geq$ **37 but < 44**, and relatively "plethoric" if $\geq$ **44**. The boundary of **37** is a conventional clinical demarcation for anemia; the value of **44**, although not alone indicative of polycythemia, was chosen to divide the remaining range into zones of normal and relatively plethoric. In another partition, *percentage weight loss* was demarcated as **none, minor if > 0 but < 10%** and **major if ≧ 10%**. These boundaries were chosen because they have been found to demarcate distinct prognostic gradients in previous research.[7] The main prognostic effects occurred at weight loss percentages beyond **10%,** and the intervening category was inserted to separate no weight loss from amounts less than 10%.

8.4.4.3. Quantiles

Another method of choosing boundaries is from quantiles of the data. If the data set is large enough, the five equal-sized groups produced by quintile demarcations are particularly popular and useful for determining trends in the target variable. For small data sets, the boundaries can be set at quartiles or even tertiles.

The main disadvantage of quantiles is that the zones, having been demarcated by the number of items in the data, have arbitrary boundaries that lack either esthetic or clinical attributes. Accordingly, investigators can begin by determining quantile boundaries for equal-sized groups, but can then adjust the boundaries for a demarcation that is more clinically or esthetically appealing. For example, if the quintile values for a particular variable are **29, 43, 86,** and **122,** the boundaries for five zones may be adjusted to **< 30, 30–44, 45–84, 85–119,** and **≧ 120.**

8.4.4.4. Theoretical statistics

The challenge of partitioning a dimensional variable had evoked many statistical proposals. One of the earliest, by Sturges,[8] was intended to produce class frequencies that approximated a binomial distribution. For example, a set of 16 observations would be divided into five classes with frequencies of 1, 4, 6, 4, 1. Dalenius[9] later described a set of boundaries that would minimize the variance of the population mean estimated with stratified sampling. In reviewing several other proposed methods, Cochran[10] seemed to prefer an approach based on f, the frequency of successive items, and on strata that would provide equal intervals for the cumulated values of $\sqrt{f}$. The partitioning challenge regularly evokes new mathematical ideas. Most recently, Pla[11] suggested that a principal component analysis would improve the "estimation of the vector mean" in stratum boundaries prepared with multivariate sampling.

Although the ingenuity of all the statistical proposals can be admired, most data analysts demarcate strata by one of the three preceding pragmatic methods, or by the correlated approach discussed in the next section.

8.4.4.5. Correlated choices

The boundaries of individual variables are often set in a purely univariate manner according to one of the foregoing methods. More commonly, however, the chosen boundaries, after being selected, receive a bivariate check to determine their correlated impact on the target variable. As long as the adjustment is reasonable, the boundaries can then be modified to improve the gradient.

For example, the three ordinalized categories of *hematocrit* **(< 37, 37–43,** and **≧ 44)** in the illustrative data set had the following gradients respectively for

6-month survival rate: 10/36 *(28%)*; 48/103 *(47%)*; and 36/61 *(59%)*. The corresponding survival rates for the three ordinalized categories of *percent weight loss* (0%, > 0–< 10%, and ≧ 10%) were 32/55 *(58%)*, 38/79 *(48%)*, and 24/66 *(36%)*. The prognostic gradients thus "confirmed" the value of the clinical boundaries selected for the demarcations.

Some analysts do not like this type of correlated appraisal, and claim it may improperly "rig the data". The procedure is no more improper, however, than the rigging that occurs when best-fitting coefficients are chosen in an algebraic model, or when diverse splits are checked (as discussed in Chapters 20 and 21) to find the best binary partition. A data analyst is always allowed to find the best fits, best splits, or best trends in the data under analysis. After these best arrangements have been selected, however, they should be "validated" by various methods that are discussed in the next section.

8.5. Validation of results

Regardless of whether an analysis is done with algebraic or stratified cluster methods, the results will have been "driven by the data". The algebraic model or the system of strata will have been deliberately chosen to provide an optimal arrangement. For the results to be applied beyond the immediate set of data, some sort of validation becomes necessary. Three main strategies can be used for this purpose.

8.5.1. Concept of validation

Validation is one of those words—like *health, normal, probability,* and *disease*—that is constantly used and seldom defined. We can easily define it as a process used to confirm validity, but we are then left with the multitudinous meanings of *validity*.[12] To avoid the long digression required to discuss *validity*, we can simply say that, in data analysis, validation consists of efforts made to confirm the accuracy, precision, or effectiveness of the results.

Because the results of either an algebraic or cluster analysis have been fitted to a particular group of data, the best and most obvious form of validation is to see what happens in an entirely different group of pertinent data. Because such groups are seldom available at the time of the original analysis, most forms of validation are done with various arrangements of the original group. The next few sections describe the diverse forms of "external" and "internal" validation.

8.5.2. External validation

The most convincing test of validity is an *external* validation, which checks the analytic accomplishments in an entirely different data set. The new data set is usually acquired from pertinent persons at an institution, region, or calendar time different from that of the original group whose data were analyzed. Because of the considerable time and effort required to obtain the new group of data, however, this type of validation is seldom possible until long after the original analysis.

8.5.3. Resubstitution

Of the different strategies that internally use the original group of data, the first is really a resubstitution rather than validation. The data for each member of the original group are reentered into the algebraic model or targeted clusters, and the accomplishments are calculated for fit and trend by the methods discussed here and in Chapter 7. Since the results merely confirm the analytic process itself, the only real validation here is achieved by "face validity" if what emerges seems plausibly "sensible".

8.5.4. Cross-validation

Because a completely external data set is seldom available and the resubstitution process is relatively trivial, a third approach, called *cross-validation*, is the most commonly used validation procedure. The strategy avoids resubstitution because a part of the available data set is set aside, excluded from the analysis, and then used as a "challenge" to test what was found in the part that was analyzed.

The part of the data used to form the algebraic model or targeted clusters is called the "training", "generating", or "development" set. The excluded rest of the data, called the "challenge" or "test" set, is then used to check the accomplishment. The several strategies of cross-validation differ according to the methods of choosing the training and challenge sets, and of denoting the accomplishment.

8.5.4.1. Jackknife method

In the jackknife method, the first member of the group is set aside, and the algebraic or clustered arrangement is developed from all the rest of the data. The accuracy (or fit) of the estimate is then noted when data for the first member are entered into the arrangement generated from the other $N - 1$ members. After the first member is restored to the data, the second member of the group is set aside; the new analytic arrangement is formed; and its corresponding fit is noted. The process then continues to reiterate successively through all N members of the data set. Each time the analytic arrangement is derived from the remaining $N - 1$ members of the group after one member is removed. The overall accomplishment is determined from the average of the N individual sets of results.

Although the jackknife procedure offers an excellent account of how well the data might be fitted, a single arrangement does not emerge for future application. Such an arrangement might be obtained, however, by determining the average of the coefficients in an algebraic model or by using the particular clustered arrangement that occurred most frequently among the individual jackknife formulations.

8.5.4.2. Split-group method

In a relatively common strategy, the group of data is split into two distinct parts, rather than appraised with a series of jackknife removals of one member. The part used for the generating set is usually larger (in a ratio of about 2:1) than the part used for the challenge set. The generating set can be chosen randomly from the total group; but if the group was assembled in a secular (calendar) chronology, the "earlier" persons can be the basis of estimations that will be tested in the "later" persons. (The latter tactic will resemble the way that information is used in clinical practice.)

If the generating estimates are sufficiently well "validated" in the challenge set, the entire group can then be recombined for development of a "final" arrangement that uses all the data. For example, when Justice et al.[13] formed a prognostic stratification to predict outcome of patients with AIDS, the generating set consisted of 76 patients seen in an AIDS clinic before April 1, 1987; the results were confirmed in a challenge set that contained 41 patients seen afterward during April 1–July 31, 1987; and the final stratification was based on all 117 patients.

8.5.4.3. V-fold validation

V-fold validation uses a strategy that is between the jackknife and split-group approaches. The original group is divided randomly in a ratio such as 9/10 to 1/10. The analysis is done in the larger group and then tested in the smaller group. The process is repeated ten times. Validation is obtained if the ten sets of results seem consistent.

8.5.5. Bootstrapping

The bootstrap method is also a form of cross-validation, but differs from the other methods, because no part of the original data is specifically set aside for later use as a "challenge set". Instead, the bootstrap method is a resampling process in which all members of the data are eligible to be chosen at each step of the sampling with replacement.

For example, suppose we do a resampling from the data set {1, 3, 5, 8}. The first item randomly chosen might be **3**. It is noted and then replaced. The second item randomly chosen might be **3** again. After it is replaced, the third item might be **8**; and the fourth item might be another **3**. For a set of data that contains N different members, the total number of possible bootstrap samples is N^k if each sample contains k members. Thus, for bootstrap samples of size 4 from the foregoing data set, there are $4 \times 4 \times 4 \times 4 = 256$ possible samples. If the samples are restricted to 3 members, there are $4 \times 4 \times 4 = 64$ possible samples. In the illustrative data set, with 200 members, a "complete" bootstrap would contain 200^{200} samples, which is a huge number, approximately 1.61×10^{460}. A complete bootstrap would contain 8.88×10^{84} samples for a data set of 50 members, and 1.05×10^{26} samples for a data set of 20 members.

To avoid the massive computation, the complete bootstrap is usually replaced by a designated smaller set of 100–500 bootstrap samplings. Each of those samples is then subjected to an algebraic or cluster analysis, and the results can be checked for their range of values in goodness of fit, coefficients of trend, etc. The range of bootstrapped results can provide "confidence intervals" for the original results, and can also be averaged, if desired, to offer a single analytic result.

Bootstrapping procedures can also be used for other forms of cross-validation. For example, in one tactic noted later in the discussion of logistic regression (Section 14.2.2.1.2), all of the observed values of Y_i can be bootstrapped into pairs [forming a total of $(N)(N-1)/2$ pairs]. The direction of the increment in each pair is then compared with the incremental direction found in the corresponding pair of estimates for $\hat{Y}_i$.

Chapter References

1. Feinstein, 1972; 2. Burnand, 1990; 3. Fleiss, 1981; 4. Armitage, 1955; 5. Cochran, 1968; 6. Shannon, 1948; 7. Feinstein, 1990; 8. Sturges, 1926; 9. Dalenius, 1950; 10. Cochran, 1961; 11. Pla, 1991; 12. Feinstein, 1987; 13. Justice, 1989.

9 Bivariate Evaluations of Illustrative Data Set

Outline . . . .
9.1. Examination of correlation coefficients
   9.1.1. Symbols and nomenclature
      9.1.1.1. Variables and persons
      9.1.1.2. Categories
      9.1.1.3. Variance and covariances
      9.1.1.4. Correlation coefficients
      9.1.1.5. Display of covariate results
   9.1.2. Correlations in illustrative data set
      9.1.2.1. Targeted bivariate relationships
      9.1.2.2. Intervariable relationships
      9.1.2.3. Other calculations for correlation coefficients
9.2. Bivariate tabulations
   9.2.1. Modicum values of categories
   9.2.2. Inspection of gradients
   9.2.3. 'Significance' for binary variables
   9.2.4. Linear trends in ordinal variables
   9.2.5. Stochastic significance for ordinal variables
   9.2.6. Demarcation and linear trends in dimensional variables
9.3. Graphical portraits
9.4. Bivariate 'conclusions'
. . . .

The principles presented throughout Chapters 7 and 8 were discussed in reasonably extensive detail, because they will appear recurrently in later chapters when used for multivariate analyses. The main job of those principles now, however, is to help guide the bivariate analyses of the illustrative data set.

9.1. Examination of correlation coefficients

The simplest and easiest form of bivariate screening is to inspect the matrix of correlation coefficients for the independent and dependent variables.

9.1.1. Symbols and nomenclature

Certain symbols and terms that do not appear in computer printouts are needed for a verbal description of the ideas.

9.1.1.1. Variables and persons

As noted in Section 7.1, any one of the k independent variables is denoted as X_j, and the entire set is $\{X_j\}$. With i as the subscript for each person, the individual values of the target variable are $Y_1, Y_2, \ldots, Y_i, \ldots, Y_N$. The value of X_j for each person, i, becomes doubly subscripted as X_{ij}. Thus, X_{96} would be the value of the

6th variable for the 9th person. The entire collection of data for the independent variables would be $X_{11}, X_{21}, X_{31}, \ldots, X_{i1}, \ldots, X_{N1}$ for the first variable: $X_{12}, X_{22}, \ldots, X_{i2}, \ldots, X_{N2}$ for the second variable; and so on through $X_{1j}, \ldots, X_{ij}, \ldots, X_{Nj}$ for the jth variable; and eventually $X_{1k}, \ldots, X_{ik}, \ldots, X_{Nk}$ for the kth (last) variable.

Fortunately, the double subscripts can generally be avoided without causing confusion, and the different independent variables can usually be cited simply as $X_1, \ldots, X_j, \ldots, X_k$.

9.1.1.2. Categories

The symbols just cited are satisfactory when the variables are algebraically analyzed as variables. With categorical-cluster methods, however, the analysis uses individual categories of each variable. Although symbols such as X_{ij} can identify the value of the ith person's category for the jth variable, X_{ij} will not represent the particular category itself. For example, if variable 7 has the ordinal categories **mild, moderate, severe** or **1, 2, 3, 4**, the values might be X_{47} = **moderate** for the fourth person or X_{57} = **3** for the fifth person. Consequently, another symbol is needed to identify the available list of categories as **mild, moderate, severe** or **1, 2, 3, 4**. For this purpose, each variable's categories can be labelled with additional lowercase letters and numerical subscripts. Thus, the possible categories might be symbolized as follows:

Variable	Categories	Example
X_1	a_1, a_2, a_3, a_4	none, mild, moderate, severe
X_2	b_1, b_2	male, female
X_3	c_1, c_2, c_3	Stage I, Stage II, Stage III
X_4	d_1, d_2	absent, present

The subscript h can be used (as noted in Chapter 8) in a manner analogous to i and j for identifying a general category as a_h, b_h, c_h, etc. The number of members in category h would be n_h. Thus, if the foregoing variable X_1 contains 23 persons with a rating of **moderate,** the symbols for that category are a_3 = moderate and $n_3 = 23$.

9.1.1.3. Variances and covariances

For only two variables, X and Y, the group variances and covariances were easily designated as S_{xx}, S_{yy}, and S_{xy}. With multiple variables, the group variances will be $S_{11}, S_{22}, S_{33}, \ldots, S_{jj}, \ldots, S_{kk}$ for each of the X_j independent variables. The group covariances among these variables can be shown as $S_{12}, S_{13}, S_{23}, \ldots, S_{1k}, \ldots, S_{jk}$, and so on. The targeted covariances between each independent variable and the target variable will be $S_{y1}, S_{y2}, S_{y3}, \ldots, S_{yj}, \ldots, S_{yk}$. (For the intervariable group covariances, the order of subscripts is unimportant since $S_{37} = S_{73}$. For the targeted covariances, $S_{yj} = S_{jy}$, but the y is usually cited first to indicate that y is being regressed on j.)

9.1.1.4. Correlation coefficients

In simple linear regression for Y vs. X, the correlation coefficient was calculated as $r = S_{xy}/\sqrt{S_{xx}S_{yy}}$. When multiple variables are involved, the bivariate correlation coefficients are calculated in the same way; and they are marked with the same subscripts used for covariances. Thus, r_{37} is the correlation coefficient between independent variables 3 and 7, calculated as $r_{37} = S_{37}/\sqrt{S_{33}S_{77}}$; and r_{y9} is the targeted correlation coefficient between Y and X_9, calculated as $r_{y9} = S_{y9}/\sqrt{S_{yy}S_{99}}$.

9.1.1.5. Display of covariate results

The results for variance and covariance are presented in a matrix that has a symmetrical outer skeleton, with the rows and columns each labelled with the Y variable, followed by $X_1, X_2, \ldots, X_k$. The interior cells of the matrix show the group variances for each variable, and its group covariances with the target variable and with the other independent variables. The arrangement is as follows:

	Y	X_1	X_2	$\cdots$	X_k
Y	S_{yy}	S_{y1}	S_{y2}	$\cdots$	S_{yk}
X_1	S_{y1}	S_{11}	S_{12}	$\cdots$	S_{1k}
X_2	S_{y2}	S_{12}	S_{22}	$\cdots$	S_{2k}
$\vdots$	$\vdots$	$\vdots$	$\vdots$	$\cdots$	$\vdots$
X_k	S_{yk}	S_{1k}	S_{2k}	$\cdots$	S_{kk}

The *group variances* are in the top-left-to-bottom-right diagonal. The other cells in the top row or first column show the *targeted group covariances* between each variable and the target variable, Y. All the remaining cells show the *intervariable group covariances* among the $\{X_j\}$ variables.

The variance-covariance data seldom require careful inspection, since they are seldom needed for analytic interpretations. The main role of the variance-covariance data is in black-box calculations that produce the correlation coefficients. (In multivariable analysis, these calculations ultimately produce the "partial" regression coefficients discussed later.)

The correlation matrix has the same outer skeleton as the variance-covariance matrix, but the cells contain the correlation coefficients. In the left-to-right downward diagonal of the correlation matrix, the coefficients are all 1, representing the correlation of a variable with itself. In the other cells, the variance-covariance terms are converted to correlation terms by appropriate formulas. Thus,

$$r_{y1} = S_{y1}/\sqrt{S_{yy}S_{11}}, \quad r_{y2} = S_{y2}/\sqrt{S_{yy}S_{22}}, \quad r_{12} = S_{12}/\sqrt{S_{11}S_{22}}, \text{ etc.}$$

Fig. 9-1 shows the location of constituents in a correlation matrix for Y and four independent variables, $X_1, \ldots, X_4$. The main diagonal shows the values of 1 for the intravariable correlations. The remaining entries in the first row and first column show the targeted correlations. The remaining upper and lower "wedges" show the intervariable correlations.

Because the variance-covariance and the correlation matrixes are symmetrical on the two sides of the left-to-right downward diagonal, the printed summary is often shown as a wedge or triangle for half of the total matrix. Thus, the correlation matrix might be displayed as

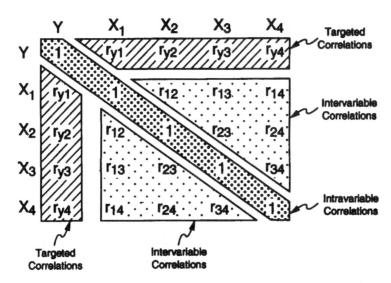

Fig. 9-1 Location of intravariable, intervariable, and targeted coefficients in a correlation matrix. In strict rules for symbolism, row numbers are listed before column numbers. To indicate the identity of coefficients, however, the order of citation is reversed here in the wedge below the main diagonal. Thus, in the strict row-then-column symbolism, r_{y1} in column 1 should be labelled r_{1y}, and r_{23}, in the X_2 column and X_3 row, should be labelled r_{32}.

$$\begin{array}{ccccc} 1 & r_{y1} & r_{y2} & \cdots & r_{yk} \\ & 1 & r_{12} & \cdots & r_{1k} \\ & & 1 & \cdots & r_{2k} \\ & & & \cdots & \cdots \\ & & & & 1 \end{array}$$

or in an L-shaped-wedge transposed form as

$$\begin{array}{ccccc} 1 & & & & \\ r_{y1} & 1 & & & \\ r_{y2} & r_{12} & 1 & & \\ \vdots & \vdots & \vdots & \vdots & \\ r_{yk} & r_{1k} & r_{2k} & \cdots & 1 \end{array}$$

The latter arrangement is an easier way to find the targeted correlation coefficients. The multiple variables can all be easily listed downward as rows in the matrix, but the page of printout may not be wide enough to show all of the corresponding columns. The "overflow" will then appear as a lower second matrix, below the upper first one. With the L-shaped arrangement, all of the *targeted* values will be completely shown in the first column.

9.1.2. Correlations in illustrative data set

For the illustrative data set here, the matrix of correlation coefficients appeared on two pages of printout, and is shown here in two parts. To avoid repetition in subsequent chapters, the arrangement of the matrix includes three expressions for the target variable, and cites each of the 4 dummy binary variables

Correlation Analysis

Pearson Correlation Coefficients / Prob > |R| under Ho: Rho=0 / N = 200

	SURV6	SURVIVE	LNSURV	HCT	PCTWTLOS	AGE	PROGIN	TNMSTAGE
SURV6 SIX MONTH SURVIVAL	1.00000 0.0	0.54540 0.0001	0.75805 0.0001	0.16432 0.0201	-0.13446 0.0577	-0.01482 0.8350	0.08453 0.2340	-0.38293 0.0001
SURVIVE SURVIVAL TIME (MOS)	0.54540 0.0001	1.00000 0.0	0.70692 0.0001	0.14058 0.0471	-0.23624 0.0008	-0.14955 0.0345	0.03842 0.5891	-0.41464 0.0001
LNSURV NAT LOG SURVIVAL TIME	0.75805 0.0001	0.70692 0.0001	1.00000 0.0	0.16442 0.0200	-0.26355 0.0002	-0.07260 0.3069	0.05954 0.4023	-0.42299 0.0001
HCT HEMATOCRIT	0.16432 0.0201	0.14058 0.0471	0.16442 0.0200	1.00000 0.0	-0.28850 0.0001	-0.13141 0.0636	0.04499 0.5270	0.02065 0.7716
PCTWTLOS PERCENT WT LOSS	-0.13446 0.0577	-0.23624 0.0008	-0.26355 0.0002	-0.28850 0.0001	1.00000 0.0	0.06651 0.3494	-0.09895 0.1633	0.12536 0.0769
AGE AGE (YEARS)	-0.01482 0.8350	-0.14955 0.0345	-0.07260 0.3069	-0.13141 0.0636	0.06651 0.3494	1.00000 0.0	-0.02325 0.7438	0.02902 0.6833
PROGIN PROGRESSION INTERVAL (MOS)	0.08453 0.2340	0.03842 0.5891	0.05954 0.4023	0.04499 0.5270	-0.09895 0.1633	-0.02325 0.7438	1.00000 0.0	-0.07210 0.3103
TNMSTAGE TNM STAGE	-0.38293 0.0001	-0.41464 0.0001	-0.42299 0.0001	0.02065 0.7716	0.12536 0.0769	0.02902 0.6833	-0.07210 0.3103	1.00000 0.0
SXSTAGE SYMPTOM STAGE	-0.30021 0.0001	-0.31460 0.0001	-0.41692 0.0001	0.07269 0.3063	0.27185 0.0001	-0.06547 0.3570	-0.10279 0.1475	0.44442 0.0001
SMKAMT RANK CUSTOMARY PPD CIGS	-0.03827 0.5905	-0.02579 0.7170	0.03062 0.6669	0.01930 0.7862	-0.18063 0.0105	-0.19329 0.0061	-0.08853 0.2126	-0.03793 0.5938
MALE MALE GENDER	-0.08891 0.2106	-0.14420 0.0416	-0.08980 0.2060	0.07852 0.2690	0.01489 0.8343	0.08965 0.2068	-0.03690 0.6039	0.04154 0.5592
WELL WELL DIFF HISTOLOGY	0.18279 0.0096	0.26977 0.0001	0.21511 0.0022	-0.06260 0.3785	-0.05268 0.4588	-0.00220 0.9753	0.09842 0.1656	-0.33573 0.0001
SMALL SMALL CELL HISTOLOGY	-0.19360 0.0060	-0.16582 0.0189	-0.17251 0.0146	0.03616 0.6113	0.13830 0.0508	0.07477 0.2927	0.02863 0.6874	0.22538 0.0013
ANAP ANAPLASTIC HISTOLOGY	-0.16143 0.0224	-0.12270 0.0835	-0.14892 0.0353	0.02413 0.7344	-0.04118 0.5626	-0.12515 0.0774	-0.08923 0.2089	0.14056 0.0471
CYTOL CYTOLOGIC PROOF ONLY	0.17322 0.0142	-0.06751 0.3422	0.07494 0.2916	0.02538 0.7213	0.00305 0.9658	0.12691 0.0733	-0.05018 0.4804	0.08293 0.2430

Fig. 9-2 Upper part of matrix of bivariate correlations for illustrative data set. (SAS PROC CORR program.)

used for *cell type*. The target variable is listed as the binary SURV6 for 6-month survival (used in categorical analyses and logistic regression), as the untransformed SURVIVE for the actual survival time (used in Cox regression), and as LNSURV or the natural logarithm of survival time, which was used in multiple linear regression.

Bivariate Evaluations of Data Set 193

The first part of the total matrix is displayed in Fig. 9-2, with the 15 variables extending downward in the rows. Because the width of the computer page allows only 8 variables to appear in the columns, the remaining 7 columns of variables,

Correlation Analysis

Pearson Correlation Coefficients / Prob > |R| under Ho: Rho=0 / N = 200

	SXSTAGE	SMKAMT	MALE	WELL	SMALL	ANAP	CYTOL
SURV6 SIX MONTH SURVIVAL	-0.30021 0.0001	-0.03827 0.5905	-0.08891 0.2106	0.18279 0.0096	-0.19360 0.0060	-0.16143 0.0224	0.17322 0.0142
SURVIVE SURVIVAL TIME (MOS)	-0.31460 0.0001	-0.02579 0.7170	-0.14420 0.0416	0.26977 0.0001	-0.16582 0.0189	-0.12270 0.0835	-0.06751 0.3422
LNSURV NAT LOG SURVIVAL TIME	-0.41692 0.0001	0.03062 0.6669	-0.08980 0.2060	0.21511 0.0022	-0.17251 0.0146	-0.14892 0.0353	0.07494 0.2916
HCT HEMATOCRIT	0.07269 0.3063	0.01930 0.7862	0.07852 0.2690	-0.06260 0.3785	0.03616 0.6113	0.02413 0.7344	0.02538 0.7213
PCTWTLOS PERCENT WT LOSS	0.27185 0.0001	-0.18063 0.0105	0.01489 0.8343	-0.05268 0.4588	0.13830 0.0508	-0.04118 0.5626	0.00305 0.9658
AGE AGE (YEARS)	-0.06547 0.3570	-0.19329 0.0061	0.08965 0.2068	-0.00220 0.9753	0.07477 0.2927	-0.12515 0.0774	0.12691 0.0733
PROGIN PROGRESSION INTERVAL (MOS)	-0.10279 0.1475	-0.08853 0.2126	-0.03690 0.6039	0.09842 0.1656	0.02863 0.6874	-0.08923 0.2089	-0.05018 0.4804
TNMSTAGE TNM STAGE	0.44442 0.0001	-0.03793 0.5938	0.04154 0.5592	-0.33573 0.0001	0.22538 0.0013	0.14056 0.0471	0.08293 0.2430
SXSTAGE SYMPTOM STAGE	1.00000 0.0	-0.03949 0.5788	0.00115 0.9871	-0.31590 0.0001	0.23540 0.0008	0.13161 0.0632	0.05325 0.4539
SMKAMT RANK CUSTOMARY PPD CIGS	-0.03949 0.5788	1.00000 0.0	0.29027 0.0001	0.04005 0.5734	-0.03591 0.6137	0.02999 0.6734	-0.07688 0.2792
MALE MALE GENDER	0.00115 0.9871	0.29027 0.0001	1.00000 0.0	0.01667 0.8148	0.06237 0.3803	0.02388 0.7371	-0.13645 0.0540
WELL WELL DIFF HISTOLOGY	-0.31590 0.0001	0.04005 0.5734	0.01667 0.8148	1.00000 0.0	-0.31103 0.0001	-0.64547 0.0001	-0.27289 0.0001
SMALL SMALL CELL HISTOLOGY	0.23540 0.0008	-0.03591 0.6137	0.06237 0.3803	-0.31103 0.0001	1.00000 0.0	-0.28300 0.0001	-0.11964 0.0915
ANAP ANAPLASTIC HISTOLOGY	0.13161 0.0632	0.02999 0.6734	0.02388 0.7371	-0.64547 0.0001	-0.28300 0.0001	1.00000 0.0	-0.24830 0.0004
CYTOL CYTOLOGIC PROOF ONLY	0.05325 0.4539	-0.07688 0.2792	-0.13645 0.0540	-0.27289 0.0001	-0.11964 0.0915	-0.24830 0.0004	1.00000 0.0

Fig. 9-3 Lower part of matrix of bivariate correlations shown in Fig. 9-2. (SAS PROC CORR program.)

including the cell types, are shown in the second part of the correlation matrix, which appears in Fig. 9-3.

Each cell in the tabular matrixes of Figs. 9-2 and 9-3 contains two numbers. The upper is the value of r, the Pearson correlation coefficient for the cited pair of variables; the lower is the associated P value for numerical stability (or stochastic significance) of the correlation. (The P value is labelled "Prob > |R|" under Ho: Rho = 0".) Whenever a correlation is calculated from something smaller than the complete N members of data, the smaller value of N would also be listed. None of the "third" numbers is shown here because no values of data were missing.

9.1.2.1. Targeted bivariate relationships

The first three tabular columns in Fig. 9-2 show the targeted correlation coefficients for the three expressions of survival. The top three rows, when intersected with the corresponding columns in the upper left corner of the matrix, show the relatively high correlations that would be expected among three forms of the same variable.

The main targeted correlations discussed in the rest of this section refer to LNSURV (in column 3), which is the target variable for multiple linear regression. If 0.3 is regarded as the minimum absolute boundary for a quantitatively impressive correlation coefficient, only two of the correlations between LNSURV and independent variables are impressive: the −.42 coefficients for symptom stage and for TNM stage. Nevertheless, because 200 is a reasonably large sample size, the relatively small .16 and −.26 correlation coefficients for hematocrit and percent weight loss emerge with P values below 0.05. The results suggest (as expected clinically) that survival is lowered with increases in symptom stage, TNM stage, or weight loss, and is raised with higher (i.e., non-anemic) hematocrits. Three of the cell-type variables also have coefficients that are quantitatively unimpressive, i.e., below |.3|, but the P values are < .05 for their correlations with LNSURV. These coefficients are .22 for well-differentiated, −.17 for small cell, and −.15 for anaplastic histology. These coefficients also have the directions expected from clinical experience.

9.1.2.2. Intervariable relationships

The lower diagonal wedge in Fig. 9-2 contains the intervariable correlation coefficients, beginning at the row and column junction for hematocrit, and extending to the bottom right corner of the matrix in Fig. 9-3. A moderately high coefficient (.44) appears for symptom stage and TNM stage, but not for any of the other pairs of independent variables. Nevertheless, despite quantitative values below 0.3, some of the intervariable correlation coefficients had P values ≦ .05. The quantitative magnitude of the stochastically significant intervariable correlations is shown in the summary arrangement of Table 9-1.

Because the dummy variables for cell type were mutually exclusive, the correlations among those variables arise from the data and coding system, not from biologic relationships. In the "significant" coefficients among the cell types themselves in Fig. 9-3, the well-differentiated group had a correlation of −0.31 with small cell, −.65 with anaplastic, and −.27 with cytology only. The anaplastic group had correlations of −.28 with the small group and −.24 with cytology only.

Table 9-1 Values of correlation coefficients with P ≤ .05 among independent variables in Figs. 9-2 and 9-3

	TNM stage	Symptom stage	Hematocrit	Smoking	Male	Percent weight loss
Symptom stage	.44	(1.00)	—	—	—	.27
Percent weight loss	—	.27	−.29	−.18	—	(1.00)
Age	—	—	—	−.19	—	—
Male	—	—	—	.29	(1.00)	—
Well-differentiated	−.34	−.32	—	—	—	—
Small-cell	.23	.24	—	—	—	.14
Anaplastic	.14	—	—	—	—	—

Almost all of the cited correlations are consistent with what is known in clinical biology. We would clinically expect that increasing symptoms would be correlated with increasing anatomic extensiveness (TNM stage) and weight loss, that well-differentiated cancers would be more likely than undifferentiated cancers to produce the lower TNM and symptom stages, that higher hematocrits would occur with less weight loss, that men are more likely to smoke than women, and that small-cell cancers are associated with weight loss. Two of the smoking correlations were somewhat surprising, however: increased smoking seemed to occur more in younger than older patients, and seemed *inversely* related to weight loss.

9.1.2.3. Other calculations for correlation coefficients

All of the coefficients cited in Figs. 9-2 and 9-3 and in Table 9-1 were calculated with the Pearson's r method cited for simple linear regression in Chapter 7. Because this form of calculation is not mathematically appropriate unless both variables are expressed in dimensional scales, several other coefficients can be used for variables cited in nondimensional scales. The additional coefficients include Spearman's rho, Kendall's tau, and φ. Nevertheless, for suitably coded nondimensional data, the other mathematical coefficients usually have only minor differences from what is found with the less appropriate computation of Pearson's r. Accordingly, since the correlation matrix is used only for screening, the crude values of r will usually suffice, and the additional coefficients need seldom be examined.

9.2. Bivariate tabulations

Although suitable for screening purposes, the correlation coefficients are not a fully satisfactory way to examine the effects of nondimensional variables on the target. The categorical effects are best displayed in tables showing the target result for each category.

When each category becomes a row in the table, the target results can be summarized as medians or means for dimensional data, or as proportions for binary events. Unfortunately, none of the currently available computer programs routinely shows the associated *median* values for a target variable. Furthermore, none of the programs routinely shows proportions of binary events (such as 6-

month survival) in juxtaposition with results of means for each category. To avoid the barrage of printout when several computer programs are used, the desired information has been consolidated and typed here in Table 9-2.

Table 9-2 Survival results for categorical variables in illustrative data set

Variable	Name of Category	Code of Category	Number of members	Mean ln survival	No. and (%) of 6-month survivors
TNM stage	I	1	51	2.43	38 (75%)
	II	2	27	2.30	15 (56%)
	III A	3	18	1.40	10 (56%)
	III B	4	36	1.11	13 (36%)
	IV	5	68	0.67	18 (26%)
Symptom stage	Asymptomatic	1	28	2.21	18 (64%)
	Pulmonic/systemic	2	82	2.16	48 (59%)
	Regional/mediastinal	3	29	0.96	12 (41%)
	Distant	4	61	0.46	16 (26%)
Sex	Male	1	182	1.43	83 (46%)
	Female	0	18	1.99	11 (61%)
Customary packs per day of cigarette smoking	None	0	9	0.65	4 (44%)
	< 1/2	1	4	1.60	1 (25%)
	1/2 to < 1	2	16	2.53	12 (75%)
	1 to < 2	3	119	1.36	54 (45%)
	≧ 2	4	52	1.58	23 (44%)
Well-differentiated cell type	Present	1	83	1.93	48 (58%)
	Absent	0	117	1.16	46 (39%)
Anaplastic cell type	Present	1	74	1.14	27 (36%)
	Absent	0	126	1.68	67 (53%)
Small-cell	Present	1	24	0.66	5 (21%)
	Absent	0	176	1.60	89 (51%)
Cytology only	Present	1	19	1.89	14 (74%)
	Absent	0	181	1.44	80 (44%)
Total	—	—	200	1.48	94 (47%)

9.2.1. Modicum values of categories

Among the individual categories in Table 9-2, three (TNM Stage IIIA, Female, and presence of Cytology only) have a size close to but below the modicum value of 20 for number of members. The last two categories (**female** and **cytology only**) were allowed to enter the multivariable analyses intact, since no consolidation was possible for the binary variables. Although the ordinal TNM Stage IIIA could have been combined with TNM Stage II because the two categories had similar 6-month survival rates, Stage IIIA was retained because of its different biologic connotation (regarding spread of the cancer) and also because the two categories had a distinctive gradient in the mean of *ln survival*.

The first three categories of the ordinal cigarette smoking variable could have been collapsed because they had small membership (9, 4, and 16 persons). On the other hand, the cigarette smoking variable had no distinctive effects on survival. The correlation with the target variable was low in Fig. 9-2, and a monotonic intrasystem gradient did not appear either in the mean of ln survival or in the associated 6-month survival rates in Table 9-2. Accordingly, the cigarette smoking variable was omitted from the multivariable analyses.

9.2.2. Inspection of gradients

In Table 9-2, TNM stage shows a progressive monotonic gradient in the mean of *ln survival*, although the second and third categories had similar 6-month survival rates. To allow the gradient in means to be preserved, *TNM stage* was kept in its 5-category coding. *Symptom stage* also shows a progressive monotonic gradient in both the log means and the proportions of 6-month survivors. Survival seemed distinctly better in women than in men, but did not show a consistent gradient in relation to the amount of customary cigarette smoking. Among the cell types, the well-differentiated group had better survival than the poorly differentiated anaplastic or small-cell groups. The 19 patients in the cytology-only group had generally higher survival than patients in the other three cell groups.

9.2.3. 'Significance' for binary variables

Despite the 15% gradient for *Sex* in Table 9-2, the distinction was not stochastically significant ($X^2 = 1.6$; $P = .2$). Nevertheless, for both esthetic and clinical reasons, the variable was retained for the multivariable appraisals.

The four binary variables for cell types in Table 6-2 were all retained because they all showed distinctive gradients in mean of *ln survival*; and the two-category gradients in 6-month survival rate were all $\geq 17\%$, with stochastically significant values of $P < .05$ by X^2 testing.

9.2.4. Linear trends in ordinal variables

Of the seven ranked independent variables under analysis, four were dimensional. As shown by the correlation coefficients in Fig. 9-2, none of the trends was quantitatively impressive, because the r values were below |.3|. Nevertheless, the trends in *ln survival* were stochastically significant for *hematocrit* ($r = .16$) and for *percent weight loss* ($r = -.26$). Neither the r values nor the P values were significant for *age* or for *progression interval*.

For the three ranked independent variables that were expressed in ordinal categories, none produced an unequivocal *rectilinear* gradient in the target variable results of Table 9-2. The variable *customary packs per day of cigarette smoking* was eliminated from further analysis, for reasons discussed in Section 9.2.1. The survival gradients are monotonic for *TNM stage* and *Symptom stage*, but the intercategory gradients have distinct inequalities in both the mean of *ln survival* values and in the *percentage of 6-month survivors*. Consequently, a rectilinear regression or correlation coefficient will give a correct average result for the impact of these two variables, but may not be accurate in the individual zones.

The intrasystem gradients, found by substraction of 6-month survival rates in Table 9-2, are 49% (= 75% − 26%) for TNM stage and 38% (= 64% − 26%) for Symptom stage. The average intrasystem gradients are 12.25% (= 49%/4) for TNM stage and 12.67% (= 38%/3) for Symptom stage. The gradients could be calculated with an algebraic model for linear trend, as discussed in Section 8.2.3, but a short cut is available from the correlation coefficients in Fig. 9-2.

When the Pearson correlation coefficients are calculated, the algebraic procedure assigns arbitrary values of **1, 2, 3,** . . . to each of the ordinal categories. Consequently, the value of Pearson's r is the same as what would be found with the standardized slope calculations in Section 8.3.4. Looking under the column marked SURV 6 in FIg. 9-2, we promptly find that r = −.38293 for TNM stage and r = −.30021 for Symptom stage. Both of these values are quantitatively significant.

If we want to convert the standardized slopes into "ordinary" algebraic slopes, we can use the formula $b = b_s(s_y/s_x)$. In Fig. 6-2, we note that s_y = .50035 for SURV 6, and that s_x is respectively 1.6316 for TNM stage and 1.0642 for Symptom stage. Therefore, the algebraic slopes are (−.38293)(.50035)/1.6316 = −.117 for TNM stage and (−.30021)(.50035)/1.0642 = −.141 for symptom stage. Both these values are reasonably close to the crude average intrasystem gradients determined directly from Table 9-2.

9.2.5. Stochastic significance for ordinal variables

The stochastic significance of the gradients produced in 6-month survival rates by *TNM stage* and *Symptom stage* was shown by the low P values for the corresponding correlation coefficients in Fig. 9-2. These results could be reproduced by doing a formal X_L^2 test for the linear trend on the data in Table 9-2, but a shortcut is again available. As noted in Chapter 8, the value of X_L^2 is Nr^2 when r is calculated with the algebraic model. Therefore, using the appropriate r coefficients in the correlation matrix, we can easily determine that X_L^2 is $(200)(-.38293)^2$ = 29.33 for TNM stage and $(200)(-.30021)^2$ = 18.03 for Symptom stage. For both these values, P is even smaller than the .0001 listed in Fig. 9-2.

The overall X^2 values for 6-month survival rates for these two variables can be determined with formula [8.3] as $[\Sigma t_h^2/n_h - T^2/N][N^2/T(N - T)]$. For 6-month survival in Table 9-2, $T^2/N = 94^2/200$ = 44.18, and $N^2/(T)(N - T)$ = 4.0145. $\Sigma t_h^2/n_h$ = 51.66 for TNM stage and 48.83 for Symptom stage. Therefore X^2 is (51.66 − 44.18)(4.0145) = 30.03 for TNM stage and (48.83 − 44.18)(4.0145) = 18.67 for Symptom stage. Both of these values are only slightly higher than the corresponding values of X_L^2, thus suggesting that X_R^2 (see Section 8.3.6.2) is quite small for the residual nonlinear effect.

Although the values of X^2 will obviously not be stochastically significant for each of the *inter*categorical gradients, the TNM and Symptom stage variables had intrasystem gradients that were linearly monotonic and significant both quantitatively and stochastically. Both variables were therefore retained for the multivariable analyses.

9.2.6. Demarcation and linear trends in dimensional variables

Of the four ranked independent variables that were expressed in dimensional scales, *progression interval* did not show a significant correlation with the target variable in FIg. 9-2. Another variable, *age*, had a stochastically significant correlation with the untransformed survival time (SURVIVE) but not with the logarithm of survival or the 6-month survival event.

The decisions about quantitative nonsignificance can be confirmed with confidence intervals, as discussed in Section 7.6.4. For example, Fig. 9-2 shows r = .05954, with P = .4023, for the bivariate relationship of *progression interval* and the *logarithm of survival time*. To be more stochastically certain that progression interval indeed has a nonsignificant effect, we could first calculate the standard error of r as $\sqrt{(1- .05954)^2/(200 - 2)} = \sqrt{0.99645/198} = .0709$, and then form a 95% confidence interval. It would be .05954 ± (1.96)(.0709), and would extend from −.0795 to +.1986. Since this interval does not include the "quantitatively significant" value of .3, we can feel reasonably confident that r = .05954 is stochastically stable in its nonsignificance.

The two remaining dimensional variables, *hematocrit* and *percent weight loss*, had stochastically significant correlations with survival, although neither correlation coefficient was quantitatively larger than |0.3|. These two variables had been partitioned into arbitrary ordinal zones for the examination of linear conformity, shown in Section 7.12.2. For the multivariable categorical analyses described later, however, the two variables were each divided into three ordinal zones that had the monotonic gradients discussed in Section 8.4.4.5.

9.3. Graphical portraits

Another way to get ideas about bivariate relationships is to inspect the graph plotted for the points of each pair of variables. Unfortunately, as noted in Section 7.10.4, the bivariate graphs produced by the packaged computer programs are not particularly helpful here because neither the BMDP nor the SAS programs have a mechanism for showing clusters of points that occur at the same location in the graph. The BMDP graphs mark single points as 1 and multiple points as 2, 3, 4, . . . , according to the number of points. The SAS graphs show single and multiple points with letters as A, B, C, D, etc. Neither format produces the clustered cannonball grouping (shown in Fig. 2-18) that gives a strong visual emphasis to the density of multiple points. Even when multiple points are not abundant, the numbers or letters produce unfamiliar visual images, particularly for relatively large amounts of data.

Fig. 9-4 shows a collection of BMDP bivariate graphs for the relationships between logarithm of survival time and the dimensional variables for hematocrit, percent weight loss, and age. Just below each graph, the printout shows the associated correlation coefficient (cited as "r"), its P value, and the equation of the corresponding bivariate regression line. Visual inspection of the highly polyvalent, widely scattered data in the three graphs in FIg. 9-4 shows no suggestion of rectilinear relationships. Nevertheless, despite the diffusely dispersed points, the corre-

200 Preparing for Multivariable Analysis

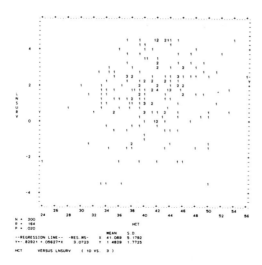
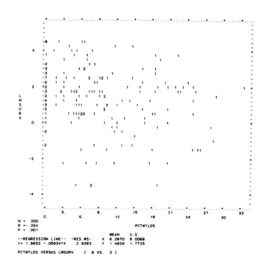
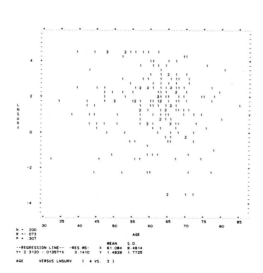

Fig. 9-4 Plots of ln survival vs. hematocrit, percent weight loss, and age. (Produced by BMDP6D–Bivariate [Scatter] Plots program.)

lation coefficient are stochastically significant at P = .02 for an upward trend (r = .164) for hematocrit and at P <.001 for a downward trend (r = −.264) for percent weight loss. The r value of −.07 indicates no impressive statistical trend for age.

The SAS and BMDP plotting programs can each be instructed to produce graphs that show points as points (or with any other special character such as *, $, @, . . .), but overlapping points are not printed. In an SAS graph, they are reported merely as the number of "hidden" observations, without indication of where they may be hiding; and BMDP graphs can be misleading because they do not mention the existence of hidden observations.

The (SAS) graph in Fig. 9-5 shows individual points of data (rather than number or letter symbols) for ln survival vs. age, but the graph does not include 15 hidden observations where multiple points occur at the same location.

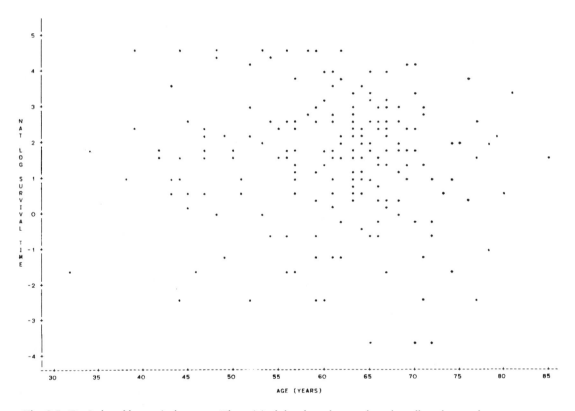

Fig. 9-5 "Dot" plot of ln survival vs. age. (The original dots have been enlarged to allow them to be distinguished on this photograph.) Although easy to read, the graph is relatively useless because, as noted at the bottom left, 15 observations are not shown. (Option in SAS PROC PLOT program.)

Fig. 9-6 shows the corresponding "point graph" produced by BMDP for ln survival vs. hematocrit in the complete set of 200 patients. The number of hidden observations is not indicated; and the dots used as symbols by BMDP do not stand out (for visual effect) any better than the hard-to-see dots in the SAS printout.

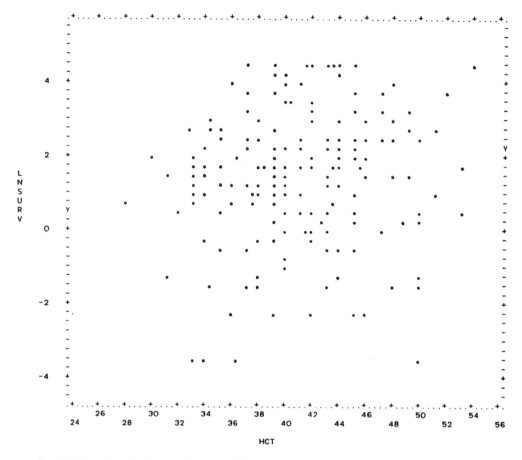

Fig. 9-6 "Dot" plot of full sample for ln survival vs. hematocrit. (The original dots have been enlarged to allow them to be distinguished on this photograph.) (Option in BMDP6D–Bivariate Plots program.)

Because the computer programs that do all the statistical calculations are not intended to be "graphics software", the cited limitations arise primarily from the output format available on standard printers. If you want visually splendid graphs, other mechanisms should be used. The main value of the automated graphical portraits here is their demonstration of polyvalent, widely scattered points for which a really close fit cannot be obtained with *any* kind of ordinary line. These displays will warn you not to expect high values for the fit of the multivariable algebraic analyses. To get ideas about general relationships for dimensional data that can readily be depicted in graphs, and also for nondimensional variables, the correlation coefficients will often offer a much simpler and more trustworthy *screening* mechanism than visual inspection of the computer-drawn graphs.

9.4. Bivariate 'conclusions'

The simple bivariate appraisals provide some useful hints about what to anticipate when the illustrative data set receives multivariable analyses in subsequent chapters. The *TNM stage* and *Symptom stage* variables seem to have the strongest impact on survival, which is also affected, in a lesser way, by *weight loss, hematocrit,* and certain cell types. Because of relatively high intervariable correlations, however, several of these independent variables may have a reduced impact in a multivariable context. For a few variables that have relatively poor distributions, precautions will be needed to ensure that important effects are neither disguised nor distorted by the maldistributions.

The multivariable analyses that occur in the rest of the text will be aimed at checking these suspicions, confirming what can be confirmed, and augmenting the existing bivariate results with whatever useful new information is produced.

Chapter Reference

1. Siegel, 1988.

Part III Basic Strategies of Targeted Algebraic Methods

After the long preparation of Parts I and II, the stage is now set for targeted multivariable analysis. Its operations are derived from the simple bivariate methods discussed in Chapters 7 and 8, and then applied in Chapter 9 for screening evaluations of the illustrative data set. The multivariable activities derived from algebraic linear regression methods are described in Parts III and IV of the text, and the targeted-cluster methods, derived from stratified rates, occupy Part V.

The first two of the three chapters in Part III are concerned with two of the main sources of complexity when simple linear regression is expanded to its multivariable counterpart. One of these sources, described in Chapter 10, is the extra variables themselves. They produce the complications of covariance relationships, as well as potential problems in collinearity and interactions. The second source of complexity, discussed in Chapter 11, is the challenge of exploring and reducing the multiple independent variables. Chapter 12 then shows the application of multiple linear regression in the illustrative data set and in published medical literature.

The subsequent discussion, in Part IV of the text, is devoted to yet another source of algebraic complexity: the different formats used for target variables that are *not* dimensional. These formats produce the three other members of the big four regression family: logistic regression, proportional hazards (Cox) regression, and discriminant function analysis.

10 Complexity of Additional Independent Variables

Outline . . . .

10.1. Additional sources of complexity
    10.1.1. Different targets
    10.1.2. Explorations and reductions
    10.1.3. Additional independent variables

10.2. Summaries of basic data and bivariate relationships
    10.2.1. Basic mean and group variance
    10.2.2. Graphic display of data
    10.2.3. Covariances and bivariate correlations
        10.2.3.1. Calculation of group variances and covariances
        10.2.3.2. Calculation of correlation coefficients
    10.2.4. Other bivariate displays

10.3. The 'general linear model'
    10.3.1. Basic structure
    10.3.2. Linearity of coefficients
    10.3.3. Partial regression coefficients
    10.3.4. Symbols in different models
    10.3.5. Parametric estimates

10.4. Fitting coefficients to the model
    10.4.1. Extra sources of covariance
    10.4.2. Connotation of the b_j coefficients
    10.4.3. Calculation of the coefficients
    10.4.4. Alternative expressions
    10.4.5. Example of calculations
    10.4.6. Application of matrix algebra

10.5. Indexes of accomplishment
    10.5.1. Sums of squares
    10.5.2. Mean square residual error (MSE)
    10.5.3. R^2 and 'explained variance'
        10.5.3.1. Advantages of R^2
        10.5.3.2. Disadvantages of R^2
    10.5.4. F ratio of mean variances
    10.5.5. Adjusted R^2
    10.5.6. Standardized regression coefficients

10.6. Stability of indexes
10.7. Stability of 'events' and independent variables
10.8 Collinearity and interactions

. . . .

Each of the big four targeted multivariable algebraic procedures uses the main operating principles of simple linear regression. As in simple regression, each multivariable procedure begins by giving the target variable a basic fit, which is then compared with the fit achieved by the independent variables; each multivariable procedure uses a specific strategy to find the best coefficients for those variables; the fit of the algebraic model and the impact of the included variables are expressed with special indexes of accomplishment; and the efficacy of the model can be appraised with regression diagnostics before final conclusions are drawn.

Despite the similarity of basic strategies, however, the tactics that carry out the strategy must often be substantially altered to deal with three main sources of complexity in multivariable methods.

10.1. Additional sources of complexity

The additional sources of complexity arise from different targets, from the methods used for explorations and reductions, and from the additional variables themselves.

10.1.1. Different targets

The most prominent source of complexity is the different formats for expressing the target variable, Y. Unlike the dimensional target of simple regression, the target of multivariable analyses can often be nondimensional, referring to events or categories; and in one instance, the target is a bivariate indication of an "exit time" and "exit state". For these nondimensional target variables, the algebraic model must have a different format, regardless of whether Y is related to a single independent variable, or to multiple variables.

The complex formats for these different targets require special discussion not because they are required by the multiple variables, but because they almost never appear in simple linear regression. If the target variable were always dimensional, the multivariable procedures would still have the complexity of multiple variables, but the main operating strategy would be similar to what is used in simple linear regression. With nondimensional targets, however, different algebraic methods are needed to express individual discrepancies between actual and estimated values; different operating principles are used to find the best-fitting coefficients; and different statistical indexes are cited to express the model's accomplishments.

The complexities caused by nondimensional target variables will be discussed in Part IV of the text when we reach the logistic, proportional hazards, and discriminant function methods that aim at such targets.

10.1.2. Explorations and reductions

A second source of complexity is the exploration and reduction of independent variables. Because not all the candidate variables may be effective, separate efforts are usually made to find the smaller set of variables that have the important impacts. These efforts involve exploring various combinations of the variables and deciding which ones to keep or to eliminate in a best final model.

During these explorations, new expressions are introduced for the algebraic accomplishments. One set of expressions contains incremental indexes that describe what happens when an existing algebraic model is augmented by addition, or depleted by removal, of a particular variable. A second set of new expressions contains special "penalty" indexes. They give "credit" for what an algebraic model does in fitting the data, and "debit" for the number of variables used to achieve the fit.

The methods of exploring and reducing variables will be discussed in Chapter 11.

10.1.3. Additional independent variables

The rest of this chapter is devoted to the most obvious source of complexity: the multiple independent variables themselves. Aside from requiring new symbols, the multiple variables will co-vary not only with the target variable but also with one another. Consequently, the coefficients attached to each variable in multiple regression do not have the same meaning and cannot be determined in the same easy way as in simple linear regression. The calculations require special computational attention to covariances; and the results produce "partial" regression coefficients that indicate the effect of each variable in the concomitant presence of the others.

The interrelationships of the multiple independent variables also create considerable opportunity for misleading results due to the collinearity or interactions discussed in Section 3.2.2. The collinearity among pairs of variables can often be suitably considered and managed when the partial regression coefficients are calculated, but the detection and management of interactions is a particularly thorny problem that involves special scientific and mathematical strategies.

The discussion that follows contains a great deal of mathematics to indicate what is happening and how it occurs. Nevertheless, the mathematics should be quite easy to follow, and each section is illustrated with results that can be shown completely for a 6-item data set having a target variable, Y, and two independent variables, X_1 and X_2.

10.2. Summaries of basic data and bivariate relationships

To avoid the algebra for k independent variables in N persons, the basic strategies of multiple linear regression are illustrated with a simple data set for six persons in Table 10-1, where the dependent Y variable is serum cholesterol (in mg/100 ml) and the two independent variables are X_1 = age (in years) and X_2 = weight (in kg).

Table 10-1 Example of data for two independent variables (age and weight) in relation to a target variable (serum cholesterol) in 6 persons

Person	Y= serum cholesterol	X_1 = age	X_2= weight
1	263	31	69
2	189	19	72
3	372	52	80
4	287	29	62
5	344	51	64
6	208	23	38
Total	$\Sigma Y_i = 1663$	$\Sigma X_1 = 205$	$\Sigma X_2 = 385$
	$\Sigma Y^2 = 487{,}243$	$\Sigma X_1^2 = 7997$	$\Sigma X_2^2 = 25{,}729$
	$\Sigma YX_1 = 61{,}739$; $\Sigma YX_2 = 109{,}229$; $\Sigma X_1 X_2 = 13{,}603$		

10.2.1. Basic mean and group variance

In multiple linear regression, the target variable, Y, has the same basic summary as in simple regression. The mean of the group is $\bar{Y} = \Sigma Y_i/N$, and the basic total discrepancy is the group's variance, $S_{yy} = \Sigma(Y_i - \bar{Y})^2$. In the second column of Table 10-1, $\Sigma Y_i = 1663$, and $\bar{Y} = 277.17$. Thus, $S_{yy} = \Sigma Y_i^2 - [(\Sigma Y_i)^2/N] = 487243 - [(1663)^2/6] = 26314.83$.

10.2.2. Graphic display of data

For two independent variables, X_1 and X_2, the relationship to Y is sometimes portrayed in the type of three-dimensional graph shown in Fig. 10-1. These three-dimensional graphs are seldom used for routine inspection of the data, however, because many people have difficulty visualizing the plane that might pass through the points; and the relationship is particularly hard to show if the points are not easily fitted with a plane. Besides, for three or more independent variables, a single graphical plot is impossible for the multiple dimensions.

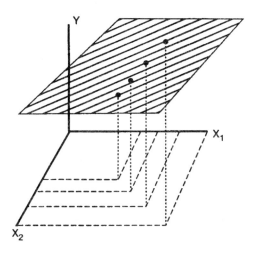

Fig. 10-1 Two-dimensional picture of three-dimensional plane passing through four points, each having coordinates in the three dimensions of X_1, X_2, and Y.

The co-relationship of any *pair* of variables can always be drawn, however, in a two-dimensional graph. For two independent variables, X_1 and X_2, the plots would show Y vs. X_1, Y vs. X_2, and X_1 vs. X_2 (or X_2 vs. X_1). Fig. 10-2 shows these bivariate graphs for the data of Table 10-1. In this instance, the graphs immediately reveal things that were not promptly apparent in the table: X_1 has a quite good linear relationship to Y; but X_2 does not seem closely related to Y or to X_1.

The bivariate graphs can help indicate the shape of individual pairs of co-relationships, but the number of graphs may become cumbersome if more than two independent variables are under appraisal. Consequently, most analysts will regularly examine statistical indexes of correlation for the bivariate graphs. After the multivariable analysis is completed, however, certain bivariate graphs (described later) may be examined during the regression diagnostics.

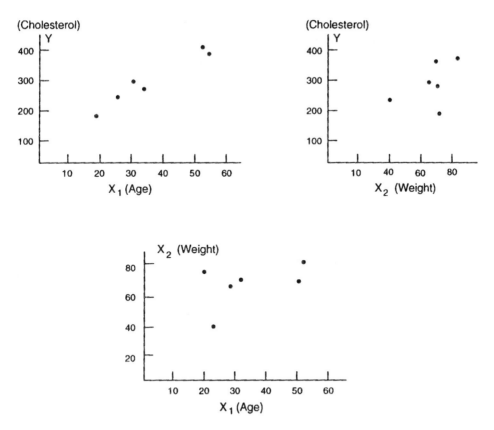

Fig. 10-2 Graphs of three sets of bivariate relationships for the data shown in Table 10-1.

10.2.3. Covariances and bivariate correlations

Before the multivariable analysis begins, the multiple bivariate relationships, discussed in Chapter 7, are shown as group covariances and coefficients of correlation for simple regression analyses of each possible pair of available variables. The calculation of these entities for Table 10-1 is shown in the next two subsections.

10.2.3.1. Calculation of group variances and covariances

Using symbols discussed in Section 9.1.1.3, the group variance-covariance results for the data in Table 10-1 would be as follows:

$S_{yy} = \Sigma(Y)^2 - (\Sigma Y)^2/N = 487{,}243 - (1663)^2/6 = 26314.834$,
$S_{11} = \Sigma(X_1)^2 - (\Sigma X_1)^2/N = 7997 - (205)^2/6 = 992.833$,
$S_{22} = \Sigma(X_2)^2 - (\Sigma X_2)^2/N = 25729 - (385)^2/6 = 1024.833$,
$S_{y1} = \Sigma(YX_1) - (\Sigma Y \Sigma X_1)/N = 61739 - (205)(1663)/6 = 4919.833$,
$S_{y2} = \Sigma(YX_2) - (\Sigma Y \Sigma X_2)/N = 109229 - (385)(1663)/6 = 2519.833$,

and

$S_{12} = \Sigma(X_1 X_2) - (\Sigma X_1)(\Sigma X_2)/N = 13603 - (205)(385)/6 = 448.833$.

10.2.3.2. Calculation of correlation coefficients

The corresponding correlation coefficients would be

$$r_{y1} = \frac{S_{y1}}{\sqrt{S_{yy}S_{11}}} = \frac{4919.833}{\sqrt{(26314.834)(992.833)}} = .963,$$

$$r_{y2} = \frac{S_{y2}}{\sqrt{S_{yy}S_{22}}} = \frac{2519.833}{\sqrt{(26314.834)(1024.833)}} = .485,$$

and

$$r_{12} = \frac{S_{12}}{\sqrt{S_{11}S_{22}}} = \frac{488.833}{\sqrt{(992.833)(1204.833)}} = .445$$

When the bivariate correlation coefficients are checked to get ideas about which variables may have close relationships with the target variable and with one another, the coefficients here confirm what was seen in the graphs of Fig. 10-2: Y seems closely related to X_1, but not to X_2; and X_1 is not impressively related to X_2.

Because the bivariate correlation coefficients are usually examined only for descriptive purposes, their stability need not be checked with tests of stochastic significance. Nevertheless, if calculated for r_{y1}, r_{y2}, and r_{12}, the respective values of t are 7.17, 1.11, and 0.99. At four degrees of freedom (for N = 6), the first of these r values is stochastically significant at P < .01. The other two values are not.

10.2.4. Other bivariate displays

Variances, covariances, and correlation coefficients can always be calculated, regardless of the binary, ordinal, or dimensional scales used for expressing the independent variables. With binary and ordinal variables, however, the results may not be as clear as tabular displays of the corresponding data. Appropriate bivariate tables or charts can therefore be arranged (see Chapter 8) to show the mean or other central values of Y in each binary or ordinal category of the pertinent independent X_j variables. Although not needed for Table 10-1, the sequential central values will often be more useful than bivariate correlation coefficients in showing the individual effect of categorical independent variables.

10.3. The 'general linear model'

The term *general linear model*, which is constantly used for algebraic combinations of independent variables, involves several important concepts about its structure, linearity, partial regression coefficients, symbols in different models, and parametric estimates.

10.3.1. Basic structure

For a set of variables, $\{X_j\}$, with coefficients $\{b_j\}$, the basic structure of the general linear model is

$$G = b_0 + b_1X_1 + b_2X_2 + b_3X_3 + \ldots + b_jX_j + \ldots \quad [10.1]$$

A particular model can contain some, all, or none of the available independent variables. When desirable (as discussed later), a subscript can be added to G to indicate the number of independent variables in the model. With no variables in the model, $G_0 = b_0$ (which is usually the mean, $\bar{Y}$). For one variable, G_1 would be $b_0 + b_1X_1$; and for two variables, G_2 would be $b_0 + b_1X_1 + b_2X_2$. If k independent variables are available, the "full" model is $G_k = b_0 + b_1X_1 + \ldots + b_kX_k$. When different models are explored, a particular "working" model will be marked G_p. The "p" subscript represents "parameters"; and in any exploration, p is $\leq$ k. Thus,

$$G_p = b_0 + b_1X_1 + b_2X_2 + \ldots + b_jX_j + \ldots + b_pX_p. \qquad [10.2]$$

For example, seven independent variables might be available, but a "working" model might contain six variables, and a "final" model might contain three. In all these arrangements, the values of $\{b_j\}$ usually differ, but b_0 is the "intercept" in any model when the values of all the $\{X_j\}$ variables are zero.

10.3.2. Linearity of coefficients

A common misconception about the general linear model is the belief that it is "linear" in its format for each X_j variable, and that it cannot contain values such as X_j^2 or $\sqrt{X_j}$. What is *linear* about the model, however, is the pattern of the b_j coefficients, not the X_j variables. As long as the b_j coefficients are retained in simple linear form, the X_j variables can have nonsimple expressions. Thus, the model would still be linear if expressed as

$$G = b_0 + b_1X_1^2 + b_2\sqrt{X_2} + b_3\ln(X_3) + b_4e^{X_4} + \ldots.$$

The model would *not* be linear, however, if it were cited as

$$G = b_0 + \sqrt{b_1}X_1 + [\ln(b_2)]X_2.$$

As long as the coefficients remain linear, the model is still linear even if it contains interaction terms such as X_1X_2 or $X_3X_4X_5$. Thus, the linear model might be

$$G = b_0 + b_1X_1 + b_2\ln X_2 + b_3X_1X_2 + b_4X_3 + b_5\sqrt{X_4} + b_6e^{X_5} + b_7X_3X_4X_5.$$

What most people think of as a linear model is the *rectilinear* (or straight-line) arrangement in which each X variable is present in its own "linear" format. Thus, $G = b_0 + b_1X_1$ is the model of a straight line; $G = b_0 + b_1X_1 + b_2X_2$ is the model of a plane; $G = b_0 + b_1X_1 + b_2X_2 + \ldots + b_pX_p$ is the model of a "rectilinear surface" in multidimensional space. On the other hand, $G = b_0 + b_1X_1 + b_2X_1^2 + b_3X_2$ is still a linear model, but the quadratic term (in X_1^2) will produce a curved surface.

10.3.3. Partial regression coefficients

As discussed later, the $b_1, b_2, \ldots$ values in the general linear model are partial regression coefficients indicating the relative weight or effect of each of the corresponding variables, $X_1, X_2, \ldots$ in the multivariable context. Because the magnitude of each b_j coefficient will be affected by the measurement scales for the corresponding variables, the b_j values are usually best compared after they have been standardized.

Using the same principle as in bivariate analysis, the value of b_j for each variable, X_j, can be converted to a standardized regression coefficient, b_{j_s}, by finding s_j as the ordinary standard deviation for the variable, and then using the formula b_{j_s}

$= (b_j s_y)/s_j$, where s_y is the standard deviation for the target variable. The procedure is illustrated later in this chapter.

10.3.4. Symbols in different models

After the candidate independent variables are examined for individual effectiveness, all of them might be eliminated except for X_2, X_5, and X_7. A different model would then be constructed, containing only those three variables. The new model would probably have different coefficients for the three variables, and could be cited as

$$G = a_0 + a_1 X_2 + a_2 X_5 + a_3 X_7. \qquad [10.3]$$

To avoid the dissociation of numerical subscripts in a written discussion, the variables X_2, X_5, and X_7 in expression [10.3] are usually renumbered to become X_1, X_2, and X_3 in the final model. To avoid the confusion of switching from b_j to a_j coefficients, the a_j coefficients are usually redesignated as b_j—with the reader warned to recognize that they differ from those cited as b_j in other formulas. The final model here would be expressed as

$$G = b_0 + b_1 X_1 + b_2 X_2 + b_3 X_3.$$

If some of the k independent variables have been eliminated, the final model will contain a total of p variables, where $p < k$. The general symbol for these reduced models is

$$G = b_0 + b_1 X_1 + \ldots + b_p X_p.$$

10.3.5. Parametric estimates

In the mathematical activities, the algebraic model offers an estimate not only for Y_i, but also for the "correct" parametric coefficient of each of the included variables. For this estimate, the N observations under analysis are assumed to be a random sample of a parent population for which the "correct" parameters of the relationship would be cited with Greek letters as

$$G_p = \beta_0 + \beta_1 X_1 + \beta_2 X_2 + \ldots + \beta_p X_p.$$

For stochastic inference in multivariable analyses, the b_j coefficients calculated during the analyses are assumed to be estimates of the correct β_j parameters. This distinction is usually ignored except in two circumstances. First, the term *beta* or *beta weight* is sometimes used in reference to the b_j coefficients. Second, in subsequent tests of stochastic significance on the b_j coefficient themselves, we assume the null hypothesis that the corresponding β_j parameter equals 0. (In stochastic tests for the entire model, rather than for individual coefficients of variables, the null hypothesis is that each of the β_j parameters equals 0.)

The total number of parameters becomes important later on when they are included in penalty indexes or when the number of degrees of freedom is used in tests of stochastic significance.

10.4. Fitting coefficients to the model

Regardless of how many X_j variables are included, the corresponding b_j coefficients in a model are determined by the same principle for each algebraic format. For the dimensional target of multiple linear regression, the principle is the same "least squares procedure" that was used in simple regression to minimize the residual variance, $S_R = \Sigma(Y_i - \hat{Y}_i)^2$. The calculations and meaning of the b_j coefficients, however, are much more complex than in simple regression.

10.4.1. Extra sources of covariance

The prime source of complexity for determining each b_j is the covariances that occur among the independent variables.

The additional effects can be shown algebraically if we start with the group residual variance in simple regression. When Y was regressed against X in formula [7.9] of Section 7.5.1, we found that

$$S_R = \Sigma(Y_i - \hat{Y}_i)^2 = S_{yy} - 2bS_{xy} + b^2 S_{xx}. \qquad [10.4]$$

This expression contains two terms for group variance (S_{yy} and S_{xx}) and one term (S_{xy}) for the group covariance of X and Y.

In multiple regression for two variables X_1 and X_2, each of the squared residual deviations would be

$$(Y_i - \hat{Y}i)^2 = (Y_i - b_0 - b_1 X_1 - b_2 X_2)^2. \qquad [10.5]$$

In multiple regression, the intercept value of b_0 is calculated in the same way as the intercept, a, in simple regression. Thus, for two variables,

$$b_0 = \overline{Y} - b_1 \overline{X}_1 - b_2 \overline{X}_2, \qquad [10.6]$$

and for k variables,

$$b_0 = \overline{Y} - b_1 \overline{X}_1 - b_2 \overline{X}_2 - \ldots - b_k \overline{X}_k.$$

Substituting the appropriate value of b_0 into formula [10.5], and rearranging terms, we get

$$(Y_i - \hat{Y}_i)^2 = [Y_i - \overline{Y}) - b_1(X_1 - \overline{X}_1) - b_2(X_2 - \overline{X}_2)]^2. \qquad [10.7]$$

When the right side of formula [10.7] is squared, and when the previous symbols for variances and covariances are substituted, the expression becomes

$$S_R = \Sigma(Y_i - \hat{Y}_i)^2 = S_{yy} + b_1^2 S_{11} + b_2^2 S_{22} - 2b_1 S_{y1} - 2b_2 S_{y2} + 2b_1 b_2 S_{12}. \qquad [10.8]$$

Formula [10.8] seems at first to correspond directly to formula [10.4]. Each formula contains the term S_{yy}. The $b^2 S_{xx}$ term for group variance of X in [10.4] corresponds to the $b_1^2 S_{11} + b_2^2 S_{22}$ terms for group variance of X_1 and X_2 in [10.8]. The term $-2bS_{xy}$ for the group covariance of Y and X in [10.4] corresponds to the $-2b_1 S_{y1}$ and $-2b_2 S_{y2}$ terms for the group covariances of Y with X_1 and X_2 in [10.8].

Formula [10.8], however, contains one glaring exception that is not present in [10.4]. The exception is the additional term $2b_1 b_2 S_{12}$, which refers to S_{12}, the group covariance of the variables X_1 and X_2. This intervariable group covariance was not (and could not be) considered in simple regression for only one variable, X.

If $\hat{Y}_1$ were estimated from four independent variables, X_1, X_2, X_3, X_4, the expression for residual deviations would become

$$(Y_i - \hat{Y}_i) = Y_i - \overline{Y} - b_1(X_1 - \overline{X}_1) - b_2(X_2 - \overline{X}_2) - b_3(X_3 - \overline{X}_3) - b_4(X_4 - \overline{X}_4).$$
[10.9]

When this expression is squared and summed, S_R would contain the following terms:

for Variances: S_{yy}, $b_1^2 S_{11}$, $b_2^2 S_{22}$, $b_3^2 S_{33}$, and $b_4^2 S_{44}$;
for Targeted Covariances of Y and X_j: $-2b_1 S_{y1}$, $-2b_2 S_{y2}$, $-2b_3 S_{y3}$, and $-2b_4 S_{y4}$;
and for Covariances among the X_j Variables:
$2b_1 b_2 S_{12}$, $2b_1 b_3 S_{13}$, $2b_1 b_4 S_{14}$, $2b_2 b_3 S_{23}$, $2b_2 b_4 S_{24}$, and $2b_3 b_4 S_{34}$.

These algebraic distinctions show that the value of S_R in multiple regression depends on more than just an appropriate addition of variances for the Y and X variables, and subtraction of the XY covariances. S_R also contains an added covariance for each pair of independent variables.

10.4.2. Connotation of the b_j coefficients

These extra sources of covariance change the meaning of the b_j coefficients and also make them difficult to calculate.

In simple regression, b is easily determined as $b = S_{xy}/S_{xx}$. It represents the direct slope of the straight-line relationship between Y and the corresponding X variable. For multiple variables, the corresponding *direct* bivariate targeted regression coefficients would be S_{y1}/S_{11} for X_1, S_{y2}/S_{22} for X_2, and so on.

In multiple regression, however, each b_j coefficient is a *partial regression coefficient*. It is calculated for formula [10.8] (or other multivariable formulas) in a manner that accounts for not only the direct variance of each X_j and the direct covariance of X_j with Y, but also the covariance of each X_j with each of the other independent variables.

This process complicates the calculations, but also gives each b_j a special new connotation. Because the calculation is done with a "derivative" calculus equation, each b_j indicates the amount by which Y_i is altered for a unit of change in X_j *within the context of the other variables, while they are held constant*. In regression for four independent variables, if X_1 were fixed at a value of **13,** with X_3 fixed at **23,** and X_4 fixed at **59,** the value of $b_2 = 7.2$ would indicate that Y_i increases by 7.2 units for a one-unit change in X_2. The mathematical relationship would pertain regardless of the particular values at which X_1, X_3, and X_4 are held constant, and regardless of whether the one-unit change in X_2 is from **1** to **2** or from **26.3** to **27.3**.

Thus, the ordinary *direct bivariate* regression coefficient, S_{yj}/S_{jj}, represents what happens between variable Y and X_j no matter what goes on in any other variables. When other independent variables are included, however, the change in X and its impact on Y are altered by the context of the other variables. The *partial* regression coefficients thus enable the multivariable analysis to fulfill its mathematical task of showing the impact of each variable in the simultaneous presence of the others.

On the other hand, the fulfillment is an act of mathematics, produced by the derivative equations, using the idea that each variable is held constant when the coefficients are calculated. Despite the mathematical splendor of this idea, the results may not always coincide with reality. As in simple regression, each coefficient will correctly represent an "average" constant value for the "linear" expression of each variable, but may not be correct in different zones of the data.

10.4.3. Calculation of the coefficients

With the least-squares principle, the b_j coefficients are aimed at minimizing the magnitude of S_R. Their values therefore emerge as solutions to the equations produced when S_R is partially differentiated with respect to each of the b_j coefficients. For example, in simple regression, the differentiation of [10.4] with respect to b produces.

$$\delta S_R/\delta b = -2S_{xy} + 2bS_{xx} \qquad [10.10]$$

When set equal to 0, equation [10.10] is solved to yield $b = S_{xy}/S_{xx}$. For formula [10.8], differentiation with respect to b_1 and b_2 produces

$$\delta S_R/\delta b_1 = 2b_1 S_{11} - 2S_{y1} + 2b_2 S_{12}, \qquad [10.11]$$

and

$$\delta S_R/\delta b2 = 2b_2 S_{22} - 2S_{y2} + 2b_1 S_{12}. \qquad [10.12]$$

These two simultaneous equations are then set equal to zero, and solved to find the values of b_1 and b_2.

If no other tools are available, the most obvious way to solve these two equations is to go through a cumbersome algebraic process. We first eliminate b_2 by multiplying [10.11] by S_{22}, and by multiplying [10.12] by S_{12}. The result is a pair of equations that can be subtracted, and then solved for b_1 to yield

$$b_1 = \frac{S_{y1} S_{22} - S_{y2} S_{12}}{S_{11} S_{22} - S_{12}^2}. \qquad [10.13]$$

Using an analogous process to eliminate b_1 in equations [10.11] and [10.12], we would eventually get

$$b_2 = \frac{S_{y2} S_{11} - S_{y1} S_{12}}{S_{11} S_{22} - S_{12}^2}. \qquad [10.14]$$

These unfamiliar formulas for b_1 and b_2 would promptly be recognized if variables X_1 and X_2 were wholly unrelated, having an intervariable covariance of zero. In the latter situation, $S_{12} = 0$, and formula [10.13] would become $b_1 = S_{y1} S_{22}/S_{11} S_{22} = S_{y1}/S_{11}$, which is the direct bivariate regression coefficient between Y and X_1. Similarly, formula [10.14] would become $b_2 = S_{y2}/S_{22}$, which is the direct bivariate regression coefficient for Y and X_2.

10.4.4. Alternative expressions

The effect of the intervariable covariance may be easier to perceive if the results are expressed using r_{12}, the correlation coefficient between X_1 and X_2. With appropriate algebraic maneuvers that are not shown here, the formulas for b_1 and b_2 can be converted to

$$b_1 = \frac{S_{y1}[1 - r_{12}(r_{y2}/r_{y1})]}{S_{11}[1 - r_{12}^2]}, \qquad [10.15]$$

and

$$b_2 = \frac{S_{y2}[1 - r_{12}(r_{y1}/r_{y2})]}{S_{22}[1 - r_{12}^2]}. \qquad [10.16]$$

Formulas [10.15] and [10.16] immediately show that partial regression coefficients for b_1 and b_2 are produced by adjusting the direct regression coefficients (S_{y1}/S_{11} and S_{y2}/S_{22}) for the correlation, r_{12}, between variables X_1 and X_2. When $r_{12} = 0$, the partial and direct regression coefficients are identical.

10.4.5. Example of calculations

The partial regression coefficients for the data in Table 10-1 can be calculated using the numerical values listed in Section 10.2.3.1. With formulas [10.13] and [10.14], the results are

$$b_1 = \frac{(4919.833)(1024.833) - (2519.833)(488.833)}{(992.833)(1024.833) - (488.833)^2} = 4.793$$

and

$$b_2 = \frac{(2519.833)(992.833) - (4919.833)(448.833)}{(992.833)(1024.833) - (448.833)^2} = .3598.$$

With formulas [10.15] and [10.16], and the data cited in Section 10.2.3.2, the corresponding calculations are

$$b_1 = \frac{4919.833[1 - .445(.485/.963)]}{992.833[1 - (.445)^2]} = 4.794$$

and

$$b_2 = \frac{2519.833[1 - .445(.963/.485)]}{1024.833[1 - (.445)^2]} = .3569.$$

(The minor differences in the two sets of results are due to rounding in calculation of the correlation coefficients.)

After the values of b_1 and b_2 are determined, we can find b_0 from formula [10.6] as

$$b_0 = 277.167 - (4.793)(34.167) - (.3598)(64.167) = 90.317.$$

The regression equation for the data in Table 10-1 can now be written as

$$\hat{Y}_i = 90.317 + 4.793 X_1 + .3598 X_2. \qquad [10.17]$$

10.4.6. Application of matrix algebra

With three or more independent variables, the formulas for the b_j partial regression coefficients become much more complicated, and the algebraic calculations become formidable. With k independent variables, the partial differentiation of S_R would produce k simultaneous equations to be solved for $b_1, \ldots, b_k$. The solutions would require intricate algebra for the formulas; and the actual calculations, in the days before automated computation, would be horrendous. Fortunately, with modern digital computers and packaged programs, all this work is done inside the "black box", which produces the results promptly and easily.

To simplify both the notation and the computational pathways, the formulas and calculations are usually expressed in matrix algebra, which can be even more formidable than everything else. Matrix algebra requires new sets of symbols, new types of operations for multiplication and division, and unfamiliar new words, such as *eigenvectors* and *eigenvalues*. If you will acknowledge that matrix algebra offers a splendid way to calculate partial regression coefficients (and many other useful things) from the diverse variances and covariances that occur among multiple variables, we can omit the complexity of the procedures.

Your multivariable education would be incomplete, however, without at least one exposure to the matrix operations. Here is a small example of how the process would be used to calculate the denominators for b_1 and b_2 in an analysis containing two independent variables:

A basic principle of matrix algebra is that four terms appearing in a 2 × 2 "square" matrix as

$$\begin{Bmatrix} a & b \\ c & d \end{Bmatrix}$$

have a *determinant*, which is written as

$$\begin{vmatrix} a & b \\ c & d \end{vmatrix}.$$

The determinant has an algebraic value, which is calculated as $ad - bc$.

The variance-covariance matrix for two variables X_1 and X_2, can be written as

$$\begin{Bmatrix} S_{11} & S_{12} \\ S_{12} & S_{22} \end{Bmatrix}.$$

According to the cited principle for a 2 × 2 square matrix, the determinant is $S_{11}S_{22} - S_{12}^2$. This expression should look familiar. It is the denominator that appeared for the partial regression coefficients calculated in formulas [10.13] and [10.14].

10.5. Indexes of accomplishment

The values of $\hat{Y}_i$ can be estimated for each person when the corresponding values of the X_j variables are inserted into the formula

$$\hat{Y}_i = b_0 + b_1X_1 + b_2X_2 + \ldots .$$

For example, for the first person in Table 10-1, $X_1 = 31$, and $X_2 = 69$. The estimated value of $\hat{Y}_i$ from formula [10.17] is $90.317 + (4.793)(31) + (.3598)(69) = 263.73$. This is a particularly good estimate, since the actual value of Y for the first person is 263. (For the second person, with $X_1 = 19$ and $X_2 = 72$, the estimate is 207.29 for the observed value of 189.)

10.5.1. Sums of squares

When each $\hat{Y}_i$ is compared with the actual value of Y_i, the process in multiple linear regression is the same as in simple linear regression. The individual indexes of discrepancy will be the deviations, $Y_i - \hat{Y}_i$; and the sum of the squared individual deviations will be the group residual variance, or "error variance",

$$S_R = \Sigma(Y_i - \hat{Y}_i)^2.$$

As in simple regression, the basic group variance in multiple regression will be partitioned as

$$S_{yy} = S_M + S_R.$$

S_M, the *group model variance*, is sometimes called the *variance due to regression, regression variance*, or *regression sum of squares*. It is expressed, as in simple regression, as

$$S_M = \Sigma(\hat{Y}_i - \bar{Y})^2.$$

The data in Table 10-1 produce $S_R = 1828.965$. Since $S_{yy} = 26314.833$, the value of $S_M = S_{yy} - S_R = 24485.868$.

The values of S_{yy}, S_R, and S_M can then be used, along with the partial regression coefficients, in indexes that express the accomplishments of the model. Because of the way that multiple variables can affect the fit of a model, and because of the need to appraise the effects of eliminating some of the variables, the indexes of accomplishment are more complicated and extensive in multiple than in simple regression,

10.5.2. Mean square residual error (MSE)

The mean value of S_R, which is regarded as "error variance" around the regression surface for the group, is sometimes used as an index of accomplishment. The smaller its value, the better the accomplishment. As noted earlier (Section 7.5.2.2), Y_i has N degrees of freedom; and $\hat{Y}_i$ has $p + 1$ degrees of freedom if the model contains p variables and an intercept. The degrees of freedom in S_R will therefore be $N - (p + 1) = N - p - 1$. The mean value of S_R, written as MSE, will be

$$\text{MSE} = S_R/(N - p - 1). \qquad [10.18]$$

For the data in Table 10-1, $S_R = 1828.965$ and $N - p - 1 = 6 - 2 - 1 = 3$. Thus, MSE = $1828.965/3 = 609.655$. Note that "p" in formula [7.12] referred to a total of p parameters. The "p" in formula [10.18], however, refers to p independent variables. They are augmented by 1, for the intercept, to form a total of $p + 1$ parameters.

10.5.3. R^2 and 'explained variance'

In multiple linear regression, as in simple regression, the proportion of basic group variance that has been reduced or "explained" by the model is

$$R^2 = \frac{S_{yy} - S_R}{S_{yy}}. \qquad [10.19]$$

Like r^2 in simple regression, R^2 will range from 0 to 1, and higher values indicate a better fit for the model.

Because of the multiple variables, the formula for calculating R^2 cannot be cited with the simple expression used to show that $r^2 = S_{xy}^2/(S_{xx}S_{yy})$ for two variables, X and Y. Because the formula for R^2 will change according to the number of included variables, R^2 is usually determined in multivariable computations by first finding S_{yy} and then calculating S_R according to formulas such as [10.8]. The values of S_{yy} and S_R are then entered into formula [10.19] to get R^2.

For the data in Table 10-1, R^2 is $(26314.833 - 1828.965)/26314.833 = .930$. In packaged programs, the computer does all the work immediately and R^2 appears as one of the first items in the printout.

10.5.2.1. Advantages of R^2

R^2 has two major advantages as an index of accomplishment. First, as a direct counterpart of r^2 in simple regression, R^2 is easily understood as representing the model's proportionate reduction in basic group variance. Second, when a particular variable is added or deleted for an existing model, the previous and subsequent values of R^2 can be compared to indicate the incremental effect of that variable. If R_1^2 is the previous value of R^2, and R_2^2 is the subsequent value, the effect can be noted as a direct increment, $R_1^2 - R_2^2$, or as a proportional increment $(R_1^2 - R_2^2)/R_1^2$. The change in R^2 thus offers a good way of choosing the best candidate when one of several candidate variables is to be added or deleted for the model.

For example, suppose we want the regression model to retain variables X_1 and X_2, but to include the other candidate variables $(X_3, X_4, \ldots, X_k)$ only if they make a significant additional contribution. To decide which candidate variable makes the most cogent contribution, we could first calculate regression equations for trios of variables that contain X_1, X_2, and one of the other possibilities from X_3 to X_k. Each of those regression equations would have the form

$$\hat{Y}_i = b_0 + b_1 X_1 + b_2 X_2 + b_p X_p.$$

Each equation would have different values for the coefficients b_0, b_1, b_2, b_p, and also for R^2. The additional variable that fits best would be the one that yields the largest value of R^2.

10.5.3.2. Disadvantages of R^2

The first disadvantage of R^2 is that it always increases merely if more variables are included in the model. Because the value of S_{yy} is fixed for any set of data, and because the value of S_R will become smaller with more variables, the value of $R^2 = (S_{yy} - S_R)/S_{yy}$ will always increase. Consequently, R^2 is satisfactory for comparing models that contain the same number of variables, but not for comparing one model with another having more or fewer variables.

A second disadvantage is that R^2 contains no provision for giving credits or debits for the number of variables in the model. More variables will enlarge R^2, but will also increase the complexity of interpreting the results. Thus, if a relatively simple two-variable model does almost as good a job as a more complex model with many more variables, we currently have no way of "adjusting" R^2 to acknowledge the distinction.

A third disadvantage is that R^2 cannot be tested for stochastic significance to indicate the numerical stability of the analysis. Because the number of constituent variables will change in different models, R^2 does not have a specific sampling distribution, and is not itself amenable to stochastic tests.

For all these reasons, R^2 offers an excellent single descriptive index of accomplishment. It shows the proportionate reduction in group variance after the final model is chosen, but is often not particularly useful in the pathway that leads to choosing the model.

10.5.4. F ratio of mean variances

The algebraic structure of the F ratio was discussed in Section 7.5.2.3. To form the ratio, we first find the degrees of freedom for S_R and S_M. As noted earlier, if a model contains p independent variables, it will estimate p + 1 parameters. (The simplest explanation for the extra parameter is that it represents the b_0 intercept term in $b_0 + b_1 X_1 + \ldots + b_p X_p$). The degrees of freedom in S_R will be $N - (p + 1) = N - p - 1$. The degrees of freedom in S_M will be p, which is calculated as p + 1 for the $\hat{Y}_i$ component minus 1 for the $\overline{Y}$ component. As a general expression for p independent variables, therefore,

$$F = \frac{S_M/p}{S_R/(N - p - 1)}.$$

In the specific instance of a two-variable model,

$$F = \frac{S_M/2}{S_R/(N - 3)};$$

and for a three-variable model

$$F = \frac{S_M/3}{S_R/(N - 4)}.$$

The structure of the F ratio eliminates two of the disadvantages of R^2. Having a specific sampling distribution, the F ratio can be interpreted for stochastic significance according to the number of degrees of freedom in S_M and S_R. The P value can be determined from an appropriate table, checked at the appropriate pair of degrees of freedom; but in packaged programs, the computer immediately presents the corresponding P value for each F value. In the data of Table 10-1, where N = 6, the value of F is (24485.868/2)(1828.965/3) = 20.08, which has P = .018 at 2, 3 degrees of freedom.

The degrees of freedom used in calculating the F ratio also produce an "adjustment" for the number of variables in the model. The adjustment occurs because S_M/S_R is multiplied by the factor $(N - p - 1)/p$. Since N is constant, this factor becomes a "penalty" value that gets smaller as p increases. For example, if we do a one-variable regression for the six persons in the data of Table 10-1, the factor is $(6 - 1 - 1)/1 = 4$. In a two-variable regression, the factor is $(6 - 2 - 1)/2 = 3/2$. For three variables, it is $(6 - 3 - 1)/3 = 2/3$. Therefore, for a three-variable model to get the same (or larger) F value obtained with a two-variable model, the three variables must do a substantially better job in reducing S_R and thereby increasing $S_M = S_{yy} - S_R$.

As noted in Section 7.5.2.4, however, the F ratio has the major disadvantage of having a wide range, from 0 to infinity, which makes a descriptive interpretation difficult. Thus, F has a penalty factor and can be used stochastically, but is difficult to interpret, whereas R^2 is easy to interpret but has no penalty factor and cannot be used stochastically.

10.5.5. Adjusted R^2

The solution to the problem just cited is to preserve F for its stochastic merits, and to develop a "penalty index" for R^2. The job of a penalty index is to offer a constrained range or criterion value for interpreting the model's goodness of fit, while giving suitable credit for reduction in variance and suitable debit for the number of variables.

From the formula of $(S_{yy} - S_R)/S_{yy}$, R^2 can be expressed as

$$R^2 = 1 - \frac{S_R}{S_{yy}},$$

which shows that R^2 is formed when the S_R/S_{yy} ratio for *group* variance is subtracted from 1. If the subtracted element is changed to the ratio for *mean* variance, rather than for *group* variance, the result can be a penalty factor that accounts for the number of variables in the model. The new ratio would contain $S_R/(N - p - 1)$ in the numerator and $S_{yy}/(N - 1)$ in the denominator.

The result is an adjusted value of R, which is often marked adj R^2 or $\overline{R}^2$ (pronounced R-bar-squared). The formula is

$$\text{adj } R^2 = \overline{R}^2 = 1 - \frac{S_R/(N - p - 1)}{S_{yy}/(N - 1)}, \quad [10.20]$$

which can be rewritten as

$$\overline{R}^2 = 1 - \left[\frac{S_R}{S_{yy}} \times \frac{N - 1}{(N - p - 1)}\right].$$

The $(N - 1)/(N - p - 1)$ term is the penalty factor. As p increases with more variables, the denominator gets smaller, so that the penalty increases. For $\overline{R}^2$ to enlarge when more variables are added, the value of S_R must become substantially smaller, indicating substantially better fit.

Like F, the value of $\overline{R}^2$ gives credit for the reduction in S_R, and debit for the increase in p. For the data of Table 10-1, we can calculate $\overline{R}^2$ from formula [10.20] as $1 - [(1828.965/3)/(26314.834/5)] = .884$, which is smaller than the unpenalized $R^2 = .930$.

The main advantage of $\overline{R}^2$ is that it has a standard range from 0 to 1. It can therefore be used to compare results in models with different numbers of variables, and in analyses for different sets of data.

10.5.6. Standardized regression coefficients

As noted in section 7.5.4, the relative effectiveness of an independent variable is best denoted by its standardized regression coefficient. If many noneffective variables are eliminated when different models are explored (as discussed in Chapter 11), only a few b_j coefficients remain to be appraised when the final model emerges as the apparent best candidate. The elimination process thus offers an additional mechanism for evaluating effective variables in multiple regression. The ineffective variables can be appraised and eliminated during sequential decisions about which ones to retain.

Regardless of whether the final model contains a complete or reduced set of variables, however, the regression coefficients are standardized the same way in multiple linear regression as in simple regression. Any partial regression coefficient, b_j, becomes a standardized as

$$b_{j_s} = b_j \times \frac{s_j}{s_y},$$

where s_y is the standard deviation for Y and s_j is the standard deviation for variable X_j.

For the data of Table 10-1, $s_y = 72.546$, $s_1 = 14.091$, and $s_2 = 14.317$. The standardized regression coefficients will be

$$b_{1_s} = (4.793)(14.091)/72.546 = .93 \text{ and } b_{2_s} = (.3598)(14.317)/72.546 = .07.$$

The relative magnitudes of these two standardized coefficients immediately confirm that variable X_1 (for age) is much more effective in these data than X_2 (for weight).

In listing the ordinary partial regression coefficients as b_j, computer programs usually also show the standard error for each coefficient, the accompanying value of t in the test for stochastic significance. and the corresponding P value—but may not list the standardized coefficient. If not offered, its value must be calculated by using s_j and s_y, which can be found in the printout of univariate data for each variable.

A different approach, not involving a full calculation of b_{j_s}, can be used if you recall that the b_{j_s} coefficients are standardized to allow comparison of their relative magnitudes. Since each value of $b_j s_j$ within the same model becomes standardized when divided by the same value of s_y, only the values of $b_j s_j$ need be calculated and compared. Things are much easier, of course, if the computer printout is well behaved and shows the standardized coefficients directly.

Some analysts may ignore the standardized partial regression coefficient, preferring to assess the impact of each variable according to the corresponding magnitude of t, or the inverse magnitude of the accompanying P value. This assessment will indicate the stability of the coefficient for a probabilistic decision about *stochastic* significance, but does *not* correctly denote the *quantitative* effectiveness noted from either $b_j s_j$ or the standardized b_{j_s}.

Because of this distinction, P values can often give misleading ideas about the importance of a variable. A P value low enough to be stochastically significant will indicate only that the corresponding b_j coefficient is numerically stable. After the P value confirms the stability, however, the interpretation of the variable's quantitative effectiveness depends on the value of b_j, not on the value of P. The P value is important only if it is too high. If P exceeds the boundary for α, an apparently important b_j coefficient cannot be taken seriously because it has *not* been demonstrated stochastically to differ from 0.

10.6. Stability of indexes

In tests of stability in multiple linear regression, the F index is used for the stochastic significance of either F or R^2. The result is interpreted with p and $N - p - 1$ degrees of freedom. The t test is used for the partial regression coefficients, b_j. The value of t is calculated as b_j/(standard error of b_j) and the result is interpreted at $N - p - 1$ degrees of freedom. Because multivariable analysis is seldom done with small groups, $N - p - 1$ is almost always large enough for the t ratio to be interpreted as though it were a Z test.

Stochastic significance is declared when P is less than an established boundary value, called α. Although customarily set at .05, the choice of α is completely arbitrary. It can be raised or lowered according to the justifying features of the analysis. The α boundary is sometimes made more lenient, i.e., higher than .05, if the data set is small or if certain sequential processing activities, discussed later, will continue or stop according to P values associated with the F test. For the data of Table 10-1, the P values are .01 for coefficient b_1, which refers to age, and .70 for b_2, referring to weight.

Arguments can be offered[1,2] that the α levels in multivariable analyses should receive stricter, i.e., lower, boundaries to account for the multiple comparisons that occur during the activities. Since a substantially stricter value of α might thwart the progress of the analysis (particularly during sequential selection of variables), most data analysts tend to ignore the multiple-comparison problem. A compromise solution is to let the α values be lenient enough for suitably processing all the data; but afterward, when the analysis is finished, the α levels can be "toughened" for interpreting the effect of the selected variables. A demand for stricter values of α seems particularly appropriate, however, if the apparently effective variables were not anticipated in advance, do not have cogent biologic plausibility, and may have emerged mainly as an artefact of statistical analysis.

10.7. Stability of 'events' and independent variables

Regardless of the values obtained for stochastic significance in the F ratio and in the b_j coefficients, a separate problem can occur for stability of the independent variables.

This issue is particularly important later in logistic and Cox regression, where the outcome consists of binary events that are present or absent. If e represents the smaller number of the two possible events, and if p represents the number of independent variables in the model, a "rough rule of thumb"[3] demands e/p be ≥ 20 for the results of b_j coefficients to be trustworthy. In other words, the multiple regression should preferably contain at least 20 outcome events per independent variable. Concato, Peduzzi, et al.[4] have recently shown that this boundary can often be lowered to e/p $\geq$ 10 for Cox regression.

With a dimensional target variable, each of the available outcomes can be regarded as an event. Consequently, in multiple linear regression, this ratio should be N/p $\geq$ 20, where N is the effective sample size for the p included variables. In the data of Table 10-1, where N = 6, this ratio has the too-small value of 6/2 = 3; and so the accuracy of the b_1 and b_2 values is uncertain, despite the apparent stochastic accolade for b_1.

10.8. Collinearity and interactions

The multiple independent variables also create many opportunities for collinearity and interactions. These phenomena can be checked within any particular algebraic model, but some of the examinations can occur while different models are being constructed and explored. Accordingly (and to keep this chapter from getting too long) the examination of collinearity and interactions will be discussed in Chapter 11.

Chapter References

1. Kleinbaum, 1988; 2. Glantz, 1990; 3. Harrell, 1985; 4. Concato, 1993.

11 Multivariable explorations and reductions

Outline . . . .

11.1. Indexes of accomplishment for multivariable models
    11.1.1. Symbols and nomenclature
    11.1.2. Residual sums of squares
    11.1.3. Sums of squares for models
    11.1.4. R^2 and adjusted R^2
    11.1.5. Mean group variances

11.2. Full regression model
    11.2.1. Printout for SAS full-regression model
    11.2.2. Interpretation of coefficients
    11.2.3. Interpretation of results
    11.2.4. BMDP printout for full regression

11.3. Incremental explorations
    11.3.1. Sequential examinations
    11.3.2. Incremental sums of squares
    11.3.3. 'Unique' or 'distinctive' variance
    11.3.4. Partial F values
    11.3.5. Incremental changes in R^2
    11.3.6. Partial correlation coefficients
    11.3.7. Penalty indexes
        11.3.7.1. Mallow's C_p
        11.3.7.2. Other credit-debit indexes
    11.3.8. Sequential stepped procedures
        11.3.8.1. Forward (step-up) regression
        11.3.8.2. Backward (step-down) regression
        11.3.8.3. Stepwise regression

11.4. Tests for collinearity
    11.4.1. Previous illustrations
    11.4.2. Algebraic problems produced by collinearity
    11.4.3. Implications of collinearity
    11.4.4. Bivariate events in collinearity
        11.4.4.1. Correlation matrix
        11.4.4.2. Weighted coefficients
        11.4.4.3. Example of weightings
        11.4.4.4. Problems of maldistributed variables
        11.4.4.5. Hazards in sequential eliminations
    11.4.5. Multivariate assessment of 'tolerance'

11.5. Stepwise regression in the illustrative data set
    11.5.1. SAS program
    11.5.2. BMDP program

11.6. Forced regression
    11.6.1. 'Closed' approach
    11.6.2. 'Open' approach

11.7. All possible regressions
   11.7.1. Criteria for "best" regression
   11.7.2. SAS program
   11.7.3. BMDP program
11.8. Inclusion of nominal variables
   11.8.1. Dummy binary variables
   11.8.2. Compound nominal group
11.9. Importance of operational specifications
   11.9.1. Entry-removal criteria
   11.9.2. Tolerance criteria
   11.9.3. Precision of calculations
11.10. Choice of exploratory procedures
11.11. Additional evaluations

. . . .

The fewer the variables entered into a targeted multivariable model, the more understandable will be the results. Consequently, an important analytic goal is to find and eliminate the ineffective variables.

A variable may be ineffective for several reasons: It may have no impact on the target; it may be essentially redundant, offering the same information as some other variable, such as the "synonyms" of *hematocrit* and *hemoglobin* in routine measurement of red blood status; or it may masquerade as effective because of high correlation with a truly effective variable, as shown earlier (Table 3-3) for the association of *age* and *anatomic extensiveness*.

To check for effectiveness, all the available candidate variables can be analyzed simultaneously, but additional separate strategies are often used to check each independent variable alone, and to explore different combinations of variables. For example, suppose age, sex, systolic blood pressure, and serum cholesterol level are available as independent variables for predicting the subsequent development of cardiovascular disease in healthy persons. We can examine a model that includes all four independent variables, but we could also check the predictions made by trios of the four variables, by pairs, or by each variable alone. If the model performs just as well with three variables rather than four, with two variables rather than three, or with one particular pair rather than some other pair, we would prefer the model with the fewest variables that gave the best of the acceptably good results.

11.1. Indexes of accomplishment for multivariable models

Additional indexes of accomplishment are needed to compare different multivariable models, particularly for explorations in which an existing model is sequentially augmented or depleted by a single variable. One type of index would compare accomplishment for two models having the same number of different variables, such as $\{X_1, X_3, X_7\}$ vs. $\{X_2, X_4, X_7\}$. Another index would compare models with different numbers of variables, such as a three-variable model $\{X_1, X_3, X_7\}$ vs. a five-variable model $\{X_2, X_3, X_5, X_6, X_7\}$. A third index could be used to check the incremental changes that occur when candidate variables are systematically added or removed sequentially, one step at a time.

Although applicable for all these purposes, the conventional indexes of accomplishment, R^2 and F, are regularly supplemented with additional indexes for the number of variables included in a model, for incremental changes from one model to another, and for credit-debit penalties. This section is concerned with indexing accomplishments for the number of variables in a model. The other additional indexes will be discussed later in the chapter when we examine changes in candidate models.

11.1.1. Symbols and nomenclature

In simple linear regression, an S symbol was used for two kinds of sums of squares. In referring to variables, S_{xx} and S_{xy} indicated a particular variable's basic group variance or covariance. In referring to models, S_R and S_M indicated the model's accomplishments in fitting the data.

To avoid ambiguity in denoting the diverse sums of incremental squares, the S symbol will hereafter be reserved for relationships among variables. With this convention, S_{23} would be the group covariance for independent variables 2 and 3; S_{55} is the basic group variance for variable 5; and S_{y4} is the targeted covariance between Y and variable 4.

A D symbol will be used hereafter for discrepancies in the fit of a model. In linear regression, these discrepancies are calculated as sums of squared deviations. In logistic and Cox regression, as discussed later, the discrepancies can be calculated as likelihood values. Because the discrepancies have analogous roles in all three multivariable methods, a change to the D symbols will allow the same ideas to be shown with similar symbols. With a subscript j, which can range from 0 to k, D_j will denote the number of variables included in the model.

11.1.2. Residual sums of squares

With the foregoing conventions, D_0 in linear regression will represent the basic "residual" sum of squares with no independent variables in the model, which is $G_0 = \bar{Y}$. Thus,
$$D_0 = S_{yy} = \Sigma(Y_i - G_0)^2 = \Sigma(Y_i - \bar{Y})^2.$$
When independent variables are added, the residual sums of squares previously called S_R will have new symbols: $D_1, D_2, \ldots$. With one variable in the model,
$$D_1 = \Sigma(Y_i - G_1)^2 = \Sigma[Y_i - (b_0 + b_1X_1)]^2.$$
With p variables in the model,
$$D_p = \Sigma(Y_i - G_p)^2 = \Sigma[Y_i - (b_0 + b_1X_1 + b_2X_2 + \ldots + b_pX_p)]^2.$$
If a single variable, X_j, is removed from a model that contains p variables, the residual sum of squares becomes
$$D_{p-1} = \Sigma(Y_i - G_{p-1})^2 = \Sigma[Y_i - (b_0 + b_1X_1 + b_2X_2 + \ldots + b_{p-1}X_{p-1})]^2.$$
Note that the D_{p-1} symbol indicates merely that one variable has been removed. The identity of the removed variable is not shown in the symbol.

11.1.3. Sums of squares for models

With D_0 representing the basic group variance, and D_p representing the residual group variance of a model containing p variables, the symbol D_M will represent the model group variance, calculated as $\sum(\hat{Y}_i - \overline{Y})^2$. Note that D_M does not indicate the number of variables in the model.

In simple regression, these sums of squares were related as $S_M = S_{yy} - S_R$ or $D_M = D_0 - D_1$. In multiple regression, the counterpart expression is

$$D_M = D_0 - D_p.$$

11.1.4. R² and adjusted R²

With the new symbols just introduced, a model's accomplishment in proportionately reducing (or "explaining") basic variance is

$$R^2 = \frac{D_0 - D_p}{D_0}. \qquad [11.1]$$

If R^2 is known, the magnitude of D_p can be determined from formula [11.1] as

$$D_p = (1 - R^2)D_0. \qquad [11.2]$$

Formula [10.20] for adjusted R^2 becomes

$$\text{adj } R^2 = \overline{R}^2 = 1 - \frac{D_p/(N - p - 1)}{D_0/(N - 1)}. \qquad [11.3]$$

Note that p here indicates the number of independent variables in the model, not the number of estimated parameters (which is p + 1 when the b_0 intercept is included). When desired, p can be used as a subscript for R^2 or adj R^2. Thus, R_1^2 has one variable in the model, R_2^2 has two variables, and R_p^2 has p variables.

11.1.5. Mean group variances

The mean of the residual sum of squares is called *mean square error* in SAS printouts and *mean square residual* in BMDP printouts. It can be symbolized as M_p, and is calculated as

$$M_p = D_p/(N - p - 1).$$

Since $M_0 = D_0/(N - 1)$, adj R_p^2 can be expressed as $1 - (M_p/M_0)$. The smaller the value of M_p, the better the fit of the model. The square root of M_p is the standard error of the estimated values of $\hat{Y}_i$; it is called "Root MSE" in SAS printouts and "std. error of est." in BMDP printouts.

11.2. Full regression model

The most obvious and simple approach in multivariable analysis is to do a *full regression*. It includes all of the available k independent variables in the model $b_0 + b_1X_1 + b_2X_2 + \ldots + b_kX_k$. When the analytic process is completed, the "ineffective" variables will be those whose partial regression coefficients have very small standardized values or are not stochastically significant.

11.2.1. Printout for SAS full-regression model

The main regression printouts in both the SAS and BMDP programs begin with an "analysis-of-variance" format, which shows labels for the group variances together with their degrees of freedom (DF), sums of squares values, mean square values, F ratio, and P value for the F ratio. (The SAS printouts often show eight decimal places, but only two or three will be used here since they are all that is needed in most civilized discourse.)

Analysis of Variance

Source	DF	Sum of Squares	Mean Square	F Value	Prob>F
Model	7	188.06349	26.86621	11.799	0.0001
Error	192	437.17620	2.27696		
C Total	199	625.23970			

Root MSE	1.50896	R-square	0.3008
Dep Mean	1.48291	Adj R-sq	0.2753
C.V.	101.75663		

Parameter Estimates

Variable	DF	Parameter Estimate	Standard Error	T for H0: Parameter=0	Prob > |T|	Standardized Estimate	Tolerance
INTERCEP	1	2.631136	1.25503585	2.096	0.0374	0.00000000	
TNMSTAGE	1	-0.310586	0.07342761	-4.230	0.0001	-0.28588842	0.79718936
SXSTAGE	1	-0.463129	0.11783868	-3.930	0.0001	-0.27804295	0.72763293
HCT	1	0.055472	0.02210709	2.509	0.0129	0.16208729	0.87276182
AGE	1	-0.008797	0.01149127	-0.766	0.4449	-0.04705755	0.96387217
MALE	1	-0.525225	0.37675152	-1.394	0.1649	-0.08501179	0.97933520
PCTWTLOS	1	-0.022599	0.01471251	-1.536	0.1262	-0.10208455	0.82449521
PROGIN	1	-0.001095	0.00591959	-0.185	0.8534	-0.01127624	0.98061482

Fig. 11-1 Full regression of ln survival on seven independent variables in illustrative data set. (SAS PROC REG program.) The two columns for Standardized estimate and tolerance are requested options.

Fig. 11-1 shows an excerpt of the SAS printout produced when the illustrative data set was exposed to a full regression procedure using seven independent variables (without the cell-type variable). With ln survival as the target variable, the row marked MODEL shows the DF as k = 7 for seven independent variables plus an intercept, for which D_M = 188.06, with a "mean square" of $D_M/7$ = 26.87. The row marked ERROR shows the corresponding values for the residual group variance, which is D_k, i.e., D_7, for the seven independent variables in this full model. Since D_k is 437.18 and N − k − 1 = 200 − 7 − 1 = 192, the mean square in that row is M_k = 437.17/192 = 2.28. The row marked C TOTAL shows 199 for N − 1 and 625.24 for D_0. The F value is 11.799 for $(D_M/7)(D_k/192)$. The P value of F, which is marked PROB > F, is 0.0001.

The square root of the mean square error of the estimate, marked ROOT MSE, is calculated as $\sqrt{2.28}$ = 1.51. The value of R^2 (marked R-SQUARE) is 0.30 and $\overline{R}^2$ (marked ADJ R-SQ) is 0.28. Adjusted R^2 here could be calculated, using formula [11.3], as 1 − [(437.18/192)/(625.24/199)] = 1 − .72 = .28. The term DEP MEAN refers to the mean of the dependent variable, which in this instance is *ln survival*. It is cited here as 1.48, which is the same value noted earlier in the univariate results of Fig. 6-2. The C.V. (coefficient of variation) is usually calculated as 100 times the standard deviation of the data divided by the mean. In this instance, 100 ROOT

MSE/DEP MEAN = $(100)(1.51)/1.48 = 102.02$. (The value in the printout, 101.76, was calculated with less rounding.)

The rest of Fig. 11-1 shows the name of each X_j variable, its 1 degree of freedom, its b_j coefficient (marked PARAMETER ESTIMATE), and the associated standard error. The value of each t test, marked T for H0: PARAMETER = 0, is calculated as [(coefficient)/(standard error)]. Thus, for SXSTAGE, the t value of -3.93 equals $-0.463/0.118$. With the null hypothesis assumption that each $ß_j = 0$, the t values are interpreted as though they were Gaussian Z values, and the associated P values are labelled PROB > |T|. Only three variables in the list have P values below .05.

The last two columns on the right appear as requested options in the SAS PROC REG program. The "standardized estimate" column shows the b_{s_j} standardized value for each coefficient. [To demonstrate the method of calculation, recall from Fig. 6-2 that s_y was 1.773 for ln survival and s_x was 1.63 for TNM stage. With b_j for TNM stage being $-.3106$, the formula $b_{s_j} = b_j(s_x/s_y)$ will produce $(-.3106)(1.632/1.773) = -.2859$ as the "standardized estimate" for the TNM stage coefficient.] The last column, marked TOLERANCE, represents a special check for multivariate correlation. It is discussed in Section 11.4.5.

From the information in the "parameter estimate" column, we could now write the full-regression equation (using only two or three decimal points) as

ln survival time = 2.63 − .31(TNM stage) − .46(symptom stage) + .055(hematocrit) − .009(age) − .53(male sex) − .02(percentage weight loss) − .001(progression interval).

Each negative coefficient would denote a reduction in survival as the corresponding variable increases. In examining the relative effect of the different variables, we can note that MALE has the largest b_j coefficient and HCT (hematocrit) has one of the smallest. We could not conclude, however, that *male* is the most effective variable. First, it is not stochastically significant (P = .16). Second, the effects of variables are best compared when the coefficients are standardized, as discussed in the next section.

11.2.2. Interpretation of coefficients

In the far right of the printout in Fig. 11-1, the highest standardized coefficients are $-.286$, $-.278$, and .162, respectively, for TNM stage, symptom stage, and hematocrit. These three variables, therefore, have the largest effects; and their negative and positive signs indicate their corresponding impacts on duration of survival. Of the remaining standardized values, which are all below |.15|, the highest is -0.10 for percent weight loss.

Before writing the final regression equation or reaching any conclusions about effects of variables, however, we first examine the stability of the coefficients by checking their t and corresponding P values. The associated P values show that only three of the coefficients—for TNM stage, symptom stage, and hematocrit— were "numerically stable," with P values below .05.

11.2.3. Interpretation of results

The three most effective variables in the full regression model are biologically plausible as well as stochastically significant. From previous research[1,2] and from the bivariate evaluations in Chapter 9, as well as from general clinical experience, the two most important prognostic predictors would have been expected to be (as they were) *TNM stage* and *symptom stage*. From previous experience, we also would have expected *hematocrit*, as a reflection of general severity of disease, to be important—and so it was.

In previous research,[2] percent weight loss had been an important predictor for the binary event of survival at six months, but had not been checked for effect on overall survival *duration*. Progression interval had previously been found important[1] for patients with pulmonic symptoms only. (The patients who had remained in the pulmonic symptom stage despite long durations of those symptoms had better survival rates—presumably due to slow-growing tumors—than patients with short durations.) The *progression interval* variable had not previously been checked, however, in the concomitant presence of all the other variables; and the special effect in the pulmonic symptom group was a type of possible interaction that had not been algebraically tested.

Accordingly, with the potential interaction temporarily ignored, a model that was biologically plausible and statistically effective could be obtained with only three independent variables.

11.2.4. BMDP printout for full regression

Fig. 11-2 shows the way that this same set of data was handled by the program called BMDP1R—linear regression by groups. The program can contrast linear regressions across subgroups of the data set, but the statement on line three indicates that this option was not invoked here.

```
DEPENDENT VARIABLE.  . . . . . . . . . . . . . . .      1 LNSURV
TOLERANCE  . . . . . . . . . . . . . . . . .   0.0100
ALL DATA CONSIDERED AS A SINGLE GROUP

MULTIPLE R              0.5484           STD. ERROR OF EST.        1.5089
MULTIPLE R-SQUARE       0.3008

ANALYSIS OF VARIANCE
              SUM OF SQUARES      DF      MEAN SQUARE     F RATIO    P(TAIL)
 REGRESSION       188.0620         7         26.8660       11.800     0.0000
 RESIDUAL         437.1597       192          2.2769

                              STD.      STD. REG
 VARIABLE      COEFFICIENT     ERROR      COEFF       T     P(2 TAIL)  TOLERANCE

 INTERCEPT        2.63108
 AGE       2    -0.0088       0.0115     -0.05     -0.77      0.44      0.9639
 MALE      3    -0.5252       0.3767     -0.09     -1.39      0.16      0.9793
 TNMSTAGE  4    -0.3106       0.0734     -0.29     -4.23      0.00      0.7972
 SXSTAGE   5    -0.4631       0.1178     -0.28     -3.93      0.00      0.7276
 PCTWTLOS  6    -0.0226       0.0147     -0.10     -1.54      0.13      0.8245
 HCT       7     0.0555       0.0221      0.16      2.51      0.01      0.8728
 PROGIN    8    -0.0011       0.0059     -0.01     -0.19      0.85      0.9806
```

Fig. 11-2 Full regression of ln survival on seven independent variables in illustrative data set. (BMDP1R program. For futher details, see text.)

The BMDP printout offers a relatively useless value for MULTIPLE R as the square root of the useful R², which is labelled MULTIPLE R-SQUARE. The entry marked STD. ERROR OF EST. (1.51) is the square root of the residual mean square (2.28). The sums of squares are cited in a two-line table labelled ANALYSIS OF VARIANCE, which does not show D_0 for the original basic (or "total") sum of squares. The BMDP labels use REGRESSION, where SAS uses MODEL, for D_M; and RESIDUAL, instead of the SAS term ERROR, for D_p.

The rest of the BMDP printout shows the same things as the SAS printout; and the b_j coefficients have values identical to those found with SAS, including the associated P values, which are labelled P (2 TAIL). In the BMDP program, the values of *std. reg coeff* and *tolerance* are produced as part of the no frills output, but they must be requested as options in the corresponding SAS procedure. The TOLERANCE results, identical to those of the SAS program, will be explained in Section 11.4.5.

11.3. Incremental explorations

Although a full regression is the most direct way to analyze the data, other models may be better because they contain fewer variables, while producing (almost) equal statistical accomplishments. These models can be sought in several ways, of which the most popular is a stepped sequential procedure whereby individual variables are added or removed in a systematic manner, one at a time.

This section discusses the strategy and concepts used for those sequential explorations; and Section 11.4 describes the tests that are simultaneously conducted for collinearity. The application of these ideas in the illustrative data set will be shown in Section 11.5.

11.3.1. Sequential examinations

At each step, the incremental process goes in a particular direction, beginning with a model containing p variables. When an individual variable, X_j, is added or removed, the new model will contain either p + 1 or p - 1 variables. Before variables are added, the incremental effect is explored for each of the remaining X_j candidate variables that are not yet contained in the model. When variables are deleted, the X_j to be removed is explored among each of the candidates that have already been included.

The names *forward* and *backward*, or *step-up* and *step-down*, are used for these two directions of exploration. The initial value of p is either 0 with a step-up direction that begins with no variables in the model, or k with a step-down direction that begins with all k variables. To illustrate an intermediate phase of the process, suppose seven independent variables are available, designated as $X_1, X_2, \ldots, X_7$. If a model already contains variables X_1, X_2, and X_3, the candidate variables for addition in a forward direction would be X_4, X_5, X_6, and X_7. In a backward direction, the candidates considered for deletion would be X_1, X_2, and X_3.

To make decisions about changes from one step to the next, a new set of incremental indexes is needed.

11.3.2. Incremental sums of squares

When variable X_j is removed from a model containing p variables, the incremental difference between D_{p-1} and D_p is

$$I_{p-1} = D_{p-1} - D_p.$$

This same result is obtained, and the same symbols are used, if variable X_j is *added* to a model containing $p - 1$ variables. For example, suppose a model contains two variables, X_1 and X_2, and we consider adding a third variable, X_3. The value of D_2 is $\Sigma[Y_i - G_2]^2$. The value of D_3 is $\Sigma[Y_i - G_3]^2$. The incremental index will be $I_{3-2} = D_2 - D_3$, which, in words, is: (Residual sum of squares for the two-variable model) − (Residual sum of squares for the three-variable model).

Because D_p gets smaller as more variables are added, the D values seem to be subtracted in a reverse direction of their subscripts. Intuitively, the expected subtraction would be $D_p - D_{p-1}$, but the other direction ($D_{p-1} - D_p$) is used because D_p has smaller values.

The process and symbols can be illustrated with data in Section 10.2.3 from the information in Table 10-1. With no independent variables in the model, $D_0 = S_{yy} = 26314.834$. With only one of the two independent variables in the model, formula [11.2] can be used to find each value of D_1 from D_0. Since R_1^2 is the same as r^2, the calculation will be $D_1 = (1 - r^2)D_0$. For X_1, $r_{y1} = .963$ and so $D_1 = (1 - .963^2)(26314.834) = 1911.27$. For X_2, $r_{y2} = .485$ and its $D_1 = 20124.93$. In Section 10.5.1, with both X_1 and X_2 in the model, the value of $S_R = D_2$ was 1828.97.

Table 11-1 Sums of squares and symbols and data of Table 10-1

Model	Sum of squares	Symbol or designation
Y alone	26314.83	D_0
Y regressed on X_1 alone	1911.27	D_1 (for X_1 alone)
Y regressed on X_2 alone	20124.93	D_1 (for X_2 alone)
Y regressed on X_1 and X_2	1828.97	D_2 (for X_1 and X_2)
$D_1 - D_2$ for removal of X_2 from two-variable model	82.30	I_{2-1} (for incremental effect of X_2, added to X_1 in model)
$D_1 - D_2$ for removal of X_1 from two-variable model	18295.96	I_{2-1} (for incremental effect of X_1, added to X_2 in model)

Table 11-1 indicates the results and corresponding symbols. Note that when both variables are in the model, D_2 represents D_1 for X_1 alone minus the I_{2-1} increment for X_1. The I_{2-1} incremental sum of squares for X_1 (age) is much much larger than the corresponding sum of squares for X_2 (weight), thus indicating that age has a much greater impact than weight in the total model.

11.3.3. 'Unique' or 'distinctive' variance

Whenever a single variable, X_j, is removed from or added to *any* model, the incremental difference is called a *Type I sum of squares*. If the model is the "final" G_p for the data set, the incremental difference when X_j is appropriately added or removed is called a *Type II sum of squares*. In any final model, the Type II sum of squares for each X_j will represent the distinctive contribution of X_j beyond whatever other variables are in the model. For this reason, the Type II sum of squares for variable X_j is sometimes called its "unique" or "distinctive" group variance in the final model. In Table 11-1, the unique group variance is 18295.96 for Variable X_1 and 82.30 for X_2.

11.3.4. Partial F values

Because an F ratio is constructed as a mean residual variance divided by a mean model variance, "partial F" ratios can be prepared for increments in residual variance.

The numerator of each ratio is the value for I_{p-1}. This value is itself a mean, since the incremental sum of squares has only one degree of freedom when a single variable is added or removed. The denominator of the partial F ratio is the residual mean variance of the larger model, which is D_p, divided by $N - p - 1$ (for the degrees of freedom). Thus, if X_j is added to a model containing $p - 1$ variables, the partial F value will be

$$F_{p-1} = \frac{D_{p-1} - D_p}{(D_p)/(N - p - 1)} \quad . \tag{11.4}$$

With the additional symbols introduced previously, the formula for the partial F ratio can also be written as

$$F_{p-1} = I_{p-1}/M_p.$$

The result is called a *partial F* ratio because it refers to the increment of an intermediate model. Interpreted at the appropriate degrees of freedom, partial F values can be converted to stochastic P values.

As a small example of the principle, suppose X_1 is the sole variable in the regression model for the data of Tables 10-1 and 11-1. At this point in the analysis, D_1 is 1911.27. If X_2 is considered as a second variable to be added to the model, the value of D_2 would become 1828.97. With $D_2 = 1828.97$, we can find the partial F value for X_2 by substituting in formula [11.4] to get

$$F_{2-1} = \frac{1911.27 - 1828.97}{(1828.97)/(6 - 2 - 1)} = \frac{82.30}{609.657} = .125.$$

At 1, 3 degrees of freedom, this F value is not stochastically significant. Consequently, variable X_2 would not be added to the model if we insist that the incremental contribution be "significant."

The partial F values have crucial roles in choosing among candidate variables to be added or deleted for a particular model. Before a variable is added, a partial F_{p-1} is first calculated for each of the candidates. The selected candidate is the one with largest F_{p-1} value that also meets whatever criterion has been set for stochastic significance. To delete a variable, the process is reversed. The variable chosen for removal has the *smallest* F_{p-1} that is also *not* stochastically significant. Criteria called *F-to-enter* and *F-to-delete*, or appropriate levels of P, are established for the desired stochastic boundaries during these decisions.

11.3.5. Incremental changes in R^2

In multiple linear regression, the multiple correlation coefficient for a model containing p variables is

$$R_p^2 = \frac{D_0 - D_p}{D_0}.$$

For the data in Table 11-1, $R_2^2 = (26314.83 - 1828.97)/26314.83 = .930$.

With D_0 as the denominator of the calculations, an incremental R^2 coefficient can be calculated for each successive variable's distinctive contribution as

$$R_{p-1}^2 = \frac{D_{p-1} - D_p}{D_0} = \frac{I_{p-1}}{D_0}.$$

Because the same D_0 is used in the denominator, the incremental changes in R^2 from one model to the next will be proportionately identical to the sequential D_{p-1} increments. Thus

$$I_{p-1} = D_{p-1} - D_p = (R_{p-1}^2)D_0.$$

In Table 11-1, after X_1 is added to the model, $R^2 = (26314.83 - 1911.27)/26314.83 = .927$. If X_2 were added next, the incremental increase in R^2 would be $(1911.27 - 1828.97)/26314.83 = .003$.

11.3.6. Partial correlation coefficients

The incremental indexes can also be used to form partial correlation coefficients for the contributions of individual variables.

In simple linear regression, the squared partial correlation coefficient, when a single new variable was added, was obtained as

$$r^2 = \frac{D_{\text{"old model"}} - D_{\text{"new model"}}}{D_{\text{"old model"}}}.$$

In multiple regression, when a new variable X_j is added to a model containing $p - 1$ variables, the analogous squared partial correlation coefficient will be

$$r_p^2 = \frac{D_{p-1} - D_p}{D_{p-1}} = \frac{I_{p-1}}{D_{p-1}}. \qquad [11.5]$$

The result indicates the proportion of "explainable" variance—i.e., the amount contained in the residual after $p - 1$ variables—that was explained when the next variable was entered. For example, when X_2 is added after X_1 has been included in the model for the data of Table 11-1, the squared partial correlation coefficient for X_2 is $(1911.27 - 1828.97)/1911.27 = .04306$. (The square root of this value is .2075.)

For choosing or deleting variables in sequential regression explorations, the partial correlation coefficient does not offer much help beyond the partial F ratio, since both calculations have $I_{p-1} = D_{p-1} - D_p$ in their numerator. The partial correlation coefficient, however, can be a useful approximation of the standardized partial regression coefficient if the latter is not immediately available. You may recall, from Chapter 10, that $b_s^2 = r^2 = S_{xy}^2/(S_{xx}S_{yy}) = (D_0 - D_1)/D_1$. Since $b =$

S_{xy}/S_{xx}, and $b_s = bs_x/x_y$, some further algebra will show that in multiple linear regression,

$$b_{j_s} = \frac{(\text{Partial correlation for } X_j)(\text{Root mean square for } D_{p-1})}{(\text{Root mean square for } D_0)}. \quad [11.6]$$

Alternatively,

$$b_{j_s} = (\text{Partial correlation for } X_j)\sqrt{M_{p-1}/M_0}.$$

Since the root mean square for D_{p-1} will be smaller than the root mean square for D_0, the standardized partial regression coefficient for b_j will be slightly smaller than the partial correlation coefficient. (The calculation is illustrated in Section 11.5.5.) This concept, although not especially needed in linear regression, will later be valuable in logistic regression.

Because division by D_{p-1} converts incremental sums of squares to a "standard" expression that does not depend on scales of measurement, the partial correlation coefficients at any step in the procedure have a valuable role in denoting the incremental impact of individual variables. Although partial *correlation* coefficients are not standardized in the same way as standardized partial *regression* coefficients, the values of the two sets of coefficients are reasonably similar. As noted in the illustrations later, the partial correlation coefficient for variable X_j, before it is entered into the model, is quite close to the standardized regression coefficient after X_j is in the model. In simple linear regression, of course, there is only one partial correlation coefficient. It is calculated, according to the same formula cited here, as $r^2 = (D_0 - D_1)/D_0$; and the value of r is the standardized regression coefficient, b_s.

11.3.7. Penalty indexes

Although adjusted R^2 can be used to compare results in models with different numbers of variables, other penalty indexes are sometimes preferred mathematically for this purpose, and particularly for comparing different models with the same number of variables.

11.3.7.1. Mallows' C_p

A frequently used penalty index proposed by Mallows[3] is written in text as C_p and in computer printouts as C(P).

Mallows set up a standardized ratio between the D_p observed in a particular model, and the mean value of residual variance (M_k) in the "best estimate," D_k, that is provided by the full model containing all of the available k variables. The formula for this standardized ratio is

$$\frac{D_p \text{ in observed model}}{(M_k \text{ in full model})}. \quad [11.7]$$

Using inferential reasoning not further discussed here, Mallows determined that the standardized variance for the estimated true parametric value of Y_i would be $N - 2(p + 1)$. He then calculated C_p as the difference between the observed standardized ratio and the estimated standardized variance. The formula is

$$C_p = \frac{D_p \text{ in observed model}}{(M_k \text{ in full model})} - [N - 2(p + 1)] . \qquad [11.8]$$

If the mean value for a particular D_p is "perfect," i.e., the same as the best estimate offered by the full model, the value of $D_p/(N - p - 1)$ will equal $M_k = D_k/(n - k - 1)$. The standardized ratio in formula [11.7] would be $N - p - 1$. The value of C_p would then become

$$C_p = N - p - 1 - [N - 2(p + 1)] = p + 1.$$

Mallows chose $p + 1$ as the ideal criterion value for C_p. (If the algebraic model does *not* contain an intercept term for b_0, the criterion value will be p rather than $p + 1$.)

To illustrate the calculations, $M_k = 1828.97/(6 - 2 - 1) = 609.66$ for the two-variable model in Table 11-1. Suppose a one-variable model is to be chosen from either X_1 or X_2 in that table. With only X_1 in the model, $D_p = 1911.27$. The value of C_p will be $[1911.27/609.66] - [6 - 2(1 + 1)] = 1.13$. With only X_2 in the model, $D_p = 20124.19$, and C_p will be $[20124.19/609.66] - [6 - 2(1 + 1)] = 31.00$. The single-variable model with X_1 would therefore be preferred because the corresponding C_p is much closer to the criterion value of 2.

As just noted, the diverse D_p values produced by different combinations of variables in the model will yield values of C_p that can be above or below the criterion value of $p + 1$. To evaluate different models, a graph can be drawn showing values of $p + 1$ as abscissas and C_p values as ordinates. The line that shows the optimal values of C_p will appear at a 45° diagonal on this graph. An excellent reduced model is indicated whenever the point for $p + 1$ and C_p is on or very near this line. The process will be discussed further in Section 11.7.2.

11.3.7.2. Other credit-debit indexes

Many imaginative mathematical efforts have been evoked by the challenge of crediting the fit of a model while simultaneously debiting its number of constituent variables. The additional indexes, which are usually labelled eponymically according to their proponents, bear such names as Amemiya, Hocking, Judge, Sawa, and Stein. The original references are listed when the indexes appear as options in computer manuals. Credit-debit indexes proposed by Akaike and by Schwartz will be discussed when we reach logistic regression.

11.3.8. Sequential stepped procedures

When individual variables are added to (or deleted from) the model in separate steps, one variable at a time, sequential stepped regression can be conducted in one direction, in the other direction, or in a "zigzag" manner.

11.3.8.1. Forward (step-up) regression

In a *forward* or *step-up regression*, the process begins with no independent variables in the model. To find the best-fitting first variable, each of the available k candidates is checked to see what would happen if it were entered as the only variable in the model. An R^2 or F value is calculated for each of these individual regressions. The candidate that provides the best fit or the highest F value is entered in step 1 as the first variable. It is renumbered as X_1 and the model becomes

$$a_0 + a_1 X_1.$$

In the second step, X_1 is forced in the model, and the remaining $k - 1$ independent variables are each checked in separate two-variable regression equations to see what would happen if the candidate were individually entered as the second variable. The candidate with the highest incremental F value during these explorations is chosen and entered in step 2. The second variable is conceptually renumbered and the model becomes

$$c_0 + c_1 X_1 + c_2 X_2.$$

(The coefficients here are altered from a_j to c_j because they will change as different combinations of variables are included.)

In the next step, with variables X_1 and X_2 forced in the model, each of the remaining k-2 variables is examined to see its potential effects as an additional third variable. When the third variable is added in step 3, the new arrangement would become

$$d_0 + d_1 X_1 + d_2 X_2 + d_3 X_3.$$

The process would continue iterating until no additional variable passes the F-to-enter boundary, which is set at a selected level of probability, such as .05. If the corresponding P value does not satisfy the requirement, a new variable cannot be added and the analytic process will stop. The p variables that have been selected for inclusion would then be cited in the final model as

$$b_0 + b_1 X_1 + b_2 X_2 + b_3 X_3 + \ldots + b_p X_p.$$

11.3.8.2. Backward (step-down) regression

In *backward* or *step-down regression*, the sequential stepping is reversed. The model begins with a full regression, containing all k of the available variables, as

$$a_0 + a_1 X_1 + a_2 X_2 + \ldots + a_k X_k.$$

The variables are then eliminated (or "deleted") one at a time—using selected criteria for F or some other appropriate index—according to their *ineffectiveness* in altering the fit of the model. (In BMDP computer programs, the criterion is usually called "F to remove.").

For the first step of the exploration, results are checked for removal of each variable individually, with the regression model recalculated for the remaining $k - 1$ variables. The variable that had the most trivial F value or effect on fit is eliminated first, and the new model—with variables appropriately renumbered—would be

$$c_0 + c_1 X_1 + c_2 X_2 + \ldots + c_{k-1} X_{k-1}.$$

The process would then iterate, at each step eliminating the particular variable that had the least substantial effect in the preceding model. The process stops when an F value becomes impressive or when the fit would be significantly altered by eliminating any of the included variables. The p variables that remain would then form the model

$$b_0 + b_1X_1 + b_2X_2 + \ldots + b_pX_p.$$

The final models that emerge from the step-up or step-down process might be expected to be similar, but they may sometimes differ because of vagaries in the data. The step-up process is more popular than the step-down technique, because the computations seem simpler, and the results are somewhat easier to follow in the associated printout of the computer's iterations. The step-down process, however, has been preferred[4] as a "safer" approach to avoid losing important variables due to stochastic idiosyncrasies in the data, but the recommendation has been disputed.[5] According to Cohen,[6] another reason for preferring the step-down direction is that the step-up process may lead to misinterpretation of "significance" for the decomposed dummy codes of categorical variables.

11.3.8.3. Stepwise regression

Although often used for either step-up or step-down forms of operation, the name *stepwise* actually refers to a zigzag process, using both directions of stepping. The activity usually goes in a step-up direction, but the possibility of backward elimination is checked at each incremental step. For example, suppose the sequential process has reached a model with three variables, expressed as

$$a_0 + a_1X_2 + a_2X_2 + a_3X_3.$$

When a fourth variable X_4 is added, this model becomes

$$c_0 + c_1X_1 + c_2X_2 + c_3X_3 + c_4X_4.$$

In the new model, however, the variable X_2 may no longer have a significant effect. At this point in a *stepwise* procedure, the variable X_2 would be eliminated. The model would be reformulated for three variables as

$$d_0 + d_1X_1 + d_2X_3 + d_3X_4.$$

The process would then continue to iterate, determining whether any other variable can be added to improve the fit, and then redetermining whether any previous variables should be deleted after that new variable is added.

The *stepwise* method thus sequentially adds variables at each step, while testing for deletion of previously added variables. The stepwise process is now probably the most popular form of the sequential options.

11.4. Tests for collinearity

A substantial correlation among independent variables is called *collinearity*. It can produce problems in any type of regression analysis, and its possibility receives special tests.

11.4.1. Previous illustrations

A problem produced by collinearity was shown in Section 3.2.3.2 and Table 3-3. When examined alone in that set of data, *age* seemed to have an important effect on *survival*. When *age* and *anatomic extensiveness* were examined simultaneously, however, the effect of age vanished. Its originally deceptive impact arose because *age* was highly correlated with a really effective variable, *anatomic extensiveness*. A similar problem was shown in Section 3.4.3.3 and Table 3-9. The apparently different effects of the two similar treatments arose from a correlation: the susceptibility bias produced when "good prognosis" patients were preferentially assigned to one of the compared treatments.

In tabular arrangements such as Table 3-3, the high collinearity of the *age* variable was manifested by its failure to show a gradient in survival when *anatomic extensiveness* was examined simultaneously. In a successful multivariable analysis (such as Table 3-3), the collinear ineffectiveness of *age* would be recognized and it could be eliminated from inclusion as an important variable. The simple, clear results of Table 3-3 are not always easily evident with multivariate algebraic models, however, because the models do not show the results directly in categorical tables.

11.4.2. Algebraic problems produced by collinearity

Collinear independent variables can create two types of statistical problems in algebraic procedures. The main difficulty is that the correlation between variables can alter their combined effect on the target, leading to distorted or misleading results if both independent variables are retained in the regression model.

A second, equally important problem is that extreme instances of collinearity may make the regression coefficients difficult or impossible to calculate. For example, in formulas [10.15] and [10.16], the denominator for b_1 and b_2 contains the term $(1 - r_{12}^2)$. If X_1 and X_2 are very highly correlated, r_{12}^2 is close to 1, and $1 - r_{12}^2$ is close to zero. Since division by zero is unacceptable as well as incalculable, b_1 and b_2 cannot be determined. In computer printouts, this problem is responsible for remarks that sometimes appear about "singularity" of matrices. Because a singular matrix has zero as its determinant, the computer program regularly checks for singularity to avoid being asked to divide by something close to zero.

11.4.3. Implications of collinearity

The existence of a high correlation between two variables does not necessarily imply that one of them should be eliminated. Each may still be effective. For example, consider variable A with categories a_1 and a_2, and variable B, with categories b_1 and b_2. The bivariate distribution of frequency counts for these two variables is as follows:

	Variable A		
Variable B	a_1	a_2	Total
b_1	80	20	100
b_2	20	80	100
Total	100	100	200

Because $\phi^2 = [(80 \times 80) - (20 \times 20)]^2 / (100 \times 100 \times 100 \times 100) = .36$, these two variables have a relatively high correlation: 0.6.

Despite the high correlation, however, the two variables may have substantially separate individual effects on the target variable, as shown by the "double gradient" in Table 11-2. In Table 11-3, however, variable A has no effect within the context of variable B. Thus, a correlation between two variables does not necessarily imply that one of them has no effect on the target variable. Despite the correlation, both variables may be individually effective (or ineffective).

Table 11-2 Two highly correlated variables that nevertheless exert individual effects on success rates

Variable B	Variable A		Total
	a_1	a_2	
b_1	80/80 (100%)	10/20 (50%)	90/100 (90%)
b_2	10/20 (50%)	0/80 (0%)	10/100 (10%)
Total	90/100 (90%)	10/100 (10%)	100/200 (50%)

Table 11-3 Two highly correlated variables with same denominator distributions as in Table 11-2, but Variable A has no effect on success rates

Variable B	Variable A		Total
	a_1	a_2	
b_1	48/80 (60%)	12/20 (60%)	60/100 (60%)
b_2	6/20 (30%)	24/80 (30%)	30/100 (30%)
Total	54/100 (54%)	36/100 (36%)	90/200 (45%)

11.4.4. Bivariate events in collinearity

In multivariable statistical operations, the intervariable correlation coefficients do not account for higher-order collinearity among multiple variables. The bivariate coefficients can be checked, however, for possible warnings; and certain effects of bivariate collinearity are often "adjusted" with suitable weightings when the partial regression coefficients are calculated.

11.4.4.1. Correlation matrix

High coefficients in the bivariate correlation matrix will indicate the particular pairs of variables that can immediately be regarded with suspicion, but the bivariate correlations are not an infallible diagnostic guide to problems, for reasons just discussed. The bivariate correlations are a useful "screening test" for problems of collinearity, but the actual diagnoses (and therapy) of those problems require further procedures.

11.4.4.2. Weighted coefficients

If everything works well in a full-regression process, the standardized partial regression coefficients should be relatively high for the effective member of a pair of collinear variables, and low for the ineffective member. The mechanism for this process can be noted from the arrangement of numerators in formulas [10.15] and [10.16], when b_1 and b_2 were calculated for independent variables X_1 and X_2 in Table 10-1. The coefficients for both variables are affected by the intervariable correlation of r_{12}, but the rest of the effect will depend on the relative magnitudes of r_{y2} and r_{y1}. If $r_{y1} > r_{y2}$, the value of b_1 will be larger than the value of b_2, and vice versa. The effect will increase with higher values of r_{12} and with greater disparities in the relative magnitudes of r_{y1} and r_{y2}.

11.4.4.3. Example of weightings

Although multiple linear regression is not strictly appropriate for analyzing the binary target of Table 3-3, the calculations will help illustrate the foregoing distinction. The results will also show that the regression procedure, even when possibly abused by being aimed at a binary target, is robust enough to work quite well.

To be submitted to regression analysis, the data of Table 3-3 must be put into an appropriate form, which is shown here in Table 11-4. Instead of listing 200 rows of data to cite the three variables for each person, Table 11-4 contains a shortcut that shows how data can be arranged for the numbers of people who had each of the eight possible combinations of data for X_1, X_2, and Y.

Table 11-4 Coding of data previously shown in Table 3-3*

X_1 = age	Anatomic X_2 = extensiveness	Y = survival	Number of patients
0	0	0	14
0	0	1	46
1	0	0	7
1	0	1	23
0	1	0	34
0	1	1	6
1	1	0	57
1	1	1	13

*For age: young = 0; old = 1
For anatomic extensiveness: localized = 0; spread = 1
For survival at 5 years: dead = 0; alive = 1.

From the earlier Table 3-3, we already know the univariate distributions of X_1, X_2, and Y. The *bivariate* relationships of Y with each of the other two variables alone would be expressed in the following linear regression equations: For X_1 alone, $\hat{Y}_i = .52 - .16X_1$, with $r^2 = .03$ (P = .02). For X_2 alone, $\hat{Y}_i = .77 - .59X_2$, with $r^2 = .35$ (P = .0001). For the correlation of X_1 and X_2, r = .30 (P < .0001).

All of this information simply confirms what was already discerned in Table 3-3. The foregoing linear regression equation for Y vs. X_1 shows a value of .52 when $X_1 = 0$, and .36 when $X_1 = 1$. This result is identical to the marginal totals for X_1 in Table 3-3. Similarly, the linear regression for Y vs. X_2 produces results almost identical to the marginal totals for X_2 in Table 3-3. The correlation between X_1 and X_2 was apparent (and discussed in Section 3.2.3.2) when the denominators of cells in

Table 3-3 were distributed in the fourfold pattern of $\{^{60}_{30}\ ^{40}_{70}\}$. The correlation of X_1 and X_2 can be calculated from the latter table, for which $X^2 = 18.18$, as $\phi^2 = 18.18/200 = .0909$, and $\phi = .30$. Perhaps the most striking aspect of these results is their demonstration that the linear regression principle, despite the use of binary data in the target variable and in each independent variable, produced highly accurate models in fitting the data.

In this example, however, the main thing to be examined is the multivariable analysis when X_1 and X_2 are entered in the model simultaneously. When carried out, the process produces

$$\hat{Y}_i = .76 + .02X_1 - .60X_2.$$

Because the respective P values are .74 and .0001 for the b_1 and b_2 coefficients, the X_1 variable could be eliminated. The model would then be recalculated, with only X_2 included, to produce the earlier equation, $\hat{Y}_i = .77 - .59X_2$.

The result demonstrates how the analytic process, during its own internal operations, will recognize important bivariate collinearities and deal with them appropriately. As shown probably best in formula [10.15], the X_1 variable for *age* has a chance to produce its solo impact on Y by virtue of the S_{y1} component of the numerator. To yield a substantial partial regression coefficient, however, the impact of S_{y1} must substantially exceed the effect contributed by r_{12} (S_{y2}/S_{22}), which reflects the covariance transmitted through variable X_2. In this instance, the covariance effect overwhelmed the solo effect, and b_1 turned out to have the almost negligible value of .02 when $r_{12}(S_{y2}/S_{22})$ was subtracted from S_{y1}.

The explanation becomes more complex when more variables are involved, with results that are not as easy to show as in Table 3-3. If you understand what happens to the data in Table 3-3 (or Table 11-4), however, the same basic principle pertains in more complex situations.

11.4.4.4. Problems of maldistributed variables

If collinearities are to be recognized and managed only during the internal operations of the regression process, a major problem can arise if each of the two variables actually has a substantial individual effect, but if one of the variables is maldistributed. Because of the problem in distribution, a relatively high coefficient for b_{j_s} may not be stochastically significant. To dismiss the variable as ineffectual would be the counterpart of a Type II error in inferential statistics (where a true distinction exists, but the null hypothesis is conceded).

The main way to avoid this problem is to look for maldistributions when reviewing the univariate data for each variable. If a particular variable is maldistributed, and if it emerges with a relatively high but "nonsignificant" b_{j_s} coefficient, the risk of a Type II error should be suspected.

11.4.4.5. Hazards in sequential eliminations

If the various steps of a sequential process work successfully, the effective member of a pair of collinear variables should be retained, and the ineffective member should be eliminated.

This elimination process has two hazards. The first was noted in the preceding section: a variable may be deleted not because it is actually ineffective, but because it failed to achieve a stochistically significant b_j coefficient. The second hazard is that the two variables may have a large amount of closely "shared variance," which renders each variable ineffectual when considered separately after the other is included in the model. During the sequential processing, one member of the closely correlated pair will be chosen first in a step-up process (or second in a step-down process). The decision about which variable to choose, however, may depend more on caprices in the distribution of data than on the true effect of the variables. A variable that was eliminated as "unimportant" may sometimes be just as powerful as the variable kept as "important."

There is no good way to avoid this problem except to recognize its potential existence in a set of data, and to draw conclusions appropriately. For example, because hemoglobin and hematocrit values are so closely correlated, the candidates in a multivariable analysis should include one or the other variable, but not both. A data analyst who does not know this biologic distinction may put both variables into the analysis, note their high bivariate correlation, and then be pleased that the unimportant member of the pair was eliminated during the analysis. The subsequent conclusion that hematocrit is important but hemoglobin is not (or vice versa) would be biologically silly, although possibly helpful for purely *statistical* decisions about which variable to use in future analyses.

11.4.5. Multivariate assessment of 'tolerance'

No matter how well the bivariate adjustments may work, they cannot account for the higher-order collinearity that can occur between a single variable and a combination of two or more other variables. Consequently, in sequential procedures, a separate search is done to check for high correlation between an additional candidate variable and those that have already been entered into a model. This additional search is called a check for "tolerance." It is regularly used to detect multivariable collinearity during sequential operations, but can also be applied in a full regression model.

The process is easiest to illustrate with forward regression. Before being added to a model, a new variable is first screened for collinearity with the *combination* of all other variables already entered in the regression model. For this procedure, the candidate variable X_j is regressed not against Y, but against all the other variables $X_1, X_2, X_3, \ldots$ that currently constitute the model.

Suppose a model already includes three variables, arranged as $a_0 + a_1X_1 + a_2X_2 + a_3X_3$, and suppose five more variables (X_4, X_5, X_6, X_7, X_8) are available as candidates for entry as a fourth variable in the model. Before any candidacy is advanced, each of those five available variables is tested for its regression *against the variables that already exist in the model*. Thus, if X_4 is the candidate, it is regressed against $X_1, X_2,$ and X_3 as the independent variables. The equation would have the form $\hat{X}_4 = c_0 + c_1X_1 + c_2X_2 + c_3X_3$; and a corresponding value of R^2 would be calculated. Similar procedures would separately regress $X_5, X_6, X_7,$ and X_8 individually against the combination of variables $X_1, X_2,$ and X_3.

If any of the R² values is too high for each of these five "regressions within a regression," the associated variable (X_4, X_5, ..., X_8) would be regarded as too collinear with the others, and omitted from further consideration. In the usual process, the *tolerance value* is calculated as $1 - R^2$ for each of these special regressions. The tolerance should be relatively close to 1 (i.e., $R^2 \simeq 0$) for an additional variable to be accepted as noncollinear with the others.

In Figs. 11-1 and 11-2, the tolerance of each independent variable indicates the value of $1 - R^2$ when that variable is regressed against the other six *independent* variables. In the full-regression model, rather than in the stepping process, all of the tolerances are calculated as a regression of one independent variable on the others. The lowest tolerances in the full regression here are for TNMSTAGE and SXSTAGE, as might be expected from their relatively high bivariate correlation, because each of these two variables is present in the group used to calculate tolerance for the other.

Checks for tolerance are prominently displayed in computer printouts, thereby giving the procedure an apparently important status. The importance is more often mathematical than substantive, however, Mathematically, tolerance is important because of the problems previously discussed for matrix computations when variables are collinear. Substantively, however, tolerance is not particularly important if the collinear variable has important effects. Therefore, if a substantively important variable is eliminated because some "quirk" in the data made its tolerance unsatisfactory, an alternative method should be sought to "force" the variable into the regression model.

11.5. Stepwise regression in the illustrative data set

To keep the illustrations from becoming excessively long, sequential linear regression will be displayed here only for the stepwise procedures.

11.5.1. SAS program

Stepwise regression was done with the STEPWISE model selection in the SAS program PROC REG. With this mode of operation, the default levels (used here) are P = 0.15 for both entry and removal of variables. Several items in the printout were requested options. They are the values of Tolerance, Model R**2, and F in the "Statistics for Entry" before each step, and also the values of Partial R**2 and Model R**2 in the "Statistics for Removal" in each step after step 2.

The first step shown in Fig. 11-3 is a prototype for all successive steps. The SAS program begins by considering all independent variables—seven, in this instance—that are not yet in the model. The tolerance for each variable is initially 1, because no variables have yet been entered. Values of R^2 (marked MODEL R**2), F, and the corresponding P value (marked PROB > F) are calculated for what would happen if each available variable were added individually to the model.

After this inspection of possibilities, the first step occurs; and the variable with the highest stochastically significant F value is added to the model. In this instance,

```
                    Stepwise Procedure for Dependent Variable LNSURV
                           Statistics for Entry: Step 1
                                    DF = 1,198
                                        Model
                    Variable    Tolerance    R**2         F        Prob>F

                    TNMSTAGE    1.000000    0.1789     43.1462    0.0001
                    SXSTAGE     1.000000    0.1738     41.6590    0.0001
                    HCT         1.000000    0.0270      5.5017    0.0200
                    AGE         1.000000    0.0053      1.0492    0.3069
                    MALE        1.000000    0.0081      1.6098    0.2060
                    PCTWTLOS    1.000000    0.0695     14.7798    0.0002
                    PROGIN      1.000000    0.0035      0.7045    0.4023

Step 1    Variable TNMSTAGE Entered    R-square = 0.17892124    C(p) = 29.46341223

                         DF        Sum of Squares      Mean Square          F      Prob>F
          Regression      1          111.86866382      111.86866382      43.15      0.0001
          Error         198          513.37103220        2.59278299
          Total         199          625.23969602

                         Parameter        Standard          Type II
          Variable       Estimate         Error          Sum of Squares      F      Prob>F

          INTERCEP        2.96030900      0.25209633      357.52564765    137.89    0.0001
          TNMSTAGE       -0.45953299      0.06995932      111.86866382     43.15    0.0001

Bounds on condition number:      1,          1
```

Fig. 11-3 Printout through step 1 of stepwise regression for seven independent variables in illustrative data set. (Produced by STEPWISE model selection option in SAS PROC REG. For further details, see text.)

the first entered variable is TNM stage, whose F of 43.15 and P of .0001 were better than the corresponding F = 41.66 and P = .0001 of SXSTAGE, the nearest contender. With the variable TNMSTAGE added to the model, the corresponding results are first shown for R^2 (now called R-square) and Mallows C(p). The partition of variance is shown for D_M (called REGRESSION), D_1 (called ERROR), D_0 (now called TOTAL), and the associated F and P values. The PARAMETER ESTIMATE shows each b_j coefficient, followed by its standard error, the distinctive (TYPE II) sums of squares for that variable, and the corresponding F and P values. (Note that although the same PROC REG program is used, the SAS stepwise and full regression procedures use inconsistent labels for the analysis-of-variance components. In the printout of the stepwise program, the terms *regression* and *total* correspond respectively to *model* and *C Total* in the "full" program.)

The expression BOUNDS ON CONDITION NUMBER, which appears at the end of each step, refers to a check for phenomena that might lead to inaccurate computation of the regression coefficients. If the constituent variables are too collinear, the variance-covariance matrixes can have internal problems that make their "determinants" difficult to compute, producing unstable numerical estimators. The CONDITION idea refers to numerical instabilities that might occur because of subtle collinearities in variables that have successfully passed the *tolerance* test. Berk's proposal[7] of the procedure suggests that the "bound for the condition . . . be monitored instead of tolerance to assure (numerical) stability." For most pragmatic purposes, however, this examination is unnecessary, and it will not be further discussed here.

After TNMSTAGE is added as the first variable in step 1, the entire process is repeated in step 2 with TNMSTAGE becoming a forced variable. In the preliminary appraisals of step 2, shown in Fig. 11-4, SXSTAGE had a relatively low value (.802) for tolerance, because of its relatively high bivariate correlation (.44) noted earlier

```
                       Statistics for Entry: Step 2
                              DF = 1,197
                                    Model
              Variable    Tolerance   R**2        F         Prob>F

              SXSTAGE     0.802492    0.2442    17.0247     0.0001
              HCT         0.999573    0.2089     7.4701     0.0068
              AGE         0.999158    0.1826     0.8777     0.3500
              MALE        0.998274    0.1841     1.2620     0.2626
              PCTWTLOS    0.984284    0.2239    11.4305     0.0009
              PROGIN      0.994801    0.1798     0.2037     0.6523

Step 2  Variable SXSTAGE Entered   R-square = 0.24423432   C(p) = 13.52882499

                         DF      Sum of Squares    Mean Square      F        Prob>F
              Regression  2       152.70499053     76.35249526    31.83      0.0001
              Error     197       472.53470549      2.39865333
              Total     199       625.23969602

                         Parameter      Standard        Type II
              Variable   Estimate        Error      Sum of Squares     F       Prob>F

              INTERCEP   3.76010478    0.31043108    351.91434930   146.71    0.0001
              TNMSTAGE  -0.32179352    0.07511488     44.02203652    18.35    0.0001
              SXSTAGE   -0.47519241    0.11516749     40.83632671    17.02    0.0001

Bounds on condition number:    1.246119,   4.984476
```

Fig. 11-4 Printout for step 2 of SAS stepwise regression begun in Fig. 11-3.

with TNM stage. Nevertheless, among the seven candidates, symptom stage had the highest incremental F value (17.02) with P = .0001, and was entered as the second variable.

In the rest of step 2, as shown in Fig. 11-4, the results for TYPE II SS reflect the value I_{2-1} if either TNMSTAGE or SXSTAGE were removed from the model containing two independent variables. To illustrate the calculations, we know (from Figs. 11-3 and 11-4) that D_0 = 625.239 with no variables in the model. From Fig. 11-3, we know that D_1 = 513.371 when TNM stage is in the model. The associated F value when TNM stage was the only independent variable in the model was calculated as $[(D_0 - D_1)/1] \div [D_1/(N - p - 1)] = [(625.239 - 513.371)]/[513.371/198] = 111.868/2.593 = 43.15$. We do not know what the value of D_1 would have been if SXSTAGE had been added alone, but we can determine it from the cited F value of 41.659 for SXSTAGE in Fig. 11-3. From formula [11.4], $F = (625.239 - D_1)/(D_1/198) = 41.659$, which becomes $625.239 - D_1 = .2104 D_1$; and so D_1 for SXSTAGE would have been 516.556. In Fig. 11-4, D_2 is 472.535 with both variables in the model. Consequently, I_{2-1} for the incremental addition of SXSTAGE is 513.371 − 472.535 = 40.836. If SXSTAGE had been entered first, the value of I_{2-1} for the incremental addition of TNM stage would be 516.556 − 472.535 = 44.021.

In the column to the right of TYPE II SS, the values of F are the partial F ratios discussed in Section 11.3.4. They are calculated by noting that at this point, with two variables in the model, D_p is D_2 = 472.535, and the value of N − p − 1 is 200 − 2 − 1 = 197. For the partial F ratio of TNMSTAGE, the value of $D_1 - D_2 = I_{2-1}$ = 44.021, and the value of $D_2/(N - p - 1) = 472.535/197 = 2.399$. Substituting in formula [11.4] produces the partial F_{2-1} for TNMSTAGE as 44.021/2.399 = 18.35. For SXSTAGE, the corresponding calculation is 40.836/2.399 = 17.02. Because the associated P values are < .05, neither variable would be removed at this point in the stepwise procedure.

Among the six remaining candidates after step 2, hematocrit had the highest F value (10.03) with P = .0018. It was entered in step 3 to produce a three-variable model shown later in Fig. 11-8. In this particular sequential regression, with a lax default mechanism that allowed continued operation as long as P was ≤ .15 for the partial F ratios, percent weight loss was entered in step 4 with an F value of 2.66 (P = .105); and male was entered in step 5 with F = 2.17 (P = .142). Fig. 11-5 shows the rest of the process for step 5 and thereafter. Another nothing could be entered

```
Step 5    Variable MALE Entered      R-square = 0.29854250    C(p) = 4.61611781

                              DF        Sum of Squares      Mean Square         F       Prob>F
          Regression           5         186.66061910       37.33212382       16.51     0.0001
          Error              194         438.57907692        2.26071689
          Total              199         625.23969602

                          Parameter         Standard            Type II
          Variable        Estimate           Error         Sum of Squares       F       Prob>F
          INTERCEP        2.01643339       0.97446543         9.68012733      4.28      0.0398
          TNMSTAGE       -0.31356714       0.07300998        41.70064848     18.45      0.0001
          SXSTAGE        -0.45392345       0.11672421        34.18924561     15.12      0.0001
          HCT             0.05724677       0.02187601        15.48145873      6.85      0.0096
          MALE           -0.55042084       0.37341486         4.91193621      2.17      0.1421
          PCTWTLOS       -0.02295654       0.01461966         5.57422541      2.47      0.1180

Bounds on condition number:      1.358135,       29.76877
-------------------------------------------------------------------------------------------------
                                       Statistics for Removal: Step 6
                                                DF = 1,194

                                                    Partial        Model
                                  Variable           R**2          R**2

                                  TNMSTAGE          0.0667        0.2318
                                  SXSTAGE           0.0547        0.2439
                                  HCT               0.0248        0.2738
                                  MALE              0.0079        0.2907
                                  PCTWTLOS          0.0089        0.2896

                                       Statistics for Entry: Step 6
                                                DF = 1,193

                                                              Model
                          Variable        Tolerance           R**2              F          Prob>F

                          AGE             0.964101           0.3007           0.5848        0.4454
                          PROGIN          0.980848           0.2987           0.0301        0.8625

All variables left in the model are significant at the 0.1500 level.
No other variable met the 0.1500 significance level for entry into the model.

          Summary of Stepwise Procedure for Dependent Variable LNSURV

       Variable         Number   Partial    Model
Step   Entered Removed    In     R**2       R**2      C(p)        F       Prob>F   Label
 1     TNMSTAGE            1     0.1789    0.1789   29.4634    43.1462    0.0001   TNM STAGE
 2     SXSTAGE             2     0.0653    0.2442   13.5288    17.0247    0.0001   SYMPTOM STAGE
 3     HCT                 3     0.0368    0.2810    5.4269    10.0289    0.0018   HEMATOCRIT
 4     PCTWTLOS            4     0.0097    0.2907    4.7734     2.6567    0.1047   PERCENT WT LOSS
 5     MALE                5     0.0079    0.2985    4.6161     2.1727    0.1421   MALE GENDER
```

Fig. 11-5 Conclusion of SAS stepwise regression begun in Figs. 11-3 and 11-4.

or removed for step 6, the procedure ended with the statement that none of the remaining variables met the 0.15 significance criterion for entry. The last few rows of the printout then display a summary of the sequential results, but do not show the coefficients of the variables, which must be sought in an earlier segment of the printout, immediately after step 5. The summary indicates the sequence in which each variable entered, and the incremental value of R^2 (labelled PARTIAL R**2) that was added by each successive variable to the previous total R^2 for the model. The other columns show Mallows' C_p for each model, and the successive values of partial F and the corresponding P value for each additional variable.

If a probability value of .05 (rather than .15) had been demanded for the F that allows entry into the model, the stepwise process would have terminated after step 3, leaving the model with the same three "significant" variables noted earlier in the full regression.

Note that the selected five variables did not enter the stepwise equations in the same order as the ranked magnitude of their bivariate correlations with *ln survival* in Fig. 9-2. The two first-entered variables did have the highest absolute magnitudes for the targeted bivariate coefficients, .423 for TNMSTAGE and .417 for SXSTAGE; and the last entered variable (MALE) had the lowest coefficient in this group: .090. According to the rank order of the bivariate targeted correlation coefficients, however, PCTWTLOSS (with r = .264) would have been expected to enter the multivariable equation before HCT (with r = .164). The impact of PCTWTLOSS in the multivariable context was reduced, however, by its having much higher intervariable correlations with both TNMSTAGE (r = .125) and SXSTAGE (r = .272) than HCT, for which the corresponding correlation coefficients were .021 and .073.

11.5.2. BMDP program

For stepwise regression on the same set of data, the BMDP2R—stepwise regression program was also allowed to use its own default values, which are much stricter than the analogous "laxity" of the SAS program. Left to its own default devices, the BMDP program demarcates 4.0 and 3.9, respectively, as the values for Minimum F to enter and Maximum F to remove. These boundaries will often be higher than needed for P values of .05. For example, at 3 and 200 degrees of freedom, F = 3.88 for $P \leq .01$, and F = 2.65 for $P \leq .05$. At 8 and 200 degrees of freedom, the corresponding values of F are 2.60 and 1.98. These default boundaries would make the BMDP sequential process stop well before the five steps of the foregoing SAS program.

Fig. 11-6 shows the printout of steps 1 and 2 in the BMDP2R stepwise regression programs. Although operated here in a no-frills fashion, i.e., with no requested extra options, BMDP prints several things not shown in the corresponding SAS program. One of them, MULTIPLE R, as the square root of R^2, has no apparent value beyond R^2; but adjusted R-square can be a useful index. PARTIAL CORR., another BMDP value that is not in the SAS printout, is the partial correlation coefficient for each potential variable that has not yet been included in the model. As noted in formula [11.5], PARTIAL CORR. is calculated as $\sqrt{[(D_{p-1}) - D_p]/D_{p-1}}$ for each variable. For example, in the printout of Step 1, the partial correlation for SXSTAGE is listed as −0.28204. In step 1, with one variable in the model, D_1 is 513.35352. In step 2, after symptom stage is added to the model, D_2 is 472.51831. The value of $\sqrt{(513.35352 - 472.51831)/513.35352}$ is .28204. Its sign is chosen to coincide with the sign of the corresponding coefficient.

For evaluating variables to enter or delete, the value of PARTIAL CORR. adds relatively little that cannot be discerned from the partial F-to-enter value, which is calculated as I_{p-1}/M_p at each step, with $M_p = D_p/(N - p - 1)$ for the variable to be entered in the next step. In this instance, where we go from step 1 to step 2, p will

```
STEP NO.            1
-----------------
VARIABLE ENTERED    4 TNMSTAGE

MULTIPLE R           0.4230
MULTIPLE R-SQUARE    0.1789
ADJUSTED R-SQUARE    0.1748

STD. ERROR OF EST.   1.6102

ANALYSIS OF VARIANCE
                SUM OF SQUARES    DF    MEAN SQUARE    F RATIO
    REGRESSION     111.86803       1      111.8680      43.15
    RESIDUAL       513.35352     198        2.592694
```

	VARIABLES IN EQUATION FOR LNSURV						VARIABLES NOT IN EQUATION			
VARIABLE	COEFFICIENT	STD. ERROR OF COEFF	STD REG COEFF	TOLERANCE	F TO REMOVE	LEVEL	VARIABLE	PARTIAL CORR.	TOLERANCE	F TO ENTER
(Y-INTERCEPT	2.96026)									
TNMSTAGE 4	-0.45954	0.0700	-0.423	1.00000	43.15	1	AGE 2	-0.06660	0.99916	0.88
							MALE 3	-0.07979	0.99827	1.26
							SXSTAGE 5	-0.28204	0.80248	17.02
							PCTWTLOS 6	-0.23418	0.98428	11.43
							HCT 7	0.19114	0.99957	7.47
							PROGIN 8	0.03213	0.99480	0.20

```
STEP NO.            2
-----------------
VARIABLE ENTERED    5 SXSTAGE

MULTIPLE R           0.4942
MULTIPLE R-SQUARE    0.2442
ADJUSTED R-SQUARE    0.2366

STD. ERROR OF EST.   1.5487

ANALYSIS OF VARIANCE
                SUM OF SQUARES    DF    MEAN SQUARE    F RATIO
    REGRESSION     152.70325       2       76.35162      31.83
    RESIDUAL       472.51831     197        2.398570
```

	VARIABLES IN EQUATION FOR LNSURV						VARIABLES NOT IN EQUATION			
VARIABLE	COEFFICIENT	STD. ERROR OF COEFF	STD REG COEFF	TOLERANCE	F TO REMOVE	LEVEL	VARIABLE	PARTIAL CORR.	TOLERANCE	F TO ENTER
(Y-INTERCEPT	3.76002)									
TNMSTAGE 4	-0.32179	0.0751	-0.296	0.80248	18.35	1	AGE 2	-0.09551	0.99150	1.80
SXSTAGE 5	-0.47519	0.1152	-0.285	0.80248	17.02	1	MALE 3	-0.08886	0.99790	1.56
							PCTWTLOS 6	-0.17794	0.92607	6.41
							HCT 7	0.22063	0.99455	10.03
							PROGIN 8	0.01025	0.98856	0.02

Fig. 11-6 Steps 1 and 2 of BMDP2R program for stepwise regression in illustrative data set.

be 2, and M_2 would be the *residual mean square* in the printout of step 2. Thus, F-to-enter for symptom stage in step 2 was calculated at the end of step 1 as $(513.35352 - 472.51831)/2.398570 = 17.02$. It can be shown algebraically that if c is the value of the partial correlation coefficient, the corresponding F can be determined as $F = (N - p - 1)c^2/(1 - c^2)$, or conversely, $c = \sqrt{F/(F + N - p - 1)}$.

Fig. 11-7 shows step 3 and the subsequent conclusion of the BMDP stepwise regression program. At the end of step 2, among the variables not yet in the model, HCT had the highest F-to-enter and partial correlation coefficient. When HCT was entered in step 3, D_3 became 449.517 and M_3, its root mean square, was 2.293. Since D_2, without HCT, was 472.518 in Fig. 11-6, the F-to-enter value for HCT was $(472.518 - 449.517)/2.293 = 10.03$; and the partial correlation coefficient was $\sqrt{(472.518 - 449.517)/472.518} = .22063$. As noted in Section 11.3.6, the partial correlation coefficient is somewhat larger than the standardized regression coefficient for HCT. To use formula [11.6], we first note that the root mean square for D_0 is $\sqrt{625.23940882/199} = 1.7725$ and that the root mean square for D_{p-1} is found in step 2 as $\sqrt{2.398570} = 1.5487$. According to formula [11.6], the standardized partial regression coefficient for HCT will be $(.22063)(1.54873)/1.7725 = .192$, which is the value that appears in step 3.

```
STEP NO.   3
---------------
VARIABLE ENTERED      7 HCT

MULTIPLE R              0.5301
MULTIPLE R-SQUARE       0.2810
ADJUSTED R-SQUARE       0.2700

STD. ERROR OF EST.      1.5144

ANALYSIS OF VARIANCE
                   SUM OF SQUARES    DF     MEAN SQUARE    F RATIO
    REGRESSION       175.70493        3      58.56830       25.54
    RESIDUAL         449.51685      196       2.293453

                 VARIABLES IN EQUATION FOR LNSURV                              VARIABLES NOT IN EQUATION
                        STD. ERROR   STD REG                 F                                 PARTIAL                      F
    VARIABLE  COEFFICIENT  OF COEFF   COEFF    TOLERANCE  TO REMOVE  LEVEL.   VARIABLE         CORR.    TOLERANCE       TO ENTER
(Y-INTERCEPT     1.11198 )
 TNMSTAGE   4   -0.31876    0.0735   -0.293    0.80235     18.83       1 .   AGE      2      -0.06983   0.97559           0.96
 SXSTAGE    5   -0.50054    0.1129   -0.301    0.79845     19.66       1 .   MALE     3      -0.10940   0.99162           2.36
 HCT        7    0.06582    0.0208    0.192    0.99455     10.03       1 .   PCTWTLOS 6      -0.11594   0.83056           2.66
                                                                         .   PROGIN   8      -0.00137   0.98584           0.00

***** F LEVELS(  4.000,  3.900) OR TOLERANCE INSUFFICIENT FOR FURTHER STEPPING


          STEPWISE REGRESSION COEFFICIENTS

    VARIABLES   0 Y-INTCPT   2 AGE    3 MALE   4 TNMSTAGE  5 SXSTAGE  6 PCTWTLOS  7 HCT     8 PROGIN
STEP
  0           1.4829*     -0.0136   -0.5548   -0.4595    -0.6945    -0.0583     0.0563     0.0058
  1           2.9603*     -0.0113   -0.4470   -0.4595*   -0.4752    -0.0473     0.0593     0.0028
  2           3.7600*     -0.0156   -0.4778   -0.3218*   -0.4752*   -0.0356     0.0658     0.0009
  3           1.1120*     -0.0112   -0.5755   -0.3188*   -0.5005*   -0.0239     0.0658*   -0.0001

*** NOTE   *** 1) REGRESSION COEFFICIENTS FOR VARIABLES IN
                 THE EQUATION ARE INDICATED BY AN ASTERISK.
               2) THE REMAINING COEFFICIENTS ARE THOSE WHICH WOULD BE
                  OBTAINED IF THAT VARIABLE WERE TO ENTER IN THE NEXT STEP.

SUMMARY TABLE

STEP         VARIABLE          MULTIPLE    CHANGE    F TO     F TO    NO.OF VAR
 NO.    ENTERED   REMOVED       R    RSQ   IN RSQ   ENTER   REMOVE   INCLUDED
  1    4 TNMSTAGE            0.4230 0.1789  0.1789   43.15                 1
  2    5 SXSTAGE             0.4942 0.2442  0.0653   17.02                 2
  3    7 HCT                 0.5301 0.2810  0.0368   10.03                 3
```

Fig. 11-7 Step 3 and Conclusion of BMDP2R Stepwise Regression program.

At about the middle row of Fig. 11-7, the printout produces five asterisks and announces that the process stopped because it could not find a sufficiently high F-to-enter, a low enough F-to-remove, or an "intolerant" tolerance. Immediately thereafter the printout offers two summary tables that have dubious value. One table, labelled STEPWISE REGRESSION COEFFICIENTS, indicates what might have happened if all of the excluded variables had been included at each step. The coefficients marked with asterisks in the row designated as step 3 are the same as those produced at the end of step 3, just before the process concluded. The second SUMMARY TABLE, at the bottom, shows mainly the successive and incremental values of R^2 at each step. Neither of these additional summaries seems to offer information that cannot be perceived more effectively in the preceding printout and at the end of step 3.

Several optional approaches, not shown here, can get the BMDP program to use alternative tactics for its stepwise operations. The iterative zigzagging can be done in a backward then forward manner, rather than in the reverse direction shown here. The stepwise work can also be done in a batch arrangement, in which all of the forward steps are taken first, so that all the variables have been entered before the backward process begins its iterative attempts at removal. For these activities, the default criteria for entry of variables are "relaxed" to less stringent values.

11.6. Forced regression

In *forced regression,* the model is constrained to include certain variables that are specified in advance. They can be chosen from substantive knowledge of the phenomena, or from results of previous analyses. For example, a forced regression can be done with the set of variables that emerged as most "effective" in the full regression.

Regardless of how the included variables are chosen, the forced regression can be "closed" or "open." If closed, the procedure is confined to only a specified set of variables, such as X_1, X_2, and X_3. If open, the algebraic model is forced to retain the selected X_1, X_2, and X_3, but additional available variables can be explored and possibly included. For the operations here, the SAS program was used for a closed-forced regression, and both the SAS and BMDP stepwise programs were used for an open-forced stepwise regression, which was assigned to include (i.e., force) the three main variables before subsequent explorations began.

11.6.1. 'Closed' approach

Fig. 11-8 shows the SAS results for forced regression of the three independent variables previously noted as important in the full regression of Fig. 11-1.

Source	DF	Sum of Squares	Mean Square	F Value	Prob>F
Model	3	175.70657	58.56886	25.536	0.0001
Error	196	449.53312	2.29354		
C Total	199	625.23970			

Root MSE	1.51444	R-square	0.2810	
Dep Mean	1.48291	Adj R-sq	0.2700	
C.V.	102.12637			

Parameter Estimates

Variable	DF	Parameter Estimate	Standard Error	T for H0: Parameter=0	Prob > |T|	Standardized Estimate	Tolerance
INTERCEP	1	1.112046	0.88957699	1.250	0.2128	0.00000000	.
TNMSTAGE	1	-0.318759	0.07345680	-4.339	0.0001	-0.29341148	0.80235503
SXSTAGE	1	-0.500547	0.11289996	-4.434	0.0001	-0.30050766	0.79845578
HCT	1	0.065822	0.02078462	3.167	0.0018	0.19232823	0.99454656

Fig. 11-8 "Closed" forced regression for three main variables of Fig. 11-1 (SAS PROG REG program for "full model" containing only those three variables. Standardized Estimate and Tolerance columns were requested options.)

The results indicate, as expected, that all three variables were stochastically significant. From the data of Fig. 11-8, the final regression equation could be written as ln (survival time)

= 1.11 − 0.32 (TNM stage) − 0.50(symptom stage) + 0.066 (hematocrit).

These coefficients are different—some in small amounts, others larger—from their corresponding values in the full regression equation of seven variables in Fig. 11-1. Despite having only three variables, however, the model produced an R^2 of 0.281 and an F value of 25.54, whereas the seven-variable model in Fig. 11-1 raised R^2 only to 0.301, and had a lower F value of 10.27 (because of the "penalty" exacted for the extra variables). The value of adjusted R^2 is only trivially lower, 0.2700, in the three-variable model than the corresponding 0.2753 for seven variables.

The respective standardized coefficients in the three-variable model are -0.29 for TNM stage, -0.30 for symptom stage, and 0.19 for hematocrit. Thus, each variable had a modestly high standardized coefficient for its impact. This impact occurred even though the three-variable model had a barely good fit, with $R^2 = .28$. (Lest you mistakenly think that the $-.29$ standardized coefficient for TNM stage is higher than the .28 achieved with the entire model, recall that .28 is a value for R^2. Its square root, which is .53, is the appropriate value to be compared with standardized coefficients such as $-.29$. Alternatively, you can square the standardized coefficients, so that $-.29$ becomes .084. You can then compare these squared values with the R^2 of .28.)

Because the R^2 values are not impressively high, we might begin to worry about the suitability of both the three-variable and the full seven-variable regression models. Despite the biologic appeal of having identified plausible variables as effective, neither model provides a really impressive fit. This difficulty does not seem due to a scarcity of variables, since the fit with seven variables was not much better than with three. The question thus arises about whether the models are being used in the best format of the variables. Should they contain interaction terms, or transformed expressions such as X^2, $\sqrt{X}$, or log X for some of the variables? To avoid making this chapter too long, the regression diagnostics will be saved for Chapter 12.

11.6.2. 'Open' approach

When asked to force the three main variables and then do an open stepwise regression with remaining candidates, the BMDP2R program produced exactly the same results as those shown in Figs. 11-6 and 11-7. The three main variables were entered in these consecutive steps and the procedure could then not continue further. In an additional BMDP option, not shown here, the three forced variables can be entered as a single batch before the stepping begins.

With the entry criteria set at .05, the SAS forced-then-open stepwise program entered the three forced variables immediately in step 0, and indicated that nothing further could be done. The final printout was identical to what appeared in Fig. 11-8.

11.7. All possible regressions

In an era of inexpensive digital computation, a special program can produce all possible regressions for all possible combinations of variables. The data analyst can then choose whatever regression seems best. For k available variables, 2^k combinations will be explored, with each of the k variables being either in or out of each combination. (In one combination, none of the variables is included; and $\hat{Y}_i$ is estimated merely as $\overline{Y}$.)

For example, suppose four independent variables are available. Regression equations and the associated coefficients can be determined for models in which the variables are included one at a time (as X_1 alone; X_2 alone; . . .), two at a time (X_1 and X_2; and X_1 and X_3; X_1 and X_4; X_2 and X_3; . . .), three at a time (X_1, X_2, and X_3; X_1, X_2, and X_4; . . .) and all four at once. Each of the 15 possibilities (4 singles, 6 pairs, 4 triplets, and 1 quadruplet) can be examined with appropriate criteria to determine which combination does the best job.

11.7.1. Criteria for 'best' regression

At least four indexes can be examined to decide which regression is "best." Since R^2 should always increase with more variables, its most effective use is to compare models with the same number of variables. For comparing models with different numbers of variables, adj R^2 is a preferable index.

The value of D_p for residual group variance (or "error") should always decrease with more variables, but its mean value, which is $D_p/(N - p - 1)$, may increase as $N - p - 1$ becomes smaller with increasing values of p. Thus, the smallest mean square error might be an effective index of the best model.

Many statisticians, however, prefer to use Mallows' C_p index as a guide to the best model. The best C_p should be the one that most closely approximates the criterion value of p + 1, as discussed in Section 11.3.7.1.

Thus, as fit improves, R^2 and adj R^2 should increase, D_p and its mean value should decrease, and C_p should approach the criterion value of p + 1. If all these indexes have concordant ranks, the choice of the best-fitting model is easy. If the ranks do not agree (e.g., one model is best for adj R^2, another is best for mean square error, and a third is best for C_p), the analyst can make a "dealer's choice," which might be affected by substantive biologic issues.

11.7.2. SAS program

The best fit among all possible combinations of the available variables involves a particularly complex search. For the seven independent variables of the illustrative data set, 127 (= $2^7 - 1$) combinations had to be checked.

The appropriate SAS computer program produced these 127 combinations in a printout that occupied three consecutive pages of single-spaced type. The combinations were reported in groups containing one, two three, . . . , six, or all seven variables. Within each group, the combinations were arranged according to descending values of R^2. With optional specification, the combinations can be arranged according to values of adj R^2 or C_p.

According to the R^2 and adj R^2 values, whose ranks were always in concordance here, TNM stage is the best solo variable; it is joined by symptom stage in the best pair; they are joined (as expected from the previous results) by hematocrit in the best trio; percent weight loss is added for the best quartet; male joins those four in the best quintet; age comes in for the sextet; and progression interval enters at the last (septet) step. This pattern of augmenting rather than substantially changing the contents of the successive best combinations is not unexpected. In general, as each "good" variable is successively identified, it tends to remain in all subsequent good combinations.

Table 11-5 Indexes of accomplishment for models with highest R^2 in all possible regressions shown in Fig. 11-7.

$p =$ no. of variables	R^2	ADJ R^2	C(P)	MSE	Number of alternative combinations with C(P) closer to $p + 1$ in complete printout
1	.18	.17	29.5	2.59	0
2	.24	.24	13.5	2.40	0
3	.28	.270	5.4	2.29	0
4	.29	.276	4.8	2.27	1
5	.299	.280	4.6	2.26	4
6	.3007	.279	6.0	2.27	1
7	.3008	.275	8.0	2.28	—

Table 11-5 shows the relationship of the indexes of accomplishment for the regression model having the highest R^2 in each of the successive series of combinations in Fig. 11-10. Although the indexes generally agree within the same series, concordance does not extend across series. R^2 has its highest and C_p its best values, as expected, in the model that contains all seven variables. Adj R^2 has its highest value (.280) for five variables, and MSE also has its lowest value (2.26) for five variables. After combinations of three variables have been tested, the value of C_p becomes relatively dissociated from the values of R^2 and adj R^2. At four or more variables, the complete printout showed alternative combinations of variables having C_p closer to $p + 1$ than the value listed in Table 11-5. The number of such combinations is shown in the last column of Table 11-5.

Table 11-5 illustrates the difficulty of trying to use just one of these indexes as a sole criterion of accomplishment, particularly when the indexes have discordant ranks in the results. For example, the best C_p cited in Table 11-5 occurs with four variables, where the value of 4.8 is close to the "perfect" value of 5. For three variables, the C_p of 5.4 is farther from the optimal value of 4; but the values for R^2 and adj R^2 in the three-variable model are quite close to the higher values produced by much larger numbers of variables; and the mean square error is at the same level—about 2.3—found in all the best combinations containing ≥ 3 variables. If we ignore C_p, an argument might be offered in favor of using the three-variable model. The argument might even prefer the two-variable model, because the simplicity of two variables might compensate for the lowered values in indexes of accomplishment.

11.7.3. BMDP program

For all possible regressions, the BMDP9R printout groups the results in arrays containing 1, 2, 3, 4, . . . , 7 variables, but the coefficients are shown (together with the associated t-statistics) only for good models. At the end of the printout, the BMDP program chooses a best subset according to the analyst's specified criterion using R^2, adj R^2, or C_p. In this instance, the program was allowed to use C_p, which is the default criterion prepared for the program's own built-in specifications. The result is an interesting contradiction. Contrary to the recommendation of Mallows, as discussed in Section 11.3.7.1, the BMDP program chooses the best subset according to the lowest value of C_p, regardless of the number of variables in the model. In Fig. 11-9, which shows the end of the BMDP printout, a five-variable model was erroneously chosen as the best subset because it had the *smallest* value

```
STATISTICS FOR 'BEST' SUBSET
-------------------------------
MALLOWS' CP                            4.62
SQUARED MULTIPLE CORRELATION        0.29854
MULTIPLE CORRELATION                0.54639
ADJUSTED SQUARED MULT. CORR.        0.28046
RESIDUAL MEAN SQUARE               2.260716
STANDARD ERROR OF EST.             1.503568
F-STATISTIC                           16.51
NUMERATOR DEGREES OF FREEDOM              5
DENOMINATOR DEGREES OF FREEDOM          194
SIGNIFICANCE (TAIL PROB.)            0.0000

*** NOTE    *** THE ABOVE F-STATISTIC AND ASSOCIATED SIGNIFICANCE
                TEND TO BE LIBERAL WHENEVER A SUBSET OF VARIABLES IS
                SELECTED BY THE CP OR ADJUSTED R-SQUARED CRITERIA.

-----------------------------------------
                                                                      CONTRI-
     VARIABLE      REGRESSION    STANDARD   STAND.     T-    2TAIL      TOL-  BUTION
  NO.   NAME       COEFFICIENT     ERROR    COEF.     STAT.  SIG.      ERANCE TO R-SQ

        INTERCEPT    2.01643     0.974465    1.138    2.07   0.040
   3    MALE        -0.550421    0.373415   -0.089   -1.47   0.142  0.989804  0.00786
   4    TNMSTAGE    -0.313567    0.0730100  -0.289   -4.29   0.000  0.800584  0.06670
   5    SXSTAGE     -0.453923    0.116724   -0.273   -3.89   0.000  0.736304  0.05468
   6    PCTWTLOS    -0.0229566   0.0146197  -0.104   -1.57   0.118  0.829045  0.00892
   7    HCT          0.0572468   0.0218760   0.167    2.62   0.010  0.884939  0.02476

THE CONTRIBUTION TO R-SQUARED FOR EACH VARIABLE IS THE AMOUNT
BY WHICH R-SQUARED WOULD BE REDUCED IF THAT VARIABLE WERE
REMOVED FROM THE REGRESSION EQUATION.
```

Fig. 11-9 Conclusion of BMDP printout (BMDP9R) for all possible regressions.

(4.62) for C_p in all of the examined models. The optimal value, according to Mallows, would have been 5 + 1 = 6 for this five-variable model. If C_p had been chosen for an optimal value, the preferred choice would have the top-listed four-variable model, whose C_p of 4.8 (in Table 11-5) was quite close to the ideal value of 5 for such models.

11.7.4. Problems of interpretation

To be able to show all possible regressions is a remarkable technologic feat that is marred only by the problem of interpreting the results. If the analyst's goal is merely to achieve a best *fit* for the data, it can be found somewhere in the plethora of printout—as long as the results are consistent for the different indexes of fit. If they do not all point to the same arrangement, however, an arbitrary judgment must be made on purely statistical grounds.

Rather than making the judgment according to an isolated mathematical doctrine, the analyst may prefer to use substantive knowledge—i.e., "enlightened common sense"—about the scientific connotations of the data. On the other hand, if the latter source of thought is allowed to enter the reasoning, the analyst may promptly recall that the main goal is usually to find the impact of important variables, not just to achieve a best fit for the data. With the latter goal, the analyst will rely on other regression procedures, as discussed in Section 11.10, that give better indications of important variables.

Consequently, the all-possible-regression procedure can be admired for its technologic prowess and acknowledged for its possible role as a screening mechanism if the analyst has absolutely no substantive knowledge or ideas about the data. Beyond the admiration and acknowledgment, however, the procedure can usually be ignored.

11.8. Inclusion of nominal variables

In all of the regressions considered thus far, the candidate independent variables did not include *cell type*, which is expressed with a nominal scale. The nominal variable was omitted to avoid confusion, because nominal variables can be coded with various scales (as noted in Section 5.4.3) and, regardless of the coding scale, they can be analyzed in at least two ways.

In one analytic strategy, the nominal categories are decomposed and reconstituted as a set of dummy binary variables. In a second analytic strategy, the nominal categories, maintained as a single "compound" group, are examined with a special analysis-of-variance approach that does the counterpart of a separate regression analysis within each category. The latter approach is intended to discern whether all the other independent variables have the same impact within the nominal groups.

Both of these strategies were checked with both SAS and BMDP procedures, but only the BMDP illustrations will be shown here.

11.8.1. Dummy binary variables

As discussed in Chapter 5, the four nominal cell-type categories were initially coded as four dummy binary variables: well-differentiated (called WELL), anaplastic (called ANAP), small cell (called SMALL), and cytology only (called CYTOL).

When these four dummy variables and the seven other independent variables were entered in a full regression that contained eleven independent variables, the BMDP1R program produced the results shown in Fig. 11-10. As expected, the regression program used its tolerance procedure to recognize promptly (and to state at the bottom of the printout) that one of the dummy variables was excluded because of its high correlation with the others. In this instance, the anaplastic category was chosen and eliminated probably because *anaplastic* had either the worst score for tolerance or the worst "singularity" when the appropriate covariance matrixes were checked. The anaplastic variable did *not* have either the smallest number of members or the highest number in the identifying digits for variables. (In the SAS program, the excluded dummy variable is usually the one with the highest identifying digits.) With the elimination of the anaplastic dummy variable, the degrees of freedom for the model dropped to 10, since estimates were made for an intercept and for the coefficients of ten independent variables.

260 Strategies of Targeted Algebraic Methods

```
DEPENDENT VARIABLE. . . . . . . . . . . . .        1 LNSURV
TOLERANCE . . . . . . . . . . . . . . . . . 0.0100
ALL DATA CONSIDERED AS A SINGLE GROUP

MULTIPLE R              0.5633          STD. ERROR OF EST.           1.5028
MULTIPLE R-SQUARE       0.3173

ANALYSIS OF VARIANCE
              SUM OF SQUARES    DF      MEAN SQUARE      F RATIO     P(TAIL)
REGRESSION         198.4050     10         19.8405         8.786      0.0000
RESIDUAL           426.8167    189          2.2583

                                   STD.     STD. REG
VARIABLE        COEFFICIENT       ERROR       COEFF        T      P(2 TAIL)  TOLERANCE

INTERCEPT           2.51510
AGE         2      -0.0123        0.0116      -0.07      -1.06      0.29       0.9304
MALE        3      -0.4178        0.3801      -0.07      -1.10      0.27       0.9541
TNMSTAGE    4      -0.3004        0.0754      -0.28      -3.98      0.00       0.7502
SXSTAGE     5      -0.4427        0.1200      -0.27      -3.69      0.00       0.6957
PCTWTLOS    6      -0.0231        0.0147      -0.10      -1.57      0.12       0.8153
HCT         7       0.0541        0.0221       0.16       2.45      0.02       0.8672
PROGIN      8      -0.0011        0.0059      -0.01      -0.18      0.85       0.9682
WELL        9       0.3036        0.2542       0.08       1.19      0.23       0.7198
SMALL      10       0.0687        0.3640       0.01       0.19      0.85       0.8070
ANAP       11                        VARIABLE NOT USED.   TOLERANCE =         -0.000008
CYTOL      12       0.7930        0.3957       0.13       2.00      0.05       0.8389

*** NOTE   *** ONE OR MORE VARIABLES WERE NOT INCLUDED BECAUSE
               THEIR INCLUSION WOULD REDUCE THE TOLERANCE OF AN
               ALREADY INCLUDED VARIABLE BELOW THE TOLERANCE LIMIT
```

Fig. 11-10 Full regression with four dummy cell-type variables in addition to seven other independent variables. (BMDP P1R program for Linear Regression by Groups. In this instance, no subgroups were demarcated.)

The results for the full ten-variable model in Fig. 11-10 show that the three previously significant variables had essentially the same quantitative and stochastic indexes as in the previous full regression in Fig. 11-2, and that R^2 rose only slightly to .3173. The WELL and SMALL dummy variables were not stochastically significant, but CYTOL was borderline, with t = 2.00 and P = .05. The other regression coefficients were changed only slightly from the values obtained in the previous full model with seven variables. A comparison of stochastically significant coefficients in Figs. 11-2 and 11-10 suggests that relatively little effect was produced by adding or omitting the cell-type variables.

11.8.2. Compound nominal group

In the second approach, the four cell-type categories were maintained as a single compound nominal group, but analyzed separately with a full regression that contained seven independent variables within each cell-type category. Of the four such regressions that were done, the top part of Fig. 11-11 shows the results obtained for the 83 patients in the well-differentiated group. The remaining three regressions (for the other three cell types) are not shown, but the bottom of Fig. 11-11 (below the solid line) shows the analysis of variance table that appeared after all four regressions were completed.

The analysis of variance calculations depend on the F ratio of means for two sums of squares. One of them is the regression sum of squares over (or across) the

```
REGRESSION FOR GROUP   1  WELL

    NUMBER OF CASES READ. . . . . . . . . . . . .       200
        CASES WITH GROUPING VALUES NOT USED. . . . .     117
            REMAINING NUMBER OF CASES . . . . . . . .     83


    DEPENDENT VARIABLE. . . . . . . . . . . . . . . .         1 LNSURV
    TOLERANCE . . . . . . . . . . . . . . . . . . .    0.0100

  MULTIPLE R              0.5761          STD. ERROR OF EST.          1.6425
  MULTIPLE R-SQUARE       0.3318

  ANALYSIS OF VARIANCE
                 SUM OF SQUARES    DF    MEAN SQUARE    F RATIO    P(TAIL)
  REGRESSION        100.4948        7      14.3564       5.321     0.0001
  RESIDUAL          202.3396       75       2.6979

                                  STD.    STD. REG
  VARIABLE        COEFFICIENT    ERROR     COEFF      T     P(2 TAIL)  TOLERANCE
  INTERCEPT           1.29926
  AGE         2       0.0017     0.0212     0.01    0.08      0.94      0.9386
  MALE        3      -0.7771     0.6716    -0.11   -1.16      0.25      0.9333
  TNMSTAGE    4      -0.5135     0.1346    -0.40   -3.81      0.00      0.7899
  SXSTAGE     5      -0.0802     0.2036    -0.04   -0.39      0.69      0.7742
  PCTWTLOS    6      -0.0423     0.0250    -0.18   -1.69      0.10      0.8087
  HCT         7       0.0761     0.0374     0.20    2.04      0.05      0.8862
  PROGIN      8      -0.0019     0.0095    -0.02   -0.21      0.84      0.9119
```

```
ANALYSIS OF VARIANCE OF REGRESSION COEFFICIENTS OVER GROUPS
      REDUCTION OF RESIDUALS DUE TO GROUPING

                         SUM OF SQUARES   DF   MEAN SQUARE   F RATIO    P(TAIL)
                         ---------------------  -----------------------------
REGRESSION OVER GROUPS      51.187        24      2.133       0.928     0.56372
RESIDUAL WITHIN GROUPS     385.972       168      2.297

A SIGNIFICANT F RATIO INDICATES THAT THE SLOPES OR INTERCEPTS DIFFER
BEYOND CHANCE BETWEEN THE GROUPS.
```

Fig. 11-11 Results for regression for seven independent variables in well-differentiated cell group, augmented by analysis-of-variance results in histologic subgroups for four dummy histologic variables. (BMDP P1R program for Linear Regression by Groups. For further details, see text.)

groups. The other is the residual sum of squares within groups. If the same p independent variables are used throughout, the degrees of freedom for the first sum of squares are determined as gp + g − p − 1 = (g − 1)(p + 1), where g is the number of groups. In this instance, the first DF = (4 − 1)(7 + 1) = 24. For the second sum of squares, DF = N − g − gp = N − g(p + 1). In this instance, the second DF = 200 − 4(8) = 168.

In the example here, the residual sum of squares was 202.3396 for the *well* group. When this value was added to the counterpart residuals for the other three histologic groups, not shown here, the total was 385.972. Its mean became 385.972/168 = 2.297. For the seven independent variables alone, as shown in Fig.

11-2, the residual sum of squares was 437.1597. Therefore, the regression sum of squares across the four groups was 437.1597 − 385.972 = 51.877. Its mean square was 51.187/24 = 2.133; and the F ratio (shown near the bottom of Fig. 11-11) was 0.928. The unimpressive F-ratio value indicates that the four regressions did not have substantially different results.

11.9. Importance of operational specifications

Although the computer programs are automated to work according to their own default values for certain operational specifications, a user should beware that these values are different in BMDP and SAS systems, and that the difference can lead to apparently disparate results. The main source of disparities is in the entry-removal criteria, the boundary of "tolerance", and the precision of calculations.

11.9.1. Entry-removal criteria

In the entry-removal criteria of sequential regression, the default P values of the SAS programs are .50 to enter in forward regression, .10 to remain in backward regression, and .15 to enter or stay in stepwise regression.

The BMDP sequential programs use F values rather than P values; and the default criterion is F = 4.0 to enter, regardless of the direction of stepping. The corresponding F value to remove (or "delete") is 3.9. Although the associated P values will change for different degrees of freedom in the number of variables, the BMDP programs use the same F criteria throughout, because of "ease in computation."

The BMDP program reports the values of the F criteria just before the beginning of the main regression printout. The SAS program shows these values near the end of the printout, when the stepping stops. The effects of the different criteria were shown earlier when the SAS stepwise program, with more "relaxed" boundaries, entered five variables in Fig. 11-5, but the BMDP counterpart program in Fig. 11-7 entered only three variables.

11.9.2. Tolerance criteria

The SAS default for tolerance is 1×10^{-7} in all linear regression programs. Although not shown in the routine printout, the value can made to appear as an option. The corresponding default for tolerance, shown just before the main printout begins, is .01 in the BMDP1R and BMDP2R (stepwise) programs, but changes to .0001 in the BMDP9R (all possible regressions) program.

The different default values did not affect any of the results shown in this chapter, but will alter some of the interactions explored in Chapter 12.

11.9.3. Precision of calculations

All SAS calculations are done with double precision, i.e., carried out to 16 decimal places. The BMDP calculations are done with single precision (i.e., about 7 decimal places) for the 1R and 2R programs, but in double precision for the 9R program. The differences in precision did not change any of the results in this chapter, but can alter the calculations of tolerance, and its subsequent effects, as shown for the interactions in Chapter. 12.

11.10. Choice of exploratory procedures

Different analysts have different preferences for methods of exploring combinations of variables. The sequential step-up procedures originally became popular because they saved time and costs of computation. With the current availability of personal computers that involve no fee for usage, however, this advantage is no longer pertinent, and may sometimes be outweighed by the disadvantage of using arbitrary algorithms in the stepping process.

The all-possible-regressions method, which allows all possibilities to be inspected, is valuable in finding a best-fitting equation for the data, particularly if the analyst has no advance idea of what variables should be important. If the main goal is to find (or confirm) effective variables, however, the extensive printout of results may require excessive amounts of time for evaluation, and the large number of possibilities may be confusing.

A particularly useful approach is a two-phase procedure. In the first phase, a full regression indicates the accomplishments of each available variable. The advantage of examining a full regression is that no variables are omitted; the substantive importance of each variable can be considered when its statistical effectiveness is appraised; and the statistical results are unaltered by any of the arbitrary algorithms of a sequential stepping procedure.

In the second phase, the variables that seem both substantively and statistically worthwhile in the full regression are entered into a forced regression, which shows their distinctive interrelationships. The forced regression in the second phase is probably best conducted in a sequentially "open" manner, retaining the selected forced variables, while giving the remaining variables a chance to be included. This two-phase process lets the analyst use substantive as well as statistical judgment in choosing a final model, and avoids the need for relying on arbitrary values of penalty indexes such as F, adj R^2, or C_p.

A quite different approach also uses full regression, but it is applied only as a screening procedure for detecting the effective variables. After being identified, they are rearranged into either a score or a clustered format arranged with decision-rule techniques. The methods are described in Chapter 12 and in Part V of the text.

11.11. Additional evaluations

Before any conclusions are reached for *any* of the exploratory models just discussed, two additional features should be evaluated: interactions and regression diagnostics. These two evaluations for the illustrative data set will be presented in Chapter 12, together with demonstrations of the use of multiple linear regression in published literature.

Chapter References

1. Feinstein, 1968; 2. Feinstein, 1990; 3. Mallows, 1973; 4. Mantel, 1970; 5. Beale, 1970; 6. Cohen, 1991; 7. Berk, 1977.

12 Evaluations and Illustrations of Multiple Linear Regression

Outline . . . .

12.1. Appraising impact of variables
   12.1.1. Concepts of 'stability'
   12.1.2. Relative magnitude of coefficients
      12.1.2.1. Stochastic values of t/Z/P ratios
      12.1.2.2. Standardized regression coefficients
      12.1.2.3. Boundaries for quantitative significance of coefficients
   12.1.3. Flexible boundaries for stochastic significance
   12.1.4. Confidence intervals for 'small' coefficients
   12.1.5. Problems in events per variable
   12.1.6. Additional appraisals

12.2. Assessment of interactions
   12.2.1. Method of citation
   12.2.2. Illustration of procedures for Table 3-5
      12.2.2.1. Additional codings
      12.2.2.2. Summary of results
   12.2.3. Choosing the interaction terms
   12.2.4. Problems of interpretation
      12.2.4.1. Magnitude of effects
      12.2.4.2. Zones of effect
      12.2.4.3. Instability of coefficients
   12.2.5. Recognizing potential interactions
   12.2.6. Alternative strategy
   12.2.7. Examples in published literature

12.3. Interactions in illustrative data set
   12.3.1. Two-way interactions
      12.3.1.1. Full regression
      12.3.1.2. Forced stepwise regression
   12.3.2. Three-way interaction
   12.3.3. 'Saturated' interaction model
   12.3.4. 'Reverse' codings
   12.3.5. Interpretation of results

12.4. Regression diagnostics
   12.4.1. Examination of simple residuals
      12.4.1.1. Display of univariate residuals
      12.4.1.2. Display of "Residual for YHAT"
      12.4.1.3. Display of residuals vs. independent variables
   12.4.2. Check for linear conformity
   12.4.3. Effect of 'outliers'
      12.4.3.1. Definition of an outlier
      12.4.3.2. Scientific concept of outliers
      12.4.3.3. Indexes of measurement for outliers

12.4.3.4. Indexes for residuals
12.4.3.5. Measuring 'leverage'
12.4.3.6. Indexes of 'influence'
12.4.4. Display of results
12.5. Therapy for 'lesions'
12.5.1. Discarding observations
12.5.2. Repairing inadequately measured variables
12.5.3. Altered formats
12.5.4. Subsequent procedures
12.6. Subsequent decisions for illustrative data set
12.7. Adjustments and scoring systems
12.8. Illustrations from medical literature
12.8.1. Goals
12.8.2. Programs
12.8.3. Operational tactics
12.8.4. Transformations
12.8.5. Fit, stability, and conformity
12.8.6. Regression diagnostics
12.8.7. Impact of variables
12.8.8. Interactions
12.8.9. Quantitative conclusions
12.8.10. Summary and recommendations
12.9. Commendations and caveats

....

As a direct expansion of the simple bivariate procedure, multiple linear regression gets its main operational complexities from the extra variables, explorations, and reductions that were considered in Chapters 10 and 11. This chapter discusses the complexities of evaluating results for impact of variables, effects of interactions, and regression diagnostics; and the chapter concludes with illustrations of published examples of multiple linear regression and with some cautionary comments.

12.1. Appraising impact of variables

The statistical penalty indexes (F, adjusted R^2, C_p, etc.) are used mainly for comparing available candidate variables, before a final model is chosen. Thereafter, however, these indexes have little or no role in evaluating its accomplishment.

If the prime analytic goal is fitting a model to the data, the main index of accomplishment is R^2, which denotes a poor fit if close to 0, and an excellent fit if close to 1. In most medical analyses, however, the main goal is to identify and quantify the effective variables. The exactness of the model's fit may therefore be relatively unimportant, as long as it has not distorted the impact of either the variables themselves, or of categorical zones within the variables. In the absence of such distortions, a model that has a relatively poor fit may sometimes still do a good job in identifying cogent variables.

Consequently, the first step in interpreting accomplishments is to check that the "important" variables and their partial regression coefficients have been identified without distortion. Because the possibility of distortion is not shown by the customary indexes of accomplishment, the results are appraised for the interactions discussed in Sections 12.2. and 12.3, and for the regression diagnostics illustrated in Section 12.4.

Assuming that no serious distortions have occurred, the impact can be appraised for each variable that has a stable coefficient. If the coefficient is too unstable, the variable itself may be important but is difficult to appraise.

12.1.1. Concepts of 'stability'

The scientific concept of "stability" implies that a particular numerical value would not be substantially altered by the loss of individual members from the data set. For example, the proportion .25 is stable if derived as 300/1200 from a set of 1200 members, but not if derived as 1/4 from a set of four.

The concept of stability is applied directly when "jackknife" procedures[1] are used in modern "computer-intensive" statistics. For the jackknife activity, as discussed in Section 8.5.4.1, each member of the set of data is successively removed, one at a time, and a regression equation is calculated from the other $N - 1$ members. The removed member is replaced and the next member is removed before each calculation. The distribution of iterated jackknife results will show the range of variation and counterpart of a "standard error" for each regression coefficient.

In conventional statistical approaches today, however, questions about stability have been converted to issues in probability, as expressed with P values and confidence intervals. Although generally quite successful, the probabilistic approach has three main problems that have been mentioned before but that are worth briefly reviewing now. The first problem is the frequently erroneous belief that a variable's importance is denoted by the minuteness of the associated P value, rather than by the quantitative magnitude of the actual impact. If large sample sizes produce small P values, variables that have relatively trivial effects may emerge as "significant." A second problem is that variables with important effects may be erroneously dismissed as "nonsignificant" because, with maldistributed categories or a relatively small sample size, the P value did not quite get below the arbitrary threshold for α. The third problem occurs when a variable that is scientifically important because it shows a small effect (or none) is fallaciously dismissed as too unstable for its nonsignificant result to be given scientific credibility. The management of these problems is discussed in the next three subsections.

12.1.1. Relative magnitude of coefficients

The relative magnitudes of partial regression coefficients can be appraised stochastically or quantitatively.

12.1.2.1. Stochastic values of t/Z/P ratios

The conventional method of getting P values for individual variables is to examine the ratio of t (or Z) formed when each regression coefficient, b_j, is divided by its associated standard error, s_{b_j}. The ratio of b_j/s_{b_j} forms a "standardized normal deviate" that can be interpreted as a t or Z value in tests of stochastic significance.

In addition to its stochastic role, this ratio, being expressed in dimension-free units that seem to remove the effects of arbitrary measurements and codings, becomes an attractive way of ranking the effects of the variables. Within a particular regression model, the sample size and total number of variables will be the same for each regression coefficient. Accordingly, the relative intra-model effect of the variables could be ranked simply by comparing the t or Z values of these ratios. The larger the value, the greater is the relative stochastic effect.

Since the corresponding P values are usually calculated and displayed for each t (or Z) ratio, this ranking can promptly be done from the P values alone. The smaller the P value, the more stochastically effective is the variable within the model. This tactic works well except when the P values become very tiny. In some computer printouts, the value of .0000 is displayed whenever the P value is below .00005. Other printouts show a • when the P value is "infinitesimal," at levels below 1×10^{-16}. If two variables each receive P values of .0000 or •, their relative effect cannot be ranked without examining the t or Z ratio directly.

Although useful for comparing the stochastic *rank* of variables within the same model and same sample size, the t/Z/P tactic does not supply an absolute *rating* for the actual impact of the variable. For example, with a very large sample size, many variables may emerge as stochastically significant, and one of them may have the highest rank in the group; but none of the variables may have a really substantial impact. The stochastic tactic also cannot be used to compare the impact of variables *across models* containing different sample sizes or different numbers of variables for the same set of data.

12.1.2.2. Standardized regression coefficients

The best way to get a standardized *rating* for the quantitative effect of an individual variable—regardless of sample size and the number of variables in the model—is to inspect the standardized partial regression coefficient, b_{j_s}, for the variable. Being free of dimensional units of measurement for the variable, this rating can be compared across models and used as a general index of an independent variable's impact on the target.

The main problem with the standardized coefficient is that it is not always automatically provided in the computer printout. The calculation will then require the nuisance of finding s_j and s_y in the univariate statistics tabulated for each variable, and then determining either $b_j \times s_j$ to compare coefficients within the same study, or $(b_j \times s_j)/s_y$ for comparisons across studies.

12.1.2.3. Boundaries for quantitative significance of coefficients

A particularly difficult choice in multivariable analysis is to demarcate a boundary for "quantitative significance" of standardized partial regression coefficients. In simple linear regression for two variables, the boundary of $b_{j_s} \geq \sqrt{.10} = .32$ was easy to set because, as discussed in Chapter 7, it represented a 10% reduction in basic variance. In multivariable regression, this boundary may be much too high. For example, in multiple linear regression of the illustrative data set here, TNM stage and symptom stage are each clearly important, yet neither variable had a b_{j_s} exceeding .32.

Furthermore, as noted earlier, variables that produce important gradients in trend may sometimes fail to get good scores for their accomplishments, reflected by b_{j_s}, in improving fit and reducing variance. Consequently, realistic boundaries for quantitative significance of b_{j_s} may often be set at much lower levels than .32, and sometimes the absolute value of b_{j_s} might be ignored. It can always be used for *relative* rankings of important variables, but may not be suitable for definitive ratings.

12.1.3. Flexible boundaries for stochastic significance

To avoid missing an important variable, the actual P value should be checked for each partial regression coefficient, without peremptorily discarding any variable whose coefficient is marked as NS (nonsignificant) when P exceeds an arbitrary α boundary such as .05. If sample size is small for the total group, or for pertinent subgroups, the P value of the coefficient may exceed .05, even though the variable is substantively important and has a relatively high standardized b_{j_s}. In such situations, the best analytic approach is usually to raise the level of α and to keep the important variable, rather than reject it because of rigid stochastic boundaries.

The main disadvantage of sequential stepping procedures is that they eliminate variables according to inflexible boundaries of α; and a major advantage of *full* regression analysis is that the P value can be checked for each variable. The inflexibility of sequential stepping can be mitigated if the P-value criteria for entry or removal are set at very lenient boundaries, such as .15 or even .5. The sequential procedure will then allow many variables to be included, and they can be ranked or rated both according to their final coefficients and their order of entry or removal. The lenience that allows a liberal entry of variables can avoid omitting some that may be quantitatively important although stochastically borderline.

12.1.4. Confidence intervals for 'small' coefficients

Although P values are used for almost all the stochastic decisions of multivariable analysis, confidence intervals are particularly important and worth examining when small values of b_j (or b_{j_s}) suggest that a variable be dismissed as nonsignificant.

If b_{j_s} is impressively large, the failure to achieve stochastic significance is obviously due to a too-small sample size. In this situation, a cautious data analyst will be reluctant to reject the associated variable, X_j, as being unimportant. If b_{j_s} is small, however, two explanations are possible. The first is that the numerical value of b_{j_s} is unstable; it may be a random stochastic variation from a truly larger value. The second explanation is that b_{j_s} indeed has a stable value, and that the associated X_j variable is important for its nonsignificant impact. For example, if serum cholesterol has a small b_{j_s} that suggests the variable has little or no effect on atherosclerosis, the unexpected "negative" result can be regarded either as a chance variation or as important evidence against the pathogenetic role attributed to cholesterol.

The choice of explanations can usually be resolved if the standard error of b_j is converted to a confidence interval as $|b_j| \pm Z_\alpha(s_{b_j})$, with Z_α selected for the desired magnitude of "confidence." Because b_j was not stochastically significant, the lower border of this interval will always include 0, and so attention is concentrated on the upper border. If it exceeds a quantitatively significant value of b_j (or b_{j_s}), the observed value of b_j can be regarded as a stochastic variation from the larger value. If the upper border does not exceed the quantitatively significant value, b_j can be regarded as truly small.

The choice of magnitude for Z_α is controversial. Most analysts usually let $Z_\alpha = 1.96$ to get a two-tailed 95% confidence interval. Other analysts argue, however, that the upper end of the 95% interval is being checked for a one-sided hypothesis, and that Z_α should be set at the one-tailed level of 1.645. If sample sizes are small, the statistics can use t rather than Z values.

12.1.5. Problems in events per variable

As noted in Section 10.7, the coefficients for the variables may be unstable[2] if the regression analyses contain too few events per independent variable included in the calculations. The coefficients may be either "overfitted," i.e., deemed important when they are not, or "underfitted," i.e., erroneously deemed as unimportant, despite whatever "confirmation" emerges from P values and confidence intervals. In a recent documentation of this problem, Concato, Peduzzi, et al.[3] showed empirically that the coefficients start to become unstable when the events per variable (EPV) ratio gets below 20, and are distinctly unstable when the ratio is below 10.

The problem is particularly troublesome when the target variable is a binary event, since the counted number of events is the *smaller* of the two numbers in the **alive/dead** or **success/failure** dichotomy. In multiple linear regression, however, each outcome of the dimensional target variable is an "event," so that the full value of N can be used in calculating the N/p ratio for p independent variables. Thus, when seven independent variables were used in multiple linear regression of the illustrative data set here, the N/p calculation was 200/7 = 28.6—a "safe" value for the EPV ratio.

12.1.6. Additional appraisals

Several other approaches can be used to assess the quantitative importance of individual variables. In a forward stepwise regression, the variables can be ranked according to their sequential order of entry into the model, but the ranking does not provide a quantitative rating. In another approach, the partial correlation coefficients shown in BMDP printouts of stepwise regression can avoid the calculation of standardized regression coefficients while offering essentially the same results.

Hauck and Miike,[4] fearing the possible distortions of a stepwise procedure when variables are correlated, suggest that results be tabulated to show the variables that were "close alternative" contenders at each step in a backward stepwise process. A close alternative is defined as "a variable whose removal causes another variable to switch from non-significant to significant."

12.2. Assessment of interactions

The conjunctive effects called *interactions* (discussed throughout Section 3.2.3.3 and illustrated in several tables of Chapter 3) are a major source of unresolved problems in algebraic forms of multivariable analysis. If the analysis is aimed mainly at fitting an algebraic equation to the data, the use of interaction terms can sometimes be a valuable method of improving fit. This potential improvement usually leads to discussions having considerable enthusiasm and verve when the statistical orientation is aimed exclusively at getting a well-fitting model. The enthusiasm for interactions may vanish, however, in a substantively realistic discussion, which recognizes that the investigator's prime goal is usually to determine the impact of individual variables. The interaction terms create major difficulties in efforts to interpret the results: what is gained in fit is lost in interpretation, and vice versa.

After various statistical accounts of how best to explore interactions, a realistically oriented discussion often concludes with the suggestion that they are perhaps best checked by examining stratified clusters rather than algebraic models. The rest of this section will provide a similar discussion and conclusion, but will also offer specific illustrations of some of the difficulties in arranging, coding, and interpreting interaction data even for relatively simple challenges.

12.2.1. Method of citation

Although routinely displayed in cluster formats such as Table 3-4, interactions are not checked in *routine* arrangements for algebraic models. To be examined in the model, an interaction must be specified for inclusion as a separate "variable." The interaction of variables X_1 and X_2 would be entered as an additional "variable," X_1X_2; and an interaction of three variables, X_3, X_7, and X_8, would be cited as $X_3X_7X_8$. Values for the interaction terms are obtained by direct multiplication of the constituent values. Thus, if a person's *age* is **27** and *systolic blood pressure* is **125,** the corresponding value for the interaction of *age* and *systolic blood pressure* would be 27 × 125 = 3375.

12.2.2. Illustration of procedures for Table 3-5

The algebraic problems and challenges of interactions can be shown with the data of Table 3-5, where an interaction in the middle cell led to a "valley" reversal of survival rates. For algebraic analysis, the three variables in Table 3-5 can be coded as follows: **1** = young, **2** = middle, and **3** = old for X_1 = age; **1** = thin, **2** = medium, and **3** = fat for X_2 = weight; and **0** = dead and **1** alive for Y = 5-year survival. When *bivariate* linear regression equations are suitably computed, the results for the 320 people in Table 3-5 are

$$\hat{Y}_i = 0.940 - 0.159X_1 \text{ for age alone, and}$$
$$\hat{Y}_i = 0.986 - 0.179X_2 \text{ for weight alone.}$$

For both independent variables, the multiple linear regression equation is

$$\hat{Y}_i = 1.206 - 0.133X_1 - 0.157X_2.$$

In the foregoing three equations, each coefficient and each "model" is significant at P < .0001. The three corresponding values of R^2 are .07, .09, and .14. Nevertheless, the coefficients for b_0, b_1, and b_2 in the last equation do not suitably reflect the actual phenomena in Table 3-5, because a major interaction occurs when $X_1 = 2$ *and* $X_2 = 2$.

If an interaction term is inserted as X_1X_2, the multivariable algebraic model becomes

$$\hat{Y}_i = 1.003 - 0.030X_1 - 0.056X_2 - 0.050X_1X_2.$$

The fit for the new model has P < .0001 and the value of R^2 rises to .15, but none of the b_j values is stochastically significant. Furthermore, the standardized values for the partial regression coefficients were previously −.23 for b_1 and −.27 for b_2 without an interaction term; in the interaction model, these values fall to −.05 for b_1 and −.10 for b_2, but the interaction term has a standardized coefficient of −.27.

Since none of the b_j coefficients was stochastically significant, the model containing an interaction term would not identify the distinctive interaction seen in Table 3-5.

12.2.2.1. Additional codings

To see whether some other system of algebraic coding might be more informative, the codes of **0, 1, 2** could be tried instead of **1, 2, 3** for variables X_1 and X_2. (The Y variable would keep a **0/1** coding.) With this approach, the regression equation would become

$$\hat{Y}_i = .867 - .080X_1 - .106X_2 - .050X_1X_2.$$

The standardized coefficients here are −.136 for X_1 (age), −.182 for X_2 (weight), and −.153 for the X_1X_2 interaction. The fit of the model would have $R^2 = .15$ with P < .0001, but the X_2 variable (at P = .03) has the only stochastically significant coefficient. A simple change in coding digits would thus have two dismaying effects: it would alter the results of the previous analysis; and it would still not clearly identify the distinctive interaction in Table 3-5.

In a different effort to achieve an informative model, the *age* and *weight* variables could be transferred to a reference-cell, dummy variable coding arrangement. For this purpose, the original variables could be arranged as follows:

Original codes for age	New dummy variables	
	X_3 = age DM-1	X_4 = age DM-2
Young	0	0
Middle	1	0
Old	0	1

Original codes for weight	New dummy variables	
	X_5 = weight DM-1	X_6 = weight DM-2
Thin	0	0
Medium	1	0
Fat	0	1

This arrangement would replace each of the previous two variables by two new dummy terms, and would produce four interaction terms: X_3X_5, X_3X_6, X_4X_5, and X_4X_6.

In a regression analysis without interaction terms, the equation for the four dummy variables is

$$\hat{Y}_i = 1.00 - .255X_3 - .276X_4 - .290X_5 - .328X_6.$$

With this equation, fit of the model is $R^2 = .17$ and $P < .0001$; and each of the partial regression coefficients has $P < .0001$. Their respective standardized values are $-.244$, $-.270$, $-.277$, and $-.324$.

In the model with the four dummy and four interaction terms, the regression equation is

$$\hat{Y}_i = .900 - .054X_3 - .159X_4 - .100X_5 - .200X_6 - .546X_3X_5 \\ -.098X_3X_6 - .054X_4X_5 - .226X_4X_6.$$

The value of R^2 is .23 with $P < .0001$ for fit, but all of the partial regression coefficients are unstable (i.e., $P > .05$) except for variable X_6 (for which $P = .056$) and for the X_3X_5 interaction (for which $P < .0001$).

The last model, although not perfect, would at least indicate that an important interaction was occurring. Furthermore, when appropriate values of the variables are substituted, the last model would also allow discernment of the valley in the center cell of Table 3-5.

12.2.2.2. Summary of results

Because linear regression is not strictly applicable to the categorical data of Table 3-5, the exploration here is intended only to illustrate the process, and should not be regarded as a conclusive test of the capacity of algebraic interaction models. (The challenge of Table 3-5 will also be explored in Chapter 14 with the more appropriate format of *logistic* regression.) Nevertheless, the current results suffice to demonstrate three main points: (1) inserting an interaction term in the model does not always promptly resolve the algebraic problem of interactions, particularly for categorical data; (2) the results produced with the interaction model may vary according to the coding system; and (3) despite the algebraic difficulties, the interaction is clearly and easily seen in the "gold-standard" stratified clusters of Table 3-5.

12.2.3. Choosing the interaction terms

In most multivariable analyses, the main challenge is to choose the structure and analytic process for the interaction terms. If a model contains six independent variables, each variable is always entered separately for its "main effect" as X_1, X_2, ..., X_6. To examine all possible interactions among those six variables would require many additional terms. They would include 15 terms ($=[6 \times 5]/2$) for the pairs of interactions X_1X_2, X_1X_3, ..., X_5X_6, and 20 ($=[6 \times 5 \times 4]/[3 \times 2]$) terms for the interaction trios $X_1X_2X_3$, $X_1X_2X_4$, ..., $X_4X_5X_6$. An additional 15 terms would be needed for the quartets, plus 6 for the quintets, and 1 for the grand sextet interaction of $X_1X_2X_3X_4X_5X_6$. Instead of a relatively simple analysis of terms for six variables, the procedure would become an analysis of 63 ($= 6 + 15 + 20 + 15 + 6 + 1$) terms.

To avoid this excessive complexity, most multivariable analyses are done *without* the initial insertion of any interaction terms, although one or two can sometimes be included if the analyst has good reason believe they are needed. Otherwise, interactions are usually contemplated after results are evaluated for the simpler, interaction-free model.

A substantive knowledge of the underlying subject matter is particularly important in anticipating the types of interactions to be considered. Lacking this knowledge, the data analyst has no idea of how to manage the potential for interactions. Most analysts agree that the main-effect terms must be retained (and therefore forced) in any model that contains those terms as components of interactions. In the absence of substantive guidance, a purely statistical approach evokes different preferences from different analysts. Some prefer an "upward" approach, examining all pairs first and then going to higher order interactions only if significant terms emerge. Other analysts prefer a "downward" approach, starting with the grand complete interaction first and then adding lower-order terms. Yet other analysts use a stepwise procedure that gives all possible regression terms a chance to enter the model after the main effects are forced. A full regression that contains the entire array of possible interactions could be routinely examined, but the process would increase the magnitude of the computations while making the results particularly awkward to interpret when b_j coefficients are compared for individual variables, pairs of interactions, triple interactions, etc. Results of these comparisons are difficult to evaluate, particularly in the "contradictory" circumstances when one of the interactions seems effectual for a variable that itself seems ineffectual.

12.2.4. Problems of interpretation

Even if the interaction itself has been suitably discerned by the algebraic model, however, the interaction terms can create confusion in interpreting the magnitude of effects for individual variables, and in determining the zones where the interactive effects are most prominent.

12.2.4.1. Magnitude of effects

Suppose an algebraic model contains X_1 = age, X_2 = blood pressure, and their interaction term, X_1X_2, with the coefficients being b_1 for age, b_2 for blood pressure, and b_3 for their interaction. If the coefficients are significant for b_3, but not for b_1 or for b_2, the individual effects of age alone and blood pressure alone might be dismissed because the principal effect of the two variables occurs exclusively in their interaction. On the other hand, if b_1 or b_2 (or both) are also significant, the result shows that all three terms are needed for the best fit of the model. The individual effect of the two variables and of their interaction, however, will be difficult to evaluate because the effect is diffused among the three terms.

The evaluation process could be relatively easy for only two independent variables and one paired interaction, but would become extremely difficult and perhaps impossible if the model contained multiple variables, with multiple pairs of interactions, as well as triplet, quadruplet, and even higher-order interactions. The model might fit the data beautifully, but the magnitude of impact for the individual variables would be obscured.

12.2.4.2. Zones of effect

A separate problem is that interaction terms do not indicate where the interaction occurs. For example, in the first and last cells of each row and column of Table 3-5, each variable seems to do "its own thing" happily until the striking reversal suddenly occurs in the middle cell of the table. In algebraic equations for these data, however, this zone is more difficult to discern. In Section 12.2.2, the "interaction equations" that were constructed in the customary manner, i.e., without dummy-variable coding, did not indicate that a significant interaction occurred. When the usual approaches are avoided and the analysis was transferred to a dummy-variable regression equation, however, a significant effect was noted for the interaction variable X_3X_5. Since this coding denotes the middle zones of each variable, the location of the interaction is clearly identified.

Consequently, for an analyst who wants to know about the gradients and zones that denote the impact of a variable, the interaction terms may sometimes act as useful screening tests, but even when successful in screening, they do not offer accurate diagnoses. If the terms raise a flag of warning, they still do not indicate whether the problem is diffuse or focal, or the zones where it is located. The zones may be identified if the customary algebraic arrangement is supplanted by a special dummy-variable approach, but an optimal arrangement is difficult to anticipate unless hints are previously available from a stratification such as Table 3-5. Consequently, a search for the actual gradients and zones of cogent interactions will almost always require a cluster analysis rather than an algebraic procedure.

12.2.4.3. Instability of coefficients

Because each interaction term adds a separate independent variable to the number of variables under analysis, the ratio of events per variable may drop precipitously, increasing the possibility that the calculated regression coefficients will have unreliable values.

For example, as noted in Section 12.1.5, the N/p ratio for events per variable was 28.6 with seven independent variables in the model for *ln survival* as the dimensional outcome in the illustrative data set. If 21 (= 7 × 6/2) pairs of interaction terms were routinely added for these seven variables, however, the N/p ratio would drop to 200/28 = 7.1, possibly threatening the stability of the results. If 6-month survival is the binary outcome, the number of events becomes 94, and the initial N/p ratio becomes 94/7 = 13.4, suggesting that the regression coefficients are borderline unstable even before any interaction terms are added. With 21 additional interaction terms, however, the E/p ratio of 94/28 = 3.3 would immediately threaten the credibility of the calculated coefficients.

In Table 3-5, with a survival rate of 199/320, the numerator of E/p would have 121 (= 320 − 199) events. The event-per-variable ratio stayed high in the regression equations that contained three terms (X_1, X_2, and X_1X_2), but was lowered to 15.1 (= 121/8) in the dummy-variable regression equation that had eight terms.

12.2.5. Recognizing potential interactions

Because interaction terms are not included in the ordinary "linear" algebraic model, the decision that they are needed requires a special strategy, which is not easy to create. If we decide to look routinely for all possible interactions, the ensuing computer printout will be bulky and cumbersome to examine and interpret. If we decide to examine only all possible pairs, we might worry about important interactions that are missed because they occur in trios, quartets, etc.

The decision is not eased by delaying it until the results of an ordinarily linear model have been examined. The model may not fit well for many reasons other than interactions: important variables may have been omitted; the individual variables may have been specified in suboptimal formats (such as X rather than X^2, log X, or $\sqrt{X}$); or the entire model may have been put in a wrong format (such as linear rather than logistic regression). These alternative sources will usually cause a "poor-fit lesion" much more often than the omission of interaction terms.

As noted later, the examination of regression diagnostics can help discover poorly specified variables or formats, but is not particularly effective for suggesting interactions. Consequently, probably the only way to avoid missing important interactions is to use the following approach before or after the routine linear results are obtained:

1. Check that all the important variables have been included in the basic analysis. This tactic is easy to suggest, but difficult to carry out. By the time the data have reached the analytic phase, the research is usually completed. If important variables have been omitted in the research, they are not available for the analysis.

2. Check for important "lurking" variables that can be contained or derived in the collected data, but that may not have been hitherto regarded as important enough to be included in the analysis. Joiner[5] has pointed out that such variables are particularly likely to arise from features of timing or spatial location in the collection of data. For example, in a long-term study, the methods of measuring certain variables may have been altered so that the calendar time of measurement may become important; in addition, the measurement process or characteristics of patients at various collaborating institutions may be sufficiently different so that institutional location becomes important.

3. Check that an appropriate algebraic method has been chosen for the analysis. This suggestion is also theoretically splendid, but pragmatically ineffectual. As noted when different algebraic methods are compared (see Chapter 22), relatively similar results often emerge for the same data set even when a particular method is mathematically suboptimal.

4. Check each individual variable to be sure it has been optimally specified. Transform it to log X, $\sqrt{X}$, or whatever other conversion seems best.

5. Steps 1–4 can sometimes remove suspicions that a poor fit is caused by poorly specified variables, by omission of important variables, or by an inadequate model format. The next step is to examine the bivariate correlation matrix, and

find the independent variables that have important bivariate effects, *in the same direction,* on the target variable. Then check the pairwise *intervariable* correlation coefficients for those variables. An interaction can be suspected if the corresponding coefficients have signs that go in a reverse direction. Thus, an important interaction can be suspected for X_1 and X_2 if they each have a distinctly positive correlation with Y, but a distinctly negative correlation with one another.

6. With or without any hints from step 5, the next step is to add specific interaction terms to the existing model. Different analysts use different tactics for this procedure. Some analysts begin with pairwise terms for all the important variables; other analysts begin with a single grand interaction term for all the important variables. Thus, if X_1, X_2, and X_3 are the important variables, we might begin either by adding a single grand interaction term $X_1X_2X_3$ or the three pairs X_1X_2, X_1X_3, and X_2X_3. The grand term has the advantage of adding fewer variables; the three paired terms add more variables but yield more precise results for localizing the interaction.

The additional interaction terms may help improve the fit of the new model, but will usually obscure the interpretation of impact for variables. To explain and understand impact, a different strategy is needed.

12.2.6. Alternative strategy

The alternative strategy is to use a targeted cluster method in which interactions are easily assessed, because the clusters are formed by combining categories that have been specifically examined for their conjoint effects.

As discussed in Part V of the text, a multicategorical, rather than multivariable, analytic format is particularly well suited for dealing with interactions. All of the operational algorithms in the stratified cluster methods are designed to find important effects by routinely examining conjunctions of multivariate categories, identifying their locations, and showing their results. The final product of the cluster procedures, however, may show impact for the categorical combinations, not for the individual original variables.

The "Catch-22" of the algebraic-model strategy, therefore, is that it is often used to avoid the apparent complexity and mathematical inelegance of targeted-cluster analyses, but the latter analyses cannot be escaped if the results are to be checked for zones of interaction and conformity of gradients. The categorical strategy, on the other hand, will show the effects of interactions but will not offer single coefficients to quantify the impact of each variable. Many enlightened analysts now avoid this dilemma by doing *both* forms of analysis. The algebraic methods are used to screen and to suggest average quantitative impacts for important variables; the targeted clusters are used to produce the definitive results.

12.2.7. Examples in published literature

Significant interactions were found in three reports that used algebraic regression models. Searching for factors influencing blood concentrations of chlordiazepoxide (such as Librium), Greenblatt et al.[6] deliberately checked "the importance of various interactions among the several independent variables, . . . which were multiplied in all possible combinations to produce unique interactive components." The "component" interaction terms were allowed to enter a stepped regression only after all the main variables were forced. Among the important interactions noted in the multivariable equations, one was a two-way age-weight interaction and another was a four-way age-sex-weight-time interaction. In a world survey of determinants of fertility, Little[7] detected a distinctive effect for the paired interactions between education and two variables: age at marriage and years since first marriage. Exploring the alleged risk of exposure to aluminum and the cognitive impairment of Alzheimer's disease, Jacqmin et al.[8] found no direct effect for the aluminum in drinking water, although calcium concentration was protective, and the interaction of aluminum and pH had a weak but distinct effect whose direction changed in different zones of pH.

12.3. Interactions in illustrative data set

To save space, interactions in the illustrative data set are reported here only for the three main variables (TNMSTAGE, SXSTAGE, and HCT) used in the forced regressions of Section 11.6.

Each of the three interaction models contained the three main variables as X_1, X_2, X_3. One model was augmented by their paired interaction terms as X_1X_2, X_1X_3, and X_2X_3; the second model included the grand interaction of $X_1X_2X_3$; and the third (or "saturated") model included the three pairs of interaction terms as well as the grand trio.

12.3.1. Two-way interactions

For *ln survival* as the dimensional target, the three main variables and the three pairs of two-way interactions were examined for a full regression and as candidates in stepwise regression where the SAS and BMDP programs were deliberately allowed to use their own default options. In the full regression, all six variables were available, without forcing. In the stepwise programs, the three main variables were forced first.

12.3.1.1. Full regression

The results of the SAS and BMDP full-regression programs for the standardized regression coefficients and their associated P values are shown in Table 12-1. In the SAS calculations, none of the single variable or interaction coefficients were stochastically significant. In the BMDP system, TNM stage was excluded for "poor tolerance," and the TNM × hematocrit interaction then became the only term with a stochastically significant coefficient.

Table 12-1 Comparison of results in full regression for pairwise (2-way) interactions analyzed with different default boundaries of SAS and BMDP programs

Variable	SAS program		BMDP program	
	b_s	P value	b_s	P value
TNM stage	.44	.47	not used	("Poor tolerance")
Symptom stage	−.93	.11	−.80	.15
Hematocrit	.24	.19	.19	.27
TNM × symptom stage (TS)	.31	.24	.34	.19
TNM × hematocrit (TH)	−.99	.12	−.55	.00
Symptom stage × hematocrit (SH)	.55	.37	.38	.50

The SAS and BMDP systems were allowed to use their own default criteria, thus demonstrating yet another problem of interactions: different results will appear according to the boundaries set for entry/removal of variables and for tolerance. As noted in Section 11.9, P = .15 is the SAS criterion for both entry and removal, whereas the BMDP criteria are F = 4.0 to enter and F = 3.9 to remove; and the default values for tolerance are .0000001 in SAS but higher in BMDP. Thus, the "tougher" demands of the BMDP default criteria kept·certain terms from being included.

12.3.1.2. Forced stepwise regression

In the forced-then-open stepwise regression, after the three main variables were forced into the model, none of the pairwise interactions entered. The final result in both SAS and BMDP programs was the same as what was found earlier (Section 11.6.1) in forced regression of the three main variables.

12.3.2. Three-way interaction

For the full-regression model containing the three main variables and a single three-way interaction term (TNM × symptom stage × hematocrit), the SAS and BMDP programs gave identical results. The b_s coefficients were $-.35$ for TNM and symptom stage and $.17$ for hematocrit; and all three were stochastically significant at $P < .05$. The three-way interaction term had a standardized coefficient of 0.10, with $P = .69$.

In stepwise regression, the three-way interaction term did not enter the model; and the result was the same as if the three main variables had been included alone.

12.3.3. 'Saturated' interaction model

A model is called "saturated" if it includes all possible interaction terms, as well as the main variables. In this instance, the model would contain seven terms: X_1, X_2, X_3, X_1X_2, X_1X_3, X_2X_3, and $X_1X_2X_3$. This model was examined in full and forced-open stepwise regressions for both the SAS and the BMDP programs.

In the full regression model with the SAS program, all seven variables were included, but none emerged with a stochastically significant coefficient. In the same model with the BMDP program, the three-way interaction and the TNM stage were excluded because of poor tolerance. For the remaining five variables that were included, only the TNM-hematocrit (TH) interaction (with $b_s = -0.55$) had $P < .05$.

When the three main variables were forced into a forced-open stepwise regression model for both SAS and BMDP, no further variables were entered.

12.3.4. 'Reverse codings'

The prognostic implications of increased values are reversed for hematocrit vs. either TNM stage or symptom stage. This reversal, as noted in Section 11.2.2, could lead to peculiar implications in the interaction terms. Consequently, two additional sets of evaluations were done for interactions. In one set, the impact of increasing hematocrit values was made consistent with that of the TNM and symptom stages by coding hematocrit as (100−hematocrit). In the other set of evaluations, the ordinal rankings of the TNM and symptom stages were reversed so that the original values of **1, 2, 3, 4, 5** were coded as **5, 4, 3, 2, 1** for TNM stage, and correspondingly **4, 3, 2, 1** for symptom stage.

These reverse codings led to some interesting minor variations in results, but no major changes or discoveries. Whenever the three main terms were forced into the model, none of the interaction terms entered.

12.3.5. Interpretation of results

The results suggest that a three-way interaction need not be considered for the three main variables in the illustrative data set. The role of a two-way interaction, however, is uncertain. It appeared to be nonsignificant when the three main variables were forced into the model; but nothing seemed significant when the SAS model included all six terms; and the TNM-hematocrit interaction was the only significant term when all six variables were candidates in the BMDP program. With these conflicting results, you can either choose whatever interaction analysis you like, or you can examine targeted clusters to see how the three variables are actually related.

12.4. Regression diagnostics

The methods of regression diagnostics consist of simple, conformal, and "outlier" examinations. The simple methods are visual checks of residual increments between the estimated $\hat{Y}_i$ and actual Y_i values. The conformal methods deal with issues in linear conformity. The outlier methods use sophisticated mathematical tactics to detect the effect of outlier values among the individual observations. To avoid excess complexity, the regression diagnostics are discussed (and shown) here for the three-main-variable model of Section 11.6.

12.4.1. Examination of simple residuals

The simple residuals can be displayed as a univariate distribution, and as plots of standardized residuals versus either the estimated $\hat{Y}_i$ values or the independent variables.

12.4.1.1. Display of univariate residuals

If everything has worked well, the univariate residual values of $Y_i - \hat{Y}_i$ should have a Gaussian distribution with a mean of 0. In Fig. 12-1, the mean is printed as zero and the stem-leaf plot looks reasonably Gaussian, but the box plot is asymmetrical. The lower quartile, at $-.85$, is farther away from the median of 0.26 than the upper quartile, at .95.

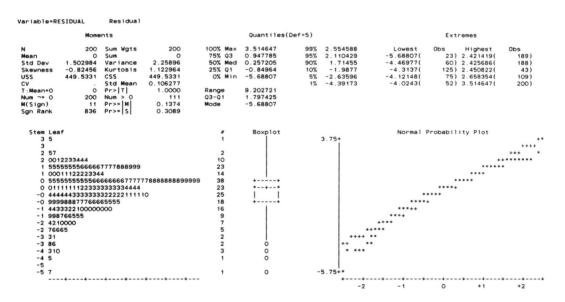

Fig. 12-1 Univariate display of residual values for $Y_i - \hat{Y}_i$, using three-main-variable regression model, with Y = ln survival. (SAS PROC UNIVARIATE program applied to residuals from PROC REG Program.)

(If you are not familiar with stem-leaf plots, the one in Fig. 12-1 looks somewhat odd. If contains two stem entries for 0 and two for −0, as well as two stem entries for each of the other integers. Examining the "leaves" on each stem, however, will make things clear. One stem for −0 contains all decimal entries from 0.0 ... to −0.4 .... The other stem for −0 contains decimals beginning with −0.5 to −0.9. Thus, the stem that shows −0 999988 ..., contains 4 entries that begin with −0.9, three beginning with −0.8, three with −0.7, and four each with −0.6 and −0.5. Other stems and leaves are organized analogously.)

The normal probability plot in Fig. 12-1 also shows nothing that is strikingly egregious, although the residual values seem too low at the extremes and slightly too high in the middle. All of this inspection shows that the fit of the model is imperfect, but nothing obvious emerges as a prime or major diagnostic "lesion."

12.4.1.2. Display of "Residual for YHAT"

Fig. 12-2 shows the standardized residuals of $Y_i - \hat{Y}_i$ plotted against the values of $\hat{Y}_i$ (often called YHAT) for each of the 200 persons. The standardized residual values are calculated as $(Y_i - \hat{Y}_i)/s_R$, where s_R is the square root, i.e., standard error, of the mean residual variance around the regression line, i.e., $s_R^2 = \Sigma (Y_i - \hat{Y}_i)^2/(N - 2)$.

The results immediately demonstrate the relatively poor fit of the model, which has a wide range of discrepancies occurring at similar values of $\hat{Y}_i$. Because an unbiased model should have residual values that are randomly scattered

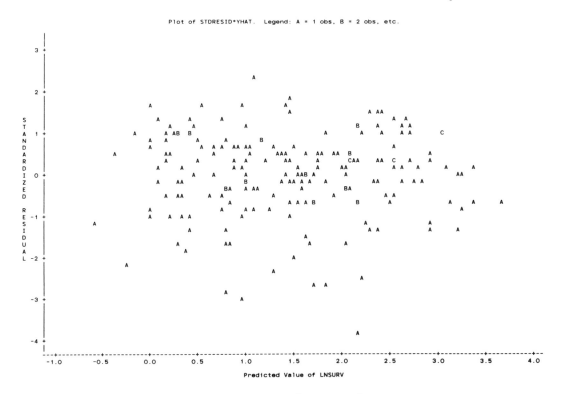

Fig. 12-2 Standardized residual values of $Y_i - \hat{Y}_i$ plotted vs. $\hat{Y}_i$ (SAS PROC PLOT program.)

around 0 as $\hat{Y}_i$ increases, the graph can be inspected to find strikingly discrepant points occurring in a consistent pattern. Such patterns would suggest a linear nonconformity that can grossly distort the results for trends. No such discrepancies are immediately evident in Fig. 12-2, although an asymmetry appears if you draw a horizontal line through the ordinate of 0 and a vertical line through the YHAT, i.e., predicted LNSURV, value of 1.8. The lower right quadrant formed by those two lines has a substantially different pattern of points from that of the upper right-hand quadrant.

If we were interested mainly in improving fit, the asymmetry suggests a possible change in format of the variables. In a set of explorations not shown here, the three main independent variables were included together with their squared values. The results did not effectively improve either the R^2 achievements or the configuration of the residual estimates.

12.4.1.3. Display of residuals vs. independent variables

A plot of the standardized residuals against the corresponding values of each independent variable can offer hints about variables that were poorly fit and that might be transformed to produce better fit.

Fig. 12-3 shows this plot for hematocrit. In an unbiased model, the residuals should randomly vary around their mean at zero and around the mean of the

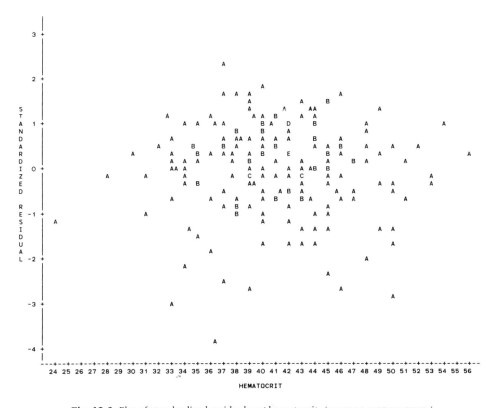

Fig. 12-3 Plot of standardized residuals vs. hematocrit. (SAS PROC PLOT program.)

independent variable (which in this instance is 41 for hematocrit). No striking pattern emerges in the four quadrants although the residual deviations usually seem larger in their below-zero than in their above-zero values.

Because the residuals here are expressed in natural logarithms for survival time, the results are difficult to interpret. With 1.503 as the standard deviation of ln Y_i, a standardized residual deviation of -3 represents $(-3)(1.503) = -4.51$ units. This becomes $e^{-4.51} = .01$ months, i.e., less than a day. Conversely, a standardized residual deviation of $+2.5$ units becomes $e^{(2.5)(1.503)} = e^{3.76} = 42.8$ months, or more than 3 1/2 years.

The standardized residual displays for the other two important independent variables are unfortunately not helpful, because the variables have ordinal categories that do not offer useful "portraits" unless something is grossly awry. Within each category, the displays merely showed polyvalent points scattered around 0.

12.4.2. Check for linear conformity

As discussed in Chapter 7, the ranked independent variables of the illustrative data set were not linear in their bivariate relationships to the target variable. They would therefore not be expected to become linear in a multivariable context.

Nevertheless, to demonstrate the test procedure, the linearity of TNM stage was examined in a regression model that contained sxstage and hct as the additional independent variables. For checking linearity, the five TNM stages were decomposed and expressed in four dummy variables marked TNMD1, TNMD2, TNMD3, and TNMD4. In a linear relationship, as noted in Section 7.12.2.2, the coefficients for the dummy variables should increase consistently. This consistent increase did not appear for this model, which showed the following regression result:

$$\text{ln survival} = 1.063 - .258\ \text{TNMD1} - .930\ \text{TNMD2} - 1.16\ \text{TNMD3} - 1.18\ \text{TNMD4} - .536\ \text{SXSTAGE} + .062\ \text{HCT}.$$

The model's fit was $R^2 = .286$ with $P < .0001$. Each of the partial regression coefficients had $P < .05$ except for TNMD1, which had $P = .48$.

Thus, the exploration demonstrated (as previously known) that TNM stage did not have a distinctly linear relationship to *ln survival*. Since no quantitative *linear* conclusions were being drawn about TNM stage (or other cogent variables), however, no efforts were needed to transform or otherwise alter those variables.

12.4.3. Effect of 'outliers'

In most statistical approaches to regression diagnostics, the focus is on outliers. The mathematical assumption is that the collection of data can be fitted well, but that outlier points will either prevent a good fit or lead to a wrong model, and that the lesion can be repaired after the outlier values are identified and removed. In this approach, the good observations are the ones that are consistent with the selected mathematical model, in contrast to the bad or "contaminant" observations, which are not.

12.4.3.1. Definition of an outlier

An immediate problem is to define what is meant by an outlier. According to Fieller,[9] it is a "Humpty-Dumpty word; it means just what the user chooses it to mean." With this fundamental ambiguity allowing unrestrained expression, about seventeen hundred statistical publications (according to Barnett)[10] have reflected "the highly diverse range of interests and emphases in outlier concepts and methodology."

In simple univariate distributions of dimensional data, the outliers are observations occurring in the "tail areas," far beyond the central indexes and near-central zones of the distribution. The main disputes for univariate distributions are the choice of a mechanism for setting boundaries (standard deviations? percentiles? multiples of percentiles?) and for locating the separation (1.5 times the quartile deviation? 3 times the standard deviation?) of bad outliers from good observations. After an algebraic model has been fitted, however, an outlier observation can be identified not from its univariate location, but from the magnitude of its discrepancy from the estimated value of $\hat{Y}_i$.

The extensive mathematical attention given to outliers has been evoked by the almost limitless opportunity offered by multiple independent variables, by multiple formats for expressing the estimates of $\hat{Y}_i$, and by "dealer's choice" in demarcating the outliers.

12.4.3.2. Scientific concept of outliers

Most scientific investigators, however, have relatively little interest in outliers. If a particular observation is erroneous (wrong patient, wrong measurement, poor technician, etc.), its discovery is valuable, and it should of course be eliminated from the data. Sometimes the detection of outliers has a pragmatic rather than error-finding or model-fitting goal. For example, outlying countries were deliberately sought in a regression analysis[11], aimed at finding unusual effects on infant mortality rates; and multivariate outliers were identified in a study[12] done to aid "future advertising promotions" by letting a bank obtain "a better 'feel' for traits of families who write a large number of cheques per month."

Assuming that all the observations are correct and accurate as data, however, most scientists would not want to discard observations arbitrarily because they fail to comply with a mathematical model. For many scientists, the peremptory exclusion of legitimate observations for any reason—mathematical, conceptual, commercial, or political—is almost tantamount to fraud.

The statistical concern with outliers, therefore, is another manifestation of the disparity between mathematical and scientific goals in data analysis. Using an algebraic model whose main goal is to get good fit, the mathematician wants to make that fit as good as possible by finding and removing outliers, or by changing the configuration of the model. The investigator, however, seldom expects the true biologic configuration of a relationship to emerge from computerized number crunching, and has relatively little interest in perfecting the fit of a mathematical model. The investigator's main goal is to discern the trends and importance of variables. Most medical scientists would be horrified at the idea of discarding or otherwise manipulating legitimate *data* for the sole purpose of making a mathematical model "feel better."

Sometimes an unscrupulous investigator may have "shady" reasons for wanting to remove outliers. This idea was conveyed in a recent vignette published[13] as an "ethics case" in which professional statisticians were asked to state what response they would give if the investigator demanded removal of outliers from the analysis of a data set.

Accordingly, for most scientists, the regression diagnostics are usually finished after the foregoing checks for interactions, residuals, and linear conformity. For many statisticians, however, all these checks are relatively unimportant because the regression diagnostic process has not yet begun its main work in detecting and potentially fixing the problems of outliers.

The next few sections describe what is done during those statistical activities. You can read about them if you want to know what the statistical consultant is contemplating and what appears in the corresponding sections of the computer printouts. Otherwise, skip directly to Section 12.5.

12.4.3.3. Indexes of measurement for outliers

The potential outlier effect of each observation can be appraised either directly from its result in the processed data, or theoretically, from the changes that would occur if that observation were removed from the data. The statistical indexes can focus on three types of effects. The first type, which is already familiar, examines various expressions of *residuals* to identify outliers in the dependent Y variable. The other two types of index are labelled with new jargon. An index of *leverage* can reflect outlier observations in the collection of the independent X_j variables; and an index of *influence* reflects the effect of individual observations on the values calculated for the b_j regression coefficients.

12.4.3.4. Indexes for residuals

In addition to the simple and standardized residuals discussed throughout Section 12.4.1, several other expressions can be calculated for residuals. They can be *weighted*, if a weight has previously been specified for the observation. They can also be converted to a *studentized residual*, which resembles the standardized residual (in Section 12.4.1.2), but is calculated with a different value of s_R in the denominator. The altered s_R is determined from the mean square error in the regression equation formed after removal of the ith observation. Another index, called DFFITS (or *dffits*), denotes the number of standard errors by which $\hat{Y}_i$ would change if the ith observation were removed.

12.4.3.5. Measuring 'leverage'

An index of *leverage*, which denotes the outlier effect of the $\{X_j\}$ variables for a particular observation, indicates the way that an observed value of Y_i can affect the estimated $\hat{Y}_i$. The principle is easy to show in simple linear regression where $\hat{Y}_i = \overline{Y} + b(X_i - \overline{X})$. Replacing b by S_{xy}/S_{xx}, rewriting S_{xy} as $\Sigma(X_i - \overline{X})Y_i$, and replacing $\overline{Y}$ by $\Sigma Y_i/N$, we can then write $\hat{Y}_i = \Sigma Y_i/N + [\Sigma(X_i - \overline{X})Y_i(X_i - \overline{X})/S_{xx}]$. This expression becomes $\hat{Y}i = \Sigma \{(1/N) + [(X_i - \overline{X})^2/S_{xx}]\}Y_i$. The expressions shows how $\hat{Y}_i$ is calculated from the observed values of X_i and Y_i. The term in braces involves only values of X, and can be symbolized as h_i. In simple regression, at point i,

$$h_i = 1/N + [(X_i - \overline{X})^2/S_{xx}].$$

The value of h_i is called the *leverage*. Because 1/N is fixed, the magnitude of leverage for a particular observation depends on the proportion of S_{xx} that is occupied by the corresponding $(X_i - \bar{X})^2$. The farther the value of X_i from the mean, the greater is the leverage.

In multivariable analysis, h_i is determined as a much more complex entity, called the *hat diagonal*, from a data structure called the *hat matrix*. For k independent variables, the average leverage is $\bar{h} + (k + 1)/N$. Hoaglin and Welsch[14] suggest that an observation is an outlier if $h_i > 2(k + 1)/N$.

Another expression of multivariate "outlierness" is *Mahalanobis distance*, which is based on the deviation of each observation's $\{X_j\}$ values from the centroid formed by the mean of each of the $\{X_j\}$ collections. The larger the value of Mahalanobis distance, the more likely is an individual observation to be an outlier.

12.4.3.6. Indexes of 'influence'

Observations with large values for residuals or leverage are regarded as potentially *influential* in affecting the regression coefficients and their variances. The indexes of influence are intended to combine the effects of both residuals and leverage.

The index called COVRATIO, which appears in SAS printouts, reflects certain changes that occur in the covariance matrix when the ith observation is removed. According to Belsley, Kuh, and Welsch,[15] an observation is "worth investigation" as a possible outlier if

$$|\text{COVRATIO} - 1| \geq 3p/N,$$

where p is the number of parameters in the model and N is the number of observations used to fit the model.

Cook's distance,[16] an index that has received substantial attention, determines a counterpart of the average amount of change in the regression coefficient(s) when the ith observation is removed from the data set. An observation is suspected as an outlier if Cook's distance exceeds 1, and as a probably serious outlier if the distance exceeds 4.

The indexes called DFBETAS are determined for each regression coefficient, b_j, when the ith observation is removed. They show the corresponding change in the number of standard errors of b_j.

Obs	Residual	Rstudent	Hat Diag H	Cov Ratio	Dffits	INTERCEP Dfbetas	TNMSTAGE Dfbetas	SXSTAGE Dfbetas	HCT Dfbetas
1	-0.2705	-0.1799	0.0191	1.0398	-0.0251	0.0062	0.0157	0.0005	-0.0126
2	1.5469	1.0319	0.0199	1.0189	0.1470	0.0296	-0.0561	-0.0767	0.0205
3	1.4860	0.9911	0.0199	1.0207	0.1412	0.0284	-0.0539	-0.0736	0.0196
4	-0.9824	-0.6611	0.0400	1.0537	-0.1349	0.0592	0.0352	0.0555	-0.0966
5	0.8162	0.5440	0.0221	1.0374	0.0818	0.0628	-0.0485	0.0046	-0.0487
6	-0.7988	-0.5304	0.0145	1.0297	-0.0643	-0.0093	0.0464	-0.0005	-0.0077
7	1.9487	1.2991	0.0155	1.0016	0.1629	-0.0024	-0.1134	-0.0006	0.0457
8	0.0736	0.0494	0.0370	1.0598	0.0097	-0.0002	-0.0077	0.0076	0.0002
9	0.2643	0.1765	0.0273	1.0486	0.0296	-0.0065	-0.0095	-0.0140	0.0158
10	1.7220	1.1472	0.0160	1.0097	0.1463	-0.0096	-0.1001	-0.0011	0.0483
11	1.5764	1.0630	0.0404	1.0394	0.2182	-0.1559	-0.0393	-0.0350	0.1950
12	1.8167	1.2100	0.0148	1.0054	0.1482	0.0638	-0.1065	0.0044	-0.0277
13	0.3148	0.2096	0.0209	1.0416	0.0306	0.0175	-0.0116	-0.0147	-0.0081
14	0.00312	0.0021	0.0200	1.0414	0.0003	0.0003	-0.0001	-0.0000	-0.0002
15	0.5875	0.3899	0.0145	1.0324	0.0472	0.0069	-0.0341	0.0004	0.0056
16	-0.1515	-0.1006	0.0173	1.0385	-0.0134	0.0022	0.0088	0.0002	-0.0056
17	-0.8072	-0.5353	0.0123	1.0273	-0.0598	0.0190	0.0203	0.0130	-0.0347
18	2.1688	1.4442	0.0113	0.9893	0.1545	-0.0400	-0.0550	-0.0341	0.0815
19	1.5673	1.0387	0.0068	1.0052	0.0860	0.0252	0.0102	-0.0424	-0.0052
20	0.6970	0.4645	0.0221	1.0391	0.0698	0.0537	-0.0415	0.0039	-0.0416
21	0.0818	0.0553	0.0504	1.0747	0.0127	0.0061	-0.0088	0.0090	-0.0066
22	0.3718	0.2514	0.0513	1.0745	0.0585	0.0289	-0.0399	0.0410	-0.0309
23	-5.6881	-3.9279	0.0183	0.7667	-0.5370	-0.3609	0.3486	-0.0259	0.2533
24	0.6254	0.4164	0.0209	1.0388	0.0609	0.0454	-0.0371	0.0033	-0.0344
25	2.0521	1.3679	0.0144	0.9967	0.1652	0.0294	-0.1198	0.0018	0.0141
26	1.4830	0.9871	0.0160	1.0168	0.1259	-0.0083	-0.0861	-0.0010	0.0416
27	-0.2798	-0.1857	0.0150	1.0355	-0.0229	-0.0105	0.0164	-0.0007	0.0050
28	-0.0120	-0.0080	0.0373	1.0602	-0.0016	0.0001	0.0013	-0.0012	-0.0001
29	0.4539	0.3012	0.0145	1.0337	0.0365	0.0053	-0.0264	0.0003	0.0044
30	1.5805	1.0515	0.0144	1.0125	0.1273	0.0459	-0.0924	0.0031	-0.0139

Fig. 12-4 Printout of "observational residuals" for 30 members of illustrative data set, with three main independent variables. ("Influence" option in SAS PROC REG program. For further details, see text.)

12.4.4. Display of results

The packaged computer programs can optionally produce a massive matrix of "observational residuals" for each member of the data set. Fig. 12-4 shows the SAS printout of these results for the first 30 members, i.e., observations, of the 200-member illustrative data set, after analysis with the model containing the three main variables. The first six columns identify, respectively, the particular observation, its residual value ($Y_i - \hat{Y}_i$), its studentized residual, its leverage or "hat diagonal" (h_i) value, covariance ratio, and Dffits value. The remaining columns show the Dfbetas values for the intercept and each of the three main independent variables.

The analogous BMDP printout (not shown here) is even more elaborate. It displays nineteen statistical indexes that can express the individual effects of each observation in the data.

12.5. Therapy for 'lesions'

The main treatments available for the lesions detected during regression diagnostics are to discard observations, fix variables, or change formats.

12.5.1. Discarding observations

Amputation is a standard treatment for outlier observations: after they are removed, the model is re-fitted. This approach is justified by the mathematical idea that the outlier observations are somehow bad, or erroneous contaminants. Not really belonging to the data, they can then readily be discarded to get a better fit.

This idea is happily received by the algebraic model, which always wants to do its best job in producing a good fit. The idea may not be equally attractive to investigators who recognize that outlier observations (when correct) are often a crucial clue to important phenomena. For example, the era of molecular biology was inaugurated by attempts to explain the unusual outlier shape of "sickled" red blood cells. To discard such cells as outliers might have been desirable for mathematics, but not for science.

Accordingly, the attempts to "clean up" the data by eliminating outlying observations can be a highly useful tactic if the observations are indeed erroneous. If the observations are correct, however, a valuable scientific baby may sometimes be thrown out along with the apparently dirty mathematical bathwater.

12.5.2. Repairing inadequately measured variables

Another method of eliminating or correcting wrong information is to aim at individual variables, rather than individual people. The strategy uses statistical methods to determine when an entire variable has been poorly or inadequately measured. In one approach, the results for a particular variable are compared with analogous measurements obtained by other methods.[17] In two other approaches, a set of multivariable results is checked either for alterations when the suspected variable is removed,[18] or for diverse intervariable correlations that may lead to distorted results.[19]

12.5.3. Altered formats

In the third main "therapeutic" strategy, the existing observations and variables are accepted, with or without influential outliers and mismeasured variables.

With the third approach, the format of the existing algebraic model is altered to provide a better fit for the data.

One obvious method, discussed earlier, is to convert either Y or X to logarithms. Another approach is to change from a rectilinear to a curvilinear or polynomial format for the existing independent variable(s). Thus, instead of $Y = a + bX$, for a single independent variable, the format of the model might become $Y = a + bX + cX^2 + dX^3 + \ldots$.

Yet another strategy is to abandon the use of a *single* model and to apply one or more "spline" functions that are interpolated for fitting different zones of the data.[20, 21] For variables that have different impacts in different zones, the spline approach can produce better fits than a single monolithic model and can avoid the possible deception of using a single coefficient for the impact of each variable throughout all zones. On the other hand, because the spline functions are themselves polynomials, they will yield multiple coefficients for each variable—one coefficient for each "power" included in the polynomial. The result may improve the fit and predictive accuracy of the model in each zone, but the impact of a variable cited in polynomials for different zones will be difficult or impossible to interpret, for reasons discussed in Section 2.4.2.1.

12.5.4. Subsequent procedures

If the regression diagnostics indicate that the model is "unhealthy," a satisfactory treatment may or may not be available. If the model fits poorly and the residuals show no specific pattern, nothing can be done about the model because the independent variable X is unrelated to Y. If something is to be related to Y, some other independent variable must be sought.

If a poor-fitting model, however, shows distinctive patterns in the residuals or nonconforming linear zones, the fit can often be improved by changing the format of X. Whatever the new format may be, X becomes transformed into a different variable. An entirely new regression analysis must be done, and the subsequent process of evaluation resumes at a "new" beginning.

12.6. Subsequent decisions for illustrative data set

Since the regression diagnostics for the illustrative data set do not indicate any particular lesion other than generally suboptimal fit, further treatment is difficult to prescribe on a purely statistical basis.

The omission of important predictor variables is the main source of the problem here, and is the most common reason for poor fit of any statistical model. The important variables may be omitted because they are unknown, i.e., we do not know what else affects the outcome of lung cancer; or because additional effective variables are known, but were not included in the available data.

One reason for the omission may be absence of a suitable taxonomy for classifying and coding the information. For example, observant clinicians have long known that the prognosis of cancer is affected by the pattern of symptoms, but the information—although used here—has seldom been specifically classified and deliberately included in formal statistical analyses of prognosis for lung cancer.[22]

Another reason for the omissions is that the investigators may be reluctant to deal with too many variables. For example, to keep the current illustrations relatively simple, two important ordinal variables—severity of symptoms and severity of concomitant co-morbid disease—were not included in the analyses, although those two variables have previously been shown[23-25] to have a significant impact on survival in lung cancer. (Omitted as an independent variable in most of the displayed analyses, *cell type* has previously been shown[25] to be relatively unimportant within the context of the other significant variables.)

Probably the main reason for the omission of important predictors, however, is that they are simply unknown. Even when all of the recognized and suspected predictor variables have been included in multivariable arrangements of prognostic or risk factors, the arrangement will seldom yield perfect predictions or really close fits to the data despite significant improvements in the general accuracy of prediction.

In all of these explanations, poor fit of the algebraic model is caused by either biologic ignorance or taxonomic defects. The main problem is scientific and substantive, not statistical. Nevertheless, since the available data are all that is available, the only approach that can be used in statistical analysis is to consider mathematical inadequacy: the existing independent variables may not have been statistically well utilized. For example, the heavily skewed distributions noted earlier for percent weight loss and progression interval might have been more effective if transformed into a Gaussian pattern via logarithms or some other conversion. An extra term might have been added for the interaction of symptom stage and progression interval. The ordinal variables of TNM stage and symptom stage might have produced better fits if converted into the binary coded ordinalized patterns discussed in Chapter 5.

These and many other mathematical possibilities could be considered and tried if we insisted on getting a better-fitting algebraic model and if we were unsatisfied with finding three important variables. For illustrative purposes, however, the current three-variable model has done its work and can rest, as we go on to other topics.

12.7. Adjustments and scoring systems

After the selected multivariable model has been chosen and checked, the analysis is finished if its main goal was merely to confirm (or refute) the importance of variables previously identified, in simple bivariate analysis, as being important. In many instances, however, an additional goal may be to use the multivariable results for "adjusting" covariate imbalances in comparisons of therapeutic, etiologic, or other maneuvers. For example, in Table 3-9, the stratification into **good** and **poor** prognostic groups provided an adjustment for the susceptibility bias that occurred when treatments E and F were allocated to groups of patients with imbalances in prognostic expectations.

With algebraic models, these adjustments can be done in several ways. In one way, the treatment (or other main independent variable) is entered into the algebraic analysis along with the other covariates. With this "intra-equation" approach, the coefficient that emerges for treatment has been adjusted for the concomitant effect of the covariates. Although statisticians find this approach attractive, most thoughtful scientists dislike it intensely for at least three reasons: (1) the different timing of pre-therapeutic and therapeutic variables is combined into a single equation; (2) the "average" result does not show the effects of treatment in pertinent clinical subgroups; and (3) the differing results in different subgroups may not be effectively detected as interactions. Thus, the intra-equation procedure adjusts, but does not illuminate, and may distort.

For an "extra-equation" adjustment, the therapy variable is omitted from the multivariable analysis. The model that emerges for the covariates can be maintained as an index or converted into a score. If the model is maintained, a prognostic or "susceptibility" index can be calculated for each person when that person's corresponding values of $\{X_j\}$ are entered into the model, $b_0 + b_1X_1 + b_2X_2 + \ldots + b_pX_p$. The effects of treatment can then be compared in groups of patients with similar prognostic indexes.

Alternatively, to make things simpler and easier, the b_j coefficients and X_j values can be arranged arbitrarily into an additive score. Patients are then classified and analyzed according to the arbitrary score. The scores can be simple, such as giving 1 point for each binary risk factor, or much more complex, with ascending or descending values of points for corresponding levels of the risk factor.[26] Sometimes the values of several variables can be converted into a predictive nomogram.[27] Scoring systems are commonly constructed after logistic and Cox regression, and will be further discussed when those topics are reached.

12.8. Illustrations from medical literature

A focus on dimensional target variables, such as blood pressure or health care costs, makes multiple linear regression particularly likely to be used in specialized studies of pathophysiology or health services research. Nevertheless, the results of multiple linear regression regularly appear in general medical literature,[28] although less frequently than those of logistic or Cox regression, for which the target variable is a binary event, such as death, rather than a dimensional measurement.

Table 12-2 cites fifteen publications,[29-43] appearing in diverse medical journals during 1987–92, in which data were analyzed with multiple linear regression. Although chosen haphazardly, without any systematic search, these publications were intended to cover a wide spectrum of topics and journals.

In the nine subsections that follow, the fifteen reports are summarized for the way in which they managed different principles, operations, and analyses of the multiple linear regression procedure.

12.8.1. Goals

None of the fifteen cited publications was aimed at making estimates. In all fifteen, the goal of the multiple regression analyses was to identify or confirm

Table 12-2 Target and independent variables in fifteen reports that used multiple linear regression

First author	Journal	Year	Target variable(s)	Independent variables
Kawazoe[29]	Clin. Cardiol.	1987	Plasma renin activity	Blood pressure, serum chemical factors
Green[30]	Int. J. Epid.	1987	Systolic blood pressure	Age, albumin, calcium, total cholesterol, Quetelet index
Heckerling[31]	Am J. Med.	1988	Carboxyhemoglobin level	Cigarette smoking, gas stoves as heaters, symptomatic cohabitants
Vollertsen[32]	Mayo Clin. Proc.	1988	Duration of stay; hospital charges	Age, osteoporosis, therapeutic injections, spondylosis, special procedures
Lane[33]	J. Gen. Int. Med.	1988	Discomfort score; Days home from work	Extent of physical disease; scores for somatization, anxiety, depression, hostility
Solvoll[34]	Am. J. Epid.	1989	Coffee; Serum cholesterol	Age, body mass index, leisure activity, intake of various foods
Lindberg[35]	Brit. Med. J.	1991	Cardiovascular mortality	Sialic acid concentration, blood pressure, cholesterol, body mass
Carlberg[36]	Stroke	1991	Admission blood pressure in stroke	History of hypertension, age, sex, impaired consciousness
Sclar[37]	Am. J. Med.	1991	Medicaid expenditure	Age, gender, prior expenditures, no. of medications, individual medications
Barker[38]	Brit. Med. J.	1992	Plasma fibrinogen, factor VII	Age, smoking, alcohol, body-mass, waist-hip ratio
Lauer[39]	Ann. Int. Med.	1992	L. ventricular mass	Age, systolic BP, body mass, sex
Barzilay[40]	Ann Int. Med.	1992	Variation in RR intervals on ECG	HLA-DR 3/4, 3/X, or 4/X; hemoglobin 1AC
Nobuyoshi[41]	Am. Heart. J.	1992	Coronary artery spasm	Age, sex, smoking, alcohol, hypertension, diabetes mellitus, cholesterol, uric acid
Koenig[42]	Circulation	1992	Plasma visocsity	Sex, age, smoking, alcohol, hypertension, apoprotein levels
Feskens[43]	J. Clin. Epid.	1992	Area under glucose tolerance curve	Systolic blood pressure, serum lipids

"predictive" factors. The procedure was done either for screening or to check effects that had already been noted in simple bivariate analyses (which were often called "univariate"). In some instances, the bivariate trend was confirmed; in others, when confirmation did not occur, the investigators concluded that the independent variable was ineffectual in a multivariate context. In many instances, the investigators regarded the multivariable results as an adjustment for the effect of the covariate or "confounder" variables.

12.8.2. Programs

The source of the computer program (as BMDP, GLIM, SAS, SPSS, or SYSTAT) was identified in only eight reports; and the process of regression (as full, stepwise, etc.) was mentioned in only six.

12.8.3. Operational tactics

In the six instances of sequential stepping, five were forward stepwise and one was backward; the incremental constraints (F to enter or delete; or incremental P values) were cited in only two reports. In one instance,[34] the investigators said they first did a full regression and thereafter excluded the apparently noncontributory variables. In another instance,[33] the investigators used a model that first forced four independent variables; the contribution of additional variables was then appraised from their incremental change in R^2.

12.8.4. Transformations

The target variable was transformed by Kawazoe et al.[29] as a logarithm of plasma renin activity, and by Lane et al.[33] as the square root of number of days staying home from work. (The latter variable was changed because it "approximated a Poisson distribution.") When Solvoll et al.[34] tested the logarithmic transformation of cholesterol, the results were no different from the untransformed variable. Heckerling et al.[31] also found no change from the untransformed results when they "normalized" their outcome variable, carboxyhemoglobin (COHb), by using its square root and also the transformation log (1 + COHb).

For two outcome variables (duration of stay and hospital charges) that had skewed distributions, Vollertsen et al.[32] reported individual results in geometric means, and used natural log transformations in the regression analyses. Koenig et al.[42] used the log of all variables that had skew univariate distributions; and Lindberg et al.[35] and Barker et al.[38] used log transformations for target variables.

12.8.5. Fit, stability, and conformity

The fit of the multiple regression model was reported in only nine papers, usually with an R^2 value for the total model or for sequential components. The R^2 values ranged from 0.09 to 0.34. In one instance,[36] fit was reported with an "r" value and P value for the model; and in another instance, only the F and P values were cited. The stability ratio of events per variable was never cited, however, and the linear conformity of the model was mentioned in only one report.[35]

12.8.6. Regression diagnostics

In one instance,[42] the investigators tested models with "2nd and 3rd order polynomial terms," but otherwise a specific examination of regression diagnostics was not mentioned in any of the reports. The investigators generally seemed to accept the multivariable results without major doubts about adequacy of the model.

12.8.7. Impact of variables

The specific impact of individual variables was not cited in two reports. In eight instances, the investigators reported partial regression coefficients (b_j) values for each coefficient, together with P, F, or t values that would indicate stochastic significance. Four investigative groups[30,36,41,43] reported standardized regression coefficients. Two groups[41,42] reported partial correlation coefficients for each independent variable. In one instance,[31] only the F values were cited for each coefficient. In two reports[35,38] no coefficients were listed for the variables; instead, results were either cited according to the adjusted values[38] or listed in adjusted quartiles for a variable whose linearity was in doubt.

12.8.8. Interactions

Only four reports[30,33,42,43] mentioned a specific exploration for interaction terms, but none of the four cited quantitative results. A distinct interaction for two variables was noted in only one report.[42]

12.8.9. Quantitative conclusions

Because the multiple linear regressions were done mainly to identify important variables, the authors usually concluded merely that the "significant" independent variables had a definite impact on the target variable. With one exception, no effort was made to quantify impact with a specific "risk" estimate for the variable. In the exception,[35] the authors constructed indexes of relative risk, but did so from results in quartile ordinalized zones, not from the partial regression coefficients.

12.8.10. Summary and recommendations

Although the foregoing fifteen papers were not randomly chosen, they seem reasonably representative of the way that multiple linear regression is used in medical literature.

The most obvious feature of these reports is that none of them used the regression procedure for its main mathematical purpose in fitting a model to the data. In many instances, the investigators did not bother to mention the fit; and many of the reported fits had unimpressive R^2 values below .1. None had an R^2 higher than .34.

Instead, the regression analyses were focused on identifying, confirming, and/or making adjustments for important independent variables. The quantitative magnitude of the importance was generally ignored, however. Standardized regression coefficients were listed in only three reports, and partial correlation coefficients in two others.

The investigators seemed satisfied, perhaps excessively so, with the results of the regression models. Reports of regression diagnostic procedures were seldom mentioned, interactions were checked (or reported as checked) in only four instances, and unstable ratios of events per variable were not considered.

A striking phenomenon that seemed to be accepted by reviewers and editors was the frequent absence of reported basic information about the regression program that had been used, its mode of operation, and the boundaries set for sequential decisions.

Although investigators can apply regression analysis for any appropriate purpose they want, certain guidelines might be recommended for the way the results are reported. The requirements could demand suitable identification of the operation of the regression procedures, an indication of the fit of the model, checks for suitability of the model, and standard quantifications for the partial regression coefficients.

12.9. Commendations and caveats

The analysis of the illustrative data set and the review of published reports indicate that multiple linear regression is being medically used to get results that seem reasonably satisfactory, despite the apparent neglect of many mathematical principles. The investigators, reviewers, editors, and subsequent readers generally seem to ignore the statistical "machinery" and recommended precautions.

Instead, the published reports focus on the main investigative goals—to identify or confirm important variables. The results then seem to be accepted without reservation if they are consistent with the original *scientific* hypotheses. The evaluators apparently give little or no attention to Gaussian assumptions, covariance matrixes, sequential directions of elimination, fit of the models, assessment of interactions, regression diagnostics, or other basic issues in statistical theory and practice. This inattention to statistical principles can be either lauded or denounced, according to the viewpoint of the commentator.

Praise might be given for the maintenance of a "scientific common sense" that refuses to be distracted by all the "regulations" available in statistical doctrines and procedures. Conversely, the investigator-analysts might be chastised for overconfidence and under-cautiousness in so readily accepting what agrees with their preconceived beliefs.

Courageous "common sense" may also sometimes represent "invincible ignorance" by users who are impressed with what they do not comprehend. As pointed out by Roweth[44] in 1980,

> The widespread respect for mathematical models amongst those who have difficulty in understanding them and those with a smattering of statistical know-how can result in inadequate examination of the validity of a particular model.

A prominent statistician whose warnings have often been ignored is *Frank Yates*[45]:

> The interpretation of the results of multiple regression analysis requires the greatest care. Nothing is easier to reach than false conclusions. The first point to remember is that all regression and correlation analysis merely deals with associations. By itself it tells us little of the causative factors that are operating.

David Finney, a prominent member of today's statistical Pantheon, is an "equal opportunity denouncer" in lamenting the peccadillos of both statisticians and investigators. According to Finney, [46]

> [T]he inadequacies of many statisticians . . . [occur because of] the present academic trend . . . [that] is likely to reduce opportunities for statisticians to spare time for the arduous but fascinating task of familiarizing themselves and their students with situations in the real world. Too often the result is that a biologist submits data, which may have been collected with little attention to design or plan, to analysis by software that he does not understand, and that he has chosen because glossy advertisements claim for it unlimited virtues.

In view of the possible uncertainties in both data and algebraic models, it is not surprising that controversies sometimes arise about the results of multiple regression analysis. (What is more surprising, perhaps, is that the controversies are not more frequent.) For example, in a topic that is itself controversial, the results of a multiple regression analysis relating lead exposure to cognitive development in children[47] were attacked[48] because (among other things) "linear representation of curvilinear relations may have resulted in spurious 'significant' results due to overprediction of mental development in the group with high lead levels." The original authors responded[49] that they had taken care of the problem by "examination of the residuals of fitted models" and by checking results "in an earlier paper" according to three ordinalized zones.

In another controversy, one group of epidemiologists[50] complained vigorously and received a vigorous rejoinder from another group of epidemiologists[51] about the effects of different methods for "categorizing fat intake" in a multiple regression analysis of the association between dietary fat and postmenopausal breast cancer. In yet another example, a multiple-regression indictment of cadmium as a cause of lung cancer[52] was rejected by R. Doll[53] because the mathematical models did not have "biologic justification," and also by Lamm et al.,[54] who found, in another study, that the "excess cancer risk . . . [could be attributed] to arsenic exposure" rather than cadium. The original authors[55] then complained about erroneous assumptions and "flawed" analyses by their critics.

Chapter References

1. Efron, 1983; 2. Harrell, 1985; 3. Concato, 1993 (Clin. Res.); 4. Hauck, 1991; 5. Joiner, 1981; 6. Greenblatt, 1977; 7. Little, 1988; 8. Jacqmin, 1994; 9. Fieller, 1993; 10. Barnett, 1993; 11. Helfenstein, 1990; 12. Gillespie, 1993; 14. Mann, 1993; 14. Hoaglin, 1978; 15. Belsley, 1980; 16. Cook, 1982; 17. Sacree, 1988; 18. Herbert, 1988; 19. Glasbey, 1988; 20. Harrell, 1988; 21. Gray, 1992; 22. Feinstein, 1985; 23. Feinstein, 1966; 24. Feinstein, 1968; 25. Feinstein, 1990; 26. Kannel, 1992; 27. Cease, 1986; 28. Concato, 1993 (Ann. Intern. Med.); 29. Kawazoe, 1987; 30. Green, 1987; 31. Heckerling, 1988; 32. Vollertsen, 1988; 33. Lane, 1988; 34. Solvoll, 1989; 35. Lindberg, 1991; 36. Carlberg, 1991; 37. Sclar, 1991; 38. Barker, 1992; 39. Lauer, 1992; 40. Barzilay, 1992; 41. Nobuyoshi, 1992; 42. Koenig, 1992; 43. Feskens, 1992; 44. Roweth, 1980; 45. Yates, 1971, pg. 327; 46. Finney, 1992; 47. Bellinger, 1987 (N. Engl. J. Med.; 316); 48. Marler, 1987; 49. Bellinger, 1987 (N. Engl. J. Med.; 317); 50. Pike, 1992; 51. Kushi, 1992; 52. Stayner, 1992; 53. Doll, 1992; 54. Lamm, 1992; 55. Stayner, 1993.

Part IV Regression for Nondimensional Targets

For the next three regression methods, the target variables are expressed as categories, not in the dimensions used for multiple linear regression. In *logistic regression*, the target categories are usually binary events, having two complementary states—such as **absent/present or dead/alive**—that can be coded with a **0/1** citation. In the proportional hazards method called *Cox regression*, the target is a moving binary event. Its occurrence (or nonoccurrence) at different points in time is summarized with displays such as a survival curve. In *discriminant function analysis*, the target is an unranked category in a nominal variable, such as ethnic group, occupation, or diagnosis. The nominal categories can receive coding numbers—such as **1 = hepatitis, 2 = cirrhosis, 3 = extrahepatic biliary obstruction**—but the numbers have no quantitative meaning.

As in multiple linear regression, these additional analytic methods examine intervariable relationships and use special procedures to explore and eliminate candidate variables, but the nondimensional targets create a major new complexity: the different formats needed to express and evaluate the estimated probabilities for the target categories.

In the new algebraic formats, the discrepancies between estimated and observed values can be cited directly as categorical errors, but can also be expressed with new entities such as *congruences* and *likelihoods*; the best-fitting coefficients are found with new procedures that search for maximum values in likelihoods or in generalized distances rather than least-square deviations; and the results are stated with different indexes of accomplishment. Despite many fundamental alterations in the formats of "interior decoration," however, the basic "architectural strategy" of the three major multivariable methods for categorical targets is similar to that used in multiple linear regression.

The six chapters that follow are presented in pairs for each of the logistic, Cox, and discriminant function procedures. In each pair, the first chapter discusses the main principles; and the second offers further details and pragmatic illustrations.

13 Multiple Logistic Regression

Outline . . . .
13.1. Concepts, symbols, and transformations
    13.1.1. Definition of odds
    13.1.2. Definition of logit
    13.1.3. Logistic transformation and linear model
13.2. Historical development
    13.2.1. Shape of curve
    13.2.2. Original biodemographic applications
    13.2.3. Biostatistical attractions
    13.2.4. Subsequent mathematical background
13.3. Basic format for multiple logistic regression
13.4. Basic structure of data
    13.4.1. Basic fit
    13.4.2. Index of individual discrepancy
    13.4.3. Alternative expressions for discrepancy
        13.4.3.1. Absolute 'distance'
        13.4.3.2. Squared 'distance'
        13.4.3.3. Calculation of 'residuals'
    13.4.4. Index of total discrepancy
13.5. Principles and applications of likelihood
    13.5.1. Calculation of 'basic' likelihood
    13.5.2. Application of likelihood principle in a 2×2 table
    13.5.3. The log likelihood ratio
        13.5.3.1. Uses of the log likelihood ratio
        13.5.3.2. LLR as index of accomplishment
        13.5.3.3. LLR as stochastic test
    13.5.4. Application of likelihood in logistic regression
13.6. Finding the coefficients
    13.6.1. Easing the trial-and-error process
    13.6.2. Choosing a best model
13.7. Quantitative fit of the model
    13.7.1. 'Explained likelihood'
    13.7.2. Penalty indexes
        13.7.2.1. Akaike (AIC) index
        13.7.2.2. Schwartz (SC) index
        13.7.2.3. R_H^2 penalty index for explained likelihood
    13.7.3. 'Goodness of fit'
    13.7.4. Accuracy of classification
        13.7.4.1. Choice of boundaries
        13.7.4.2. Weighting of errors
    13.7.5. Index of relative concordance
        13.7.5.1. General principles and symbols
        13.7.5.2. Construction of C index
        13.7.5.3. Disadvantages of C index
    13.7.6. Indexes of correlation

13.8. Stochastic appraisals
   13.8.1 'Goodness' and 'independence'
   13.8.2. Calculating chi-square for 'independence' of 'model'
      13.8.2.1. Log likelihood ratio
      13.8.2.2. Score index
      13.8.2.3. Wald index
   13.8.3. Goodness of fit of model
      13.8.3.1. Hosmer-Lemeshow test
      13.8.3.2. C. C. Brown's test
   13.8.4. Evaluation of b_j coefficients
      13.8.4.1. Incremental tests
      13.8.4.2. Direct test
      13.8.4.3. Confidence intervals
   13.8.5. Additional chi-square tests
13.9. Quantitative interpretation of b_j coefficients
   13.9.1. Relationship to odds ratio
   13.9.2. Multiplicative feature of b_j coefficients
   13.9.3. 'Standardized' coefficients and coding
      13.9.3.1. Disadvantages of 'standardized' coefficients
      13.9.3.2. 'Standardized' coding
   13.9.4. R index for partial correlation coefficients
13.10. Regression diagnostics

Appendix for Chapter 13
Mathematical Background and Derivation for Logistic Regression
Outline:

A.13.1. Role of discriminant function analysis

A.13.2. Unification of Bayes, Odds, and Gauss

A.13.3. Application for Bayes Theorem

A.13.4. Odds for a probability

A.13.5. Gaussian probability density

A.13.6. Conversion to logistic expression

A.13.7. Logistic arrangement

A.13.8. Logistic and discriminant coefficients

A.13.9. Standard deviation of Y in logistic distribution

. . . .

Logistic regression uses the same symbols that were discussed earlier for linear regression. Y and $\hat{Y}$ represent observed and estimated values of the target variable; and the symbols $\{X_j\}$ and $\{b_j\}$ represent the independent variables and their corresponding weighting coefficients in an algebraic linear model, $G = b_0 + b_1X_1 + b_2X_2 + \ldots + b_pX_p$. In logistic regression, however, the target variable is the occurrence of a binary event, and the value that emerges for each $\hat{Y}_i$ is the estimated probability of that event for person i.

    The use and fitting of the linear logistic model require some special mathematical concepts and transformations, which have had a long historical background of development.

13.1. Concepts, symbols, and transformations

The logistic principles for estimating probability begin with a concept called *odds*.

13.1.1. Definition of odds

If P is the probability of occurrence for an event, the probability of its non-occurrence is $Q = 1 - P$. The ratio of P/Q is the odds.

For example, the chance of drawing a spade from a well-shuffled deck of 52 cards is $13/52 = 1/4$. The chance of not drawing a spade is $(52 - 13)/52 = 39/52 = 3/4$. The odds will be $(1/4)/(3/4) = 1/3$. In gambling, the odds would be stated as a 3 to 1 bet against drawing a spade. If $\hat{Y}_i$ is the estimated probability for a binary occurrence, the odds will be $\hat{Y}_i/(1 - \hat{Y}_i)$.

13.1.2. Definition of logit

The mathematical activities are eased if we work with the natural logarithm of the odds, expressed as $\ln (P/Q)$, and called the *logit* of P. If $\hat{Y}_i$ is an estimated probability, the *logit* is

$$\ln [\hat{Y}_i/(1 - \hat{Y}_i)].$$

For the odds of 1/3, the logit is $\ln (.3333) = -1.099$.

13.1.3. Logistic transformation and linear model

The reason for estimating probability with a logit is that the general algebraic linear model, G, is used to fit the logit as

$$\ln [\hat{Y}_i/(1 - \hat{Y}_i)] = G.$$

Since $\ln u = w$ becomes $u = e^w$, the foregoing expression becomes

$$\hat{Y}_i/(1 - \hat{Y}_i) = e^G.$$

Solving this equation produces $\hat{Y}_i = e^G - \hat{Y}_i e^G$, then $\hat{Y}_i (1 + e^G) = e^G$, and finally

$$\hat{Y}_i = \frac{e^G}{1 + e^G}.$$

The expression $e^G/(1 + e^G)$ is also called the *logistic transformation* of G. If e^G is factored out of the numerator and denominator, the expression becomes

$$\hat{Y}_i = \frac{1}{1 + e^{-G}}.$$

If $\hat{Y}_i$ is being estimated with a single variable, X, so that $G = a + bX$, the logistic transformation is

$$\hat{Y}_i = \frac{1}{1 + e^{-(a + bX)}}.$$

For p independent variables in the general linear model, $G_p = b_0 + b_1 X_1 + \ldots + b_p X_p$, the expression is

$$\hat{Y}_i = \frac{1}{1 + e^{-(b_0 + b_1 X_1 + \ldots + b_p X_p)}}.$$

The main statistical work of logistic regression consists of finding the best-fitting coefficients for $b_0, b_1, \ldots, b_p$ when appropriate data are analyzed for N members of a group. The fit of the model is then appraised, and the coefficients are evaluated to indicate the impact of the individual variables. The activity uses the same basic approach as in linear regression, but—as discussed in the rest of this chapter—the mathematical maneuvers are substantially different.

13.2. Historical development

The original appeal of the logistic procedure was the *shape* of the curve, which was first used in 1838 by the Belgian mathematician P. F. Verhulst[1,2] as a "growth function" in demographic studies of populations.

Table 13-1 Table of values of G and $1/(1+e^{-G})$ in the logistic transformation

G	e^{-G}	$1+e^{-G}$	$1/(1+e^{-G})$
−10	22026.47	22027.47	.0000454
−5	148.41	149.41	.00669
−2	7.39	8.39	0.135
−1	2.72	3.72	0.269
0	1	2	0.500
+1	0.368	1.368	0.731
+2	0.135	1.135	0.881
+5	0.0067	1.0067	0.993
+10	0.0000454	1.0000454	0.999...

13.2.1. Shape of curve

The effect of the logistic transformation is shown in Table 13-1, where arbitrary values of G, extending from −10 to +10, are accompanied by the values of $1/(1+e^{-G})$. In Fig. 13-1, the points in Table 13-1 are graphed to form a sigmoidal (or S) shape that flattens at the low end toward values of 0 as G becomes increasingly negative, and at the high end toward values of 1 as G becomes increasingly positive. (This shape appeared in Figs. 2-5 and 2-15.)

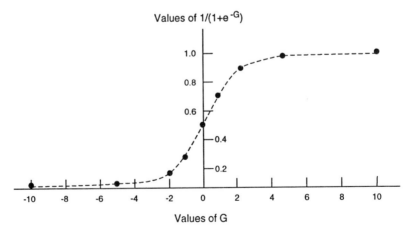

Fig. 13-1 Graph showing values of G and logistic transformation of G for data in Table 13-1.

The logistic shape also resembles the *ogive* curve of the cumulative distribution of frequencies in a Gaussian set of data. The Gaussian curve itself is obtained by plotting the frequency of items, f_i, occurring at any value, X_i. The cumulative frequency ogive is a plot of Σf_i vs. X_i, as values of f_i accumulate with increases in X_i. (This Gaussian connection is later used in the Appendix, which shows the mathematical background for the logistic-regression strategy.)

13.2.2. Original biodemographic applications

Verhulst regarded the sigmoidal curve as a "law" that showed how a population increases over time, beginning at low levels, going through a period of accelerating growth, and then tapering off at full size. Almost a century later, in 1920, R. Pearl and L. J. Reed independently "discovered" the same mathematical function and gave it the *logistic* name.[3] It was subsequently applied both in laboratory work[4] and in demographic epidemiology, to demonstrate the "integrated correlation" of "the biologic forces of natality and mortality" in human populations.[5] Fig.

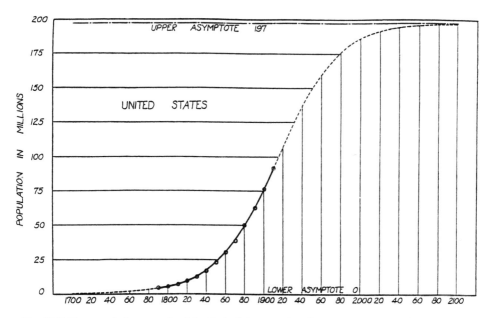

Fig. 13-2 The population growth of the United States. Points for 1790 to 1910 are from actual data. Values thereafter are estimates from the logistic equation. (Source: p. 591 in Chapter reference 5.)

13-2 illustrates Pearl's fitting of a logistic equation to the pattern of growth of the population of the United States based on census data from 1790 to 1910. From this curve, Pearl estimated that an "upper asymptote" of about 197 million would be reached by 2100 and that the 1980 population would be about 175 million. Like many mathematical extrapolations, however, the estimate was inaccurate. The U.S. population[6] in 1940 was close to the curve, at 132 million, but then rose far above, at 178 million in 1960, 203 million in 1970, and 222 million in 1980.

In describing the growth of a population, the logistic equation acts as a type of dose-response curve, with time as the "dose" and population size as the response. The response has two clear boundary levels as limits: the initial nadir size of the population and the zenith at which growth seems to stop. The attractive sigmoidal shape of the logistic curve seemed to offer a good fit for the early, ascending pattern of population growth, before the zenith boundaries were reached; and the curve could be used even when no upper boundaries had been attained.

13.2.3. Biostatistical attractions

Although the original attraction was the sigmoidal shape, the upper and lower boundaries of the curve became particularly appealing when multivariable data were studied for the risks affecting a binary event, such as death or development of a particular disease. To estimate a probability for that event, the algebraic model should constrain the values of $\hat{Y}_i$ to a range between 0 and 1. This constraint does not occur in multiple linear regression, where data for individual values of the variables $\{X_j\}$ can sometimes make $\hat{Y}_i$ yield impossible probability estimates that are negative (i.e., below 0) or greater than 1.

The 0/1 algebraic constraint of $1/(1 + e^{-G})$ in the logistic transformation is easy to show. When $G = 0$, $e^{-G} = 1$, and $1/(1 + e^{-G}) = 1/2$. When G approaches negative infinity, e^{-G} approaches infinity, and the transformed value approaches 0. When G approaches positive infinity, e^{-G} approaches zero, and the transformation approaches 1.

13.2.4. Subsequent mathematical background

Although the shape and boundaries of the logistic curve are appealing, the strategy itself requires a suitable mathematical provenance to justify the "legitimacy" of estimating probabilities with the logistic transformation. A "function" that has boundaries between 0 and 1 could be prepared with many other mathematical structures. Why should the logistic curve be the preferred choice?

The answer to this question involves a long tour through the history of statistical methods. The tour begins with the temporary role of the discriminant function procedure for separating binary groups, and culminates with the formation of logistic regression as a unification of three basic mathematical strategies. They involve Bayesian conditional probabilities, odds ratios, and Gaussian distributions. This tour, if you want to take it, is available in the Appendix to this chapter. Otherwise, go directly to the next section.

13.3. Basic format for multiple logistic regression

With the target variable being a binary event, coded as **0** or **1**, the odds of the estimated probability are $\hat{Y}_i/(1 - \hat{Y}_i)$; and the logistic model is formatted as the logit or natural logarithm (ln) of the odds. Despite the name "logistic," the procedure uses the general linear model, G.

In the ordinary format of a linear model, the logistic expression is

$$\ln[\hat{Y}_i/(1 - \hat{Y}_i)] = G = b_0 + b_1X_1 + b_2X_2 + \ldots + b_kX_k \qquad [13.1]$$

for a full regression on all k variables. After a reduction to p variables, where p < k, the last term in [13.1] would be $b_p X_p$.

In addition to constraining $\hat{Y}_i$ to values between 0 and 1, the logistic transformation also removes a mathematical problem in ordinary linear regression, where the target variable is assumed to have a Gaussian distribution. This assumption, which is mathematically inappropriate for the 0/1 data of a binary target event, is not required in logistic regression. The coefficients for $\{b_j\}$ can be determined and the results evaluated without any demands for Gaussian distributions.

The main disadvantage of the multivariable logistic method for many years, however, was computational. The best-fitting coefficients for a logistic model cannot easily be calculated with the relatively simple mathematical methods used for least squares procedures. Logistic coefficients are usually found through trial-and-error computations that require an extensive amount of number crunching for even small data sets. When the calculational problem was removed in the past few decades by the availability of modern digital computers, multiple logistic regression became (and has remained) one of the most popular analytic procedures in epidemiologic research.

Although the original appeal of the logistic transformation was an accurate fit when monovalent data had a sigmoidal shape, this shape is essentially ignored in multivariable applications, where the data are almost always highly polyvalent. The multivariable advantages of the logistic method are that it offers boundaries, not shape; it can process any type of independent variable; and it produces an extra bonus (to be explained later): the coefficients can be interpreted as odds ratios.

These virtues have made multiple logistic regression a commonly used procedure when the target variable is a single binary event described as either a current state—such as **yes/no, present/absent, case/control**—or a future state, such as **alive/dead.** The independent variables act as "predictors" or "explainers" for this event, whose probability of occurrence (or nonoccurrence) is estimated by the algebraic model.

The rest of this chapter describes the many new ideas and statistical indexes of logistic regression. Their application is shown in the next chapter with printouts for the illustrative data set and with examples from published literature. Although the heaviest doses of mathematics have been relegated to the Appendix, the discussion that follows contains many statistical symbols and ideas. Since they are all illustrated with specific data in Chapter 14, you may want now to browse through the rest of this chapter rapidly, and then return for another reading (and better understanding) after the specific examples of printout appear in the next chapter.

13.4. Basic structure of data

In logistic regression, the target variable for each of N people is cited as $Y_1, Y_2, \ldots, Y_N$, and each Y_i is customarily coded as either **0** or **1**. The k *independent* variables are $X_1, X_2, \ldots, X_j, \ldots, X_k$. For each of the array of N individual persons, the data for each variable can be cited as $X_{11}, X_{12}, \ldots, X_{ij}, \ldots, X_{Nk}$. In the discussion that follows, we shall assume that n_1 persons have a value of **0** in the target variable, that n_2 persons have a value of **1**, and that $N = n_1 + n_2$.

13.4.1. Basic fit

With no independent variables in the model, the basic summary for Y is $\overline{Y} = \Sigma Y_i/N$. Because each Y_i is either 0 or 1, $\overline{Y}$ will be a proportion, expressed as either n_1/N or n_2/N. For example, if 94 of 200 people survive, $n_2 = 94$ and $n_1 = 106$. The value of $\overline{Y}$ will be $94/200 = .47$ and $1 - \overline{Y}$ will be $106/200 = .53$, which could also be calculated as $1 - .47$.

The logistic model can be aimed at estimating either survival or mortality. For survival, Y is coded as **0** for *dead* and **1** for *alive*. For estimating mortality, *alive* is coded as **0** and *dead* as **1**. In either direction of estimation, the process and eventual results will be similar, but complementary.

To prove the latter point, note that $\hat{Y}_i = e^G/(1 + e^G) = 1/(1 + e^{-G})$, and so the value of $1 - \hat{Y}_i$ will be $1 - [e^G/(1 + e^G)] = 1/(1 + e^G)$. Thus, if the logistic model aims at estimating $\hat{Y}_i$, the coefficients that emerge are for G, which is transferred to $-G$ when $1 - \hat{Y}_i$ is calculated. If the estimate is made directly for $1 - \hat{Y}_i$, the coefficients that emerge are for $-G$, which will have identical values but opposite signs.

When no X_j variables are in the model, the logistic result is

$$\ln [\overline{Y}/(1-\overline{Y})] = G_0, \qquad [13.2]$$

calculated from the mean of the Y values. If aimed at survival in the foregoing example, $\overline{Y}/(1 - \overline{Y})$ will be $.47/.53 = .8868$, which could be alternatively found as $n_2/n_1 = 94/106$. G_0 will then be $\ln .8868 = -.12$. Thus, with no variables in the model, $\overline{Y} = .47$ and $G_0 = -.12$.

13.4.2. Index of individual discrepancy

The actual value of a binary event, Y_i, will be either **1** or **0**, according to whether the event did or did not occur. The estimated value, $\hat{Y}_i$, will be a probability, such as **.17**, **.46**, or **.83**, that lies between **0** and **1**. Consequently, the discrepancy between the actual and estimated values of probability is not the same type of increment as the distance between two dimensional measurements. A mathematically preferable way of expressing the probabilistic discrepancy is to cite it as a *congruence*, which is

$$1 - |Y_i - \hat{Y}_i|.$$

This idea was previously described and briefly illustrated in Section 8.1.2.3. Since Y_i will always be either 0 or 1, the value of congruence will be $\hat{Y}_i$ if the actual Y_i is 1, and $1 - \hat{Y}_i$ if Y_i is 0. As shown in the earlier example, the method of calculating congruence produces its higher value when the estimated value is closer to what actually happened.

Although formula [13.3] is correct and easy for nonmathematicians to understand, it is a simplification of a more formal mathematical expression. In the formal expression, which is shown here to keep statistical readers comfortable, the congruence for person i is the complex product of

$$\left\{(\hat{Y}_i)^{Y_i}\right\} \left\{(1 - \hat{Y}_i)\right\}^{(1-Y_i)}. \quad [13.4]$$

This expression yields exactly the same result as formula [13.3]. It produces $\hat{Y}_i$ if Y_i = 1, and $(1 - \hat{Y}_i)$ if $Y_i = 0$. The simplified expression in [13.3] is preferred here because it is easier to display and understand.

13.4.3. Alternative expressions for discrepancy

Despite the mathematical impropriety of regarding $Y_i - \hat{Y}_i$ as a dimensional distance, it can be used for several alternative expressions of discrepancy.

13.4.3.1. Absolute 'distance'

In one alternative approach, each discrepancy would be expressed as a "distance" for absolute values of $|Y_i - \hat{Y}_i|$. The sum of the discrepancies would be $\Sigma |Y_i - \hat{Y}_i|$, and higher values would represent *worse* fit, as in linear regression. Although intuitively appealing, the use of absolute deviations is not statistically popular, and the tactic is seldom employed.

13.4.3.2. Squared 'distance'

In another approach, which avoids the absolute values, the squared discrepancies are cited, with the same expression used in linear regression, as $(Y_i - \hat{Y}_i)^2$. The sum of these discrepancies would be $\Sigma(Y_i - \hat{Y}_i)^2$. Thus, if the discrepancies are regarded as "legitimate" distances, logistic regression analyses might be done with the same least-squares principle used in linear regression. As noted in Appendix Section A.13.8, this tactic was used in the days before the complex computations of likelihoods became well automated. The same tactic is applied today, as noted in Section 13.8.2.2, for rapid calculation of a model chi-square *score* test. In addition, squared discrepancies have been used in an index proposed by Brier[7] for quantifying the accuracy of probabilistic predictions for 0/1 events.

13.4.3.3. Calculation of 'residuals'

In almost all of the available computer packages for logistic regression, the model is fitted and the b_j coefficients are determined by expressing the discrepancies as congruences, which are then multiplied to form the *likelihoods* discussed in the next section. After the model has been fitted, however, the directly subtracted values of $Y_i - \hat{Y}_i$ become regarded as *residuals* that are examined during the regression diagnostics. For the latter purpose, the "distances" noted in the residuals are converted into several "standard" indexes that will be discussed when they are illustrated in Section 14.7.

13.4.4. Index of total discrepancy

In the most mathematically "proper" activities, the individual discrepancies are expressed as congruences. Representing the closeness of a probability rather than a dimensional distance, the individual congruences for each $\hat{Y}_i$ can be multiplied to form an entity called *likelihood*. Likelihood is usually defined as the joint probability of occurrence for the observed data under the stated model. When applied to binary events, the product of the set of congruences would indicate the combined likelihood of finding that the first estimate (for $\hat{Y}_i$) was correct, *and* that the second estimate (for $\hat{Y}_2$) was correct, *and* that the third estimate (for $\hat{Y}_3$) was correct, and so on.

For multiplication, the Π symbol has the same role that the Σ symbol has in addition. Thus, the product of congruences that forms the *likelihood* is symbolized as

$$L = \Pi (1 - |Y_i - \hat{Y}_i|). \qquad [13.5]$$

For example, suppose the congruences for a series of five estimates are .87, .36, .72, .62, and .23. The value of likelihood will be

$$L = (.87)(.36)(.72)(.64)(.23) = .003.$$

For a perfect set of estimates, each value of congruence will be 1, and the likelihood will be 1.

The multiplicative product of a large series of probabilities, each of which is < 1, will usually be very small. The tiny values, however, are easily managed when converted to logarithms. With this conversion, the *log likelihood* becomes an addition of the natural logarithms for congruences. Thus,

$$\ln L = \Sigma \ln (1 - |Y_i - \hat{Y}_i|). \qquad [13.6]$$

When the appropriate values are substituted in the foregoing example,

$$\ln L = \ln (.87) + \ln(.36) + \ln(.72) + \ln(.64) + \ln(.23)$$
$$= -.139 - 1.02 - .329 - .446 - 1.470 = -3.405.$$

This is the same result obtained from the original product likelihood of .033, since $\ln (.033) = -3.405$.

For probability values, which always lie between 0 and 1, the logarithms are always negative. Furthermore, the absolute magnitudes of the logarithms *increase* as the probabilities become *smaller*. Thus, the larger the value of likelihood, the smaller is the value of log likelihood. For example, $\ln (.87) = -.139$, but $\ln (.23) = -1.470$. To avoid the negative values, the log likelihood expression is regularly cited as $-\ln L$. For a perfect fit, $L = 1$ and $-\ln L = 0$.

The value of the negative log likelihood, cited as -ln L, thus has two obvious resemblances to D_p (or S_R), which is residual group variance in a linear regression. Negative log likelihood is a sum of individual indexes of discrepancy; and decreasing values indicate an improved fit for the model.

13.5. Principles and applications of likelihood

The likelihood strategy did not become statistically popular (or feasible) until digital computers became commonly available to do the calculations. Now that computation is easy, the strategy is regularly used in multivariable analysis, and has even begun to appear in elementary statistics. The likelihood method has many applications, some of which are illustrated in the subsections that follow, beginning with the "basic" likelihood of a total group where $\overline{Y} = n_2/N$, then extending to a 2 x 2 table, and ultimately concluding with logistic regression.

13.5.1. Calculation of 'basic' likelihood

Before any X variables are entered in the model, the basic likelihood L_0, is determined using $\bar{Y}$, the mean of the Y_i values, as

$$L_0 = [1 - |Y_1 - \bar{Y}|][1 - |Y_2 - \bar{Y}|][1 - |Y_3 - \bar{Y}|] \ldots [1 - |Y_N - \bar{Y}|],$$

which can be written as

$$L_0 = \prod_{i=1}^{N} [1 - |Y_i - \bar{Y}|]. \qquad [13.7]$$

If $Y_i = 1$, the congruence is $1 - |1 - \bar{Y}| = \bar{Y}$; and if $Y_i = 0$, the congruence is $1 - |0 - \bar{Y}| = 1 - \bar{Y}$. In the basic summary, n_2 values of Y_i will be **1**, and n_1 will be **0**. Therefore, the basic likelihood value will be

$$L_0 = (\bar{Y})^{n_2} (1 - \bar{Y})^{n_1}$$

or

$$L_0 = \left(\frac{n_2}{N}\right)^{n_2} \left(\frac{n_1}{N}\right)^{n_1}. \qquad [13.8]$$

Expressed in natural logarithms, formula [13.8] becomes $\ln L_0 = n_2 \ln(n_2/N) + n_1 \ln(n_1/N)$. When terms are collected and rearranged, and then cited as a negative for $\ln L_0$, the customary expression will be

$$-\ln L_0 = N \ln(N) - n_1 \ln(n_1) - n_2 \ln(n_2). \qquad [13.9]$$

For example, suppose a group of 49 people contains $n_1 = 35$, who are alive, and $n_2 = 14$, who are dead. If survival is being predicted, the summary proportion for n_1/N will be $35/49 = .714$. To make advance or "prior" estimates of the probability of survival for each member of this group of 49 people, we would guess .714 for each person.

With the likelihood principle, the congruences for these estimates would be scored as .714 for each of the 35 living persons, and as .286 ($= n_2/N$) for each of the 14 dead persons. The basic likelihood would be calculated as

$$L_0 = \underbrace{(.714)(.714)\ldots(.714)}_{35 \text{ times}} \underbrace{(.286)(.286)\ldots(.286)}_{14 \text{ times}}$$

which is $L_0 = (.714)^{35}(.286)^{14}$. If obtained from formula [13.8], the value would be

$$L_0 = \left(\frac{35}{49}\right)^{35}\left(\frac{14}{49}\right)^{14} = [7.6828 \times 10^{-6}][2.4157 \times 10^{-8}] = 1.8559 \times 10^{-13},$$

which becomes converted to $-\ln L_0 = 29.315$.

With the alternative logarithmic calculation in formula [13.9],

$$-\ln L_0 = 49 \ln(49) - 35 \ln(35) - 14 \ln(14)$$
$$= 49(3.892) - 35(3.555) - 14 \ln(2.639) = 29.315.$$

The value of $-\ln L_0$ is the basic total discrepancy, analogous to S_{yy} (or D_0), that is to be reduced by the algebraic model.

13.5.2. Application of likelihood principle in a 2 × 2 table

Suppose that the 49 people in the previous section are divided into two groups, M_1 and M_2, by something such as treatment or a prognostic staging system. The survival results for these two groups are shown in the four cells of Table 13-2.

Table 13-2 Survival results in two groups

Group	Alive	Dead	Total
M_1	30	6	36
M_2	5	8	13
Total	35	14	49

In group M_1 of Table 13-2, the estimated probability of survival would be 30/36 = .833. The congruent discrepancies of these estimated probabilities in the two cells of M_1 would be .833 for each of the 30 patients who lived, and (1 − .833) = .167 for each of the 6 patients who died. In group M_2, the corresponding results would be 5/13 = .385 for the overall estimated probability, and the congruences would be .385 for 5 living people and .615 for 8 dead people.

The new value of the "residual" likelihood formed by this partition (or "model") would be

$$L_R = \left(\frac{30}{36}\right)^{30} \left(\frac{6}{36}\right)^{6} \left(\frac{5}{13}\right)^{5} \left(\frac{8}{13}\right)^{8} = 1.563 \times 10^{-11}.$$

When appropriately expanded and converted to a negative logarithm, this expression would be

$$-\ln L_R = 36\ln 36 + 13\ln 13 - 30\ln 30 - 6\ln 6 - 5\ln 5 - 8\ln 8$$
$$= 129.007 + 33.344 - 102.036 - 10.751 - 8.047 - 16.635$$
$$= 24.882.$$

The value of $e^{-24.882}$ is $L_R = 1.563 \times 10^{-11}$.

The application of the model that formed the two groups M_1 and M_2 has thus enlarged the baseline likelihood from 1.8558×10^{-13} to 1.5635×10^{-11}. The value of -ln L has been reduced from 29.315 to 24.882.

In more general terms, the data for a 2 × 2 table can be shown in the form of Table 13-3. The baseline likelihood in this table will be the same as formula [13.9].

Table 13-3 Frequencies in a general 2 x 2 table

Group	Target Variable 1	2	Total
M_1	a	b	m_1
M_2	c	d	m_2
Total	n_1	n_2	N

The value for the residual likelihood produced by the M_1/M_2 "model" will be

$$-\ln L_R = m_1 \ln(m_1) + m_2 \ln(m_2) - a \ln(a) - b \ln(b) - c \ln(c) - d \ln(d).$$

13.5.3. Log likelihood ratio

The improvement produced in L_0 by a fitted likelihood, L_R, can be expressed as a ratio, L_R/L_0. The natural logarithm of this ratio will be the *log likelihood ratio*, which is

$$\text{LLR} = \ln (L_R/L_0) = \ln L_R - \ln L_0. \qquad [13.10]$$

This statement can be rewritten as $\text{LLR} = -\ln L_0 + \ln L_R$, which reflects the structure of

$$\text{LLR} = -\ln L_0 - (-\ln L_R). \qquad [13.11]$$

In the cited example for Table 13-2, $\text{LLR} = 29.315 - 24.882 = 4.433$.

13.5.3.1. Uses of log likelihood ratio

With the application of an algebraic model in linear regression, group variance is divided into two parts as

$$S_{yy} = S_M + S_R$$

for simple regression, and as

$$D_0 = D_M + D_p$$

for multiple regression. In logistic regression, the original likelihood, expressed as $-\ln L_0$, is similarly divided into two parts as

$$-\ln L_0 = \text{LLR} - \ln L_R. \qquad [13.12]$$

Each of the three terms in expression [13.12] has a positive value. The value of $-\ln L_0$ is analogous to the baseline group variance, D_0 (or S_{yy}). The value of LLR is analogous to the model group variance, D_M (or S_M); and the value of $-\ln L_R$ is analogous to the residual group variance, D_p (or S_R).

13.5.3.2. LLR as index of accomplishment

In models that use the likelihood principle, the goal is to achieve a best fit by maximizing the value of L_R. Because LLR is the difference of two positive values ($-\ln L_0$ and $-\ln L_R$ in equation [13.11]), and because the value of $-\ln L_0$ is fixed, a minimal value for $-\ln L_R$ will maximize the value of LLR.

This partitioning tactic allows LLR to be used, in logistic regression, as a direct counterpart of D_M [$= \Sigma(\hat{Y}_i - \bar{Y})^2$] in linear regression. In sequential stepping procedures, LLR can be checked to determine what has been accomplished at the end by the completed model, and also at each step, when variables are added or deleted. Like D_M in variance reduction, the log likelihood ratio shows the model's effect in dividing basic likelihood, L_0, into two groups (i.e., L_R and LLR). The larger the value of LLR, the better the accomplishment.

13.5.3.3. LLR as stochastic test

A splendid mathematical property of the log likelihood ratio is that twice its value (i.e., 2 LLR) has a chi-square type of distribution that can be used directly for tests of stochastic significance. With 2 LLR interpreted as though it were the value of X^2 in a chi-square test, the result is readily converted, with appropriate degrees of freedom, to a P value.

The value of 2 LLR is also reasonably close to the value of X^2 that would be obtained with the customary statistical calculations. For example, in Tables 13-2 and 13-3, the value of X^2 would ordinarily be calculated as $(ad - bc)^2 N/(n_1 n_2 m_1 m_2)$. In this instance, the result is $[(30 \times 8) - (5 \times 6)]^2 \times 49/(35 \times 14 \times 36 \times 13) = 9.42$. The value of LLR for this table was previously noted to be 4.433, and 2LLR = 8.866, which is not far from $X^2 = 9.42$. Either stochastic result, interpreted with 1 degree of freedom for a 2×2 table, would be highly significant at $P < .005$.

If $\hat{Y}_i$ is estimated from an algebraic model (rather than in the example just cited for a 2×2 table), the log likelihood ratios can be regarded as *chi-square test statistics*, which are interpreted as P values at the appropriate degrees of freedom. Because LLR corresponds to the model group variance, D_M (or S_M), in linear regression, the value associated with 2 LLR is often called *model chi square*.

13.5.4. Application of likelihood in logistic regression

If a single variable, X, is entered into the algebraic model, G, each value of Y_i will be estimated from the corresponding X_i as

$$\hat{Y}_i = \frac{e^{a+bX_i}}{1 + e^{a+bX_i}}.$$

For calculations of congruence, the estimated probabilities will be the corresponding $\hat{Y}_i$ for each of the n_1 people coded as **1**, and $1-\hat{Y}_i$ for the n_2 people coded as **0**. The latter estimates will be calculated as

$$1 - \hat{Y}_i = \frac{1}{1 + e^{a+bX_i}}.$$

The residual likelihood for the group will be calculated as

$$L_R = \underbrace{\left(\frac{e^{a+bX_1}}{1 + e^{a+bX_1}}\right) \cdots \left(\frac{e^{a+bX_{n_1}}}{1 + e^{a+bX_{n_1}}}\right)}_{\substack{\text{for the } n_1 \text{ people} \\ \text{coded as } \mathbf{1}}} \underbrace{\left(\frac{1}{1 + e^{a+bX_{n_1+1}}}\right) \cdots \left(\frac{1}{1 + e^{a+bX_N}}\right)}_{\substack{\text{for the } n_2 \text{ people, extending from} \\ n_1 + 1 \text{ to N, who are coded as } \mathbf{0}}}.$$

When converted to logarithms, and when appropriate terms are collected, this expression becomes

$$-\ln L_R = \sum_{1}^{N} \ln (1 + e^{a+bX_i}) - \sum_{1}^{n_1} (a + bX_i). \qquad [13.13]$$

To maximize L_R, expression [13.13] becomes differentiated with respect to both *a* and *b*. The subsequent equations are set equal to zero, and then solved. The process is the same but more complex for more than one independent variable, where the expression $a + bX_i$ is replaced by $b_0 + b_1 X_{i1} + b_2 X_{i2} + \ldots + b_j X_{ij} + \ldots$.

13.6. Finding the coefficients

When expression [13.13] is differentiated, the equation derived from *a* turns out to be $\Sigma[e^{a+bX_i}/(1 + e^{a+bX_i})] = N$; and the equation derived from *b* is $\Sigma[X_i e^{a+bX_i}/(1 + e^{a+bX_i})] = \Sigma X_i$. The latter two simultaneous equations are then solved to yield the optimum coefficients for *a* and *b*. This solution is not too difficult for only 1 independent variable X, but the nonlinear structure of the algebra for likelihood does not allow application of the relatively simple mathematical procedures used for variance in linear regression.

If the algebraic model were expressed for p variables as $G = b_0 + b_1X_2 + b_2X_2 + \ldots + b_pX_p$, instead of $G = a + bX$ for one X variable, we would need to work with p + 1 coefficients in the formula

$$-\ln L_R = \sum_1^N \ln(1 + e^{b_0 + b_1X_1 + \ldots + b_pX_p}) - \sum_1^{n_1}(b_0 + b_1X_1 + \ldots + b_pX_p).$$

The process of solving the p + 1 simultaneous equations would become enormously more complex than before. In the absence of a well-defined algebraic strategy for solving these equations, the solution is obtained by trial and error, an approach that can be done promptly and without complaint by a capable digital computer.

13.6.1. Easing the trial-and-error process

Several algorithms have been developed to ease the computer's work in the trial-and-error process. Seeking the coefficients that yield a maximum value for likelihood, the computer program essentially "climbs a hill" and tries to reach the top, where the maximum value is located. At each step, the location is checked to see whether L_R has ascended or descended from its prior status. For the check, the computational process examines the slope between consecutive values of L_R at different choices of b_j coefficients. If this slope, expressed as a derivative of the basic computational equation, becomes very close to 0 on any side of a particular point, the point can be regarded as marking the top of the hill, where "convergence" is obtained.

In one trial-and-error strategy, sometimes called the *Newton-Raphson procedure*, the activity begins by assigning certain arbitrary values to each bj coefficient. With every other coefficient held constant, a particular b_j is enlarged (or reduced) by one unit. If the change does not increase L_R, b_j is enlarged by a half unit. If the progressive half-steps do not increase L_R, the changes in b_j are made to go in the opposite direction for the next step. A "full step" is completed when the process has been checked for each b_j in each variable X_j.

With this (or some other) searching algorithm, the computer printouts may contain statements indicating the number of trial-and-error steps needed to achieve convergence in the search process, and citing the value of the derivative (i.e., the slope) at the point chosen to represent the maximum likelihood.

Sometimes, if the criterion is too strict (or the data are not "well behaved"), the process may fail to converge. For example, as noted in Chapter 14, the SAS procedure did not converge for one variable when the default criterion required a derivative slope of < .0001, but convergence promptly occurred when the requirement was lowered to < .005. Thus, before concluding that nonconverging data are grossly misbehaved, the analyst can "loosen the lid" on the criterion.

13.6.2. Choosing a best model

The trial-and-error process produces the best fit of b_j coefficients for a particular model, but does *not* necessarily offer the best model. The role of the b_j coefficients is to provide a maximum likelihood value for the p variables *that have been included* in the model as $b_0 + b_1X_1 + b_2X_2 + \ldots + b_pX_p$. The search for a best model requires an exploratory process in which models are altered to include different variables.

The decision about a best model will depend on assessments of accomplishment, using (among other things) *log likelihood ratios* rather than *likelihoods*. This same principle is used when b_j coefficients are fitted in multiple linear regression. For any set of p variables, the b_j coefficients will minimize the group variance of D_p for those variables, but D_p does not determine a best model. It is chosen in linear regression by appraisal of the F variance ratio, the R^2 proportionate reduction in variance, or other appropriate indexes for *different* models. In logistic regression, $-\ln L_R$ corresponds to D_p, but decisions about a best-fitting logistic model will depend on evaluating the effect on the log likelihood ratio (LLR) and on other appropriate indexes. In the sequential operations, shown in Chapter 14, values of LLR will also be used in logistic regression, in a manner analogous to the linear-regression F and R^2, for making decisions about when to enter or delete variables, and when to stop the sequential process.

13.7. Quantitative fit of the model

The fit of the model can be evaluated for its quantitative and stochastic accomplishments. The quantitative accomplishments, discussed in this section, are cited with indexes of "explained likelihood," "goodness of fit," and accuracy in classification. The stochastic tests, discussed in Section 13.8, use a variety of chi-square procedures. The principles of the activities are briefly outlined here, and their application in the illustrative data set is shown in Chapter 14.

13.7.1. 'Explained likelihood'

In linear regression, the proportion of variance reduced or "explained" by the model G_p is $R^2 = (D_0 - D_p)/D_0 = 1 - (D_p/D_0)$. The logistic counterpart of this expression is $R_L^2 = 1 - [(-\ln L_R)/(-\ln L_0)]$. Also, since $R^2 = D_M/D_0$, and since the log likelihood ratio is a counterpart of D_M, a proportion of explained likelihood for any model could be calculated as

$$R_L^2 = \frac{LLR}{-\ln L_0} , \qquad [13.14]$$

where L_0 is the basic likelihood, with no variables in the model.

For the same set of data, two models with the same number of variables can readily be compared using the ratio in formula [13.14]. In fact, since $\ln L_0$ will be the same for both models, the comparison can be done merely with each model's value of LLR.

As in multiple linear regression, the inclusion of more variables will almost always improve the fit of a logistic regression model. Consequently, if two models have different numbers of variables, they will be unfairly compared with the explained likelihood construction of R_L^2 in formula [13.14]. Several penalty indexes have therefore been developed to allow comparison of models with different numbers of variables and also to provide a relatively "standardized" expression for the accomplishment of any logistic model. With these other indexes available, R_L^2 is seldom used.

13.7.2. Penalty indexes

The penalty indexes that have recently become popular are eponymically named for Akaike and for Schwartz. Another useful penalty index, which adjusts R_L^2, was developed by Harrell.

13.7.2.1. Akaike (AIC) index

The Akaike information criterion,[8] abbreviated as AIC, produces its penalty by augmenting the value of -2 log likelihood. The calculation is AIC = -2 (log likelihood of model) + 2 (number of free parameters in model). The formula is

$$\text{AIC} = -2 \ln L + 2p_j. \qquad [13.15]$$

In this formula, $p_j = t + p$, where p is the number of "explanatory" independent variables, and t is the number of ordered values for the target variable. In most circumstances, the target variable is binary, so that $t = 1$. When logistic regression is applied (as discussed later) to polytomous ordinal targets, t can be 2, 3, or more. In the customary binary situation, $p_j = 1 + p$ for p independent variables.

13.7.2.2. Schwartz (SC) index

Abbreviated as SC, the Schwartz information criterion[9] modifies AIC to incorporate N, the total number of observations. The formula is

$$\text{SC} = -2 \ln L + [(t + p)(\ln N)], \qquad [13.15]$$

where t and p have the same values as in the AIC index. The value of SC will always be larger than AIC except when $\ln N \leq 2$, which occurs only when $N < 8$. According to Koehler and Murphree,[10] "the AIC will overfit the data and ... SC is a better criterion for applications." Nevertheless, AIC seems to be more popular, and has had an entire textbook[11] devoted to it.

13.7.2.3. R_H^2 penalty index for explained likelihood

Being analogous to a penalized D_p (or S_R) value, the AIC and SC indexes are useful for comparing models within the same set of data. They do not, however, have a "proportionate reduction" or standardized explanatory value, like R^2 or adj R^2 in linear regression, that ranges from 0 to 1 to allow an interpretation across different sets of data in different studies.

A useful penalty index, analogous to adjusted R^2, was contained in earlier versions[12] of the SAS logistic regression procedure, but has been omitted in the latest version.[13] Labelled as R^2 in the computer printouts, the index was devised by Frank Harrell, and will be designated here as R_H^2. It is calculated for p estimated parameters as

$$R_H^2 = \frac{\text{model chi-square} - 2p}{-2 \log \text{baseline likelihood}}. \quad [13.17]$$

The subtraction of 2p in the numerator is the penalty for the number of p independent variables in the model, excluding intercepts. Since the value of model chi-square is -2 log likelihood ratio, the result is $R_H^2 = (-2LLR - 2p)/(-2 \ln L_0)$. It is a counterpart of $(D_M - p)/D_0$, and thus resembles the value of the R^2 calculated as D_M/D_0 in linear regression.

If $-2LLR < 2p$, R_H^2 is set to 0, indicating that the model has no explanatory value. Otherwise, R_H^2 could be regarded as approximating the proportion of explained log likelihood. In a perfect explanatory model (before 2p is subtracted), R_H^2 will be 1, since model chi-square will equal -2 log baseline likelihood. Because of the penalty, R_H^2 does not necessarily rise as more variables are added to the model.

The reason for omitting the R_H^2 index from the latest SAS logistic procedure is not clear. One possibility is that the simple subtraction of 2p in the numerator of formula [13.17] does not acknowledge the "high-powered" mathematical theory that led to the AIC and SC indexes. Nevertheless, if you work out all the algebra, it turns out that

$$\frac{AIC}{-2 \ln L_0} \sim 1 - R_H^2,$$

and that $SC/(-2 \ln L_0)$ is often also reasonably close to $1 - R_H^2$.

Consequently, if the computer program does not produce an R_H^2 index for explained likelihood, its value can be approximated as

$$\frac{\text{Explained}}{\text{likelihood}} \simeq 1 - \frac{AIC}{-2 \ln L_0} \simeq 1 - \frac{SC}{-2 \ln L_0}. \quad [13.18]$$

Since $-2 \ln L_0$ has a fixed value in each set of data, the proportion of explained likelihood will increase as AIC and SC become smaller.

13.7.3. 'Goodness of fit'

The observed values of Y_i and those estimated as $\hat{Y}_i$ by the model can be arranged into the observed-minus-expected pattern of the chi-square goodness-of-fit test. The method, described by Cochran[14] in 1955 and proposed for logistic regression by Cox[15] in 1970, is now often associated with Hosmer and Lemeshow.[16] It is described, along with other stochastic uses of chi-square, in Section 13.8.

If the results of $\hat{Y}_i$ are always divided into deciles for the goodness-of-fit chi-square test, the value calculated for X_{H-L}^2 (see Section 13.8.3.1) need not be converted to a P value, and can be used descriptively as a quantitative index of fit. The smaller the value of X_{H-L}^2, the better the fit.

13.7.4. Accuracy of classification

If a boundary such as .50 is chosen, each estimate of Y_i can be converted to a specific estimate of **0** or **1**. For example, if .50 is the boundary, the estimate would be **0** if $\hat{Y}_i < .50$ and **1** if $\hat{Y}_i \geq .50$. Each estimate can then be counted as **right** or

wrong according to the actual value of the corresponding Y_i. The total number of correct and incorrect estimates can then be cited in a table showing accuracy of classification.

These tables are an effective way of showing the accomplishments of a final model, but are seldom used when models are explored sequentially for intermediate decisions, which are made almost exclusively from the examination of incremental likelihood values.

13.7.4.1. Choice of boundaries

The boundary value for classifying each $\hat{Y}_i$ estimate as **0** or **1** is commonly set at .50, but can be chosen in at least two other ways. In the first way, the boundary depends on the "prior probability" found in the data set, without a model. Thus, in the illustrative data set of 200 persons with 94 6-month survivors, the prior probability of survival is 94/200 = .47. If $\hat{Y}_i$ is aimed at estimating survival, the selected boundary might be set at .47. The second approach constructs the type of receiver-operating-characteristic (ROC) curve proposed by Metz et al.[17] and by McNeil et al.[18] for choosing an analogous boundary for diagnostic marker tests. The boundary for $\hat{Y}_i$ is then chosen according to the best result found in the ROC curve.

13.7.4.2. Weighting of errors

The total count of right and wrong estimates is usually expressed as the proportion correct for all estimates, but separate proportions can be calculated (like indexes of sensitivity and specificity) for accuracy in groups with **alive** and **dead** predictions, or with **1** and **0** actual outcomes.

For more complex analyses, a "loss function" can be established to give different scoring weights to different types of right or wrong estimates. For example, a correct estimate might be scored as **+1**, a false negative estimate as **0**, and a false positive as **−1**. The predictive result can then be expressed with the complex total score for the estimates.

13.7.5. Index of relative concordance

Using a special type of resampling, a C index of concordance[19] can be applied to the results of logistic regression.

13.7.5.1. General principles and symbols

If each of the N observations of Y_i is paired with each of the other observations, a total of $N(N − 1)/2$ pairs will be formed. In some of these pairs, the two values of Y_i will be similar—i.e., **1,1** or **0,0**. In other pairs, however, the two values of Y_i will differ, i.e., **1,0** or **0,1**. These dissimilar pairs become the focus of the appraisal.

In each dissimilar pair, the corresponding estimates of Y_i are called *concordant* if their rank goes in the opposite direction as the observed values of **1,0** or **0,1**. If the rank goes in the opposite direction, the paired estimates are *discordant*. For example, suppose $\hat{Y}_6 = .823$ and $\hat{Y}_9 = .637$. The estimates for the pair are concordant if the actual observed values were $Y_6 = 1$ and $Y_9 = 0$, and discordant if $Y_6 = 0$

and $Y_9 = \mathbf{1}$. The pair is tied if the two estimates are identical or very close, such as $\hat{Y}_6 = .823$ and $\hat{Y}_9 = .824$.

The pertinent entities are expressed in six symbols used for calculating not only the C index, but also the indexes of correlation discussed in Section 13.7.6. The symbols are as follows: N = total number of observations; u = number of paired Y_i observations that are similar; and t = number of paired Y_i observations that are dissimilar. Among the dissimilar pairs of observations, r = number of paired estimates that are concordant; s = number of paired estimates that are discordant; and v = number of paired estimates that are tied, i.e., $\hat{Y}_1 - \hat{Y}_2 < |.002|$. Expressed in the cited symbols, the total number of paired observations, N(N − 1)/2, will be composed of *u* that are similar and *t* that are dissimilar. For the *t* dissimilar pairs, i.e., those that are not tied on the observed values of Y_i, the corresponding paired estimates of $\hat{Y}_i$ will contain r concordances, s discordances, and v ties.

13.7.5.2. Construction of C index

The *C index* is constructed as a ratio in which the denominator contains the total number of observed dissimilar pairs, and the numerator contains the number of paired estimates that are concordant plus one-half the number of tied estimated pairs. In the most recent SAS manual,[13] the formula is cited as

$$C = [r + (v/2]/t. \qquad [13.19]$$

13.7.5.3. Disadvantages of C index

The C index has three disadvantages. The first, which is a problem in any "retrospective" appraisal of a fitted model, is that C is sometimes claimed to be an index of "validity" for the model. This claim is inappropriate because C offers merely an "internal" check of the analyzed data. A true "validation" would require "external" examination in a different group of data. The second problem is that the C index, being restricted to only the t pairs of Y_i values that are dissimilar, does not indicate the relative accuracy of estimates for the u pairs of Y_i that are similar. For example, suppose $\hat{Y}_5 = .015$ and $\hat{Y}_7 = .024$. If Y_5 and Y_7 each have observed values of **1**, both of the corresponding estimates are poor, but they will be unblemished if Y_5 and Y_7 are classified as a similar pair and omitted from the calculations. Conversely, if Y_5 and Y_7 are each **0**, the two corresponding estimates are quite good, but they will be uncredited. An additional problem is that if $Y_5 = \mathbf{0}$ and $Y_7 = \mathbf{1}$, the two estimates will be credited with concordance although $\hat{Y}_5$ is a good estimate and $\hat{Y}_7$ is poor. If $Y_5 = \mathbf{1}$ and $Y_7 = \mathbf{0}$, however, the estimates will be debited as discordant, although $\hat{Y}_5$ is poor and $\hat{Y}_7$ is good.

A third problem arises if the C index is interpreted as though it were analogous to the area under the receiver operating curve in a diagnostic marker test.[20] Since a useless marker test will nevertheless get a score of .5 for the area under the curve, the accomplishment of the C index might be better reflected if .5 were subtracted from its value. With this "adjustment," an apparently impressive value of C = .762 would become the more modest result of .262.

13.7.6. Indexes of correlation

The SAS program also produces three indexes that are ordinarily used to denote statistical correlation in two-way tabulations of bivariate categorical data. Of these indexes, *Kendall's tau* requires ranked data for both variables. The other two indexes, *gamma* and *Somers D*, can be applied to either ordinal or nominal variables.

All three of the cited indexes would ordinarily be calculated from the results shown in the pertinent two-way table. If the selected table, however, is the 2 × 2 agreement matrix of estimated and observed results in logistic regression for a binary target variable, the indexes become "underachievers." They are not challenged by the many other patterns that can occur in larger-size tables. Accordingly, the SAS program does not calculate these indexes in the customary statistical manner, from results of the published 2 × 2 table. Instead, the calculations use the values of r, s, and v determined for the contrived pairings described in Section 13.7.5.1.

The formulations and examples of results for these "correlation indexes" will be shown in Chapter 14 for the illustrative data set.

13.8. Stochastic appraisals

The relatively simple chi-square stochastic procedure can become bewildering in logistic regression, because the test can be calculated in at least three different ways, applied for at least four different purposes, and used to evaluate both goodness of fit and independence.

13.8.1. 'Goodness' and 'independence'

In elementary statistics, the chi-square procedure relied on a test statistic, X^2, which was calculated as

$$\sum \frac{(\text{observed value} - \text{expected value})^2}{\text{expected value}}$$

for each of the observed and expected frequencies.

The procedure was used in two ways, according to the mechanisms that produced the expected values. If they came from a theoretical external model, the procedure was a test of the model's goodness of fit. For example, suppose we have external reasons for anticipating that a particular group will contain 50% women. If the group actually has 39 women and 61 men, a chi-square test can be used to determine how well this group is fit by the expected model of 50% women. The smaller the value of X^2, the better is the goodness of fit for the 50% anticipation of the model. (Because high values of X^2 lead to rejection of the model, the test would be better labelled as indicating "poorness of fit.")[21]

In the other (more familiar) usage, the X^2 procedure offers a test of independence. For example, in Table 13-2, a single group was divided into two or more groups that seem different. The X^2 test of independence is done to check the null hypothesis that the two groups are really similar. Thus, when a sufficiently high value of X^2 is found (9.42, with .001 < P < .005) for the data of Table 13-2, we would reject the null hypothesis that division into two groups, M_1 and M_2, had no effect on the target variable.

In multiple logistic regression, X^2 scores can be calculated for both independence and goodness of fit.

13.8.2. Calculating chi-square for 'independence' of 'model'

X^2 scores of independence can be calculated in at least three ways: *log likelihood, score,* and *Wald*. The procedures, which can be used to test independence for the entire model or just for the individual coefficients, are first described here for the model.

In the *model chi-square* test, the combined effectiveness of all the independent variables in the model is examined under the null hypothesis that the logistic model has exerted essentially no effect on the target, i.e., the parametric value of $ß_j$ is assumed to be 0 for all of the b_j coefficients calculated for the independent variables. The "null" idea here is that fitting the model has not accomplished anything more than what would be achieved with no model. With rejection of this hypothesis, we would conclude that the model has had a distinctive, independent effect. The *model chi-square* value for this null hypothesis can be calculated in three ways.

13.8.2.1. Log likelihood ratio

The easiest and most straightforward test of independence is the log likelihood ratio, which is the counterpart of $D_M = D_0 - D_p$ in linear regression. When doubled as 2 LLR, this ratio has a chi-square interpretation at the appropriate degrees of freedom.

13.8.2.2. Score index

In the days when computation was expensive, Rao[22] devised a simpler, quicker way of determining a model chi-square *score* index by using derivatives of the log likelihood at the hypothesized parameter values. Being computationally easier and faster than the customary likelihood calculations, the *score* index is sometimes preferred for certain screening procedures. The model X^2 score calculation is also recommended for the rare situations in which the likelihood computations fail to converge.

The likelihood and score calculation of chi-square for the model usually have reasonably similar results, as shown for the 2×2 table in Section 13.5.3.3.

13.8.2.3. Wald index

Yet another method of determining chi-square is with the Wald index, which does its calculations from the variance-covariance matrix under the assumption that the regression coefficients have "asymptotic normality," i.e., a Gaussian distribution in very large samples. The Wald index is usually relatively close to the other two-chi square results; and all three indexes are regularly cited in computer printouts.

13.8.3. Goodness of fit of model

A second purpose of chi-square tests is to check the goodness of fit of the logistic model. Two different tests can be done.

13.8.3.1. Hosmer-Lemeshow test

In a conventional goodness-of-fit test, now often labelled as *Hosmer-Lemeshow*,[16] the ascending values of the estimated probabilities for Y_i are divided into 10 groups, partitioned at the decile values. The average estimated $\hat{Y}_i$ in each of the 10 groups is then compared with the average value of Y_i observed for that group. With the hypothesis that the estimated mean value of $\hat{Y}_i$ equals the observed mean value of $\overline{Y}_i$ in each decile zone, the value of X^2_{H-L} becomes a sum of observed-minus-expected-squared-divided-by-expected calculations interpreted with 8 (= 10 − 2) degrees of freedom. Because the test is aimed at measuring goodness of fit, a well-fitting logistic model will be indicated by a small X^2_{H-L} value and a correspondingly large P value, i.e. > .05.

Because the chi-square test of independence for the model and the Hosmer-Lemeshow chi-square test of its goodness of fit are examining very different things, an excellent logistic model for variables having high impact should get a large model chi-square value (for independence) and a low Hosmer-Lemeshow chi-square value (for goodness of fit).

Unless something drastic has gone wrong, however, the Hosmer-Lemeshow result should usually show a good fit, because the test poses a relatively easy challenge. The observed and expected results are compared not for each individual value of Y_i, but for the means of values in batches of 10 groups. The challenge becomes progressively easier with smaller numbers of groups. The proponents of the test[23] have stated that when "calculated from fewer than 6 groups, it will always indicate that the model fits."

13.8.3.2. C. C. Brown's test

A different goodness-of-fit chi-square test, proposed by C.C. Brown,[24] is intended to check the "logisticness" of the model by comparing its estimates with those made by a more complex mathematical formulation. The comparison is really between estimates made by two models, rather than between observed and expected values. A low value of the Brown chi-square, with a correspondingly high P value, suggests an appropriately good fit for the logistic model. The test has relatively little pragmatic utility, but its result constantly appears as a routine feature of BMDP printouts.

13.8.4. Evaluation of b_j coefficients

The first two sets of X^2 indexes were aimed at independence and goodness of fit of the entire model. All the other chi-squares indexes are used for tests of independence. The indexes in this section are employed to check the individual b_j coefficients. In logistic regression, they have stochastic and descriptive connotations that differ substantially from the corresponding coefficients in linear regression.

The stochastic significance of individual variables can be determined incrementally during the steps of sequential regression procedures, or appraised from the corresponding b_j coefficient after a final model is chosen.

13.8.4.1. Incremental tests

As in multiple linear regression, the logistic contribution of a particular independent variable is determined as the increment of D with and without that variable in the model. For a forward-regression model containing $p - 1$ variables or a backward-regression model containing p variables, the incremental contribution is $D_{p-1} - D_p$. When used for likelihoods, this result is a log likelihood ratio, which is doubled to get a chi-square value, interpreted with 1 degree of freedom. The procedure, illustrated in Chapter 14, is analogous to the incremental F value for a variable in multiple linear regression.

13.8.4.2. Direct test

After each bj has been calculated as an estimate of the parametric ß$_j$, a "standard normal deviate" can be formed as the ratio

$$\frac{\text{estimated value of } b_j}{\text{standard error of the estimate}}.$$

This ratio is interpreted as a Z test, referred to a Gaussian distribution of probability under the null hypothesis that ß$_j$ = 0. The standard error of b_j is estimated (with mathematics not discussed here) from the matrix of variances and covariances of second partial derivatives of the log likelihood function.[25]

The Z ratio that forms the standard normal deviate for b_j is sometimes called the *Wald statistic,* but the Wald eponym is usually applied to the square of the ratio, which is a chi-square statistic. Since the value of the squared Z ratio is identical to a value for X^2, either index can be used to test b_j. (The results are reported for the Z ratio in BMDP printouts, and for X^2 in the SAS system.)

13.8.4.3. Confidence intervals

The stochastic range of variability for each b_j coefficient is shown with a confidence interval, calculated in the customary manner by forming the product (Z_α)(s.e. b_j), where s.e. b_j is the standard error of b_j; and Z_α is chosen to produce a confidence interval of size $1 - \alpha$. For the usual 95% confidence interval, $1 - \alpha = .95$; $\alpha = .05$; and $Z_{.05}$ (from conventional Gaussian tables) is 1.96. For example, suppose b_j = .5248 and s.e. b_j = .0991. The direct 95% confidence interval, obtained as .5248 ± (1.96)(.0991) = .5248 ± .1942, would extend around b_j from .3306 to .7190.

This result reflects the confidence interval around b_j. As noted in Section 13.9, however, the b_j coefficients are interpreted as odds ratios when expressed as e^{b_j}; and the confidence intervals for the odds ratios are produced with similar exponentiation. Thus, in the example just cited, the estimated b_j coefficient produces $e^{.5248}$ = 1.690 as the odds ratio. The lower bound of the 95% confidence interval for this odds ratio is $e^{.3306}$ = 1.392, and the upper bound is $e^{.7190}$ = 2.054. Because of the exponentiation, the 95% confidence interval from 1.392 to 2.054 is *not* symmetrical around the estimated odds ratio of 1.690.

13.8.5. Additional chi-square tests

Beyond the three uses just described (for the model's independence and goodness of fit and for stochastic evaluation of the b_j coefficients) chi-square tests are also used, as shown in Chapter 14, as intermediate indexes in sequential stepping procedures to screen the effects of variables that have not yet been included in a model.

13.9. Quantitative interpretation of b_j coefficients

A particularly striking feature of the logistic model is that the b_j coefficients are quantitatively interpreted as odds ratios.

13.9.1. Relationships to odds ratios

By definition of the logistic formula, the logarithm of the odds for Y is

$$\ln [\hat{Y}_i/(1 - \hat{Y}_i)] = G_p = b_0 + b_1 X_1 + \ldots + b_j X_j + \ldots + b_p X_p,$$

and the odds itself is

$$\hat{Y}_i/(1 - \hat{Y}_i) = e^{b_0 + b_1 X_1 + b_2 X_2 + \ldots + b_j X_j + \ldots + b_p X_p}.$$

Now suppose that X_1 advances from 0 to 1 while all other variables are held constant. With $X_1 = 1$, the odds will be

$$\hat{Y}_1/(1 - \hat{Y}_1) = e^{b_0 + b_1 + b_2 X_2 + \ldots + b_j X_j + \ldots + b_p X_p}.$$

With $X_1 = 0$, the b_1 term vanishes and the odds will be

$$\hat{Y}_0/(1 - \hat{Y}_0) = e^{b_0 + b_2 X_2 + \ldots + b_j X_j + \ldots + b_p X_p}.$$

The two expressions are identical except that the first contains b_1 and the second does not. If we now construct the odds ratio for Y_1 and Y_0, we get

$$\frac{\hat{Y}_1/(1 - \dot{Y}_1)}{\hat{Y}_0/(1 - \hat{Y}_0)} = \frac{e^{b_0 + b_1 + b_2 X_2 + \ldots + b_j X_j + \ldots}}{e^{b_0 + b_2 X_2 + \ldots + b_j X_j + \ldots}} = e^{b_1}.$$

Thus, e^{b_1} is the odds ratio for the change in the probability, Y, when the value of X_1 advances one unit, from 0 to 1, while all other variables are held constant. A similar result would be obtained if X_1 advanced one unit from 6 to 7 or from 23 to 24, while all other variables remained unchanged.

In more general terms, each b_j coefficient in logistic regression is the natural logarithm of the odds ratio for the variable X_j. The odds ratio itself will be e^{b_j}.

13.9.2. Multiplicative feature of b_j coefficients

Another striking distinction of the logistic regression format is that when variables are combined, their coefficients—as odds ratios—have a multiplicative, rather than additive, effect.

For example, consider the odds when $X_1 = 1$ and $X_2 = 1$ versus the odds when $X_1 = 0$ and $X_2 = 0$, while all other variables are held constant in both situations. In the first instance,

$$\hat{Y}_1/(1 - \hat{Y}_i) = e^{b_0 + b_1 + b_2 + b_3 X_3 + \ldots + b_j X_j + \ldots + b_p X_p}.$$

In the second instance,

$$\hat{Y}_0/(1 - \hat{Y}_0) = e^{b_0 + b_3 X_3 + \ldots + b_j X_j + \ldots + b_p X_p}.$$

The two expressions are identical, except that the first contains $b_1 + b_2$ and the second does not. When the two expressions are divided to form the odds ratio, the result will be

$$\text{Odds ratio} = e^{b_1 + b_2} = e^{b_1} e^{b_2}.$$

Since e^{b_1} is the odds ratio for X_1 and e^{b_2} is the odds ratio for X_2, their combined effect is multiplicative rather than additive.

For example, if $b_1 = 1$ and $b_2 = 1$, the odds ratio for each variable is $e^1 = 2.72$, but their combined odds ratio is

$$\text{OR} = e^{1+1} = e^2 = 7.39.$$

If $b_1 = 3$ and $b_2 = 3$, each of the individual odds ratios is $e^3 = 20.09$. Their combined odds ratio is e^6, which is $(20.09)^2 = 403.43$.

This multiplicative attribute of odds ratios for combined variables must be carefully evaluated when the coefficients of a logistic regression are considered simultaneously. In the reported results, the individual coefficients, when suitably exponentiated, are usually each cited as an "odds ratio." If the associated variables act concomitantly, however, the effect of the odds ratios will be multiplicative.

13.9.3. 'Standardized' coefficients and coding

Despite all of these statistical activities, the values of the estimated odds ratio and the associated confidence interval (see Section 13.8.4.3) are *not* standardized. Consequently, they can be substantially affected by either the units of measurement or the coding system in which each X_j variable is expressed.

To avoid this problem, the b_j logistic coefficients can be standardized in a process similar to what is done in linear regression. The standardized coefficients are obtained by multiplying each b_j by the variable's standard deviation, s_j, and dividing by s_y, the standard deviation of the Y variable. The procedure will be illustrated in the next chapter.

13.9.3.1. Disadvantages of 'standardized' coefficients

Despite the many apparent advantages of standardized coefficients and the frequency with which they are advocated, their usage has recently been vigorously denounced.[26] The crux of the argument is that the standard deviation, s_j, of a categorical variable depends on its distribution, and that the same value of b_j may produce different standardized results when multiplied by different values of s_j.

Regardless of how this argument is resolved, the standardized coefficient is a reasonable screening test for comparing the impact of variables within the same model and across different models. The result becomes a problem only if you decide (improperly perhaps) to use the standardized coefficient to calculate a "standardized odds ratio."

13.9.3.2. 'Standardized' coding

Logistic regression is regularly used to identify the magnitude of risk factor variables, such as "exposure," "severity," or "relative dosage," that are expressed in binary or ordinal categories. As an alternative approach to the standardizing problem, a uniform set of digits can be used for coding these categories. The easiest and most obvious tactic is to use 1-unit increments between the codes. *Nonexposed/ exposed* or *absent/present* can be coded as **0/1**; and grades of severity or magnitude can be coded either as **0, 1, 2, 3** ..., or as **1, 2, 3, 4,** .... If this convention were always used, the values of b_j or e^{b_j} could always be promptly interpreted.

Instead, however, different investigators may use different patterns of coding, such as **−1/+1,** for binary variables, and other arrangements for ordinal variables. Unless the reader knows what codes have been used, the published results can be highly confusing or impossible to interpret. In some publications, for example, the odds ratios listed in the multivariable tabulations seem to be twice the result that would be expected from the bivariate tables. The puzzle can be solved only by discovering that *exposure* was coded as −1/+1 for the multivariable analysis. With this coding, the odds ratio becomes doubled, because the change from *no* to *yes* would represent an increment of 2 units rather than 1.

If a uniform coding system is not adopted, authors should always publish (and editors should always demand citation of) the system used for coding the important variables.

13.9.4. R index for partial correlation coefficients

The SAS logistic program now routinely shows standardized regression coefficients, but the BMDP program does not, either routinely or as an option. An earlier version of the SAS logistic program[12] displayed the analog of a partial *correlation* coefficient, which could approximate a standardized value for b_j in the manner discussed for linear regression in Section 11.3.6. The "partial correlation coefficient," designated as *index R*, is calculated as

$$\text{index } R^2 = \frac{\text{Wald chi-square} - 2}{-2 \ln L_0}. \qquad [13.20]$$

For example, if $-2 \ln L_0$ is 276.54 and Wald chi-square for b_j is 15.76, the value of index R is $\sqrt{(15.76 - 2)/276.54} = .223$. The sign of b_j is attached to index R. Its values range, like partial correlation coefficients, from −1 to +1. The subtraction of 2 in the numerator is a penalty imposed to insist that the Wald chi-square value be high enough to exceed 2. If the Wald value is ≤ 2, R is recorded as 0. Note that the value of index R applies only to a coefficient, and has nothing to do with the R^2 term that describes proportionate reductions in variance or likelihood.

The main virtue of the index R coefficient is that it is essentially standardized within the algebraic model, thereby avoiding the antistandardization argument discussed in Section 13.9.3.1. The standard format also allows index R, being independent of sample size, to be used to rank the impact of variables across models.

13.10. Regression diagnostics

For regression diagnostics, the computer printouts in both the SAS and BMPD programs can show the estimated probability values of $\hat{Y}_i$ either for each person in the actual data or for each "covariate cell," i.e., persons having the same collection of values for the X_j variables. These estimates can be accompanied by standard errors, confidence intervals, and other calculations for each estimate. Histograms can also be shown for the estimated predictions made for each of the binary target groups.

All of these results, however, deal with the *fit* of the models rather than with the conformity of its linear spatial configuration. There are currently no readily available *visual* strategies for examining conformity of the polyvalent or oscillating values of Y_i in relation to the linearity of the $\ln[\hat{Y}_i(1-\hat{Y}_i)]$ transformation used in the logistic model. The nonvisual methods of checking conformity of the regression model are discussed (and illustrated) in Chapter 14.

Appendix for Chapter 13
Mathematical Background and Derivation for Logistic Regression

The first part of the mathematical discussion in this appendix is concerned with discriminant function analysis, which was often used before logistic regression became popular, and which is still sometimes applied today (as described in Chapter 17). In the subsequent discussion, the logistic regression equation is derived for the relationship of a binary target variable, Y, to a *single* independent variable X. (An analogous strategy, with much more complex symbols and ideas, is used if Y is related to a set of multiple independent variables $\{X_1, X_2, \ldots, X_j, \ldots\}$.)

A.13.1. Role of discriminant function analysis

For many years, the mathematical problems of using linear regression for estimates of a 0/1 event were avoided by resorting to a different algebraic method: two-group discriminant function analysis. The method had been introduced by R. A. Fisher[27] in 1936 as a strategy for separating (or discriminating between) two groups, designated as 1 and 2, using a series of variables, $X_1, X_2, X_3, \ldots$, that described the members of the groups. Fisher created a linear discriminant function

$$L = b_1 X_1 + b_2 X_2 + b_3 X_3 + \ldots,$$

that would have the value L_i when each member's values of the $\{X_j\}$ variables were inserted. After a boundary value, L_0, was established, each member would be classified as being in Group 1 if the result for L_i was $< L_0$, and in Group 2 if $L_i \geq L_0$. The mean values of the discriminant function would be $\bar{L}_1$ and $\bar{L}_2$ in each group; and half the distance between them, i.e., $(\bar{L}_1 + \bar{L}_2)/2$, was the value usually chosen for L_0.

Using principles of the analysis of variance, Fisher determined that the groups would be best separated if the coefficients $b_1, b_2, b_3, \ldots$ for L were chosen to maximize the ratio B/W for the sums of squares of two group variances: B, between the groups, and W, within the groups. With n_1 and n_2 as the numbers of members in each group, the grand mean of $\bar{L}$ would be $(n_1 \bar{L}_1 + n_2 \bar{L}_2)/(n_1 + n_2)$. The two sum-of-square variances, for appropriate values of L_i in each group, would be $B = n_1(\bar{L}_1 - \bar{L})^2 + n_2(\bar{L}_2 - \bar{L})^2$; and $W = \Sigma(L_i - \bar{L}_1)^2 + \Sigma(L_i - \bar{L}_2)^2$. Substituting appropriately for $\bar{L}$, the value of B becomes $[n_1 n_2/(n_1 + n_2)](\bar{L}_1 - \bar{L}_2)^2$, and is at a maximum when $\bar{L}_1 - \bar{L}_2$ is as large as possible. The values of the $\{b_j\}$ coefficients are then chosen to maximize the discriminant-function distance, $\bar{L}_1 - \bar{L}_2$, between the means of the two groups.

In the days before electronic computation, a great deal of calculation was required to do the matrix algebra that found the appropriate values of $\{b_j\}$ for an array of multiple variables. When the computational process became eased, the discriminant procedure became attractive. Because it does not produce a probability estimate, it avoids the need to constrain results in the range from 0 to 1. This advantage gave two-group discriminant analysis a long period of statistical popularity despite an intriguing mathematical irony: As discussed in Chapters 17 and 18, if the same binary outcome variable is analyzed with multiple *linear* regression, the b_j coefficients for the X_j variables are essentially similar; and the estimated value for $\hat{Y}_i$ is identical to what emerges with discriminant function analysis. Nevertheless, despite the similarity, the results could be regarded as statistically "legitimate" if obtained with the discriminant method, but not with the linear regression method.

On the other hand, despite the apparent advantage of avoiding the demand for a Gaussian target variable (as in linear regression), discriminant analysis makes a more difficult demand: the multiple independent variables are assumed to have a multivariate Gaussian distribution and reasonably similar values for individual variances. This demand can seldom be fully satisfied in the pragmatic realities of data analysis.

A.13.2. Unification of Bayes, Odds, and Gauss

Logistic regression had the advantage of avoiding the need for a Gaussian multivariate distribution, but its greatest mathematical attraction was a conceptual unification of three basic ideas in the armamentarium of statistical strategies. One idea is the use of Bayes Theorem for conditional probability; the second is the expression of probabilities as odds; and the third is the common statistical assumption that data are distributed in a Gaussian pattern of probability.

A.13.3. Application of Bayes Theorem

An intriguing mathematical "provenance" of the logistic procedure is that it can be derived by using Bayes Theorem, which you have probably met or heard about because of its frequent appearance in the literature of diagnostic marker tests.

To understand how the theorem works, recall that in the usual clinical situation, we begin with a positive result, X, for a diagnostic marker test, and we want to know the probability that the patient has disease D. This conditional probability, symbolized as $P(D|X)$, is called "the probability of D, given X." This type of conditional probability is sought whenever binary values of a variable such as X are used to separate two groups designated as **1/0, yes/no,** or **D/$\bar{D}$**. (The symbols D and $\bar{D}$ would represent **diseased** and **nondiseased,** or **dead** and **alive.**)

In ordinary clinical activities, X is the positive result of a binary diagnostic marker test. With direct clinical reasoning, the clinician contemplates the number of patients who have been encountered with a positive value of X, recalls the number of those patients who also had disease D, and determines $P(D|X)$ directly. In the strategy advocated by "Bayesophilic" diagnosticians, however, the value of $P(D|X)$ is obtained indirectly by applying Bayes Theorem as an algebraic transformation of three other items of information. They are: $P(D)$, which is the prevalence of the disease D in the group under study; $P(X|D)$, which is the "sensitivity" of the marker test, representing the proportion of positive test results in diseased patients; and $P(X)$, which is the prevalence of positive test results. These three entities are transformed into $P(D|X)$ by using Bayes Theorem, which was developed as a simple algebraic truism. The procedure is easily illustrated with the Venn diagram in Fig. A.13-1, which shows the relationship between occurrence of disease D and occurrence of a positive

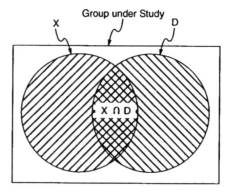

Fig. A13-1 Illustration of sets and subsets for Bayes Theorem. D = persons with the cited disease. X = persons with a positive diagnostic marker test. X ∩ D = persons with a "true positive" test.

test result X in the group under study. The zone of true positive results for the test is shown as the intersection, $X \cap D$.

If N represents the total number of people in the population of Fig. A.13-1, and if X, D, and $X \cap D$ represent the corresponding numbers of people with the cited attributes, then $P(X) = X/N$, $P(D) = D/N$, and $P(X \cap D) = (X \cap D)/N$. By definition, the value of $P(X|D) = P(X \cap D)P(D)$, and the value of $P(D|X) = P(X \cap D)/P(X)$. The latter two equations can then each be solved for $P(X \cap D)$, and the results can be equated to produce

$$P(X|D)P(D) = P(D|X)P(X). \quad [A13.1]$$

The truism stated in formula [A13.1] is Bayes Theorem. It will produce the desired conditional probability result as

$$P(D|X) = \frac{P(X|D)P(D)}{P(X)}. \quad [A13.2]$$

In formula [A13.2], the value of $P(X)$ is usually not as readily available as the values for $P(D)$ and $P(X|D)$. Another set of easy algebraic calculations is therefore used to obtain $P(X)$. For those calculations, people who do not have the disease are designated as $\bar{D}$; people with a negative test result are marked $\bar{X}$; and the value of $P(\bar{X}|\bar{D})$ represents the *specificity* of the diagnostic marker test. Because the value of $P(X|\bar{D})$ will be $1 -$ specificity, $P(X)$ can be determined as

$$P(X) = P(X|D)P(D) + P(X|\bar{D})P(\bar{D}). \quad [A13.3]$$

Substituting the result of [A13.3] into [A13.2] leads to the standard diagnostic-marker expression of Bayes Theorem as

$$P(D|X) = \frac{P(X|D)P(D)}{P(X|D)P(D) + P(X|\bar{D})P(\bar{D})}. \quad [A13.4]$$

With this Bayesian conversion, the values of prevalence, sensitivity, and specificity for a diagnostic marker test are transformed to an expression of its "predictive diagnostic accuracy," $P(D|X)$.

All of the foregoing algebra was used merely to explain the way Bayes Theorem is applied for diagnostic-marker tests. For eventual use in the logistic procedure, the numerator term of formula [A13.4] is factored out of both the numerator and denominator to produce

$$P(D|X) = \frac{1}{1 + \frac{P(X|\bar{D})P(\bar{D})}{P(X|D)P(D)}}. \quad [A13.5]$$

We now rearrange the complex denominator of expression [A13.5] by applying concepts of odds and Gaussian distributions for probabilities.

A.13.4. Odds for a probability

As noted in Section 13.1.1, if p is the probability of an event such as D, the odds against occurrence of the event are $(1 - p)/p$, which is $P(\bar{D})/P(D)$. Thus, the expression $P(\bar{D})/P(D)$ in the denominator of formula [A13.5] can be replaced by the inverse odds ratio $(1 - p)/p$ where p is the probability (or prevalence) for the occurrence of event D. Before any calculations begin, this information will be known (or estimated) as the proportion of group D members in the population under analysis.

We are now left to deal with the ratio of $P(X|\bar{D})/P(X|D)$ in formula [A13.5]. For diagnostic marker tests, X is the positive result of a binary marker test that can be either positive or negative. In other analytic activities, however, X is a dimensional variable, and the values of $P(X|\bar{D})$ and $P(X|D)$ cannot be as easily determined as results for sensitivity and specificity.

Consequently, some other approach will be needed to get values of $P(X|\overline{D})$ and $P(X|D)$. This is where Gaussian distributions get into the act.

A.13.5. Gaussian probability density

The univariate frequency distribution of the variable X in any group shows the proportionate occurrence of each of the observed values of X. In decisions about stochastic significance, the accrued proportions become cited as P values for cumulative relative frequencies at or beyond a selected boundary of X. In other appraisals, however, the individual relative frequencies at each value of X are called the *probability density* of X. In a Gaussian distribution, with mean X and standard deviation s, the probability density for the curve at any point X is

$$\frac{1}{\sqrt{2\pi}\, s} e^{-\frac{1}{2}\left[\frac{X - \overline{X}}{s}\right]^2}.$$

The expression of the algebra can be simplified by use of the convention that exp u = e^u. Consequently, the foregoing expression can be written as $[1/(\sqrt{2\pi}\, s)]\exp[(-1/2)\{(X - \overline{X})/s\}^2]$. If X is a member of Group D, with mean $\overline{X}_D$ and standard deviation s_D, the probability density is

$$P(X|D) = [1/(\sqrt{2\pi}\, s_D)]\exp[(-1/2)\{(X - \overline{X}_D)s_D\}^2. \qquad [A13.6]$$

The corresponding expression for the probability of X as a possible member of Group $\overline{D}$, with mean $\overline{X}_{\overline{D}}$ and standard deviation $s_{\overline{D}}$, is

$$P(X|\overline{D}) = [1/(\sqrt{2\pi}\, s_{\overline{D}})]\exp[(-1/2)\{X - \overline{X}_{\overline{D}})/s_{\overline{D}}\}^2. \qquad [A13.7]$$

A.13.6. Conversion to logistic expression

We now assume that the two groups D and $\overline{D}$ are samples having essentially equal variances, so that $s_D = s_{\overline{D}} = s$. Substituting this value appropriately and dividing expression [A13.7] by [A13.6], we can form

$$\frac{P(X|\overline{D})}{P(X|D)} = \exp\{[-1/2s^2][(X - \overline{X}_{\overline{D}})^2 - (X - \overline{X}_D)^2]\}.$$

Because $\overline{X}_D$, $\overline{X}_{\overline{D}}$, and s are all constant values in the two groups D and $\overline{D}$, the foregoing expression can be expanded and further simplified into

$$\frac{P(X|\overline{D})}{P(X|D)} = \exp\left\{-\left[\frac{\overline{X}_D^2 - \overline{X}_{\overline{D}}^2}{2s^2} + \left(\frac{\overline{X}_D - \overline{X}_{\overline{D}}}{s^2}\right)X\right]\right\}. \qquad [A13.8]$$

If we let a' = $(\overline{X}_D^2 - \overline{X}_{\overline{D}}^2)/2s^2$ and b = $(\overline{X}_D - \overline{X}_{\overline{D}})/s^2$, the right side of formula [A13.8] becomes $e^{-(a' + bX)}$. The latter expression can be usefully substituted for $P(X|\overline{D})/P(X|D)$ in formula [A13.5], but it is multiplied by $P(\overline{D})/P(D)$, which is the $(1 - p)/p$ expression of odds that was derived in Section A.13.4. For mathematical convenience in simplifying matters, the $(1 - p)/p$ term can be converted to a power of e. Recalling that if $e^c = d$, then ln d = c, we can express $(1 - p)/p$ as $e^{\ln[(1-p)/p]}$. We thus can substitute to get

$$\frac{P(X|\overline{D})P(\overline{D})}{P(X|D)P(D)} = \left\{e^{-(a'+bX)}\, e^{\ln[(1-p)/p]}\right\},$$

which becomes e raised to the power $\ln[(1 - p)/p] - a' - bX$.

Because $\ln(1 - p)/p$ is a constant in a fixed group of data, the foregoing expression has the form

$$e^{-(a+bX)},$$

where

$$a = \ln[p/(1-p)] + (\overline{X}_D^2 - \overline{X}_{\overline{D}}^2)/2s^2, \qquad [A13.9]$$

and

$$b = (\overline{X}_D - \overline{X}_{\overline{D}})/s^2. \qquad [A13.10]$$

With suitable substitution, therefore, the formidable-looking expression in [A13.5] can be cited relatively simply as

$$P(D|X) = 1/[1 + e^{-(a+bX)}]. \qquad [13.11]$$

A.13.7. Logistic arrangement

If we let $\hat{Y}_i$ represent the estimated probability of D for a given value of X_i, then $\hat{Y}_i = P(D|X_i)$. In formula [A13.11], the expression has been placed into a logistic format, so that

$$\hat{Y}_i = \frac{1}{1 + e^{-(a+bX_i)}} = \frac{e^{a+bX_i}}{1 + e^{a+bX_i}}. \qquad [A13.12]$$

Since $\hat{Y}_i$ is a probability, the value of $1 - \hat{Y}_i$ will be $1/(1 + e^{a+bX_i})$, and the odds for $\hat{Y}_i/(1 - \hat{Y}_i)$ will be e^{a+bX_i}. The natural logarithm of this odds will be

$$\ln[\hat{Y}_i/(1 - \hat{Y}_i)] = a + bX_i. \qquad [A13.13]$$

The latter expression represents the culmination of the long mathematical tour used to justify the estimation of probabilities with the format of logistic regression.

With more ambitious and complex mathematical operations for more than one X variable, it could be shown that the format of the latter two expressions holds true for the general linear model, G_p, containing the p variables designated as $\{X_j\}$. Thus, the basic format of logistic regression using a linear model is

$$\ln[\hat{Y}_i/(1 - \hat{Y}_i)] = G_p = b_0 + b_1X_1 + b_2X_2 + \ldots + b_pX_p. \qquad [A13.14]$$

The mathematical tour that has just ended began with Bayes formula [A13.2] for estimating the probability of event D, given the result for variable X. After travels through odds and Gaussian distributions, the tour culminated in the logistic expressions in [A13.12] through [A13.14].

A.13.8. Logistic and discriminant coefficients

As noted in Chapter 17, the structures of discriminant analysis also use the same mathematical background of Bayes, odds, and Gauss that was developed to produce the coefficients *a* and *b* expressed in formulas [A13.9] and [A13.10].

The calculations of *a* and *b* are particularly simple when only one X variable is involved. For multiple independent variables $(X_1, X_2, \ldots, X_p)$, however, things get much more complicated, since they involve not only matrix algebra for the computations, but also the exploration of different combinations of variables to find maximum values for distances between the group centroids of the selected variables.

For discriminant function analysis, the b_j coefficients can be calculated with matrix versions of the least-squares regression principle. For logistic regression, however, the b_j coefficients are best calculated with the likelihood methods discussed in the main text. Because the likelihood calculations are particularly difficult and computer-intensive, the least-squares discriminant method was used as a first estimate of the logistic coefficients before the likelihood procedure became well automated.

Therefore, in the early days of logistic regression, the $\{b_j\}$ coefficients were calculated with the methods used for discriminant function analysis. The coefficients were then inserted into suitable formulas to produce the logistic regression equation. This sharing of coefficients is responsible for the occasional use of the now somewhat archaic term *logistic discriminant analysis*.

A.13.9. Standard deviation of Y in logistic distribution

When the b_j coefficients are standardized in logistic regression, a value is needed for s_y. Although the observed Y is a 0/1 outcome variable for n_1 persons coded as **0** and n_2 persons coded as **1**, with $n_1 + n_2 = N$, the standard deviation of $\hat{Y}_i$ is not calculated as $s_y = \sqrt{pq} = \sqrt{n_1 n_2 / N^2}$, with the approach that would be used in ordinary linear regression. Instead, for logistic regression, s_y is $\pi/\sqrt{3} = 1.8138$.

The reason for this formula for s_y is that if X is a logistic variable with mean $\bar{X}$ and standard deviation s, and if X_i is expressed in the standardized form of $Z_i = (X_i - \bar{X})/s$, it can be mathematically shown[28] that the probability density for the logistic function of X is

$$\frac{\pi}{s\sqrt{3}} \frac{e^{-Z_i(\pi/\sqrt{3})}}{[1 + e^{-Z_i(\pi/\sqrt{3})}]^2}.$$

This formula is not particularly worth knowing or remembering, but its $\pi\sqrt{3}$ constant accounts for the standard deviation of the logistic $\hat{Y}_i$ when standardized coefficients are calculated in Chapter 14.

Chapter References

1. Verhulst, 1838; 2. Verhulst, 1845; 3. Pearl, 1920; 4. Reed, 1929; 5. Pearl, 1924; 6. U.S. Bureau of the Census, 1980; 7. Brier, 1950; 8. Akaike, 1974; 9. Schwartz, 1978; 10. Koehler, 1988; 11. Sakamoto, 1986; 12. Harrell, 1986, p. 271; 13. SAS Technical Report P-200, 1990, pp. 175-230; 14. Cochran, 1955; 15. Cox, 1970, p. 96; 16. Hosmer, 1980; 17. Metz, 1973; 18. McNeil, 1975; 19. Harrell, 1986, p. 273; 20. Hanley, 1982; 21. Bradley, 1968; 22. Rao, 1973, pp. 418-19; 23. Hosmer, 1989, p. 144; 24. Brown, 1982; 25. Breslow, 1980; 26. Greenland, 1986; 27. Fisher, 1925; 28. Balakrishan, 1992.

14 Evaluations and Illustrations of Logistic Regression

Outline ....

14.1. Strategies for finding a best model
14.2. Full regression
    14.2.1. SAS procedure
        14.2.1.1. Results of model
        14.2.1.2. Appraisal of coefficients
        14.2.1.3. Standardized coefficients and Index R values
        14.2.1.4. Correlation and covariance matrixes
    14.2.2. Indexes of concordance, correlation, and accuracy
        14.2.2.1. C index of concordance
        14.2.2.2. Indexes of correlation
            14.2.2.2.1. Gamma
            14.2.2.2.2. Somers D
            14.2.2.2.3. Kendall's tau
        14.2.2.3. Accuracy of classification
        14.2.2.4. Conclusions about fit of model
    14.2.3. BMDP procedure
14.3. Indexes for sequential decisions
    14.3.1. Joint effect of all remaining variables
    14.3.2. Tests for inclusion of individual variables
    14.3.3. Deletion of variables
14.4. Illustrations of stepwise regression
    14.4.1. BMDP procedure
        14.4.1.1. Steps 1 and 2
        14.4.1.2. Step 3 and conclusion
    14.4.2. SAS procedure
    14.4.3. Checking results
14.5. Forced regression
14.6. All possible regressions
14.7. Regression diagnostics
    14.7.1. Covariate patterns
    14.7.2. Indexes for individual 'misfits'
        14.7.2.1. 'Hat matrix diagonal'
        14.7.2.2. 'Pearson residuals'
        14.7.2.3. Standardized residual
        14.7.2.4. 'Deviance residuals'
    14.7.3. Indexes of postdeletion change
    14.7.4. Tabular printouts
    14.7.5. Additional graphic displays
    14.7.6. Tests for interaction
        14.7.6.1. Two-way interactions only
        14.7.6.2. Three-way interactions
        14.7.6.3. Saturated model
        14.7.6.4. Interactions after 'reversed' recoding

14.7.6.5. Conclusions
14.7.7. Examination for linear conformity
   14.7.7.1. Linearity of log odds ratios for individual variables
   14.7.7.2. Constancy of odds ratios across conjoint strata
14.7.8. Events per variable
14.7.9. Summary of 'diagnostic' examinations
14.8. Challenges to interaction models
14.9. Scores and scoring systems
14.10. Additional topics
14.10.1. Use of 'design variables'
   14.10.1.1. Multiple dummy variables
   14.10.1.2. Coding system for design variables
   14.10.1.3. Results of design coding
14.10.2. 'Conditional' regression
   14.10.2.1. Matched and unmatched arrangements
   14.10.2.2. Strategy of conditional analysis
   14.10.2.3. Applications and reassurance
14.10.3. Quadratic logistic regression
14.10.4. Polytomous target variables
14.10.5. Application to diagnostic-marker tests
14.11. Use of logistic regression in published literature
14.11.1. Goals of the analyses
14.11.2. Checks for fit
14.11.3. Regression diagnostics
14.11.4. Checks for linear conformity
14.11.5. Transformations
14.11.6. Interactions
14.11.7. Standardization procedures
14.11.8. Reporting of results
14.12. Concluding remarks
14.12.1. Hazards of quantitative coefficients
14.12.2. Pragmatic usage and scoring systems
14.12.3. Caveats, precautions, and opportunities

....

This chapter has four components. The first, comprising printouts produced when multiple logistic regression was applied to the illustrative data set, shows results for full regression, various sequential explorations, and some additional statistical indexes of accomplishment. The second part of the chapter is devoted to regression diagnostics. The third part describes several additional topics and methods in logistic regression. The chapter concludes with a review of published reports in which logistic regression was used, and with some additional remarks.

14.1. Strategies for finding a best model

As in linear regression, the best subset of independent variables for a logistic regression can be chosen in several ways. Most analysts, troubled by the arbitrary results that can emerge from sequential stepping, prefer to do a full regression that includes all independent variables. The apparently most effective candidates can then be retained and examined in a forced-then-open stepped regression. Another approach is to use a sequential stepping procedure, which begins with all candidate variables and reduces them to those that seem most effective. In yet another approach, all possible regression arrangements for all the variables are checked to determine which is regarded as best according to a stipulated criterion.

All of these approaches in the illustrative data set were examined with pertinent programs in both the SAS and the BMDP systems. The target variable for these analyses was a binary outcome event: **dead** (or **alive**) at 6 months after zero time.

14.2. Full regression

The first group of illustrations show the results of full logistic regression for the same seven independent variables that were examined earlier (in Section 11.2) for linear regression.

14.2.1. SAS procedure

The first part of the SAS PROC LOGISTIC printout, not shown here, begins by stating that the dependent variable is SURV6 (6-month survival), with 200 observations, i.e., patients, of whom 106 are dead and 94 are alive. The printout then optionally shows the means, standard deviations, and minimum and maximum values for each of the seven independent variables in the total group and in each of the dead and alive groups. (This information offers a quick way to check the univariate distributions if you have not already done so.)

14.2.1.1. Results of model

The top part of Fig. 14-1 shows the statistical indexes used as "Criteria for Assessing Model Fit." The column marked "Intercept Only" indicates the basic values calculated before any independent variables were entered into the model. Thus, the value of 276.538 for the basic likelihood, $-2 \ln L_0$ (labelled -2 LOG L), was obtained by substituting the basic summary values into formula [13.9] and getting $-\ln L_0 = 200 \ln(200) - 94 \ln(94) - 106 \ln(106) = 138.269$, which becomes doubled to 276.538.

After the model was fitted for the seven independent "covariates," the residual likelihood, L_R, had 230.780 for "-2 LOG L." The difference in the L_0 and L_R log values is 2LLR = 45.758, which has P = .0001 when interpreted as a chi-square value at 7 degrees of freedom. (The additional model chi-square *score* statistic, as calculated in Section 13.8.2.2, is 42.174.)

The AIC index, described in formula [13.15], adds $2p_j$ to the value of -2 LOG L. With no variables in the model, $p_j = 1 + 0 = 1$, so the AIC value for intercept only is -2 LOG L + 2. Later, with seven independent variables, $p_j = 1 + 7 = 8$, and $2p_j = 16$, so the corresponding AIC (= 246.780) is 16 units higher than -2 LOG L in the full model. The SC index, according to formula [13.16], adds $[(t + p)(\ln N)]$ to the value of -2 LOG L. With no variables in the model, the increment is $(1 + 0)(\ln 200) = 5.298$; and with seven independent variables, the increment is $(1 + 7)(\ln 200) = 42.387$, which is the amount by which SC, at 273.167, exceeds the corresponding -2 LOG L, at 230.780.

```
                        Criteria for Assessing Model Fit

                                      Intercept
                          Intercept      and
            Criterion       Only      Covariates    Chi-Square for Covariates

            AIC            278.538     246.780
            SC             281.837     273.167
            -2 LOG L       276.538     230.780      45.758 with 7 DF (p=0.0001)
            Score             .           .         42.174 with 7 DF (p=0.0001)


                          Analysis of Maximum Likelihood Estimates

                 Parameter   Standard     Wald        Pr >      Standardized    Odds     Variable
Variable   DF    Estimate     Error    Chi-Square  Chi-Square    Estimate      Ratio      Label

INTERCPT    1      1.2734    1.8811     0.4582      0.4984                     3.573     Intercept
AGE         1     -0.00495   0.0175     0.0801      0.7771      -0.025850      0.995     AGE (YEARS)
MALE        1      0.7867    0.5622     1.9581      0.1617       0.124443      2.196     MALE GENDER
TNMSTAGE    1      0.4253    0.1074    15.6922      0.0001       0.382619      1.530     TNM STAGE
SXSTAGE     1      0.4161    0.1746     5.6804      0.0172       0.244109      1.516     SYMPTOM STAGE
PCTWTLOS    1     -0.00364   0.0217     0.0282      0.8667      -0.016058      0.996     PERCENT WT LOSS
HCT         1     -0.0948    0.0345     7.5578      0.0060      -0.270564      0.910     HEMATOCRIT
PROGIN      1     -0.00543   0.00940    0.3331      0.5638      -0.054604      0.995     PROGRESSION INTERVAL (MOS)


              Association of Predicted Probabilities and Observed Responses

                  Concordant = 76.1%          Somers' D = 0.523
                  Discordant = 23.7%          Gamma     = 0.524
                  Tied       =  0.2%          Tau-a     = 0.262
                  (9964 pairs)                c         = 0.762
```

Fig. 14-1 Main results of full logistic regression for illustrative data set. (SAS PROC LOGISTIC program.)

The SAS printout offers no indexes for appraising the explanatory effect of the model. If we use formula [13.17], the R_H^2 value for "proportionate explanation" is $\{45.758 - [(2)(7)]\}/276.538 = .115$. With the relative AIC structure of formula [13.18], $R_H^2 \simeq 1 - (246.780/276.538) = .108$.

14.2.1.2. Appraisal of coefficients

The middle part of Fig. 14-1, marked "Analysis of Maximum Likelihood Estimates," shows the label for each variable, its corresponding degrees of freedom, the b_j coefficient (called *Parameter Estimate*), and the coefficient's corresponding values for standard error, Wald chi-square, P (marked *Pr > Chi-Square*), standardized estimate, and odds ratio calculated as e^{b_j}.

The four largest values of the "raw" b_j coefficients are for male (.79), TNM stage (.43), symptom stage (.42), and hematocrit (−.09); but only the values for TNM stage, symptom stage, and hematocrit are stochastically significant at P < .05. As noted in Section 13.8.4.2, the Wald chi-square value for each coefficient is calculated as $[b_j/(\text{std. error of } b_j)]^2$. Thus, for TNM stage, the b_j coefficient is 0.4253, with a standard error of 0.1074, so that $(0.4253)/(0.1074)$ is 3.960, which becomes 15.681 when squared, corresponding to 15.69 (calculated without rounding) in the printout.

The standardized estimates are formed as $b_j(s_j/s_y)$ where s_y for the logistic distribution is $\pi/\sqrt{3} = 1.8138$, as explained in Appendix section A.13.9. For example, for TNM stage, the standard deviation, s_j, is known from previous data (see Fig. 6-2) to be 1.6316. Since $b_j = .4253$, b_{j_s} is calculated as $(.4253)(1.6316)/1.8138 = .3826$.

The three variables in Fig. 14-1 that emerge with P values below .05, and with standardized estimates above .2, have the same counterpart ranks in magnitude for their chi-square and P values. According to those ranks, the most effective variable here is TNM stage (which has a positive effect on death), followed by hematocrit (which has a negative effect), followed by (the positive effect of) symptom stage. These are the same three variables that emerged as effective in the earlier linear regression analysis, but hematocrit here has a higher rank for impact than symptom stage. The transposition is not surprising, since the level of hematocrit might have a stronger effect on 6-month survival than on the longer durations of survival that would have been considered in the earlier linear regression.

Calculated as e^{b_j}, the odds ratios for each of these three variables in affecting mortality are $1.530 (= e^{.4253})$ for TNM stage, 1.516 for symptom stage, and .910 for hematocrit. Because the standardized estimates of b_j seem much higher for *TNM stage* (.383) than for *symptom stage* (.244), the relatively similar odds ratios (1.53 and 1.52) for these two variables may seem surprising until you think about the associated scales in the coding: *TNM stage* had five ordinal grades, whereas *symptom stage* had only four. Thus, a one-unit change in symptom stage could be expected to have a somewhat greater effect on survival than a one-unit change in TNM stage. The relatively small odds ratio (0.91) for *hematocrit* is also not surprising, since the variable was coded in dimensional units as . . . , **37, 38, 39,** . . . . The ratio is below 1, since increasing hematocrit would have a "protective" effect against mortality.

14.2.1.3. Standardized coefficients and Index R values

The standardized values of the b_j coefficients can help rate the importance of variables within a model and across different models, but cannot be converted into explanatory indexes. Thus, the squared value of b_{j_s} in *linear* regression will roughly denote the proportion of basic group variance explained by that variable. In logistic regression, however, the corresponding squared value of b_{j_s} does *not* denote the explained proportion of basic likelihood.

The latter explanation would be better provided by the Index R value for each variable, which can be calculated as shown in formula [13.20]. Thus, for TNM stage, Index R^2 would be $(15.6922 - 2)/276.538 = .0495$. Its square root, .223, would roughly correspond to a partial correlation coefficient, which could be used, if desired, instead of the standardized estimates to rate the impact of each variable. Index R also turns out to be .115 for symptom stage and $-.142$ for hematocrit. The magnitudes and ranks for these three Index R values correspond to those of the standardized coefficients. (The same results cited here for R_H^2 and for the three Index R coefficients appeared when these indexes were printed in an analysis of the same data, using the previous [Version 5.18] SAS PROC LOGIST program.)

The standardized coefficients in logistic regression are valuable mainly for preventing delusions about the magnitude of odds ratios calculated without standardized values. The Index R and Index R^2 values (which are unfortunately not available in either the BMDP or the most recent SAS programs) are most helpful for denoting, respectively, the relative impact of each variable and its effect in explaining the basic likelihood.

14.2.1.4. Correlation and covariance matrixes

After the main results are presented, the SAS logistic program can optionally print estimated correlation and covariance matrixes (not shown here) for the intercept and seven independent variables. These matrixes, which will not be further discussed, are used in the regression diagnostics when variance and confidence intervals are calculated for the individual $\hat{Y}_i$ estimates.

14.2.2. Indexes of concordance, correlation, and accuracy

The results at the bottom of the SAS printout in Fig. 14-1 show the indexes of concordance and correlation for the individual estimates. They are discussed here with the concepts and symbols previously defined in Sections 13.7.5 and 13.7.6.

14.2.2.1. C index of concordance

The total number of simulated pairs in the 200-member illustrative data set would be (200)(199)/2 = 19900. As noted at the bottom of Fig. 14-1, however, 9964 was the value of t for the dissimilar pairs considered in the concordance calculations. [This number can be obtained if you recall that 94 patients survived and 106 died. They will form 94 × 106 = 9964 dissimilar pairs. There will be (106 × 105/2) + (94 × 93/2) = 9936 similar pairs.] Among the dissimilar pairs, the corresponding estimates of $\hat{Y}_i$ were concordant in 76.1%, discordant in 23.7%, and tied in 0.2%. Thus, the value of C is 76.1% + (0.2%/2) = 76.2%, which is cited as .762 in the printout. If adjusted for the .5 value that would be expected if the model did not improve the estimates, C would drop to .762 − .5 = .262.

14.2.2.2. Indexes of correlation

The SAS printout also produces three indexes of correlation—gamma, Somers D, and Kendall's tau—that are seldom helpful, but available if you want them.

14.2.2.2.1. Gamma

Attributed to Goodman and Kruskal,[1] gamma can be regarded as an index of correlation, or as a straightforward index of accuracy for agreement in a 2 × 2 classification table. Gamma would ordinarily be calculated as (concordances − discordances)/(concordances + discordances) in an observed table. Thus, if we used the results that will be presented shortly in Table 14-1 (which has not yet been discussed), gamma would be (138 − 62)/(138 + 62) = .38.

In the SAS procedure,[2] however, gamma = (r − s)/(r + s), and its calculation is confined to results of the dissimilar pairs. From data shown at the bottom of Fig. 14-1, the latter formula for gamma would produce the cited value of (.761 − .237)/(.761 + .237) = .525, which is more impressive than .38.

14.2.2.2.2. Somers D

Somers D resembles gamma but takes tied rankings into account. The SAS manual[2] indicates D = (r − s)/t. Since the tied values of v are included in t = r + s + v, D can be expected to be slightly smaller than gamma.

For the data cited at the bottom of Fig. 14-1, the calculation would be (.761 − .237)/(.761 + .237 + .002) = .524. (The printout values for gamma and Somers D differ slightly from those determined here because they were probably calculated without the rounded percentages.)

14.2.2.2.3. Kendall's tau

Kendall's tau, another index of rank correlation, is probably better known and more generally used than the two others. The index has three versions. One, called tau-a, pertains to bivariate correlations among ordinal ranks that have no ties. It is generally not applicable to contingency tables, where multiple frequencies occur as "ties" at the same rank. Tau-b, which contains provision for ties, is the version pertinent for contingency tables that are "square," i.e., have equal numbers of rows and columns (as in Table 14-1 below). Tau-c is a version of the index pertinent for "rectangular," i.e., nonsquare, contingency tables.

The SAS manual calls its index TAU-A (rather than the more appropriate TAU-B), and cites the formula as $[r - 2]/[0.5N(N - 1)]$. To use this formula, we would have to determine from the data at the bottom of Fig. 14-1 that $r = (.761)(9964) = 7583$ and that $s = (.237)(9964) = 2361$. The calculated result for tau would then be $(7583 - 2361)/19900 = .262$, which is the value listed in the printout.

The SAS formulas seem inconsistent in using the dissimilar pairs as the denominator for C, gamma, and Somers D, but then transferring to N for the calculation of Kendall's tau. Besides, if the all-pairs denominator is used for the Kendall coefficient, its numerator might be expected to come from Table 14-1 as $[(78 \times 60) - (34 \times 28)] = 3728$, so that tau would be $3728/19900 = .19$.

14.2.2.3. Accuracy of classification

To choose an optimal boundary, the SAS procedure works with the optional printout, shown in Fig. 14-2, of a "classification table." It displays the correct and incorrect "batting averages" that would occur for each possible level of demarcation or cutpoint for the probability, $\hat{Y}_i$. The two highest percentages for total "correct" in that table are the values of 69.0 at probability boundaries of .48 and .50. The 69% overall accuracy of estimation is a distinct improvement over the 53% accuracy (= 106/200) that would have been obtained if everyone were estimated as **dead** without the regression model. The lambda proportion of error reduction (see Section 8.2.2.2) is not listed among the diverse SAS indexes, but would be an impressive $(.47 - .31)/(.47) = 34\%$.

If .50 is chosen as the cutpoint, the classification results can be summarized as shown in Table 14-1. The estimates would have a sensitivity of .736 (= 78/106) in

Table 14-1 Estimated and observed results after full logistic regression (boundary of .50 used for results in Fig. 14-2)

	Observed		
Estimated	Dead	Alive	Total
Dead	78	34	112
Alive	28	60	88
Total	106	94	200

dead patients, a specificity of .638 (= 60/94) in alive patients, and a total accuracy of .69 (= 138/200). The false positive rate is .304 (= 34/112) when the estimate is *dead*; and the false negative rate is .318 (= 28/88) when the estimate is *alive*. (All of these results can easily be determined from Table 14-1, but are also cited in the appropriate line of the classification table in Fig. 14-2.)

Classification Table

Prob Level	Correct Event	Correct Non-Event	Incorrect Event	Incorrect Non-Event	Percentages Correct	Sensitivity	Specificity	False POS	False NEG
0.080	106	0	94	0	53.0	100.0	0.0	47.0	.
0.100	106	1	93	0	53.5	100.0	1.1	46.7	0.0
0.120	105	2	92	1	53.5	99.1	2.1	46.7	33.3
0.140	105	6	88	1	55.5	99.1	6.4	45.6	14.3
0.160	105	10	84	1	57.5	99.1	10.6	44.4	9.1
0.180	105	12	82	1	58.5	99.1	12.8	43.9	7.7
0.200	105	15	79	1	60.0	99.1	16.0	42.9	6.3
0.220	101	19	75	5	60.0	95.3	20.2	42.6	20.8
0.240	100	23	71	6	61.5	94.3	24.5	41.5	20.7
0.260	97	28	66	9	62.5	91.5	29.8	40.5	24.3
0.280	97	33	61	9	65.0	91.5	35.1	38.6	21.4
0.300	96	36	58	10	66.0	90.6	38.3	37.7	21.7
0.320	96	37	57	10	66.5	90.6	39.4	37.3	21.3
0.340	92	39	55	14	65.5	86.8	41.5	37.4	26.4
0.360	89	42	52	17	65.5	84.0	44.7	36.9	28.8
0.380	87	44	50	19	65.5	82.1	46.8	36.5	30.2
0.400	85	49	45	21	67.0	80.2	52.1	34.6	30.0
0.420	83	53	41	23	68.0	78.3	56.4	33.1	30.3
0.440	81	53	41	25	67.0	76.4	56.4	33.6	32.1
0.460	81	56	38	25	68.5	76.4	59.6	31.9	30.9
0.480	80	58	36	26	69.0	75.5	61.7	31.0	31.0
0.500	78	60	34	28	69.0	73.6	63.8	30.4	31.8
0.520	76	60	34	30	68.0	71.7	63.8	30.9	33.3
0.540	76	60	34	30	68.0	71.7	63.8	30.9	33.3
0.560	71	62	32	35	66.5	67.0	66.0	31.1	36.1
0.580	68	63	31	38	65.5	64.2	67.0	31.3	37.6
0.600	62	65	29	44	63.5	58.5	69.1	31.9	40.4
0.620	58	68	26	48	63.0	54.7	72.3	31.0	41.4
0.640	54	71	23	52	62.5	50.9	75.5	29.9	42.3
0.660	50	74	20	56	62.0	47.2	78.7	28.6	43.1
0.680	46	74	20	60	60.0	43.4	78.7	30.3	44.8
0.700	42	76	18	64	59.0	39.6	80.9	30.0	45.7
0.720	36	80	14	70	58.0	34.0	85.1	28.0	46.7
0.740	33	83	11	73	58.0	31.1	88.3	25.0	46.8
0.760	31	84	10	75	57.5	29.2	89.4	24.4	47.2
0.780	28	87	7	78	57.5	26.4	92.6	20.0	47.3
0.800	24	88	6	82	56.0	22.6	93.6	20.0	48.2
0.820	17	89	5	89	53.0	16.0	94.7	22.7	50.0
0.840	12	89	5	94	50.5	11.3	94.7	29.4	51.4
0.860	6	91	3	100	48.5	5.7	96.8	33.3	52.4
0.880	4	92	2	102	48.0	3.8	97.9	33.3	52.6
0.900	3	94	0	103	48.5	2.8	100.0	0.0	52.3
0.920	1	94	0	105	47.5	0.9	100.0	0.0	52.8
0.940	1	94	0	105	47.5	0.9	100.0	0.0	52.8
0.960	0	94	0	106	47.0	0.0	100.0	.	53.0

Fig. 14-2 "Classification table" for choosing optimal boundary for estimations. (Option in SAS PROC LOGISTIC program. For further details, see text.)

14.2.2.4. Conclusions about fit of model

From the diverse available indexes of likelihood, concordance, correlation, and accuracy, the results for fit of the model can be given almost any evaluation you would like. On the one hand, an explained likelihood proportion of about .11 is unimpressive; on the other hand, a classification accuracy of 69% and a concordance C index (unadjusted) of .76 seem rather good; and a lambda proportionate reduction index of 34% in rate of errors seems impressive. Nevertheless, 31% of the estimates were wrong.

Regardless of how well the model fits, however, it has clearly selected three effective variables; they are the same ones that were chosen in the linear regression model; and they are biologically sensible.

14.2.3. BMDP procedure

Because the BMDP package has a program for stepwise but not full logistic regression, the latter is obtained by getting the stepwise program to force all seven independent variables. Details of the BMDP stepwise procedure will be illustrated in Section 14.4.1.

A noteworthy feature of the BMDP printout, not shown here, is a statement that the data for the seven-variable regression model had 200 "distinct covariate patterns." In other words, each of the 200 patients under study had different values for the collection of seven independent variables. (The number of distinct covariate patterns is used later when the BMDP [but not the SAS] regression diagnostics appraise the observed and estimated results.)

```
PREDICTED PROBABILITIES CAN BE USED TO CLASSIFY CASES INTO GROUPS.

A CASE IS 'PREDICTED' TO BE IN GROUP
      ALIVE    IF PROBABILITY LESS THAN OR EQUAL TO CUTPOINT, OR IN GROUP
      DEAD     IF PROBABILITY GREATER THAN CUTPOINT.

EACH CUTPOINT YIELDS A CLASSIFICATION MATRIX, CONTAINING COUNTS OF

                        PREDICTED AS DEAD            PREDICTED AS ALIVE

    CASES IN DEAD        (A BELOW)         I          (B BELOW)
                        ---------------------+------------------------
    CASES IN ALIVE       (C BELOW)         I          (D BELOW)

THE CROSS PRODUCT RATIO (COL. 11 BELOW) IS COMPUTED FROM THE CLASSIFICATION
MATRIX AS THE PRODUCT OF THE COUNTS OF CORRECT CLASSIFICATIONS
DIVIDED BY THE PRODUCT OF THE COUNTS OF INCORRECT CLASSIFICATIONS  ( AD/BC )

THE 'COST' MATRIX TIMES THE CLASSIFICATION MATRIX YIELDS THE GAIN/LOSS FUNCTION.

THE COST MATRIX                        P R E D I C T E D
                                       DEAD       ALIVE
                         DEAD          0.00       -1.00
                         ALIVE        -1.00        0.00

LOSS = + 0.0000 A - 1.0000 B - 1.0000 C + 0.0000 D
```

Fig. 14-3 Explanation of BMDP classification table after full regression of seven variables. The statement "Col. 11 below" refers to the table shown in Fig. 14-4. (BMDP LR program.)

The only BMDP display for full regression here consists of the classification tables produced as an optional request. The BMDP classifications are both more and less informative than the SAS approach. As shown in the printed explanation offered in Fig. 14-3, BMDP allows the formation of a "cost matrix" and the calculation of a "gain/loss function" for the accuracy of the classifications. When (as in the situation here) costs are not deliberately specified, the program's default option assigns values of **0** for correct estimates and -1 for incorrect estimates. With this convention, the best demarcation in the classification table will produce the smallest negative number for the loss function.

The BMDP classification table in Fig. 14-4 differs from the corresponding SAS table (Fig. 14-2). The BMDP cutpoints are cited for values differing by increments of .016 or .017; and results are shown both for the AD/BC cross product ratio of correct and incorrect predictions, and also for the gain/loss function. In this instance, a cutpoint at .492 or .508 produces 71% total accuracy for the predictions, a relatively high AD/BC ratio (5.989), and the lowest loss function value (-58.00) for the 58 wrong predictions in the group.

In all three respects, the BMDP tabulation is more informative than the SAS approach. The BMDP procedure, however, does not produce a value for lambda or

340 Regression for Nondimensional Targets

CUT-POINT	CORRECT PREDICTIONS			PERCENT CORRECT			INCORRECT PREDICTIONS			CR PROD RATIO	GAIN OR LOSS
	DEAD	ALIVE	TOTAL	DEAD	ALIVE	TOTAL	DEAD	ALIVE	TOTAL		
	A	D	E=A+D	A/(A+B)	D/(C+D)	E/(E+F)	B	C	F=B+C	AD/BC	
0.108	106	2	108	100.00	2.13	54.00	0	92	92	UNDEFINED	92.00
0.125	106	4	110	100.00	4.26	55.00	0	90	90	UNDEFINED	-90.00
0.142	105	8	113	99.06	8.51	56.50	1	86	87	9.767	-87.00
0.158	105	11	116	99.06	11.70	58.00	1	83	84	13.916	-84.00
0.175	105	13	118	99.06	13.83	59.00	1	81	82	16.852	-82.00
0.192	105	16	121	99.06	17.02	60.50	1	78	79	21.538	-79.00
0.208	105	18	123	99.06	19.15	61.50	1	76	77	24.868	77.00
0.225	105	22	127	99.06	23.40	63.50	1	72	73	32.083	-73.00
0.242	102	26	128	96.23	27.66	64.00	4	68	72	9.750	-72.00
0.258	102	29	131	96.23	30.85	65.50	4	65	69	11.377	-69.00
0.275	100	34	134	94.34	36.17	67.00	6	60	66	9.444	-66.00
0.292	98	36	134	92.45	38.30	67.00	8	58	66	7.603	-66.00
0.308	97	37	134	91.51	39.36	67.00	9	57	66	6.996	-66.00
0.325	97	39	136	91.51	41.49	68.00	9	55	64	7.642	-64.00
0.342	96	41	137	90.57	43.62	68.50	10	53	63	7.426	-63.00
0.358	92	43	135	86.79	45.74	67.50	14	51	65	5.541	-65.00
0.375	90	48	138	84.91	51.06	69.00	16	46	62	5.870	-62.00
0.392	89	51	140	83.96	54.26	70.00	17	43	60	6.209	-60.00
0.408	87	53	140	82.08	56.38	70.00	19	41	60	5.919	60.00
0.425	83	53	136	78.30	56.38	68.00	23	41	64	4.665	-64.00
0.442	83	55	138	78.30	58.51	69.00	23	39	62	5.089	-62.00
0.458	82	57	139	77.36	60.64	69.50	24	37	61	5.264	-61.00
0.475	81	59	140	76.42	62.77	70.00	25	35	60	5.462	-60.00
0.492	81	61	142	76.42	64.89	71.00	25	33	58	5.989	-58.00
0.508	81	61	142	76.42	64.89	71.00	25	33	58	5.989	-58.00
0.525	79	61	140	74.53	64.89	70.00	27	33	60	5.409	-60.00
0.542	77	62	139	72.64	65.96	69.50	29	32	61	5.144	-61.00
0.558	75	64	139	70.75	68.09	69.50	31	30	61	5.161	-61.00
0.575	72	66	138	67.92	70.21	69.00	34	28	62	4.992	-62.00
0.592	70	70	140	66.04	74.47	70.00	36	24	60	5.671	-60.00
0.608	67	73	140	63.21	77.66	70.00	39	21	60	5.972	-60.00
0.625	60	74	134	56.60	78.72	67.00	46	20	66	4.826	-66.00
0.642	56	76	132	52.83	80.85	66.00	50	18	68	4.729	-68.00
0.658	55	76	131	51.89	80.85	65.50	51	18	69	4.553	-69.00
0.675	51	76	127	48.11	80.85	63.50	55	18	73	3.915	-73.00
0.692	47	80	127	44.34	85.11	63.50	59	14	73	4.552	-73.00
0.708	43	82	125	40.57	87.23	62.50	63	12	75	4.664	-75.00
0.725	39	83	122	36.79	88.30	61.00	67	11	78	4.392	-78.00
0.742	34	84	118	32.08	89.36	59.00	72	10	82	3.967	-82.00
0.758	31	88	119	29.25	93.62	59.50	75	6	81	6.062	-81.00
0.775	29	88	117	27.36	93.62	58.50	77	6	83	5.524	-83.00
0.792	25	89	114	23.58	94.68	57.00	81	5	86	5.494	-86.00
0.808	23	89	112	21.70	94.68	56.00	83	5	88	4.933	-88.00
0.825	18	91	109	16.98	96.81	54.50	88	3	91	6.205	-91.00
0.842	12	91	103	11.32	96.81	51.50	94	3	97	3.872	-97.00
0.858	9	93	102	8.49	98.94	51.00	97	1	98	8.629	-98.00
0.875	4	94	98	3.77	100.00	49.00	102	0	102	UNDEFINED	-102.00
0.892	4	94	98	3.77	100.00	49.00	102	0	102	UNDEFINED	-102.00
0.908	1	94	95	0.94	100.00	47.50	105	0	105	UNDEFINED	-105.00
0.925	1	94	95	0.94	100.00	47.50	105	0	105	UNDEFINED	-105.00
0.942	1	94	95	0.94	100.00	47.50	105	0	105	UNDEFINED	-105.00
0.958	0	94	94	0.00	100.00	47.00	106	0	106	UNDEFINED	-106.00

Fig. 14-4 BMDP 'Classification table' (BMDP LR program) for choosing optimal boundary for estimates after full regression of seven variables.

for any of the additional indexes (C, Kendall's tau, etc.). that are offered by SAS. According to your opinion of the latter indexes, their absence here may be an advantage or disadvantage.

Perhaps to compensate for the absence of the additional indexes, the BMDP program also produces a spidery graph that plots, for each cutpoint, the proportion of correct estimates in each group and in the totals. The graph is not shown here because it offers nothing that cannot be recognized easily, and more precisely, in the tabulation of Fig. 14-4.

14.3. Indexes for sequential decisions

Because sequential logistic regression is usually conducted in a stepwise—i.e., step up and then maybe down—manner, the customary decisions at each step involve an appraisal of candidate variables that have not yet entered the model. Several additional indexes can be used during these activities.

14.3.1. Joint effect of all remaining variables

At the end of each sequential step, a special chi-square score statistic can be calculated to show what would happen if *all* of the remaining candidate variables were entered into the model to produce a full regression. Labelled as "residual chi-square," it is essentially the difference between the current value for model chi-square and the corresponding value that would be obtained if *all* the remaining variables were in the model. The procedure will be illustrated in Section 14.4.1.1.

14.3.2. Tests for inclusion of individual variables

At each step, each variable not yet in the model can be checked in an individual regression that allows an appraisal of the effect of the candidate.

In SAS procedures, this effect is shown with an associated score chi-square, which approximates the Wald chi-square value that would be obtained for the selected variable if it alone were the next entry into the model. In BMDP procedures, a new log likelihood value is displayed and used incrementally to show what would occur in the new model if the selected variable were entered alone. To illustrate this tactic, suppose the existing model, with p − 1 variables, has ln L_{p-1} = −138.267 and suppose the potential new model, with p variables, has ln L_p = −123.1075 for variable X_p. The doubled incremental value of −2[138.267 − 123.1075] = 30.319 would be the approximate chi-square to enter (marked APPROX. CHI-SQ. ENTER) for that variable.

The variable selected to enter is chosen in the SAS procedure as the one with the smallest P value for its chi-square score statistic, and in the BMDP procedure as the one with the highest increment in log likelihood or "approximate chi-square."

14.3.3. Deletion of variables

If the SAS stepwise logistic procedure does not remove any variables after each sequential step, the removal efforts are not reported until the end of the procedure. The BMDP printout, however, shows each of the stepwise phases. At the end of each step, the program determines a value of chi-square-to-enter for each remaining candidate variable, but also checks APPROX. CHI-SQ. REMOVE for taking out variables that are already in the model. The latter tactic is the reverse of what was just described for *entering* a variable: the increment in log likelihoods is examined for what would happen if a variable were *removed*. For example, suppose that variable 3 has just entered the model with an incremental chi-square of 6.20, that the new log likelihood is −116.607, and that the log likelihood would become −120.645 if variable 2 were removed at this point. The doubled increment in the two log likelihoods is 2(120.645 − 116.607) = 8.08, which is the approximate chi-square for removal of variable 2. Since the associated P value (at 1 D.F.) is .0045, variable 2 would *not* be removed.

14.4. Illustrations of stepwise regression

The printouts for stepwise logistic regression analyses of the illustrative data set are shown in the subsections that follow.

14.4.1. BMDP procedure

In the BMDP stepwise procedure, the maximum likelihood estimates for the *age* variable did not converge (for unknown reasons) when the convergence criterion (called PCONV) had the default value of .0001. Convergence promptly occurred however—with no change in the pertinent statistical indexes—when PCONV was relaxed to .005. The results that follow were obtained with the latter criterion.

```
STEP NUMBER    0
---------------
                       LOG LIKELIHOOD  =   -138.268
GOODNESS OF FIT CHI-SQ (2*O*LN(O/E))   =    276.535   D.F.= 199   P-VALUE= 0.000
GOODNESS OF FIT CHI-SQ ( C.C.BROWN )   =      0.000   D.F.=   0   P-VALUE= 1.000
                         STANDARD                      95% C.I. OF EXP(COEF)
    TERM       COEFFICIENT   ERROR   COEF/SE   EXP(COEF)   LOWER-BND UPPER-BND

CONSTANT          0.1201     0.142    0.848      1.13        0.853     1.49


STATISTICS TO ENTER OR REMOVE TERMS
-----------------------------------
                APPROX.            APPROX.
    TERM       CHI-SQ. D.F.       CHI-SQ. D.F.                    LOG
                ENTER              REMOVE         P-VALUE      LIKELIHOOD

AGE              0.04    1                         0.8340      -138.2458
MALE             1.58    1                         0.2081      -137.4755
TNMSTAGE        30.32    1                         0.0000      -123.1075
SXSTAGE         18.54    1                         0.0000      -128.9966
PCTWTLOS         3.66    1                         0.0556      -136.4357
HCT              5.48    1                         0.0192      -135.5283
PROGIN           1.43    1                         0.2315      -137.5518
CONSTANT                            0.72    1      0.3952      -138.6292
CONSTANT                            IS IN              MAY NOT BE REMOVED.
```

Fig. 14-5 Beginning of stepwise logistic regression in BMDP LR program.

Fig. 14-5 shows step 0 in the BMDP LR program. The intercept (called CONSTANT) is entered as a coefficient of .1201, with ln likelihood = −138.268. This value is negatively doubled to become 276.535 as a goodness of fit chi-square, marked "2*0*LN (O/E)". The C. C. Brown chi-square test, which appears somewhat incongruously at step 0, has nothing to fit at this point, and receives a value of 0. When each candidate variable is tested in the section marked STATISTICS TO ENTER OR REMOVE TERMS, the smallest absolute value for log likelihood is produced at −123.1075 by TNM stage. The difference between this value and the ln L_0 of −138.268 is −15.16, which becomes negatively doubled to yield 30.32, the value cited for the "Approx. chi-sq. enter" of TNM stage.

14.4.1.1. Steps 1 and 2

Fig. 14-6 shows what happens when TNM stage is entered in step 1. The log likelihood falls to −128.108, which is negatively doubled to produce the first goodness of fit chi-square as 246.216. It is marked "2*0*LN (O/E)" and has an associated P value of .011 at 198 degrees of freedom. The latter value, calculated as N − (p + 1), is 200 − (1 + 1). (This chi-square result is really mislabelled, since it tests independence for variables in the model, not goodness of fit.) The improve-

```
STEP NUMBER    1              TNMSTAGE              IS ENTERED
--------------
                    LOG LIKELIHOOD =   -123.108
IMPROVEMENT CHI-SQUARE  ( 2*(LN(MLR) ) =    30.320  D.F.=   1  P-VALUE= 0.000
GOODNESS OF FIT CHI-SQ  (2*O*LN(O/E)) =   246.216  D.F.= 198  P-VALUE= 0.011
GOODNESS OF FIT CHI-SQ (HOSMER-LEMESHOW)=   0.808  D.F.=   3  P-VALUE= 0.848
GOODNESS OF FIT CHI-SQ  ( C.C.BROWN ) =     0.186  D.F.=   2  P-VALUE= 0.911
                        STANDARD                      95% C.I. OF EXP(COEF)
     TERM      COEFFICIENT  ERROR    COEF/SE   EXP(COEF)   LOWER-BND UPPER-BND

TNMSTAGE        0.5056    0.971E-01    5.21     1.66         1.37      2.01
CONSTANT       -1.495     0.346       -4.32     0.224        0.113     0.443


CORRELATION MATRIX OF COEFFICIENTS
----------------------------------

                TNMSTAGE   CONSTANT

TNMSTAGE         1.000
CONSTANT        -0.896     1.000


STATISTICS TO ENTER OR REMOVE TERMS
-----------------------------------
               APPROX.              APPROX.
     TERM      CHI-SQ. D.F.         CHI-SQ. D.F.                     LOG
               ENTER                REMOVE          P-VALUE      LIKELIHOOD

AGE             0.00    1                            0.9528      -123.1058
MALE            1.23    1                            0.2681      -122.4944
TNMSTAGE                              30.32   1      0.0000      -138.2678
SXSTAGE         4.93    1                            0.0265      -120.6446
PCTWTLOS        1.80    1                            0.1794      -122.2062
HCT             6.80    1                            0.0091      -119.7064
PROGIN          0.79    1                            0.3741      -122.7125
CONSTANT                              20.67   1      0.0000      -133.4414
CONSTANT                              IS IN              MAY NOT BE REMOVED.
```

Fig. 14-6 Step 1 of BMDP stepwise logistic regression procedure begun in Fig. 14-5.

ment chi-square of 30.320 (P = ".000") is the increment between 246.216 and the previous value of 276.535 in Fig. 14-5. The Hosmer-Lemeshow and C. C. Brown goodness of fit chi-square tests show good fit, with P values above .8. The printout in Fig. 14-6 then shows the estimated coefficients for TNM stage and the intercept (marked CONSTANT). Each entry of the b_j coefficient is followed by the associated STANDARD ERROR, the Z ratio of COEF/SE, the value of e^{b_j} [marked EXP(COEF)] for the corresponding odds ratio, and the lower and upper boundaries of the 95% confidence interval around the odds ratio. (The BMDP program prints these boundaries routinely. They must be requested as an option in SAS.) As an example of the calculations, b_j for TNMSTAGE is .5056, and e^{b_j} will be $e^{.5056} = 1.66$. The standard error is .0971, so that the 95% interval *for the coefficient* will be .5056 ± (1.96)(.097), which extends from .3153 to .6959. The values of $e^{.3153}$ and $e^{.6959}$ are, respectively, 1.37 and 2.01, which are cited boundaries for 95% C.I. of EXP(COEF).

The section marked CORRELATION MATRIX OF COEFFICIENTS can be ignored, for reasons discussed earlier (Section 14.2.1.4). In the list of statistics to enter or remove

344 *Regression for Nondimensional Targets*

terms for step 2, Hematocrit (HCT) has the lowest log likelihood and highest chi-square value (with P = .0091).

```
STEP NUMBER    2              HCT                    IS ENTERED
---------------
                    LOG LIKELIHOOD =   -119.706
IMPROVEMENT CHI-SQUARE    ( 2*(LN(MLR) ) =      6.802  D.F.=   1   P-VALUE= 0.009
GOODNESS OF FIT CHI-SQ    (2*O*LN(O/E)) =    239.414  D.F.= 197   P-VALUE= 0.021
GOODNESS OF FIT CHI-SQ (HOSMER-LEMESHOW)=      9.697  D.F.=   8   P-VALUE= 0.287
GOODNESS OF FIT CHI-SQ    ( C.C.BROWN ) =      3.736  D.F.=   2   P-VALUE= 0.154
                               STANDARD                       95% C.I. OF EXP(COEF)
     TERM        COEFFICIENT   ERROR   COEF/SE   EXP(COEF)   LOWER-BND  UPPER-BND

TNMSTAGE          0.5248     0.991E-01   5.30      1.69        1.39       2.05
HCT              -0.7855E-01 0.309E-01  -2.54      0.924       0.870      0.983
CONSTANT          1.673      1.28        1.31      5.33        0.428     66.2


CORRELATION MATRIX OF COEFFICIENTS
----------------------------------

                TNMSTAGE    HCT      CONSTANT

TNMSTAGE         1.000
HCT             -0.138     1.000
CONSTANT        -0.111    -0.961     1.000


STATISTICS TO ENTER OR REMOVE TERMS
-----------------------------------
                APPROX.              APPROX.
     TERM       CHI-SQ.  D.F.        CHI-SQ.  D.F.                     LOG
                ENTER                REMOVE             P-VALUE     LIKELIHOOD

AGE              0.09     1                             0.7695      -119.6635
MALE             1.81     1                             0.1786      -118.8017
TNMSTAGE                              31.64     1       0.0000      -135.5282
SXSTAGE          6.20     1                             0.0128      -116.6068
PCTWTLOS         0.36     1                             0.5508      -119.5284
HCT                                    6.80     1       0.0091      -123.1075
PROGIN           0.62     1                             0.4318      -119.3974
CONSTANT                               1.74     1       0.1873      -120.5757
CONSTANT                              IS IN                      MAY NOT BE REMOVED.
```

Fig. 14-7 Step 2 of BMDP stepwise logistic regression begun in Figs. 14-5 and 14-6.

In step 2, shown in Fig. 14-7, hematocrit is entered and the improvement chi-square becomes 6.802 (as might be expected from the previous chi-square-to-enter value for hematocrit). The respective values for the intercept, TNM stage, and hematocrit coefficients become 1.673, 0.5248, and −0.0786. (The last has been printed as −0.7855 E − 01.) Among variables that are candidates for entry at the next step, symptom stage has the highest chi-square value, 6.20, together with a log likelihood of −116.6068, and P = .0128.

In the explorations for removal, nothing is deleted after step 2, because the constant (intercept) must remain, and the TNMSTAGE and HCT variables continue to have significant chi-square results.

14.4.1.2. Step 3 and conclusion

The rest of the BMDP stepwise process is shown in Fig. 14-8. The respective coefficients for intercept, TNM stage, hematocrit, and symptom stage become 1.314,

```
STEP NUMBER    3               SXSTAGE              IS ENTERED
---------------

                      LOG LIKELIHOOD  =    -116.607
IMPROVEMENT CHI-SQUARE   ( 2*(LN(MLR) ) =       6.199   D.F.=   1   P-VALUE= 0.013
GOODNESS OF FIT CHI-SQ   (2*O*LN(O/E)) =     233.215   D.F.= 196   P-VALUE= 0.035
GOODNESS OF FIT CHI-SQ (HOSMER-LEMESHOW)=      4.403   D.F.=   8   P-VALUE= 0.819
GOODNESS OF FIT CHI-SQ   ( C.C.BROWN ) =       1.811   D.F.=   2   P-VALUE= 0.404
                         STANDARD                      95% C.I. OF EXP(COEF)
     TERM       COEFFICIENT   ERROR   COEF/SE  EXP(COEF)   LOWER-BND  UPPER-BND

TNMSTAGE         0.4258       0.106     4.01    1.53        1.24       1.89
SXSTAGE          0.4106       0.166     2.47    1.51        1.09       2.09
HCT             -0.8804E-01   0.320E-01 -2.75   0.916       0.860      0.975
CONSTANT         1.314        1.30      1.01    3.72        0.284     48.8


CORRELATION MATRIX OF COEFFICIENTS
----------------------------------

               TNMSTAGE   SXSTAGE    HCT       CONSTANT

TNMSTAGE        1.000
SXSTAGE        -0.315     1.000
HCT            -0.078    -0.166     1.000
CONSTANT       -0.078    -0.079    -0.932     1.000


STATISTICS TO ENTER OR REMOVE TERMS
-----------------------------------
               APPROX.             APPROX.
    TERM       CHI-SQ. D.F.        CHI-SQ. D.F.                  LOG
               ENTER               REMOVE          P-VALUE    LIKELIHOOD

AGE             0.02    1                          0.8847     -116.5963
MALE            2.01    1                          0.1567     -115.6037
TNMSTAGE                            16.82    1     0.0000     -125.0174
SXSTAGE                              6.20    1     0.0128     -119.7063
PCTWTLOS        0.01    1                          0.9179     -116.6015
HCT                                  8.08    1     0.0045     -120.6449
PROGIN          0.38    1                          0.5370     -116.4162
CONSTANT                             1.02    1     0.3115     -117.1189
CONSTANT                            IS IN                     MAY NOT BE REMOVED.


NO TERM PASSES THE REMOVE AND ENTER LIMITS (  0.1500  0.1000 ) .


SUMMARY OF STEPWISE RESULTS

STEP      TERM                   LOG       IMPROVEMENT      GOODNESS OF FIT
NO.   ENTERED  REMOVED    DF  LIKELIHOOD  CHI-SQUARE P-VAL  CHI-SQUARE P-VAL
---   ---------------    ---  ----------  ----------------  ----------------
  0                              -138.268                    276.535  0.000
  1   TNMSTAGE            1      -123.108   30.320  0.000    246.216  0.011
  2   HCT                 1      -119.706    6.802  0.009    239.414  0.021
  3   SXSTAGE             1      -116.607    6.199  0.013    233.215  0.035
```

Fig. 14-8 Remainder of BMDP stepwise logistic regression begun in Figs. 14-5 through 14-7.

0.4258, −0.0880, and 0.4106. Since P values for the coefficients are not listed in the BMDP printout, they can be obtained by interpreting COEF/SE as a Gaussian Z test, for which $P \leq .05$ when $|Z| \geq 1.96$. The TNMSTAGE, SXSTAGE, and HCT variables all fulfill this requirement. Alternatively, stochastic significance can be declared at $P \leq .05$ whenever the 95% confidence interval for the odds ratio, i.e., EXP(COEF), excludes the value of 1. This exclusion also occurs for the three main variables.

After the correlation matrix and the statistics showing that no additional variable has a high enough APPROX. CHI-SQ. ENTER and that existing variables are too high in APPROX. CHI-SQ. REMOVE, the stepping process ends with the statement NO TERM PASSES THE REMOVE AND ENTER LIMITS (0.1500 0.1000). Those limits were set by the default selection mechanism of the BMDP program.

The last entry in Fig. 14-8 is a SUMMARY OF STEPWISE RESULTS. It shows, for the variables entered at each step, the corresponding degrees of freedom, log likelihood, improvement chi-square, its P value, a model chi-square that is erroneously labelled "goodness of fit," and its P value. The "goodness of fit" chi-square values here are twice the negative log likelihood values, and the improvement chi-square at each step is the increment between the two adjacent "goodness-of-fit" chi-square values.

14.4.2. SAS procedure

To save space, details of the stepwise SAS logistic procedure are not shown here. The printout at each step resembles what appeared in Fig. 14-1. The main difference from the BMDP sequence is that before doing step 1, after the intercept is entered for a model containing no variables in step 0, the SAS program does essentially a full regression with all variables that are *not* yet in the model. The residual chi-square for that full regression is 42.1739 with 7 D.F. (The result is the same as the score value of chi-square, shown in Fig. 14-1, when all seven variables were in the model.) Immediately afterward, but still before step 1, individual logistic regressions are done, one variable at a time, for each variable that is *not* in the model. For each of these regressions, the printout shows the associated model chi-square score and the corresponding P value that would occur if that variable were added alone to the model. From those results, the highest ranking variable (in this case, TNMSTAGE) is then entered in step 1.

At the end of step 1, the model chi-square score statistic is 29.327 and the new residual chi-square is 14.9078 with 6 D.F. and P = .02. The latter value represents the maximum improvement that could occur if all the remaining six variables were added to the model.

After hematocrit was entered in step 2 and symptom stage in step 3, the model chi-square (as -2 log likelihood) rose to 43.322, which had P = .0001 at 3 D.F. The corresponding score chi-square was 40.240. The residual chi-square (2.39) and the score chi-square values for the remaining four variables showed no significant P values. The stepping process then stopped because no additional variables met the criterion for entry.

When compared with the full regression in Fig. 14-1, the three-variable model that emerged from the stepwise regression had a desirably lower AIC (241.216 vs. 246.780) and a lower SC (254.409 vs. 273.167). Although the P values were listed as .0001 for both models, the P value was probably lower for the three-variable model, whose slightly smaller model chi-square values would have been interpreted with 3 rather than 7 degrees of freedom. The three-variable model has slightly smaller results for the various indexes of accuracy and concordance, and the crude indexes of explained likelihood are $(43.322 - 6)/276.538 = .135$ with the R_H^2 tactic and $1 - (241.216/276.538) = .128$ with the relative AIC tactic (discussed in Section 13.7.2.3).

In the next phase of the printout (not shown here), the classification table showed a total accuracy of 71.5% with the cutoff placed at a probability level of .48.

14.4.3. Checking results

The logistic maximum likelihood estimates for $\hat{Y}_i$ have the property that the average value of $\hat{Y}_i$ equals $\bar{Y}$. This property could be checked directly, but a simpler, although cruder, approach for checking the regression results is to insert mean values for the variables into the regression formula. For the variables whose coefficients are shown in Fig. 14-8, the algebraic model would be expressed as

G = 1.314 + .4258 (TNM stage) + .4106 (symptom stage) − .08804 (hematocrit).
The mean values for these three variables (see Fig. 6-2) were TNM stage = 3.2150, symptom stage = 2.6150, and hematocrit = 41.089. When they are inserted into the model, the results for the average prediction would be
G = 1.314 + (.4258)(3.2150) + (.4106)(2.6150) − (.08804)(41.089) = 0.1392.

When expressed as $\hat{Y} = e^{0.1392}/(1 + e^{0.1392})$, the probability for the average estimate for death would be $1.1493/2.1493 = .5347$, which is quite close to the overall observed value of $106/200 = .53$.

14.5. Forced regression

The full and stepwise regression procedures have both indicated that the illustrative data set has three important independent variables. An optimal model, therefore, could be constructed as a forced regression of these three variables. The construction is really unnecessary now, however, since it appeared as the last of the three steps in the stepwise procedures. Nevertheless, to check the consistency of the computer package systems, a forced regression was checked with TNMSTAGE, SXSTAGE, and HCT as the three main variables. The results are discussed here, but printouts are not shown.

For the SAS program, the forced results were identical in every respect to what emerged after step 3 of the stepwise procedure. For the BMDP program, almost all the forced results were analogously identical, but one small quirk was noted. At the end of step 3 in the stepwise BMDP procedure in Fig. 14-8, the goodness of fit chi-square for the model was 233.215, with D.F. = 196 and P = .035. After the three variables were entered in step 0 of the forced regression, however, the corresponding value for chi-square was 171.749 with D.F. = 140 and P = .035. Since all other statistical aspects of the two printouts were otherwise identical, the reason for the disparity is uncertain. The D.F. = 140 is mystifying, but probably arises because (as noted in Section 14.7.1) the three-variable BMDP model contains 144 unique "covariate patterns," and because DF was calculated here (for obscure reasons) with N = 144 rather than 200. The value of N − p − 1 is then 144 − 3 − 1 = 140. On the other hand, since BMDP does not use the AIC or SC scores, and since everything else was similar, the disparity appears to be inconsequential.

14.6. All possible regressions

The latest version of the SAS system has a program for doing all possible regressions with a logistic format. To increase rapidity of computation, the procedure uses the model chi-square score test as the criterion for decisions. Like the all-possible-regression strategy for linear regression, the logistic procedure (whose results are not shown) is possibly attractive for screening, and may be helpful for an analyst who has no substantive knowledge of the topic under investigation and whose sole interest is a best model.

14.7. Regression diagnostics

As discussed earlier (see Section 12.4.3), the formal activities called regression diagnostics are directed solely at finding excessively influential observations, manifested by poorly fitting estimates. Consequently, the printouts produced under the diagnostics title are not particularly helpful for discerning more cogent problems either in interactions or in linear conformity of the model. The discussion that follows in Sections 14.7.1 through 14.7.5 can therefore be omitted for most practical purposes, but is available for readers who might like to know what is displayed in the diagnostics printouts. The interaction problems are discussed in Section 14.7.6 and linear conformity in Section 14.7.7.

14.7.1. Covariate patterns

In the jargon of regression, the independent variables $X_1, X_2, \ldots, X_k$ are called *covariates*. Two persons (or "observations") have the same covariate pattern if their collection of individual values is identical for each of the X_j variables. Thus, persons 2 and 7 have the same covariate pattern if $X_{21} = X_{71}$, $X_{22} = X_{72}$, $X_{23} = X_{73}$, and so on. Such persons will have identical estimates for $\hat{Y}_i$.

With all the possibilities permitted by seven independent variables, it is not surprising that no 2 persons in the 200 members of the illustrative data set had the same covariate pattern. On the other hand, with a smaller number of independent variables, some of which have a limited number of categories, duplicate covariate patterns can occur. Thus, with only the three main independent variables in the model, similar covariate patterns for 2 or more persons occurred in 38 instances, involving 94 persons. The remaining 106 persons in the group of 200 had unique covariate patterns, so that the total number of distinct covariate patterns was 144.

The covariate patterns are used in diverse ways during the regression-diagnostics process.

14.7.2. Indexes for individual 'misfits'

Both the SAS and BMDP programs use counterparts of the indexes described earlier (throughout Section 12.4) to express the magnitude and effect of individual discrepancies between observed and estimated values. The first four indexes of individual "misfit" all refer to the residual values of $Y_i - \hat{Y}_i$.

14.7.2.1. 'Hat matrix diagonal'

The "hat matrix diagonal," h_i, is a counterpart of the same entity (see Section 12.4.1.2) in linear regression, converted by Pregibon[3] into a logistic-regression format. The value of h_i represents, in essence, the "leverage" of covariate pattern i as determined by the multivariate distance of the observation's values from the centroid of the independent variables.

14.7.2.2. 'Pearson residuals'

Being somewhat analogous to the entities used in a chi-square calculation, the *Pearson residuals* are labeled as such in the SAS program, but are called (OBS-PRED)/(SQRT PQ/N) in the BMDP program.

Each Pearson residual, symbolized as r_i for covariate pattern i, is calculated as

$$r_i = \frac{Y_i - m_i \hat{Y}_i}{\sqrt{m_i \hat{Y}_i (1 - \hat{Y}_i)}}.$$

In this expression, m_i = number of persons with a particular covariate pattern; Y_i = number of those persons observed to have the coded value of **1** (e.g., **dead** or **alive**); and $\hat{Y}_i$ = estimated probability produced by the covariates in pattern i. According to Hosmer and Lemeshow,[4] the Pearson chi-square statistic for independence (rather than goodness of fit) of the model can be calculated as $X^2 = \Sigma r_i^2$.

To illustrate the calculation, 2 persons (both **alive**) had the same covariate pattern, for which the estimated $\hat{Y}_i$ for death was .3832. Since the prediction was made for death, the alive patients will be cited as **0**. Accordingly, r_i is calculated as $\{0 - [2(.38732)]\}/\sqrt{2(.3832)(.6168)} = -1.1147$.

14.7.2.3. Standardized residual

This index appears only in BMDP printouts, where it is marked (OBS-PRED)/S.E. RES. For the denominator, the variance of the residual for covariate pattern i is calculated as $m_i \hat{Y}_i (1 - \hat{Y}_i)(1 - h_i)$. In this formula, h_i and m_i have the meanings discussed in Sections 14.7.2.1 and 14.7.2.2. If r_i is the Pearson residual, the standardized residual is $r_i/\sqrt{1 - h_i}$.

14.7.2.4. 'Deviance residuals'

In the expression for congruence, the value of $1 - |Y_i - \hat{Y}_i|$ is $(1 - \hat{Y}_i)$ if $Y_i = 0$, and $\hat{Y}_i$ if $Y_i = 1$. The natural logarithm of the estimated probability for either $\ln(1 - \hat{Y}_i)$ or $\ln \hat{Y}_i$ will be analogous to a distance.

The square root of twice this distance is called a *deviance residual,* symbolized as d_i. Thus,

$$d_i = \sqrt{2|\ln(1 - |Y_i - \hat{Y}_i|)|}.$$

If $Y_i = 0$, $d_i = \sqrt{2|\ln(1 - \hat{Y}_i)|}$ and if $Y_i = 1$, $d_i = \sqrt{2|\ln \hat{Y}_i|}$. Because the logarithms will always be negative for values of $\hat{Y}_i$ that lie between 0 and 1, the absolute values are used so that square roots can be derived. For example, suppose the $\hat{Y}_i$ estimate for death was .5856 for someone who was alive, so that Y_i would be coded as 0. The value of $d_i = \sqrt{2|\ln(1 - \hat{Y}_i)|}$ will be $-\sqrt{2|\ln .4144|} = -1.327$.

The value of each d_i^2 will be $2|\ln(1 - |Y_i - \hat{Y}_i|)|$; and the sum of d_i^2 values will be $\Sigma d_i^2 = 2[-\Sigma \ln(1 - |Y_i - \hat{Y}_i|)]$, which is twice the value of $-\ln L$. Thus, the "deviance," Σd_i^2, is equal to twice the negative log likelihood, symbolized as $\Sigma d_i^2 = -2\ln L$.

Each deviance residual is calculated with the foregoing formula when $m_i = 1$, i.e., when the covariate pattern has only one member. A more complex formula, cited on page 138 of the Hosmer and Lemeshow book,[4] is used when $m_i > 1$.

14.7.3. Indexes of postdeletion change

Five additional diagnostic indexes refer to the changes produced when the logistic regression is repeated after removal (or deletion) of a cited observation (or covariate pattern). One of these indexes, *Dfbeta,* is printed as a set of multiple values, representing the standardized difference in the estimated value of b_j for the intercept and for each independent variable.

The other four indexes represent the influence of the individual observations. Two of the indexes, called C and CBAR in the SAS printouts, represent "confidence interval displacement diagnostics" that are "based on the same idea as the Cook distance in linear regression theory." (In the BMDP program, a "measure similar to the Cook distance" is called INFLUENCE.)

350 Regression for Nondimensional Targets

The indexes called DIFDEV and DIFCHISQ in the SAS printouts represent, respectively, the change in deviance and the change in the Pearson chi-square statistic for the individual observation.

14.7.4. Tabular printouts

The SAS printout of diagnostic residuals for the logistic regression program must be ordered as an option. For the model containing only the three main independent variables, the complete printout occupied eighteen pages. It showed all the observed covariate values in the X_j variables for each observation (without combining them into similar covariate patterns), together with the Pearson and deviance residuals, the hat matrix diagonals, the Dfbeta values for the intercept and each main variable, and the C, CBAR, DIFDEV, and DIFCHISQ indexes. Each of these indexes and Dfbeta values for each observation is accompanied by an "asterisk graph" indicating its magnitude. (The largest values can be sought as the asterisks with greatest magnitudes.)

The option that produces the corresponding BMDP printout has more compact results, since they do not graphically display the magnitude of the misfit indexes and may not have a line for each observation. For the diagnostics of the three-main-variable logistic model, the BMDP printout for the three-variable model contained entries for 144 cells, consisting of 106 solitary cells (having only one person) and 38 cells whose covariates were shared by two or more persons. For each row (or covariate cell) the tabulation shows the number of dead and alive persons, the observed proportion dead, the predicted probability of death, the standard error of the predicted probability, the standardized residual, the predicted log odds, the Pearson residual, the deviance residual, the hat matrix diagonal, and the influence. The far right columns of the printout begin to show the individual values for each associated covariate, but horizontal space becomes exceeded, and values for some of the covariates are "wrapped around" to appear as a "double entry" beneath the previous columns.

14.7.5. Additional graphic displays

In addition to the asterisk graphics cited in Section 14.7.4, the SAS program can be asked to show plots of each of the diverse misfit indexes (as ordinate) versus the case number of each observation (as abscissa). The plots enable prompt identification of the observations with the largest indexes for each misfit category.

Hosmer and Lemeshow Goodness-of-Fit Test

Group	Total	SURV6 = DEAD		SURV6 = ALIVE	
		Observed	Expected	Observed	Expected
1	20	3	3.32	17	16.68
2	20	4	4.93	16	15.07
3	20	6	6.72	14	13.28
4	20	9	8.02	11	11.98
5	20	13	10.49	7	9.51
6	20	13	11.85	7	8.15
7	20	13	13.32	7	6.68
8	20	13	14.53	7	5.47
9	20	14	15.82	6	4.18
10	20	18	17.01	2	2.99

Goodness-of-fit Statistic = 4.1294 with 8 DF (p=0.8453)

Fig. 14-9 Display of SAS results used to calculate Hosmer-Lemeshow test after three main variables were forced. (Option in SAS PROC LOGISTIC program.)

Illustrations of Logistic Regression 351

Although the SAS programs, unlike their BMDP counterparts, do not routinely show the result of the Hosmer-Lemeshow goodness-of-fit test, an option allows the calculations to be completely displayed. Fig. 14-9 shows these results for the three-variable forced model in the illustrative data set. The 200 patients are divided into 10 groups of 20 each, ranked according to the estimated values of $\hat{Y}_i$. The number of observed dead and alive persons is shown for each group, together with the corresponding "expected" number estimated by $\hat{Y}_i$ from the forced regression model. When calculated as $\Sigma[\text{(observed-expected)}^2/\text{expected}]$ for each of the 20 cells of the table, the X^2 result here (4.1294) differs slightly from the corresponding value of 4.403 shown near the top of Fig. 14-8 when the same test was done for the same model with the BMDP program, although the P values are in the same vicinity (~.8). The slight disparity in X^2 probably arises from the way that tied values are divided in the decile partitions.

The BMDP program can optionally convert the diagnostics information into graphs showing the observed proportions of the target event versus its predicted probabilities or predicted log odds in each of the covariate cells. For seven independent variables in the illustrative data set, no two patients had identical covariate values; and so the graphic results produced useless patterns. All entries on the ordinate for observed proportions were at either 0 or 1 values along the top and bottom of the graphs. When the logistic model contained only the three main independent variables, some of the 200 patients could cluster in observed cells to form 144 distinct covariate patterns, thus allowing a few points to appear in the body of the graph, not just across the bottom and top. Despite the increased clustering, however, the spread of items was too diffuse for the graph to be helpful.

14.7.6. Tests for interaction

Since the examination of residuals and misfits is not particularly useful here, the next phase of a more complete regression diagnostics is to check for interactions. With three independent variables as the main effects, the additional entries can include three pairwise terms as two-way interactions and one term for a three-way interaction. If the three main variables are always entered as "main effects," the interactions can be tested for a six-term model, which adds only the three paired interactions; a four-term model, which adds only the single three-way interaction; and a seven-term saturated model, which adds all four types of interaction term.

All of these possibilities were explored for the illustrative data set. The results are discussed but are not displayed in the subsections that follow.

14.7.6.1. Two-way interactions only

For two-way interactions in a six-term model that contained three main terms and three paired interaction terms, the SAS and BMDP procedures gave identical results. In a full regression that allowed all terms to be included, none of the b_j coefficients was stochastically significant. After the three main terms were forced, none of the two-way paired interaction terms entered a stepwise model.

14.7.6.2. Three-way interaction

When the three main variables and the single three-way interaction term received a full regression that included all four terms, the three-way interaction term did not have a stochastically significant coefficient, but the coefficients changed for the three main terms. The respective values of b_j were .49 (with P =

.04) for TNM stage, $-.22$ (with P = .059) for hematocrit, and .32 for symptom stage (with P = .093). The three-way interaction term had $b_j = -.17$ with P = .620. When the three main terms were forced in a stepwise regression, the interaction term was not retained, and the model included only the three main-effect terms.

14.7.6.3. Saturated model

In the saturated model that contained terms for each of the main effects and for each possible interaction, none of the seven terms emerged as stochastically significant in full regression. After a forced entry of the three main variables, no other terms entered the stepwise model.

14.7.6.4. Interactions after 'reversed' recoding

Because the impacts on outcome were reversed for increases in hematocrit vs. increases in TNM stage and symptom stage, the interactions were re-explored after a 'reverse' recoding, previously discussed in Section 12.3.4, that would allow all effects for these variables to go in the same direction. In one recoding, hematocrit was expressed as 100-hematocrit, with the other two variables remaining intact. In a second recoding, hematocrit remained intact, but TNM stage and symptom stage had their ordinal codes reversed.

All of the interactions described in the previous three subsections were then reexamined for both of these revised coding systems. The results showed minor variations from what was noted before, but no striking discrepancies or dramatic revelations.

14.7.6.5. Conclusions

The results of the tests for interaction suggest that the previous conclusions about the main effects need not be altered. Neither the three-way nor any of the two-way interaction terms was significant when the main-effect terms were retained in the model.

14.7.7. Examination for linear conformity

Since neither the SAS nor BMDP programs offer a distinctive method for examining linear conformity of the logistic regression model, the check can be done in two simple ways, using multicategorical stratification. One approach is to examine the linearity of successive target proportions in categories of the ordinal variables. Another approach, if data are sufficiently abundant, is to examine odds ratios across appropriate conjoint strata. Both methods are displayed here.

To get enough members in all the cells, stages 2–3 and 4–5 of the TNM stage variable were combined in pairs so that the variable was condensed to three categories. A paired combination of stages 3–4 of the symptom stage variable also condensed it to three categories. To allow odds and odds ratios to be calculated for mortality, the results for these two variables were converted to rates of mortality rather than survival. With the cited arrangements, Table 14-2 shows the 6-month mortality rates for the 3 × 3 array of cells in the two-way conjunction of TNM stage and symptom stage.

Table 14-2 6-month mortality rates for compressed TNM and symptom stages in illustrative data set

Coding for TNM stage	Coding for symptom stage			Total
	1	2	3–4	
1	2/13 (15.4%)	4/27 (14.8%)	7/11 (63.6%)	13/51 (25.5%)
2–3	2/7 (28.6%)	14/29 (48.3%)	4/9 (44.4%)	20/45 (44.4%)
4–5	6/8 (75.0%)	16/26 (61.5%)	51/70 (72.9%)	73/104 (70.2%)
Total	10/28 (35.7%)	34/82 (41.5%)	62/90 (68.9%)	106/200 (53.0%)

14.7.7.1. Linearity of log odds ratios for individual variables

The odds ratios in successive categories of each individual variable can be determined from the row and column *totals* in Table 14-2. For the column totals of symptom stage alone, the odds ratios are $[34/(82 - 34)]/[10/(28 - 10)] = 1.275$ for the second column vs. the first, and $(62/28)/(34/48) = 3.13$ for the third column vs. the second. For TNM stage, the odds ratios are $(20/25)(13/38) = 2.34$ for row 2 vs. 1, and $(73/31)/(20/25) = 2.94$ for row 3 vs. 2. These results clearly show that the odds ratios are *not* similar from one cell to the next in totals for the individual variables. The average odds ratio for the symptom stage variable can be calculated as an arithmetic mean of 2.20 or as a geometric mean, $[(1.275)(3.13)]^{1/2} = 2.00$. The corresponding average odds ratios for TNM stage are 2.64, or 2.62 for the geometric mean. (These results are higher than the corresponding values of about 1.5 shown for both variables in Fig. 14-8, but the latter calculations were based on more categories for each variable.)

In the logistic formula, when $\ln(\text{odds}) = a + bX$, the increment in $\ln(\text{odds})$ should have a constant value as bX moves from one category to the next. The results for the successive increments in ln odds for the symptom stage columns are $\ln(34/48) - \ln(10/18) = .242$ for column 2 vs. 1 and $\ln(62/28) - \ln(34/48) = 1.140$ for column 3 vs. 2. The corresponding successive increments in ln odds for the TNM stage rows are .849 and 1.080.

Since the increments in *ln odds* are obviously not constant, the *single* coefficients that emerge in algebraic models for symptom stage and TNM stage will not give an accurate portrait of the impact of these variables in different zones.

14.7.7.2. Constancy of odds ratios across conjoint strata

The constancy of the successive odds ratios can also be checked by examining results within the rows and columns of the two-variable conjunction in Table 14-2. Thus, for columns 2 vs. 1 and columns 3 vs. 2 of symptom stage, the respective odds ratios are 0.95 and 10.06 in TNM row 1, 2.33 and 0.85 in TNM row 2, and 0.53 and 1.67 in TNM row 3. Going in the other direction, for rows 2 vs. 1, and rows 3 vs. 2 of TNM stage, the respective odds ratios are 2.2 and 7.5 in symptom stage column 1, 5.36 and 1.71 in column 2, and .45 and 3.36 in column 3.

Because the odds ratios for one variable have striking inconsistencies within the individual zones of the second variable, these results further demonstrate the inevitable difficulty of logistic efforts to model the real world. For the six odds-ratio values of symptom stage, the mean is 2.73 and the geometric mean is 1.59. For the corresponding results in TNM stage, the mean is 3.43 and the geometric mean is 2.47. Thus, the average values for odd ratios of each variable in the bivariate cells are also different from the average values found when each variable was examined alone.

14.7.8. Events per variable

As noted in Section 12.1.5, the regression coefficients (despite stochastic significance) may become unstable or misleading when the analysis contains too few events per variable (*epv*). In the full regression models of Section 14.2, the *epv* ratio was $94/7 = 13.4$, for 94 events and 7 independent variables. This ratio is below the "desirable" level of $\geqq 20$, but above the "dangerous" level of 10. In the procedures containing a three-variable model, however, the *epv* ratio of $94/3 = 31.3$ was well above the threshold of concern.

14.7.9. Summary of 'diagnostic' examinations

The complete "diagnostic workup" for logistic regression analysis of the illustrative data set has included an examination of the possible interaction effects, the linear conformity of the model, and the ratio of events per variable, as well as the search for outliers conducted with statistical regression diagnostics. The main diagnostic conclusion or "take-home message" is that the logistic procedure did an excellent job of identifying the three main independent variables, although the linear model was not really appropriate for the pattern of the data, and did not offer a particularly good fit. Thus, if the main goal was to find the important predictors (or to confirm those identified in previous analyses), the logistic procedure was splendid. The results could be used effectively to create a prognostic staging system either from a point score assigned to the main predictors, or from combining them into a conjunctive consolidation. On the other hand, if the goal was to quantify the effects of the *individual* predictors, the single values of odds ratios for each variable would be unsatisfactory. The inconsistent variations within zones of the same variable, and across zones of other variables, would prevent a single odds ratio value from being more than an imprecise and possibly distorted average.

14.8. Challenges to interaction models

To determine the efficacy of interactions in algebraic models, they were challenged with data from the four major "interaction" tables that appeared in Chapter 3. Because each of the four tables contained two independent variables and an outcome variable, each of the algebraic interaction models was constructed with a term for each of the two independent variables, and with a third term for their interaction. The algebraic models were explored with both full and stepwise regression procedures. An algebraic analysis was regarded as successful if it correctly identified the existence of a major interaction in the corresponding table.

Table 3-5, for which the interaction produced an "internal valley" in the middle cell, had previously been algebraically examined with multiple *linear* regression in Section 12.2.2. When subjected to multiple *logistic* regression in the full three-term model, none of the three terms was stochastically significant. In stepwise logistic regression, the interaction term did not enter the model after the two main-effect terms were forced. Thus, the algebraic logistic models did not successfully identify the interaction valley problem of Table 3-5.

A different challenge occurred in the conjunctive subgroup effects of Table 3-7, where treatment B was superior in the totals and in patients with good clinical condition, but was worse than treatment A for patients in poor clinical condition. In the three-term full-regression model, all three terms were significant; and the sign of the interaction coefficient was opposite to that of the other two terms. The same results were obtained in stepwise regression. In this situation, therefore, the algebraic interaction model was successful.

Table 3-8, which somewhat resembled 3-7, had conjunctive subgroup effects that were better in the good group for treatment C and in the poor group for treatment D. In the overall totals, however, the disparate effects were balanced, so that the results were similar for the two treatments. Algebraically, all three terms in a full logistic model were significant, and the sign of the interaction coefficient was opposite to that of the other two. In stepwise regression, however, the *clinical condition* variable entered first, and neither of the other two terms then met the entry criterion of P = .05. The algebraic model was thus successful in a full but not in a stepwise regression.

In Table 3-10, the clinical trial had an inefficient design because the two treatments had similar results in the mild and severely ill groups, although treatment G was significantly better than placebo in the subgroup with moderate illness. The problem in this table might be regarded as an interaction because the striking effects of the compared treatments in the moderate subgroup would not be expected from the marginal totals, where treatment G was not significantly better than placebo. The algebraic interaction model, however, was unsuccessful in identifying this phenomenon. In full regression, none of the three terms had a stochastically significant coefficient. In stepwise regression, after the main-effect terms for clinical severity and therapy entered the model, the interaction term did not meet the P = .05 entrance criterion. The results of these (and several other) algebraic explorations of Table 3-10 gave no hint of what was readily perceived in the cells of the tabular arrangement.

An interesting feature of the algebraic analysis for Table 3-10 is that when the model was restricted to the two main variables—*without* an interaction term—the treatment variable emerged as stochastically significant, although it did not previously achieve that distinction in the totals of Table 3-10. Thus, the algebraic model, by adjusting for the results of the ordinal clinical-condition variable, showed that treatment was significant, but did not indicate *where* the significance occurred. An attempt to find this location via an ordinary interaction term, however, would have obscured the significance itself.

The results thus show that the algebraic interaction models were easily able to manage only one of the four challenging interaction tables of Chapter 3. The model was clearly successful for Table 3-7, where the conjunctive effects were most obvious. The model was "half" successful—being effective with full but not with forced-stepwise regression—in Table 3-8, where the opposite subgroup effects led to balanced totals. The model failed for the central valley in Table 3-5, and for the disguised significance in Table 3-10. These demonstrations indicate that algebraic interaction models cannot be relied upon consistently to detect interactions that are clearly shown with clustered tabular arrangements.

14.9. Scores and scoring systems

The results of a logistic regression analysis are often converted into scores or scoring systems, in a manner analogous to what was considered in Section 12.7. To keep this chapter from becoming too long, the discussion of the scoring process, which is similar for both logistic and Cox regression, will be relegated to Chapter 16.

14.10. Additional topics

This section is devoted to five additional uses of logistic regression. They involve design variables, conditional regression, "quadratic logistic regression," polytomous target variables, and diagnostic-marker evaluations.

14.10.1. Use of 'design variables'

When entered into regression analyses, nominal variables must be decomposed into a series of dummy binary variables for each of the nominal categories. In the subsequent analysis, the regression coefficient obtained for each dummy variable may indicate the effects of the corresponding category, but the decomposition does not produce a single coefficient to denote the effect of the entire nominal variable itself. Thus, with the multiple dummy binary variables, an effect might be noted for **Catholic** but not for **religion,** or for **coal miner** but not for **occupation.** In the tactic called *design variables,* which was developed to deal with this problem, the dummy variables are still analyzed individually, but a separate result is reported for the *overall* effect of the original nominal variable. The tactic resembles the group approach discussed for linear regression in Section 11.8.2.

To show this tactic in the illustrative data set, the four cell types constituted a single independent nominal variable, called **histology,** that was deliberately omitted in all the previous examples of logistic regression. This nominal independent variable could be managed in several different ways, as shown in the illustrations that follow. All the illustrations contain the three main variables (TNMSTAGE, HCT, and SXSTAGE), but various arrangements (including design variables) are used to include the nominal categories of *histology.*

14.10.1.1. Multiple dummy variables

As noted in Section 5.4.3.1, a nominal variable with four categories should be converted to three dummy variables. Nevertheless, to see what the computer programs would do, each of the four dummy cell–type binary variables for histology was made separately available for the model. When given this collection of variables, the SAS program for full logistic regression promptly recognized their interrelationship. As shown near the top of Fig. 14-10, the printout stated that the cytology variable was given a value equal to the values of *intercept − well − small − anaplastic*. The last line of the printout shows that *cytol* was then discarded from the analysis.

```
                          Intercept
               Intercept    and
Criterion        Only     Covariates    Chi-Square for Covariates

AIC             278.538   234.421
SC              281.837   257.509
-2 LOG L        276.538   220.421       56.118 with 6 DF (p=0.0001)
Score              .         .          50.673 with 6 DF (p=0.0001)
```

NOTE: The following parameters have been set to 0, since the variables are a linear combination of other variables as shown.
CYTOL = 1 * INTERCPT - 1 * WELL - 1 * SMALL - 1 * ANAP

Analysis of Maximum Likelihood Estimates

							Conditional Odds Ratio and 95% Confidence Limits			
Variable	DF	Parameter Estimate	Standard Error	Wald Chi-Square	Pr > Chi-Square	Standardized Estimate	Odds Ratio	Lower	Upper	Variable Label
INTERCPT	1	0.0625	1.4961	0.0017	0.9667	.	1.065	0.057	19.982	Intercept
HCT	1	-0.0938	0.0339	7.6680	0.0056	-0.267942	0.910	0.852	0.973	HEMATOCRIT
TNMSTAGE	1	0.4366	0.1138	14.7198	0.0001	0.392744	1.547	1.238	1.934	TNM STAGE
SXSTAGE	1	0.3689	0.1755	4.4169	0.0356	0.216442	1.446	1.025	2.040	SYMPTOM STAGE
WELL	1	1.4585	0.6298	5.3639	0.0206	0.397208	4.300	1.251	14.774	WELL DIFF HISTOLOGY
SMALL	1	2.3107	0.7794	8.7904	0.0030	0.415024	10.081	2.188	46.442	SMALL CELL HISTOLOGY
ANAP	1	1.8860	0.6198	9.2584	0.0023	0.503283	6.593	1.956	22.217	ANAPLASTIC HISTOLOGY
CYTOL	0	0	.	.	.	.	.	.	.	CYTOLOGIC PROOF ONLY

Fig. 14-10 Full logistic regression for three main variables and four dummy histology variables. (SAS PROC LOGISTIC program; the lower and upper boundaries of the confidence limits are obtained as options.)

For this same challenge, the BMDP program responded somewhat differently. To obtain a full logistic regression, which is not offered by the BMDP programs, all of the independent variables must be forced into the BMDP stepwise procedure. After this forcing of the four histologic dummy variables, which could not be removed, the BMDP program discarded the intercept (marked CONSTANT) as shown in Fig. 14-11.

The results in both the SAS and BMDP printouts show exactly the same b_j coefficients for the three main variables, but striking differences for the histologic cell types. In the SAS procedure, all three coefficients for the dummy cell types emerged with P < .05, but none of them was significant in the BMDP printout. Furthermore, the *anaplastic* variable had a BMDP coefficient that was vanishingly small (multiplied by a factor of 10^{-8}).

The disparities show the problems that can arise when one tries to "fool mother nature" by giving the program all four binary dummy variables in a situation where only three should be entered.

14.10.1.2. Coding system for design variables

As noted in Section 5.4.3, at least two different codings can be used for multiple dummy binary variables. In the dummy variables of the "reference" or "partial" system, one of the binary categories is coded with all zeros, and other categories receive staggered codes of 1. In the "effect" or "marginal" system, the

```
STEP NUMBER    0
---------------
                      LOG LIKELIHOOD  =   -110.209
GOODNESS OF FIT CHI-SQ  (2*O*LN(O/E)) =    198.918   D.F.= 171   P-VALUE= 0.071
GOODNESS OF FIT CHI-SQ (HOSMER-LEMESHOW)=    7.087   D.F.=   8   P-VALUE= 0.527
GOODNESS OF FIT CHI-SQ  ( C.C.BROWN )  =     2.010   D.F.=   2   P-VALUE= 0.366
                     STANDARD                         95% C.I. OF EXP(COEF)
       TERM   COEFFICIENT   ERROR    COEF/SE    EXP(COEF)    LOWER-BND  UPPER-BND

   TNMSTAGE      0.4366     0.114      3.84        1.55        1.24       1.94
   SXSTAGE       0.3689     0.176      2.10        1.45        1.02       2.04
   HCT          -0.9383E-01 0.339E-01 -2.77        0.910       0.852      0.973
   WELL         -0.4275     1.39      -0.309       0.652       0.424E-01 10.0
   SMALL         0.4247     1.55       0.274       1.53        0.720E-01 32.5
   ANAP         -0.1758E-08 1.44      -0.122E-08   1.00        0.584E-01 17.1
   CYTOL        -1.886      1.50      -1.26        0.152       0.793E-02  2.90
   CONSTANT      1.949      0.000E+00  0.000E+00   7.02        7.02       7.02
            THE ABOVE TERM DID NOT PASS THE TOLERANCE TEST.


STATISTICS TO ENTER OR REMOVE TERMS
-----------------------------------
            APPROX.              APPROX.
   TERM     CHI-SQ.  D.F.        CHI-SQ.  D.F.                      LOG
            ENTER                REMOVE           P-VALUE        LIKELIHOOD

TNMSTAGE                          15.45    1      0.0001         -117.9333
TNMSTAGE                          IS IN           MAY NOT BE REMOVED.
SXSTAGE                            4.46    1      0.0346         -112.4414
SXSTAGE                           IS IN           MAY NOT BE REMOVED.
HCT                                8.27    1      0.0040         -114.3419
HCT                               IS IN           MAY NOT BE REMOVED.
WELL                               0.00    1      1.0000         -110.2092
WELL                              IS IN           MAY NOT BE REMOVED.
SMALL                              0.00    1      1.0000         -110.2092
SMALL                             IS IN           MAY NOT BE REMOVED.
ANAP                               0.00    1      1.0000         -110.2092
ANAP                              IS IN           MAY NOT BE REMOVED.
CYTOL                              0.00    1      1.0000         -110.2092
CYTOL                             IS IN           MAY NOT BE REMOVED.
CONSTANT                           0.00    1      1.0000         -110.2092
```

Fig. 14-11 "Full" (forced) logistic regression for three main variables and four dummy histology variables. (BMDP LR program.)

reference category receives values of −1 in all the dummy variables, and the other categories receive staggered codes of 1.

Results were examined here for both systems of coding. The single design variable, called HISTOLOGY, contained three dummy variables, HISTOL (1), HISTOL (2), and HISTOL (3), which were coded in the partial system as follows:

	Histology dummy design variable		
Original variable	1	2	3
Well-differentiated	0	0	0
Small cell	1	0	0
Anaplastic	0	1	0
Cytology only	0	0	1

(With the marginal system, all entries in the foregoing *well-differentiated* row were coded as −1.)

14.10.1.3. Results of design coding

Because the SAS program does not have a design strategy for nominal variables, all of the discussion that follows refers to the BMDP procedure, which was used with stepwise forced regression for both coding systems.

```
STEP NUMBER     0
---------------
                         LOG LIKELIHOOD  =   -110.209
GOODNESS OF FIT CHI-SQ    (2*O*LN(O/E))  =    198.918   D.F.= 171   P-VALUE= 0.071
GOODNESS OF FIT CHI-SQ (HOSMER-LEMESHOW)=      7.087    D.F.=   8   P-VALUE= 0.527
GOODNESS OF FIT CHI-SQ    ( C.C.BROWN  ) =     2.010    D.F.=   2   P-VALUE= 0.366
                           STANDARD                           95% C.I. OF EXP(COEF)
     TERM        COEFFICIENT   ERROR    COEF/SE   EXP(COEF)   LOWER-BND  UPPER-BND

TNMSTAGE          0.4366       0.114     3.84      1.55        1.24       1.94
SXSTAGE           0.3689       0.176     2.10      1.45        1.02       2.04
HCT              -0.9383E-01   0.339E-01 -2.77     0.910       0.852      0.973
HISTOL      (1)   0.8522       0.623     1.37      2.34        0.687      8.01
            (2)   0.4275       0.377     1.13      1.53        0.729      3.23
            (3)  -1.459        0.630    -2.32      0.233       0.672E-01  0.805
CONSTANT          1.521        1.39      1.10      4.58        0.298     70.3


STATISTICS TO ENTER OR REMOVE TERMS
-----------------------------------
                  APPROX.           APPROX.
     TERM         CHI-SQ.  D.F.     CHI-SQ.  D.F.                      LOG
                  ENTER             REMOVE             P-VALUE      LIKELIHOOD

TNMSTAGE          15.45     1                          0.0001       -117.9333
TNMSTAGE                             IS IN             MAY NOT BE REMOVED.
SXSTAGE            4.46     1                          0.0346       -112.4414
SXSTAGE                              IS IN             MAY NOT BE REMOVED.
HCT                8.27     1                          0.0040       -114.3419
HCT                                  IS IN             MAY NOT BE REMOVED.
HISTOL            12.80     3                          0.0051       -116.6068
HISTOL                               IS IN             MAY NOT BE REMOVED.
CONSTANT           1.22     1                          0.2692       -110.8195
CONSTANT                             IS IN             MAY NOT BE REMOVED.
```

Fig. 14-12 Results for design variable with "partial" (or "reference") coding in BMDP LR logistic regression program. (For further details, see text.)

Results for the partial system are shown in Fig. 14-12. The results of the marginal system were identical to those in the upper part of Fig. 14-12 for the likelihood and other indexes that cite fit of the model, and also for the three main variables. As might be expected, however, the coefficients in the two coding systems differed for the three components of the design variable and also for the constant. As shown in Fig. 14-12, only the third design variable (i.e., **cytology only**) was stochastically significant (according to COEF/SE ratios) in the partial system. With the marginal system, however, both the third and the first design variable **(small cell)** were stochastically significant.

In the lower half of Fig. 14-12, in the section marked STATISTICS TO ENTER OR REMOVE TERMS, the main result is shown for the entire *single* design variable, HISTOL. The result was the same in both coding systems: an APPROX. CHI-SQ. REMOVE of 12.80 with 3 degrees of freedom, a P value of .0051, and an incremental log likelihood of

−116.6068. The only variable that was more "powerful" in the total array was TNM stage, for which the corresponding respective values were 15.45, .0001, and −117.9333.

Since the overall effect of the histology variable was *not* significant when previously examined in the linear regression analyses of Chapter 12, the results suggest that histologic cell type may be an important variable for 6-month survival, but not for the longer survival periods examined with the linear regression analyses. The results also indicate that the quantitative impact of *individual* histologic categories cannot be suitably recognized from the algebraic models, because the effects will depend on vicissitudes of the coding arrangement.

14.10.2. 'Conditional' regression

In many epidemiologic studies of risk factors for disease, the diseased cases are chosen as one group, and the nondiseased controls as another. In some studies, however, the controls are individually matched to the cases, using such features as age, sex, etc. Although more than one control may be matched to each case, the discussion that follows is easier to understand if confined to a 1-to-1 matching.

14.10.2.1. Matched and unmatched arrangements

When presented in matched form, the frequency counts for the N matched pairs would appear in the following 2 × 2 "agreement matrix":

Cases	Controls		Total
	Exposed	Nonexposed	
Exposed	a	b	f_1
Nonexposed	c	d	f_2
Total	n_1	n_2	N

The analysis of the matched table makes use of the stratification concept that is regularly employed for Mantel-Haenszel combinations of 2 × 2 tables. Each particular case-control pair will be a member of one of the following four "strata" in the 2 × 2 agreement matrix:

$$\begin{Bmatrix} 1 & 0 \\ 0 & 0 \end{Bmatrix}, \begin{Bmatrix} 0 & 1 \\ 0 & 0 \end{Bmatrix}, \begin{Bmatrix} 0 & 0 \\ 1 & 0 \end{Bmatrix}, \text{ and } \begin{Bmatrix} 0 & 0 \\ 0 & 1 \end{Bmatrix}.$$

These four strata will contain, respectively, *a, b, c,* and *d* members. In the *a* and *d* cells (or strata), the members have no differences in ratings for exposure (or nonexposure) but the *b* and *c* cells (or strata) the ratings differ. Consequently, the *a* and *d* cells become "noninformative" in the "stratified" analysis, and the odds ratio becomes calculated as *b/c*.

The matched arrangement is opposed by many data analysts, however, who argue that matching is often an act of administrative convenience rather than specific attention to "confounding variables." Accordingly, the matched results would be "unpacked" and analyzed in an unmatched format for the 2N persons who are the cases and controls. The unmatched format would be

	Cases	Controls	Total
Exposed	f_1	n_1	f_1+n_1
Nonexposed	f_2	n_2	f_2+n_2
Total	f_1+f_2	n_1+n_2	2N

In the unmatched format, the odds ratio would be $(f_1 n_2)/(f_2 n_1) = [(a + b)(b + d)]/[(c + d)(a + c)]$. When algebraically expanded, with terms collected and rearranged, this expression becomes $[b + \{(ad - bc)/N\}]/[c + \{(ad - bc)/N\}]$, and will equal b/c whenever $ad = bc$. Otherwise, the matched and unmatched odds ratios, as well as the corresponding chi-square test values, will differ.[5]

A strong argument against the matched analysis is that it discards the information in the a and d cells of the agreement matrix. If ad is substantially different from bc, the two odds ratios can be dramatically different. For example, consider the matched agreement matrix $\{{}^{200}_{\ \ 8}\ \ {}^{40}_{100}\}$. Its b/c odds ratio is $40/8 = 5.0$. When these results are converted to an unmatched format, however, the table becomes $\{{}^{240}_{108}\ {}^{208}_{104}\}$, and the odds ratio drops to a much lower value of 1.5. By ignoring the many paired matchings that had similar results for either exposure or nonexposure, the matched odds ratio produced a major distortion of the total situation. (This same argument can be invoked against the C index of concordance [see Sections 13.7.5 and 14.2.2.1], which uses only the results in the dissimilar observed pairs.)

An analogous opportunity and potential problem can occur when logistic regression is applied in case-control studies. In the ordinary (unmatched) circumstances, the outcome event is each person's status as a case or control. The independent variables include the risk factors, as well as other pertinent features that are to be adjusted or balanced in the multivariable analysis. This type of analysis is called *unconditional*, to distinguish it from conditional logistic regression, which maintains the case-control matching.

14.10.2.2. Strategy of conditional analysis

The strategy of the conditional logistic analysis is not easy to understand. It will be described here without further attempts at justification, for which you can look, and perhaps find clarity of explanation, in references elsewhere.[4,6,7] For the conditional procedure, with a 1-to-1 matching, each independent variable, X_j, is replaced by a separate variable X'_j, which represents the increment in X_j between each matched case and control. In addition, the value of Y_i for each matched case-control pair is replaced by their increment in Y values, which will always be $1 - 0 = 1$. Thus, in variable X_j, if the first matched pair has $X_j = 17$ for the case and 5 for the control, then $X'_j = 12$ and $Y_1 = 1$. The linear format for the algebraic model becomes $G' = b'_0 + b'_1 X'_1 + b'_2 X'_2 + b'_3 X'_3 + \ldots$, and the logistic equation is arranged as

$$1 = e^{G'}/[1 + e^{G'}].$$

The maximum likelihood method is then used to find the best-fitting values of b'_1, b'_2, b'_3, ... for the data in this format, containing all pertinent pairs of persons. The values of b'_1, b'_2, b'_3, ... are then exponentiated to form the conditional (or matched) odds ratios. The results are intended to allow appraisal of "the effect of several variables simultaneously in the analysis while allowing for the matched design."[6]

14.10.2.3. Applications and reassurance

The conditional form of logistic regression is particularly likely to appear in reports of epidemiologic studies of risk factors, which are often constructed with a matched case-control arrangement. If the raw frequency-count data are not published, readers have no way of checking to determine whether the results are affected by the general problems of small data sets or by the imbalances discussed in Section 14.10.2.1.

By showing that both the conditional and unconditional analyses produced similar findings, the investigators can reassure themselves, as well as reviewers and readers, that such problems have not occurred. The similarity of results was demonstrated for data sets that were used to illustrate proposals of the matched methods[8] and computational techniques.[9] Checks for similarity have seldom been reported, however, when the conditional method is employed in new studies. When such reports appear, however, the investigators often state that "the results of the matched and unmatched procedures were similar, [and] only the unmatched analyses are presented."[10]

14.10.3. Quadratic logistic regression

If the linear arrangement of

$$G_p = b_0 + b_1X_1 + b_2X_2 + \ldots + b_pX_p$$

does not seem to fit well, the investigator may include certain terms in which an X_j variable is squared. The tactic, called *quadratic regression,* is seldom used for more than two independent variables. The format of the model might be

$$G = b_0 + b_1X_1 + b_2X_1^2 + b_3X_2.$$

In one published report,[11] the authors say they used quadratic logistic regression, but offer no explanation why the quadratic format was chosen, what variables were entered in quadratic form, or what results were produced other than good fits.

14.10.4. Polytomous target variables

Although logistic regression was deliberately prepared to aim at a binary target variable, the method has been extended to manage polytomous targets that have three or more categories. The categories can be ordinal or nominal.

The problems of using such targets and interpreting the results will receive detailed discussion in Section 17.1.1 for discriminant function analysis, and, in Part V of the text, in Sections 19.5.3 and 21.6.3 for multicategorical stratification. To avoid these problems in algebraic analysis, an investigator can convert the polytomous targets into various binary splits, and then do a separate regression analysis for each split. Thus, if the outcome variable is coded as **none, some,** and **many,** the binary splits for targets in the two regressions might be **none** vs. **at least some,** and **many** vs. **not many.**

If the investigator wants to preserve the polytomous format in a single logistic procedure, however, an "authorized" analysis can be arranged. Letting the three outcome categories be **0, 1,** and **2,** the logistic model will work out appropriate coefficients to find "proportionate odds" for the X_j variables. In essence, the outcome is regarded as a dummy variable; b_j regression coefficients for each variable are calculated for the comparisons of outcome category 2 vs. category 0, and for category 1 vs. 0. The regression coefficients for comparing category 2 vs. category 1 are then obtained as the difference between the two values for each b_j. The methods and results of algebraic analyses for polytomous targets are not easy to understand and interpret. They demonstrate, however, that logistic regression, when asked to do what it was not intended to do, can perform at least as gracefully as the famous Russian dancing bear.

If you insist on using a polytomous approach, it is described in the Hosmer-Lemeshow book[4] and in several pertinent references.[12-15] The method was well illustrated by Ashby et al.[16] to relate alcohol consumption to serum biochemistry and hematology, and by Brazer et al.[17] to estimate the likelihood of colorectal neoplasia. In the illustrative example by Armstrong and Sloan,[14] the investigators, after doing a polytomous regression, concluded that ordinary (binary) logistic regression gave "results quite similar . . . after dichotomizing outcome in the conventional way."

14.10.5. Application to diagnostic-marker tests

Because diverse features of the patient's medical status will create problems of bias or confounding in using "unadorned" Bayes Theorem and likelihood ratios for evaluating diagnostic-marker tests, logistic regression has been proposed[18,19] as a strategy of adjusting covariates to deal with the problems.

The outcome variable is presence or absence of the disease under consideration. The independent variables include the result of the diagnostic-marker test, plus diverse additional data—such as age, sex, other laboratory tests, clinical severity of illness, and co-morbidity—that become the adjusted covariates. The effectiveness of the process has been disputed,[20] but it has many attractions that are being further explored.

In an interesting report of patients with gastrointestinal symptoms, Talley et al.[21] applied the logistic procedure to make (i.e., estimate) the diagnosis of *three* clinical conditions: essential dyspepsia, peptic ulcer, and gallstones. Three logistic equations were derived—one for each clinical condition as the target, with the other two conditions being used as the control groups in that equation. The analytic results were not particularly dramatic or different from what would be described in most textbooks of physical diagnosis: i.e., food often relieved ulcer pain but provoked dyspepsia. The investigators did not describe how the trio of equations would be applied to make a diagnosis in a new patient with GI symptoms.

14.11. Use of logistic regression in published literature

Table 14-3 lists twenty papers, all appearing (with one exception) in chronologic order in published literature during the years 1988–93, in which multiple logistic regression was used for analysis of the data. The first paper was chosen as a neurosurgical example; all the others were selected with the same "haphazard" technique described earlier (Section 12.7) for examples of multiple linear regression. Despite the absence of random selection, the choices were made without any goal other than to note recently published instances of logistic regression in a reasonably wide scope of journals. Consequently, the results can reasonably indicate the way this analytic procedure is generally being used in clinical literature.

Table 14-3 Twenty publications in which logistic regression was used

First author	Journal	Year	Target variable(s)	Independent variables
Stablein[22]	Neurosurg.	1980	Good vs. poor recovery after head injury	Age, sex, physiological status, and nine neurologic variables
Makuch[23]	Stat. in Med.	1988	Liver metastases in small-cell lung cancer	Age, four liver function disease scores
Eskenazi[24]	JAMA	1991	Pre-eclampsia	Parity, body mass, race, family history of hypertension, smoking, worked during pregnancy, events in previous pregnancies
Holt[25]	JAMA	1991	Ectopic pregnancy	Demographic, reproductive, sexual and contraceptive variables, previous tubal sterilization
Lockwood[26]	NEJM	1991	Preterm contractions or rupture of membranes	Fibronectin, gestational age, cervical dilation, rate of uterine contractions, birth weight
Dougados[27]	Arth & Rheum	1991	Diagnosis of spondylarthropathy	Twenty-five "clinical" variables
Aube[28]	Am. J. Med.	1992	Septic shock	Sex, age, creatinine, prothrombin time, blood counts, other chemical factors
Berkman[29]	Ann. Int. Med.	1992	6-month mortality after myocardial infarction	Age, gender, emotional support, co-morbidity, disability, cardiac status
Reiber[30]	Ann. Int. Med.	1992	Amputation in patients with diabetes mellitus	Age, sex, education, income, duration of diabetes, distance to hosp., disease severity, Hx, PVD, Percut. O_2 tension
Hubbard[31]	Arch. Int. Med.	1992	Dx of coronorary disease	Demographic, clinical, and ECG variables
Daneshmend[32]	BMJ	1992	Mortality, operation, re-bleeding	Age, blood pressure, bleeding site, centre, treatment group

O'Connor[33]	Circulation	1992	In-hospital death after coronary graft surgery	Age, sex, body surface area, co-morbidity score, prior operation, several cardiac status variables
Higgins[34]	JAMA	1992	Morbidity and mortality (separately) after coronary graft surgery	Age, weight, emergency case, prior operation, several co-morbid diseases, LV dysfunction
Von Korff[35]	J. Clin. Epid.	1992	Hospitalization or death	Age, gender, chronic disease score
Classen[36]	NEJM	1992	Surgical wound infection	Age, sex, hospital service, attending M.D., underlying disease, nursing service, surgeon, types and duration of surgery, postoperative procedures, timing of antibiotics
Garrison[37]	Am. J. Epid.	1993	Post-traumatic stress disorder after a hurricane	Severity of exposure, other traumatic events, race, sex, social class
Schubert[38]	Am. J. Med.	1993	Gastric or duodenal ulcer in patients referred for endoscopy	H. pylori infection, history of ulcer, use of alcohol, aspirin, and/or nonsteroidal anti-inflammatory drugs, symptoms, sex, age, smoking
Landesberg[39]	Lancet	1993	Cardiac ischemic complications after major noncardiac vascular surgery	Cardiac risk index, features of previous cardiac history and co-morbid diseases, age, timing and duration of peri-operative ischemia
Pahor[40]	J. Clin. Epid.	1993	Adverse drug reactions to digoxin	Daily dose, age, sex, co-morbidity, concomitant therapy, various laboratory tests
Fieselmann[41]	J. Gen. Int. Med.	1993	Cardiopulmonary arrest in intensive care unit	Vital signs, age, sex, discharge diagnoses, abnormalities in pulse rate, respiratory rate, and/or blood pressure

14.11.1. Goals of the analyses

All of the logistic analyses in the twenty reports were aimed exclusively at determining, or adjusting for, the impact of the independent variables. In no instance did the analysis produce a specific equation that was then used alone to offer an accurate substitute fit for the target variable.

The impact of the variables was usually checked for purposes of confirmation. The investigators had almost always first checked the variables for their direct individual bivariate effect (often called "univariate") on the target event. The apparently effective variables, and sometimes others that seemed interesting, were then entered into the multiple logistic analysis. The latter result was often confirmatory, showing that the individual effect was retained in the multivariate context.

In six instances,[23,24,27,28,30,31] the goal was to find the factors that were important predictors in the multivariable context, and in five other instances,[29,36–39] the aim was to confirm that a variable retained its previously noted bivariate impact. In one instance,[25] the aim was to "adjust for confounders" and in another,[35] the impact of certain variables was quantified as an odds ratio. In three instances,[27,31,34] the "explanatory" factors were converted into a simpler score for pragmatic usage. In one instance[33] the regression equation gave results that were assigned to one of five ordinalized risk categories. One paper[22] was intended mainly to show that logistic regression was predictively superior to Bayesian methods.

The infrequency of specific quantification for individual odds ratios probably reflects the predominant origin of these twenty reports in clinical literature, where the "risks" under evaluation are usually prognostic factors for outcome of an illness, rather than "risk factors" for development of a disease. For the clinical decisions, the investigators are almost always concerned with the overall predictive effect of the combined variables, rather than with specific citations for individual variables. In epidemiologic literature, however, analysts often emphasize[4,7] the odds ratios associated with individual etiologic risk factors.

14.11.2. Checks for fit

Only three[23,27,33] of the twenty reports mentioned the results of a check for *fit* of the model.

14.11.3. Regression diagnostics

Except for one report,[23] which gave specific examples and results of efforts to improve the fit of the model, regression diagnostics were not mentioned by any of the investigators.

14.11.4. Checks for linear conformity

The results of checks for linear *conformity* were mentioned in only four reports,[33,35,36,38] none of which relied on the customary algebraic model. Instead, the investigators deliberately decomposed pertinent ordinal or dimensional variables into binary variables or into a set of dummy variables. Odds ratios were then determined for the binary or dummy categories rather than for the entire original variable.

14.11.5. Transformations

The product of *age* × *log age* was explored in one report[23] "to detect suspected curvilinearity," and in another report,[33] body surface area was transformed to its square-root value. Otherwise, no transformations were cited. (In one illustrative example, in a textbook,[42] the linear model contains variables for both *Age* and *Age*². The results are then discussed for their goodness of fit, but not for the problem of interpreting the impact of the *age* variable when it is expressed in two formats simultaneously.)

14.11.6. Interactions

In two instances[24,34] the investigators added an interaction term for a variable of special interest (such as maternal *parity*), and also did separate regressions within groups of patients defined by categories of that variable. Stablein et al.[22] and Reiber et al.[30] entered and gave quantified results for certain interaction terms that were kept in the final model. Schubert et al.[38] investigated and found some significant results with two-way and three-way interaction terms for variables such as age, drug use, race, and history of ulcer. The investigators did not discuss, however, how the interactions were used when the "final" odds ratios were calculated for individual variables.

14.11.7. Standardization procedures

None of the cited reports used standardized variables or made efforts to convert the results of b_j coefficients to a standardized citation.

14.11.8. Reporting of results

No particular method was used consistently for citing the impact of the individual variables. They were sometimes reported merely as being significant, or with their associated P values. In many instances, the regression coefficient and/or odds ratio were listed, with or without an associated standard error for the coefficient. In several instances, 95% confidence intervals were offered for the odds ratios. Sometimes the result for a b_j coefficient was accompanied by a chi-square value.[31]

14.12. Concluding remarks

The automation of logistic regression has made it an attractive procedure for the many clinical and epidemiologic studies in which a plethora of predictor variables are related to a categorical (usually binary) outcome event. Consequently, investigators have an unprecedented opportunity to analyze complex data for multiple variables. The data are inserted, the program is operated, and the significant results promptly emerge.

14.12.1. Hazards of quantitative coefficients

In exchange for this simplicity, the procedure has many hazards, most of which have already been discussed in this chapter. Can we really assume that the average odds ratio for a variable is constant throughout the entire scope of that variable, and that it remains constant in the interaction presence of other variables? When expressed as odds ratios, do the combined variables really have a multiplicative rather than additive effect? Is a conditional regression really necessary or desirable when the cases and controls were matched for convenience rather than for scientific reasons?

14.12.2. Pragmatic usage and scoring systems

The foregoing questions are particularly pertinent when the investigators quantify the analysis of a risk factor in conclusions such as, "You live 2 years longer for every 3-point drop in substance X." In many pragmatic activities, however, the investigators are less ambitious. The goals may be merely to verify that a particular variable maintains its impact in the multivariate context, or to adjust for covariate imbalances by establishing a prognostic scoring or staging system. For both of these goals—confirmation and adjustment via prognostic staging—the mathematical hazards are much less frightening. Furthermore, when the logistic coefficients are converted into points that form a scoring system, the performance of the scoring system is checked separately, and the nuances of its logistic derivation become unimportant if the results are sensible and if the system performs well.

14.12.3. Caveats, precautions, and opportunities

To enhance commercial sales appeal, the packaged programs for personal computers have been made relatively easy to use. After a few adventures in wading through the extensive printouts, investigators can readily find the most pertinent results.

Euphoric and sometimes intellectually anesthetized by the results (particularly if cherished beliefs are confirmed), the investigators may then report the findings with no further concern for the many problems and complexities that require intensive evaluation. Beyond some of the difficulties already cited, at least three more infelicities may occur:

1. The investigators may claim that suitable adjustments have been made for confounding variables, even though some of the most important confounders have *not* been identified for inclusion in the data. Furthermore, many of the adjusted covariates, although impressive as a list of features that have been "taken care of," were neither tested nor demonstrated to have a role as confounders.

2. Some of the variables noted as risks may actually indicate effects rather than causes.[43]

3. The research may be so unsuitable in many features of its scientific architecture that the mathematical splendor of the analysis becomes essentially irrelevant.

Perhaps the main reason for understanding what happens in logistic regression is to surmount the intellectual obstacles it offers to the uninitiated. If you do not know what is really happening and what is represented by all the symbols, terms, and numbers, they can be an overwhelming barrier to the more fundamental scientific appraisals. If you can get past the mathematical barrier, however, the really important scientific evaluations can become, as they should be, the main focus of attention.

Chapter References

1. Goodman, 1979; 2. SAS Technical Report P-200, 1990; 3. Pregibon, 1981; 4. Hosmer, 1989, p. 144; 5. Feinstein, 1987; 6. Holford, 1978; 7. Breslow, 1980; 8. Breslow, 1978; 9. Campos-Filho, 1989; 10. Harvey, 1985; 11. Giles, 1988; 12. McCullagh, 1980; 13. Begg, 1984; 14. Armstrong, 1989; 15. Peterson, 1990; 16. Ashby, 1986; 17. Brazer, 1991; 18. Hlatky, 1984; 19. Coughlin, 1992; 20. Diamond, 1992; 21. Talley, 1987; 22. Stablein, 1980; 23. Makuch, 1988; 24. Eskenazi, 1991; 25. Holt, 1991; 26. Lockwood, 1991; 27. Dougados, 1991; 28. Aube, 1992; 29. Berkman, 1992; 30. Reiber, 1992; 31. Hubbard, 1992; 32. Daneshmend, 1992; 33. O'Connor, 1992; 34. Higgins, 1992; 35. Von Korff, 1992; 36. Classen, 1992; 37. Garrison, 1993; 38. Schubert, 1993; 39. Landesberg, 1993; 40. Pahor, 1993; 41. Fieselmann, 1993; 42. Kleinbaum, 1988; 43. Rubin, 1992.

15 Proportional Hazards Analysis (Cox Regression)

Outline ....

15.1. Principles of survival analysis
    15.1.1. Fitting a moving target
    15.1.2. Sources of censoring
        15.1.2.1. Timing of follow-up process
        15.1.2.2. Insufficient duration of follow-up
        15.1.2.3. Intermediate losses
        15.1.2.4. Competing events
    15.1.3. Classification of exit state
    15.1.4. Choice of survival curves
        15.1.4.1. Parametric models
        15.1.4.2. Nonparametric life-table structure
        15.1.4.3. Berkson-Gage actuarial method
        15.1.4.4. Kaplan-Meier method
    15.1.5. General strategy of Cox regression

15.2. Concept of a hazard function
    15.2.1. Radioactive decay curve
    15.2.2. Corresponding ideas in survival curve
    15.2.3. Algebraic and arithmetical features of hazard function
    15.2.4. Modification of hazard function
    15.2.5. Illustration of modification

15.3. Basic structure and summary of data
    15.3.1. Basic summary curve, $S_o(t)$
    15.3.2. Subsequent calculations and complexities

15.4. Fitting the coefficients
    15.4.1. Tactics in calculation
    15.4.2. Partial likelihood method
    15.4.3. Assumption about proportional hazards
    15.4.4. Centered and noncentered models
    15.4.5. 'Score' methods of calculation

15.5. Indexes of accomplishment
    15.5.1. Problems in expression and comprehension
    15.5.2. Indexes for models
        15.5.2.1. −Log likelihood
        15.5.2.2. Log likelihood ratio
        15.5.2.3. Chi-square global score
        15.5.2.4. 'R_H^2 statistic'
        15.5.2.5. Additional expressions
    15.5.3. Incremental indexes
        15.5.3.1. Maximum partial likelihood ratio (MPLR)
        15.5.3.2. Improvement chi-square
        15.5.3.3. Alternative PHH method
    15.5.4. Indexes for variables
        15.5.4.1. 'Parameter estimates'

15.5.4.2. Partial 'Index-R'
15.5.4.3. Standardized coefficients
15.5.4.4. Risk ratios and hazard ratios
15.6. Sequential exploration procedures
15.7. Stochastic appraisals
15.8. Regression diagnostics
15.8.1. Conventional procedures
15.8.2. Shape of S(t)
15.8.3. Constancy of e^G
15.8.4. Likelihood estimates
15.8.5. 'Linearity' of the model
15.9. Additional applications
15.9.1. 'Time-dependent covariates'
15.9.1.1. Longitudinal analysis
15.9.1.2. Survival analysis
15.9.1.2.1. Availability of data
15.9.1.2.2. Interpretation of results
15.9.1.2.3. Basic scientific goals
15.9.2. 'Conditional' methods

By using a multivariable model to modify the summary of a group, Sir David R. Cox became an eponym in the analysis of survival data.[1]

The concept itself was not particularly new. It was previously applied, without being so labeled, in both linear and logistic regression. When those procedures made estimates for individual persons, the summary of the group was modified by a general linear model containing data for the person's independent variables. In multiple linear regression, as discussed in Chapter 11, the estimate for an individual person is

$$\hat{Y}_i = G,$$

where G is the general linear model, $G = b_0 + b_1X_1 + b_2X_2 + \ldots b_jX_j + \ldots$ for the independent variables $X_1, X_2, \ldots, X_j, \ldots$. The value of b_0 in that model uses the summary mean of the Y and $X_1, X_2, \ldots, X_j, \ldots$ variables, expressed as $b_0 = \bar{Y} - b_1\bar{X}_1 - b_2\bar{X}_2 - \ldots - b_j\bar{X}_j - \ldots$. The estimate for an individual person, i, can then be algebraically rearranged as

$$\hat{Y}_i = G_i = \bar{Y} + b_1(X_1 - \bar{X}_1) + b_2(X_2 - \bar{X}_2) + \ldots + b_j(X_j - \bar{X}_j) + \ldots.$$

In this expression, the basic summary value, $\bar{Y}$, and the summary for the mean values of the group's $\bar{X}_j$ variables, are modified by the additional terms containing each coefficient for $b_1, b_2, \ldots, b_j, \ldots$ and the values of $X_1, X_2 \ldots, X_j, \ldots$ for each person's independent variables. A similar arrangement and modification of the group summary, $\bar{Y}$, occurs in logistic regression when an individual estimate is expressed as $\ln[\hat{Y}_i/(1 - \hat{Y}_i)] = G_i$.

Cox used this same basic approach in survival analysis, but he added several ingenious strategies that will be discussed shortly. What eventually emerges, however, is an estimate in which the general linear model modifies the group's summary survival *curve*. The curve is constructed from the summary values of the group's survival proportion, $\overline{Y}_t$, at each point, t, as time progresses. The entire curve is commonly denoted by the symbol S(t), and often has the shape shown earlier in Fig. 7-8. The S(t) curve begins with a value of 1 at zero time, when everyone is alive, and progressively drops downward toward a value of 0 as time goes on.

Cox's approach had three novel features. First, in contrast to many previous mathematical efforts, he did not try to establish a "parametric" algebraic model for the S(t) curve itself. Instead, he let the curve be constructed by a "nonparametric" method. Second, he did not modify S(t) directly. Instead, he modified a different curve, closely related to S(t), that is called the *hazard function*. Third, when he determined values of the b_j coefficients for the general linear model G, he used a method of *partial likelihood*. These three distinctions will be discussed shortly, but their eventual effect is that the general linear model, G_i, for a particular person, i, becomes a modifying factor, expressed as e^{G_i}; and it exponentiates the summary value of S(t). Thus, the algebraic model for the multivariable method that is now called proportional hazards analysis or (more often) *Cox regression* has a 'double exponential" format, expressed as

$$\hat{Y}_i = S(t)^{e^{G_i}}.$$

The value of $\hat{Y}_i$ denotes the probability that person i will survive at least until time t.

15.1. Principles of survival analysis

The basic background for understanding what happens in Cox regression is a nonparametric method, called *life-table* or *actuarial analysis*, which has been used for several centuries to summarize a group's survival. Although usually aimed at analyzing life or death, the life-table method can also be applied to other targets. They can be any "failure event" that concludes the period of "risk" for a particular person. Thus, a "survival analysis" might be aimed at such nonfatal failure events as development of a particular disease, occurrence of an unwanted pregnancy, or malfunction of an intracardiac prosthetic valve. For the discussion in this chapter, the failure event will be death attributed to a particular disease, such as cancer or coronary disease. In statistical jargon, the failure event is often called a *response*, and the survival analysis is based on each person's *time to response*. For application to a clinical cohort, the analytic technique requires special adjustments to account for patients who have not "responded", i.e., they were not dead when last observed.

The construction of a survival curve is not an inherent part of Cox regression, but is conceptually important for understanding what is done, why it is done, and how to interpret the results.

15.1.1. Fitting a moving target

In *survival analysis,* the dependent outcome variable is a moving target, rather than a stationary state at a single point in time. In linear regression, the single state is expressed with a dimensional variable, such as *blood pressure,* and in logistic regression, the target state is a binary event, such as **success/failure** or **alive/dead.** In survival analysis, the target is also a binary event, but it is dynamic rather than static. Each person's "bivariate" outcome is the state of being alive or dead at a specified point in time. The analysis itself, however, is aimed at a single dependent variable: the duration of time until each person responds with the failure event, e.g., death. The analytic summary of individual results produces the type of "survival curve" shown in Fig. 15-1 for a cohort of patients with cancer.

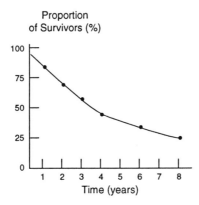

Fig. 15-1 Survival curve for a cohort with cancer.

For an individual person, of course, survival is a "step function", not a "curve". Each person's life is mathematically in a coded state of **1** until it suddenly drops to **0** at death. For summarizing the results of a group of these step functions, the dynamic time component creates two new challenges that do not occur when the target variable is a single state. The challenges require managing unknown or "censored" events, and choosing an appropriate curve to show the summary.

15.1.2. Sources of censoring

If we knew each person's survival time, no new analytic formats would be needed. The selected independent variables could be entered into a conventional linear (or other pertinent) regression model that estimates each person's duration of survival, while assuring that each estimated value would be ≥ 0. (The linear regression tactic was used with *ln survival* as the dependent variable for the illustrative data set in Chapter 12.)

In the realities of medical research with human life, however, survival duration is seldom known for everyone. When it is unknown, the patient is called *censored*. The censoring can be produced by three main mechanisms: insufficient duration of follow-up, intermediate loss-to-follow-up, or competing failure events. (These three forms of censoring are called *right-censored* because they occur after each person's period of observation begins at zero time. In another form of censoring, called *left-censored*, the persons enter observation long after a zero time that occurred previously.[2] Left-censored analyses are rare, and will not be further discussed here.)

15.1.2.1. Timing of follow-up process

Every cohort study begins and ends at selected dates in secular (or calendar) time. Thus, the study itself might extend for eight years from January 1, 1989, to December 31, 1996. After the study begins, different people enter the cohort successively at different calendar dates, which usually become the start of *serial* time for observing each person. If the individual people do not die, their follow-up continues until either the study ends or they become withdrawn by censoring.

Regardless of the associated calendar dates, the assembled data are organized according to each person's serial duration, which begins with zero time on admission to the cohort and continues until the person departs at the exit time of the last observation. Fig. 15-2 gives an example of what might occur as points of data for ten people in a cohort followed for eight time intervals.

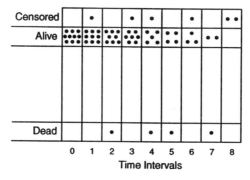

Fig. 15-2 Location of individual points of data for status of 10 persons at regular intervals of follow-up in a cohort. All 10 people are alive at the zero-time onset. One person each is censored in intervals 1, 3, 4, and 6. One person each dies in intervals 2, 4, 5, and 7. The two people who are alive in the last interval are censored in that interval. Note that the sum of deaths and censorings = 10.

15.1.2.2. Insufficient duration of follow-up

Unless everyone is enrolled on the first calendar day of a study, relatively few living members of the cohort will have a chance to be followed alive for the study's full duration, such as eight years. In many cohorts, therefore, the maximum serial duration chosen for each person's follow-up time is usually much shorter than the study's actual calendar time. If the study itself lasts eight years, the maximum duration of serial time might be set at five years. With this maximum duration, people who entered the study relatively late in calendar time may not have had the chance to be observed for five years when the study ends. For example, someone who enters on July 1, 1994, can be followed for only 2 1/2 years if still alive when the study ends on December 31, 1996. On the last day of follow-up in the study, such a patient will be "terminally" censored because of insufficient duration.

This phenomenon occurred in the illustrative data set of Chapter 12, where some of the lung cancer patients were still alive at 99.9 months, the end of the maximum coded duration of survival. For the multiple linear regression analysis, these patients were regarded as having died at 99.9 months. The foreshortening of their true survival slightly lowered the mean survival duration for the cohort, but was preferable to the alternative option of classifying the survival times as *unknown*, and removing them from the data set. The complete removal of the long-term survivors would have lowered the group's mean survival much more drastically.

In a preferable form of survival analysis, therefore, the censored patients should be allowed to contribute their full period of follow-up, without being either omitted entirely from the data or cited inappropriately as dead.

15.1.2.3. Intermediate losses

For many censored patients, the duration of survival is unknown because they have become "lost-to-follow-up" at an intermediate time before the study ends. They were last observed and "withdrawn" alive at a particular serial duration, but nothing is known of their fate thereafter. Because the intermediately censored durations of follow-up may be short or long, the summary results could be seriously distorted if such patients are analyzed as though they either had wholly unknown survivals or were dead on their last observation date.

15.1.2.4. Competing events

A different mechanism of censored withdrawal is a "competing event", which is usually death due to a presumably unrelated cause. Thus, patients being followed for survival after treatment of cancer would be competingly censored if they died of an unrelated event, such as an unexpected myocardial infarction or automobile accident. The clinical decision that a particular event is *unrelated* rather than *related* is not always easy, particularly if someone with far-advanced cancer commits suicide or dies of supervening infection. The problem of classifying competing deaths is avoided if any death, regardless of cause, is regarded as the failure event.

A different type of competing event can occur in randomized trials when patients significantly alter or violate the originally assigned treatment. For example, in a trial of medical vs. surgical therapy, some of the patients who were randomized to medical treatment may later decide to have the surgical operation. They cannot be regarded as part of the original surgical cohort, but they also have not had only medical therapy. In an intention-to-treat analysis,[3] they would be counted as having had medical treatment throughout—a tactic that can be justified statistically, but that clinically seems inappropriate. An alternative approach would regard such patients as "withdrawn alive" and censored at the time of surgery. The approach is controversial, because of the potential bias introduced if the change of treatment is not a random event, and is caused by success or failure of the preceding therapy. Nevertheless, if this approach is used, a fourth type of censoring would be produced by a substantial change of treatment as the competing event.

15.1.3. Classification of exit state

For the binary exit state in survival analysis, the complement of **dead** is **censored.** Because of insufficient duration of observation, loss to follow-up, or a competing event, the censored persons exit **alive** before having developed the response (e.g., an appropriate death).

15.1.4. Choice of survival curves

With different persons having different exit times and states, the choice of a single algebraic model for the bivariate outcome becomes a formidable challenge, sometimes leading to inscrutable mathematical complexity.

In an alternative approach, the modelling activity is divided into two phases. The first is directed at getting a model for the group's summary survival curve, $S(t)$. The second model then arranges the covariates (X_1, X_2, X_3, . . .) in a format that modifies $S(t)$ and converts it into estimates for individual persons. The two-phase modelling process, although not simple, produces a more "tractable" approach than a single, highly complex model for the bivariate outcomes.

The search for a suitable algebraic model for $S(t)$ is difficult because survival curves often have different shapes that are not easily fitted with the same approach. When an algebraic model is used for a set of data in linear or logistic regression, the problem of nonconformity, as discussed earlier, seldom receives much mathematical attention. The problem is important in survival analysis, however, where we want particularly close fit and conformity for a model that expresses the summary values $\bar{Y}_t$, not the individual items of data, $Y_{i,t}$, and that also becomes modified to give estimates for individual persons.

15.1.4.1. Parametric models

In search of a closely conforming fit for $S(t)$, many algebraic models have been proposed.[4-6] They reflect parametric distributions or functions that have such names as Exponential, Weibull, Lognormal, Gamma, Log-logistic, Gompertz, and Generalized F. (In BMPD programs for "survival analysis with covariates", one of these parametric functions can be chosen as an "accelerated failure time model," rather than the nonparametric approach used in Cox regression.)

The parametric models for the summary survival curve all have their advantages and disadvantages, but none of them can offer the reassurance of routinely providing a close, well-conforming fit. In fact, if such a fit is always desired, it cannot be routinely produced by an "off the rack" mathematical model. A "custom-tailored" approach is needed.

15.1.4.2. Nonparametric life-table structure

The custom-tailored approach to survival curves has been available for several centuries. It was originally proposed in 1662 in England by John Graunt,[7] who worked with "Bills of Mortality" to derive estimates of rates of death at different ages. He then arranged the rates into a life-table, which could be used for estimates of life expectancy. The procedure was later adapted and applied by the actuaries of life insurance companies to help set premiums that would be profitable for the companies while simultaneously attractive to the purchasers of policies. Because of this usage, the life-table method of analysis is also called *actuarial analysis*.

The actuarial method is today called nonparametric, because it has no mathematical algebraic model. The curve was prepared simply by connecting the survival rate values at different annual intervals of life. In actuarial work, the curves were easy to construct because everyone's date of death becomes known when the insurance policy is paid. For medical activities, however, a modification was needed to deal with the problem of censored patients, whose final survival durations are unknown.

The desired modification for censored durations was provided by successively decrementing the denominators. In the decrementing principle, each censored person is included for the pertinent length of follow-up time, and is thereafter eliminated from the denominator of people under continuing observation. The decrementations can take place at equal regular intervals of serial (follow-up) time, or at irregular ad hoc intervals that are demarcated whenever a person dies.

15.1.4.3. Berkson-Gage actuarial method

In the general-population method developed and still used by the actuaries of life insurance companies, the life-table curve shows survival rates at equal intervals of time, such as every six months or one year. This same equal-interval tactic was applied about fifty years ago by Berkson and Gage[8] in the survival method proposed for medical cohorts and used thereafter for many decades. At the end of each interval, survival rates are calculated for an adjusted denominator, which consists of all persons who died in the interval, plus all who survived, plus half the number of persons censored during the interval. (The assumption is made that censored persons, on average, survive half the interval.) After contributing half of their numbers to the denominator of the interval in which they are censored, the censored persons are omitted from the cohort thereafter.

When the number of deaths in each interval, ending at time t, is divided by the "adjusted" denominator of people at risk during that interval, the result is an interval mortality rate, q_t, which represents the risk or *hazard* of death during that interval. (In a general population, this type of interval hazard is often called the "force of mortality" when cited as a one-year incidence rate of death, or annual mortality rate.) The interval survival rate, p_t, calculated as $1 - q_t$, is the complement of the interval mortality rate. For example, if 100 people enter an interval in which 5 die, 4 are censored, and 91 survive, the interval mortality rate is $5/[91 + 5 + (4/2)] = 5/98 = .051$. The interval survival rate is $1 - .051 = .949$.

The symbol S(t) is used for the entire summary survival curve. It consists of a set of *cumulative* survival rates, for which the summary, $\overline{Y}_t$, at each time point, t, is calculated as the product of all the successive preceding interval survival rates. At the beginning, $\overline{Y}_0 = p_0 = 1$. If the survival rate is p_1 for the first interval, the cumulative value of $\overline{Y}_1 = p_0 \times p_1 = 1 \times p_1 = p_1$. After the second interval, the cumulative survival rate is $\overline{Y}_2 = p_1 \times p_2$. After the third interval, $\overline{Y}_3 = p_1 \times p_2 \times p_3$, and so on. Thus, if the interval survival rates are 80% from years 0 to 1, 70% from years 1 to 2, and 93% from years 2 to 3, the value of S(t) at three years will be the cumulative three-year survival rate, $.80 \times .70 \times .92 = .52$, or 52%.

For the Berkson-Gage life-table analysis, the data for the ten persons in Fig. 15-2 would be arranged as shown in Table 15-1. If the life-table analysis were *not* used, and data were taken directly from Fig. 15-2, four persons would have been censored so that six of the original ten persons could have been followed as long as the end of the seventh time interval. At that point, since four were dead, the survival rate would have been $2/6 = .33$.

Table 15-1 Life-table (Berkson-Gage) arrangement of data in Fig. 15-2

Interval	Interval mortality rate	Interval survival rate	Cumulative survival rate
1	$q_1 = 0/[9+(1/2)] = 0$	$p_1 = 1$	$S_1 = 1$
2	$q_2 = 1/9 = .111$	$p_2 = .889$	$S_2 = .889$
3	$q_3 = 0/[7+(1/2)] = 0$	$p_3 = 1$	$S_3 = .889$
4	$q_4 = 1/[6+(1/2)] = .154$	$p_4 = .846$	$S_4 = .752$
5	$q_5 = 1/5 = .2$	$p_5 = .8$	$S_5 = .602$
6	$q_6 = 0/[3+(1/2)] = 0$	$p_6 = 1$	$S_6 = .602$
7	$q_7 = 1/3 = .333$	$p_7 = .667$	$S_7 = .401$
8	$q_8 = 0/1 = 0$	$p_8 = 1$	$S_8 = .401$

The use of equal time intervals in the acturial or Berkson-Gage method produces a set of points that often form a relatively smooth-looking curve, such as Fig. 15-1, for S(t).

15.1.4.4. Kaplan-Meier method

About thirty-five years ago, investigating the "life-history" of vacuum tubes at Bell Laboratories, E. L. Kaplan worked with Paul Meier (then a biostatistician at Johns Hopkins University medical school) to develop a different approach to intervals. In the Kaplan-Meier method,[9] the interval durations for the adjusted denominators are determined by the deaths (or failure events). Whenever someone dies (due to an appropriate event), the periodic mortality and survival proportions are calculated for the antecedent interval, however short or long it may have been. This tactic gives the Kaplan-Meier method the advantage, to be discussed later,

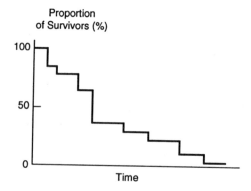

Fig. 15-3 Kaplan-Meier summary survival curve for a cohort. (Strictly speaking, the Kaplan-Meier curve is a series of step functions, which are disconnected horizontal lines. The vertical lines are usually added for esthetic reasons to avoid a ghostly staircase.)

that the interval mortality rates, q_t, always have non-zero values. Censored persons are included in the entire denominator for the interval in which they became censored, and are eliminated thereafter. As in Berkson-Gage analyses, the cumulative survival at each interval is the product of survival proportions for all preceding intervals.

In the Kaplan-Meier method, each interval's cumulative survival proportion is drawn as a flat line, and the series of lines forms a "step function". Because the intervals are demarcated in an ad hoc manner at the time of deaths, the lines have unequal lengths, and the survival curve looks like an irregular staircase, as shown in Fig. 15-3.

For the data in Fig. 15-2, the Kaplan-Meier tabulation would require using the exact durations before death, rather an arbitrary set of intervals. For simplicity of illustration, however, we can assume that each death occurred at the renumbered durations shown in Table 15-2.

Table 15-2 Kaplan-Meier arrangement of data in Fig. 15-2

Interval in Fig. 15-2	Number censored	Number dead	Renumbered duration	Interval mortality rate	Interval survival rate	Cumulative survival rate
1	1	0	—	No deaths; no calculations		
2	0	1	1	$q_1 = 1/9 = .1111$	$p_1 = .8889$	$S_1 = .8889$
3	1	0	—	No deaths; no calculations		
4	1	1	2	$q_2 = 1/7 = .1429$	$p_2 = .8571$	$S_2 = .7619$
5	0	1	3	$q_3 = 1/5 = .2$	$p_3 = .8$	$S_3 = .6095$
6	1	0	—	No deaths; no calculations		
7	0	1	4	$q_4 = 1/3 = .3333$	$p_4 = .6667$	$S_4 = .4063$
8	2	0	—	No deaths; no calculations		

The Berkson-Gage and Kaplan-Meier curves for the data under discussion are shown in Fig. 15-4. The Kaplan-Meier lines always maintain the cumulative survival noted at the previous death until each step occurs at the new death. In this instance, despite very small groups, and with approximations for the times of death, the Berkson-Gage and Kaplan-Meier methods still yield relatively similar results. By giving credit for the durations lived by censored persons, both methods raise the survival rate that would otherwise have been calculated with a "direct" approach, based only on those persons whose alive or dead state is always known.[10]

The Kaplan-Meier method has become particularly popular in medical statistics today because arbitrary durations need not be chosen for time intervals, and because the log-rank test[11,12] can easily be applied to check for stochastically significant differences in survival curves.

15.1.5. General strategy of Cox regression

Regardless of a Berkson-Gage or Kaplan-Meier construction, the summary curve of a survival analysis can be modified by the multiple variables that exist at baseline. Good prognostic factors should move the curve upward, and poor ones, downward. To discern and arrange these factors, Cox decided to modify a curve called the *hazard function*, which is closely related to S(t).

380 *Regression for Nondimensional Targets*

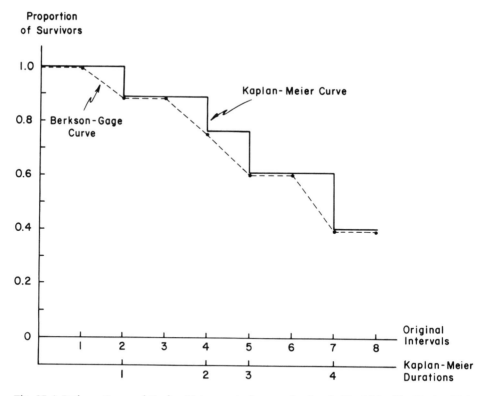

Fig. 15-4 Berkson-Gage and Kaplan-Meier survival curves for data in Fig. 15-2. (The Kaplan-Meier curve is always plotted as a step function. The Berkson-Gage points are usually connected directly, but can also be regarded, in strict construction, as a step function.)

The next three sections (15.2 through 15.4) describe the mathematical concepts and statistical principles that are used in Cox's proportional hazards method. The ideas, which should not be too difficult to digest, are worth knowing if you want to understand the basic reasoning that underlies the Cox procedure. If the reading becomes too hazardous for your intellectual survival, however, advance promptly to Section 15.5, which discusses the indexes of accomplishment that appear in computer printouts. The text has been arranged so that it can be followed from Section 15.5 onward, even if you have skipped Sections 15.2–15.4. Nevertheless, you may later want to review some of those sections when pertinent clarifications seem needed.

15.2. Concept of a hazard function

The "radioactive decay" curves of physics offer a relatively easy way to understand the role of the hazard function in survival analysis.

15.2.1. Radioactive decay curve

In a curve that shows the decay of radioactivity, the change at a given instant in time, t, is proportional to the amount of radioactivity, y, that exists at that instant. The differential-calculus expression for this change is $dy/dt = -cy$, where c is a constant proportion. When integrated, this expression produces $\ln y = -ct$, which becomes converted to $y = e^{-ct}$. When plotted against time, the descending exponential curve for y resembles the survival curve shown in Fig. 15-1.

If expressed in the ideas of survival analysis, the corresponding entities would be: y = cumulative survival function, and $-\ln y = ct$ = cumulative hazard function. The time derivative of the cumulative hazard function would be

$$d(\ln y) = c = \text{hazard function}.$$

In a radioactive decay curve, the hazard function has a constant value of c, and the cumulative hazard function, $-\ln y = ct$, forms a rising straight line as time progresses. The cumulative survival curve, $y = e^{-ct}$, is the exponential of the negative cumulative hazard function.

15.2.2. Corresponding ideas in survival curve

The corresponding principles in survival analysis are shown in Table 15-3. The cumulative survival function is $S(t)$. Its negative logarithm is the cumulative hazard function, $H(t)$. The time derivative of the latter is the hazard function, $h(t)$. The symbols for the two hazard terms are sometimes written with upper- and lowercase Greek lambda letters as shown in the far right of Table 15-3.

Table 15-3 Correspondence of concepts and symbols in survival analysis and in radioactive decay curve

Concept	Radioactive decay curve	Survival analysis	Alternative statistical symbol
Cumulative survival	$y = e^{-ct}$	$S(t)$	—
Cumulative hazard function	$ct = -\ln y$	$H(t) = -\ln S(t)$	$\Lambda(t)$
Hazard function	$c = d(-\ln y)$	$h(t) = d[H(t)]$ $= d[-\ln S(t)]$	$\lambda(t)$

In a smooth radioactive decay curve, the constant value of the hazard function indicates the same force of mortality at each instant. In survival analysis, which seldom produces a smooth curve, the hazard function is approximated by the values of the successive interval mortality rates. For example, in the Kaplan-Meier Table 15-2, the successive interval mortality rates that form the hazard function would be $q_1 = .1111$, $q_2 = .1429$, $q_3 = .2$ and $q_4 = .3333$. The negative logarithms of the cumulative survival rates that form the cumulative hazard function would be $H_1 = -\ln S_1 = -\ln .8889 = 1178$; $H_2 = -\ln S_2 = -\ln .7619 = .2719$; $H_3 = -\ln .6095 = .4951$; and $H_4 = -\ln .4063 = .9007$.

15.2.3. Algebraic and arithmetical features of hazard function

In radioactive decay, the hazard function for the amount of change at each instant is proportional to the existing amount of radioactivity, y. For survival analysis, the corresponding change at time t is the proportional increment, $[S(t-1) - S(t)]/S(t-1)$. With the cumulative calculation, $S(t) = S(t-1) \times P_t$, and so the change is $[S(t-1) - S(t)]/S(t-1) = 1 - p_t = q_t$. Thus, the successive values of q_t are the counterpart of a hazard function.

Since $S(t) = S(t-1) \times p_t$, another interesting algebraic distinction is that $\ln S(t) = \ln [S(t-1)] + \ln p_t$, so that $\ln p_t = \ln S(t) - \ln [S(t-1)]$. For example, since $S_4 = .4063$ and $S_3 = .6095$ in the Kaplan-Meier tabulation in Table 15-2, $\ln p_4 = \ln (.4063) - \ln (.6095) = -.9007 - (-.4951) = -.4056$ and $p_4 = e^{-.4056} = .6665$ (which differs from .6667 due to rounding).

15.2.4. Modification of hazard function

Cox's strategic decision was to use the linear model, G, to modify the hazard function. Instead of the direct value of G, however, he used its exponentiated value, e^G. This structure had at least two advantages. First, the value of e^G would always be positive, no matter what G might become for an individual person. Second, when an individual e^G value is divided by the sum of e^G values for a group (as discussed later), the quotient can be regarded as a type of probability, analogous to what was done in logistic regression. Since an intercept was not needed, the linear model would be expressed as $G = b_1X_1 + b_2X_2 + \ldots + b_jX_j + \ldots$.

If $h_o(t)$ was the underlying summary hazard function for the group, the modification that estimated the hazard function for an individual person would be

$$\hat{h}(t) = h_o(t) \times e^G.$$

Since e^G is a constant for each person, the cumulative hazard for that person would be

$$\hat{H}(t) = H_o(t) \times e^G.$$

Since $H_o(t) = -\ln S_o(t)$, the estimated cumulative survival function will be constructed as $\exp[-H_o(t)]^{e^G}$ to form the doubly exponentiated

$$\hat{S}(t) = S_o(t)^{e^G}.$$

At each point in time, the summary curve, $S_o(t)$, expresses a probability value that ranges from 1 at the beginning, when everyone in the cohort is alive, to 0 when everyone has died. When the general linear model, $G = b_1X_1 + \ldots + b_pX_p$, is used to modify the probability value of $S_o(t)$, the modified estimate must also lie between 0 and 1. If $S_o(t)$ were merely multiplied by G, the product of $S_o(t) \times G$ might be negative or might exceed 1 if G was sufficiently large. If $S_o(t)$ were multiplied by e^G, the values of e^G would never be negative, but they might still become large enough to make the product of $S_o(t) \times e^G$ exceed 1. For example, if $S_o(t) = .4$ at time t and if $e^G = 6$, the value of $S_o(t) \times e^G$ would be an impossible 2.4 for estimated probability of survival.

The great advantage of exponentiating $S_o(t)$, which always lies between 0 and 1, by the always positive value of e^G, is that the value of $S_o(t)^{e^G}$ will always lie between 0 and 1. Because exponential powers have a reverse impact on numbers in the range between 0 and 1, the exponential expression will make survival decrease as e^G enlarges, and increase as e^G gets smaller. Thus, if $S_o(t) = .4$ and $G = .693$, the value of $e^G = 2$, and $S_o(t)^{e^G} = (.4)^2 = .16$. If $G = -.693$, so that $e^G = .5$, $S_o(t)^{e^G}$ becomes $(.4)^{.5} = .632$. Another way to remember the idea is that e^G modifies the hazard of death. As the hazard increases, survival decreases, and vice versa.

15.2.5. Illustration of modification

To illustrate the general modification procedure, suppose that the longitudinal results for nonoccurrence of the failure event in a cohort are summarized in a survival curve, such as Fig. 15-1. This curve could be used to estimate the summary probability that a member of the cohort would be alive at any particular point in serial time after the start of the observation period. Thus, in Fig. 15-1, the chance of surviving to three years is about 50%.

In the proportional hazards analytic procedure, the independent baseline variables modify the summary curve for the group; and the result is an estimated probability of survival for individual persons (or subgroups) as time progresses. For example, in Fig. 15-5, the summary curve of Fig. 15-1 is modified so that the upper part shows the survival of young patients with localized cancer, and the lower part shows the survival of old patients with disseminated cancer.

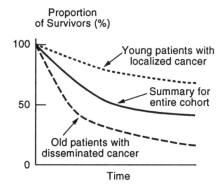

Fig. 15-5 Summary survival curve of Fig. 15-1 modified to show survival in two subgroups.

15.3. Basic structure and summary of data

In proportional hazards analysis, the outcome for the ith person is expressed as $Y_{i,t}$ for both a time and a state. The outcome data for the entire group can be summarized with the cumulative *survival curve*, $S_o(t)$. Because the analytic target is a response event, such as death, it is usually coded as **1**, and the complementary censored state is coded as **0**. Thus, if **1** = dead, and the 53rd person dies at 28 months, the observed value for $Y_{i,t}$ is $Y_{53,28} = 1$. In the analytic process, the estimated individual values for a particular person refer not to a single state, such as $\hat{Y}_i$, but to both a state and a specified time, $\hat{Y}_{i,t}$.

In the analysis of data, the selected value of t for each person is the observed duration of serial time before that person's *exit*. The person's binary *outcome state* at the exit date is either **death,** or its converse, the state of **censored.** For each person, the expression of $Y_{i,t}$ will have three components: the i that identifies the person; the t that shows the serial duration elapsed before the exit; and the **0** or **1** citation for Y itself as the exit state.

If T is the selected maximum duration (such as 5-year survival) for the group's observation period, each person's exit state and duration can have four possible citations: dead before or at T; intermediately censored as withdrawn alive (by loss to follow up) at a duration shorter than T; terminally censored as withdrawn alive at T; or censored at or before T by a competing event. In most analyses (because of the assumption that the censorings and outcomes are unrelated), the reason for censoring is ignored; and each person is classified as either **dead** or **censored** at the exit time.

15.3.1. Basic summary curve, $S_o(t)$

The underlying cumulative survival curve, $S_o(t)$, is the basic summary for the $Y_{i,t}$ values. It includes no X_j variables, and depends only on the observed states and times of the $Y_{i,t}$ values. For the commonly used Kaplan-Meier method, $S_o(t)$ is cited as a step function, with different constant values for different intervals of time. If the centerpoints were joined for each "stair" in the "steps", however, the connected line for $S_o(t)$ might resemble the smoother type of curve usually obtained with a Berkson-Gage actuarial analysis.

Although $S_o(t)$ is a fundamental concept in the general reasoning, the Berkson-Gage and Kaplan-Meier methods need not be used for determining a summary survival curve in Cox regression. Instead, a summary hazard function can be calculated, and the summary survival curve, $S_h(t)$, is then formed by appropriately integrating the values of the hazard function. As noted in Chapter 16, the summary survival curve is $S_h(t)$ in the BMDP program, but is the Kaplan-Meier $S_o(t)$ in the SAS program.

15.3.2. Subsequent calculations and complexities

In the linear and logistic methods of regression, the discrepancies between the observed Y_i and the estimated value of $\hat{Y}_i$ for each person were used to form a total discrepancy. The coefficients of the general linear model, G, were calculated to give the total discrepancy an optimum value, which was a minimum for the sum of squared differences in linear regression, and a maximum likelihood for the product of congruences in logistic regression.

This approach might have been used in survival analysis to determine the congruence between each person's actual exit value at time t, and the corresponding estimated value of S(t). The product of the congruences could form a likelihood, and the b_j coefficients for the X_j variables could be chosen to maximize this likelihood.

Cox used a different approach, however, based on determining a probability for each person's hazard function at the time of death. The probabilities were multiplied to form a likelihood, which was maximized to produce the coefficients for the b_j values. A summary hazard function could then be determined when the b_j values were set to 0. Integration of the summary hazard function would produce the cumulative hazard function, which is the negative logarithm of the cumulative survival function, expressed as $S_h(t)$.

Fortunately for the user of computer programs for Cox regression, the complexity of these activities need not be understood and is not displayed in the printouts. A possibly confusing consequence of the activities, however, is that the summary cumulative survival curve, $S_h(t)$, determined by the hazard-function method, may regularly show differences from the corresponding values of $S_o(t)$, determined by the customary Kaplan-Meier (or Berkson-Gage) method. Another possible source of confusion is that the Cox likelihoods are not calculated in the same way as in logistic regression. Nevertheless, the likelihood results are eventually partitioned and interpreted in almost exactly the same way for Cox as for logistic regression.

The main mathematical complexity of the proportional hazards regression method for survival analysis arises because the hazard function, which is the derivative of a logarithm, is multiplied by an appropriate linear combination of multiple variables. The result is then integrated and exponentiated.

15.4. Fitting the coefficients

At any point in time, t, the summary hazard function for the group is $h_o(t)$. Whenever a person dies at time t, the modified hazard function will be $h_i(t) = h_o(t) \times e^{G_i}$, where G_i contains the result of $b_1X_1 + b_2X_2 + \ldots$ for that person's values of $X_1, X_2, \ldots$. This value of $h_i(t)$ becomes a probability when divided by the sum of all the corresponding $h_i(t)$ values for all persons who were at risk just before time t. In somewhat oversimplified notation, the probability would be $[h_o(t) \times e^{G_i}]/\Sigma[h_o(t) \times e^{G_i}]$. Since the same constant, $h_o(t)$, appears in both numerator and denominator of this expression, the probability can be cited as $e^{G_i}/\Sigma e^{G_i}$ for each death. The likelihood is the product of these probabilities, expressed as $\Pi[e^{G_i}/\Sigma e^{G_i}]$. The best b_j coefficients for the model, $G = b_1X_1 + b_2X_2 + \ldots$, are found, by trial and error, as the values that will maximize the likelihood.

The basic likelihood, L_0, is obtained with no variables in the model. For a model containing the independent variables, $X_1, X_2, \ldots, X_p$, different values of the coefficients $b_1, b_2, \ldots, b_p$ are examined to find the combination that produces a maximum value of the likelihood, L_R. Since the log likelihood ratio (as noted in Chapter 13) is

$$\text{LLR} = -\ln L_0 + L_R,$$

and since $-\ln L_0$ has a fixed value, a maximum for L_R will minimize the value of $-\ln L_R$, thereby maximizing LLR. For example, if the original value of $\ln L_0$ is -1678, and if $\ln L_R$ is -1597, the value of LLR is $-(-1678) + (-1597) = 81$.

15.4.1. Tactics in calculation

As in multiple logistic regression, the maximum likelihood is determined by trial and error. In the first step, all the b_j values are set at zero, and the "basic" likelihood, L_0, is determined. Each b_j coefficient is then advanced backward or forward in unitary steps and successive half-steps to reach a value that maximizes the residual likelihood, L_R, for that coefficient. The search process is repeated over and over until the huge number of appropriate possibilities is exhausted. The best choice of coefficients is the particular collection that "converges" in giving the best, i.e., lowest, value for the negative log of the residual likelihood, L_R, and the highest value for the log likelihood ratio, LLR.

Fortunately, various strategies are used inside the "black box" to enhance the efficiency of the arbitrary choices of coefficients, and to reduce the enormous amount of time required for the computations. Another fortunate thing is that the intermediate computations are usually not displayed in the computer printout, which shows only the final results, and (if asked) a statement of the number of iterative steps needed to reach them.

15.4.2. Partial likelihood method

The name "partial likelihood" is applied in Cox regression because the likelihoods are calculated from numerators for only the people who died. The total group at risk appear in the denominator of each probability ratio, $e^{G_i}/\Sigma e^{G_i}$, but the e^{G_i} numerators depend solely on the deaths. With this method of "conditioning on the deaths", the reliability of the process may be impaired if too few deaths are available for the calculations.

The term *partial maximum likelihood* is regularly used for the method of finding the $\{b_j\}$ coefficients; and the acronym PMLR in computer printouts refers to the *partial maximum likelihood ratios* that become statistical indexes of accomplishment.

15.4.3. Assumption about proportional hazards

An important assumption in the Cox regression method is that e^G for each person or each group has a *constant* value over time. When multiplied by e^G, the hazard function, $h(t)$, should be modified consistently throughout its temporal course, being proportionately displaced upward or downward, without substantially altering the basic shape.

When survival results are compared for two groups, or for two categories within a single variable, X_j, the assumption about a constant modification will be correct if the corresponding curves for the cumulative survival, $S(t)$ [or for the hazard, $h(t)$], are proportionately separated and do not cross throughout their course. If the curves cross, however, the hazard cannot be constant in both. When the proportional hazards assumption is violated, the analytic results may be incorrect.

For this reason, the diagnostics process in Cox regression includes special tests, discussed in Chapter 16, that are aimed at checking the validity of the assumption about constant proportional hazards.

15.4.4. Centered and noncentered models

The value of G in the Cox regression model can be calculated in a centered or noncentered manner. The noncentered model is the customary

$$G = b_1X_1 + b_2X_2 + b_3X_3 + \ldots.$$

In the centered model, the means of the X_j variables are subtracted to form

$$G_c = b_1(X_1 - \bar{X}_1) + b_2(X_2 - \bar{X}_2) + b_3(X3 - \bar{X}_3) + \ldots.$$

Either model can be used, but the differences must be kept in mind when the eventual results are used for estimates of $\hat{S}(t)$. As noted later, certain estimates are made using mean values of the X_j variables. For such estimates, the value of $G_c = 0$, and $e^{G_c} = 1$. Consequently, when the computer printout shows individual mean values of the $S_h(t)$ curve at different time intervals, the results can be used directly for individual estimates if $S_h(t)$ was obtained with a centered model. If not obtained with a centered model, an additional conversion is needed.

The two procedures eventually yield the same results, as discussed in Section 16.1.1.2.1, but the individual calculations are done differently.

15.4.5. 'Score' methods of calculation

Although all of the basic interpretations continue to depend on the likelihood strategy, the calculations for various chi-square indexes in the Cox model are often done with the "score" method (see Section 13.8.2.2.) used in logistic regression. The score method saves time in computation, is particularly valuable in the uncommon situations where the "conventional" trial-and-error likelihood calculations fail to converge, and generally gets results quite close to those obtained with the "gold-standard" trial-and-error procedure.

15.5. Indexes of accomplishment

As in other forms of regression, the indexes of accomplishment for the Cox method refer to achievements of the survival model, and to the effects of the independent variables. The indexes are discussed in this section and in Sections 15.6 and 15.7. Their application in the illustrative data set is shown with printouts in Chapter 16.

15.5.1. Problems in expression and comprehension

As noted in Chapters 2, 7, and 12, an algebraic model of the survival curve cannot possibly give a good fit to the *individual* items of data. They will appear, as shown in Fig. 15-2, as cannonball-like batches of points at $Y = 0$ or $Y = 1$ for corresponding values of time; and the individual points will usually be too polyvalent for a close direct fit.

With appropriate selection, the summary curve for $S(t)$ will give an excellent fit to the *summary proportions*, $\bar{Y}_t$. Nevertheless, when each person's estimate is made after the person's values of X_j are entered into the formula for $\hat{S}(t)$, the result is an estimate for probability of the event at the selected time for that person, not for an entire curve.

In addition to this problem, the *summary* proportion estimated by $\hat{S}(t)$ at each time point is not readily calculated. The values of t and the b_j coefficients are easy to use, but not the values of X_j, which will differ for each person. Accordingly, the summary survival proportions for $\hat{S}(t)$ are usually estimated by entering the *mean* values of the X_j variables into the formula for $S_h(t)^{e^G}$. In a noncentered model, if $G = 0.0135X_1 - 0.5946X_2 + .8181X_3$, and if the mean values are 59.7 for $\bar{X}_1$, .255 for $\bar{X}_2$, and .210 for $\bar{X}_3$, the mean value of G will be $(.0135)(59.7) - (.5946)(.255)$

+ (.8181)(.210) = .826. The summary value of e^G will be $e^{.826}$ = 2.28, and the summary value for $\hat{S}(t)$ will be $S_h(t)^{2.28}$ at each time point.

Since all post-zero-time values of S(t) are $\leq$ 1, the modified values will also be $\leq$ 1, but the modification will raise or lower S(t) according to the value of e^G. Thus, if $S_h(t)$ at a particular time point is .56, the value of $\hat{S}(t)$ when e^G = 2.23 would be $.56^{2.28}$ = .27. If the values of G are negative, $\hat{S}(t)$ can be increased but will still not exceed 1. For example, if G is negative rather than positive at a value of $-.826$, e^G will be .438, and the foregoing modification would be $S_h(t)^{.438}$ = $56^{.438}$ = .78.

15.5.2. Indexes for models

As in logistic regression, the main indexes of accomplishment for a Cox model refer to its achievements in improving residual likelihood. They can be expressed as values for $\ln L_R$ and for the log likelihood ratio, LLR.

15.5.2.1. $-$Log likelihood

The residual likelihood, L_R, can always be calculated for whatever variables have been included in a model, $G = b_1X_1 + \ldots + b_pX_p$. As L_R gets larger, i.e., closer to 1, its negative logarithm will get smaller. Thus, if the baseline likelihood, L_0, is doubled from .0035853 to a value of L_R = .0071706, the value of negative log likelihood decreases from 5.6309 to 4.9378.

The negative log likelihood value, $-\ln L_R$, is analogous to the value of the residual (or error) sum of squares, symbolized earlier as D_p in ordinary linear regression; and $-\ln L_R$ also gets smaller, like D_p, as the fit of the model improves.

15.5.2.2. Log likelihood ratio

The log likelihood ratio, LLR, has the same role in Cox regression as in ordinary logistic regression:

$$\text{LLR} = \ln L_R - \ln L_0 = -\ln L_0 - (-\ln L_R).$$

The value of LLR will always be positive because both original logarithms will be negative, but the absolute value of $|\ln L_0|$ will exceed that of $|\ln L_R|$. In the preceding example, if $\ln L_R = -4.9378$ and $\ln L_0 = 5.6309$, the value of LLR will be $-4.9378 + 5.6309 = .6931$. As might be expected from the earlier statement that the likelihood ratio was doubled, the actual value associated with this LLR is $e^{.6931} = 2$.

The value of 2 LLR is called the *model likelihood-ratio chi-square,* and is interpreted with a chi-square distribution.

15.5.2.3. Chi-square global score

The global chi-square "score", as calculated in Section 15.4.5, compares the fit of the model with the fit that would have been obtained with no variables in the model. Both the likelihood ratio and global chi-square score values for the model are converted to P values, according to the appropriate degrees of freedom (for the number of included variables).

The score results for chi-square are usually quite close to the values obtained with the likelihood method for the same model.

15.5.2.4. 'R_H^2 statistic'

This index is analogous to the R_H^2 statistic (described in Section 13.7.2.3), proposed by Harrell[13] for multiple logistic regression, with a "penalty" correction inserted for the number of parameters, p. The result resembles a quantitative proportion of explained likelihood, analogous to the adjusted R^2 results obtained for explained variance in least-squares linear regression.

The calculation is done the same way as in logistic regression

$$R_H^2 = \frac{(\text{model chi-square} - 2p)}{-2 \ln L_0}.$$

If model chi-square is less than 2p, the value of 0 is assigned to R_H^2.

15.5.2.5. Additional expressions

Although a useful "crude" index of explained likelihood, R_H^2 is not regarded as mathematically optimum. Several other (more complicated) indexes have therefore been proposed by Schemper,[14] Korn and Simon,[15] and Verweij and van Houwelingen.[16] These indexes have not yet begun to appear in packaged computer programs, but (given programmers' avidity for adding new expressions) will probably do so in the foreseeable future.

15.5.3. Incremental indexes

The two main incremental indexes in Cox regression show the changes when a sequential exploration goes from one model to the next.

15.5.3.1. Maximum partial likelihood ratio (MPLR)

The maximum partial likelihood ratio is particularly useful in sequential explorations when a current model is either augmented (or depleted) incrementally with the addition (or removal) of a new variable. This ratio in a step-up regression process is calculated as

$$\text{MPLR} = \frac{\text{current likelihood before candidate variable}}{\text{new likelihood after candidate is entered}}.$$

It is usually transformed to

$$\ln \text{MPLR} = [\ln \text{current likelihood}] - [\ln \text{new likelihood}].$$

For example, suppose ln likelihood is -846.3585 for an existing model, and becomes -820.5814 after a potential candidate variable is added. The value of ln MPLR will be $-846.3585 - (-820.5814) = -25.7771$. The MPLR index is a direct counterpart of the increment $D_{p-1} - D_p$ in logistic regression, where D_{p-1} is analogous to the log of the current likelihood and D_p, to the log of the new likelihood.

15.5.3.2. Improvement chi-square

The value of $-2 \ln$ MPLR is an exact counterpart of the incremental chi-square formed from -2LLR in logistic regression. The result, often called "improvement chi-square", is also analogous to the incremental value of R^2 between two successive models in multiple linear regression.

Calculated as $-2 \ln$ MPLR, the value of improvement chi-square will be 51.554 if ln MPLR is -25.777. The multiplication by -2 makes the negative logs of MPLR become positive, and allows the result to have a chi-square interpretation, converting the value of $-2 \ln$ MPLR to a P value. The degrees of freedom depend on how the MPLR was obtained. In a stepwise procedure, when one new variable is added or deleted at a time, each successive MPLR has one degree of freedom.

15.5.3.3. Alternative PHH method

In BMDP programs, the improvement chi-square can also be determined with a score rather than MPLR calculation, using a PHH method that is named after its proponents: Peduzzi, Hardy, and Holford.[17] In the PHH method, a score-type procedure is used to calculate improvement chi-square. It is essentially the ratio of $b_j^2/$(variance of b_j), which is often called *Wald chi-square*, that would occur for each candidate variable if it individually entered into an existing model. In forward stepping, the variable with the largest potential chi-square score is chosen for entry. In backward stepping, the process is reversed, and the removed variable has the lowest chi-square score.

15.5.4. Indexes for variables

Indexes for the "impact" of the component variables can be determined form the final regression model, but are also often calculated incrementally as individual candidate variables are sequentially entered or removed.

15.5.4.1. 'Parameter estimates'

The corresponding b_j coefficient is the "parameter estimate" for each variable included in the model. The quotient formed when the coefficient is divided by its associated standard error is analogous to a Z-test statistic, which can be converted to a P value for stochastic significance. Since the coefficient will appear in the model as $e^{b_j X_j}$, rather than as $b_j X_j$, a term labelled "exp(coefficient)" can be calculated as e^{b_j} to show the actual effect. Thus, if a particular coefficient is .6804, the value of exp(coefficient) will be $e^{.6804} = 1.9747$. It is interpreted as the "hazard ratio" for a one-unit change, as discussed in Section 15.5.4.4.

15.5.4.2. Partial 'Index-R'

The partial-correlation Index-R statistic that Harrell proposed for logistic regression (see Section 13.9.4) can also be applied to each variable in Cox regression. The numerator of the ratio for R_j^2 is a penalized value of the improvement chi-square for the variable; and the denominator is twice the negative log of the baseline likelihood. The formula is

$$R_j^2 = \frac{(\text{improvement chi-square} - 2)}{-2 \ln L_0}.$$

The coefficient of b_j is attached to R_j, which will have values ranging from -1 to $+1$. If the improvement chi-square is less than 2, R_j is set at 0.

Values of the partial Index-R can avoid some of the problems of distribution when standardized coefficients are calculated for categorical variables, and are

15.5.4.3. Standardized coefficients

The main value of *standardized* coefficients is to help indicate the relative rank of importance of the variables, unaffected by the arbitrary coding or measurement units that will alter the values of b_j or e^{b_j}. Although not offered by either the BMDP or SAS programs for Cox regression, standardized regression coefficients can be obtained if each independent variable in the regression process is entered in its standard Z-score format as $Z_j = (X_j - \bar{X}_j)/s_j$.

A much simpler approach, if the partial regression coefficients are supplied as b_j in the conventional printout, is to find each variable's standard deviation, s_j, in the univariate data. Each standardized coefficient can then be calculated as

$$b_{s_j} = b_j \times s_j.$$

Because an intercept, b_0, is not included in Cox models, division by s_y is not necessary for the standardized adjustment.

The problem noted earlier (see Section 13.9.3.1) for standardized coefficients of categorical variables will usually make the partial Index-R a better way to rate and rank the relative impact of individual variables.

15.5.4.4. Risk ratios and hazard ratios

In Cox regression, the estimated probability of survival when $\hat{S}(t)$ is calculated as $S_h(t)^{e^G}$ can be converted to an estimate of mortality, $\hat{F}(t) = 1 - \hat{S}(t)$. The estimated value of $\hat{F}(t)$ would represent a "direct" risk, rather than an odds. The ratio of two such estimates would therefore form a *risk ratio*, which is also sometimes called a *relative risk* or *rate ratio*. Using the same reasoning discussed in Section 13.9, the value of e^{b_j} for each b_j coefficient is sometimes regarded as a risk ratio for the effect of variable X_j in Cox regression.

The idea that $\exp(b_j)$ is a risk ratio is not quite correct, however, because it exponentiates rather than multiplies $S_h(t)$. As a demonstration of this point, consider the effect of variable X_3 with all other variables held constant. When $X_3 = 0$, the model is $G = b_1X_1 + b_2X_2 + b_4X_4 + \ldots$. With a unitary advance that lets $X_3 = 1$, the model is $G' = b_1X_1 + b_2X_2 + b_3X_3 + b_4X_4 + \ldots$. The risk ratio for the two estimates of F(t) will be $[1 - S_h(t)^{e^{G'}}]/[1 - S_h(t)^{e^G}]$, rather than the expected value of e^{b_3} for the analogous ratio in logistic regression. For example, suppose $S_h(t) = .72$ at a particular time point for a model in which b = .4 for a single variable X. When X = 0, G = bX = 0, $e^G = 1$, $\hat{S}(t) = .72^1 = .72$, and $\hat{F}(t) = .28$. When X = 1, G = .4, $e^G = 1.493$, $\hat{S}(t) = .72^{1.492} = .61$, and $\hat{F}(t) = .39$. Tha ratio of the two $\hat{F}(t)$ risks is $.39/.28 = 1.39$, not the value of $e^{.4} = 1.49$.

On the other hand, since e^G was originally constructed to multiply the hazard function, h(t), and since the estimated hazard function will be $h_0(t)e^G$, the algebraic structure allows e^{b_j} to be a *hazard ratio*. For example, when the model changes from G to G', the ratio of the two corresponding hazard functions will be $[h_0(t) \times e^{G'}]/[h_0(t) \times e^G]$. The $h_0(t)$ factors will cancel, and the result will be

$e^{G'}/e^G = e^{(G'-G)}$. If all other variables are held constant, $G'-G$ for variable X_3 will be b_3, and the hazard ratio will be e^{b_3} or $\exp(b_3)$. In the foregoing example with just one X variable, where the risk ratio turned out to be 1.39 for the lowered survival when X went from 0 to 1, the hazard ratio would have been $e^{.4} = 1.49$.

The structure that allows the exponentiated b_j coefficients to act as hazard ratios is another reason the Cox regression model is arranged to modify the hazard function rather than the cumulative survival. Because the exponents are added, the combined hazard ratios for more than one variable have a multiplicative effect. Thus, $e^{b_3+b_4} = e^{b_3}e^{b_4}$.

The values of b_j will depend on the units of measurement and coding for X_j. Consequently, the best way to compare the hazard impact of different variables is to examine their partial R indexes or the standardized regression coefficients, b_{j_s}. As noted in Section 13.9.3.1, however, standardized coefficients can be misleading for variables that have a peculiar distribution of frequencies. To allow a relatively easy and consistent direct interpretation for the e^{b_j} results, the coding for categorical variables should be maintained at one-unit intervals (such as **0/1**).

15.6. Sequential exploration procedures

The sequential exploration process in Cox regression can go only forward, only backward, or in a stepwise (forth and back) direction. Except for some different indexes of accomplishment, the processes are identical to those in linear and logistic regression. The discussion here will outline the forward stepwise method, which seems most popular today.

The process begins with no variables in the model. Each of the candidate X_j variables is evaluated individually for its possible accomplishments if entered. The candidate with a stochastically significant coefficient that has the highest value for improvement chi-square (e.g., $-2 \ln$ MPLR) is selected and entered into the model.

In the second step, with this first variable retained in the model, the process is iterated. All the remaining variables are examined as candidates for entry as the second variable, with all of the coefficients recalculated afterward for each variable. The second chosen variable is the one with a stochastically significant coefficient that also has the highest value for improvement chi-square.

During the iterative process, each variable that is already in the model is examined after each step, along with those that have just been entered. The aim is to find variables entered earlier that may later become nonsignificant after additional variables have been included. Variables that have become nonsignificant will then be removed.

The stepwise iterations continue until stopped by one of two events: (1) All available variables have been entered and retained in the model; or (2) No available new candidate variable produces a sufficient improvement in chi-square, and no entered variable can be removed.

15.7. Stochastic appraisals

Almost all the stochastic appraisals of significance in Cox regression make use of chi-square statistics, interpreted at the appropriate degrees of freedom. For the model chi-square, the degrees of freedom are usually p for the number of independent variables in the model. For the Wald chi-square of individual variables, the D.F. is 1. [The ratio of b_j/(standard error of b_j) is the square root of the corresponding Wald chi-square statistic, and, when used, is interpreted as a Z-test.]

The 95% confidence intervals for e^{b_j}, which is usually labelled as "exp(b_j)" in computer printouts, are calculated in exactly the same way as in logistic regression, using the standard error for b_j.

The Cox regression curves for two groups are regularly compared with the log rank statistic, another variation of the chi-square procedure, which is discussed further in Chapter 16.

15.8. Regression diagnostics

With the ingenuity of creative statisticians and computer programmers, many procedures have been developed for displaying the output of Cox regression analysis and for certain checks of regression diagnostics. Unfortunately, the procedures do not routinely produce some important diagnostic data, and the few helpful diagnostic tests do not indicate what is needed for prophylactic or remedial therapy. The procedures will be further described and illustrated in Chapter 16, but are briefly outlined here, particularly for readers who may have skipped Sections 15.2–15.4.

15.8.1. Conventional procedures

The conventional activities formally called regression diagnostics are usually conducted as "search-and destroy" missions, aimed at getting a better fit for the model by finding and removing outlier observations. As discussed earlier, this strategy has much more mathematical than scientific appeal. For scientific purposes, the key diagnostic examinations are not detection of outliers, but rather a check of four problems that can arise from the following features of Cox regression: (1) The model uses two components [h(t) and e^G] rather than one; (2) the constant value for e^G; (3) the partial likelihood calculations; and (4) the linearity of the algebraic model. Some of these features were discussed in Sections 15.2–15.4, and their importance is now emphasized for readers who may not have read those sections.

15.8.2. Shape of S(t)

The summary shape of S(t), which is derived from h(t), represents all of the data for the entire group. This shape is retained and becomes the basis for all modifications produced by e^G. The modifications may alter the location of S(t), but if it has substantially different shapes for different subgroups (such as persons in different clinical "stages"), the differences are *not* taken into account by the model.

15.8.3. Constancy of e^G

In the customary method, without any "time-dependent" covariates (discussed shortly), the values of e^G depend exclusively on the values of X_j that were present at zero time, t_0, for each person. At a later time, such as t_4, a particular person's covariate values of hematocrit, weight, or functional severity might be substantially different from what they were originally at the baseline t_0. Nevertheless, the e^G expression will contain only the values present at t_0. Accordingly, the individual predictions for any point in subsequent time always depend on the person's baseline state at t_0.

For this reason, the value of an estimated $\hat{S}(t)$ at t_4 should *not* be used as a *baseline* for estimating what may happen later at t_9. If values of b_j change over time, a time-dependent model might be used to account for the nonconstant factors, but the latter model has many disadvantages (see Section 15.9.1.2). Rather than estimating outcome at t_9 based on status at t_4, a preferable approach is to calculate an entirely different Cox regression, using the t_4 values as a baseline state for persons who have survived to t_4.

15.8.4. Likelihood estimates

As noted earlier, the maximum likelihood in Cox regression is determined only partially, from the hazard values in patients who died. All the patients who were censored appear in the denominators of the calculations, but do not have the opportunity to contribute to the numerators. In various statistical discussions and explorations, the partial approach seems to have been validated, and it is not regarded as a significant problem. Nevertheless, the partial-likelihood calculations may become unstable if based on only a few deaths, or if applied with relatively few events (deaths) per variable, as discussed in earlier chapters and also in Chapter 16.

15.8.5. 'Linearity' of the model

Two types of constancy are assumed for the b_j values in Cox regression. One of them, discussed previously, is the proportional hazards assumption of constancy over time. The other assumption—common to all algebraic models—is that each independent variable has the constant linear impact denoted by the regression coefficient, b_j. The second assumption implies that the hazard ratio (e^{b_j}) remains constant as the variable moves through its own range of values and within the ranges of other variables. Suitable checks for linear conformity and for interactions are therefore a crucial part of the total array of regression diagnostics to be examined before any final conclusions are drawn about each variable's quantitative impact. This problem is particularly important for deciding whether to believe such quantitative claims as, "You live 2 years longer for every 3-unit drop in substance X."

The four types of diagnostic examinations just cited are not included in the usual computer printouts of regression diagnostics and must be conducted separately, as shown in Chapter 16.

15.9. Additional applications

The Cox regression method can also be applied in two other ways that are briefly outlined here.

15.9.1. 'Time-dependent covariates'

Throughout the discussion until now, the individual independent variables, $X_1, X_2, X_3, \ldots$ each had a single value that represented the patient's condition in the baseline state at zero time. The prognostic estimates may have changed for the outcomes as time progressed, but the "predictors" did not. Nevertheless, because a patient's values for $X_1, X_2, X_3 \ldots$ can also change as time progresses, the Cox method can be modified to include these changes by adding "time-dependent covariates". With this tactic, a particular model will include "fixed covariates" as the baseline values of $\{X_j\}$ and time-dependent covariates for those $\{X_j\}$ variables that have data entered for different serial time points.

The time-dependent-covariate approach has different merits for the different challenges to which it can be applied in longitudinal and in survival analysis.

15.9.1.1. Longitudinal analysis

The data for each *survival analysis* are usually arranged in a similar manner and the analyses are aimed at a similar goal. The dependent variable is the interval of time until a single failure event, such as death, which ends each person's duration of being at risk; and the independent variables are analyzed to discern which ones affect the occurrence of the failure event.

Longitudinal analyses, however, need not be aimed at a single failure event, and can have many different arrangements and purposes. As the cohort is followed over time, the longitudinal analytic goals may include the following: "tracking" the independent variables for relative changes with time in location at high, middle, or low ranks in the cohort; discerning correlations in the temporal changes among covariates; determining whether the temporal changes and correlations affect the outcome events; and evaluating the independent variables that are predictors for recurrent outcome events (such as repeated streptococcal infections) or for temporal changes in a dimensional outcome variable (such as blood pressure).

Longitudinal analysis is a relatively young and underdeveloped field, but it is being given increasing attention as more longitudinal research is done with cohort groups. The time-dependent covariate approach seems quite appropriate when a cohort is being followed closely and examined repeatedly for longitudinal changes in both independent and dependent variables. Altman and De Stavola[18] have written a thoughtful account of the benefits and problems of using "updated" covariates—a linguistic substitute for the ambiguity of "time-dependent" covariates—in longitudinal analysis of post-randomization measurements in a clinical trial.

15.9.1.2. Survival analysis

In survival analysis, however, the time-dependent-covariate approach has three important disadvantages for pragmatic application in predicting a single outcome event.

15.9.1.2.1. Availability of data

The first problem is that the data needed for a suitable time-dependent analysis are seldom available. We would have to know each person's values of the $\{X_j\}$ variables (hematocrit, cholesterol, white blood count, clinical stage, etc.) not just at the baseline zero time, but also at each appropriate serial point *after* zero time. In most survival follow-up studies, however, the investigator is usually delighted if information is obtained merely to denote each person's outcome. The data about intervening changes in baseline variables are seldom fully available, except for change in age.

15.9.1.2.2. Interpretation of results

Even if all the intervening information were available for all members of the cohort under analysis, the second problem is that the analytic results are particularly difficult to interpret or apply. The results could not be used for future predictions in new patients who are being seen at zero time, and whose future changes in the time-dependent covariates are unknown. Another problem in interpretation, however, is the question of how to evaluate the ambiguous impact of a variable when it receives *two* partial regression coefficients—one for the variable itself, and the other for its time-dependent formulation.

15.9.1.2.3. Basic scientific goals

A third problem in the time-dependent approach is its conflict with the basic predictive goals of most survival analyses, where we want to use the results to forecast outcomes for individual persons, according to their baseline condition at zero time. We may also want the analyses to help demonstrate (when appropriate) the best treatment or other intervention to apply at zero time. These decisions, which are all made with only the data available at zero time, would not be aided by analytic results based on unavailable data obtained afterward.

Furthermore, if the patient's baseline state later changes, and if at that later time we want to reestimate a new prognosis, reevaluate the treatment, and alter it if necessary, an entirely new analysis would be done at that time—based on what has actually happened during the interval. A thoughtful clinician would *not* try to make the subsequent decisions, after things have changed, using only the baseline information and statistical estimates of what might happen thereafter. Serious clinical attention would seldom (if ever) be given to a *single* prediction, made at baseline, that never needs to be altered because it allegedly provides for all the subsequent changes in the covariates.

Yet another problem that makes the time-dependent covariate method scientifically unsatisfactory is that the analysis does not recognize either that temporal changes in the covariates may be caused by treatment, or that changes in treatment may be caused by changes in the covariates (or by other factors not included in the model).

For all these reasons, the time-dependent models may be useful for the different goals of *longitudinal analysis*, although unattractive for *survival analysis*. The strategies and tactics of longitudinal analysis are beyond the scope of the discussion here.

15.9.2. 'Conditional' methods

Published reports of 1-to-k matched case-control studies, in which each case is matched to a set of k controls, sometimes contain a surprising statement that the analysis was done with the Cox proportional hazards method. The confused reader may then wonder why and how the forward-directed cohort approach of Cox regression was used to analyze the backward-directed data of a case-control study. The explanation for the apparently peculiar strategy is that with a 1-to-k matching in conditional logistic regression (see Section 14.10.2), the likelihood function becomes the same as that of Cox regression if the matched control sets are regarded as strata. (The use of strata in Cox regression is discussed in Section 16.4.) According to Harrell,[13] when Cox regression is used this way, "a dummy time variable must be created so that the case appears to be the only failure in each stratum and the controls appear to be censored at a time later than the cases".

Chapter References

1. Cox, 1972; 2. Kurtzke, 1989; 3. Cramer, 1991; 4. Kalbfleisch, 1980; 5. Lee, 1980; 6. Harris, 1991; 7. Greenwood, 1935; 8. Berkson, 1950; 9. Kaplan, 1958; 10. Feinstein, 1985; 11. Mantel, 1966; 12. Peto, 1972; 13. Harrell, 1986; 14. Schemper, 1990; 15. Korn, 1990; 16. Verweij, 1993; 17. Peduzzi, 1980; 18. Altman, 1994.

16 Evaluations and Illustrations of Cox Regression

Outline ....

16.1. Full regression
    16.1.1. BMDP program
        16.1.1.1. Achievements of model and variables
        16.1.1.2. Comparison of observed and estimated results
            16.1.1.2.1. Individual estimates
            16.1.1.2.2. Illustration of calculations
            16.1.1.2.3. Additional estimates
    16.1.2. SAS program
        16.1.2.1. Differences from BMDP displays
        16.1.2.2. Additional options
        16.1.2.3. Differences from previous SAS displays
            16.1.2.3.1. R_H^2 for the model
            16.1.2.3.2. Utility of Z:PH test
            16.1.2.3.3. Utility of individual Index-R values
            16.1.2.3.4. Standardized coefficients
        16.1.2.4. Interpretation of results
        16.1.2.5. Individual estimates
    16.1.3. Stratified-cluster method for impact of variables

16.2. Sequential stepped regression
    16.2.1. SAS program
    16.2.2. BMDP program
    16.2.3. BMDP/SAS differences in entry/removal criteria

16.3. Three-main-variable model

16.4. Regression for groups
    16.4.1. Descriptive comparisons
        16.4.1.1. Display of curves
        16.4.1.2. Median survival time
        16.4.1.3. Relative hazard
        16.4.1.4. 'Interaction effect' of compared variables
    16.4.2. Log-rank and other stochastic tests
    16.4.3. 'Adjusted' effects of interventions
        16.4.3.1. 'Intra-' and 'extra-model' approaches
        16.4.3.2. Illustration with cell-type variables
    16.4.4. Group curves for constant proportional hazards

16.5. 'Diagnostic examinations'
    16.5.1. Routine graphs for actual and estimated survivals
    16.5.2. Checks for constant proportional hazards
        16.5.2.1. Insertion of time-interaction terms
        16.5.2.2. Graph of estimated LOG MINUS LOG SURVIVOR FUNCTION
    16.5.3. Checks for linear conformity
    16.5.4. Tests for interactions
    16.5.5. Events per variable
    16.5.6. Examination of residuals

16.6. Remedial therapy
16.7. Conversion of algebraic models to scores
16.8. Illustrations from medical literature
    16.8.1. Purpose
    16.8.2. Identification of program and operation
    16.8.3. 'Fit' of model
    16.8.4. Transformation of variables
    16.8.5. Coding for variables
    16.8.6. Impact of variables
    16.8.7. Group curves
    16.8.8. Interactions
    16.8.9. Constant proportional hazards
    16.8.10. Checks for linear conformity
    16.8.11. Events per variable
    16.8.12. Other comments

....

Like Chapter 14, this chapter has several parts. The first shows excerpts of printout for BMDP and SAS programs of Cox regression in the illustrative data set. The second part discusses and displays the diagnostic examination of results. The rest of the chapter contains a review of published reports in which Cox regression was used.

For the illustrative data set, the printouts show full regression, stepped regression, and then regression within groups, before the diagnostic activities appear. The BMDP displays for Cox regression are from the mainframe printout of the program[1] BMDP2L, Version 1990. The SAS displays, which were not available for personal computers when this chapter was written, come mostly from PROC PHREG in the currently newest 6.07 version,[2] but several displays are shown from a precursor[3] PROC PHGLM version 5.18.

Both the BMDP and SAS programs can produce a bare bones or no frills printout that shows the fundamental information for fit of the model and impact of the independent variables. Both programs also have a plethora of options that can yield many additional pages of display for various ancillary indexes, tests, and graphs that usually have limited value. The illustrations here show mainly the no frills program, with an occasional useful option.

16.1. Full regression

For most of the BMDP and SAS programs for Cox regression in the lung cancer illustrative cohort, the maximum duration of survival was truncated at 60 months. This is the conventional 5-year survival time often used in analyzing outcome for patients with cancer. Because no patients had been lost to follow-up, all censorings occurred at 60 months.

16.1.1. BMDP program

The first portion (not shown here) of the BMDP2L printout gives descriptive univariate statistics for the "fixed covariates," which are the seven independent variables used in the regression. The rest of the main printout is shown in Fig. 16-1.

```
INDEPENDENT VARIABLES
   3 AGE          4 MALE         5 TNMSTAGE     6 SXSTAGE      7 PCTWTLOS
   8 HCT          9 PROGIN


   LOG LIKELIHOOD =      -798.5963
GLOBAL CHI-SQUARE =        83.03   D.F.=    7    P-VALUE =0.0000
NORM OF THE SCORE VECTOR=    0.188E-04

                                      STANDARD
        VARIABLE       COEFFICIENT      ERROR       COEFF./S.E.    EXP(COEFF.)
        --------       -----------    ---------     -----------    -----------
     3  AGE               0.0043       0.0084         0.5084         1.0043
     4  MALE              0.5042       0.2970         1.6974         1.6557
     5  TNMSTAGE          0.2463       0.0530         4.6509         1.2793
     6  SXSTAGE           0.3480       0.0893         3.8950         1.4162
     7  PCTWTLOS          0.0141       0.0095         1.4840         1.0142
     8  HCT              -0.0465       0.0166        -2.8001         0.9546
     9  PROGIN            0.0023       0.0045         0.5064         1.0023
```

Fig. 16-1 BMDP printout for full regression in proportional hazards seven-variable analysis of illustrative data set. (BMDP²L program.)

16.1.1.1. Achievements of model and variables

After naming the seven independent variables and indicating that the full model achieved a log likelihood (ln L_R) of -798.5963, the printout shows that the global chi-square for the model, calculated with the rapid score method, is 83.03, which has (at 7 degrees of freedom), a P value of 0.0000 (i.e., < .00005). This global chi-square, which represents the improved difference in log likelihoods for the model vs. no model, is usually calculated as twice the log likelihood ratio, i.e., $2LLR = 2(\ln L_R - \ln L_o)$. The BMDP printout indicates that $2LLR \simeq 83.03$ and that $\ln L_R = -798.5963$, but does not show $-2 \ln L_o$. It can be approximated as $-2 \ln L_o \simeq (2 \times 798.5963) + 83.03 = 1680.22$. The entity called "norm of the score vector" is the approximate increment between the score calculation for the model's global chi-square and what would have been obtained with twice the log likelihood ratio. The tiny value (.188 × 10^{-4}) shows that the two results were almost identical.

The BMDP printout gives no indication of the model's accomplishment in explaining a proportion of likelihood. If desired, such an index can be calculated (see Section 15.5.2.4) as $R_H^2 = $ (model chi-square $-$ 2p)/($-2 \ln L_o$). In this instance, the printout shows model chi-square, and we know that p = 7, but $-2 \ln L_o$ is not stated. From the previous approximation of $-2 \ln L_o$, the value of R_H^2 can be calculated as $(83.03 - 14)/1680.22 = .041$, thus indicating that only about 4% of the original likelihood has been explained by the model.

Of the b_j coefficients for the seven independent variables, only three have absolute Z scores (i.e., COEFF./S.E.) that exceed 1.96 for stochastic significance. The three variables are (as we have come to expect) TNMSTAGE, SXSTAGE, and HCT. The corresponding values of e^{b_j} (labelled EXP(COEFF.)) are 1.28, 1.42, and 0.95, respectively. These are the estimated hazard ratios for each variable's impact on mortality. Because survival increases as hematocrit increases, the hazard ratio is below 1 for hematocrit, indicating its "protective" effect.

The no frills printout for the BMDP program of Cox regression can be augmented, if desired, by options (not shown here) that display (1) the postregression correlation matrix of coefficients; (2) an estimated asymptotic covariance matrix; (3) the "records" for each case, i.e., a list of all values of the independent variables; (4) an indication of the number of iterations (and corresponding results) that were needed to estimate the b_j parameters; and (5) a list of "tested effects," which denote the effects on each model if the component variables are removed one at a time, two at a time, three at a time, etc. These effects will be examined during the sequential explorations in Section 16.2.

16.1.1.2. Comparison of observed and estimated results

The subsequent printout in Fig. 16-2 is a useful option that compares the observed and certain estimated results for each person, arranged according to their exit times (marked SURVIVAL TIME), and exit state (marked STATUS). The display of printout has been "cut and pasted" with a horizontal gap in the middle, so that

CASE LABEL	CASE NUMBER	SURVIVAL TIME	STATUS	CUM EVENTS	CUM INCMPL	REMAIN AT RISK	KAPLAN MEIER SURVIVAL	PROPORTIONAL HAZARDS MODEL			
								BASELINE SURVIVAL	BASELINE HAZARD	BASELINE CUM HAZARD	COX-SNELL RESIDUAL
1821	199	0.03	DEAD	1	0	199		0.9847		0.0154	0.0567
1050	125	0.03	DEAD	2	0	198		0.9847		0.0154	0.0254
400	60	0.03	DEAD	3	0	197		0.9847		0.0154	0.0262
132	23	0.03	DEAD	4	0	196	0.9800	0.9847	0.5134	0.0154	0.0096
1765	193	0.10	DEAD	5	0	195		0.9616		0.0392	0.1922
1429	162	0.10	DEAD	6	0	194		0.9616		0.0392	0.1008
572	82	0.10	DEAD	7	0	193		0.9616		0.0392	0.0447
513	75	0.10	DEAD	8	0	192		0.9616		0.0392	0.0279
509	74	0.10	DEAD	9	0	191		0.9616		0.0392	0.1152
353	52	0.10	DEAD	10	0	190	0.9500	0.9616	0.3400	0.0392	0.0343
1200	141	0.20	DEAD	11	0	189		0.9377		0.0643	0.2071
1155	135	0.20	DEAD	12	0	188		0.9377		0.0643	0.0427
1038	120	0.20	DEAD	13	0	187		0.9377		0.0643	0.1322
998	117	0.20	DEAD	14	0	186		0.9377		0.0643	0.0728
678	96	0.20	DEAD	15	0	185		0.9377		0.0643	0.1305
387	59	0.20	DEAD	16	0	184	0.9200	0.9377	0.2511	0.0643	0.1167
1302	147	0.30	DEAD	17	0	183		0.9174		0.0862	0.1205
764	97	0.30	DEAD	18	0	182		0.9174		0.0862	0.2268
568	81	0.30	DEAD	19	0	181		0.9174		0.0862	0.2045
440	66	0.30	DEAD	20	0	180		0.9174		0.0862	0.2396
331	47	0.30	DEAD	21	0	179	0.8950	0.9174	0.2193	0.0862	0.1371
872	107	0.40	DEAD	22	0	178	0.8900	0.9132	0.0460	0.0908	0.0776
547	77	0.50	DEAD	23	0	177	0.8850	0.9089	0.0462	0.0955	0.3030
1062	129	0.60	DEAD	24	0	176		0.8879		0.1189	0.2331
1054	127	0.60	DEAD	25	0	175		0.8879		0.1189	0.1002
1039	121	0.60	DEAD	26	0	174		0.8879		0.1189	0.1682
501	72	0.60	DEAD	27	0	173		0.8879		0.1189	0.3405
261	38	0.60	DEAD	28	0	172	0.8600	0.8879	0.2345	0.1189	0.0711
1453	165	0.70	DEAD	29	0	171	0.8550	0.8836	0.0487	0.1238	0.4080
1556	172	0.80	DEAD	30	0	170		0.8706		0.1386	0.4487
99	15	23.20	DEAD	170	0	30	0.1500	0.0820	0.6013	2.5008	1.1478
1702	187	23.90	DEAD	171	0	29	0.1450	0.0771	0.0883	2.5626	0.9848
1228	145	27.90	DEAD	172	0	28	0.1400	0.0724	0.0158	2.6260	0.7976
302	44	28.00	DEAD	173	0	27	0.1350	0.0678	0.6459	2.6906	0.8068
1510	167	28.60	DEAD	174	0	26	0.1300	0.0635	0.1098	2.7564	4.2947
110	19	29.60	DEAD	175	0	25	0.1250	0.0590	0.0734	2.8298	2.6829
1051	126	31.70	DEAD	176	0	24	0.1200	0.0545	0.0376	2.9087	1.1107
1393	159	33.90	DEAD	177	0	23	0.1150	0.0503	0.0370	2.9901	0.7497
58	9	38.50	DEAD	178	0	22	0.1100	0.0463	0.0180	3.0731	0.7442
815	102	39.90	DEAD	179	0	21	0.1050	0.0425	0.0605	3.1578	1.7796
1232	146	41.30	DEAD	180	0	20	0.1000	0.0389	0.0636	3.2468	1.5005
1710	188	46.00	DEAD	181	0	19		0.0323		3.4324	3.9529
1666	186	46.00	DEAD	182	0	18	0.0900	0.0323	0.0395	3.4324	4.4932
1048	124	50.90	DEAD	183	0	17	0.0850	0.0286	0.0245	3.5527	2.1631
163	30	54.90	DEAD	184	0	16	0.0800	0.0252	0.0324	3.6824	1.8527
1444	164	55.90	DEAD	185	0	15	0.0750	0.0219	0.1388	3.8213	1.7621
1030	119	56.60	DEAD	186	0	14	0.0700	0.0189	0.2119	3.9696	2.2831
892	109	62.00	CENSORED	186	1	13					
149	26	64.80	CENSORED	186	2	12					
80	12	66.40	CENSORED	186	3	11					
1515	168	79.70	CENSORED	186	4	10					
62	10	82.30	CENSORED	186	5	9					
10	3	94.00	CENSORED	186	6	8					
143	25	98.40	CENSORED	186	7	7					
1825	200	99.90	CENSORED	186	8	6					
1309	149	99.90	CENSORED	186	9	5					
1191	138	99.90	CENSORED	186	10	4					
109	18	99.90	CENSORED	186	11	3					
72	11	99.90	CENSORED	186	12	2					
44	7	99.90	CENSORED	186	13	1					
8	2	99.90	CENSORED	186	14	0					

Fig. 16-2 BMDP printout for actual and estimated results in Cox regression with seven independent variables. (Option in BMDP2L program.)

results are shown here for the first 30 and last 30 patients. The column marked CUM EVENTS maintains a successive numerical tally of the failure events, i.e., deaths. The CUM INCMPL column tallies the censorings and REMAIN AT RISK shows the number of people retained in the cohort after each death or censoring.

The KAPLAN-MEIER SURVIVAL column shows the cumulative summary survival rate, $S_o(t)$, for the group at each date of death(s). The first value, .9800, is [1 − (4/200)] when four deaths occur "simultaneously" at .03 months, i.e., 1 day. At 0.1 month, six deaths occur, making the interval mortality 6/196 and the interval survival [1 − (6/196)] = .9644. Consequently, just before the deaths at 0.2 months, the Kaplan-Meier statistic is (.9800)(.9694) = .9500. The third listed Kaplan-Meier value, .9200, is [1 − (4/200)] [1 − (6/196)] [1 − (6/190)]. These are the overt summary values of $S_o(t)$, but a separate set of summary values, $S_h(t)$, was determined for the unmodified survival curve, using the hazard function method described in Sections 15.3.1 and 15.3.2. This $S_h(t)$ value is used in subsequent calculations for individual estimates.

16.1.1.2.1. Individual estimates

Making estimates for individual persons is not an easy job in Cox regression because the estimate will differ according to the selected exit time. If a specific time, t, is chosen, the corresponding value of $S(t) = S_h(t)^{e^G}$ estimates the probability that the person will survive at least as long as t. In the set of four columns marked PROPORTIONAL HAZARDS MODEL in the BMDP printout, the value of BASELINE SURVIVAL can be converted to an estimate for individual persons at that time point. The conversion process is somewhat convoluted because BASELINE SURVIVAL represents the calculated value of $S_h(t)$ for each successive exit time *in the group*, using the mean of each variable, X_j, for the entire cohort, rather than the X_j values for individual persons. Thus, if $\overline{G}$ represents the mean value of $\overline{G} = b_1\overline{X}_1 + b_2\overline{X}_2 + \ldots$ in the cohort, the value of BASELINE SURVIVAL is $\overline{S}(t) = S_h(t)^{e^{\overline{G}}}$. If this result is converted to get the actual value of $S_h(t)$, we can then estimate survival for individual persons as $S_h(t)^{e^G}$, using their distinctive values of $X_1, X_2, \ldots$.

The choice of the corresponding value of $S_h(t)^{e^G}$ is also tricky, because G can be calculated (see Section 15.4.4) either with a direct model, $G_d = b_1X_1 + b_2X_2 + b_3X_3 + \ldots$, or with a centered model, $G_c = b_1(X_1 - \overline{X}_1) + b_2(X_2 - \overline{X}_2) + b_3(X_3 - \overline{X}_3) + \ldots$, in which the mean, $\overline{X}_j$, is subtracted from X_j for each variable. The conversion process using G will depend on the way the model for G has been constructed. In a direct model, G becomes $G_d = b_1X_1 + b_2X_2 + b_3X_3 + \ldots$. In a centered model, $G = G_c = 0$.

Although the tactic is not specifically stated, the BMDP program in this instance uses a centered model. Its usage can be discerned at the end of the printout (not shown here), where the program indicates the mean values of the variables used to form a "conversion factor." Because this conversion factor is listed as 1, the program must have used a centered model, which makes the value associated with BASELINE SURVIVAL become $e^{\overline{G}_c} = e^0 = 1$. Therefore, since $\hat{S}(t) = S_h(t)$ exponentiated by 1, the results printed in the BASELINE SURVIVAL column for $\hat{S}(t)$ are also $S_h(t)$.

To make estimates for individual persons, we insert the appropriate values of X_j to calculate $G_c = b_1(X_1 - \bar{X}_1) + b_2(X_2 - \bar{X}_2) + b_3(X_3 - \bar{X}_3) + \ldots$ The estimated survival will be $S_h(t)^{e^{G_c}}$. If a direct model had been used, we would first calculate $\bar{G}_d$ and then determine $S_h(t)$ as the $(1/\bar{G}_d)$ root of BASELINE SURVIVAL. The estimated $S(t)$ for an individual person would then be $S_h(t)^{e^{G_d}}$ where $G_d = b_1X_1 + b_2X_2 + b_3X_3 + \ldots$. The two processes will produce similar results. To prove this point, note that if G_d is the direct model and $\bar{G}_d$ is its mean, the centered model is $G_c = G_d - \bar{G}_d$. The value of $S_h(t)^{e^{(G_d - \bar{G}_d)}} = [S_h(t)^{e^{G_d}}][S_h(t)^{e^{-\bar{G}_d}}]$. Since $e^{-\bar{G}_d}$ is $1/e^{\bar{G}_d}$, the result of $S_h(t)^{e^{G_d}}$ is the $(1/e^{\bar{G}_d})$ root of $S_h(t)$. Thus, the two approaches will reach the same estimate for $\hat{S}(t)$ for individual persons.

16.1.1.2.2. Illustration of calculations

To provide a relatively simple illustration of the estimation procedure for individual persons, a separate Cox regression was done with the three-main-variable model. It showed that the b_j coefficients were .2598 for TNMSTAGE, .3381 for SXSTAGE, and −0.0489 for HCT. The printout also indicated that the conversion factor was 1, (i.e., the regression was done with a centered model) for the mean values of 3.215 for TNMSTAGE, 2.615 for SXSTAGE, and 41.090 for HCT. This result could be used with either a noncentered or centered approach to estimate survival for the first case 199, with case label 1821, who died at 0.03 months. In Fig. 16-2, this person's BASELINE SURVIVAL was listed as .9847 for the seven-variable model. In the printout (not shown here) for the three-variable model, the corresponding BASELINE SURVIVAL is .9842.

In a noncentered approach, we would calculate $\bar{G}_d = (.2598)(3.215) + (.3381)(2.615) - (0.0489)(41.090) = -.2899$. The value of $e^{\bar{G}_d} = e^{-.2899} = .7483$. We then take the $(1/.7483)^{th}$ root of BASELINE SURVIVAL, so that $S_h(t)$ is determined as $(.9842)^{1.336} = .9789$. For case 199, the actual values of the $\{X_j\}$ variables were TNMSTAGE = 5, SXSTAGE = 4, and HCT = 34.0. Therefore, $G_d = (.2598)(5) + (.3381)(4) - (.0489)(34.0) = .9888$, for which $e^{.9888} = 2.688$. The estimated $S(t)$ for that person will be $S_h(t)^{2.688} = (.9789)^{2.688} = .9442$. With the centered approach, the BASELINE SURVIVAL value is $S_h(t)$. We would then calculate $G_c = (.2598)(5 - 3.215) + (.3381)(4 - 2.615) - (.0489)(34.0 - 41.090) = 1.2787$, and $e^{G_c} = 3.592$. The value of $\hat{S}(t)$ will be $(.9842)^{3.592} = .9444$. The two estimates are quite close, and differ only because of rounding in the calculations.

16.1.1.2.3. Additional estimates

The column called BASELINE HAZARD in Fig. 16-2 shows the summary value of the hazard function, $h_o(t)$, at each time point. It is essentially the quotient of two increments. The numerator is the difference in the natural logarithms of two consecutive estimated (BASELINE) survivals. The denominator is the difference in the time for each estimate. The formula is (ln survival estimate at time $(t - 1)$ − ln survival estimate at time t] ÷ [time t − time $(t - 1)$]. For example, in Fig. 16-2 for case number 23 having cumulative event 4 at .03 month, the BASELINE SURVIVAL is listed at .9847, for which the logarithm is −0.015418. The prior survival estimate (i.e., at the beginning) was 1; its ln is 0, with $t - 1 = 0$. The "baseline hazard" becomes [0 − (−0.015418)]/.03 = .5139. For case number 52, which has cumulative event 10 at .10 month, the estimated BASELINE HAZARD is .3400. It can be obtained from (ln .9847 − ln .9616)/(.10 − .03) = .339. At 16 lines from the bottom of Fig. 16-2, for case 164 with event 185 at 55.90 months, the estimated BASELINE HAZARD is listed as .1388. It comes from (ln .0252 − ln .0219)/ (55.9 − 54.9) = (.1404)/(1) = .1404. The estimates listed in the printout differ somewhat from those just calculated, because the estimates are prepared from the mathematical construction of the instantaneous hazard at each time point, whereas the calculations here were based on "jumps" from one time point to another.

The column marked BASELINE CUM HAZARD is the negative logarithmic value for each survival estimate, $\hat{S}(t)$. For example when BASELINE SURVIVAL = .9847, its natural logarithm is −.0154, which becomes BASELINE CUM HAZARD for the first four cases. For the next six cases, −ln (BASELINE SURVIVAL) = −ln (.9616) = .0392 = BASELINE CUM HAZARD.

The column marked COX-SNELL RESIDUAL is one of many ways[4] to express the "residual" difference between the observed and estimated values for each person. The approach, discussed in Section 16.4, has relatively little value unless you are looking for "improper" observations that are denoted by large outlier residuals.

The bottom of the printout in Fig. 16-2 does not indicate values of Kaplan-Meier survival or any proportional hazards model estimates for the 14 persons who were still alive and terminally censored at the end of 60 months. The absence of this information should not be surprising, since none of the cited entities can be estimated until deaths occur.

16.1.2. SAS program

The main results of the current SAS program, PROC PHREG, for a full Cox regression with seven variables are shown here in Fig. 16-3.

16.1.2.1. Differences from BMDP displays

The SAS printout contains several pertinent results that are absent in the BMDP

Testing Global Null Hypothesis: BETA=0

Criterion	Without Covariates	With Covariates	Model Chi-Square
-2 LOG L	1678.032	1597.193	80.839 with 7 DF (p=0.0001)
Score			83.029 with 7 DF (p=0.0001)
Wald			78.189 with 7 DF (p=0.0001)

Analysis of Maximum Likelihood Estimates

Variable	DF	Parameter Estimate	Standard Error	Wald Chi-Square	Pr > Chi-Square	Risk Ratio	Lower	Upper	Label
AGE	1	0.004273	0.00841	0.25844	0.6112	1.004	0.988	1.021	AGE IN YEARS
MALE	1	0.504205	0.29705	2.88108	0.0896	1.656	0.925	2.964	PRESENCE OF MALE GENDER
TNMSTAGE	1	0.246331	0.05296	21.63084	0.0001	1.279	1.153	1.419	TNM STAGE
SXSTAGE	1	0.348001	0.08934	15.17124	0.0001	1.416	1.189	1.687	SYMPTOM TYPE
HCT	1	-0.046492	0.01660	7.84046	0.0051	0.955	0.924	0.986	HEMATOCRIT OR HEMOGLOBIN
PCTWTLOS	1	0.014132	0.00952	2.20536	0.1378	1.014	0.995	1.033	PERCENT WT LOSS

(Conditional Risk Ratio and 95% Confidence Limits)

Fig. 16-3 Main results of Cox regression analysis for "full" seven-variable model. (SAS PROC PHREG program, Version 6.07.)

printout. Labelled −2 LOG L, they are: the value of −2 ln likelihood (1678.032) for the basic likelihood with no variables in the model, and the corresponding value (1597.193) for −2 ln L_R when all seven variables are entered. The difference in the latter two values is the source of the model chi-square, listed as 80.839, which was *not* shown in the BMDP program. (Half of the value of −2 ln L_R was shown in Fig. 16-1 as −798.5963.)

The SAS printout also shows the results of two additional ways to calculate a chi-square value for the model. One of them is the *score* statistic (83.029), which was the model value reported in the BMDP printout. The other is a Wald value of 78.189. Publishing all three of these chi-square tests for the model is probably redundant, since they usually have relatively similar results.

For each independent variable, the SAS program shows Wald chi-square, which is the square of the BMDP's COEFF./S.E., and also gives the associated P value, marked "Pr > Chi-Square." The value of e^{b_j} for each parameter estimate, b_j, is listed as "Risk Ratio," which is an erroneous label for the hazard ratio; and lower and upper boundaries (an option) are cited for the 95% confidence interval around the "risk ratio."

16.1.2.2. Additional options

In additional options not shown here, the SAS PHREG procedure can print (1) the "iteration history" of steps and results explored before the final coefficients were achieved; (2) estimated covariance and correlation matrices; (3) results of "linear hypothesis testing," showing changes in the Wald model-chi-square values when variables are removed from the model in single, double, triple, or other increments. These options are analogous to those offered by the BMDP program, and have the same limited general value.

The SAS program also has a set of four options, not shown here, for managing "ties in the failure time," i.e., situations in which two or more patients die at exactly the same time duration. These four options are generically called DISCRETE and EXACT, and eponymously cited as BRESLOW and EFRON. They refer to four different ways—some shorter and some longer—of calculating the likelihood values. The default BRESLOW method was used throughout the programs here.

Other options of the PHREG program are discussed in Section 16.1.2.5 and also as part of regression diagnostics in Section 16.4.

16.1.2.3. Differences from previous SAS displays

The Cox regression program has been significantly changed in its most recent version (PHREG, Version 6.07) from a precursor program, called PHGLM, in the previous version 5.18. No explanation was offered for the change, which yields similar results for all of the calculated direct statistics.

Several valuable "interpretive" statistics, however, have been eliminated in the new version. (You may want to save and use the old version if it is available.) The useful PHGLM statistics that have been eliminated in the new PHREG program are R_H^2 for the model, an Index-R for each variable, and a "z:PH" index for each variable.

16.1.2.3.1. R_H^2 for the model

The AIC and SC indexes (see Section 13.7.2), which describe accomplishments of likelihood changes in a logistic model, do not appear in the printouts of either the BMDP or current SAS versions of Cox regression. Consequently, if you want an index that penalizes for number of variables, and that offers an idea about proportion of explained likelihood, the only proposed index is R_H^2 for the model. The R_H^2 index was included in the previous (PROC PHGLM) version of the SAS Cox regression procedure, but is not presented in the current version.

In Fig. 16-4, the old PROC PHGLM program was applied to the same data. The printout shows the R_H^2 index as "R = 0.200," to the far right of the statement about

```
                    DEPENDENT VARIABLE: T

                    EVENT INDICATOR: STATUS


            200 OBSERVATIONS
            186 UNCENSORED OBSERVATIONS
              0 OBSERVATIONS DELETED DUE TO MISSING VALUES


       -2 LOG LIKELIHOOD FOR MODEL CONTAINING NO VARIABLES= 1678.03

       MODEL CHI-SQUARE=   83.03 WITH   7 D.F.   (SCORE STAT.) P=0.0    .
       CONVERGENCE IN  5 ITERATIONS WITH  0 STEP HALVINGS     R= 0.200.
       MAX ABSOLUTE DERIVATIVE=0.8082D-13.               -2 LOG L= 1597.19.
       MODEL CHI-SQUARE=   80.84 WITH   7 D.F.   (-2 LOG L.R.) P=0.0    .
```

VARIABLE	BETA	STD. ERROR	CHI-SQUARE	P	R	Z:PH
TNMSTAGE	0.24633135	0.05296427	21.63	0.0000	0.108	-1.74
SXSTAGE	0.34800056	0.08934482	15.17	0.0001	0.089	-0.79
MALE	0.50420514	0.29704990	2.88	0.0896	0.023	1.53
AGE	0.00427319	0.00840576	0.26	0.6112	0.000	1.34
PROGIN	0.00226040	0.00446394	0.26	0.6126	0.000	0.67
HCT	-0.04649225	0.01660389	7.84	0.0051	-0.059	0.68
PCTWTLOS	0.01413174	0.00952251	2.20	0.1378	0.011	-0.20

Fig. 16-4 Results of "old" SAS PROC PHGLM procedure for same data presented in Fig. 16-3. The distinctive items to note in this printout are the values of R = 0.200 for the model, and the values of Index-R and Z:PH for each variable.

convergence, just below the line that shows MODEL CHI-SQUARE. If not displayed, R_H^2 could be calculated from the data of Fig. 16-4 as

$$\frac{(1678.03 - 1597.19) - (2 \times 7)}{1678.03} = .0398,$$

for which the square root is $R_H = .200$. (This value for R_H^2 is almost identical to that noted in Section 16.1.1.1, when the basic log likelihood had to be approximated from the BMDP printout.)

16.1.2.3.2. Utility of Z:PH Test

Perhaps the most valuable PHGLM index that has been omitted in PHREG is the Z:PH statistic, calculated for each independent variable, which offers a check for the assumption that the variable has a constant hazard ratio over time. According to Harrell,[5] the test statistic is calculated as

$$Z = r\sqrt{N_u - 2}/\sqrt{1 - r^2}.$$

For this calculation, r is the correlation between the residuals of the estimates and the rank order of the failure times; and N_u is the number of uncensored observations. The result is interpreted as a conventional Z test and is stochastically significant (for the customary $\alpha = .05$) when $|Z| \geq 1.96$. The crucial component of Z:PH is r, which shows the correlation of residuals and order of failure times. With a constant hazard ratio, this correlation should be 0. The value of r tends to be positive if the hazard rate is increasing over time, and negative if decreasing. Since $\sqrt{1 - r^2}$ appears in the denominator, the value of Z increases as r^2 increases. If Z:PH becomes stochastically significant, the constant proportional hazards assumption is presumably being violated by that variable.

The test can be regarded as a type of screening procedure. If not significant, the result is reassuring. If significant, the result can be checked with other diagnostic methods (described later) to determine how serious the violation may be. (The violation is apparently not too devastating, since a stochastically significant Z:PH [$= -2.37$] appears repeatedly without further comment on pages 460, 462, 463, and 464 of an example in the older SAS manual[3] showing an illustrative printout from the PROC PHGLM procedure. The violation involves a variable [LOGBUN] that is cited as significant among the independent variables.)

The PROC PHGLM printout in Fig. 16-4 shows that none of the absolute values of Z:PH for the independent variables exceeded the "danger" boundary of 1.96.

16.1.2.3.3. Utility of individual Index R-values

The Index-R values for individual variables, as in logistic regression, are useful for rating the impact of the variables, because neither the BMDP nor the SAS printout produces standardized coefficients. If impact were determined only from the regression coefficient and the associated hazard ratio for each stochastically significant variable, symptom stage would seem more important (in Figs. 16-1, 16-3, and 16-4) than TNM stage because of the larger b_j value (.348 vs. .246), and the correspondingly larger "risk ratio." Having the standardizing effect of a partial correlation coefficient, the Index-R value of .108 in Fig. 16-4 shows that TNM stage actually has a greater effect than symptom stage, for which R = .089.

In the absence of a printed result, the Index-R value could be calculated from the formula shown in Section 13.9.2.5, if the improvement chi-square values were available *for each variable*. Those values, however, do not appear in SAS printouts of the *full* regression model.

16.1.2.3.4. Standardized coefficients

If Index-R values are not available, standardized coefficients can be compared to rate (or rank) the impact of the variables. Using the univariate standard deviations of 1.6316 for TNM stage and 1.0642 for SXSTAGE, the standardized b_j coefficients will be (1.6316) (.2463) = .402 for TNM stage and (1.0642) (.3481) = .370

for SXSTAGE. These two standardized coefficients now have the same rank as their Index-R values. For hematocrit, the standardized coefficient is $(5.1793)(-.0465) = -.241$, placing it in the third rank of impact, in accordance with the rank of its Index-R value in Fig. 16-4.

16.1.2.4. Interpretation of results

All of the Index-R values for individual variables, as well as the R_H of .200 (and R_H^2 of about .04) for the entire model, are quite low in Fig. 16-4. The results suggest that the fit is unimpressive, offering relatively little proportionate explanation for the original baseline likelihood. Nevertheless, if you concluded that the analysis has not accomplished very much, you would confuse the goals of estimating fit and determining impact.

None of the three effective variables, nor the model itself, can be expected to substantially improve fit in the predictive estimates for a cohort in which the 5-year survival rate is only 7%, i.e., 14/200. In such a cohort, concordant accuracy (at 5 years) will be 93% merely if everyone is predicted as dead, without further analysis. The virtue of the regression analysis here is its demonstration of important variables, not its improvement of fit. None of these virtues would be adequately expressed, however, in the customary statistical indexes of accomplishments for the model itself.

16.1.2.5. Individual estimates

Unlike the BMDP program, which requires special calculations to produce individual estimates of $\hat{S}(t)$ for individual persons, the SAS PHREG program can show those estimates directly in one of the options requested for the printout. To compare the SAS performance with the BMDP results discussed in Section 16.1.1.2.2, the PHREG program was run with only the three main independent variables in the model.

The printout, which is the SAS three-main-variable counterpart of the seven-main-variable BMDP result shown in Fig. 16-2, is presented in Fig. 16-5. Further details of this printout will be discussed in Section 16.5.6. The main item to be noted now is the value of .94374, in the column marked SURVIVAL, for the first case, identified as ID 1821. This result corresponds to the value of .9444 obtained for this same person after the BMDP calculations in Section 16.1.1.2.2. The two results differ slightly because the SAS estimate of the individual $\hat{S}(t)$ is done by exponentiating the baseline Kaplan-Meier value of $S_o(t)$, whereas the BMDP program uses $S_h(t)$, as determined from the hazard function values.

The SAS procedure has the advantage of offering the individual estimate directly, without the additional calculations needed for the BMDP printout.

16.1.3. Stratified-cluster method for impact of variables

Probably the most direct way to demonstrate the impact of individual variables is with a stratified-cluster method. It was constructed here with a strategy called *conjunctive consolidation*, which was briefly outlined in Section 4.6.2.5, and which will receive major attention in Part V of this text. The approach begins in this instance by conjoining the categories of two variables, symptom stage and

OBS	ID	SURVIVE	STATUS	TNMSTAGE	SXSTAGE	HCT	XBETA	STDXBETA	SURVIVAL	LOGSURV	LOGLOGS	RESMART	RESDEV
1	1821	0.0	0	5	4	34.0	0.98790	0.59547	0.94374	-0.05790	-2.84902	0.94210	1.95291
2	1050	0.0	0	5	4	50.0	0.20510	0.81969	0.97388	-0.02647	-3.63182	0.97353	2.30577
3	460	0.0	0	1	4	33.0	-0.00236	0.57470	0.97872	-0.02151	-3.83929	0.97849	2.39198
4	132	0.0	0	1	2	36.3	-0.84000	0.57517	0.99074	-0.00931	-4.67692	0.99069	2.71523
5	1765	0.1	0	4	4	24.0	1.21735	0.46065	0.82987	-0.18649	-1.67939	0.81351	1.31596
6	1429	0.1	0	5	3	36.0	0.55196	0.60782	0.90858	-0.09587	-2.34478	0.90413	1.69744
7	572	0.1	0	5	4	45.0	-0.14816	0.72786	0.95351	-0.04760	-3.04490	0.95240	2.04573
8	513	0.1	0	1	4	46.0	-0.63839	0.75469	0.97127	-0.02915	-3.53513	0.97085	2.26464
9	509	0.1	0	5	4	42.0	0.59650	0.70490	0.90463	-0.10023	-2.30024	0.89977	1.67360
10	353	0.1	0	3	2	39.0	-0.45250	0.62452	0.96550	-0.03511	-3.34924	0.96489	2.18374
11	1200	0.2	0	5	4	38.0	0.79220	0.64935	0.81893	-0.19976	-1.61065	0.80024	1.27311
12	1155	0.2	0	5	2	37.0	-0.87424	0.58594	0.96297	-0.03774	-3.27710	0.96226	2.15167
13	1038	0.2	0	5	4	43.0	0.20949	0.70887	0.89446	-0.11154	-2.19337	0.88846	1.61549
14	998	0.2	0	5	3	48.0	-0.29494	0.77293	0.93486	-0.06735	-2.69779	0.93265	1.87891
15	678	0.2	0	4	4	50.0	0.20510	0.81969	0.89489	-0.11105	-2.19775	0.88895	1.61790
16	387	0.2	0	5	4	34.2	0.45852	0.58460	0.86668	-0.14308	-1.94433	0.85692	1.47473
17	1302	0.3	0	4	4	38.0	0.53240	0.64082	0.81337	-0.20657	-1.57712	0.79343	1.25195
18	764	0.3	0	4	4	38.0	0.79220	0.64935	0.76502	-0.26785	-1.31733	0.73215	1.08183
19	568	0.3	0	5	4	44.0	0.49865	0.73319	0.81897	-0.19971	-1.61088	0.80029	1.27325
20	440	0.3	0	3	4	31.0	0.61508	0.54213	0.79902	-0.22437	-1.49445	0.77563	1.19902
21	331	0.3	0	5	4	50.0	0.20510	0.81969	0.86165	-0.14891	-1.90443	0.85109	1.45144
22	872	0.4	0	5	1	40.0	-0.31992	0.67620	0.91147	-0.09270	-2.37843	0.90730	1.71530
23	547	0.5	0	5	4	40.0	0.69435	0.67694	0.76459	-0.26842	-1.31520	0.73158	1.08039
24	1062	0.6	0	4	3	37.0	0.24324	0.60983	0.80833	-0.21279	-1.54745	0.78721	1.23308
25	1054	0.6	0	4	1	35.0	-0.33509	0.58285	0.88751	-0.11934	-2.12578	0.88066	1.57805
26	1039	0.6	0	5	3	45.0	0.11164	0.73832	0.82982	-0.18655	-1.67905	0.81345	1.31575
27	501	0.6	0	5	4	43.0	0.54758	0.71901	0.74940	-0.28848	-1.24311	0.71152	1.03111
28	261	0.6	0	3	2	44.0	-0.69712	0.70147	0.92027	0.08309	-2.48782	0.91691	1.77252
29	1453	0.7	0	5	4	42.0	0.59650	0.70490	0.72968	-0.31514	-1.15473	0.68486	0.96940
30	1556	0.8	0	4	4	34.0	0.72810	0.58628	0.66893	-0.40207	-0.91112	0.59793	0.79145
...													
170	99	23.2	0	1	2	42.0	-1.11887	0.66323	0.33540	-1.09244	0.08842	-0.09244	-0.08974
171	1702	23.9	0	2	2	49.0	-1.20155	0.77384	0.35708	-1.02979	0.02936	-0.02979	-0.02950
172	1228	27.9	0	2	1	47.0	-1.44179	0.74672	0.43640	-0.82919	-0.18731	0.17081	0.18164
173	302	28.0	0	1	2	45.0	-1.60373	0.71072	0.48581	-0.72194	-0.32581	0.27806	0.30904
174	1510	28.6	0	5	2	37.0	0.16495	0.62067	0.01308	-4.33698	1.46718	-3.33698	-1.93381
175	110	29.6	0	3	2	40.5	-0.52589	0.64753	0.10752	-2.23010	0.80205	-1.23010	-0.92526
176	1051	31.7	0	1	1	40.0	-1.35911	0.63138	0.36975	-0.99493	-0.00508	0.00507	0.00508
177	1393	33.9	0	1	2	42.0	-1.11887	0.66323	0.27289	-1.29848	0.26315	-0.29868	-0.27324
178	58	38.5	0	1	1	47.0	-1.70158	0.74249	0.47502	-0.74440	-0.29518	0.25560	0.28134
179	815	39.9	0	1	1	37.0	-0.87424	0.58594	0.17411	-1.74807	0.55851	-0.74807	-0.61572
180	1232	41.3	0	1	1	52.0	-1.34832	0.82081	0.32710	-1.11748	0.11108	-0.11748	-0.11317
181	1710	46.0	0	2	2	39.0	-1.19270	0.63518	0.02331	-3.75867	1.32407	-2.75867	-1.69388
182	1666	46.0	0	5	2	45.0	-0.22645	0.73962	0.02641	-3.63393	1.29031	-2.63393	-1.63928
183	1048	50.9	0	5	2	41.0	-0.81015	0.64927	0.12345	-2.09194	0.73809	-1.09194	-0.84125
184	163	54.9	0	1	1	40.0	-1.02102	0.63224	0.17378	-1.74997	0.55960	-0.74997	-0.61704
185	1444	55.9	0	2	2	48.0	-1.15262	0.75821	0.20479	-1.58575	0.46106	-0.58575	-0.49939
186	1030	56.6	0	1	1	36.0	-0.82532	0.57056	0.10269	-2.27604	0.82244	-1.27604	-0.95247
187	892	62.0	0	1	4	40.0	-0.24163	0.65029	0.01691	-4.08014	1.40613	-4.08014	-2.85662
188	149	64.8	0	1	1	44.0	-1.21672	0.69432	0.21463	-1.53885	0.43104	-1.53885	-1.75434
189	80	66.4	0	1	1	39.3	-0.98677	0.62141	0.14418	-1.93670	0.66098	-1.93670	-1.96809
190	1515	79.7	0	1	1	43.0	-1.16779	0.67877	0.19869	-1.61601	0.47996	-1.61601	-1.79778
191	62	82.3	1	1	1	44.0	-1.21672	0.69432	0.21463	-1.53885	0.43104	-1.53885	-1.75434
192	10	94.0	1	1	1	42.0	-1.45696	0.66310	0.29813	-1.21021	0.19080	-1.21021	-1.55577
193	143	98.4	1	1	1	41.7	-1.10419	0.65858	0.17868	-1.72213	0.54356	-1.72213	-1.85587
194	1825	99.9	1	3	3	37.0	-0.01656	0.60217	0.00604	-5.10999	1.63120	-5.10999	-2.19687
195	1309	99.9	1	2	2	40.0	-0.90800	0.68030	0.13302	-2.09543	0.73976	-2.09543	-2.04716
196	1191	99.9	1	2	2	39.0	-0.97209	0.61678	0.14011	-1.96533	0.67566	-1.96533	-1.98259
197	109	99.9	1	2	2	45.0	-1.00585	0.71141	0.14955	-1.90011	0.64191	-1.90011	-1.94941
198	72	99.9	1	2	2	54.0	-1.44617	0.85219	0.29425	-1.22334	0.20159	-1.22334	-1.56419
199	44	99.9	1	2	2	43.5	-1.19226	0.68654	0.20660	-1.57696	0.45550	-1.57696	-1.77593
200	8	99.9	1	1	1	42.0	-1.45696	0.66310	0.29813	-1.21021	0.19080	-1.21021	-1.55577

Fig. 16-5 SAS printout for actual and estimated results together with various additional displays of residuals in three-main-variable Cox regression. The sequence of survival times is shown for only the first 30 and last 30 patients. (Option in SAS PHREG program. For further details, see text.)

TNM stage, to form the bivariate 5-year survival rates shown in the cells of Table 16-1. The results are consistent with a double-gradient, which is difficult to discern

Table 16-1 Conjunctive consolidation of 5-year survival rates in TNM and Symptom stages

TNM stage Symptom stage	I	II	IIIA	IIIB	IV	Total
Asymptomatic	2/13 (15%) **[A]**	0/3 (0%)	0/4 (0%) **[B]**	0/6 (0%)	0/2 (0%) **[C]**	2/28 (7%)
Pulmonic/ Systemic	7/27 (26%)	3/21 (14%)	0/8 (0%)	1/17 (6%)	0/9 (0%)	11/82 (13%)
Mediastinal/ Regional	0/0 (−)	0/0 (−)	1/4 (25%)	0/7 (0%)	0/18 (0%)	1/29 (3%)
Distant	0/11 (0%)	0/3 (0%)	0/2 (0%)	0/6 (0%)	0/39 (0%)	0/61 (0%)
Total	9/51 (18%)	3/27 (11%)	1/18 (6%)	1/36 (3%)	0/68 (0%)	14/200 (7%)

Dashed lines show the pattern of consolidation; Letters **A, B, C** show the three composite ST stages; denominators in each cell show number of patients in that category; numerators show the corresponding number of 5-year survivors.

from the small numbers in many of the cells. The bivariate categories were then conjunctively consolidated, in clusters shown by the dashed lines in Table 16-1, to form a composite S-T staging system, containing three ordinal categories. The survival results in the three composite ST stages are **A,** *9/40 (23%);* **B,** *5/70 (7%);* and **C,** *0/90 (0%).*

Table 16-2 Conjunctive consolidation of 5-year survival rates for hematocrit and composite S-T stage

Hematocrit category	Composite S-T stage			Totals
	A	B	C	
≥ 44	2/11 (18%) [I]	2/21 (10%) [II]	0/29 (0%) [III]	4/61 (7%)
≥ 37 to < 44	7/21 (33%)	3/36 (8%)	0/46 (0%)	10/103 (10%)
< 37	0/8 (0%)	0/13 (0%)	0/15 (0%)	0/36 (0%)
Totals	9/40 (23%)	5/70 (7%)	0/90 (0%)	14/200 (7%)

Dashed lines show the pattern of consolidation.

Table 16-2 shows the distinctive survival effects produced when the three hematocrit categories were conjoined with the composite S-T variable. When these results were conjunctively consolidated, as shown by the dashed lines and letters in Table 16-2, the trivariate composite STH 5-year survival rates are: **I,** *9/32 (28%);* **II,** *5/57 (9%)* and **III,** *0/111 (0%).* In the next step of the conjunctive consolidation method, the composite STH stages were conjoined in separate tables with each of the remaining four candidate variables. Since none of them showed an impressive impact within the three STH stages, the process was concluded.

The two composite staging systems did not improve the accuracy of predictions because the original total survival rate was 7%, and none of the results in Tables 16-1 and 16-2 were able to span the meridian of 50% (see Section 8.2.2.2.2). Nevertheless, the two composite stagings have some useful statistical accomplishments. The composite ST stages **(A, B, C)** derived from Table 16-1 have a (regular) chi-square of 21.5, a $-\ln L_o$ of 50.73, and $-\ln L_R$ of 39.34, so that LLR = 11.35, which is 22% of the original $\ln L_o$. The composite STH stages **(I, II, III)** derived from Table 16-2 have $X^2 = 30.6$ and $-\ln L_R = 35.95$, so that LLR = 14.70, which is 29% of $-\ln L_o$. In addition to these achievements, the nadir set in the third STH stage would allow unequivocally accurate predictions for 111 of the 200 patients.

Although the three composite STH stages were constructed from 5-year survival rates alone, the Kaplan-Meier survival curves in Fig. 16-6 show the clear separations throughout the entire interval.

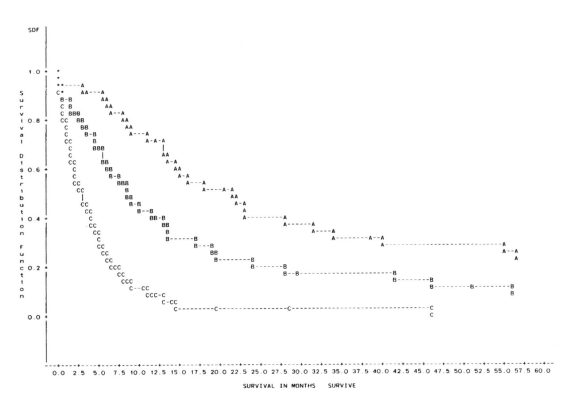

Fig. 16-6 Kaplan-Meier survival plot for three stages of the composite S-T-H variable. (For further details, see text.) The symbols are A, B, and C respectively for the first, second, and third stage. The asterisks at the beginning show values that overlap. Letters show locations of deaths. Dashes show intervals without deaths. (SAS PROC LIFETEST program.)

16.2. Sequential stepped regression

After the three important variables were identified in full regression, many analysts would do a separate regression containing only those variables. Alternatively, in a forced-then-open stepped procedure, the three main variables could be forced into the model, and the remaining candidate variables could enter thereafter sequentially if they fulfilled the admission criteria. To see how the sequential stepped procedure would work in a natural manner, a stepwise regression, starting with all seven independent variables, was done with the ordinary default criteria of both the SAS and BMDP programs. Results are shown for the first part of the SAS and the last part of the BMDP printouts. (The three-main-variable regression is discussed in Section 16.3.)

16.2.1. SAS program

As shown in Fig. 16-7, stepwise regression in the SAS PROC PHREG begins by citing the chi-square score for each candidate variable and by stating that the "residual chi-square" is 83.0294 for the joint effect of all variables (in this case, the seven candidates) that are not yet in the model. The highest individual chi-square score

```
                                    The PHREG Procedure

                              Analysis of Variables Not in the Model

                                    Score          Pr >
                   Variable      Chi-Square     Chi-Square    Label

                   SXSTAGE         40.5367        0.0001      SYMPTOM TYPE
                   TNMSTAGE        53.8472        0.0001      TNM STAGE
                   AGE              2.2183        0.1364      AGE IN YEARS
                   MALE             2.9814        0.0842      PRESENCE OF MALE GENDER
                   PROGIN           0.9806        0.3221      FIRST MANIFESTATION TO ZT
                   HCT              5.9686        0.0146      HEMATOCRIT OR HEMOGLOBIN*3
                   PCTWTLOS        14.6482        0.0001      PERCENT WT. LOSS

                        Residual Chi-square = 83.0294  with 7 DF (p=0.0001)

Step  1: Variable TNMSTAGE is entered. The model contains the following explanatory variables.
         TNMSTAGE

                                Testing Global Null Hypothesis: BETA=0
                                Without        With
                   Criterion    Covariates     Covariates    Model Chi-Square

                   -2 LOG L     1678.032       1625.761      52.271 with 1 DF (p=0.0001)
                   Score           .              .          53.847 with 1 DF (p=0.0001)
                   Wald            .              .          50.592 with 1 DF (p=0.0001)

                            Analysis of Maximum Likelihood Estimates

                                                              Conditional Risk Ratio and
                                                                95% Confidence Limits

                Parameter   Standard    Wald         Pr >       Risk
  Variable  DF  Estimate    Error    Chi-Square  Chi-Square    Ratio    Lower    Upper   Label
  TNMSTAGE   1   0.34934    0.04911   50.59204     0.0001      1.418    1.288    1.561   TNM STAGE

                              Analysis of Variables Not in the Model
                                    Score          Pr >
                   Variable      Chi-Square     Chi-Square    Label

                   SXSTAGE         11.4231        0.0007      SYMPTOM TYPE
                   AGE              0.7463        0.3876      AGE IN YEARS
                   MALE             2.0684        0.1504      PRESENCE OF MALE GENDER
                   PROGIN           0.0054        0.9413      FIRST MANIFESTATION TO ZT
                   HCT              6.1083        0.0135      HEMATOCRIT OR HEMOGLOBIN*3
                   PCTWTLOS         7.2195        0.0072      PERCENT WT. LOSS

                        Residual Chi-square = 29.4792  with 6 DF (p=0.0001)
```

Fig. 16-7 Beginning and first step of stepwise regression for illustrative data set with SAS PROC PHREG program.

was for TNMSTAGE, which was entered in step 1. After its accomplishments are displayed, the bottom of Fig. 16-7 shows that SXSTAGE was the leading candidate for entry in the second step. In the rest of the printout, not shown here, SXSTAGE was added in the second step, HCT in the third, and MALE in the fourth. The b_j coefficients for these variables were identical to those shown below in the BMDP stepwise program. After the fourth step, the SAS program concluded with the statement that "no additional variables met the 0.05 level for entry into the model." The conclusion was followed by a "Summary" that helps demonstrate the statistical focus on fit of models rather than impact of variables. The summary indicates the order of entry, the chi-square score, and the P values for each variable, but nothing about the b_j coefficients themselves. Their values have to be sought in the results for step 4.

16.2.2. BMDP program

In steps 0 and 1 (not shown here) for stepwise regression with BMDP2L, the first variable to enter in step 1 was TNM stage, which had 52.27 as the highest APPROX. CHI-SQ. ENTER in step 0. This same value appeared as the IMPROVEMENT CHI-

SQ, obtained as 2 ln MPLR after the first step, for which the log likelihood became −812.8803. The original baseline likelihood, which is not shown in BMDP printouts, can be calculated if 2 ln MPLR is added to twice the value of 812.8803, producing 2 ln L_0 = 1678.03, which is the same value shown in Fig. 16-7 for the SAS printout.

Fig. 16-8 shows steps, 2, 3, and 4 of the BMDP stepwise process before it ends because NO TERM PASSES THE REMOVE AND ENTER LIMITS (0.1500 0.1000). As expected,

```
STEP NUMBER   2              SXSTAGE  IS ENTERED
------------
                LOG LIKELIHOOD  =  -807.2037
IMPROVEMENT CHI-SQ  ( 2*(LN(MPLR) )  =   11.35  D.F. =  1   P = 0.0008
                GLOBAL CHI-SQUARE =   67.40  D.F. =  2   P = 0.0000

                              STANDARD
    VARIABLE    COEFFICIENT    ERROR    COEFF./S.E.   EXP(COEFF.)
    --------    -----------    ------   -----------   -----------
  5 TNMSTAGE      0.2657       0.0543     4.8949        1.3043
  6 SXSTAGE       0.2932       0.0870     3.3713        1.3407

STATISTICS TO ENTER OR REMOVE VARIABLES
---------------------------------------
                  APPROX.      APPROX.
    VARIABLE      CHI-SQ       CHI-SQ                     LOG
NO. N A M E       ENTER        REMOVE     P-VALUE      LIKELIHOOD
  3 AGE            2.09                    0.1481      -806.1579
  4 MALE           3.38                    0.0661      -805.5150
  5 TNMSTAGE                    25.18      0.0000      -819.7919
  6 SXSTAGE                     11.35      0.0008      -812.8803
  7 PCTWTLOS       5.96                    0.0146      -804.2232
  8 HCT            9.59                    0.0020      -802.4071
  9 PROGIN         0.01                    0.9062      -807.1968


STEP NUMBER   3              HCT  IS ENTERED
------------
                LOG LIKELIHOOD  =  -802.4071
IMPROVEMENT CHI-SQ  ( 2*(LN(MPLR) )  =    9.59  D.F. =  1   P = 0.0020
                GLOBAL CHI-SQUARE =   75.38  D.F. =  3   P = 0.0000

                              STANDARD
    VARIABLE    COEFFICIENT    ERROR    COEFF./S.E.   EXP(COEFF.)
    --------    -----------    ------   -----------   -----------
  5 TNMSTAGE      0.2598       0.0536     4.8480        1.2967
  6 SXSTAGE       0.3381       0.0880     3.8400        1.4023
  8 HCT          -0.0489       0.0160    -3.0611        0.9523

STATISTICS TO ENTER OR REMOVE VARIABLES
---------------------------------------
                  APPROX.      APPROX.
    VARIABLE      CHI-SQ       CHI-SQ                     LOG
NO. N A M E       ENTER        REMOVE     P-VALUE      LIKELIHOOD
  3 AGE            1.40                    0.2367      -801.7070
  4 MALE           4.98                    0.0257      -799.9181
  5 TNMSTAGE                    24.71      0.0000      -814.7602
  6 SXSTAGE                     14.73      0.0001      -809.7716
  7 PCTWTLOS       3.30                    0.0693      -800.7574
  8 HCT                          9.59      0.0020      -807.2037
  9 PROGIN         0.01                    0.9243      -802.4026
```

```
STEP NUMBER   4              MALE  IS ENTERED
------------
                LOG LIKELIHOOD  =  -799.9181
IMPROVEMENT CHI-SQ  ( 2*(LN(MPLR) )  =    4.98  D.F. =  1   P = 0.0257
                GLOBAL CHI-SQUARE =   80.63  D.F. =  4   P = 0.0000

                              STANDARD
    VARIABLE    COEFFICIENT    ERROR    COEFF./S.E.   EXP(COEFF.)
    --------    -----------    ------   -----------   -----------
  4 MALE          0.5826       0.2823     2.0639        1.7907
  5 TNMSTAGE      0.2482       0.0530     4.6857        1.2817
  6 SXSTAGE       0.3597       0.0881     4.0826        1.4329
  8 HCT          -0.0529       0.0160    -3.3095        0.9485

STATISTICS TO ENTER OR REMOVE VARIABLES
---------------------------------------
                  APPROX.      APPROX.
    VARIABLE      CHI-SQ       CHI-SQ                     LOG
NO. N A M E       ENTER        REMOVE     P-VALUE      LIKELIHOOD
  3 AGE            0.34                    0.5615      -799.7496
  4 MALE                         4.98      0.0257      -802.4071
  5 TNMSTAGE                    23.14      0.0000      -811.4860
  6 SXSTAGE                     16.60      0.0000      -808.2181
  7 PCTWTLOS       2.08                    0.1496      -798.8799
  8 HCT                         11.19      0.0008      -805.5150
  9 PROGIN         0.21                    0.6484      -799.8172

NO TERM PASSES THE REMOVE AND ENTER LIMITS ( 0.1500  0.1000 )

SUMMARY OF STEPWISE RESULTS

STEP  VARIABLE                  LOG           IMPROVEMENT           GLOBAL
NO.  (E)NTERED     DF       LIKELIHOOD      CHI-SQUARE P-VALUE   CHI-SQUARE P-VALUE
     (R)EMOVED
---  ---------    ---      -----------      ---------- -------   ---------- -------
  0                         -839.016
  1  E  5 TNMSTAGE  1       -812.880          52.271    0.000      53.847    0.000
  2  E  6 SXSTAGE   2       -807.204          11.353    0.001      67.398    0.000
  3  E  8 HCT       3       -802.407           9.593    0.002      75.384    0.000
  4  E  4 MALE      4       -799.918           4.978    0.026      80.625    0.000
```

Fig. 16-8 BMDP stepwise procedure for Cox regression in illustrative data set. Steps 2 and 3 are shown on the left. Step 4 and remainder of printout are on the right. (BMDP2L program.)

symptom stage entered in step 2, and *hematocrit* in step 3. In step 4, *male* entered the model, with a P value of .0257. Having had a P value of .09 in the full regression model (see Fig. 16-3), *male* seemed to improve its effect when the remaining three candidate variables (AGE, PCTWTLOS, and PROGIN) were not included.

At the end of each step, the BMDP program checks all entered variables to see what might be removed (with APPROX. CHI-SQ. REMOVE tests) and determines which candidate variable to enter next (via APPROX. CHI-SQ. ENTER) from those that have not yet been included. In this instance, no variables were removed after having been entered. At the end of step 4, PCTWTLOS was the best contender, but did not enter because its P value of .1496 was higher than the criterion level of .1000. The BMDP summary at the bottom right of Fig. 16-8, like its SAS counterpart, shows the rank of entry of the variables and their log likelihood and chi-square accomplishments for the model, but does not indicate the individual coefficients or hazard ratios.

16.2.3. BMDP/SAS differences in entry/removal criteria

As noted earlier (see Section 11.9), the default entry criteria for linear regression are different in the SAS and BMDP programs. The BMDP approach also differs for types of regression, where an F value is used for linear regression, and P values for logistic and Cox regressions. SAS uses P for all three methods.

The logistic operations here were done with the default boundary values of P, which were 0.10 to enter and 0.15 to remove in the BMDP program, and 0.05 for both entry and removal in the SAS program. Other values can be assigned if desired.

16.3. Three-main-variable model

The results of forced regression with the three main independent variables are not shown here, since they were identical to what was obtained after the first three steps of the stepwise regressions in Section 16.2. For example, in the BMDP printout of the forced three-variable regression, the residual log likelihood was -802.4071; global chi-square was 75.38, and the respective values of the b_j coefficients were 0.2598 for TNMSTAGE, 0,3381 for SXSTAGE, and -0.0489 for HEMATOCRIT. These results are identical to what is shown after step 3 in Fig. 16-8.

16.4. Regression for groups

Cox regression analyses are regularly done and shown for groups. The groups can be "external" entities—such as Treatments A and B—that might otherwise be cited as an additional binary variable. The groups can also be formed from "internal" categories of an ordinal or nominal variable, such as Stages I, II, III, ...

The survival curves in groups can be shown for several purposes: to compare the groups either directly or after the adjustment of covariates, to check the "constant hazards" assumption, or to evaluate differences.

16.4.1. Descriptive comparisons

Because dynamic survival for a group is summarized as an entire curve, the curves for two or more groups cannot be easily compared with single values of descriptive summary indexes, like the means or binary proportions of single-state target variables. The result for an entire curve might be described from the area under the curve, but two curves with substantially different shapes might nevertheless have a crossing pattern that gives them the same areas. For this reason, the area under the curve has not become established as a single, simple descriptive index.

16.4.1.1. Display of curves

In the absence of simple descriptive indexes, the summary survival curves for two or more groups are usually compared by visually examining their graphic displays. They can be shown either as the unmodified $S_o(t)$ curves, before any multivariable analysis, or as the $\hat{S}(t)$ curves produced after the Cox modifications. Although a quantitative expression will not be available for the comparison, the displays will promptly indicate obvious differences in the patterns and shapes of group survival.

16.4.1.2. Median survival time

If a single quantitative summary value is sought for the entire pattern, an excellent routine choice is the *median* survival time. Unlike a mean, the median will seldom be affected by outlier values of long-term survivors, or by unknown values for persons who are still alive at the truncation time.

16.4.1.3. Relative hazard

Another descriptive index of quantitative comparison is the *hazard ratio* or *relative hazard*. For this comparison, the two treatments, A and B, are entered into the Cox regression as a single binary variable, coded as **0/1,** and symbolized here as X_T. After the regression model is determined as $b_1X_1 + b_2X_2 + b_3X_3 + \ldots + b_TX_T$, a prognostic index (also called a risk index) can be determined for each group. The value of the prognostic index, R_A, for group A is calculated after the mean values of the other $\{X_j\}$ variables for group A, and its **0** value for X_T, are substituted into the equation. The corresponding value of R_B is then calculated for group B with its value of **1** for X_T.

The incremental value of $R_A - R_B$, when exponentiated appropriately as $e^{(R_A - R_B)}$, will then be the average *relative hazard* for the effects of the two treatments. In the usual approach, however, all the other covariates are regarded as adjusted, and the value of b_T itself in the foregoing equation is used to express the comparison as a single hazard ratio, e^{b_T}.

16.4.1.4. 'Interaction effect' of compared variables

In the foregoing approach, each of the X_j variables is assigned a single b_j coefficient. If worried that the variables have different effects for the different treatments, the analyst can insert a set of interaction terms, as $\{X_jX_T\}$, into the equation. To avoid the complexity of appraising the interaction, Christensen[6] has proposed that each variable X_j be entered into the equation as two component variables, X_{jA} and X_{jB}. The values used for X_{jA} and X_{jB} will be only those of persons receiving the corresponding Treatment A or Treatment B, and are set at 0 otherwise. Thus, recipients of Treatment A are coded 0 in X_{jB}, and recipients of Treatment B are coded 0 in X_{jA}. The increment in the two coefficients that emerge for each b_{jA} and b_{jB} can then be divided by the standard error of the difference, calculated from the appropriate variance-covariance matrix, to form a ratio interpreted as a Z statistic. If this ratio is *not* significant, an interaction need not be feared; and the X_j variable can be entered into the equation as a single variable rather than as X_{jA} and X_{jB}.

16.4.2. Log-rank and other stochastic tests

Although a single quantitative index has not been developed to summarize and compare differences in the shapes *descriptively,* they can be contrasted *stochasti-*

cally with the log-rank test. Like the New York City soft drink called an *egg cream*, which uses neither eggs nor cream, the log-rank test contains neither logs nor ranks. To compare the survival curves of two groups, it applies an expanded variation of the observed-vs.-expected comparison strategy used in chi-square tests. Although often attributed to Mantel[7] in 1966, the log-rank statistic produces the same result, when there is no censoring, as the Savage[8] statistic, introduced a decade earlier, which does use logs and ranks. The observed-minus-expected calculations of the log-rank survival procedure are well illustrated with worked examples in papers by Peto et al.[9] and by Tibshirani.[10]

A log-rank test is unique among all the stochastic procedures discussed thus far in this text. With every other test, *quantitative* significance can be appraised for each comparison expressed with a *descriptive* index—such as the magnitude of an increment, correlation coefficient, standardized regression coefficient, or proportion of reduced variance. The comparative index is then evaluated for stochastic significance, with a P value or confidence interval. For survival curves, however, no standard form of descriptive expression has been developed to allow *quantitative* comparisons that cover the dynamic scope of the curves. Consequently, in the absence of a suitable descriptive index, the only kind of significance available for evaluation is stochastic, via the log-rank test.

Survival curves for censored data can also be stochastically compared with several other tests that are all variations[11,12] of either the Savage test or the familiar Wilcoxon ranks test. The additional procedures include Gehan's generalized Wilcoxon test,[13] and Peto and Peto's versions[14] (which use logarithms of the survival function) of the log-rank test and of the Wilcoxon test.

Because the P values of the stochastic tests are so greatly influenced by the number of events, an investigator (or reader) who fails to examine the actual magnitudes and shapes shown on the graph may mistakenly conclude that a low P value implies a significant difference in survival curves for two groups. When directly inspected, however, the curves may actually show relatively little difference; but a very large sample size may have brought the P value down to a significantly low level in the log-rank (or other stochastic) test.

16.4.3. 'Adjusted' effects of interventions

Cox regression (as well as logistic regression) is commonly used to "adjust" or correct for the baseline disparities that commonly exist in comparisons of interventional maneuvers, such as therapy. The multivariable regression in this situation has a role analogous to the stratifications that adjusted the baseline states for the treatments compared in Tables 3-6 through 3-10.

16.4.3.1. 'Intra-' and 'extra-model' approaches

With Cox or logistic regression models, the adjustment process can be done in two ways, as discussed in Section 12.7. With an "intra-model" approach, the "covariates" in the regression model include the therapeutic agents along with the baseline variables; and the "covariate-adjusted" effects of the agents are indicated by their partial regression coefficients.

With an "extra-model" approach, the Cox or logistic regression model is applied to only the baseline variables, without the inclusion of treatment. The adjusted survival curves are then compared for each treatment, using "grouping" methods (or strata) to arrange the treatment groups. The extra-model approach is scientifically more attractive because it does not mix the therapeutic agents contemporaneously, as though they were just another preceding covariate, along with the baseline variables.

16.4.3.2. Illustration with cell-type variable

In the illustrative data set here, treatment was not one of the variables under analysis, but the four available cell-type variables were not included in the full regression model. They can therefore be analyzed (for this illustration) as though their effects represented the "intervention" of a nominal "therapeutic" variable, having four categories of "treatment." The four nominal categories can then be compared with extra-model adjustments for other baseline variables.

The results of the extra-model procedure, in Fig. 16-9, show that the baseline-adjusted "survivor function" curves are relatively close together, thus suggesting that the four cell-type groups do not have strikingly different effects. Nevertheless, Group D (cytology only) had higher survival rates than Group A for months 4 to 8, and substantially lower rates thereafter. (The automated display of the curve for C is somewhat misleading because it stops, when the last death occurred, at a survival proportion that seems higher than the other three curves. The display is also misleading for Curve D, which seems to stop about 31 months, although it actually overlaps Curve B for at least another 15 months.) The constant-hazard assumption in Cox analysis is not violated by the crossover of curves in Fig. 16-9, because the crossover is produced by the histological variables, which were *not* entered into the Cox model. The crossover indicates, however, that their entry might not be appropriate.

For an intra-model adjustment, the logistic-regression design-variable option (see Section 14.10.1) that could test the four subgroups simultaneously is not available in either BMDP or SAS programs for Cox regression. If the four cell-type groups are to be analyzed concomitantly as independent variables within the model, the only approach would be to enter them with dummy variables.

To save space, the printout is not shown for results obtained when the dummy histologic variables were entered along with the other seven independent variables into the SAS old PROC PHGLM procedure (used to allow a check of the Z:PH indexes). The *cytology* variable, which was Group D in Fig. 16-9, received an excessively high Z:PH of 2.39, consistent with its nonproportional-hazard pattern. All other variables had absolute Z:PH values below 1.8. Stochastic significance was achieved for the three main variables (TNMSTAGE, SXSTAGE, HCT) but not for any of the cell-type or other variables.

16.4.4. Group curves for constant proportional hazards

A fourth use of group curves is to check the proportional constant-hazards assumption for individual variables by letting their categories form the groups or strata of survival curves whose patterns are inspected over time. The categories can be constructed as ordinalized zones for a dimensional variable, or maintained in the "natural" categories of coding for binary or ordinal variables. (Fig. 16-9 showed the temporal pattern for four categories of a nominal variable.)

To illustrate this process for an ordinal variable, the four categories of *symptom stage* become the subgroups graphed in Fig. 16-10 for an adjustment that contained six independent variables, but not symptom stage, in the model. Fig. 16-10 shows that the A and B curves cross at about 3 months and again at about 19 months, and also that the C and D curves cross at about 8 months. These crossings were surprising, because SXSTAGE had received a z:PH value of −0.79 (i.e., nonsignificant) in Fig. 16-4 for the analysis with 60-month truncation. When the previous analysis was repeated with a shortened truncation at 12 months, however, z:PH for SXSTAGE rose to −2.35 (i.e., significant), perhaps because early crossings have a greater effect when follow-up has a shortened duration.

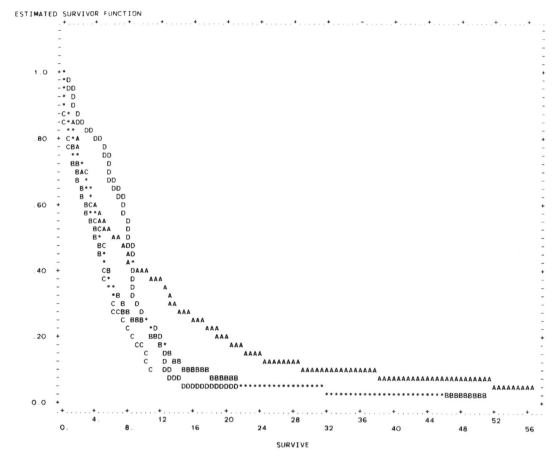

Fig. 16-9 Baseline adjusted survival curves for four cell-type groups. The symbols are as follows: A = WELL (well-differentiated); B = ANAP (anaplastic); C = SMALL (small cell); and D = CYTOL (cytologic proof only). The asterisks indicate locations of overlapping curves. The curve for "C" stops at 11.2 monts, when the last death occurs in this group. In its last segment, the curve for "D" overlaps with "B," and then stops at 46 months, at the last death in the "D" group. (Done with BMDP2L program, using the survival plot option, determined at mean values of the three main covariates.)

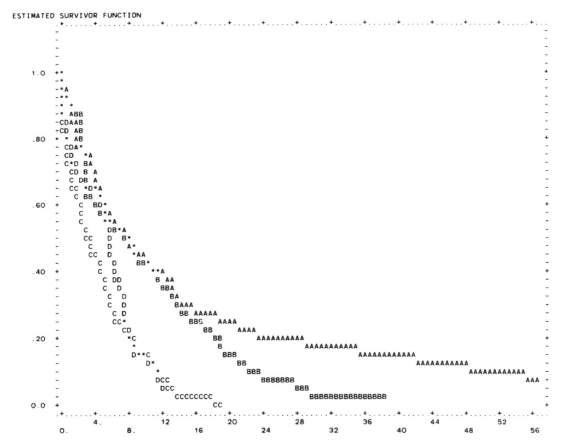

Fig. 16-10 Survivor function stratified for the four categories of symptom stage. A = pulmonic-systemic; B = asymptomatic; C = distant; D = mediastinal-regional. (BMDP2L program. Survival plot for mean value of six covariates.)

16.5. 'Diagnostic examinations'

As discussed earlier (Sections 15.4.3 and 15.8), two of the main potential "ailments" in the proportional hazards regression method are produced by nonlinear conformity in ranked variables, and by nonconstant hazards in any variables. With the first ailment, a single coefficient cannot correctly denote more than the "average" of a quantitative impact that varies in different zones of the ranked variable. The second ailment will raise doubt about including the variable in a conventional proportional hazards model. (A third ailment, noted earlier in the discussion of logistic regression, consists of having too few outcome events for the number of independent variables in the analysis. This problem, however, should have been noted and suitably managed before the analysis began.)

Before any special diagnostic tests are checked, the ordinary "physical examination" consists of inspecting graphic portraits of the results.

16.5.1. Routine graphs for actual and estimated survivals

If you have not yet inspected the actual summary survival curve, $S_o(t)$, before any variables are used for modeling, it can be displayed in BMDP or SAS printouts for both the Kaplan-Meier and the LIFE TABLE (i.e., Berkson-Gage) methods of calculation.

As an option in the BMDP2L program (but not contained in SAS PHREG), an analogous display can appear for the *estimated* summary survival curve, i.e., $S_h(t)$. Since $S_o(t)$ is calculated from the original data, and $S_h(t)$ is calculated from the estimated hazard function, an examination of the two curves offers a quick check that the model did what it was supposed to do in fitting the summary curve.

16.5.2. Checks for constant proportional hazards

Checking for violation of constant proportional hazards over time seems to be a fashionable indoor sport in Cox regression, although the violations, when found, may sometimes be ignored, as in the earlier example near the end of Section 16.1.2.3.2. The different procedures for diagnosing violations in proportional hazards have been described with reasonable clarity in reviews by Kay,[15] Andersen,[16] and Harrell and Lee.[17] The popularity of Cox regression leads to continuing statistical research, which will doubtlessly produce newer methods for this task in the future.

We have already met two methods for doing the check. One method is the Z:PH index (Section 16.1.2.3.2) and the other is a graphic plot of subgroups (Section 16.4.4). Another statistical tactic and another graphic plot are also available, however.

16.5.2.1. Insertion of time-interaction terms

Constancy of proportional hazards for a particular variable can be checked if an additional time-interaction term is inserted into the model for that variable. For example, if three main variables are being considered, three additional models would be checked. Each model would contain X_1, X_2, and X_3, but the first model would include an extra time-interaction term for X_1; the second model would include an extra time-interaction term for X_2; and so on. If the extra time-interaction term is *not* significant in these models, the corresponding variable probably does not violate the constant hazard assumption.

16.5.2.2. Graph of estimated LOG MINUS LOG SURVIVOR FUNCTION

A graph with this interesting name is yet another mechanism that can be used in checking for constancy of proportional hazards. Available as an option that is easy to obtain in BMDP but more difficult in SAS programs, the "double logarithm" function represents the values of $\ln[-\ln \hat{S}(t)]$ over time; and the graph offers a visual "isolation" for the effect of the linear model, G. To understand why this visual phenomenon occurs, recall that $\hat{S}(t) = S(t)^{e^G}$. [The "S(t)" symbol can represent either the original $S_o(t)$ summary curve or the $S_h(t)$ summary curve prepared from the summary hazard function.] The negative logarithm of this expression will be

$$-\ln \hat{S}(t) = e^G[-\ln S(t)], \qquad [16.1]$$

where $-\ln S(t)$ is the *cumulative hazard function*. The logarithm of the entities in expression [16.1] will be

$$\ln[-\ln \hat{S}(t)] = G + \ln[-\ln S(t)]. \qquad [16.2]$$

Therefore, if the value of $\ln[-\ln \hat{S}(t)]$ is plotted against time, the graphic ordinate at any time t would be $G + \ln[-\ln S(t)]$. For the entire group, or for any particular subgroup or category—for example, men or women; Stage I, Stage II, or Stage III; Treatment A or Treatment B—G will have its own distinctive value when the appropriate mean values are inserted into the expression for $G = b_1X_1 + b_2X_2 + \ldots + b_pX_p$. Each total group or subgroup, however, will have the same value of $\ln[-\ln S(t)]$, which is based on the overall unmodified summary curve. Therefore, the subgroup plots for $\ln[-\ln \hat{S}(t)]$ vs. time should show a series of parallel curves that all have the fundamental shape of $\ln[-\ln S(t)]$ while being separated, in their ordinates, by the corresponding values of G. Although S(t) becomes progressively smaller with time, $-\ln S(t)$ and $\ln[-\ln S(t)]$ become progressively larger, so that the double-logarithm graphs are a form of ascending mirror inverse of the descending shape of S(t).

Fig. 16-11 shows the double-logarithm graph of data for the symptom-stage variable, which previously showed crossings when checked for proportional hazards of the constituent categories in Fig. 16-10. Because each of the four curves contains the same $\ln[-\ln S(t)]$, adjacent curves should show a constant vertical

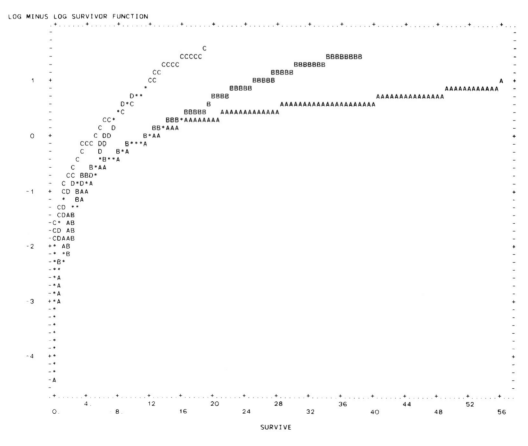

Fig. 16-11 "Double logarithm" plot for each of the four symptom stages, identified in legend of Fig. 16-10.

difference when augmented by the value of G for each category of the symptom-stage variable. The vertical differences of the four plots in Fig. 16-11, however, are obviously not constant with time, thus confirming that symptom stage does not suitably meet the assumption of constant proportional hazards.

16.5.3. Checks for linear conformity

The linear conformity of ranked (dimensional or ordinal) variables can be checked with a graphic method proposed by Harrell and Lee,[17] but the procedure is not easy to use, requiring special "macro" programs to augment the main computer program. Probably the simplest approach for checking linear conformity is to run the programs with the ranked variables converted into dummy variables for ordinalized zones, as discussed in Sections 12.4.2 and 14.7.7. With linear regression, linear conformity is checked by looking directly for a linear trend in the coefficients from one ordinalized zone to the next. With logistic regression, the linear trend should occur in the odds ratios from one zone to the next. With Cox regression, the linear trend should occur in *hazard ratios* from one ranked zone to the next.

This examination was carried out when *symptom stage* was recoded for dummy variable analysis in a Cox regression that included six other independent variables. In the reference cell coding system, the *asymptomatic* first stage was used as the reference cell and the three other stages were coded successively as SXDUMMY 1, SXDUMMY 2, etc. In the Cox regression results, the parameter estimates for the three successive *SXDUMMY* variables were, respectively, −.333, .322, and .689, which correspond to hazard ratios of .72, 1.38, and 1.99. The incremental values of about .6 between adjacent categories of these ratios suggest that the algebraic structure of symptom stage had good linear conformity, despite the nonconstant proportional hazard over time.

16.5.4. Tests for interactions

The independent variables originally entered into a Cox regression procedure seldom contain any interaction terms. Since the procedure does not look for interactions unless asked to do so, now is the time at which such explorations might be conducted (while you fervently hope that no significant interactions are found). The terms to be explored for interactions can be chosen in the manner discussed earlier throughout Section 12.3.

The interactions in Cox regression for the illustrative data set were examined in the same way as in logistic regression. The three main-effect variables—TNM stage, symptom stage, and hematocrit—were included in each model. The extra terms were the three paired interactions in one model, the single three-way interaction in a second model, and the paired and three-way interactions in a third, saturated model. Each of these models was then explored with full and forced-then-open stepwise regression procedures.

In the full regression procedures, when the model contained the three main effects and the three paired (two-way) interaction terms, *none* of the variables was stochastically significant. When the model contained three main effects and the single three-way interaction, the main effects remained stochastically significant, with b_j values respectively of .33, .44, and $-.04$ for TNM stage, symptom stage, and hematocrit, but the three-way interaction term had $b_j = -.0007$, which was not stochastically significant. When the model was saturated, none of the terms was stochastically significant.

In the stepwise procedures, after the three main effects were forced into the model, none of the interaction terms were entered, regardless of the pattern in which they were presented. The results indicate that interaction problems do not require attention in the three-main-variable model.

16.5.5. Events per variable

The events-per-variable (epv) ratio, discussed in Sections 12.1.5 and 14.7.8, should have been checked before the analyses began. If not yet determined, however, the ratio can be examined now, particularly if additional variables have been included for interactions (or for other purposes). In logistic regression, the events are counted as the *smaller* frequency of whatever has been coded as **0/1** for the binary outcome variable. By contrast, in Cox regression, the counted events are the deaths, because they are used (as discussed throughout Section 15.4) to determine the b_j coefficients for whatever X_j variables are in the model.

As noted earlier, the epv ratio should preferably exceed 20, and begins to suggest unstable results when below 10. For the 5-year survival analyses in the illustrative data set, 186 persons had died. With seven variables in the model, the epv ratio was a safe value of $186/7 = 26.6$. When three additional dummy variables were added for analyzing the histologic cell types in Section 16.4.3.2, the epv ratio dropped to $186/10 = 18.6$, but was still reasonably safe. For the eventual three-variable model, the ratio was a reassuring $186/3 = 62$.

16.5.6. Examinations of residuals

As noted in the discussion in earlier chapters, the idea of regression diagnostics has received large amounts of mathematical attention that is usually confined to a check for outliers in the residual discrepancies between observed and estimated results. Since the removal of outliers or major repairs of the basic linear model (with splines or other inserted functions) are seldom a scientific desideratum, the inspection of residuals is probably the least important pragmatic form of diagnostic examination.

The BMDP printout earlier in Fig. 16-2 showed Cox-Snell residual values for individual observations. The counterpart SAS printout, obtained as optional output and shown earlier in Fig. 16-5, displays various attributes including the value of $\Sigma b_j X_j$ (marked XBETA) calculated for each person, the standard error of that value (marked STDXBETA), the estimated value of SURVIVAL at each person's exit time; the natural logarithm (LOGSURV) of the survival function; and the natural log (LOG LOG S) of its negative value. The printout for each person also shows a "martingale residual" (RESMART), conceptualized via complex statistical reasoning,[18] which seems to be pragmatically calculated as LOGSURV for censored patients and LOGSURV + 1 for dead patients. An entity called "residual deviance" (RESDEV), is calculated with efforts to "normalize" the usually skew pattern of the martingale residuals. Various graphs can also be obtained to show diverse plots of the residuals.

All of this information can give you a pleasant sense of having complied with statistical guidelines about examining residuals. Pragmatically, however, the information has little or no real value unless you intend to try to improve the fit of the model by removing outliers, changing the formats of variables, adding splines, etc.

16.6. Remedial therapy

The main goal of conventional regression diagnostics is to improve the fit of the model by finding and removing outlier observations. This goal seems somewhat futile for survival analyses, because no algebraic model can be expected to offer really good fit to a polyvalent series of **0/1** data points that are widely dispersed over time. Even if nonfutile, however, the plan of removing major outliers creates a substantial disparity between scientific and mathematical goals in data analysis. Scientifically, the outliers should be retained unless they represent obvious or confirmed errors in the data. Mathematically, however, the outliers might be removed to improve fit of the model.

The fit of the model might also be improved by transforming variables, incorporating interconnecting splines, or adding interaction terms. These activities, however, will usually impair the scientific goal of identifying (and adjusting) the impact of important variables. Consequently, obtaining a splendid fit for the model is usually less important than being sure the important variables have received an undistorted identification. The latter goal is checked (and problems repaired), however, with methods other than those of regression diagnostics.

If the problem is too few events per variable, the analyses can be repeated (and results compared) when a smaller number of the most cogent variables are used in the model. For nonconstant proportional hazards, the offending variable can be examined in stratified format. For linear nonconformity, the variable can be transformed into linearity or converted into an ordinal array of dummy variables. Significant interactions can be checked with appropriately stratified analyses.

If these therapies for overt ailments are unsuccessful, the subsequent treatment depends on the specific scientific goal of the analysis. If the analytic goal was merely to *screen* for important variables, you can take the original (pre-interaction) results as suitable hints and then proceed accordingly with whatever you want to do with the hints thereafter. If you wanted to confirm what was previously noted in bivariate analyses, and if the expected confirmation occurs, you might want to depart without any further activity, unless the interactions are so strong that the individual bivariate results become untrustworthy.

The main requirements for remedial therapy arise if major surprises occur when you do the multivariable confirmations, or if your goal is to offer quantitative risk estimates for each variable. If the multivariable confirmations show "collinearity" that excludes an apparently important variable, you can be surprised but not dismayed. You an always check things out, as discussed earlier (see Section 12.5), to be sure that the collinear exclusion is not a capricious act of numerical fate, abetted by small numbers, arbitrary P values for decisions, and the rigidity of algebraic models.

With the surprise of a significant interaction, however, you may be dismayed and then feel obligated to do something. Reluctant to "clutter up" the interpretation of impact for individual variables, however, you may seek a way out of the problem. There is no way out, alas, if you retain an algebraic model and insist on getting a hazard ratio or other single expression of impact for each variable. If the latter goal is abandoned, however, several escapes are available. You can keep the cluttered-up interaction model; you can convert the algebraic results into a simpler array of prognostic scores; or you can shift from an algebraic to a targeted-cluster stratified analysis.

Essentially these same remedies can be used for the other two ailments (linear nonconformity or nonconstant hazards). With either of those problems, the single *average* hazard ratios for individual ranked variables will not be correct. You can try using dummy binary variables for the categories or ordinal variables or for the ordinal zones of dimensional variables, but the results will deprive you of a *single* coefficient for expressing each variable's impact. If you try transforming the variables into logarithms, square roots, squares, or other formats, you may improve the state of linear conformity, but the individual coefficients require peculiar interpretations. (For example, what is the meaning of a hazard ratio of 2.3 for the logarithm of a TNM stage?)

Therefore, if your goal is to offer a single expression of quantitative impact for each variable, the goal cannot be achieved with an algebraic model that has significant regression ailments. Since the goal also cannot be achieved with a targeted-cluster approach, the goal would have to be abandoned. A shift to a simple scoring system (described in the next section) would then allow effective estimates without the need for a complex algebraic model; and a shift to a targeted stratification system would allow the changing gradients to be observed directly as they occur.

16.7. Conversion of algebraic models to scores

The results of regression analyses, particularly with logistic and Cox methods, are commonly used for prognostication and for the types of adjustment described on many previous occasions in this test. The algebraic model can be used directly for both of these purposes. The prognostic outcome for an individual person can be predicted when that person's values of X_1, X_2, X_3, . . . are appropriately entered into the logistic model for Y_i or into the Cox model for $S(t)$. The adjustment of covariate confounders for an interventional "maneuver" (such as treatment or exposure to a "risk factor") can be done with the intra-model or extra-model approaches described in Section 16.4.3.

When used directly in this manner, however, the models often evoke major intellectual and scientific discomforts. Intellectually, the user is kept distant from the data, not knowing exactly what happened inside the "black box" that produced the coefficients, and feeling uncertain about possible caprices that can occur during the complex interplay of numbers when multiplications and additions lead to the estimates for each person. Scientifically, a clinical user may be reluctant to let a single regression model be the basis of realistic predictions for individual persons. As noted in Chapter 22, the results of the models can be quite effective for the average predictions in groups, but quite inconsistent for predictions in individuals.

For these reasons, the results found in the regression analyses are often converted into simple scoring systems that may be mathematically suboptimal, but that have the major advantage of letting the user see and know what is happening. The scoring systems can be constructed as factor counts, arbitrary weight points, or formula scores. The tactics are briefly outlined here, and discussed in greater detail in Section 19.6.

In a factor count, each important binary variable in the regression analyses becomes a "factor" that is counted as present or absent for a particular person. The sum of the factors is the person's score. With arbitrary weight points, the presence of a particular factor is given a weighted point value, such as **1, 2, 3, . . .** , usually corresponding to the "weight" of the regression coefficient for that factor. A person's score is the sum of these weighted points. In a formula score, which is the least preferred method, the multivariable equation is either preserved or somewhat simplified, so that $2.9X_1 - 0.8X_2 + 1.7X_3$ might become $3X_1 - X_2 + 2X_3$. The person's score is produced when the corresponding values of X_1, X_2, X_3 are entered into the formula.

The scores are then usually demarcated, as discussed in Sections 19.6 and 19.7, into stratified groups, which are then used for the predictions and adjustments.

16.8. Illustrations from medical literature

Table 16-3 lists twenty publications that reported the use of Cox regression analysis. These twenty reports, like those presented in Chapters 12 and 14, were

Table 16-3 Twenty published reports that used Cox regression

First author	Journal	Year	Outcome variable	Independent variables
Christensen[6]	Hepatology	1987	Survival in arbitrary data set (liver disease)	Albumin, bilirubin, alcoholism, treatment
Menotti[19]	J. Epidemoil. Community. Health	1987	All-cause mortality in rural Italian men	Demography, family, behavior, anthropometric, clinical Dx, physical exam, lab tests
Laberge[20]	Cancer	1987	Survival in small-cell cancer	CEA, perf. status, % weight loss, extension of disease
Lashner[21]	Am. J. Med.	1988	Mortality in biopsy-proven chronic hepatitis	Amniopyrine breath test, clinical biochemical & serologic data

Levy[22]	Psychosom. Med.	1988	Survival in recurrent breast cancer	Karnofsy score, age, no. of metastases, hostility and mood scores, disease free interval
Ooi[23]	Hypertension	1989	Cardiovascular morbidity and mortality in hypertension	Demographic, BP, body mass, ECG, cholesterol, smoking, treatment
Gail[24]	J. Nat. Cancer Inst.	1989	Development of breast cancer	Age at first live birth, age, age at menarche, no. of previous biopsies, no. of relatives' various "risk factors"
Grambsch[25]	Mayo Clin. Proc.	1989	Survival in binary cirrhosis	Age, edema, various lab tests
Nyboe[26]	Am. Heart J.	1991	Development of first infarction	Age, sex, family history, smoking, type of tobacco, earlobe crease, education, income, marital status, BP, alcohol, cholesterol, physical activity
Miranda[27]	Circulation	1991	Post-myocardial infarction survival	Angiographic status, Q waves on ECG, exercise response, age, medication
Skene[28]	Br. Med. J.	1992	Time to heal uncomplicated venous leg ulcer	Area and duration of ulcer, age, BP, deep vein involvement, body mass, other clinical and lab data
Merzenich[29]	Am J. Epidemiol.	1993	Age at menarche	Physical activity, energy expenditure, body weight, body mass index, fat intake, other dietary features, anthropometric data, family history
Kim[30]	Am. J. Public Health	1993	Mortality in persons using vitamin & mineral supplements	Dietary elements, various demographic features, clinical status
Klatsky[31]	Ann. Epidemiol.	1993	Mortality in persons using coffee and/or tea	Age, gender, race, education, marital status, body mass index, smoking, alcohol
McDonald[32]	Ann. Intern. Med.	1993	Veno-occlusive disease of liver after bone marrow transplant	Age, sex, diagnosis, and about 25 clinical and therapeutic variables
Goldbourt[33]	Arch. Intern. Med.	1993	6-month mortality affected by nifedipine after acute myocardial infarction	Congestive heart failure, age, sex, other clinical and laboratory data
Port[34]	J. Am. Med. Assn.	1993	Survival for renal dialysis vs. cadaveric transplant	Primary cause of renal disease, age, sex, year of first dialysis
Victora[35]	Lancet	1993	Duration of breastfeeding	Use of pacifiers, sex, birth weight, maternal education, family income

Pezzella[36]	New. Engl. J. Med.	1993	Survival in lung cancer	bcl-2 protein, sex, age, tumor differentiation, T stage, N stage
Cole[37]	Stat. Med.	1993	Quality adjusted survival in breast cancer	Treatment, age, tumor size, tumor grade, number of nodes involved

chosen from reprints haphazardly collected by the author. Each of the nine reports at the bottom came from a different journal in 1993, and the preceding eleven reports came from yet other journals in preceding years. Thus, twenty different journals are represented in the collection. The reports were reviewed to note their compliance with various principles that one might expect to find stated in published presentations. (Many of these principles are also mentioned in an excellent review by Andersen.)[38]

16.8.1. Purpose

All twenty sets of analyses were aimed at identifying or confirming the importance of the independent covariates, or adjusting for their effects. None of these analyses was intended alone to establish a predictive equation. In three instances,[6,25,28] the results were converted into a prognostic (or risk) score.

16.8.2. Identification of program and operation

The particular package program was identified in only four reports, one of which[28] used the BMDP system, and three[30,33,36] the SAS. The method of operation was identified in only six reports. One[23] used a full regression; one[19] used a stepwise procedure; two[22,25] used forward regression; and two[6,28] used all possible regressions.

16.8.3. 'Fit' of model

The "fit" (or general explanatory effect) of the model was mentioned in only five reports. The citations were an R_H^2 and P value,[6] a global chi-square and P value;[19,26] a global chi-square only;[22] and a P value only.[25]

16.8.4. Transformation of variables

Dimensional variables were transformed into their logarithms in four reports[6,21,25,27] and into squares or square roots in one report.[25] In six reports,[20,24,26,28,29,37] dimensional variables were analyzed after being given ordinal categorizations.

16.8.5. Coding for variables

In most reports, the coding of variables was identified clearly either in the text or tables, or, in one instance,[19] in an appendix deliberately prepared to show the coding. Specific details of coding could not be readily determined, however, in five reports.[22,29,31,33,35]

16.8.6 Impact of variables

In two reports,[20,32] after the previous bivariate results were confirmed, the important variables received no further descriptions. In the other eighteen reports, the multivariate-adjusted effect of the individual variables was cited in different ways. The partial regression coefficients, i.e., the b_j values, were usually cited directly; only one report[19] gave a standardized value, and none listed an Index-R value. The citations were always accompanied by stochastic indexes of "uncertainty," such as a P value, standard error variance, Z ratio, chi-square, or improvement chi-square. In seven instances,[26,27,29,30,31,33,34] the variable was quantified as a "relative risk." The correct terms *hazard rate ratio, hazard ratio,* or *relative hazard* were used instead of *relative risk* in two reports.[26,35]

16.8.7. Group curves

In three instances,[20,27,33] group curves were used to illustrate a log-rank test, and in ten instances, the curves showed survival for a particular group or categories *without* the Cox model.

16.8.8. Interactions

Tests for interactions, if done, were not mentioned in thirteen of the twenty reports. In two reports,[23,24] the results of interactions were included in the final model; and in four reports,[26,30,34,36] interactions were managed by appropriate analyses within subgroups.

16.8.9. Constant proportional hazards

A check for constant proportional hazards was not mentioned in fifteen of the twenty reports. The problem was stated without further discussion in one,[24] checked with a "log (−log survival)" plot in three,[6,21,28] and with comparison of stratified curves in one.[37] The report by Christensen,[6] which offers an excellent discussion and illustrations of these checks, also contains a lucid account of many other issues in the clinical use of Cox regression.

16.8.10. Checks for linear conformity

None of the twenty reports mentioned specific examinations to check for linear conformity between the model and the individual data points.

16.8.11. Events per variable

This potential problem was not mentioned in any of the reports. In many reports, with an abundance of outcome events, the problem did not seem pertinent. In one report,[23] the potential existence of the problem could not be checked, since the number of outcome events was not listed (and could not be determined from the reported data). In three reports, the event-per-variable ratio was doubtlessly too small, having the values of 18/4, 17/5, and 14/5, respectively.[6,21,22]

16.8.12. Other comments

In a case-control study of risk for breast cancer,[24] the authors did most of their analyses with an "unconditional logistic regression model" but then converted the odds-ratio coefficients into relative risks after determining a "baseline hazard." The authors state that "the proportional hazards assumption was relaxed to allow separate proportional hazards models for those under age 50 and those of age 50 or more." In two reports,[32,34] the authors say they used time-dependent Cox proportional hazards regression models, but the mechanism of operation and analysis was not clear.

A substantial hazard of Cox regression analysis is the problem of comprehension when results published in medical journals are not presented in a "reader-friendly" manner.

Chapter References

1. Dixon, 1990; SAS Technical Report P-229, 1992, pp. 433–80; 3. SUGI, 1986, pp. 437–66; 4. Cox, 1968; 5. SUGI, 1986, p. 443; 6. Christensen, 1987; 7. Mantel, 1966; 8. Savage, 1956; 9. Peto, 1977; 10. Tibshirani, 1982; 11. Kalbfleisch, 1980, pp. 16–19; 12. Lee, 1980, Chapter 5; 13. Gehan, 1965; 14. Peto, 1972; 15. Kay, 1977; 16. Andersen, 1982; 17. Harrell, 1986, pp. 823–8; 18. Therneau, 1990; 19. Menotti, 1987; 20. Laberge, 1987; 21. Lashner, 1988; 22. Levy, 1988; 23. Ooi, 1989; 24. Gail, 1989; 25. Grambsch, 1989; 26. Nyboe, 1991; 27. Miranda, 1991; 28. Skene, 1992; 29. Merzenich, 1993; 30. Kim, 1993; 31. Klatsky, 1993; 32. McDonald, 1993; 33. Goldbourt, 1993; 34. Port, 1993; 35. Victora, 1993; 36. Pezzella, 1993; 37. Cole, 1993; 38. Andersen, 1991.

17 Discriminant Function Analysis

Outline ....

17.1. Contemporary disadvantages
    17.1.1. Plurality estimates
    17.1.2. Precise technologic approaches
    17.1.3. Alternative methods for binary targets
    17.1.4. No direct coefficients for variables

17.2. Historical background

17.3. Basic concept of a discriminating function

17.4. Basic strategies
    17.4.1. Separation of groups
    17.4.2. Coefficients for linear combinations
    17.4.3. Conversion to classification functions
    17.4.4. Application of classification functions

17.5. Indexes of accomplishment
    17.5.1. Accomplishment of model
    17.5.2. Classification of each case
    17.5.3. Jackknife classifications
    17.5.4. Impact of variables

17.6. Conversion to discriminant functions

17.7. Application to two groups
    17.7.1. Relationship of coefficients
    17.7.2. Mathematical and pragmatic consequences

17.8. Regression diagnostics

Appendix for Chapter 17

Outline

A.17.1. Strategies of separation
    A.17.1.1. Separation for two groups and one variable
    A.17.1.2. Separation for two groups and two variables
    A.17.1.3. Role of centroids

A.17.2. Statistical concepts of 'standard' distance
    A.17.2.1. One variable; two groups
    A.17.2.2. Two or more variables; two groups
    A.17.2.3. One variable; more than two groups
    A.17.2.4. Multiple variables; multiple groups

A.17.3. Converting variables to functional expressions
    A.17.3.1. Choice of discriminating boundaries
    A.17.3.2. Concepts and symbols for classification functions
    A.17.3.3. Prior estimates of probability
    A.17.3.4. Estimates with classification functions
    A.17.3.5. Role of prevalence
    A.17.3.6. Role of conditional probability

A.17.3.7. Special new challenges in Bayesian components
A.17.4. Role of Gaussian conditional probabilities
A.17.5. Calculation of classification probabilities
  A.17.5.1. Calculation of $p_h f_h$
  A.17.5.2. Subsequent calculations
    A.17.5.2.1. Finding $P(C_h) [= P_h]$
    A.17.5.2.2. Finding $P(X) [= \Sigma P_h f_h]$
    A.17.5.2.3. Calculating $\hat{Y}_h$
A.17.6. Role of group separations
A.17.7. Algebraic expression for classification and discriminant functions
  A.17.7.1. Logarithms of $P_h f_h$
  A.17.7.2. Conversion to discriminant functions
  A.17.7.3. Resemblance to logistic regression for two groups
  A.17.7.4. Expression of classification functions
A.17.8. Extension to multiple variables
  A.17.8.1. Change from X to vector **X**
  A.17.8.2. Change from means to centroids
  A.17.8.3. Uses of D2 and F indexes
  A.17.8.4. Formation of multivariate classification functions

....

With discriminant function analysis, the target variable can contain a set of unranked nominal categories, such as **hepatitis, cirrhosis, biliary obstruction,** and **other** for *diagnosis of liver disease*. For such polytomous categories, the discriminant function method must differ from the three preceding members of the big four regression family, because the target cannot be estimated with a *single* algebraic expression. Instead, a separate *classification* function, with its own set of coefficients for the component independent variables, is prepared for *each* of the target categories. When a particular person's data are entered into the classification functions, the one that has the highest value is the estimated category for that person.

Because of the many (i.e., > 2) polytomous categories in the dependent variable, discriminant function analysis is a truly *multivariate* rather than merely *multivariable* procedure. Thus, if the polytomous target were decomposed into a set of dummy binary variables, their relationship to the independent variables would have the many-to-many structure discussed in Chapter 1. The polytomous format also makes the discriminant procedure much more mathematically complex than the other three forms of regression. The discriminant mathematical operations require special conditional probabilities that do not appear in the other forms of regression. Furthermore, unlike the other three forms of regression, the algebraic coefficients for the discriminant models are determined with methods that differ from those used to evaluate accomplishments.

17.1. Contemporary disadvantages

The strategy of aiming at a polytomous target gives the discriminant function method some substantial disadvantages, discussed in the next four subsections, that reduce its appeal in modern medical research.

17.1.1. Plurality estimates

A polytomous set of categories is often difficult to separate accurately with a single procedure. After the discriminant classification functions are determined for a particular data set, the estimate for a particular person is the category with the highest probability value. The highest value, however, may be a "plurality winner" that is not impressively or discriminatingly higher than the other contenders. The situation can be reminiscent of a political election in which none of the candidates gets a majority of the votes. If the votes are distributed as 23%, 24%, 26%, and 27% in a four-party election, and if no mechanism exists for a runoff, the party with the plurality vote of 27% will win the election. The prediction that everyone would vote for that party, however, would have been wrong for 73% of the voters.

17.1.2. Precise technologic approaches

If applied to the challenge of choosing a single diagnosis from among a group of several possible contenders, a mathematical model will seldom be preferred over the technologic precision available in modern science. Rather than relying on the imprecise statistical estimate offered by the model, most clinical diagnosticians will want to use a suitable algorithm or to get more data with an appropriate technologic test. For example, if clinical evaluation of the chemical tests of liver function does not clearly point to a specific hepatic diagnosis, most clinicians today will usually seek the precision of appropriate endoscopic or ultrasound examinations, or even liver biopsy. With precise technologic results available, very little diagnostic faith would be placed on the weighted summation of multivariable estimates that emerge when the liver function tests are entered into mathematical formulas.

17.1.3. Alternative methods for binary targets

Because of the first and second problems just cited, the categorical target of analytic estimation in medical research is regularly transformed from a polytomous to a binary variable. The analytic estimates are then dichotomous, aimed at discriminating category A vs. all other categories, or at identifying the presence vs. absence of Diagnosis D. If the target variable is converted to a binary event, however, two other algebraic methods are available for the estimating job. One of them, as discussed earlier, is logistic regression. The other method, discussed later in this chapter, is linear regression. Although mathematically inappropriate (because the target variable will not have a Gaussian distribution), linear regression produces the same result for a binary target as discriminant function analysis.

17.1.4. No direct coefficients for variables

A fourth disadvantage is that regardless of how well discriminant function analysis works in estimating either polytomous or binary target categories, it does not directly produce partial regression coefficients to denote the impact of the independent variables. Their importance can be determined from their accomplishments during sequential explorations, but their partial regression coefficients must be calculated indirectly (as discussed later) by subtracting the corresponding results in the classification functions.

Because of all these problems, discriminant function analysis is used much less frequently today than many years ago, when it was the dominant multivariable method for analyzing categorical targets in medical research. Some statisticians, in fact, now regard discriminant analysis as obsolete, urging that it be replaced by logistic regression for either polytomous or binary targets. Nevertheless, the medical literature continues to show results of the discriminant method (which is chosen for reasons seldom cited by the users). The method is therefore described here to prepare you for occasional encounters with it, and also to render homage, as the method approaches "retirement", to its long career of statistical service.

17.2. Historical background

When introduced about sixty years ago by R. A. Fisher,[1] the discriminant method was aimed at a binary separation of two groups. As discussed in Section A.13.1. of the Appendix to Chapter 13, Fisher combined the independent variables into a linear function, $L = b_0 + b_1X_1 + b_2X_2 + b_3X_3 + \ldots$. When the appropriate values were inserted into this function, a particular person, i, would be assigned to Group A or Group B according to whether the value of L_i was above or below a selected boundary value of L_0.

The tactic was first applied in biology by M. M. Barnard to study variations in Egyptian skulls.[2] As the technique was expanded to the challenge of separating polytomous rather than just binary categories, the strategy and tactics of discriminant function analysis received contributions from many prominent leaders of twentieth-century statistics: Bartlett, Hotelling, Kendall, Mahalanobis, Rao, Wilks.

Discriminant functions began to appear in medical research in the 1950s, accelerating through the 1960s and 1970s. The procedures have also been used in anthropology, economics, and political science. In the United States, a discriminant function has reportedly been used by the Internal Revenue Service to decide who should receive special audits of tax returns.

Although generally intended to offer a "diagnostic classification" among three or more polytomous categories, discriminant function analysis became particularly popular in medical research when aimed at a two-category binary target variable. Discriminant analysis was probably favored for this purpose because multiple linear regression had been deemed mathematically unsuitable, and because multiple logistic regression (at that time) required highly expensive computations. During the 1960s and 1970s, the medical literature contained discriminant function analyses in which predictors or risk factors were examined for the diverse binary events listed in Table 17-1.

The twenty-two reports cited in Table 17-1 (and those in the subsequent two tables) are mentioned because they all were conveniently available for review in the author's reprint collection. I have not done a systematic search of the literature to find other appropriate citations. In particular, I have not reviewed an additional twenty-one pertinent papers, all presumably having binary target events, which are listed in references for

Lachenbruch's excellent text,[3] published in 1975. Organized chronologically by year of publication and first authors, those references are as follows: 1949: Lewis, Rao; 1950: Harper; 1955: Charbonnier; 1959: Hollingsworth; 1960: Bulbrook; 1963: Hughes; 1965: Baron; 1966: Abernathy, Marmorsten, Morris, Weiner; 1967: Truett; 1968: Boyle, Dickey; 1969: Armitage; 1971: Afifi, Kleinbaum; 1972: Anderson, Jenden; 1973: Carpenter.

Table 17-1 Pre-1980 medical publications of discriminant function analysis for binary targets*

First author	Journal	Year	Binary target event
Radhakrishna	Statistician	1964	Successful response to treatment of tuberculosis
Norris	Lancet	1969	Hospital mortality in acute myocardial infarction
Winkel	J. Chron. Dis.	1971	Death in patients with shock
Watanabe	Am. Heart J.	1972	Electrocardiographic features of cor pulmonare vs. anterior wall myocardial infarction
Lehr	J. Chron. Dis.	1973	Development of coronary heart disease
Rothman	J. Chron. Dis.	1973	Trigeminal neuralgia vs. "controls"
Anderson	Appl. Statistics	1974	Dental patients with and without rheumatoid arthritis
Editorial	Brit. Med. J.	1975	Good (vs. poor) post-therapeutic response of breast cancer
Hedge	Brit. Med. J.	1975	Beta thalassemia trait vs. iron deficiency in pregnancy
Aoki	Jap. Circ. J.	1975	Fundoscopic features of cerebral hemorrhage vs. thrombosis
Loop	J. Thorac. Cardiovasc. Surg.	1975	Survival after coronary bypass
Luria	Ann. Intern. Med.	1976	Survival after acute myocardial infarction
Smith	Brit. Heart J.	1976	Occurrence of myocardial infarction during antihypertensive therapy
Walters	Brit. Vet. J.	1976	Pregnant vs. nonpregnant animals
Keighley	Lancet	1976	Postoperative recurrence of peptic ulcer
Goldman	New Engl. J. Med.	1977	Cardiac complications after major noncardiac operations
Hamilton	J. Clin. Pathol.	1977	Hepatic disease vs. "controls"
Ranson	J. Surg. Research	1977	Benign (vs. severe) course of acute pancreatitis
Jayakar	Ann. Hum. Biol.	1978	Obese vs. non-obese persons
Schildkraut	Arch. Gen. Psych.	1978	Bipolar vs. unipolar psychiatric depression
Drance	Arch. Opthalmol.	1978	Presence/absence of glaucomatous visual field defects
Ellis	Comp. Biomed. Research	1979	Diverse binary decisions in hepatic disease

*To save space and to avoid an unduly long list of references, the full title and pagination of these publications are not cited. The information noted here should allow them to be found.

As the computational problems were eased, multiple logistic regression gradually began replacing discriminant analysis as the preferred procedure for multivariable analyses aimed at binary targets. Nevertheless, despite the ready availability of logistic regression in the past fifteen years, two-group discriminant function analysis continues to be used. It has been applied, after 1980, for binary "separations" in the reports listed in Table 17-2.

Table 17-2 Post-1980 medical publications of discriminant function analysis for binary targets*

First author	Journal	Year	Binary target event
Bernt	Biometrics	1980	Miscarriages vs. normal births
Vlietstra	Circulation	1980	Arteriograms with or without coronary disease
Kennedy	J. Thorax. Cardiovasc. Surg.	1980	Survival after coronary bypass surgery
Klein	Science	1980	Prison vs. probation for criminals
Lafferty	Arch. Intern. Med.	1981	Primary hyperparathyroidism vs. other sources of hypercalcemia
Fisher	Circulation	1981	Anatomic condition of coronary artery disease
Dolgin	New Engl. J. Med.	1981	Cholesterol vs. pigment gall stones
Rupprecht	Am. J. Epidemiol.	1987	Terrestrial vs. nonterrestrial forms of rabies virus
Pintor	Lancet	1987	Good vs. poor responsiveness to treatment of children with short stature
Boey	Ann. Surgery	1987	Survival vs. death in perforated duodenal ulcer
Diehr	Med. Decision Making	1988	Fracture vs. no fracture after ankle trauma
Lerman	J. Gen. Intern. Med.	1989	"White-coat" vs. other forms of hypertension
Wang	Arch. Intern. Med.	1990	Alcoholic vs. nonalcoholic patients with cardiomyopathy
Seed	New Engl. J. Med.	1990	Congestive heart failure (CHF) vs. no CHF in familial hypercholesterolemia
Poungvarin	Brit. Med. J.	1991	Supratentorial hemorrhage vs. cerebral infarction in stroke
Kornreich	Circulation	1991	Normal vs. Q-wave and vs. non-Q-wave myocardial infarction
Sawle	Arch. Neurol.	1994	Parkinson's disease vs. normal controls

*See note in Table 17-1.

With logistic regression increasingly used for binary target variables, discriminant function analysis can be expected to appear less often today than before. The applications will presumably be aimed at targets that are polytomous nominal categories, for which the discriminant procedure seems most suited. Some of the applications of discriminant functions to "diagnose" polytomous targets are listed in Table 17-3. (The references in Lachenbruch's text[3] mention seven additional papers, which I have not reviewed, in which the analyses were probably aimed at polytomous targets. Those citations, by year and first author, are as follows: 1955: Eysenk; 1964: Baker, Collen, Hopkins; 1968: Anderson, Pipberger; and 1969: Burbank.) Since almost all of these polytomous-target analyses were published before 1988 and since the only recent (1993) example in Table 17-3 was commercial rather than medical, discriminant functions may either be losing their original appeal, or medical investigators may be getting disenchanted with making algebraic diagnoses for polytomous nominal targets.

Table 17-3 Polytomous targets in medical discriminant function analysis*

First author	Journal	Year	Categorical target variables
Fraser	Lancet	1971	4 groups of patients with hypercalcemia
Overall	Arch. Gen. Psych.	1972	Profiles for usage of 3 main types of psychotherapeutic drugs
Werner	Clin. Chemistry	1972	7 diagnoses made from immunoassays of electrophoretic globulin fractions
Barnett	Brit. Med. J.	1973	3 ordinal categories of thyroid function
Spielman	Am. J. Human Genetics	1976	17 villages of Yanomama Indians
Rosenblatt	Circulation	1976	3 ordinal levels of cardiac compensation in valvular heart disease
Rotte	Diagnostic Radiol.	1977	3 types of pulmonary lesions
Watson	Comp. Biomed. Research	1978	10 choices of physicians' specialty
Titterington	J. Roy. Statist. Soc. A Stat.	1981	3 ordinal categories of outcome after head injury
Ryback	J. Am. Med. Assn.	1982	4 diagnostic categories associated with alcoholism and liver disease
Brohet	Circulation	1984	4 categories of diagnosis in pediatric vectorcardiography
Thomas	Arch. Oral Biol.	1987	6 ethnic groups
Murata	Biosci. Biotechnol. Biochem.	1993	6 types of coffee

*See note in Table 17-1.

Regardless of how popular the discriminant function procedure may be or become, however, the discussion here will refer mainly to polytomous targets. A separate section will be added afterward for binary targets.

17.3. Basic concept of a discriminant function

The basic goals of discriminant function analysis are to develop linear models that can be used for separating polytomous groups in a observed set of data. The models can then be applied to assign an "unknown" new person to the most appropriate group.

The basic idea was illustrated in Chapter 2, in the graph of Fig. 2-24, showing two groups of points for two variables, X_1 and X_2. Although not easily distinguished from their values of either X_1 alone or X_2 alone, the groups could easily be separated if the values of X_1 and X_2 are considered simultaneously. When both variables are used, a line of clear separation could be drawn as shown here in Fig.

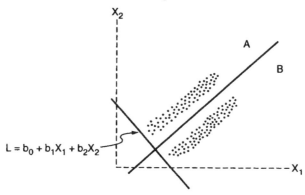

Fig. 17-1 Two-dimensional line, using variables X_1 and X_2 to separate two groups, A and B, previously shown in Fig. 2-24. The discriminant function line, L, is perpendicular to the line of separation.

17-1. If a perpendicular line is drawn as an axis orthogonal to the first line, and if we shift the axes of alignment for the graph, the orthogonal line would become the main axis, identified as

$$L = b_0 + b_1X_1 + b_2X_2.$$

When a person's data for X_1 and X_2 are inserted into this expression for L, values below 0 will indicate membership in Group A, and values above 0 will indicate membership in Group B.

This idea is the basis for creating discriminant functions. If p variables are used, the linear discriminating function that separates two groups will be

$$L = b_0 + b_1X_1 + b_2X_2 + \ldots + b_jX_j + \ldots + b_pX_p.$$

The multivariable concept is theoretically splendid, but its execution requires management of several major problems and challenges:

1. The data for the groups are seldom, if ever, as cleanly separated as the points shown in Fig. 17-1.
2. Things become much more complex if more than two groups are to be separated.
3. A strategic mechanism must be developed for achieving the separation.
4. The mechanism must be converted into the appropriate algebraic expression(s).

17.4. Basic strategies

The challenges just cited are managed with four main strategies:

17.4.1. Separation of groups

The groups are usually best separated by the set of variables whose centroids have maximum distances between them. A *centroid* is a multivariable mean; and the locations of centroids for the groups will vary according to the choice of variables. Thus, if the available variables are $X_1, X_2, X_3, \ldots$, the best set of variables is the one that will produce maximum distance in space among the group centroids $\bar{X}_1, \bar{X}_2, \bar{X}_3, \ldots$. The exploration of candidate independent variables, forming different models, is intended to find this best set.

17.4.2. Coefficients for linear combinations

For m categorical groups, denoted as $C_1, \ldots, C_h, \ldots, C_m$, there will be m classification functions, symbolized as $\hat{Y}_1, \ldots, \hat{Y}_h, \ldots, \hat{Y}_m$. In each of these functions, the selected variables receive partial regression coefficients and are combined into a linear format. For Group 1, the format for p variables would be

$$\hat{Y}_1 = a_0 + a_1 X_1 + a_2 X_2 + \ldots + a_p X_p.$$

For Group 2, the format would be

$$\hat{Y}_2 = b_0 + b_1 X_1 + b_2 X_2 + \ldots + b_p X_p.$$

For Group m, the format would be

$$\hat{Y}_m = m_0 + m_1 X_1 + m_2 X_2 + \ldots + m_p X_p.$$

The coefficients for the groups are chosen to minimize an index called Wilks lambda, which is somewhat analogous to an F statistic or a log likelihood ratio. It is calculated as

$$\frac{|\mathbf{W}|}{|\mathbf{W} + \mathbf{B}|},$$

where **W** is the dispersion (i.e., variance-covariance) matrix within groups; |**W**| is the determinant of the matrix; and **B** is the corresponding dispersion matrix between (or among) groups.

The Wilks lambda statistic resembles the F statistic used with the least-squares principle of ordinary linear regression. The F statistic has its maximum value at a counterpart ratio of |**B**|/|**W**|, when the coefficients create a suitable regression model. In that model, the mean variance (e.g., $S_M/1$) between the unmodelled mean ($\bar{Y}$) and the model values of $\hat{Y}_i$ will correspond to |**B**|, and should be as large as possible in relation to |**W**|, which corresponds to S_R/p for the mean variance values of $Y_i - \hat{Y}_i$. If the groups are well separated, the value of |**B**| will be much larger than |**W**|, and Wilks lambda will be relatively small.

17.4.3. Conversion to classification functions

The linear combinations of the variables and their coefficients are used for making decisions about an optimal separation of distances, but not for the eventual models that do the estimation of categories. The latter models are produced by converting the selected variables into another set of linear combinations. The latter combinations are the classification functions, obtained by using principles of conditional and Gaussian probability. These principles were discussed earlier (if you happened to read it) in the Appendix of Chapter 13 for logistic regression.

17.4.4. Application of classification functions

The classification functions can then be used to estimate the probability of a particular person's membership in each group. The classification functions can also be converted, if desired, into the incremental entities that are called discriminant functions.

Although logical and relatively straightforward, the strategic process is mathematically complicated because it involves multiple groups, multiple variables, measurement of multidimensional distances, and probabilistic tactics that were not required in the linear, logistic, or Cox methods of regression. The long description of that intricate process is outlined in the Appendix for this chapter.

17.5. Indexes of accomplishment

After a set of variables has been chosen to form the classification functions, the accomplishments of the "model" that comprises those functions can be described in several ways.

17.5.1. Accomplishment of model

To avoid the Greek symbol Λ for the Wilks ratio described in Section 17.4.2, the index is often called the U statistic. In an analogy to the analysis of variance, U can be regarded as a ratio in which the sums of squares within groups is divided by the total sum of squares. Its value ranges from 0 to 1, and gets smaller with better separations. If the groups all have essentially the same centroids, the value of U is close to 1. If the groups are well separated, the value of U is close to 0.

The U statistic for Wilks lambda is regularly transformed to an approximate F statistic, which can be used for both quantitative and stochastic purposes. Stochastically, it tests the hypothesis that the group means are equal for all of the variables included in the model.

Two other statistical indexes can also be used to measure accomplishment in separating groups. One index, *Pillai's trace,* is a counterpart of R^2 in linear regression; and higher values indicate better separations. Pillai's trace is close to 1 if all groups are well separated and if all or most directions show a good separation for at least two groups. The other index, the *average squared canonical correlation* (ASCC), is Pillai's trace divided by m − 1, where m is the number of groups.

In stepwise operations, the criterion for adding or deleting a particular variable depends on which variable contributes the most (forward regression) or the least (backward regression) to lowering Wilks lambda.

17.5.2. Classification of each case

The computer program for discriminant function analysis produces the values of the b_{hj} coefficients for the classification functions that estimate each group's analog of probability, $\hat{Y}_h$. The program will also determine the array of $\hat{Y}_h$ values for each person, i, having the vector of variables $X_{i1}, X_{i2}, \ldots, X_{ip}$. Each person will then be assigned to the group for which $\hat{Y}_h$ has the maximum value. Thus, the estimated $\hat{Y}_i$ for that person will be $\hat{Y}_{max}$. Alternatively, a particular case can be assigned to the group from which it has the shortest multivariate distance.

After making this estimate, the program will check the actual value of Y_i and determine whether the estimate was right or wrong. The accomplishment will be shown in a classification table, which is usually arranged to indicate the relationship of estimated and actual categories as follows:

	Estimated category				
Actual category	C_1	...	C_h	...	C_m
C_1	Correct				
⋮		...			
C_h			Correct		
⋮				⋮	
C_m					Correct

The number of correct estimates will be shown in the diagonal cells of this table. The total accuracy can be expressed as the percentage of estimates that were correct, or in various analogs of citations for sensitivity and specificity.

17.5.3. Jackknife classifications

A different set of calculations uses the jackknife procedure, in which each of the N cases is eliminated sequentially from the data, and the classification functions are calculated from all the remaining N − 1 cases. Because the group assignment for each case is estimated from data in which the case has *not* been included, the jackknife procedure is often regarded as a less "biased" method than the ordinary classification procedure. With a reasonably large group size, the results of the two approaches are similar.

17.5.4. Impact of variables

Because discriminant function analysis is aimed mainly at separating groups, the impact of individual variables is a relatively remote secondary objective in the mathematical model. In the customary printout of the packaged computer programs, which show classification rather than discriminant functions, the variables do not receive omnibus *individual* coefficients, because these coefficients will differ for each function. For more than two groups, the construction of discriminant functions (as noted shortly) will produce two or more coefficients for each variable.

Accordingly, the impact of variables in discriminant function analysis is often discerned not from their coefficients, but by other methods such as their sequential rank in stepwise procedures. These rankings can indicate whether a variable has a significant effect and whether the effect is greater than that of some other candidate variable, but will not provide the quantitative precision offered by specific omnibus coefficients in the other three regression methods. Therefore, if impact of variables is a prime goal in the analysis, the discriminant function procedure is the least satisfactory of the big four multivariable methods.

17.6. Conversion to discriminant functions

In most medical circumstances, the data analyst does *not* use discriminant function analysis to obtain estimates of $\hat{Y}_h$. The customary goal is to identify the effective variables that are selected for inclusion in the vector $X_1, X_2, \ldots, X_p$. Despite the idiosyncrasies of sequential analytic steps, they are often used to appraise the impact of individual variables. At each step in the process, the accomplishments of each added (or removed) variable are shown by incremental changes in the value of F, or in values of a standard measure of generalized distance, called Mahalanobis D^2, that is discussed in Section A.17.2.2. of the Appendix.

For a more direct assessment of effective variables, the classification functions can be subtracted and converted to discriminant functions. Thus, the discriminant function for Groups 1 and 2 will be

$$\hat{D}_{12} = \hat{Y}_1 - \hat{Y}_2$$
$$= (b_{10} - b_{20}) + (b_{11} - b_{21})X_1 + (b_{12} - b_{22})X_2$$
$$+ \ldots + (b_{1j} - b_{2j})X_j + \ldots + (b_{1p} - b_{2p})X_p.$$

If we use d_j to represent each value of $b_{1j} - b_{2j}$, the discriminant function for Groups 1 and 2 will be

$$\hat{D}_{12} = d_0 + d_1X_1 + d_2X_2 + \ldots + d_jX_j + \ldots + d_pX_p. \qquad [17.1]$$

Analogous functions can be obtained for D_{13}, D_{23}, or any other pair of groups to be separated. The results will indicate the raw magnitude of the incremental coefficients, but not their standard errors or standardized values. Furthermore, for more than two groups, a single variable can be involved in two or more incremental coefficients. Thus, for four groups, a coefficient containing X_2 might be expressed as $b_1 - b_2$, $b_2 - b_3$, or $b_2 - b_4$, where each of these b coefficients comes from the corresponding coefficient of X_2 in $\hat{Y}_1$, $\hat{Y}_2$, $\hat{Y}_3$, and $\hat{Y}_4$.

17.7. Application to two groups

An intriguing mathematical event occurs when the discriminant function process is applied to only two groups (rather than to three or more). After variables $X_1, X_2, \ldots, X_k$ have been suitably reduced to p variables during the sequential procedures, the result produces two classification functions—one for each group. Their subtraction produces the single discriminant function shown in expression [17.1]

An alternative way to analyze these same variables for two groups, however, is to use a binary code of **0** for one group and **1** for the other, so that each category becomes a value of Y_i cited as either **0** or **1**. Despite the inappropriateness of using 0/1 target variables, such data can be analyzed with *multiple linear regression*. If the same collection of p variables is included, the multiple linear regression equation would be

$$\hat{Y}_i = b_0 + b_1X_1 + b_2X_2 + \ldots + b_jX_j + \ldots + b_pX_p.$$

17.7.1. Relationship of coefficients

The coefficients produced by the two procedures will be $d_0, d_1, d_2, \ldots, d_p$ for the discriminant function, and $b_0, b_1, b_2, \ldots, b_p$ for the multiple linear regression. When the two sets of coefficients are compared, the values may differ for the intercepts b_0 and d_0. The other coefficients for each X_j variable, however, will be either identical or will have equal ratios of d_j/b_j for each value of j. Thus, either $d_1 = b_1, d_2 = b_2, \ldots, d_j = b_j, \ldots,$ and $d_p = b_p$; or else $(d_1/b_1) = (d_2/b_2) = \ldots = (d_j/b_j) = \ldots = (d_p/b_p)$.

Lachenbruch's text[3] (pages 17–19) offers a mathematical proof of the "interesting parallel" between multiple linear regression and discriminant function analysis for two groups. Formulas for the relationship of coefficients for variables have been provided by Kleinbaum, Kupper, and Muller.[4] If Group 1 has n_1 members and Group 2 has n_2 members, so that $N = n_1 + n_2$, the values of b_j in the linear regression equation correspond to the d_j discriminant coefficients as

$$b_j = cd_j, \qquad [17.2]$$

where c is a constant, calculated as

$$c = \frac{n_1 n_2}{N(N-2)}(1 - R^2), \qquad [17.3]$$

using the R^2 of multiple linear regression. The value of Mahalanobis D^2 can be determined from the discriminant analysis directly or can be calculated from the F analysis-of-variance value in the linear regression as

$$D^2 = \frac{(N)(N-2)pF}{(n_1 n_2)(N-p-1)}. \qquad [17.4]$$

(The value of p is the number of independent variables.)

Appropriate substitution and algebraic development can also show that

$$R^2 = cD^2. \qquad [17.5]$$

17.7.2. Mathematical and pragmatic consequences

The consequence of the mathematical distinction just cited is that discriminant function analysis has often been used as a substitute for multiple linear regression (or for multiple logistic regression) in the multivariable analysis of data for two groups. While producing essentially the same b_j coefficients as linear regression, the substitution of discriminant analysis avoids the likelihood ratios and other mathematical complexities of logistic regression; and it also avoids the linear-regression risk of getting $\hat{Y}_i$ values that do not lie between 0 and 1. Nevertheless, because multiple linear regression yields exactly the same estimates, it can be applied instead of discriminant function analysis whenever the target variable is binary, containing two groups or a **0/1** event, As illustrated in Chapter 18, the linear regression procedure in this situation is easier to use and simpler to interpret.

On the other hand, linear regression and discriminant function analysis have inherent mathematical contraindications that may make them both relatively inappropriate for 0/1 data. Linear regression may be inappropriate because of the risk that $\hat{Y}_i > 1$ or $\hat{Y}_i < 0$; and discriminant function analysis may be inappropriate because it requires a multivariate Gaussian distribution, which can seldom be guaranteed for the array of X_j, variables. Consequently, logistic regression is usually the preferred *mathematical* approach as an algebraic model for a target variable expressed in 0/1 data.

17.8 Regression diagnostics

Because a single model is not produced, visual graphs cannot readily be constructed for discriminant function results. There is no conventional visual or mathematical method to check whether the variables have a consistent gradient throughout their range.

The main form of regression diagnostics is another inspection of the effect of the model, not a check of individual variables. The effect is displayed with entities called *canonical variables*. The first canonical variable is the particular linear combination that best discriminates among the groups. The second canonical variable is the second-best linear combination. (In essence, the two canonical variables condense the multiple independent variables into two composite new variables.) For each of the classified persons, the values of the two canonical variables can be plotted on a graph, and if good discrimination has been obtained, the members of similar groups will cluster together in this "biplot".

A seldom-checked mathematical feature is the tenability of the assumptions about multivariate Gaussian distributions. Because various studies have claimed that the model is robust and works well even when these assumptions are violated, the assumptions themselves are seldom examined.

(The remaining discussion of discriminant function analysis is in the Appendix to this chapter, and in the text that accompanies the illustrations in Chapter 18.)

Appendix for Chapter 17
Mathematical Background and Logic of Discriminant Function Analysis

The description that follows will go step by step through the long sequential chain of mathematical reasoning that begins with one variable for two groups, and culminates with multiple variables for multiple groups. Please sign an informed consent, fasten your seat belt, and proceed.

A.17.1. Strategies of separation

In the linear, logistic, and Cox forms of regression, we needed an index to denote the accomplishments of a particular model. In linear regression, the index usually showed the proportionate reduction in variance; and in logistic and Cox regression, the index usually referred to reductions in log likelihood. With all three forms of regression, the index was used in two ways. The $\{b_j\}$ coefficients for a particular candidate model were chosen to maximize the value of the index; and the best statistical model among the diverse candidates would be chosen as the one that had the best value for the index.

An analogous process takes place in discriminant function analysis, but the goal is not to reduce variance or log likelihood. Instead, we try to separate the target groups by maximizing the "distance" between them. This new idea, which eventually involves measuring a multivariate distance, is described in this section.

A.17.1.1. Separation for two groups and one variable

For a simple challenge that will later be extended, suppose we had two groups, A and B, with data measured for a single independent variable, X. We want to separate the groups to decide how best to assign persons, using only the value of that single X variable.

The most obvious way to separate the two groups is according to their mean values of $\overline{X}$. For each person's value of X, membership would be assigned to the group whose mean is closer to X. Thus, if one group has $\overline{X} = 7$ and the other has $\overline{X} = 21$, we would almost immediately decide that a person with $X = 19$ belongs to the latter group. This decision might be altered, however, by the spread of the data in each group. If the first group is widely spread around a mean of 7, and the second group is tightly packed around 21, we might not be so confident. For example, suppose the items in the first group are {0,0,0,1,1,2,2,11,18,20,22} and the second group consists of {20.8,20.9,21.0,21.0,21.0,21.1, 21.2}. In this situation, we might want to assign a person with a value of 19 to the first group rather than the second.

Thus, the basic decision would involve thinking about not just the locations of means, but also the standard deviations for spread of the data.

A.17.1.2. Separation for two groups and two variables

If available, a second independent variable might ease the separation process. For example, groups A and B in Fig. 17-1 would have considerable overlap if examined for only the values of X_1 or only the values of X_2. Examination of the two variables together, however, shows a sharp demarcation between groups.

A.17.1.3. Role of centroids

The distinctions in Fig. 17-1 were evident just from looking at the patterns of points. Since these visual distinctions will seldom (if ever) be so clear, a more pragmatic mathematical mechanism is to note the distances between the means or centroids for any two compared groups. The farther apart the centroids, the better separated are the groups. We therefore need a good method to measure the distance between centroids.

A.17.2. Statistical concepts of 'standard' distance

When the Pythagorean theorem is used to calculate the direct or "Euclidean" value of distance between the centroids of two groups, the result would be accurate if each value for each variable were located exactly at the mean. Thus, if Group A consisted of the items {3, 3, 3, 3,} and Group B consisted of {15, 15, 15, 15, 15} for a single variable, X, the value of $\bar{X}_A - \bar{X}_B = 3 - 15 = -12$ would be a perfect expression of the distance between the means. On the other hand, if the values for Group A were {0, 1, 5, 6} and if Group B contained {2, 7, 13, 25, and 28}, the two means would still be $\bar{X}_A = 3$ and $\bar{X}_B = 15$, but the value of $\bar{X}_A - \bar{X}_B$ alone would not account for the variability occurring in the groups.

In Section A.17.1.1, when we wanted to make a decision about a two-group distance, the spread of data for the single variable was a pertinent problem. In that particular instance, we managed the problem "intuitively", after examining the data directly. For most other problems, which cannot be solved so intuitively, a more formal mechanism is needed. This mechanism—for two groups with one variable—is analogous to something taught in elementary statistics, although it was not called a measurement of distance. It was called a Z test or t test.

A.17.2.1. One variable; two groups

In the formula for a Z or t test, the numerator is $\bar{X}_1 - \bar{X}_2$ for the difference in the means of two groups for one variable. The denominator contains their common standard deviation formed as the square root of the "pooled variance". If the sums of squares for group variances are S_{xx_1} and S_{xx_2}, and the total sample size is $N = n_1 + n_2$, the common standard deviation is

$$s_c = \sqrt{(S_{xx_1} + S_{xx_2})/(N - 2)}.$$

(The divisor is $N - 2$ rather than $N - 1$ because each group has 1 degree of freedom in its data.)

The *standard distance* between two groups for one variable is then expressed as

$$\frac{\bar{X}_1 - \bar{X}_2}{s_c} = \frac{(\bar{X}_1 - \bar{X}_2)}{\sqrt{(S_{xx_1} + S_{xx_2})/(N - 2)}}.$$

In other statistical discourse, this entity is also called the *standardized increment* or the *effect size*.

Because a t or Z test requires use of the common standard error of the means rather than the common standard deviation of the data, the squared standard error of the difference in means is calculated as

$$\frac{s_c^2}{n_1} + \frac{s_c^2}{n_2} = s_c^2 \left(\frac{n_1 + n_2}{n_1 n_2}\right) = s_c^2 \left(\frac{N}{n_1 n_2}\right).$$

Thus, the customary formula for t or Z is produced by multiplying the standardized distance by the sample-size factor, $\sqrt{n_1 n_2/N}$.

A.17.2.2. Two or more variables; two groups

For two or more variables in two groups, the standardized distance between centroids is called Mahalanobis D^2. For this non-Euclidean calculation, the common multivariate standard deviation is derived from the variance-covariance matrix of the participating variables.

When the standardized multivariable distance is converted to appropriate standard errors (rather than standard deviations and codeviations), the result is an entity called Hotelling's T^2, which is a multivariable counterpart of the t or Z test for one variable in two groups.

A.17.2.3. One variable; more than two groups

For only one variable is more than two groups, the standardized distance is a counterpart of the ratio formed in an F test. From the analysis of variance (or from some of the previous discussions of linear regression analysis), you may recall that the F ratio is determined as a ratio of mean variances among and within groups. The ratio, which corresponds to the t or Z ratio for one variable in two groups, can be used to denote a counterpart of standardized distance for a single variable among two or more groups.

A.17.2.4. Multiple variables; multiple groups

For multiple variables in multiple groups, an entity called Rao's V^2 is the multi-group counterpart of Mahalanobis D^2. For stochastic decisions, the multi-group standard multivariable distance, V^2, becomes converted to an entity that corresponds to a multivariable F ratio.

The multivariate distances are used for examining models with different candidate variables. The best model is the collection of variables that maximizes the distance among the centroids.

A.17.3. Converting variables to functional expressions

In linear, logistic, and Cox regression, we began with a single functional expression that estimated $\hat{Y}_i$ from the combination of variables, $G = b_0 + b_1X_1 + b_2X_2 + b_pX_p$. The functional expression was $\hat{Y}_i = G$ in linear regression, $\hat{Y}_i = 1/(1 + e^{-G})$ in logistic regression, and $\hat{Y}_{i,t} = S_0(t)^{e^G}$ in Cox regression. This functional expression was used for calculations that determined the best coefficients for each model, and the best model for each analysis.

In discriminant function analysis, however, a single functional expression cannot be used. Because the polytomous nominal categories cannot be ranked, they cannot be estimated from the single result that would be produced if values of the $\{X_j\}$ independent variables were substituted into a *single* algebraic model. This distinction is responsible for the extra mathematical complexity of discriminant function analysis. It requires a series of mathematical models, one for each polytomous category; and the functional expression of the models is determined with a strategy different from the method used to find maximum standard distance and the best set of variables.

The procedures require a complex set of decisions to choose discriminating boundaries, to express the concepts of classification functions, to use ideas about prevalence in forming conditional probabilities, and to develop an operational mechanism for converting diverse probabilities into the classification functions.

A.17.3.1. Choice of discriminating boundaries

For the examples in Section 17.3.1. and A.17.1.1. of this Appendix, a single discriminating line was chosen to separate two groups. When three groups are available, only two lines are needed to separate them, but three lines can be drawn. We then have to choose which two of the three lines are best.

The problem is illustrated by the three groups shown in Fig. A.17-1. In this idealized set of data, the three groups can easily be separated with line A between Groups 1 and 2, line B between Groups 1 and 3, and line C between Groups 2 and 3.

An excellent separation can be obtained, however, with only two of the three lines. We can first separate Groups 1 and 2 with line A, and then separate Groups 1 and 3 with line B.

448 Regression for Nondimensional Targets

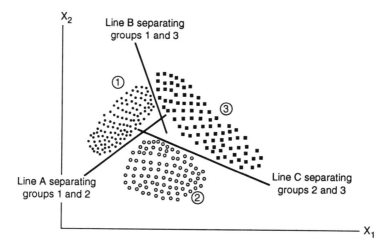

Fig. A.17-1 Locations of three groups identified with two variables, X_1 and X_2; and three lines that can be drawn to separate the three groups.

Alternatively, keeping line A, Groups 2 and 3 could be separated with line C. In yet another arrangement, we could use only line B to separate Groups 1 and 3, and line C to separate groups 2 and 3. Which pair of lines should be chosen for the separation?

The foregoing question has no immediate or standard answer. Furthermore, if four groups were being considered, at least 6 pairs [= (4 × 3)/2] of separating lines could be drawn, but a good separation should be attainable with only three boundary lines. Although the results should be the same regardless of which set of lines is chosen, the diverse possibilities preclude a simple mathematical decision about the particular lines that will be optimum discriminating functions. The examination of standardized distances will therefore identify the particular independent variables that will provide the best discrimination, but will not indicate the best set of discriminant functions.

Accordingly, when three or more categorical groups are to be separated by the analytic process, no effort is made to choose a best set of discriminant functions. Instead, a classification function is prepared for each group; and the operational strategy is aimed at finding the best classification function for that group. Afterward, as noted later, any selected pair of classification functions can be arranged, if desired, into the appropriate discriminant functions.

The foregoing discussion explains why computer programs for discriminant function analysis do not produce the coefficients of variables for discriminant functions. The coefficients that emerge are for classification functions, which can then be suitably converted into discriminant functions. Many investigators, however, may not actually try to construct the discriminant functions. Instead, as noted later, the estimates for individual persons are made directly from the classification functions; and the decisions about important variables are made from their sequential appearance in a stepped exploration.

A.17.3.2. Concepts and symbols for classification functions

A new set of symbols and coding conventions is needed for the classification functions.

Suppose we want to use liver function tests and other data as independent variables to estimate the diagnosis of liver disease. The independent variables could be X_1 = serum bilirubin, X_2 = serum glutamic oxalic transaminase (SGOT), X_3 = history of alcoholism; and so on. The four diagnostic categories in the polytomous dependent variable might be **hepatitis, cirrhosis, extrahepatic biliary obstruction,** or **other.** For coding convenience, we could designate these categories arbitrarily, like telephone numbers, as **1, 2, 3, 4** (or as $C_1, C_2, C_3,$ and C_4). What we seek is a combination of the independent variables,

$b_0 + b_1X_1 + b_2X_2 + b_3X_3 + \ldots$, that can be used to estimate the probability of each target diagnostic category.

In symbols for a set of m available categories, we can let h be a general subscript for categories (since the subscripts i and j have been reserved respectively for persons and variables). The m categories could then be designated as $C_1, \ldots, C_h, \ldots, C_m$. For any particular person, designated as i, the target variable Y_i, is membership in the group formed by one of the categories. The actual "value" of Y_i would be C_h. Thus, for a person with hepatitis, Y_i would be **hepatitis**.

A.17.3.3. Prior estimates of probability

The prior probability of a person's membership in C_h is estimated without any of the $\{X_j\}$ variables. Each of the h categories in the set of data will have n_h members, and the total group will contain $N = \Sigma n_h = n_1 + \ldots + n_h + \ldots + n_m$ members. The only basic or prior "summary" that can be used for the total group is to choose n_{max} as the largest of the n_h values. The value of $p_{max} = n_{max}/N$ would then be assigned as the best single estimate for the group. Thus, if a group of 90 people with liver disease contained 40 who had **hepatitis,** 8 with **cirrhosis,** 17 with **biliary obstruction,** and 25 with **other,** our best prior estimate for everyone would be the diagnosis of **hepatitis,** using the maximum proportion, 40/90 = .44. This prior estimate would be correct in 44% of the classifications, and wrong in 56%.

A.17.3.4. Estimates with classification functions

The classification functions use the $\{X_j\}$ variables to improve the prior estimates. If the outcome event has m categories, and if p independent variables are used, the classification function constructed at the end of the analysis would be

$$G_1 = a_0 + a_1X_1 + \ldots + a_jX_j + \ldots + a_pX_p$$
$$\vdots$$
$$G_h = h_0 + h_1X_1 + \ldots + h_jX_j + \ldots + h_pX_p$$
$$\vdots$$
$$G_m = m_0 + m_1X_1 + \ldots + m_jX_j + \ldots + m_pX_p$$

The m functions are constructed (as noted later) to make each $\hat{G}_h$ resemble a statement of probability for membership in category h. When the values of $X_1, X_2, \ldots, X_p$ for a particular person are inserted into each estimating G_h function, the corresponding values are calculated for the estimates $\hat{G}_1, \hat{G}_2, \ldots, \hat{G}_m$. One of the $\hat{G}_h$ "probabilities"—call it $\hat{G}_{max}$—will have the highest value for that person, and it becomes the best estimate of the person's correct classification. Thus, the eventual model for estimating a particular Y_i is

$$\hat{Y}_i = \hat{G}_{max}.$$

A.17.3.5. Role of prevalence

The next step is to convert the prior estimates of probability, which depended only on the frequencies of n_h, into the conditional probabilities that are the estimates for each G_h. The conversion uses the prevalence of each group, as expressed in the old medical diagnostic aphorism, "When you hear hoofbeats, think of horses before you consider zebras".

To illustrate the idea, recall the earlier challenge in Section A.17.1.1. If Group 1 is extremely common, whereas Group 2 is rare, and if the spread of data would allow the value of 19 to occur in either group, we might assign the person to Group 1 simply because of its greater prevalence, despite the shorter distance from 19 to the mean of 21 than to the mean of 6. To prepare an optimal set of classification functions, therefore, we want to use information about prevalence. This usage introduces the same concepts of conditional probability with which Bayes Theorem is applied in calculations for diagnostic marker tests.

A.17.3.6. Role of conditional probability

Suppose we had data from one variable X and wanted to find the probability of a particular category, C_h, in patients with a particular value of X. Expressed in conditional probability, the desired result would be $P(C_h|X)$. We begin by knowing the prevalence of each category, $P(C_h)$, and the prevalence, $P(X)$, for values of X. From the distribution of X among the categories, we will also know $P(X|C_h)$. Using the Bayesian formula derived in Section A.13.3, we can find the desired result as

$$P(C_h|X) = \frac{[P(X|C_h)][P(C_h)]}{P(X)} \quad [A17.1]$$

For example, in a particular set of data, $P(C_h)$ could be the probability of persons with cirrhosis; $P(X|C_h)$ could be the probability of an SGOT of 421 in persons with cirrhosis; and $P(X)$ could be the probability of persons with an SGOT of 421. $P(C_h|X)$ would then be the desired probability of cirrhosis in a person whose SGOT is 421. To illustrate the calculations, suppose the group contains 90 people, of whom 8 have cirrhosis, 40 have hepatitis, and 6 have an SGOT of 421. Among the 8 cirrhotic patients, 2 have an SGOT of 421. For these data, the value of $P(C_h)$ is 8/90 = .089; $P(X)$ is 6/90 = .067; and $P(X|C_h)$ is 2/8 = .25 Using formula (A17.1), $P(C_h|X)$ would be (.089)(.25)/.067 = .33.

A.17.3.7. Special new challenges in Bayesian components

The fundamental strategy of the preceding calculation is identical to the way Bayes Theorem is used in diagnostic marker tests, but the tactics must differ to deal with two important distinctions in the components of discriminant function analysis: (1) C_h is one of several polytomous categories, not just a dichotomous category of disease that is present or absent; and (2) X is the value of a dimensional variable, rather than the positive or negative binary result for a diagnostic marker test. The first challenge is relatively easy to manage. The second challenge, however, requires some additional mathematical strategy to find probabilities for $P(X|C_h)$ and for $P(X)$.

A.17.4. Role of Gaussian conditional probabilities

The first new challenge is to find an expression of conditional probability for $P(X|C_h)$. It will correspond to the sensitivity of a diagnostic test, but is different because C_h is one of several possible diagnostic categories and because X is a dimensional expression, not just a categorical positive or negative result. The method of solving this challenge was presented in Section A.13.5.

If variable X has a Gaussian distribution in a group, C_h, with mean $\bar{X}$ and standard deviation s, the probability of X at any point on the Gaussian curve for the C_h group is obtained from the "probability density" of the Gaussian Z-score format, which is

$$f = \frac{1}{s\sqrt{2\pi}} e^{-[(X-\bar{X})/s]^2/2} . \quad [A17.2]$$

In a standardized Gaussian curve for which $\bar{Z} = 0$ and s = 1, the complexity of [A17.2] becomes simplified to

$$f = \frac{1}{\sqrt{2\pi}} e^{-Z^2/2} .$$

(Note that f is the probability for a particular point, not the probability indicated by P values for a zone.)

In the standardized curve, when Z = 0, $e^{-(0/2)} = e^0 = 1$. Since $1/\sqrt{2\pi} = .3989$, the corresponding value of f is (.3989)(1) ≈ .4. Thus, the probability value would be .4 when Z = 0.

If Z = +1 or −1, $e^{-1/2}$ = .6065 and f = .24. If Z = +2 or −2, $e^{-4/2} = e^{-2}$ = .1353 and f = .054. These results show what you may recall about the shape of a Gaussian curve. As X gets farther and farther from the mean, $\bar{X}$, the value of Z enlarges and the value of f decreases for the probability that X "belongs" to this set of data.

To illustrate the calculations, consider the set of hepatic disease data in Section A.17.3.6. In that example, 2 of the 8 people with cirrhosis had an SGOT of 421. If X = 421, and C_h = cirrhosis, we could immediately determine that $P(X|C_h)$ was 2/8 = .25. In more extensive sets of data, however, this simplicity would not be possible. Consequently, we use the Gaussian principle to express the probability of a dimensional value in a group of data.

To apply this principle for determining $P(X|C_h)$, we assume that each of the C_h groups has a Gaussian distribution with mean $\bar{X}_h$ and standard deviation s_h. For three groups, C_1, C_2, and C_3, the respective values would be $\bar{X}_1$, s_1; $\bar{X}_2$, s_2; and $\bar{X}_3$, s_3. (These summary values would be known or promptly calculated from the data available for analysis.) We could then insert the stipulated value of X into the appropriate estimating formulas for each group h. Letting $1/\sqrt{2\pi}$ = .3989, the formula would be

$$f_h = \frac{.3989}{s_h} e^{-[(X-\bar{X}_h)/s_h]^2/2}. \qquad [A17.3]$$

When calculated for each of the three groups 1, 2, and 3, these values of f_1, f_2, and f_3 would provide the probabilities for $P(X|C_h)$.

To illustrate the latter calculations, suppose we want to estimate the probability that X = 16 occurs within each of three possible groups, each having a standard deviation of 3. The respective means for the three groups are $\bar{X}_1$ = 15, $\bar{X}_2$ = 20, and $\bar{X}_3$ = 24. When calculated with formula [A17.3], the value for each f_h will be

$$f_1 = .126, f_2 = .055, \text{ and } f_3 = .0038.$$

The values are highest in Group 1, because X = 16 is closer to $\bar{X}_1$ = 15 than to the means of the other two groups.

A.17.5. Calculation of classification probabilities

With the Gaussian mechanism available to produce P(X|Ch), we can now develop the algebraic expressions to calculate each classification function as $P(C_h|X)$. To simplify the algebraic symbols, let

$$\begin{aligned} p_h &= P(C_h), \\ f_h &= P(X|C_h), \\ p_h f_h &= P(C_h)P(X|C_h), \\ \Sigma p_h f_h &= \Sigma P(C_h)P(X|C_h) = P(X), \text{ and} \\ \hat{Y}_h &= P(C_h|X). \end{aligned}$$

Substituting the cited symbols into formula [A17.1], we get

$$\hat{Y}_h = \frac{p_h f_h}{\Sigma p_h f_h}. \qquad [A17.4]$$

For a given value of X, we could use formula [A17.4] to calculate and examine the values of $\hat{Y}_h$ for each of the h groups (which are C_1, C_2, and C_3 in this instance). With the best estimated classification, X would be assigned to the group for which $\hat{Y}_h$ has the highest probability.

The actual algebraic expressions can be formidable, but the process can be eased with some simplifying procedures.

A.17.5.1. Calculation of $p_h f_h$

We need not actually calculate the sum, $\Sigma p_h f_h$, which is the same denominator for all three groups. The maximum value of $\hat{Y}_h$ can be found simply by noting the largest $p_h f_h$ value among $p_1 f_1$, $p_2 f_2$, and $p_3 f_3$.

For estimating the location of X = 16 in the three groups under discussion, where s = 3, we can recall that f_1 = .126, f_2 = .055, and f_3 = .0038 when the respective values $\bar{X}_1$ = 15, $\bar{X}_2$ = 20, and $\bar{X}_3$ = 24 were placed into formula [A17.3] in Section A.17.4. If n_1 = 50, n_2 = 150, and n_3 = 30, the values of p_h would be p_1 = 50/230 = .217, p_2 = 150/230 = .652, and p_3 = 30/230 = .130. The values for $p_h f_h$ would be

$$p_1 f_1 = (.217)(.126) = .027$$
$$p_2 f_2 = (.652)(.055) = .036$$
$$p_3 f_3 = (.130)(.0038) = .000494.$$

From the maximum value of these expressions alone, we could estimate that $\hat{Y}_h$ is category 2. If we wanted to get a value for $\Sigma(p_h f_h)$, however, it would be .027 + .036 + .00049 = .06349

A.17.5.2. Subsequent calculations

After the Gaussian strategy produced f_h as the conditional probability of $P(X|C_h)$, the rest of the calculations become relatively straightforward. We need to find $P(C_h)$ and $P(X)$, and we can then substitute those values into formula [A17.1] to obtain the desired value of $P(C_h|X)$, which is the classification function we are looking for.

A.17.5.2.1. Finding $P(C_h)$ [=p_h]

$P(C_h)$ is easy to determine. It corresponds to $P(D)$ in a diagnostic marker test, being the prior probability of membership in the hth group. Thus, $P(C_h)$ is

$$p_h = n_h/N.$$

In the previous cited liver disease group, which contained 40 people with hepatitis, the prior probability for hepatitis was 40/90 = .44. The prior probability for cirrhosis was 8/90 = .089.

A.17.5.2.2. Finding $P(X)$ [= $\Sigma p_h f_h$]

If you read Section A.13.3, you may recall the complexity of determining $P(X)$ for its use in Bayesian formula [A13.2]. In that situation we knew $P(D)$ for the prevalence of the disease, and we knew the test's sensitivity and specificity, which were respectively $P(X|D)$ and $P(\bar{X}|\bar{D})$. Formula [A13.3] for determining $P(X)$ was really $\Sigma P(D)P(X|D)$, which became $P(D)P(X|D) + P(\bar{D})P(\bar{X}|\bar{D})$ for two groups.

A similar tactic is used to determine $P(X)$ when we know each value of $P(C_h)$ and of $P(X|C_h)$. In this instance, the appropriate sum is $P(X) = \Sigma P(C_h)P(X|C_h)$ across each of the C_h categories. In the liver disease example under discussion, the expression for $P(X|C_h)$ would be the probability of individual values of SGOT in each of the cited diagnostic groups. Accordingly, with the Gaussian formula that offers suitable expressions for each $P(X|C_h)$, and with the summary expressions for each $P(C_h)$, we can calculate $P(X) = \Sigma P(C_h)P(X|C_h)$. We would then have everything needed to solve formula [A17.1] for $P(C_h|X)$, or its counterpart in the simpler formula [A17.4] for $\hat{Y}_h$.

A.17.5.2.3. Calculating $\hat{Y}_h$

In the example under discussion, the probability for each $\hat{Y}_h$ would be estimated as

$$\hat{Y}_1 = p_1 f_1 / \Sigma(p_h f_h) = .027/.06349 = .425$$
$$\hat{Y}_2 = p_2 f_2 / \Sigma(p_h f_h) = .036/.06349 = .567$$
$$\hat{Y}_3 = p_3 f_3 / \Sigma(p_h f_h) = .000494/.06349 = .008$$

Because the maximum of these probabilities is $\hat{Y}_2$, we would assign group 2 as the estimate for X. [Note that the three values for $P(C_h|X)$ add up to a sum of 1, as they should for the total of the conditional probabilities, since X must belong to one of the three groups.] Since the expression $\Sigma(p_h f_h)$ is the same in all three calculations of $P(C_h|X)$, this same estimate could have been made using only the values of $p_h f_h$, without calculating $p_h f_h / \Sigma(p_h f_h)$.

A.17.6. Role of group separations

In calculating the values of f_h, we first determined the distance, $X - \overline{X}_h$, between X and the mean of each group, $\overline{X}_h$. The value of $X - \overline{X}_h$ was used because it is required by the Gaussian formula, and also because each group was represented by its mean, $\overline{X}_h$.

The chance of getting an accurate assignment for X will be increased if the group means are as widely separated as possible. For example, with $\overline{X}_1 = 15$, $\overline{X}_2 = 20$, and $\overline{X}_3 = 24$ in the foregoing example, a value of $X = 22$ could readily belong to either Group 2 or 3, and might even belong to Group 1. If the respective means, however, were $\overline{X}_1 = 3$, $\overline{X}_2 = 20$, and $\overline{X}_3 = 47$, the value of $X = 22$ is highly likely to belong only to Group 2.

If only a single X variable is available for the analysis, the means of the groups are what they are, and nothing can be done about them. If multiple X_j variables are available, however, the multivariate centroids will depend on the particular combination of variables that are included in the algebraic model. The best model will come from including the particular collection of variables that maximizes the distance between the centroids.

A.17.7. Algebraic expression for classification and discriminant functions

Now that the basic ideas and operations have been illustrated, we can reach the mathematical consummation that converts centroids, distances, Gaussian probabilities, and conditional probabilities into algebraic expressions for the desired classification and discriminant functions.

A.17.7.1. Logarithms of $p_h f_h$

In Section A.17.5.1, when searching for the maximum value of $p_h f_h$, we could have expressed each appropriate value as

$$p_h f_h = (p_h) \left\{ \frac{.3989}{s_h} e^{-[(X-\overline{X}_h)/s_h]^2/2} \right\}. \qquad [A17.5]$$

The calculations could be eased by taking logarithms, so that expression [A.17.5.] becomes

$$\ln (p_h f_h) = \ln (p_h) + \ln (.3989) - \ln (s_h) - \frac{[X - \overline{X}_h]^2}{2s_h^2}.$$

The last term in this expression can be expanded to form

$$\ln (p_h f_h) = \ln (p_h) + \ln (.3989) - \ln (s_h) - \frac{X^2}{2s_h^2} + \frac{\overline{X}_h X}{s_h^2} - \frac{\overline{X}_h^2}{2s_h^2}. \qquad [A17.6]$$

Since the values of p_h, s_h, and $\overline{X}_h$ are all known and constant, this expression has the form of a quadratic equation

$$\ln(p_h f_h) = a + bX + cX^2.$$

It can readily be calculated for each group C_h, and for any given value of X.

A.17.7.2. Conversion to discriminant functions

Because the largest value of $\hat{Y}_h$ depends on the largest value of $p_h f_h$, one way of deciding whether p_1f_1 is larger than p_2f_2 and p_3f_3 is to examine the ratios p_1f_1/p_2f_2, and p_1f_1/p_3f_3. This process is equivalent to subtracting the corresponding logarithms.

An important underlying mathematical assumption for discriminant function analysis is that each of the categorical groups has similar values of variance for X. In other words, s_h is assumed to be essentially the same for all three groups. With this assumption, when we subtract the logarithms for any two groups, the terms in ln (.3989), ln (s_h), and $X^2/2s_h^2$ will vanish in formula [A17.6]. We could therefore discriminate between p_1f_1 and p_2f_2 by finding the incremental function D_{12}, which is

$$D_{12} = \ln p_1f_1 - \ln p_2f_2 = \ln p_1 - \ln p_2 + \frac{(\bar{X}_1 - \bar{X}_2)X}{s_h^2} - \frac{(\bar{X}_1^2 - \bar{X}_2^2)}{2s_h^2}.$$

This expression can be rearranged into

$$D_{12} = \ln \left(\frac{p_1}{p_2}\right) - \frac{(\bar{X}_1^2 - \bar{X}_2^2)}{2s_h^2} + \frac{(\bar{X}_1 - \bar{X}_2)X}{s_h^2} \quad [A17.7]$$

Since the values are constant for $\ln(p_1/p_2)$ and for $(\bar{X}_1^2 - \bar{X}_2^2)/2s_h^2$, the expression for D_{12} is a straight line expressed in the form of a + bX for the variable X. The intercept of the line is

$$\ln \left(\frac{p_1}{p_2}\right) - \frac{(\bar{X}_1^2 - \bar{X}_2^2)}{2s_h^2}$$

and the slope, or coefficient for X, is $(\bar{X}_1 - \bar{X}_2)/s_h^2$.

An analogous expression could be calculated as D_{13} to discriminate Group 1 from Group 3. If we wanted, we could also calculate D_{23} to discriminate Group 2 from Group 3.

The expressions for D_{12}, D_{13}, and D_{23} are *discriminant functions*. We need only two of them, since the maximum value for p_hf_h could be found from evaluating any two of the comparisons for D_{12} and D_{13}, for D_{12} and D_{23}, or for D_{13} and D_{23}.

A.17.7.3.. Resemblance to logistic regression for two groups

On careful examination, equation [A17.7] for a two-group separation has a close resemblance to the values found in the Appendix of Chapter 13 when P(X|D) in equation [A13.6] was converted to the form of $e^{-(a+bX)}$. For two groups, if we let p_1 and p_2 be replaced by p and 1 − p, and D and $\bar{D}$ be identified as Groups 1 and 2, the values of a and b in equation [A13.9] and [A13.10] form an arrangement identical to that given for D_{12} in equation [A17.7].

Thus, the discriminant function D_{12} has a value identical to the logistic regression $P(D|X) = 1/[1 + e^{-(a+bX)}]$ in equation [A13.11].

A.17.7.4. Expression of classification functions

A simpler alternative for finding the maximum of p_hf_h, however, is to avoid calculating the discriminant functions and to use the logarithms of p_hf_h directly as *classification functions*. In the discriminant functions, three of the component terms of ln (p_hf_h) in expression [A17.5] will vanish. Those three terms are ln (.3989), ln (s_h), and $X^2/2s_h^2$. These terms can therefore be ignored in the classification functions, which can be derived from ln (p_hf_h), and expressed in working form, as

$$\hat{Y}_h = \ln (p_h) - \frac{\bar{X}_h^2}{2s_h^2} + \frac{\bar{X}_h X}{s_h^2}. \quad [A17.8]$$

In the previous examples for Groups 1, 2, and 3, the classification functions would be calculated with formula [A17.8] as follows:

$$\hat{Y}_1 = \ln(.217) - \frac{15^2}{2(3^2)} + \frac{15}{3^2} X = -1.528 - 12.5 + 1.667X;$$

$$= -14.028 + 1.667X;$$

$$\hat{Y}_2 = \ln(.652) - \frac{20^2}{2(3^2)} + \frac{15}{3^2} X = 22.650 + 2.222X;$$

and

$$\hat{Y}_3 = \ln(.130) - \frac{24^2}{2(3^2)} + \frac{24}{3^2} X = -34.040 + 2.667X.$$

When we insert X = 16 into these three functions, we get $\hat{Y}_1$ = 12.644, $\hat{Y}_2$ = 12.902, and $\hat{Y}_3$ = 8.632. These expressions no longer indicate probability, but their magnitudes would correspond to those found in Section A.17.5.2.3 with the $p_h f_h$ formats. The largest of the $\hat{Y}_h$ values here is $\hat{Y}_2$; we would assign X to Group 2; and the assignment would be the same as what occurred earlier.

A fundamental point to note about formula [A17.8] is its relative simplicity. If given a value of $\bar{X}$, we can promptly find $\hat{Y}_h$ from the known values of p_h, X_h, and s_h. All of the cumbersome symbols of Gaussian and conditional probabilities may have been used in the background reasoning, but they have disappeared in the final formula.

A.17.8. Extension to multiple variables

If you understand what has just transpired, you have grasped the basic strategy of discriminant function analysis. All the rest of the complexity arises when we advance from a single independent variable, X, to a set (or *vector*) of k independent variables: $X_1, X_2, \ldots, X_k$. All of the unfamiliar new terms, symbols, and eponyms are produced by the multiplicity of those independent variables.

A.17.8.1. Change from X to vector **X**

Instead of having a single value of X for the classifications, we have a vector of variables,

$$\mathbf{X} = (X_1, X_2, \ldots, X_j, \ldots, X_k).$$

For the mathematical operations, we assume that the data for each X_j variable have a Gaussian distribution, with similar variance-covariance matrixes, and that the combination of variables has a multivariate Gaussian distribution.

(This assumption is almost never true in medical reality, where a really Gaussian variable is seldom encountered, let alone a whole array of them; and even if each of the variables were individually Gaussian, their collective distribution would hardly be multivariate Gaussian. Nevertheless, to allow the discussion to proceed, the splendor of the mathematical strategy will not be interrupted by pedantic reminders of reality.)

A.17.8.2. Change from means to centroids

With a single variable, X, each of the h categorical groups has a mean $\bar{X}_h$. With a vector of k variables, **X**, each group has a multivariate centroid, expressed as

$$\bar{\mathbf{X}}_h = (\bar{X}_1, \bar{X}_2, \ldots, \bar{X}_k).$$

For different choices of p variables, the location of these centroids will vary in ways that may help or hinder the effort to separate the groups. The details of the complex formula used to calculate the inter-centroid distances as Mahalanobis D^2 (or Rao's V^2) are not needed for the discourse here. The main point to bear in mind is that the value of D^2 (or V^2) can be eventually converted to an F value that indicates the effectiveness with which the centroids separate the groups. (The conversion mechanism for two groups is discussed in Section 17.7.1.)

A.17.8.3. Uses of D² and F indexes

Sequential procedures can be used to explore the centroids for models that contain 1, 2, 3, ..., or all k of the available k variables. Since the values of D^2 and F will denote the best-separating centroids, these values (or appropriately derived other indexes) can be used to make decisions about adding or deleting individual variables. The decisions will determine whether the multivariate vector should consist of

$$\mathbf{X} = [X_1, X_2, X_3, \ldots, X_k]$$
or $\quad \mathbf{X} = [X_1, X_3, X_9, X_{14}]$
or $\quad \mathbf{X} = [X_1, X_7, X_{12}]$

or some other arrangement of p variables chosen from the available X_j variables. The process is computationally complex because the variances under consideration represent combinations of variances and covariances for the individual variables (X_j) and the individual groups (C_h).

The main point to bear in mind is that D^2 and F help determine the collection of variables for the presumably best separation of centroids. This best separation should presumably lead to the best set of discriminant and classification functions. Nevertheless, D^2 and F do not produce coefficients for the variables $X_1, X_2, \ldots, X_p$; and D^2 and F do not indicate the effectiveness of the estimates produced by those functions. The coefficients, the estimates, and the efficacy of the estimates are determined in a different way.

This distinction is another major difference between operational strategies in discriminant function analysis and in the other three (linear, logistic, and Cox) algebraic models. In the other three models, the indexes that led to F values or log likelihood ratios had an "omnibus" role. They were used to choose the included variables {X_j}, to fit the coefficients b_j, and to index the accomplishments of the model. In discriminant function procedures, the multivariate index F is used only to choose the variables. The b_j coefficients are chosen in a different way; and yet a third way is used to express the accomplishments of the estimates.

A.17.8.4. Formation of multivariate classification functions

After the appropriate set of p variables ($X_1, \ldots, X_p$) has been chosen for inclusion, the coefficients attached to the variables are selected according to the classification function procedure described in Section A.17.7.4. The procedure is now much more complicated than before, however, because it contains a vector $\mathbf{X}$, rather than a single variable X.

For a single variable X, each classification function was obtained from an expression that has been reduced to

$$\hat{Y}_h = \ln p_h - \frac{\overline{X}_h^2}{2s_h^2} + \frac{\overline{X}_h X}{s_h^2}$$

The expression had the form of a line

$$\hat{Y}_h = a_h + b_h X,$$

where a_h was the intercept for group h and b_h was the coefficient for the single variable X.

The multivariate expression for the vector of variables, $\mathbf{X} = X_1, X_2, \ldots, X_p$, will involve standard multivariate distances for each $\mathbf{X}$ from the centroid $\overline{\mathbf{X}}$ of the variables, using the total array of variances and covariances to find the multivariate standard deviation vectors, s_h, that are used in the standard adjustments. The eventual form of the multivariate classification function is a "surface", rather than a line. The probability for ($C_h | \mathbf{X}$) is estimated from the surface

$$\hat{Y}_h = b_{h0} + b_{h1} X_1 + b_{h2} X_2 + \ldots + b_{hp} X_p.$$

Chapter References

1. Fisher, 1936; 2. Barnard, 1935; 3. Lachenbruch, 1975; 4. Kleinbaum, 1988, pp. 566–68.

18 Illustrations of Discriminant Function Analysis

Outline ....
18.1. Separation of two groups
   18.1.1. Full regression
      18.1.1.1. No frills seven-variable model
      18.1.1.2. Additional options for seven-variable model
      18.1.1.3. Displays of grouped data
   18.1.2. Stepwise regression
      18.1.2.1. SAS program
      18.1.2.2. BMDP program
   18.1.3. Comparison with linear regression
      18.1.3.1. Discriminant function results
      18.1.3.2. Results of multiple linear regression
      18.1.3.3. Conversion of coefficients
18.2. Separation of four groups
   18.2.1. SAS program
   18.2.2. BMDP program
18.3. Illustrations from published literature
   18.3.1. Two-group separations
   18.3.2. Polytomous separations
   18.3.3. Additional comments
....

For reasons discussed in Section 17.1, discriminant function analysis is not common in medical research today. The main reasons for displaying the procedure in this chapter are to preserve the "paired" discussion used for regression methods, and to provide some real-world examples for readers who may use or see results from the discriminant method. As in previous chapters, the first set of displays comes from the illustrative data set, and the others come from published medical literature.

The illustrative data set is used for two types of demonstration. First, discriminant function analysis does a two-group separation for patients **dead** and **alive** at 6 months; and the results are directly compared with those obtained with multiple linear regression for the same data. Afterward, discriminant analysis is applied to a polytomous target: categories of the four histologic cell-type groups. The rest of the chapter contains examples and discussions of medically published reports of discriminant function analysis.

18.1. Separation of two groups

The management of polytomous nominal targets is a distinctive ability of discriminant function analysis, but the first challenge shown here is the separation of two groups. A binary separation is commonly done when discriminant analysis is reported in medical literature, but the task can be equally accomplished with multiple linear regression. To allow the two procedures to be compared for a binary challenge, the target variable (marked SURV6) in the illustrative data set was dichotomously divided as patients **alive** or **dead** at six months after zero time.

To avoid relatively noncontributory mathematical complexity, the SAS and BMDP programs were applied in a no-frills manner, without any of the available special options. Users who want those options can find them described and illustrated in the pertinent instruction manuals.[1,2]

18.1.1. Full regression

The "full regression" process with seven independent variables is shown here only for the SAS PROC DISCRIM method.

18.1.1.1. No frills seven-variable model

With seven independent variables in the model, the no-frills SAS PROC DISCRIM printout is shown in Fig. 18-1.

The top part of the printout indicates basic data about number of observations, variables, classes, and degrees of freedom (DF). "Class Level Information" gives the frequency of observations, their actual proportions, and the default prior probabilities of .50 assigned to each of the two outcome groups (**alive** and **dead**). The rank and natural log of the determinant of the covariance matrix are used in quadratic discrimination, and will not be further discussed here. The generalized squared distance computations show values of **0** when the observed and estimated groups are the same, and 1.062 when the two groups differ.

The bottom of Fig. 18-1 shows the coefficients of each variable in the classification functions. For the **alive** group, the function can be expressed (in no more than three decimal places) as

$\hat{Y}_{alive}$ = −68.7 + .909 symptom stage + .365 percent weight loss + 1.88 hematocrit + .047 TNM stage + .783 age + 5.43 male + .062 progression interval.

For the **dead** group, the classification function is

$\hat{Y}_{dead}$ = −67.76 + 1.33 symptom stage + .362 percent weight loss + 1.79 hematocrit + .516 TNM stage + .778 age + 6.24 male + .057 progression interval.

To get a single discriminant function, these two classification functions would be subtracted to produce

$\hat{Y}_{alive}$ − dead = −.942 − 0.422 symptom stage + .004 percent weight loss + .095 hematocrit − .469 TNM stage + .004 age − .806 male − .005 progression interval.

The printout shows no indexes (see Section 17.5.1.) for the accomplishment of the model, and no standard error or P values for the coefficients. From results of previous analyses, we can anticipate that many of the coefficients are not stochastically significant.

Fig. 18-2 shows that the estimates produced by the classification functions in Fig. 18-1 were correct for 61 members of the **alive** group and for 78 of the **dead** group.

```
                    Discriminant Analysis

        200 Observations         199 DF Total
          7 Variables            198 DF Within Classes
          2 Classes                1 DF Between Classes

                 Class Level Information

                                                    Prior
    SURV6    Frequency      Weight    Proportion   Probability

    ALIVE        94        94.0000    0.470000      0.500000
    DEAD        106       106.0000    0.530000      0.500000


    Discriminant Analysis    Pooled Covariance Matrix Information

         Covariance         Natural Log of the Determinant
         Matrix Rank        of the Covariance Matrix

              7                      15.6838767


Discriminant Analysis    Pairwise Generalized Squared Distances Between Groups

              2                  -1
           D (i|j) = (x̄ - x̄ )' COV   (x̄ - x̄ )
                       i   j            i   j

                   Generalized Squared Distance to SURV6
         From
         SURV6         ALIVE              DEAD

         ALIVE           0              1.06200
         DEAD         1.06200              0


    Discriminant Analysis     Linear Discriminant Function

                       -1                              -1
     Constant = -.5 x̄' COV  x̄      Coefficient Vector = COV  x̄
                    j        j                                j

                            SURV6
                       ALIVE         DEAD

         CONSTANT    -68.70336     -67.76121
         SXSTAGE       0.90935       1.33099
         PCTWTLOS      0.36523       0.36151
         HCT           1.88477       1.79002
         TNMSTAGE      0.04739       0.51617
         AGE           0.78253       0.77824
         MALE          5.43222       6.23828
         PROGIN        0.06152       0.05686
```

Fig. 18-1 Printout for SAS PROC DISCRIM program applied to seven variables for a 2-group separation.

The overall accuracy of estimation was thus 139/200 = 69.5% The rate of errors was 33/94 = .35 in members of the **alive** group and 28/106 = .26 in the **dead** group.

18.1.1.2. Additional options for seven-variable model

The SAS PROC DISCRIM procedure offers many additional options beyond the no-frills results just shown. Some of the options are mentioned here so that they will not be total strangers if you ever meet them.

```
Discriminant Analysis      Classification Summary for Calibration Data: WORK.TEMP1
              Resubstitution Summary using Linear Discriminant Function
```

Generalized Squared Distance Function: Posterior Probability of Membership in each SURV6:

$$D_j^2(X) = (X-\bar{X}_j)' COV^{-1} (X-\bar{X}_j)$$ $$Pr(j|X) = \exp(-.5\, D_j^2(X)) / \sum_k \exp(-.5\, D_k^2(X))$$

Number of Observations and Percent Classified into SURV6:

From SURV6	ALIVE	DEAD	Total
ALIVE	61	33	94
	64.89	35.11	100.00
DEAD	28	78	106
	26.42	73.58	100.00
Total	89	111	200
Percent	44.50	55.50	100.00
Priors	0.5000	0.5000	

Error Count Estimates for SURV6:

	ALIVE	DEAD	Total
Rate	0.3511	0.2642	0.3076
Priors	0.5000	0.5000	

Fig. 18-2 Accuracy of assignments by the classification functions developed in Fig. 18-1. (SAS PROC DISCRIM program.)

Nonparametric operations: The basic discriminant-analysis procedure is usually done with the parametric method described in Chapter 17, but nonparametric methods can be substituted if users are worried about the violation of parametric assumptions. The SAS manual[1] describes five ways of using a nonparametric "kernal" method and also a method based on "nearest neighbor" concepts. The conventional parametric technique is the default method if nothing else is specified.

Computational aim: The calculations can be aimed at a goal called canonical discriminant analysis, or can be left to be done, by default, in the conventional manner.

Specification of prior probabilities: In the default method of operation, the prior probabilities for each of the m groups are set equal to 1/m, e.g., .5 for two groups and .25 for four groups. Because each of the m groups to be separated has n_h members, with $N = \Sigma n_h$, the prior probabilities could optionally and perhaps preferably be set at the proportional values of n_h/N for each group.

Individual observations and estimates: The optional printout can produce a long list, also not shown here, that identifies each patient, the associated independent variables for that patient, the corresponding probability value calculated for each **alive** or **dead** classification, and the category estimated for the patient.

Additional matrixes: The printout can also optionally show matrixes for the sums of squares and cross products (marked SSCP), the covariances, and the correlation coefficients of the independent variables.

Additional indexes: In addition to the standard indexes for the model (discussed in Section 18.1.2), the full SAS options include two additional indexes of accomplishment: the *Hotelling-Lawley Trace* and *Roy's Greatest Root*. Although excellent acquisitions for your vocabulary of statistical eponyms, these indexes have almost no additional pragmatic value, and they are extremely difficult to introduce into ordinary social conversation.

18.1.1.3. Displays of grouped data

A useful SAS option shows displays of univariate results arranged for the total array of observations, for each class (e.g., **dead** or **alive**), for a pooled within-class group, and a pooled between-class group. In the full set of data and in each class, the univariate statistics (N, sum, mean, variance, standard deviation) are summarized for each independent variable, and the standardized class means are calculated for the total sample and for a pooled within-class arrangement.

In Fig. 18-3, the different means and standard deviations can be used to check the basic assumption that s_h is similar for each X_j variable, which has a mean $\bar{X}_{h_j}$ in each compared group, as well as a grand mean $\bar{X}_j$ in the total collection of patients. For each variable in each group, the corresponding standard deviations, s_h, for individual persons are calculated as deviations $(X_{ij} - \bar{X}_j)$ from the grand mean, $\bar{X}_j$; these are called "Total-Sample". The deviations $(X_{ij} - \bar{X}_{h_j})$ are also cal-

```
                    Discriminant Analysis      Simple Statistics

                              Total-Sample

Variable            N            Sum            Mean         Variance        Std Dev

SXSTAGE            200        523.00000        2.61500        1.13244         1.06416
PCTWTLOS           200            1653         8.26700       64.11217         8.00701
HCT                200            8218        41.09000       26.82523         5.17931
TNMSTAGE           200        643.00000        3.21500        2.66209         1.63159
AGE                200           12217        61.08500       89.89726         9.48142
MALE               200        182.00000        0.91000        0.08231         0.28690
PROGIN             200            2883        14.41400      332.98151        18.24778

-------------------------------------------------------------------------------

                              SURV6 = ALIVE

Variable            N            Sum            Mean         Variance        Std Dev

SXSTAGE             94        214.00000        2.27660        0.93342         0.96614
PCTWTLOS            94        669.90000        7.12660       64.11423         8.00714
HCT                 94            3947        41.99149       22.23606         4.71551
TNMSTAGE            94        240.00000        2.55319        2.50789         1.58363
AGE                 94            5728        60.93617       86.06040         9.27687
MALE                94         83.00000        0.88298        0.10444         0.32317
PROGIN              94            1508        16.04787      379.62639        19.48400

-------------------------------------------------------------------------------

                              SURV6 = DEAD

Variable            N            Sum            Mean         Variance        Std Dev

SXSTAGE            106        309.00000        2.91509        1.12606         1.06116
PCTWTLOS           106        983.50000        9.27830       62.52419         7.90722
HCT                106            4271        40.29057       29.77267         5.45643
TNMSTAGE           106        403.00000        3.80189        2.08419         1.44367
AGE                106            6489        61.21698       94.11438         9.70126
MALE               106         99.00000        0.93396        0.06226         0.24953
PROGIN             106            1374        12.96509      290.32953        17.03906
```

Fig. 18-3 SAS PROC DISCRIM optional printout for univariate results of each independent variable in total sample and in each group.

culated from the mean of the groups marked ALIVE and DEAD for SURV6. If the standard deviations for each variable appear reasonably close both in the total sample and within the classes, the assumption seems confirmed that the s_h values are similar. The results in Fig. 18-3 show that the standard deviations are relatively similar for the three sets of calculations for each independent variable.

```
                    Discriminant Analysis

              Total-Sample Standardized Class Means

           Variable              ALIVE              DEAD

           SXSTAGE          -.3180011969        0.2820010614
           PCTWTLOS         -.1424257728        0.1263021004
           HCT               0.1740559645       -.1543515157
           TNMSTAGE         -.4056218614        0.3597024054
           AGE              -.0156969990        0.0139199802
           MALE             -.0941836360        0.0835213376
           PROGIN            0.0895381384       -.0794017453


        Pooled Within-Class Standardized Class Means

           Variable              ALIVE              DEAD

           SXSTAGE          -.3325404629        0.2948943727
           PCTWTLOS         -.1433693719        0.1271388770
           HCT               0.1760105609       -.1560848370
           TNMSTAGE         -.4379861108        0.3884027775
           AGE              -.0156592291        0.0138864862
           MALE             -.0943202775        0.0836425103
           PROGIN            0.0896336847       -.0794864751
```

Fig. 18-4 SAS PROC DISCRIM optional printout showing standardized values for variables in each group.

In Fig. 18-4, for the Total-Sample Standardized Class Means, the value of $\bar{X}_{h_j} - \bar{X}_j$ for each variable is divided by the total-sample standard deviation. For the set of Pooled Within-Class Standardized Class Means, the divisor is the pooled standard deviation within classes. A comparison of the incremental results in the two groups of total-sample standardized class means can immediately suggest which variables might be effective. Thus, the (rounded) increment for TNM stage in the dead-alive groups is $.360 - (-.406) = .766$. The corresponding increments are $.282 - (-.318) = .600$ for symptom stage and $-.154 - .174 = -.328$ for hematocrit, whereas the corresponding increments are only $-.079 - (.090) = -.169$ for progression interval and $.014 - (-.016) = .030$ for age.

18.1.2. Stepwise regression

The stepwise regression results are shown for both the SAS and BMDP procedures.

18.1.2.1. SAS program

In steps 1 and 2 (which are not displayed) the printout of the stepwise SAS PROC STEPDISC procedure begins by testing each variable for entry in the model and determining the corresponding values of R^2, F, P, and tolerance. (When only two groups are under consideration, values of R^2 can be converted to Mahalanobis D^2 according to formula [17.5] in Section 17.7.1.)

After the best first contender, TNM stage, was entered into the model, Wilks lambda became 0.853 and Pillai's trace, 0.147. Both of these statistical indexes had the same F value, and a corresponding P value of .0001. The average squared canonical correlation (ASCC) had the same value as Pillai's trace because $m - 1 = 1$ with two groups. In step 2, after an unsuccessful attempt to remove TNMSTAGE, the SAS program entered hematocrit as the next variable, having an F value of 7.097. After entry of hematocrit, Wilks lambda dropped to .824, and Pillai's trace rose to .176.

```
                           Stepwise Discriminant Analysis
Stepwise Selection: Step 3
                          Statistics for Entry, DF = 1, 196
                            Partial
                  Variable  R**2              F           Prob > F     Tolerance

                  SXSTAGE   0.0302          6.107          0.0143       0.7985
                  PCTWTLOS  0.0018          0.349          0.5553       0.8995
                  AGE       0.0004          0.088          0.7672       0.9817
                  MALE      0.0091          1.806          0.1805       0.9922
                  PROGIN    0.0029          0.575          0.4491       0.9926

                            Variable SXSTAGE will be entered

                    The following variable(s) have been entered:
                              SXSTAGE   HCT      TNMSTAGE

                                  Multivariate Statistics

              Wilks' Lambda  = 0.79879862    F( 3, 196) =  16.456    Prob > F = 0.0001
              Pillai's Trace =  0.201201     F( 3, 196) =  16.456    Prob > F = 0.0001
                       Average Squared Canonical Correlation = 0.20120138
```

```
Stepwise Selection: Step 4
                          Statistics for Removal, DF = 1, 196
                            Partial
                  Variable  R**2              F           Prob > F

                  SXSTAGE   0.0302          6.107          0.0143
                  HCT       0.0402          8.219          0.0046
                  TNMSTAGE  0.0871         18.707          0.0001

                             No variables can be removed

                          Statistics for Entry, DF = 1, 195
                            Partial
                  Variable  R**2              F           Prob > F     Tolerance

                  PCTWTLOS  0.0000          0.008          0.9300       0.7373
                  AGE       0.0001          0.011          0.9159       0.7935
                  MALE      0.0103          2.029          0.1559       0.7980
                  PROGIN    0.0016          0.320          0.5725       0.7930

                             No variables can be entered

No further steps are possible
```

```
Stepwise Selection: Summary
                                                                                   Average
                                                                                   Squared
              Variable            Number   Partial     F       Prob >   Wilks'    Prob <   Canonical     Prob >
      Step  Entered  Removed       In      R**2      Statistic   F      Lambda    Lambda   Correlation   ASCC

       1    TNMSTAGE                1      0.1466    34.023    0.0001   0.85336366  0.0001  0.14663634   0.0001
       2    HCT                     2      0.0348     7.097    0.0084   0.82368864  0.0001  0.17631136   0.0001
       3    SXSTAGE                 3      0.0302     6.107    0.0143   0.79879862  0.0001  0.20120138   0.0001
```

Fig. 18-5 Step 3 through conclusion of printout for SAS PROC STEPDISC program.

Fig. 18-5 shows the rest of the SAS stepwise operations. After SXSTAGE is added in step 3, none of the entered variables can be removed, and no additional variables can be entered in step 4. With the entry of SXSTAGE, Wilks lambda fell to

0.799 and Pillai's trace rose to 0.201. The bottom of the printout shows a summary for the sequence of entry; and the important variables can also be ranked according to their partial R^2 values. The SAS STEPDISC printout concludes, however, without identifying any coefficients for constructing classification functions or a discriminant function. Getting the desired coefficients would require the other SAS discriminant analysis program, PROC DISCRIM. On the other hand, PROC DISCRIM will provide the coefficients, but will not show their statistical accomplishments. Thus, both programs would be needed to get a descriptive and evaluatory set of results.

18.1.2.2. BMDP program

In step 0 of the corresponding BMDP7M stepwise procedure for two groups, all seven variables were tested for F-to-enter values and tolerance. The no-frills printout in Fig. 18-6 shows the TNM stage was entered in step 1, with a U statistic (Wilks lambda) of .853 and the corresponding approximate F statistic of 34.023 for the model. Adjacent to the entered value of TNM stage in step 1 is a column of the remaining variables being tested for possible entry in step 2. When only two groups are being separated, the F-MATRIX value (34.02) is the same as the APPROXIMATE F-STATISTIC. Immediately afterward, the printout shows the coefficients for the classification functions that might be prepared at this stage. In step 2, HCT is entered since it had the largest F-to-enter after step 1.

Step 3, with entry of SXSTAGE, is shown at the top of Fig. 18-7. Each of the three selected variables is accompanied by a high value of F-to-remove, and none of the remaining variables has a high value of F to enter. The U statistic falls to 0.80, and the approximate F-statistic and F-matrix fall to 16.46. The latter value, for two groups, is an equivalent of the Hotelling T^2 test. After listing the coefficients for the classification functions, the printout shows a table of right-wrong results for the classification matrix and also a similar table for classifications obtained with the jackknife method. The last part of Fig. 18-8 shows a summary of the preceding stepwise results. Note that the BMDP7M stepwise program, unlike the corresponding SAS PROC STEPDISC, cites the coefficients for the classification functions.

After completing this work, the BMDP7M no-frills program produces five pages of printout (not shown here) that do not occur in the corresponding SAS program. For each patient in both groups, the printout shows the incorrect classifications, the patient's Mahalanobis D-square value from the **dead** and **alive** groups, each patient's posterior probability estimate for each group, and additional information (discussed in the BMDP manual[2] and usually not particularly important) regarding canonical variables, EIGENVALUES, and graphs of canonical variables.

18.1.3. Comparison with linear regression

To avoid the complexity of pairwise comparisons for seven independent variables, only the three main independent variables will be used to contrast the linear regression and discriminant function procedures for separating two groups.

466 Regression for Nondimensional Targets

```
*******************************************
STEP NUMBER   1
VARIABLE ENTERED    4 TNMSTAGE

              F TO   FORCE TOLERNCE *
VARIABLE     REMOVE  LEVEL           *      VARIABLE       F TO    FORCE TOLERNCE
             DF =  1  198            *                     ENTER   LEVEL
 4 TNMSTAGE   34.02     1  1.00000   *                DF = 1  197
                                     *      2 AGE          0.00      1     0.99936
                                     *      3 MALE         1.24      1     0.99993
                                     *      5 SXSTAGE      4.99      1     0.86021
                                     *      6 PCTWTLOS     1.77      1     0.99349
                                     *      7 HCT          7.10      1     0.99159
                                     *      8 PROGIN       0.75      1     0.99814

U-STATISTIC(WILKS' LAMBDA) 0.8533636             DEGREES OF FREEDOM  1    1   198
APPROXIMATE F-STATISTIC      34.023              DEGREES OF FREEDOM  1.00    198.00

F - MATRIX            DEGREES OF FREEDOM  =  1   198
           DEAD
ALIVE     34.02

CLASSIFICATION FUNCTIONS

VARIABLE     GROUP =    DEAD        ALIVE
 4 TNMSTAGE            1.66516      1.11825

CONSTANT              -3.85851     -2.12070

*******************************************
STEP NUMBER   2
VARIABLE ENTERED    7 HCT

              F TO   FORCE TOLERNCE *
VARIABLE     REMOVE  LEVEL           *      VARIABLE       F TO    FORCE TOLERNCE
             DF =  1  197            *                     ENTER   LEVEL
 4 TNMSTAGE   35.71     1  0.99159   *                DF = 1  196
 7 HCT         7.10     1  0.99159   *      2 AGE          0.09      1     0.98150
                                     *      3 MALE         1.81      1     0.99101
                                     *      5 SXSTAGE      6.11      1     0.85103
                                     *      6 PCTWTLOS     0.35      1     0.91445
                                     *      8 PROGIN       0.57      1     0.99686

U-STATISTIC(WILKS' LAMBDA) 0.8236886             DEGREES OF FREEDOM  2    1   198
APPROXIMATE F-STATISTIC      21.084              DEGREES OF FREEDOM  2.00    197.00

F - MATRIX            DEGREES OF FREEDOM  =  2   197
           DEAD
ALIVE     21.08

CLASSIFICATION FUNCTIONS

VARIABLE     GROUP =    DEAD        ALIVE
 4 TNMSTAGE            1.19774      0.62587
 7 HCT                 1.50348      1.58379

CONSTANT             -33.25797    -34.74500
```

Fig. 18-6 Steps 1 and 2 in BMDP7M stepwise discriminant analysis program.

```
*******************************************
STEP NUMBER   3
VARIABLE ENTERED    5 SXSTAGE

              F TO   FORCE TOLERNCE *
VARIABLE     REMOVE  LEVEL           *      VARIABLE       F TO    FORCE TOLERNCE
             DF =  1  196            *                     ENTER   LEVEL
 4 TNMSTAGE   18.71     1  0.85831   *                DF = 1  195
 5 SXSTAGE     6.11     1  0.85103   *      2 AGE          0.01      1     0.97575
 7 HCT         8.22     1  0.98101   *      3 MALE         2.03      1     0.98923
                                     *      6 PCTWTLOS     0.01      1     0.84582
                                     *      8 PROGIN       0.32      1     0.99131

U-STATISTIC(WILKS' LAMBDA) 0.7987985             DEGREES OF FREEDOM  3    1   198
APPROXIMATE F-STATISTIC      16.456              DEGREES OF FREEDOM  3.00    196.00

F - MATRIX            DEGREES OF FREEDOM  =  3   196
           DEAD
ALIVE     16.46

CLASSIFICATION FUNCTIONS

VARIABLE     GROUP =    DEAD        ALIVE
 4 TNMSTAGE            0.86012      0.38979
 5 SXSTAGE             1.37322      0.96019
 7 HCT                 1.47723      1.56544

CONSTANT             -34.08891    -35.15129

CLASSIFICATION MATRIX

GROUP        PERCENT    NUMBER OF CASES CLASSIFIED INTO GROUP  -
             CORRECT     DEAD     ALIVE
DEAD          70.8        75        31
ALIVE         63.8        34        60

TOTAL         67.5       109        91

JACKKNIFED CLASSIFICATION

GROUP        PERCENT    NUMBER OF CASES CLASSIFIED INTO GROUP  -
             CORRECT     DEAD     ALIVE
DEAD          70.8        75        31
ALIVE         63.8        34        60

TOTAL         67.5       109        91

SUMMARY TABLE
                                                                        DEGREES
        VARIABLE         F VALUE TO    NO. OF                              OF
STEP    ENTERED          ENTER OR      VARIAB.           APPROXIMATE     FREEDOM
NO.     REMOVED          REMOVE        INCLUDED  U-STATISTIC  F-STATISTIC
  1   4 TNMSTAGE          34.023          1       0.8534       34.023    1.0  198.0
  2   7 HCT                7.097          2       0.8237       21.084    2.0  197.0
  3   5 SXSTAGE            6.107          3       0.7988       16.456    3.0  196.0
```

Fig. 18-7 Step 3 and conclusion of BMDP7M stepwise program begun in Fig. 18-6.

18.1.3.1. Discriminant function results

The discriminant function results for the three main variables were shown in Fig. 18-7 for the BMDP7M program. The classification-function coefficients, and the discriminant coefficients that emerged from their subtractions, were as follows:

Variable	Classification coefficient		"Discriminant" coefficient
	Dead	Alive	= Alive − Dead
Constant	−34.08891	−35.15129	−1.06238
TNMSTAGE	0.86012	0.38979	−0.47033
SXSTAGE	1.37322	0.96019	−0.41303
HCT	1.47723	1.56544	0.08821

The estimated classifications had a sensitivity of 64% (60/94) in the living patients, and a specificity of 71% (75/106) in the dead patients. The overall accuracy, which was 53% before development of the discriminant model, was raised to 67.5% (135/200).

```
                    Analysis of Variance

                        Sum of          Mean
Source         DF       Squares         Square       F Value     Prob>F

Model           3       10.02385        3.34128      16.456      0.0001
Error         196       39.79615        0.20304
C Total       199       49.82000

     Root MSE        0.45060      R-square      0.2012
     Dep Mean        0.47000      Adj R-sq      0.1890
     C.V.           95.87263

                    Parameter Estimates

               Parameter       Standard      T for H0:
Variable  DF    Estimate         Error       Parameter=0    Prob > |T|

INTERCEP   1   0.262513       0.26468127        0.992        0.3225
TNMSTAGE   1  -0.094531       0.02185605       -4.325        0.0001
HCT        1   0.017729       0.00618417        2.867        0.0046
SXSTAGE    1  -0.083015       0.03359181       -2.471        0.0143
```

Fig. 18-8 Multiple linear regression with three independent variables and dead/alive (at 6 months) outcome variable. (SAS PROC REG program.)

18.1.3.2. Results of multiple linear regression

Fig. 18-8 shows what happened when the same three independent variables were entered into a SAS PROC REG multiple linear regression aimed at 6-month survival, coded for the target variable as **0** = dead and **1** = alive. The F value of 16.46 here is identical to the F value obtained for Wilks lambda at the end of step 3 (in Figs. 18-5 and 18-7) for the same three variables in the discriminant analysis. The value of R^2 = .2012 for the model here is also identical to the average squared canonical correlation for the same model after step 3 in the discriminant analysis of Fig. 18-5.

The intercepts and coefficients produced in the linear regression analysis of Fig. 18-8 are listed in the first column below. The corresponding results of the discriminant function analysis are shown in the second column. The ratio of the two sets of coefficients (Discriminant/Linear) is shown in the third column:

Variable	Linear regression coefficient	Discriminant function coefficient	Ratio: discriminant ÷ linear
Intercept	0.26251	−1.06236	−4.0469
TNMSTAGE	−0.09453	−0.47033	4.9755
SXSTAGE	−0.08302	−0.41303	4.9751
HCT	0.01773	0.08821	4.9752

As noted earlier (in Chapter 17), the coefficients differ for the intercepts in the two analytic methods, but the coefficients of the *independent* variables have identical ratios (except for minor rounding discrepancies). Each method would produce exactly the same set of estimates. If the variables were standardized, i.e., expressed as $(X_j - \bar{X})/s_j$ for each variable, the intercept coefficients in the two methods would remain different although the other coefficients would retain constant ratios.

The two sets of results show that if only two groups are to be analyzed as the target variable, the multiple linear regression method is simpler and easier to understand than discriminant function analysis. The latter may have a better mathematical provenance, but the final results are the same.

18.1.3.3. Conversion of coefficients

Section 17.7.1. contained several formulas that allow back and forth conversion of the coefficients for linear regression and discriminant analyses of two target groups having sizes n_1 and n_2, with $N = n_1 + n_2$. Using formula [17.2], and a factor, c, calculated as shown in formula [17.3], the b_j values of the regression coefficients can be converted into the d_j values of the discriminant function coefficients as

$$d_j = \frac{b_j}{c}. \qquad [18.1]$$

For the data in Fig. 18-8, where $R^2 = .2012$, with $n_1 = 94$, $n_2 = 106$, and $N = 200$, the value of c from formula [17.3] will be $[(94)(106)/(200)(198)] [1 - .2012] = .20099$. The value of 1/c will be 4.975. This is exactly the value noted earlier for the ratio of discriminant ÷ linear regression coefficients.

18.2. Separation of four groups

The next set of illustrations shows a more "classical" application of discriminant function analysis, with the target variable being the four polytomous histologic (cell-type) groups in the lung cancer data set. For this challenge, the seven independent variables were used to estimate each patient's cell type, rather than survival.

18.2.1. SAS program

The *stepwise* SAS discriminant analysis of the seven variables for four groups is not shown here. TNMSTAGE was entered in the first step with F = 9.703, P = .0001, and a partial R^2 of .1293. SXSTAGE was entered in the second step, with F = 3.477, P = .0170 at that step, and partial R^2 of .0508. With the default criterion set at P =

.15, the variable AGE entered in step 3, with F = 2.026, P = .166, and partial R^2 of .0304. At this point for the model, Wilks lambda was .801, Pillai's trace was .204, and ASCC, which in this instance is Pillai's trace divided by 3, was .0679. Since no variables could be removed after step 3, and none could be entered thereafter, the stepwise procedure ended before step 4. No coefficients were supplied for the three variables identified as important.

In the SAS *full* discriminant analysis, using all seven independent variables, the printout began by using its default mechanism to list equal prior probabilities of .25 for each of the WELL, SMALL, ANAP, and CYTOLOGY ONLY groups, which actually had 83, 24, 74, and 19 members, respectively. The next part of the printout (not shown here) listed the univariate statistics for the seven independent variables in each of the four groups. Fig. 18-9 shows the rest of the printout. The upper section indicates the distance from each of the four cell-type groups (marked in the four rows on the left) to each of the corresponding four groups formed in the classification functions. The bottom section of the printout in Fig. 18-9 gives the coefficients for the seven variables in each of the four groups. Unfortunately, the PROC DISCRIM printout does not provide any F values, P values, or other indications of each variable's accomplishment.

Discriminant Analysis   Pairwise Generalized Squared Distances Between Groups

$$D^2(i|j) = (\bar{X}_i - \bar{X}_j)' COV^{-1} (\bar{X}_i - \bar{X}_j)$$

Generalized Squared Distance to HISTOL

From HISTOL	ANAP	CYT ONLY	SMALL	WELL
ANAP	0	0.62337	0.68431	0.59802
CYT ONLY	0.62337	0	0.85153	1.14958
SMALL	0.68431	0.85153	0	1.82682
WELL	0.59802	1.14958	1.82682	0

Discriminant Analysis   Linear Discriminant Function

$$\text{Constant} = -.5 \bar{X}_j' COV^{-1} \bar{X}_j \qquad \text{Coefficient Vector} = COV^{-1} \bar{X}_j$$

HISTOL

	ANAP	CYT ONLY	SMALL	WELL
CONSTANT	-69.65054	-74.05787	-76.46369	-68.54374
SXSTAGE	1.77697	1.74919	2.10825	1.36730
PCTWTLOS	0.36579	0.37688	0.41317	0.38404
HCT	1.83244	1.87236	1.86996	1.82600
TNMSTAGE	0.80329	0.86205	0.99792	0.49442
AGE	0.79230	0.85962	0.83433	0.80580
MALE	5.67283	3.77372	5.94448	5.68784
PROGIN	0.05379	0.05133	0.06916	0.06438

Fig. 18-9 Full discriminant analysis, with four histologic categories being estimated from seven independent variables. (SAS PROC DISCRIM program.)

Fig. 18-10 shows accuracy in the classification table for the estimates made with the preceding coefficients. The numbers of correct estimates were 20 for ANAP, 8 for CYT ONLY, 13 for SMALL, and 48 for WELL. The total number of correct predictions is 89, an unimpressive batting average of 44.5%, hardly better than the

```
            Discriminant Analysis        Classification Summary for Calibration Data: WORK.TEMP1
                           Resubstitution Summary using Linear Discriminant Function
  Generalized Squared Distance Function:        Posterior Probability of Membership in each HISTOL:
    2           _     -1    _                                    2                    2
   D (X) = (X-X )' COV   (X-X )              Pr(j|X) = exp(-.5 D (X)) / SUM exp(-.5 D (X))
    j           j           j                                    j       k            k
```

```
                          Number of Observations and Percent Classified into HISTOL:

         From HISTOL        ANAP         CYT ONLY        SMALL          WELL          Total

              ANAP            20             14            20            20             74
                           27.03          18.92         27.03         27.03         100.00

          CYT ONLY             2              8             7             2             19
                           10.53          42.11         36.84         10.53         100.00

             SMALL             3              4            13             4             24
                           12.50          16.67         54.17         16.67         100.00

              WELL            10             11            14            48             83
                           12.05          13.25         16.87         57.83         100.00

             Total            35             37            54            74            200
           Percent         17.50          18.50         27.00         37.00         100.00

            Priors         0.2500         0.2500        0.2500        0.2500
```

```
                          Error Count Estimates for HISTOL:

                          ANAP         CYT ONLY        SMALL          WELL          Total

              Rate        0.7297        0.5789        0.4583        0.4217        0.5472

             Priors       0.2500        0.2500        0.2500        0.2500
```

Fig. 18-10 Classification accuracy for classification functions in Fig. 18-9. (SAS PROC DISCRIM program.)

41.5% accuracy obtainable if everyone had merely been estimated as having the WELL cell type. (In separate results not shown here, the discriminant analysis was repeated with the group prior probabilities set at their actual p_h values of .415, .120, .370, and .095. The total number of correct predictions rose slightly but unimpressively to 102.) Probably the best explanation for these results is that cell type cannot be effectively estimated with discriminant function analysis for the available independent variables.

A more disconcerting finding, however, is discovering again that two SAS programs are required to obtain both the coefficients and indexes of their impact. The full discriminant analysis program (PROC DISCRIM) will show coefficients but not impact. The SAS stepwise program (PROC STEPDISC) will show impact but not coefficients.

18.2.2. BMDP program

When the BMDP7M program was applied for this same challenge, the printout began with details not shown here: identifying comments; and a list of means, standard deviations, and coefficients of variation for each of the seven variables in each of the four cell-type groups.

Fig. 18-11 shows the main results of the BMDP printout. It begins at step 0 with efforts to find a first variable for entry. Only two candidates (TNMSTAGE and SXSTAGE) have high F values. When TNMSTAGE is entered in step 1, the F level for SXSTAGE drops to 3.48, and no additional variables can be entered since they do not pass

```
STEP NUMBER    0

   VARIABLE     F TO    FORCE TOLERNCE  *  VARIABLE     F TO    FORCE TOLERNCE
                REMOVE  LEVEL           *               ENTER   LEVEL
        DF =  3    197                  *        DF =  3    196
                                        *   2 AGE       1.98    1    1.00000
                                        *   3 MALE      1.39    1    1.00000
                                        *   4 TNMSTAGE  9.70    1    1.00000
                                        *   5 SXSTAGE   8.96    1    1.00000
                                        *   6 PCTWTLOS  1.30    1    1.00000
                                        *   7 HCT       0.29    1    1.00000
                                        *   8 PROGIN    0.91    1    1.00000

••••••••••••••••••••••••••••••••••••••••••••••••••••••••••••••••••••••••••••

STEP NUMBER    1
VARIABLE ENTERED    4 TNMSTAGE

   VARIABLE     F TO    FORCE TOLERNCE  *  VARIABLE     F TO    FORCE TOLERNCE
                REMOVE  LEVEL           *               ENTER   LEVEL
        DF =  3    196                  *        DF =  3    195
   4 TNMSTAGE    9.70    1    1.00000   *   2 AGE       1.93    1    0.99972
                                        *   3 MALE      1.39    1    0.99807
                                        *   5 SXSTAGE   3.48    1    0.86623
                                        *   6 PCTWTLOS  0.89    1    0.99030
                                        *   7 HCT       0.26    1    0.99999
                                        *   8 PROGIN    0.73    1    0.99745



U-STATISTIC(WILKS' LAMBDA)  0.8706881    DEGREES OF FREEDOM    1      3      196
APPROXIMATE F-STATISTIC         9.703    DEGREES OF FREEDOM    3.00   196.00

  F - MATRIX         DEGREES OF FREEDOM =    1    196

            WELL      SMALL     ANAP
  SMALL    21.33
  ANAP     14.92      3.72
  CYTOL     7.46      1.50      0.09

CLASSIFICATION FUNCTIONS

        GROUP =    WELL        SMALL        ANAP        CYTOL
  VARIABLE
   4 TNMSTAGE    1.09048      1.78825      1.49300      1.54317

  CONSTANT     -2.78553     -5.14907     -4.00913     -4.18836

CLASSIFICATION MATRIX

  GROUP     PERCENT    NUMBER OF CASES CLASSIFIED INTO GROUP -
            CORRECT
                       WELL     SMALL    ANAP     CYTOL
  WELL      67.5       56       27       0        0
  SMALL     79.2        5       19       0        0
  ANAP       0.0       29       45       0        0
  CYTOL      0.0        6       13       0        0

  TOTAL     37.5       96      104       0        0

JACKKNIFED CLASSIFICATION

  GROUP     PERCENT    NUMBER OF CASES CLASSIFIED INTO GROUP -
            CORRECT
                       WELL     SMALL    ANAP     CYTOL
  WELL      67.5       56       27       0        0
  SMALL     79.2        5       19       0        0
  ANAP       0.0       29       45       0        0
  CYTOL      0.0        6       13       0        0

  TOTAL     37.5       96      104       0        0
```

Fig. 18-11 Stepwise discriminant analysis with BMDP7M program in separating the four cell-type groups of the illustrative data set.

the BMDP default criterion for entry. (Differences in the default criterion are the reason the SAS stepwise program entered two variables whereas the BMDP corresponding program entered only one.)

The BMDP program concludes by citing the U statistic, its approximate F value, and the F matrix, which now shows the set of pairwise F values based on D^2 or Hotelling's T^2. The coefficients are next shown for the classification functions, using TNMSTAGE and a constant, for each of the four groups. This listing is followed by the classification table matrix and the jackknifed classifications—both of which arrived at 75 (37.5%) correct classifications for the total group. This result is actually worse than what would have occurred merely by assigning everyone to the WELL group.

18.3. Illustrations from published literature

As noted at the beginning of Chapter 17, discriminant function analysis has had a sharp reduction in popularity during the past decade; and a major revival or renascence of the procedure does not seem likely. Consequently, the discussion that follows contains only brief remarks about the way discriminant analysis was used in a few of the publications cited in the tables of Chapter 17.

18.3.1. Two-group separations

In a study of clonidine treatment, thirty-four children with constitutional growth delay (i.e., short stature) were classified as responders (C-R) and nonresponders. The authors[3] then applied discriminant function analysis to obtain "thresholds" for velocity (HV) and for a standard deviation score (SDS) from normal height. According to the text, discriminant analysis allowed children to "be classified with confidence of 80% as C-R" when their HV and SDS values were below certain thresholds. No additional data were reported for methods or details of the analysis.

Applying discriminant analysis to "registry data from the Interinstitutional Coronary Artery Surgery Study", Vlietstra et al.[4] wanted to determine precursor risk factors for the presence or absence of arteriographically significant coronary disease. "To minimize the effect of nonlinear interactions between variables that were considered important", the discriminant analyses were done separately in nine age-sex subgroups. The eleven independent variables consisted of dimensional variables for weight, height, and plasma cholesterol level, and eight binary variables for current, former, or never cigarette smokers, family history, and four possibly predisposing diseases, such as diabetes mellitus. The discriminant functions that predicted presence or absence of coronary disease in each of the nine subgroups are shown in the Appendix of the published paper. After presenting the effects of two frequently identified risk factors (smoking and high cholesterol) in a targeted cluster display, the authors concluded that the discriminant function analysis had produced only "a small gain . . . of limited clinical value" in diagnostic classification. Relatively few details were cited for the actual operation of the discriminant procedure.

To separate surgically proved primary hyperparathyroidism from other causes of hypercalcemia, Lafferty[5] examined serum calcium, phosphorus, and chloride levels as well as serum parathormone as independent variables in a discriminant analysis. The discriminant functions obtained for different combinations of variables were shown in a quasi-graph of discriminant scores for separating the two groups. The author contended that the predictive results have "98% accuracy if patients taking diuretics or vomiting were omitted".

Dolgin et al.[6] entered demographic and anthropometric data as well as various morphologic characteristics of gallstones to discriminate between pigmented and cholesterol gallstones. The accuracy of the discriminant classification scores for the two groups was portrayed in a quasi-graph as well as in a relatively novel graph that converted the discriminant score to a probability that each stone might have a cholesterol composition.

Using a series of measurements of monoclonal antibodies in discriminant analysis to predict whether rabies virus came from terrestrial (raccoon, skunk) or nonterrestrial (bat) sources, Rupprecht et al.[7] reported accuracy of 100% for the reference sample, 96% for a jackknifed classification, and 100% for a separate independent "validation" sample. The discriminant results were presented with parameter estimates for the classification functions and a canonical variate, as well as a histogram for the canonical variate.

18.3.2. Polytomous separations

Watson and Croft[8] analyzed ten polytomous categories of medical specialty choices in relation to 37 predictor variables noted at a time when 291 physicians had just completed medical school. The predictor variables included demographic and financial attributes of the medical students as well as MCAT (medical college admissions test) scores and certain characteristics of the students' premedical and medical education. The published text shows many details of the analytic procedure. After concluding that a "classification of physicians into 10 groups is not feasible", the authors believed that medical school admission committees might be helped in choosing "among several equally well qualified applicants" because the "classification functions may be used to predict specialty choice of the medical school applicants". You can read the paper to draw your own conclusion about the merits of this recommendation, and also about the wisdom of initially believing that a discriminant function analysis might successfully predict a single choice among at least ten different types of medical specialty.

Overall et al.[9] examined demographic attributes, psychiatric diagnoses, and symptom profile patterns in relation to three main forms of psychotropic treatment given to 661 outpatients in an adult psychiatric clinic. The analysis produced "two orthogonal discriminant functions which provide maximum separation between the three major treatment groups". The authors also formed "a geometric representation of the results which can serve as a decision model". No data were presented for the success of the discriminant function model in choosing the same treatments as the psychiatrists.

18.3.3. Additional comments

A browse through other published literature reveals some old comments that seem to have stood the test of time, and others that have not. An anthropological battle about discriminating the fossil canine teeth of humans and chimpanzees was partly resolved when two prominent statisticians[10] (F. Yates and M. J. R. Healy) obtained the raw data, repeated the analyses, and pointed out errors in the previous calculations. The statisticians concluded that the previously "published examples must be regarded more as examples of method than as practical contributions to comparative anatomy".

Responding to the paper by Fraser (see Table 17-3) in 1971, Winkel and Juhl[11] analyzed an extensive set of data on liver function tests, trying to separate cirrhosis of the liver from noncirrhosis. After noting major differences in variances of the variables, the authors said that "linear discriminant analysis at first sight appears to be perfectly suited but . . . some of the requirements of the analysis are certainly not fulfilled. Since some multivariate statistical methods require weighty assumptions, it should be emphasized that they may occasionally fail to bring about optimal solutions of differential diagnostic problems". The events of the past twenty years have shown that Winkel and Juhl were perhaps excessively optimistic in citing the frequency of failure as only *occasional*.

Another set of optimistic hopes that have not been confirmed by reality was expressed by Ryback et al.[12] after they reported accuracy rates ranging from 89% to 100% in identifying four diagnostic categories via quadratic discriminant analysis of "25 commonly ordered laboratory tests" for their paper cited in Table 17-3. Unlike the linear format of discriminant analysis, the quadratic format does not require homogeneous variances among the variables. This attribute of the quadratic method, together with the extraordinarily (and unexplained) high rate of diagnostic success, led Ryback et al. to claim that quadratic discriminant analysis "holds the most promise for differentiating among similar medical conditions . . . and can derive accurate classifications even when there are no differences between means". Nevertheless, more than a decade later, quadratic discriminant analysis is almost never used in medical diagnosis.

Chapter References

1. SAS/STAT User's Guide, 1990; 2. Jennrich, 1990, pp. 339–58; 3. Pintor, 1987; 4. Vlietstra, 1980; 5. Lafferty, 1981; 6. Dolgin, 1981; 7. Rupprecht, 1987; 8. Watson, 1978; 9. Overall, 1972; 10. Yates, 1951; 11. Winkel, 1971; 12. Ryback, 1982.

Part V Targeted-Cluster Methods

Targeted categorical-cluster methods are not particularly well known or popular today, because they lack both the mathematical cachet of algebraic models and the ready availability of automated computation. Nevertheless, many readers and investigators who were previously unfamiliar with these methods become highly enthusiastic after seeing the results, and often wonder why the methods are not taught or used more often.

One reason for the enthusiasm is that categorical output is direct and easy to understand. The results are immediately evident, requiring no conversions into odds ratios, risk ratios, or complex additive scores. Another scientific attraction is that categorical methods offer a "gold standard" for checking the results of algebraic analyses. To determine whether the linear models have suitably conformed to the contents of ranked data, whether important interactions have occurred, and whether the subsequent arrangements are an undistorted portrait of the data, the best approach is to examine the data directly. The targeted-cluster methods allow this direct examination.

The targeted-cluster procedures are also becoming increasingly easy to apply. For methods that can currently be automated, appropriate computer programs are much more generally available than in the past. For methods that use non-automated judgments, the progressively widespread availability of personal computers makes the data easy to explore, arrange, and evaluate.

The procedures are presented in the next three chapters. Chapter 19 describes basic operational principles for carrying out targeted-cluster methods and for evaluating results. Chapter 20 discusses the multicategorical procedures that allow decisions to be made with both substantive and statistical judgment. Chapter 21 presents the recursive partitioning techniques that operate with automated statistical algorithms.

19 Basic Strategies of Cluster Operations

Outline . . . .

19.1. Demarcation of multivariate categories
    19.1.1. Expression of categories
    19.1.2. Venn-diagram illustrations
    19.1.3. Nomenclature for tabulations
    19.1.4. Targeted and nontargeted tabulations
19.2. Differences in cluster and algebraic methods
    19.2.1. Illustrations of estimates
    19.2.2. Important variables and transcending categories
19.3. Mistaken beliefs about multicategorical analysis
    19.3.1. Loss of information
    19.3.2. Lack of mathematical model
    19.3.3. Absence of automation
    19.3.4. Failure to 'weight' variables
    19.3.5. Problems in sample size
    19.3.6. General misconceptions
19.4. Basic strategies of multicategorical analysis
    19.4.1. Collapsing tables (log-linear models)
    19.4.2. Aggregating categories (recursive partitioning)
    19.4.3. Segregating categories (sequential sequestration)
    19.4.4. Consolidating variables (conjunctive consolidation)
19.5. Additional methods of evaluation
    19.5.1. Collapsing categories
        19.5.1.1. Biologic 'homogeneity'
        19.5.1.2. Statistical 'homogeneity'
        19.5.1.3. Considerations of trend
        19.5.1.4. 'Terminal unions'
            19.5.1.4.1. Closed arrangement
            19.5.1.4.2. Open arrangement
            19.5.1.4.3. Criteria for terminal unions
    19.5.2. Appraising linear trend
        19.5.2.1. Linear model for slope
        19.5.2.2. Quantitative significance for slope
        19.5.2.3. Standardized slope
        19.5.2.4. Stochastic tests for 'stability'
            19.5.2.4.1. Chi-square test for linear trend
            19.5.2.4.2. Chi-square test for residual nonlinearity
    19.5.3. Polytomous targets
    19.5.4. Use of a cost matrix
        19.5.4.1. Operation of cost matrix
        19.5.4.2. Advantages and disadvantages
    19.5.5. Appraising 'survival curves'
19.6. Scores and partitions from algebraic analyses
    19.6.1. Count of factors

478 *Targeted-Cluster Methods*

    19.6.2. Arbitrary point score
    19.6.3. Multivariable formula score
    19.6.4. Uncondensed cross-partition of variables
        19.6.4.1. 'Bar-graph' arrangement
        19.6.4.2. 'Tree' arrangement
    19.6.5. Stratified arrangements
    19.6.6. Other strategies and tactics
19.7. Advantages of stratified evaluations
    19.7.1. Baseline imbalances in distribution of strata
    19.7.2. Pre- vs. postadjustment
    19.7.3. Intrinsic value of stratification

. . . .

Targeted-cluster analyses are aimed at determining the effects of multivariate categories on dependent variables.

19.1. Demarcation of multivariate categories

A *multivariate category*, such as **tall old men**, contains a composite collection of categories from two or more variables, such as *height*, *age*, and *sex*. The multivariate conjunction can be formed with Boolean logic as a simple *intersection* of several binary attributes, such as **tall** and **old** and **male**, or as a simple *union* of attributes, such as **tall** and/or **old** and/or **male**. A few distinctions of simple Boolean unions and intersections are shown in Fig. 19-1.

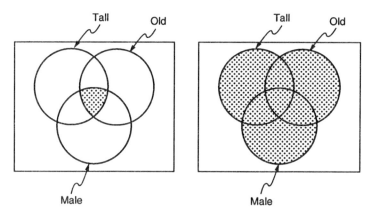

Fig. 19-1 Multivariate conjunction of three attributes. Shaded area shows the composite category formed, on the left, by their intersection, and on the right, by their union.

A multivariate category can also be formed as a complex union or intersection of categories that were previously formed as simple Boolean unions or intersections. To allow diverse forms of combination, the term *cluster* can be used for any multivariate category, regardless of how it was constructed. An example of a complex cluster is the conjunction of **old men** and/or **short women** and/or **tall young women**.

When the analysis begins, the coded values for each variable can be cited in continuous dimensions or in discrete binary, ordinal, or nominal categories. Before or during the analysis, dimensional variables such as *age* may be demarcated into categories that form either a binary scale such as **< 60/≧ 60**, or an ordinal scale, such as **< 40, 40–49, 50–59, 60–69, ≧ 70**.

The analysis produces an array of *terminal clusters*, containing locations for each of the multivariate categories that have been formed from the data. The final clusters can seem *simple*, having apparently one constituent category, or overtly *composite*, with two or more components. Thus, **age < 60** years may seem to be a simple terminal category, but it can actually be a composite cluster that includes all persons who are < 60 years of age, together with whatever their additional categories may be in any of a series of other variables such as *sex, height, weight, clinical stage, diagnosis,* etc.

19.1.1. Expression of categories

To illustrate the arrangements, consider three binary variables, with elemental categories X_1: a_1, a_2; X_2: b_1, b_2; and X_3: c_1, c_2.

The staging system for cancer described in Section 4.6.1 came from three variables, demarcated as follows:

X_1 = evidence of primary tumor: a_1 = absent; a_2 = present
X_2 = evidence of regional spread: b_1 = absent; b_2 = present
X_3 = evidence of distant spread: c_1 = absent; c_2 = present.

(Binary categories are usually coded as **0** for **absent** and **1** for **present**.)

In Boolean logic, the symbol ∩ represents an intersection, and ∪ represents a union. The stages in Section 4.6.1 were defined as the following Boolean intersections:

Stage I : $a_2 \cap c_1 \cap b_1$ [i.e., localized]
Stage II : $b_2 \cap c_1$ [i.e., regional but not distant spread]
Stage III: c_2 [i.e., distant spread]

These demarcations do not seem to include all the possible composite categories, because many of them were not pertinent for the demarcations. Thus, Stage III exists as long as c_2 is **present** in variable X_3, regardless of the status of X_1 or X_2. The same arrangement, shown in a more complete but less efficient demarcation, presents all of the possible categories as

Stage I : $a_2 \cap c_1 \cap b_1$
Stage II : $b_2 \cap c_1 \cap (a_1 \cup a_2)$
Stage III: $c_2 \cap (a_1 \cup a_2) \cap (b_1 \cup b_2)$

In this example, where all of the variables are binary, we could also demarcate categories by letting the symbols X_j represent **presence** and $\bar{X}_j$ represent **absence** of the particular attribute for binary variable X_j. The stages could then be defined as

Stage I : $X_1 \cap \bar{X}_2 \cap \bar{X}_3$
Stage II : $X_2 \cap \bar{X}_3$
Stage III: X_3

Although eight multivariate categories ($= 2^3$) would be expected, the definitions here account for only seven—one in Stage I, two in Stage II, and four in Stage III. The "missing" category represents the absence of all three attributes—a situation that does not occur when variables such as *TNM stage* are coded with ranked grades, such as **1, 2, 3,** . . . for a condition that is always present. (As discussed in Chapter 5, a *null-based* coding could be used for a variable such as *severity of pain*, in which the attribute of pain is either absent [denoted by **0**], or present in grades of magnitude [**1, 2, 3,** . . .]. In TNM classification, stage **0** is sometimes used for *carcinoma in situ*.)

19.1.2. Venn-diagram illustrations

The Venn diagrams in Figs. 4-1 and 4-2 showed the formation of seven multivariate categories from the three binary variables, and the clustering of categories to form three stages.

Because variables that contain more than two categories are not easily displayed with circles, rectangles are used in Fig. 19-2 to show the sixteen possible bivariate categories formed by conjunction of two variables, X_4 and X_5, each having four ordinal coded categories, marked, respectively, as d_1, d_2, d_3, d_4 and e_1, e_2, e_3, e_4.

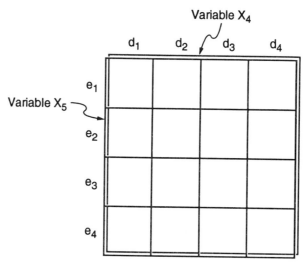

Fig. 19-2 Rectangular Venn diagram for two variables, each having four "conditionally coded" ordinal categories.

If each variable had a null-based coding such as **0, 1, 2, 3** rather than the conditional coding of **1, 2, 3, 4,** the conjunction would produce the rectangular portrait shown in Fig. 19-3. Nine cells can be formed when each variable has a positive value. Three peripheral cells in each of the rows and columns show situations where one of the variables is absent, and the upper left corner is the cell formed when both variables are absent, i.e., coded as **0**.

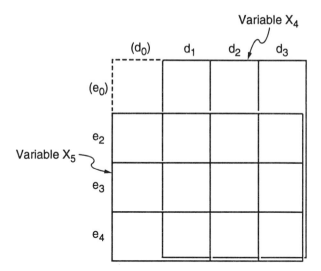

Fig. 19-3 Rectangular Venn diagram for two variables each having four null-based ordinal categories. The dotted line shows the cell where both variables are "absent".

Figs. 19-2 and 19-3 have immediate resemblances to a table. In fact, the structure of the common 2 × 2 table can be shown with a square Venn diagram in Fig. 19-4.

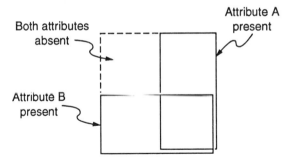

Fig. 19-4 Rectangular (square) Venn diagram for two binary variables, forming structure of the common "2 × 2 table".

19.1.3. Nomenclature for tabulations

When only two variables are included, the tabular structure is called a *two-way table*. (If each variable has only two categories, the arrangement is called a *2 × 2* or *fourfold table*.) For more than two variables, the display forms a *three-way table*, *four-way table*, etc., but the complex tables are usually called *cross-classifications* or *cross-tabulations*. The cross-tabulations show all the multivariate categories formed in conjunctions of two or more variables. Although the cross-tabulations directly correspond to rectangular Venn diagrams, the multiple categories can seldom be organized in arrangements as simple as the foregoing illustrations.

Because the horizontal rows and vertical columns resemble layers of strata, the multivariate tables are also sometimes called *cross-stratifications*. Since each variable is partitioned into categories, shown in the row and column headings, the multicategorical tables are also called *cross-partitions*.

19.1.4. Targeted and nontargeted tabulations

Many cross-tabulations are arranged in a nontargeted manner. If identified, the dependent variable is simply included among the other variables and not given any special status. For example, Table 19-1 here recapitulates the arrangement of "observed three-dimensional data" displayed in Table 3.5-4 (page 90) of the text by Bishop, Fienberg, and Holland.[1] In this arrangement, *consumer's illness* is a binary target variable, categorized as **ill** and **not ill**. The target is explored for the impact of the two binary independent variables, called *crabmeat* and *potato salad*, each categorized as **yes/no**. The table shows the eight cells formed by presence or absence of the appropriate category for each of the three binary variables.

Table 19.1 Cross-tabular arrangement of "Observed three-dimensional data" for an outbreak of food poisoning*

Consumer's illness	Food eaten			
	Crabmeat		No crabmeat	
	Potato salad		Potato salad	
	Yes	No	Yes	No
Ill	120	4	22	0
Not ill	80	31	24	23

*This arrangement is reproduced from Table 3.5-4, page 90, in Discrete Multivariate Analysis by Bishop et al.[1] Each person in the table could have eaten crabmeat, potato salad, both, or neither.

In a targeted arrangement, however, the target variable is given a special focal status. Removed from the rows and columns that indicate the independent variables, the results of the target variable are summarized in the cells. A targeted rearrangement of the data of Table 19-1 is shown in Table 19-2. The rearrangement displays the same collection of data, but is more immediately informative. The effects on the target variable are directly displayed in each bivariate categorical cell, and the effects of each independent variable are shown in the marginal totals for each row and column.

Table 19-2. Targeted cross-stratification for the data of Table 19-1 (Numerators and denominators in each cell show rates of illness)

Crabmeat eaten	Rates of consumer's illness: Potato salad eaten		Total
	Yes	No	
Yes	120/200 *(60%)*	4/35 *(11%)*	124/235 *(53%)*
No	22/46 *(48%)*	0/23 *(0%)*	22/69 *(32%)*
Total	142/246 *(58%)*	4/58 *(7%)*	146/304 *(48%)*

This type of targeted tabular arrangement was used earlier to illustrate the discussions throughout Chapter 3. For example, Table 3-10 contained six cells formed as bivariate categories from conjunctions of the three categories (**mild, moderate, severe**) of one independent variable and two categories of the other (**Treatment G** or **Placebo**). The interior of the cells showed results (as rates of success) for the dependent target variable.

This kind of cross-stratification is an excellent way to display the focus of targeted cluster analyses. Unfortunately, the targeted display is not easily obtained from packaged computer programs, and special arrangements are usually needed to get the data to appear in this format.

The term *cross-stratification* can be reserved to designate a targeted arrangement. The cells of a cross-stratification show the targeted results in each stratum, rather than merely a nontargeted *cross-tabulation* or *cross-partition* of frequency counts for the labelled rows and columns. Unfortunately, the effort to give a distinctive meaning to *cross-stratification* may not succeed, because this word and the other three terms (*cross-tabulation, cross-classification,* and *cross-partition*) are often used synonymously. If an epidemiologist is sometimes defined as someone who knows the difference between *incidence* and *prevalence,* a connoisseur of targeted-cluster analysis might be defined as someone who knows the difference between a *cross-tabulation* (such as Table 19-1) and a *cross-stratification* (in Table 19-2).

The impact of each multivariate category can easily be shown in a cross-stratification because the contents of each multicategorical cell display the corresponding summary for the dependent variable. The summary is expressed as a proportion for binary data, as a mean or median for dimensional data, etc.

19.2. Differences in cluster and algebraic methods

In targeted multivariable analysis, both the cluster and the algebraic methods begin with each person's data for a dependent (target) variable, Y, and for a series of independent variables, $X_1, X_2, \ldots, X_j, \ldots, X_k$. Algebraic methods rely on a mathematical model, G, containing a combination of variables $\{X_j\}$, each of which is weighted by a coefficient $\{b_j\}$. The arrangement is $G = b_0 + b_1X_1 + b_2X_2 + \ldots + b_kX_k$. After the algebraic analysis, each person's value of the target variable, $\hat{Y}_i$, is estimated from the score obtained as the sum of the $\{b_jX_j\}$ products when the person's values of the $\{X_j\}$ variables are entered into the algebraic model and multiplied by the weighting coefficients.

With categorical methods, however, there are no mathematical models or weighting coefficients. Instead, each $\hat{Y}_i$ is estimated from the value of Y previously found for the person's location in one of several multivariate clusters formed as diverse combinations of categories in the set of $\{X_j\}$ variables. The clustered categories, which usually form an ordinal array, are sometimes given special names, such as **Stage I, Stage II, Stage III,** ....

19.2.1. Illustrations of estimates

To illustrate the difference in the two sets of methods, suppose the lung cancer of patients in the illustrative data set here (see Chapter 5) were described with three binary variables, coded as **absent/present,** in which X_1 = no anatomic evidence of spread beyond primary cancer, X_2 = anatomic evidence of regional spread but not distant metastasis, and X_3 = evidence of distant metastasis. Of the 200 patients, 51 would be coded as 1 and all others as **0** in X_1; 27 would be coded **1** and all others **0** in X_2; 122 would have a code of **1** and 78 would have **0** in X_3. With clustered categories, the 6-month survival rate for the three groups would be expressed as

> Stage I (localized):   38/51   *(74.5%)*
> Stage II (regional):   15/27   *(55.5%)*
> Stage III (metastatic): 41/122 *(33.6%)*.

With a linear algebraic model, the 6-month survival rate for the 200 patients would be expressed as $\hat{Y} = .336 + .409X_1 + .219X_2$. (The X_3 term, being perfectly correlated with X_1 and X_2, does not appear in this equation, because someone who is coded as **0** in X_1 and **0** in X_2 will be coded as **1** in X_3.) When the appropriate values of X_1 and X_2 are substituted, the algebraic score for $\hat{Y}_i$, as shown in the table below, gives the same results just cited for the clustered categories.

Stage	Values of			$\hat{Y}$
	X_1	X_2	X_3	
I	1	0	0	.745
II	0	1	0	.555
III	0	0	1	.336

19.2.2 Important variables and transcending categories

In algebraic analyses, a prime goal is to identify the important variables and to eliminate the ones that seem unimportant, thus reducing the number of retained variables to just a few. Categorical analysis has the same goal, but it is usually achieved by reducing categories rather than variables. The aim is to get relatively simple multivariate categories that contain components from just a few variables. The result is optimal if an important impact can be shown with a single category (or variable) regardless of what happens in the others.

The term *transcending category* (or variable) is used for this type of impact. It is usually found in categories that produce the highest (or lowest) results in the target variable. The transcending effect is also apparent in symbols for the composite variables. For the example in Section 19.1.1, variable X_3 was a transcending variable, with c_2 a transcending category, because the impact on the target variable was unaffected by the conjoint simultaneous results in variables X_1 and X_2. This transcending effect was also indicated when **c_2** was used alone to identify Stage III, without any additional symbols.

In general, with a few exceptions to be noted later, the fewer symbols needed to identify a multivariate category, the more transcending is its impact.

19.3. Mistaken beliefs about multicategorical analysis

When the arrangement is possible, a cross-stratification is an effective display. The results in each cell offer an estimate of Y that is highly specific for the cited multicategorical attributes. The estimate would indicate not just what was found on average in individual results for men, or for tall people, or for thin people, or for old people, or for patients in Stage II, or for patients with anemia. Instead, results are shown for the precise conjunction of all these attributes in a multicategorical cell such as tall, thin, old, Stage II, anemic men.

Despite the appeal of precise identifications, however, multicategorical arrangements are often viewed with passive horror or active antagonism by many data analysts. The reasons for these beliefs include "loss of information," lack of a mathematical model, absence of automation, failure to "weight variables," and (most important), problems in sample size. These beliefs, however, do not reflect the realities discussed in the next few subsections.

19.3.1. Loss of information

When converted to binary or ordinal categories, dimensional variables seem to "lose information," but the loss may often be biologically irrelevant. Biologists tend to think in categorical concepts[2,3] such as **old, fat, tall, diabetes mellitus,** or **sickle cell,** rather than in precise measurements of *age, weight, height, blood sugar,* or *outer radius* and *inner radius of red blood cell.* The main biologic problems in the categorized dimensions are not the loss of information, but occasional disagreements about the boundaries chosen to demarcate binary or ordinal zones.

19.3.2. Lack of mathematical model

Categorical analyses may also be regarded as coarsely crude or esthetically inelegant because they lack the configuration (or "theory") of an algebraic mathematical model. Many years ago, scientists who respect the primacy of their own work would promptly reject this idea, but it has received respectful attention in an era when many medical biologists have been persuaded (or perhaps "brainwashed") to believe that science must satisfy the needs of mathematics, rather than vice versa. Investigators who recognize and preserve the main goals of biology know that science does indeed depend on documentary evidence and that the evidence is often quantified. Nevertheless, the particular format taken by the evidence and quantification is determined by the needs of scientific documentation, communication, and understanding, not by the arbitrary structure of mathematical models.

19.3.3 Absence of automation

Because some of the procedures are not automated, the subjective judgmental decisions about combining categories are sometimes inconsistent, and made differently by different analysts. Nevertheless, the analytic judgments are "up front" and readily apparent, whereas automated algebraic analyses (as discussed earlier) require a series of quantitative and stochastic judgments that may also be inconsistent, but that are usually hidden or obscured by either the algebraic models themselves or by the "black box" complexity of the number crunching.

Most important, however, the judgmental decisions can make a highly positive rather than negative contribution. As discussed earlier (throughout Section 5.3), these decisions create the appropriate unions and other biologic distinctions that lead to clinically sensible analyses and interpretations. To give statistical results the "face validity" of being biologically meaningful and important, judgmental decisions during the analysis engage the mind behind the face.

19.3.4. Failure to 'weight' variables

In a rectilinear algebraic model, the apparent "weight" of each independent variable is shown by a coefficient attached to the variable. Because clustered categories do not have such weighted coefficients, their impact is shown from results in the *target* variable. Consequently, clustered categories do not appear to achieve the prime analytic goal of having a single coefficient that denotes a slope, relative risk, or odds ratio to show the quantitative impact and relative importance of each independent variable.

This statement is true, but the coefficient itself is not necessarily the best or even an accurate way to show impact. In fact, the apparent precision of a single coefficient for each variable is often a delusion. In a multivariable setting, the impact will almost always change not only in different zones of the variable itself but also in the conjoined zones of impacts from all the other variables. As discussed earlier, the single coefficient for each variable is a statistical average, derived from the assumption that all the other variables are somehow being held constant. The assumption is valid mathematically, but cannot be true in pragmatic reality. Consequently, although the calculated algebraic coefficient is not wrong, its validity is limited. It represents a general value that is correct for an average impact; but the impact can vary extensively as the variable passes through its own zones and through the zones of the other variables.

In the clustered form of analysis, the impact of each multivariate category is shown directly and specifically at the site where it occurs: on the target variable. Furthermore, in certain categorical analyses, the impact of individual variables is clearly isolated either as a final product or during the operating strategy. For example, as an operating strategy in the clustering method called *conjunctive consolidation*, the impact of individual variables is specifically noted and evaluated before each "consolidation" occurs. Another example is offered by the staging system discussed in Section 19.2.1. Stage III shows the impact of variable X_3, exerting its transcending effect regardless of what happens in variables X_1 or X_2. Stage II shows the impact of X_2 when X_3 is absent; and Stage I shows the impact of X_1 when both X_2 and X_3 are absent.

In multicategorical arrangements, therefore, a single weight for each variable may often be lost, but it is replaced by the gain of showing impacts that are unequivocally specific for each of the stipulated multicategorical zones.

19.3.5. Problems in sample size

The most common source of complaints and discomforts about cross-stratification is the problem of small sample size. If only two categories are used in each scale of *sex, height, weight, age,* and *anemia,* and if only three categories are used for *stage,* the subsequent cross-tabulation of these six variables would contain $2 \times 2 \times 2 \times 2 \times 2 \times 3 = 96$ cells. A huge sample size would be required to get enough people for stable numerical results in each cell; and even with the huge sample, many of the cells might be empty. Furthermore, although specific for each cell, the multicategorical results would be difficult to understand and interpret. They could show the "trees but not the forest," offering an accurate portrait of what takes place in the cells, but not enlightenment about what is really happening biologically.

19.3.6. General misconceptions

For some or all of the foregoing reasons, many data analysts today vigorously oppose the idea of doing multicategorical analysis. They view it as an awkward cross-tabular or "multistratification" exercise that usually has inadequate sample sizes in the cells and that yields uninterpretable results, regardless of how useful the results may be for biologic explanations and predictions.

This dismissal would be fully justified if multicategorical analysis today were indeed carried out with the mega-cellular cross-tabulations that are used when algebraic models are applied to the huge tables. In newer strategies, however, the work does not require an algebraic model or the construction of giant cross-tabulations. Instead, the new methods immediately reduce the potentially vast number of multicategorical cells into a smaller array of entities that can be readily managed and easily understood.

Like algebraic procedures, the targeted-cluster methods have some distinct advantages and disadvantages. Among the major advantages, however, are the opportunities to see directly what has happened to the data, to apply (when desired) certain judgments that give clinical or biologic sensibility to the results, and to let the computer act as an agent of scientific enlightenment, not just arithmetical number crunching.

The next section contains a brief review of the old algebraic log-linear model strategy for cross-tabulations, followed by an outline of basic operational distinctions for the new strategies.

19.4. Basic strategies of multicategorical analysis

Multicategorical analyses can be done with at least four basic operating strategies. One of them *(log-linear models)* is relatively old, and depends on algebraic configurations. It will be mentioned here briefly and then dismissed, because it is not really a method of categorical clustering. The three other strategies are relatively new, and all of them use a configurationless, decision-rule approach to form clustered categories. The three new approaches are outlined here and discussed extensively in Chapters 20 and 21. One of the new procedures *(recursive partitioning)* has been automated; and the easy "computability" has made it the most popular of currently developed targeted-cluster procedures. The other two strategies *(sequential sequestration and conjunctive consolidation)* require substantive judgments that have not yet been automated, and the procedures are not well known.

The basic differences among the four operational strategies can be described as collapsing tables, aggregating categories, segregating categories, or consolidating variables.

19.4.1. Collapsing tables (log-linear models)

The strategy called *log-linear models* is one of several algebraic procedures subsumed under the title of *discrete multivariate analysis*.[1] The approach begins conceptually with a mega-cellular cross-tabulation in which each cell of the huge grid shows frequency counts for the conjoined categories of all of the candidate variables (including the target variable).

For the analytic procedure, the frequency count in each multicategorical cell is divided by the total sample size, N, and expressed as a probability, which becomes estimated as the score in an algebraic model for the structure of the table. The concept is analogous to the algebraic format used as a model for the analysis of variance. The models are called *log-linear* because they use a linear combination of variables and the natural logarithms of probabilities in the cells.

An extensive literature has been developed and can be consulted for more details of the log-linear modelling procedure,[1,4-7] which will not be discussed in this text beyond its brief overview here. Unlike the decision-rule strategy of multicategorical analysis, the log-linear approach is not intended to identify important categories or clusters of categories. The main goal of the analysis is to get an algebraic model that offers a best fit for the megacellular table.

Analogous to the "backward regression" procedure for algebraic analyses, the log-linear model starts with a big table and makes it smaller, applying mathematical criteria to remove certain variables or multivariate combinations. The big-to-small direction of the log-linear approach is reversed, however, in the small-to-big sequence of the newer multicategorical strategies, which work in a manner analogous to "forward regression." They start with single variables or categories, which are then augmented sequentially.

Because of the mathematical complexity and "elegance," log-linear models have regularly been taught during advanced statistical instruction. If they are the only "multicategorical" method to which students are exposed, however, the instruction will misrepresent the potential value of multicategorical analysis. While troubled by the paucity of data in the individual multivariate cells, the students will not have learned about the effective new decision-rule methods, outlined in the next three subsections, that can eliminate the problems.

19.4.2. Aggregating categories (recursive partitioning)

The approach that is generally called *recursive partitioning* produces a tree-like branching arrangement of targeted categories. The procedure is illustrated in Fig. 19-5, which shows the branches and combinations of categories for age, electrocardiographic ST-segment, and mental orientation that were arranged[8] to predict postdischarge survival of 202 patients with stroke.

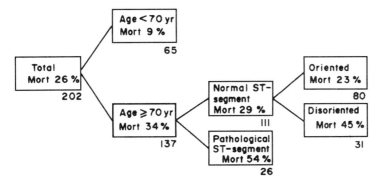

Fig. 19-5 Long-term prognosis. Predictor tree for 202 patients with cerebral haemorrhage or infarction discharged alive from hospital. Follow-up 12–39 months (mean 20 months). Number of patients and mortality in each subgroup are shown. (Legend and figure taken from Chapter reference 8.)

Beginning with the entire group of data, the partitioning process examines how the target is affected by diverse binary categorical partitions of each independent variable. The best candidate partition is chosen as the first *split*, producing two categories, as shown on the left side of Fig. 19-5. A split at **age** $\geq$ **70** divided the overall mortality of 53/202 (26%) into two groups with mortality rates 6/65 (9%) and 47/137 (34%).

Each category formed by the first split is then explored for the conjoined effects of partitioned categories from each remaining candidate variable. The best of these conjunctive partitions is selected as the second split, thus yielding a total of three categories, of which two are composite. In the example here, the older age group was split, by presence of a pathological ST-segment, into two groups with mortality rates of 33/111 (29%) and 14/26 (54%). (A second split can sometimes involve categories of the previous variable. As a hypothetical example, the older age group here might have been divided into two groups with mortality rates 15/67 [22%] for **age 70–84** and 32/70 [46%] for **age** $\geq$ **85**.)

In the next recursive iteration, the three demarcated clusters are explored for finding a best third split. At this step in Fig. 19-5, the 111 older people with normal ST-segments were divided into an "oriented" group with 18/80 (23%) mortality, and a "disoriented" group with 14/31 (45%) mortality. After producing a branching, tree-like array of simple and increasingly composite categories, the process stops when no further effective splits can be established.

Originally described by social scientists[9] under the title of *Automated Interaction Detector* (AID), recursive partitioning has been well developed, and is the most popular of the targeted-cluster methods in current use. It receives detailed discussion in Chapter 21.

19.4.3. Segregating categories (sequential sequestration)

In an approach called *sequential sequestration*,[10] multivariate categories are separated without a branching technique. Instead, the selected simple or composite categories are progressively segregated. At each step, the segregation removes the subgroup of persons who are members of the selected clustered category. The subsequent search is then directed at the persons (and categories) remaining in the available group of data.

The procedure has not been automated and involves judgmental decisions at each sequential step. The method is particularly useful not in constructing composite categories, but for arranging a series of overlapping binary attributes into a ranked, mutually exclusive set of ordinal categories. The technique, which will be further described in Section 20.1, was used to construct and evaluate the *primary symptom* union discussed in Chapter 5.

19.4.4. Consolidating variables (conjunctive consolidation)

In the two preceding methods of aggregation and segregation, the focus was on *categories* that could be joined or isolated in diverse ways, regardless of their sources in different variables. In the fourth approach, called *conjunctive consolidation*, the focus is on *variables*. The goal is to find and sequentially combine (or "consolidate") the most important variables. The conjunctive consolidation process was briefly illustrated in Section 4.6.2.5 when three composite stages were formed from the seven bivariate categories of Table 4-5. Another example appeared in Section 16.1.3 where a trivariate STH composite staging was formed from the TNM, symptom, and hematocrit variables of the illustrative data set.

In the first step of the procedure, the available k independent variables are conjoined in pairs to form $k(k-1)/2$ tables. The most effective of these tables, chosen according to substantive and statistical criteria, is then consolidated to form a new composite variable, containing conjoined categories from each of the two constituent variables.

This consolidation immediately removes 2 of the k variables, leaving $k-2$ variables as candidates for consolidation in the next step. Each of the remaining $k-2$ variables is then explored for its conjoined effect on the first composite variable. (For example, with seven independent variables, twenty-one tables are examined in the first step. For the second step, however, only five tables are explored.) When the best of the $k-2$ candidates is combined with the first composite variable, the clustered categories of the second composite variable will contain consolidations of trivariate categories from all three constituent variables. The second composite variable is then explored for conjunctions of all remaining $k-3$ variables; and the process continues iterating until no further candidate variables exert an effect within the existing clusters of the largest composite variable.

The process has not yet been automated, and has the advantage (if you so regard it) of allowing judgmental decisions about biologic coherence or sensibility for all of the newly formed consolidated clusters and composite variables. Conjunctive consolidation also has the scientific advantage of initially keeping each variable intact enough to allow evaluation of its impact. The process is further described in Section 20.2.

19.5. Additional methods of evaluation

Although taxonomic compression is the main purpose of *untargeted* analyses, a counterpart activity occurs in targeted-cluster procedures, when new composite categories are formed (or "compressed") from constituents of the independent variables. Because composite categories are not produced by algebraic analyses, special guidelines are needed to decide when a new composite category is taxonomically satisfactory, and when to separate (or combine) different categories within the scale of the same variable.

Because the compressions take place while or after the effective variables and categories are identified, the prime goals of categorical analysis are to identify trends or impacts, and to make estimates. The achievements are evaluated with the basic methods discussed in Chapter 8 for concepts of significance, fit, trend, and modicum size for categories, but some additional procedures are needed when different candidates are compared. The additional procedures involve further details of collapsing categories and testing for linear trend, but also include two new strategies that introduce a cost matrix for choosing concordant fit, and that appraise fit for polytomous targets. The additional procedures, which occur in targeted-cluster but not in algebraic analyses, are described in the next few sections before we turn, in Chapters 20 and 21, to the actual operations and illustrations of the multicategorical analyses.

19.5.1. Collapsing categories

Biologic as well as statistical judgment can be applied when categories are collapsed or otherwise clustered.

19.5.1.1. Biologic 'homogeneity'

Some of the principles of biologic "homogeneity" were discussed in Section 4.6.2.2. Two categories can often be plausibly regarded as similar if they are part of the same pathophysiologic derangement, prognostic implication, or therapeutic decision. For example, dyspnea and peripheral edema occur in different parts of the body but can both be part of congestive heart failure. Hepatic failure and uncontrolled monoblastic leukemia seem to be different derangements, but can be similar in having poor prognostic import, and also in being contraindications to a major surgical operation.

At simpler levels of decision, as discussed in Section 4.6.2.2, two adjacent categories of an ordinal variable can readily be joined, and two bivariate categories might be combined if they represent symmetric opposites such as **mild-severe** and **severe-mild** in the conjoined ranks of two component ordinal variables.

Biologic homogeneity is more difficult to discern if a cluster contains categories from three or more variables. The homogeneity is apparent if *all* of the categories come from extreme ends of the rating scale for each constituent variable. Thus, trivariate categories formed from **none-none-none, mild-mild-mild,** or **severe-severe-severe** ranked levels of the three component variables are easily justified as homogeneous. A mixture of **moderate-severe-mild** categories is more difficult to understand for biologic meaning. Consequently, except for combinations of extreme or near-extreme ranks, multivariate mixtures of three or more ordinal categories are often arranged according to statistical rather than biologic principles of homogeneity.

19.5.1.2. Statistical 'homogeneity'

Two categories can be regarded as statistically homogeneous if they have essentially similar results for the target event. This type of isometry will depend on the criteria (see Sections 4.6.2.1 and 8.2.2) used to delineate a quantitatively significant distinction. The criteria will vary with the situation under analysis. If the target refers to survival rates that can range from 0 to 100%, an increment of 10% might be set as the minimum standard for quantitative significance between two categories. If the target is development of a disease that usually occurs at a rate of .001, different criteria will obviously be needed. In the latter situation, the criterion might demand a minimum odds ratio or risk ratio, such as 2.5 or 3.0, between categories. If the target variable is expressed in means or medians, quantitative significance might require a *proportionate* increment of at least 20% between categories.

All of these examples should be regarded as merely examples rather than rigid guidelines. The boundaries will vary with different situations, but the data analyst is always obligated to explain or justify why the choices are biologically sensible.

19.5.1.3. Considerations of trend

Although two categories might be promptly collapsed if they are inadequate in size, gradient, or stochastic significance, the decision is sometimes deferred if they make a satisfactory contribution when linear trend is tested in either an algebraic or categorical analysis.

For example, if a stochastically significant result is obtained with the log-rank test, the two compared survival curves are usually regarded as distinctively different. Nevertheless, at individual points of time in the curves, the survival rates may not be significantly different, either quantitatively or stochastically. The results for the entire curve are thus given priority over the results for individual components.

Similarly, a suitably monotonic trend in an array of categories may often take precedence over the results found in pairwise comparisons of the categories. This tactic was used in Section 9.2.5 when the total categorical array was preserved for *TNM stage* and *symptom stage* even though quantitative and stochastic significance were not present in each of the intercategory comparisons.

The appraisal of linear trend in ordinal categories is discussed in Section 19.5.2.

19.5.1.4. 'Terminal unions'

After the main analysis seems to have been completed, some or many of the final clusters may be combined into *terminal unions*. They can be formed in a closed or open pattern.

19.5.1.4.1. Closed arrangement

In the closed pattern, illustrated by the conjunctive-consolidation example in Section 4.6.2.5, all of the clusters are produced by the successive conjunction and consolidation of individual variables within each table. Some of the clusters—particularly at extreme corners of the table—are easily delineated according to biologic principles of homogeneity, but the remaining composite categories in the "closed" interior of the table may be difficult to combine on biologic grounds alone. Their consolidation can then be arranged according to purely statistical guidelines, such as relatively similar results for the target variable, or concepts of spanning and meridian. These statistical tactics were used for the conjunctive consolidation in Section 4.6.2.5, and will be further illustrated in Section 20.3.6.

19.5.1.4.2. Open arrangement

Terminal unions can be formed from an open arrangement during the type of recursive partitioning shown in Fig. 19-5. After each cluster is produced by sequentially conjoining categories from binary segments (or splits) of individual variables, the remaining unused categories of each variable can then participate, via subsequent splits, in clusters formed later in the branching sequence. The locations at which the splits occur are called *nodes,* and the final clusters that receive no further splits are called *terminal nodes.*

At each branching split, the two terminal nodes formed from a "generating" cluster will always be significantly different from one another because they fulfilled criteria that allowed the split. On the other hand, some of the other terminal nodes, generated in earlier or later steps of the process, may have relatively similar results for the target event. For example, suppose an original group with 70% survival is split into two groups having 80% and 60% survival. If the group with 80% survival is then split into two groups having 90% and 60% survival, the three terminal nodes will have survival rates of 90%, 60%, and 60%.

Since a stratification does not gain increased power or precision by preserving "statistically homogeneous" terminal nodes, they can be combined into a single cluster, to make a simpler final system. The combining of terminal nodes into "terminal unions" is somewhat analogous to pruning a tree, but the terminal branches here are combined, rather than removed.

For example, in Fig. 19-5, the partitioning produced four terminal nodes, with mortality rates of 9%, 23%, 45%, and 54%. If the investigator insisted on having an intercategory gradient of $\geq 10\%$, the last two nodes could be combined into a terminal union comprising two multivariate categories: disoriented old people with normal ST-segments, and old people with pathologic ST-segments. The clustered terminal union, containing subgroups with mortality rates of 14/31 and 14/26, would have a mortality rate of $(14 + 14)/(31 + 26) = 28/57 = 49\%$.

19.5.1.4.3. Criteria for terminal unions

Although no standard guidelines have been established for forming terminal unions, the decision rule can apply the same principles used earlier throughout the components of Section 4.6.2 for joining (or collapsing) cells of a stratified table.

Unless the branched partitioning algorithm contains "interior" demands for a modicum at each split, one or more of the terminal nodes may have too few members to meet the minimum-size requirement. Such nodes may be terminally joined with other nodes that seem appropriate according to previously discussed criteria for statistical, and perhaps substantive, homogeneity. Because of the problems discussed earlier for evaluating biologic plausibility when composite categories contain components from a large number of multiple variables, most terminal-union decisions depend on statistical rather than biologic concepts of homogeneity.

Even if two terminal nodes contain ample members, they might be combined if their target gradient is too small. In the example just cited for data in Fig. 19-5, the two nodes with 45% and 54% mortality rates could be combined if the 9% internodal gradient was below the minimum of 10%. On the other hand, terminal nodes with ample members but a too-small gradient might be preserved if they span the meridan. Thus, if the nodes with 45% and 54% rates are combined, the 54% rate would be "lost" in the only terminal node that spans the meridian in that group of data.

19.5.2. Appraising linear trend

Because the impact of categories in either an individual variable or a clustered system is indicated by the gradient in the target variable, appraising this trend is an important activity. The easiest approach is to examine the direct intrasystem gradient, as discussed in Sections 8.3.1 and 8.3.2. The direct gradient, however, may sometimes be misleading if the strata have too-small denominators or substantial imbalances in size.

Table 19-3 Two staging systems for a group with total mortality rate of 50% = 40/80

Variable X_1	*Variable X_2*
a_1: 8/10 *(80%)*	b_1: 18/25 *(72%)*
a_2: 30/60 *(50%)*	b_2: 15/30 *(50%)*
a_3: 2/10 *(20%)*	b_3: 7/25 *(28%)*

For example, suppose the modicum is set at 10, and suppose two possible stratifications of the same 80 people have the results shown in Table 19-3. The first arrangement, for variable X_1, seems more impressive. The direct intrasystem gradient is larger (60% vs. 44%) than for variable X_2, and the intercategorical gradients are also larger (30% vs. 22%). Nevertheless, the results in the second system, for variable X_2, seem more stable because the sample sizes are better distributed, with substantially larger denominators in the first and third stratum. (This quantitative virtue of the second system will be confirmed shortly.)

19.5.2.1. Linear models for slope

To avoid the problem of imbalances in the size of strata, each gradient can be "adjusted" by calculating its linear slope with the algebraic mechanism discussed in Section 8.3.3. Arbitrary coding digits are assigned to the ordinal categories, which can then be regarded as a dimensional X variable, used for calculating a regression line. The slope of the regression line for the target Y variable will be the adjusted linear trend.

The numerical codes assigned as values of X for each categorical stratum should have equal-sized intervals between the coded digits; and the same coding digits should be used for comparing variables that have the same number of strata. The easiest approach, particularly if calculations are done by computer, is to assign the coding digits in a sequence of 1, 2, 3, 4, ... For an electronic hand calculator, computation is eased if the codes are "balanced" around zero, so that coding digits of -1, 0, $+1$ could be used for three strata, and -3, -1, $+1$, $+3$ for four strata. If the X variable is coded in 2-unit increments, however, the slopes will be half of their values with 1-unit incremental codes. Consequently, if a three-stratum system is to be compared with a four-stratum system, a change to codes of -2, 0, $+2$ for the three-stratum system would produce equal intervals for the codes in both systems. In a three-category partition with codes assigned as $w_1 = -1$, $w_2 = 0$, and $w_3 = +1$, the formula for the slope becomes

$$\text{Slope} = \frac{N(t_3 - t_1) - T(n_3 - n_1)}{N(n_1 + n_3) - (n_3 - n_1)^2}. \qquad [19.1]$$

Applying formula [19.1] to the stratified system for variable X_1 in Table 19-3, the slope will be $[80(2 - 8) - 40(10 - 10)]/[80(20) - (0)^2] = -480/1600 = -0.3$. For variable X_2, the slope will be $[80(7 - 18) - 40(25 - 25)]/[80(50) - (0)]^2 = -880/4000 = -0.22$. Thus, the larger intrasystem gradient in the first system is confirmed by the slopes found with the linear models for the gradient. Furthermore, the slopes here turn out to be exactly what would be anticipated from the average intrasystem gradient, which is $.60/2 = .30$ for variable X_1, and $.44/2 = .22$ for variable X_2.

For two binary categories, coded as $w_1 = 0$ and $w_2 = 1$, with the two proportions being $p_1 = t_1/n_1$ and $p_2 = t_2/n_2$, the formula for the slope reduces to

$$\text{slope} = p_2 - p_1,$$

which is simply the direct increment in the two proportions. If the binary categories have been coded as $w_1 = -1$ and $w_2 = +1$, however, the slope is halved to $(p_2 - p_1)/2$.

19.5.2.2. Quantitative significance for slope

Since the slope represents the incremental change from one category to the next, criteria for a "strong" slope can be similar to those chosen in Section 8.3.2.2 as the boundary for minimum intrasystem gradients. If the weighted codes are assigned with 1-unit intervals, and if the strata denominators are not too grossly maldistributed, the slope should generally be similar to the average intrasystem gradient, as noted in the calculations for Table 19-3.

19.5.2.3. Standardized slope

If desired, the slope b can be converted to a standardized slope, b_s, when multiplied by the conversion factor given previously as formula [8.8] in Section 8.3.4.

For three categories coded as **−1, 0, +1,** the conversion factor will be
$$\sqrt{[N(n_1 + n_3) - (n_3 - n_1)^2]/[T(N - T)]}. \qquad [19.2]$$
For variable X_1 in Table 8-3, this factor will be $\sqrt{1600/[(40)(40)]} = 1$, and for variable X_2, the factor will be $\sqrt{4000/1600} = 1.58$. Consequently, the standardized slope will be $(-0.3)(1) = -0.30$ for variable X1 and $(-.22)(1.58) = -0.348$ for variable X_2. Both standardized values equal or exceed the 0.3 boundary set for quantitative significance, but variable X_2 now has a higher value than variable X_1. The difference occurs because the standardized results will favor X_2 for its better distribution of the denominator categories.

19.5.2.4. Stochastic tests for 'stability'

Stochastic tests of the slope rest on the principle, discussed in Section 8.3.6.2, that the total value of chi-square for the system, calculated as $X^2 = \phi^2(N)$, can be partitioned into two parts so that $X^2 = X_R^2 + X_L^2$. In this partition, X_L^2 represents the fit of the linear model vs. fit of the mean $\bar{Y}$ values; and X_R^2 represents the residual of Y_i estimates around the linear "regression line," thus corresponding to S_R in ordinary linear regression.

19.5.2.4.1. Chi-square test for linear trend

A simple formula for calculating X_L^2 was shown in Section 8.3.6.2. If the standardized slope has already been calculated as the counterpart of $b_s = r$ for two variables, the formula is
$$X_L^2 = Nr^2. \qquad [19.3]$$
In a three-category system, with $w_1 = -1$, $w_2 = 0$, and $w_3 = +1$, formula [19.3] can be expressed, without calculating r, as
$$X_L^2 = \frac{N[N(t_3 - t_1) - T(n_3 - n_1)]^2}{T(N - T)[N(n_1 + n_3) - (n_3 - n_1)^2]}. \qquad [19.4]$$

For variable X_1 in Table 19-3, where each system has three categories, $X_L^2 = 80[-480]^2/[(40)(40)(1600)] = 7.20$. For variable X_2, $X_L^2 = 80(-880)^2/[(40)(40)(4000)] = 9.68$. To confirm that formulas [19.3] and [19.4] yield similar results, note that variable X_1 in Table 19-3 had $b_s = -0.30$ and so $X_L^2 = (80)(-.30)2 = 7.20$. For variable X_2, in Table 19-3, bs = $-.348$, with $X_L^2 = (80)(-.348)^2 = 9.69$.

The easiest approach to all these calculations is to do a standard simple linear regression with an appropriate computer program, using codes of **0/1** for the Y variable, and, **1, 2, 3,** . . . for the ordinal X variable. The program will produce the value of b for slope and also the value of r^2, from which the standardized slope can be determined as r, and X_L^2 and Nr^2.

19.5.2.4.2. Chi-square test for residual nonlinearity

Because the target variable has a monotonic gradient for both of the stratifications in Table 19-3, the linear model would not be suspected for offering a poor fit. With or without that suspicion, however, the residual nonlinearity in the system can be stochastically checked by examining the value of X_R^2, which can be found as $X^2 - X_L^2$.

Since X_L^2 has already been determined, the next step is to find the total value for X^2. For the two systems in Table 19-3, NPQ was previously found to be 20. The value of $\Sigma n_h p_h q_h$, using the shortcut formula $\Sigma[t_h(n_h - t_h)/n_h]$, becomes $[(8)(2)/10] + [(30)(30)/60] + [(2)(8)/10] = 18.2$ for variable X_1, and $[(18)(7)/25] + [(15)(15)/30] + [(7)(18)/25] = 17.58$ for variable X_2. The proportionate reduction in variance, expressed as ϕ^2, is $(20 - 18.2)/20 = .09$ in the first system and $\phi^2 = (20 - 17.58)/20 = .121$ in the second. The second system, as expected, has done a better job of variance reduction, and it also gets a higher value when X^2 is calculated as $N\phi^2$, so that $X^2 = (80)(.121) = 9.7$ for variable X_2 and $(80)(.09) = 7.2$ for variable X_1. The value of $X_R^2 = X^2 - X_L^2$ will be $7.2 - 7.2 = 0$ for variable X_1 and $9.7 - 9.7 = 0$ for variable X_2. Both stratifications thus achieve an essentially perfect linear fit with the linear model. This result would be expected from the equal magnitudes of the intercategory gradients within each stratification.

19.5.3. Polytomous targets

Although stratified analyses are usually aimed at two categories that are expressed as binary proportions, the target variable may sometimes have three or more polytomous categories. They can be cited in an ordinal array, such as **mild, moderate, severe,** or in a nominal group of categories for variables such as *cell type* or *principal diagnosis*. Developing a model that will correctly estimate the polytomous categories is an intriguing mathematical challenge. Algebraically, the challenge was approached with the logistic and discriminant methods discussed in Part IV. With cluster methods, the challenge is approached with the CART or KS forms of recursive partitioning discussed in Sections 21.6 and 21.7.

In most biologic research, however, the investigator will often seek a binary separation that will distinguish **mild** from **non-mild**, or **severe** from **non-severe** ordinal categories. As discussed in Section 14.10.4, a *single* analytic mechanism is seldom expected to estimate probabilities for *each* of a group of polytomous categories, but might achieve the binary aim of separating *one* category from all the others.

The main reason investigators prefer binary rather than polytomous outcomes was discussed in Sections 17.1.1 and 17.1.3. For binary estimates, the decisions usually depend on whether the probability is above or below 50%. For polytomous estimates, however, the probability values may be close to 33% for each of three categories, or close to 25% for each of four categories. A clear-cut majority decision may be impossible; and the analyst may have substantial discomfort in predicting that all members of the entire group under consideration belong to a category that "wins" by only a "modal" plurality value.

These difficulties have not prevented the development of the polytomous estimation methods discussed earlier for discriminant analysis and later in Chapter 21 for certain forms of recursive partitioning. The difficulties, however, have greatly reduced the attractiveness of polytomous estimations in pragmatic scientific research.

19.5.4. Use of a cost matrix

In all the discussion until now, independent variables were partitioned according to biologic or statistical homogeneity when categories were collapsed, and according to statistical tests of trend when the categories formed an ordinal array.

A different way to make the categorical decisions, however, is to examine concordance directly. For binary targets, it can be expressed with a simple score for accuracy, as shown in Section 8.2.2.2. A more interesting and complex scoring system for concordance, however, awards penalties for different types of errors. The complex strategy can be especially useful for scoring credits and debits in estimates of polytomous ordinal or nominal, rather than binary, categories.

The penalty strategy is most easily exemplified if we consider the binary targets of ordinary diagnostic marker tests, where a particular estimate can be a true positive, true negative, false positive, or false negative. If everything is weighted equally, the correct estimates are counted as "hits," the incorrect ones are "outs," and the simple "batting average" will be the proportion of hits per total number of estimates. If this proportion is called p_h, the proportion of outs will be $1 - p_h$.

If certain errors are regarded as more serious than others, however, appropriate weights can be used. For example, the false positive diagnosis of a particular disease might be regarded as a more serious error than a false negative. If so, we might give scores of +1 to a correct diagnosis, 0 to a false negative, and −1 to a false positive. The batting average would then be (number of correct diagnoses − number of false positive diagnoses)/(total number of diagnoses).

An operating principle can then be developed to maximize the value of batting average or "credit score." In this instance, the strategy might emphasize the aim of avoiding the penalties imposed by false positive diagnoses.

19.5.4.1. Operation of cost matrix

To assess concordant accuracy in a partitioning procedure for polytomous categories, a cost matrix can be established to give appropriate credits for correct estimates and impose different penalties for the different types of errors. A final credit score can then be calculated for the sum of correct and incorrect classifications.

For example, suppose an ordinal target variable has been expressed in the three categories of **mild, moderate, severe.** Correct prediction of each category might be credited as +1, but debit scores might vary according to clinical judgment about whether "underpredicting" a moderate or severe outcome is a more serious error than "overpredicting" it. Thus, if the severe outcome is an irreversible moribund state leading to death, the failure to predict the severe outcome, i.e., underprediction, might lead to wasted resources; but its overprediction might evoke a denial of life-saving care for patients who can be benefited. In other circumstances, where intensive care can be usefully life-saving for a severe outcome, but relatively unnecessary for a mild or moderate outcome, the reverse phenomena will occur. The overprediction of a severe outcome may lead to wasted resources, but the underprediction may be clinically disastrous if intensive care is denied.

To illustrate a cost matrix for this situation, let us assume that the underprediction of moderate or severe outcomes is not regarded as a serious error, and is given a score of 0. If someone with a **mild** outcome is predicted incorrectly, however, the error might be more serious, and debited as -1 for a **moderate** estimate, as -2 for **severe**. The cost matrix would then have the following appearance:

Predicted outcome	Actual outcome in target variable		
	Mild	Moderate	Severe
Mild	+1	0	0
Moderate	−1	+1	0
Severe	−2	0	+1

19.5.4.2. Advantages and disadvantages

As noted in Chapter 21, the cost-matrix tactic can have appealing features in choosing splits for recursive partitions of polytomous targets. The predicted estimates can be scored directly for accuracy, while avoiding the use of relatively "nonspecific" scores for variance and relative concordance. For investigators who insist on using polytomous target variables, the cost matrix can eliminate the difficulties of ranking and dichotomizing the polytomous categories.

Nevertheless, the tactic also has some prominent disadvantages. One of them, noted earlier, is that the predictions for each group are made from the largest polytomous proportion in the group's categorical estimates, thus often leading to a "plurality" rather than "majority" choice. For example, if a group's respective probability estimates for a four-category target variable are .20, .25, .35, and .20, everyone will be predicted as being in the third category although 65% of the estimates will be wrong. A second disadvantage is the problem of choosing an estimate when the categorical values are tied, particularly if they happen to occur in extreme ordinal categories. Thus, if probability estimates for a four-category ordinal target variable are .40, .10, .10, and .40, the credit scores can be dramatically different if the first or fourth category is chosen as the estimate. (The problem is less dramatic, but still substantial, if the estimates are distributed as .10, .40, .40, and .10.)

Perhaps the most obvious disadvantage, however, is the arbitrariness of the cost matrix. It allows the partitioning procedure to be automated, but it incorporates idiosyncratic judgments about the allocation and weighting of the costs. A cautious investigator can explore and examine the effects of several "plausible" cost matrixes and choose the one that seems best, but the chosen matrix (and the effects of other matrixes) may not be presented in the published report. Even if the investigator has no difficulty drawing conclusions, the reader may be unable to determine how they were made.

19.5.5. Appraising 'survival curves'

The currently developed targeted-cluster methods are aimed at a single-state outcome entity, and have not been specifically expanded to deal with dynamic phenomena, such as a survival curve over time. One way of managing the moving target problem is to convert it to a set of "stationary targets." For example, separate targeted-cluster analyses could be done for the survival rates at 6 months, 1 year, 3 years, and 5 years. The same set of independent variables may emerge as the important predictors for each occasion; if not, two (or more) "staging systems" can be established for different occasions. For stochastic testing of results at single time points, the targeted-cluster results can be evaluated with the ordinary and linear-trend chi-square procedures.

Some investigators, however, may prefer to examine and compare the entire, dynamic survival curve for each of the clusters. A generally accepted statistical index has not been developed for quantitative contrast of the survival curves, but they can be stochastically compared with the log-rank (or some other appropriate) test, as discussed earlier for Cox regression.

19.6. Scores and partitions from algebraic analyses

A quite different method of forming multicategorical scores and partitions is from algebraic analyses. The weight of each variable is first noted from its corresponding multivariable coefficient in one of the big three (linear, logistic, or Cox) multivariable algebraic methods. The results are then arranged to form factor counts, point scores, formula scores, or multicategorical partitions.

19.6.1. Count of factors

Each important variable can be cited as present or absent, and the count of these binary risk factors in each person can form an ordinal score (**0, 1, 2, 3, ...**). This tactic was used after multiple logistic analysis, as shown in Fig. 19-6, to demarcate four clinical groups having different probabilities of pneumocystis carinii pneumonia.[11]

In another example, after three binary attributes were algebraically identified as prognostic factors for cardiac death in acute myocardial infarction, Moss et al.[12]

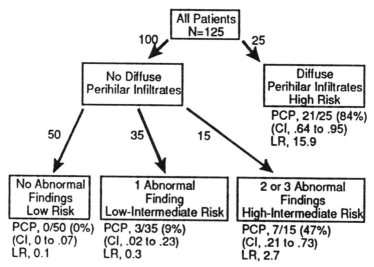

Fig. 19-6 Probability of pneumocystis carinii pneumonia (PCP) in four clinical groups. Abnormal findings were mouth lesions, presence of a lactate dehydrogenase level of more than 220 U/L, or an erythrocyte sedimentation rate of 50 mm/h or more, as identified by multiple logistic analysis. CI indicates 95% confidence interval; LR, likelihood ratio. (Figure and text of legend taken from Chapter reference 11.)

formed a simple partition as *no factor present* or *any factor present*. In a third example, after several forms of mathematical analysis for **yes/no** answers to nine questions about symptoms, Gardner and Barker[13] concluded that a simple count of symptoms was as effective in diagnosing hypothyroidism as any of the more complex mathematical methods.

19.6.2. Arbitrary point score

In a second approach, the weights noted in the multivariable coefficients are used to give each pertinent variable a point score, such as **1, 2, ...** The sum of the pertinent points is then added to form total scores, which can also be partitioned to form strata. A modification of this tactic has been used to form a set of prognostic risks for development of coronary heart disease.[14]

19.6.3. Multivariable formula score

In a third approach, the multivariable algebraic equation is kept intact, and a score is calculated when each person's data are entered into the formula of the equation. The scores can then be partitioned into a suitable number of strata. This principle was used by Miettinen[15] to perform "stratification by a multivariate confounder score," and was also applied by Smith et al.[15a] to classify discharged patients into groups with **low, medium,** or **high** risk of non-elective re-admission.

Freedman et al.[16] used this approach to construct a predictive index from a set of binary variables, but the authors then decided that the results would be "simpler to remember and use in a clinical environment," with only a small "loss in predictive power," if each variable were given "a weighting of 1 instead of the regression coefficient." The index was thus reduced to a simple count of four binary variables.

19.6.4. Uncondensed cross-partition of variables

In a fourth approach, the important predictor variables identified in an algebraic multivariable analysis are partitioned into categories. The results are displayed in a large multicategorical arrangement, without forming a more condensed system such as the stratified stagings of Chapter 20, or the "pruned trees" of Chapter 21.

19.6.4.1. 'Bar-graph' arrangement

In one uncondensed-partition approach,[17] the risk of cardiovascular disease with increasing group levels of cholesterol was demarcated, as shown in Fig. 19-7, with bar graphs formed from binary partitions of glucose intolerance, systolic blood pressure, cigarette smoking, and electrocardiographic left ventricular hypertrophy. The height of the bars shows the "8-yr. probability" of cardiovascular disease for each of the clustered categories.

Although sixteen categories ($= 2^4$) might be expected in a cross-partition of these four binary variables, the possibilities were reduced to seven basic categories, which were then conjoined to four groups of cholesterol partitions.

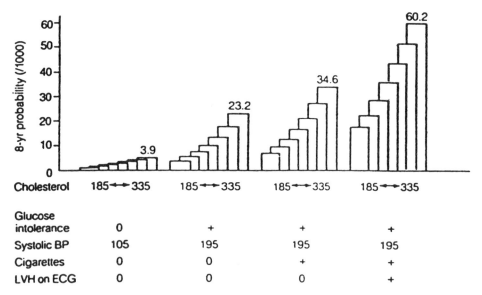

Fig. 19-7 Risk of cardiovascular disease according to serum cholesterol at specified levels of other risk factors (Framingham Study, 18-year follow-up) (men aged 35 years). BP = blood pressure; ECG = electrocardiography; LVH = left ventricular hypertrophy. (Figure and text of legend taken from Chapter reference 17.)

19.6.4.2. 'Tree' arrangement

A tree-like "layered" pattern was constructed[18] for the risk of cardiac death in the eight groups shown in Fig. 19-8. The layers depended on binary categories for three main prognostic variables that had been identified by logistic regression. The three binary variables were pulmonary congestion, maximum exercise BP < 110, and exercise duration < 9 min. The combinations of those categories formed eight groups, with mortality rates appended for each group.

19.6.5. Stratified arrangements

In a study of prognosis in a cohort with ischemic heart disease, Marantz et al.[19] used Cox regression to assess factors affecting 3-year mortality. After noting that the most important predictors were left ventricular ejection fraction (LVEF) and congestive heart (as well as age), the investigators formed a four-category stratification using presence or absence of congestive heart failure, and two levels of LVEF. The results, shown in Fig. 19-9, demonstrate a distinctive prognostic gradient in the four strata.

19.6.6. Other strategies and tactics

The challenge of forming scoring partitions has begun to stimulate the interest of statisticians, and new (mathematically complex) ideas are regularly proposed. In one method[20] of "log normal parametric analysis," the "sum of the log likelihood values" indicated the "prognostic value" of a staging system for gastric cancer. For matching and covariance adjustments, Rosenbaum and Rubin[21] offered indexes called *propensity score* and *Mahalanobis metric*, illustrated with data for effects of pre-

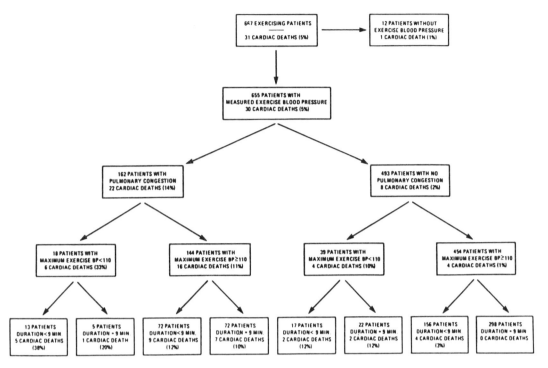

Fig. 19-8 Prognostic stratification with clinical and exercise variables. BP = blood pressure measured during the exercise test; duration < 9 min. = duration of exercise test less than 9 minutes; duration = 9 min. indicates completion of 9-minute exercise protocol. (Figure and legend taken from Chapter reference 18.)

natal exposure to barbiturates on psychological development of a cohort of Danish children. Spiegelhalter and Knill-Jones[22] proposed a system of weightings, using a combination of Bayesian estimates and logistic regression. The latter mathematical approach is summarized in the Appendix of a paper by Seymour et al.,[23] who used the method to form a complicated-looking score for predicting postoperative respiratory complications in elderly surgical patients. Seymour et al. laud the scoring method as being "simple . . . [requiring] only the ability to add, subtract, and look

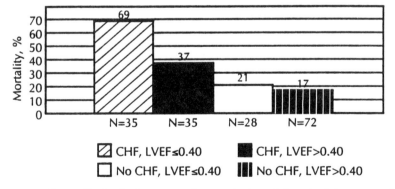

Fig. 19-9 Overall mortality for the four strata of congestive heart failure (CHF) (present or absent) and left ventricular ejection fraction (LVEF) (≤ 0.4 or > 0.4). (Figure and text of legend taken from Chapter reference 19.)

up a reference table or graph" while being able to "handle complex diagnostic concepts such as conflict of evidence and doubt."

19.7. Advantages of stratified evaluations

As shown in the foregoing discussion, the important strata in a data set can be formed directly from multicategorical analysis, or indirectly when results of an algebraic analysis are converted to scores or partitions. Regardless of how the strata are formed, however, the idea of forming strata is regularly disputed with complaints about (1) the false assurance offered when two groups have similar baseline proportions of strata, (2) the dubious merit of stratified randomization in clinical trials, and (3) the value of stratification itself.

19.7.1. Baseline imbalances in distribution of strata

An important problem in comparing results of treatment is the susceptibility bias that arises if the compared groups have prognostically different baseline states, before therapy. The problem is particularly common for comparisons of treatments given in ordinary clinical practice, where therapy is often chosen judgmentally according to each patient's anticipated prognosis. Randomized trials were developed to eliminate the "confounding" arising as susceptibility bias when the prognostic baseline state affects both the outcome and the choice of treatment. Nevertheless, despite randomized assignment of the treatment, substantial prognostic imbalances, at an α level of .05, can occur in 1 of 20 clinical trials.

One way to avoid the latter problem is to do a *stratified* randomization, which allows the assigned treatments to be "balanced" within the different baseline prognostic strata. The more common alternative strategy is to do a *complete* randomization, without regard to baseline state, but to compare the therapeutic results afterward with an appropriate "poststratified analysis." The merits of these two approaches are compared in Section 19.7.2. The rest of the discussion now is concerned with the method used for poststratified analysis.

One common approach is simply to compare the distribution of strata for the two treatments. In statistical jargon, the comparison is called a test of association between the covariates and the treatment. If the test does not show a significant association, i.e., no imbalance is found, the analyst may then not bother comparing results for each treatment within the pertinent strata. The error of this approach was shown in Table 3-8 of Section 3.4.3.2. Two treatments can be distributed equally among the strata and yet produce substantially different results in the same stratum. For this reason, a mere test of stratum (or covariate) distributions among treatments will never suffice. The therapeutic outcome should always be compared within the strata. In a lucid statistical demonstration of the problem, Begg[24] says that "significance tests of the association between covariates and treatment in randomized trials are illogical." He concludes that "analyses stratified by important covariates are recommended regardless of the observed distribution of the covariates and treatment."

19.7.2. Pre- vs. postadjustment

An unresolved dispute persists about whether to adjust for prognostic imbalances before the randomization or afterward, when the trial is finished. The adjustment commonly involves an appropriate stratification, but can be done beforehand with different methods of "equalization," such as the Taves[25] minimization technique, or other approaches proposed by Pocock and Simon,[26] and by Begg and Iglewicz.[27] For adjustments afterward, when the trial is completed, the analyses can also be done with algebraic methods rather than stratification. Regardless of whether the adjustment is done with stratification or with other methods, however, the main issue is whether to try to prevent problems prophylactically, or to fix them remedially after they have occurred.

The proponents of pre-adjustment, which usually consists of stratification followed by randomized assignment within strata, argue that the approach can avoid the prognostic imbalances arising in a complete randomization.[28] The prophylactic approach can thus substantially reduce or eliminate the need for expecting poststratification adjustment to provide successful "remedies" for imbalances that could easily have been prevented beforehand. The opponents of prestratification argue that a stratified randomization may be administratively awkward, that it increases the pragmatic difficulty of conducting a trial, and besides, an optimal arrangement of strata may be unknown.

During the statistical debates, the scientific clinical issues are generally ignored. A prime advantage of prestratification is that it will force the clinical investigators to think about and identify the prime prognostic factors. If not given suitable attention beforehand, important prognostic determinants—for example, types of symptoms, severity of symptoms, severity of co-morbidity, and responses to previous therapy—may be overlooked; and the appropriate data may not be collected. The necessary information will then be unavailable for analysis when the trial is finished.

19.7.3. Intrinsic value of stratification

Because algebraic methods can be used to adjust "covariate imbalances" either before treatment is assigned or afterward during the analysis, questions regularly arise about whether stratification itself is a worthwhile procedure. As noted throughout Section 19.3, the antistratification statistical arguments are that the stratification process may lose information, lack mathematical elegance, require cumbersome calculations, fail to weight variables, and create major problems in sample size. The first four statistical arguments can be countered as stated in the earlier discussion; and the sample size problem is avoided by stratifications constructed with the modern methods outlined here and discussed further in Chapters 20 and 21.

The most cogent value of stratification, however, is not considered or acknowledged in the statistical arguments. Those arguments are perhaps pertinent if a randomized clinical trial is regarded mainly as a statistical exercise, intended to demonstrate whether treatment A is, on average, more effective than treatment B.

This demonstration, which is the main goal of a randomized trial, has allowed the trials to have had magnificent statistical success. When practicing clinicians want to apply the results, however, the main issues are clinical, not just statistical. To treat individual patients, a good clinician wants to know the therapeutic results in pertinent clinical subgroups, not merely whether treatment A is better on average than treatment B.

Consequently, the main scientific point is usually missed in statistically oriented discussions of different methods and timing for multivariable analysis in randomized trials. If the results are to be used in the real world beyond the published document, clinicians must be able to determine what can be expected for individual patients who suitably resemble the pertinent clinical subgroups in reports of treatment. An appropriate, effective stratification is the best way to identify those subgroups.

Chapter References

1. Bishop, 1975; 2. Feinstein, 1967; 3. Kassirer, 1991; 4. Grizzle, 1969; 5. Fienberg, 1980; 6. Agresti, 1990; 7. Goodman, 1984; 8. Miah, 1983; 9. Morgan, 1963; 10. Feinstein, 1990; 11. Katz, 1991; 12. Moss, 1976; 13. Gardner, 1975; 14. Kannel, 1992; 15. Miettinen, 1976; 15a. Smith, 1988; 16. Freedman, 1987; 17. Kannel, 1990; 18. Krone, 1985; 19. Marantz, 1992; 20. Maetani, 1980; 21. Rosenbaum, 1985; 22. Spiegelhalter, 1984; 23. Seymour, 1990; 24. Begg, 1990; 25. Taves, 1974; 26. Pocock, 1975; 27. Begg, 1980; 28. Feinstein, 1976.

20 Multicategorical Stratification

Outline

20.1. Sequential sequestration
  20.1.1 Simple multivariate stratifications
  20.1.2. Formation of multibinary ordinal unions
    20.1.2.1. Ordinal stages for pattern of symptoms
    20.1.2.2. Primary-symptom union
  20.1.3. Directional application
20.2. Conjunctive consolidation
  20.2.1. Examining candidate conjunctions
  20.2.2. First step
  20.2.3. Subsequent steps
  20.2.4. Application in patients with prostatic hyperplasia
20.3. Demonstration in illustrative data set
  20.3.1. Preliminary analyses
  20.3.2. Initial explorations
  20.3.3. Special 'three-way' computer program
  20.3.4. First consolidation
  20.3.5. Second consolidation
    20.3.5.1. Sequential addition
    20.3.5.2. Parallel addition
  20.3.6. Third consolidation
  20.3.7. Subsequent examinations
  20.3.8. Statistical summaries
20.4. Illustrations in published literature
  20.4.1. Bivariate decision rules
  20.4.2. Display of bivariate stratifications
20.5. Concluding comments

In the three analytic methods called *multicategorical stratification*, a target variable is estimated from ordinal strata that contain composite multicategorical clusters. In all three methods, the clusters are formed with a sequential set of decisions, called a *decision rule*. In two of the methods, *sequential sequestration* and *conjunctive consolidation*, which are discussed in this chapter, the decision rule relies on a combination of biologic and statistical judgments. At each step of the operations, substantive principles are used to make the results sensible and biologically cohesive; and statistical principles are used to choose clustered categories that have both quantitative and stochastic significance.

For the third method, *recursive partitioning*, which is discussed in Chapter 21, the decision rule can be automated according to a selected set of statistical criteria.

20.1. Sequential sequestration

Sequential sequestration is the simplest method of multicategorical stratification. Each component of the clustered composite categories usually contains contributions from no more than two independent variables.

20.1.1. Simple multivariate stratifications

In one mode of operation, the process begins with an examination of target rates in tables showing the two-way conjunction of all pairs of the multiple independent variables. The cells in the array of tables are examined to find the particular pair of conjoined categories that exceeds modicum size while having the most strikingly extreme (high or low) result in the target rates. This cell will usually occur at the corner of one of the tables, having an extreme ordinal or binary category for both constituent variables. With a particularly effective or transcending category, an entire row or column may have the desired attribute.

The selected cell (or row or column) forms the first stratum. The subgroup of persons in that stratum is then removed (i.e., sequestered) from the total group under analysis. The pairs of tables are then reexamined in the remaining persons to find the conjoined pair of categories with the next most extreme target rate. This subgroup of persons is then sequestered, and the pairs of tables are reexamined for the remaining group. The process continues iterating until no more "effective" subgroups can be identified. After the iteration is completed, the remaining members of the nonsequestered group are joined in a terminal union. Sequestered subgroups with similar target rates can also be joined in terminal unions.

Stage	All Characteristics Must Be Present	At Least One Characteristic Must Be Present	5-Year Survival Rate (%)
Composite I	Tumor not disseminated; symptoms not deleterious; co-morbid gastrointestinal tract disease present; no periobtrusive features; no anemia; no prognostic co-morbidity	Tumor small; tumor above 8 cm	32/52 (61)
Composite II	Tumor not disseminated; symptoms not deleterious; no periobtrusive features; no anemia; no prognostic co-morbidity	Tumor low and not small; no co-morbid gastrointestinal tract disease	32/82 (39)
Composite III	Tumor not disseminated; symptoms not deleterious; no prognostic co-morbidity	Periobtrusive features; anemia	21/97 (22)
Composite IV	...	Tumor disseminated; deleterious symptoms; prognostic co-morbidity	6/87 (7)

Fig. 20-1 Results of sequential sequestration process to produce a multivariate composite staging system that can be employed preoperatively in patients with cancer of the rectum. The sequestration goes in an upward direction, beginning with demarcation of the composite IV stratum. (The figure is a reproduction of Table 2 in Chapter reference 1. For further details, see text and the reference.)

This process was used to form a prognostic staging system from preoperative morphologic, clinical, and co-morbid data for patients with cancer of the rectum.[1] The results of the sequential sequestration are shown in Fig. 20-1. In this example, the selection process went in a direction from lowest to highest survival rates.

The individual components of terminal-cluster unions are separated by semicolons in the third column of Fig. 20-1. The first three chosen categories were joined in a terminal-union stage called "Composite IV". It contained patients who had disseminated tumor and/or deleterious symptoms and/or prognostic co-morbidity. Each of these categories had been sequestered as an entire row (or column) during the first three steps of the process. They were united in a terminal union because they had similar prognostic portents and survival rates. The second main sequestration led to another terminal-union stage, called "Composite III", containing patients with anemia and/or a union of clinical manifestations called "periobtrusive features". Further details of the procedure can be found in the published report.[1]

20.1.2. Formation of multibinary ordinal unions

Although applicable to general multivariable challenges, sequential sequestration can be used in an even simpler mode of operation. The latter approach was particularly valuable for producing (or confirming) the *multibinary ordinal unions* mentioned earlier (Section 5.3.3) that formed staging systems for several major clinical variables in the illustrative data set of patients with lung cancer.

The original coded data contained binary expressions for presence or absence of such elemental clinical manifestations as hemoptysis, fever, bone pain, hoarseness, etc. The analytic goal was to arrange these multiple binary categories into three sets of hierarchical unions. The first set of unions would combine the elemental manifestations into groups of phenomena attributed to such "pathophysiologic" sources as bronchial, parenchymal, or parietal locations in the lung. The second set of unions would combine the pathophysiologic-source groups into a larger set of groups related "anatomically" to pulmonic, systemic, regional, mediastinal, or metastatic manifestations. The third step would arrange and anatomically related groups into the ordinal categories of a variable called *symptom stage*. The term *multibinary ordinal unions* describes the result of combining (and segregating) the original set of multiple-binary-variable elements into an ordinal array of stages.

Before the procedure began, all of the symptomatic, pathophysiologic, and anatomic categories had been previously selected and ranked by "clinical judgment". The role of sequential sequestration, described in the next subsection, was to separate the overlap of the multiple binary manifestations and to confirm that the clinical rankings had a distinctive prognostic gradient.

20.1.2.1. Ordinal stages for pattern of symptoms

The last phase of the procedure is shown in Fig. 20-2. The diverse categories of binary manifestations had previously been arranged[2] into a set of pathophysiologic unions, and the set of pathophysiologic unions had then been arranged into the seven anatomically oriented unions shown in the first column of Fig. 20-2.

510 Targeted-Cluster Methods

Median survival times (in months) for sequential sequestration of cohort according to categories of symptoms

(1) Sequence of Sequestered Categories	(2) Results in Original Cohort Before any Sequestration		(3) Results in Sequestered Category		(4) Results in Remaining Cohort After Removal of This Category	
	N	Median	N	Median	N	Median
ALL PATIENTS	1266	4.6	—	—	—	—
Distant symptoms	194	1.9	194	1.9	1072	5.5
Quasi-metastatic symptoms	241	1.8	197	2.2	875	6.6
Mediastinal symptoms	147	3.3	105	3.8	770	7.2
Regional symptoms	138	3.0	83	4.1	687	7.8
Systemic symptoms	494	2.9	209	5.3	478	10.6
Pulmonic symptoms	1091	4.3	394	8.8	84	16.3
Asymptomatic	84	16.3	84	16.3	—	—

Fig. 20-2 Successive results of sequential sequestration procedure discussed in Section 20.1.2.1. (The figure is a reproduction of Table 1 in Chapter reference 2.)

Before sequential sequestration of the previous multibinary unions, the top row of Fig. 20-2 shows that the entire cohort of 1266 patients had a median survival of 4.6 months; and the second column of the table shows survival medians for each of the cited six unions of symptoms and for the asymptomatic group. The trend of the medians in column 2 generally corresponds to what would be expected clinically, but the gradient is not distinctive: the distant and quasi-metastatic groups had essentially similar medians, as did the mediastinal, regional, and systemic groups. The sequential sequestration was aimed at magnifying the distinctions of this gradient by appropriately separating its constituents.

In this mode of operation, the sequestration process works in a manner analogous to TNM staging.[3] Patients with the worst manifestations are placed in the worst group. The remaining patients with the next-worst manifestations are placed in the next-worst group. The process continues iterating until patients with none of the bad manifestations are placed in the "best" group. For example, in the first step of a three-group TNM staging for cancer, the patient's information is checked to see whether any evidence exists for distant metastases. If it does, the patient is placed in Stage III. If it does not, the data are checked for evidence of regional metastases. If any is found, the patient is placed in Stage II. If not, the patient is placed in Stage I.

In the sequestration here, the category with the lowest survival results was removed first. In this instance, *distant symptoms* was chosen first because it and the next *(quasi-metastatic)* category had similar survival medians (1.9 and 1.8 months, respectively) in the total cohort, but the 6-month survival rate (not shown in Fig. 20-2) was lower for distant symptoms (16.5%) than for quasi-metastatic symptoms (22.4%).

After the 194 patients in the *distant symptom* category were sequestered in column 3, the second row of Fig. 20-2 shows, in column 4, that the remaining 1072 patients in the cohort had a median survival of 5.5 months. In the next step, median survival times were examined for each symptom category of the remaining 1072 patients. The lowest median survival, 2.2 months, occurred in the 197 patients who had quasi-metastatic but no distant symptoms. When those 197 patients were sequestered as shown in columns 3 and 4 of the third row, the remaining population of 875 patients had a median survival of 6.6 months.

The next step led to the sequestration of 105 patients, without distant or quasi-metastatic symptoms, who had mediastinal symptoms. The median survivals were 3.8 months in those patients, and 7.2 months in the remaining 770 patients. The rest of Fig. 20-2 shows the next four steps in the sequestration process, leading to successive removal of the regional symptom group (83 patients; median survival 4.1 months), the systemic symptom group (209 patients; 5.3 months), and the pulmonic symptom group (394 patients; 8.8 months). The remaining category of asymptomatic patients contained 84 people with a median survival of 16.3 months.

The sequential sequestration results in column 3 thus gave statistical confirmation to the judgmentally formed ordinal rankings. Each of the ordinal symptom categories, when examined in the sequestered absence of its predecessors, exerted a distinct monotonic effect on the survival gradient. This effect was retained regardless of whether survival was examined for medians or (in results not shown here) for proportions of 6-month, 1-year, or other survival intervals. Furthermore, the relative similarity of results in the first two categories (distant and quasi-metastatic) and in the next two categories (mediastinal and regional) would justify a subsequent judgmental decision to combine each of those two categorical pairs into a single category for the symptom stages discussed in Section 5.3.3.

20.1.2.2. Primary-symptom union

Sequential sequestration was also applied for an analogous but different purpose in an analysis that preceded the work shown in Fig. 20-2. The *pulmonic symptom* group in Fig. 20-2 contained three unranked (or "nominal") pathophysiologic unions, denoted as parenchymal, parietal, and bronchial symptoms. As binary categories, these constituent symptoms could be present or absent in various overlapping combinations for the 394 patients who had *pulmonic symptoms* alone, without more deleterious manifestations such as systemic, regional, mediastinal, quasi-metastatic, or distant symptoms. The goal of the sequential sequestration was to separate the constituent pulmonic categories into a nonoverlapping ordinal array.

Median survival times (in months) for sequential sequestration of nominal categories in the pulmonic-symptom union

(1) Sequence of Sequestered Categories	(2) Results in Original Group Before Any Sequestration		(3) Results in Sequestered Category		(4) Results in Remaining Group After Removal of This Category	
	N	Median	N	Median	N	Median
Pulmonic symptoms union	394	8.8	—	—	—	—
Parenchymal symptoms	269	7.0	269	7.0	125	11.6
Parietal symptoms	196	8.2	60	10.0	65	15.8
Bronchial symptoms	334	8.3	65	15.8	—	—

Fig. 20-3 Successive results of sequential sequestration procedure for pulmonic-symptom union discussed in Section 20.1.2.2. (The figure is a reproduction of Table 2 in Chapter reference 2.)

Column 2 of the table in Fig. 20-3 shows that before anything was sequestered for the 394 pulmonic-symptom patients, the 269 with parenchymal symptoms had the lowest median survival, 7.0 months. After sequestration of this group, as shown in column 3, the lowest median survival, 10.0 months, was in the 60 members of the parietal symptom group. With their separation, the remaining survival median was 15.8 months in the 65 members of the bronchial symptom group.

Despite the apparent statistical heterogeneity (7.0, 10.0, and 15.8 months) of the median survival times, the three component elements were eventually retained in a single *pulmonic-symptom* union. There were three reasons for this decision. First, a separation of the elements might have seemed clinically peculiar. Second, none of the component medians was either lower than the 5.3-month median shown in Fig. 20-2 for the adjacent *systemic* group or higher than the 16.3-month median for the adjacent *asymptomatic* group. Third, the *parenchymal* group, which had the lowest median survival of the three components, contained patients with dyspnea. Since severe dyspnea was a constituent of a separate variable[2] for *severity of illness*, the effects of severe dyspnea would be "captured" by the latter variable. When those effects were later removed by the *severity* variable, the survival median would rise for the remaining patients in the *parenchymal* group. Thus, after severity of illness was accounted for, the components of the pulmonic union became statistically more homogeneous than what is shown in Fig. 20-3.

The sequential sequestration principle, although not so labelled, seems to have been used when Winkel et al.[4] derived "a monotone function of the (prognostic) ranking of the classes" that produced three ordinal stages from nine clinical manifestations of patients with aortic valve regurgitation.

20.1.3. Directional application

The sequential direction of the sequestration must be followed when the results are applied. As noted earlier, the procedure resembles what is done in an ordinal TNM staging system for cancer. Persons are assigned to a particular category only if they successfully pass through the intervening cascade of categories.

20.2. Conjunctive consolidation

The conjunctive consolidation procedure is probably the most scientifically appealing form of multicategorical stratification. All the decisions involve consideration of substantive as well as statistical homogeneity, and the individual variables are completely combined (i.e., consolidated) without having "leftover" elements that may later form "heterogeneous" categories.

Heterogeneity of components is often an inevitable feature of sequential sequestration and recursive partitioning, because the different variables are split, and isolated categories from each variable can be joined at each step to produce multicategorical strata composed of diverse constituents. Category a_1 from variable X_1 may be joined with category b_2 of variable X_2. Category a_2 from X_1 may be joined with category c_1 of variable X_3; and category b_1 of variable X_2 may be joined with category d_3 of variable X_4 and category e_2 of variable X_5.

Conjunctive consolidation was developed to avoid this biologically heterogeneous mixture of constituent categories. Because an *entire* variable, with all of its categories, is included at each step, the conjunctive consolidation process is aimed at reducing variables, rather than isolating or aggregating categories. The retained variables become *composite variables* having multivariate constituent categories.

20.2.1. Examining candidate conjunctions

The process begins by examining all of the two-way stratified tables that can be formed from the candidate independent variables. If seven independent variables have "passed" the bivariate screening inspection, a total of 21 tables (= 7 × 6/2) will show outcome rates of the target event in each cell of the conjoined pairs of variables.

For the first step in the process, one of these tables is selected as best for showing the desired combination of statistical impact and biologic coherence (or clinical sensibility). Statistical impact is shown with the double gradient or "gradient-within-gradient" phenomenon (discussed in Section 3.2.4.3) that demonstrates a distinctive effect for each of two conjoined variables within the context of the other. (The phenomenon was illustrated in Tables 3-2 and 3-6). From the various tables that show suitable double gradients, and that have suitable results in the other statistical criteria discussed in Chapter 8, the pair of conjoined variables that also seems most substantively appropriate becomes the best choice for the first consolidation.

20.2.2. First step

The first step of the consolidation process is illustrated with the same data (Table 4-5) that were used in Section 4.6.2.2.2 to demonstrate the principles of biologic coherence and statistical isometry in forming a targeted decision rule.

In the research[1] that produced the cited data for patients with rectal cancer, eight other independent variables were available, beyond anatomic spread and symptom pattern. The additional variables included prognostic co-morbidity, age, sex, cell type, hematocrit, gastrointestinal co-morbidity, size of tumor, and location of tumor. The pairwise conjunction of these ten variables produced 45 tables to be examined. Many of the tables could be immediately dismissed because they did not show an impressive double gradient. Among those that did, the conjunction of anatomic spread and symptom pattern was chosen as best because these two variables had the best biologic coherence. Many of the symptoms could be directly attributed to anatomic abnormalities, but the symptom categories had produced substantial gradients within the anatomic categories, despite the biologic (and statistical) correlation.

The conjunction of the symptom stage and anatomic stage variables previously shown in Table 4-5 is consolidated with dashed lines in Table 20-1, and the three composite stages are labeled with the letters A, B, and C. The process followed the guidelines of the decision rule described in Section 4.6.2.5. The combination of endorectal anatomic and indolent symptom categories in Stage A had a

5-year survival rate of 64% (= 36/56). Stage B was formed as a cluster of two cells, one containing the conjunction of indolent symptoms and vicinal spread; and the other conjoining obtrusive symptoms and endorectal localization. Since the respective survival rates in these two cells were 6/15 (40%) and 37/79 (47%), the survival rate in the new stage B was 43/94 (46%). All the remaining cells in Table 20-1 were combined to form Stage C, with a total survival rate of 11/93 (12%). The original survival rate of 37% was thus divided into three clusters of Symptom-Anatomic strata, called Composite S-A Stages, that had respective survival rates of 64%, 46%, and 12%.

Table 20-1 Two-way table showing 5-year survival rates for patients with rectal cancer, classified according to postexcisional anatomic stage and symptom stage*

Symptom stage	Postexcisional anatomic stage			Total (1%)
	Endorectal [A]	Vicinal [B]	Distant [C]	
Indolent	36/56 (64%)	6/15 (40%)	1/18 (6%)	43/89 (48%)
Obstrusive	37/79 (47%)	6/35 (17%)	4/30 (13%)	47/144 (33%)
Deleterious	0/2 (0%)	—	0/8 (0%)	0/10 (0%)
Total	73/137 (53%)	12/50 (24%)	5/56 (9%)	90/243 (37%)

*Each denominator shows number of patients in the category. Each numerator shows number of patients who survived at least 5 years. Dashed lines show pattern of consolidation. For further details, see text and Reference 20-1.

Because so many tables must be examined, the choice of the best first table is probably the trickiest task in conjunctive consolidation. The task can be remarkably eased if a *biologic* choice is made beforehand. If the biologically best table has a statistically appropriate double gradient, this table can be chosen promptly without intense scrutiny of all the others. Regardless of how the choice is made, however, once the first composite variable is formed, the rest is relatively easy.

20.2.3. Subsequent steps

Because two of the available k variables will have been removed by the first consolidation, only k − 2 variables remain. They need no longer be examined in pairs with one another, and can be directly checked for their conjunction with the composite variable formed in step 1. Consequently, in step 2, only k − 2 tables are examined.

Sometimes, however, the previous pairwise examination in step 1 may show another pair of variables that are "good contenders". These two variables might then be consolidated in step 2 to form a second composite variable. The two pairs of composite variables can then be consolidated in step 3. This alternative approach is illustrated in Section 20.3.6.

The one-variable-at-a-time approach was used in the second step of the conjunctive consolidation process for the rectal cancer data. The Composite S-A stage

Composite S-A Stage	Patients Without Prognostic Co-morbidity (%)	Patients With Prognostic Co-morbidity (%)
A	32/47 (68)	4/9 (44)
B	41/80 (51)	2/14 (14)
C	11/82 (13)	0/11 (0)
Total	84/209 (40)	6/34 (18)

Fig. 20-4 5-year survival rates for conjunction of prognostic co-morbidity and composite symptom-anatomic (s-a) stage in 243 patients with resected cancer of rectum. (The figure is a reproduction of Table 5 in Chapter reference 1.)

variable formed in the first step was conjoined (or intersected) with each of the remaining eight independent variables, and survival rates were examined in the cells of the stratified tables. The best of those tables is shown here as Fig. 20-4, where the conjunction of prognostic co-morbidity led to another double gradient phenomenon. Each category of prognostic co-morbidity produced a distinctive survival gradient within each S-A stage; and each composite S-A stage produced a distinctive gradient within the binary categories of prognostic co-morbidity. Among patients with prognostic co-morbidity, the three cells for composite S-A stage all had membership size below a modicum of 20 (and in one instance, below 10). Nevertheless, although intercategory tests were not stochastically significant, the chi-square score for linear trend in the three cells was 6.6, with $P < .05$. Accordingly, the table was statistically acceptable, and was conjunctively consolidated.

Table 20-2 Conjoined consolidation of categories shown in Fig. 20-4, to form three new composite stages: Alpha, beta, and gamma

Composite S-A stage	Prognostic co-morbidity	
	Absent	Present
A	Alpha	Beta
B	Beta	Gamma
C	Gamma	Gamma

Table 20-2 shows the pattern of combination when the prognostic co-morbidity variable and the composite S-A variable were joined to form three new composite S-A-C stages that are labelled *Alpha, Beta,* and *Gamma*. The 5-year survival rates in these three S-A-C composite stages were Alpha, 32/47 (68%); Beta 45/89 (51%); and Gamma, 13/107 (12%). These three stages have strong quantitative gradients that are stochastically significant between adjacent stages; and the size of each stage exceeds the modicum.

In the next step, when the composite S-A-C stages were conjoined with the remaining seven candidate predictor variables, none of the two-way tables showed a significant double gradient. Accordingly, the conjunctive consolidation process was ended.

20.2.4. Application in patients with prostatic hyperplasia

After transurethral partial prostatectomy (TURP) was accused[5] of causing higher 5-year mortality than an abdominal complete prostatectomy (OPEN) in patients with benign prostatic hyperplasia, Concato et al.[6] suspected that the results, which came from 1284 cases in a medico-fiscal claims data base, had not been adequately classified for severity of co-morbidity. The Concato group then reviewed and analyzed data from 252 appropriately assembled medical records, using an improved classification of co-morbidity.

To show the concomitant effect of age and co-morbidity on postsurgical mortality, Concato et al. used three types of multivariable "adjustment". In one procedure, the three ordinal categories of age, three ordinal categories of co-morbidity, and two categories of surgery were used as independent variables in a logistic regression, for which the dependent variable was mortality at five years after surgery. In a second procedure, the nine age–co-morbidity categories received an epidemiologic "standardization"; and mortality was examined for each type of surgery in the standardized groups. The third approach, which is illustrated here, used conjunctive consolidation to form a staging system for age and co-morbidity.

Table 20-3 Conjunctive consolidation of age and co-morbidity, forming three prognostic stages for 5-year mortality after surgery for prostatic hyperplasia

Grade of co-morbidity	Age (years)						Total	
	55–64		65–74		75–84			
None	2/49	(4%)	2/39	(5%)	3/13	(23%)	7/101	(7%)
		I				II		
Intermediate	7/40	(18%)	6/62	(10%)	11/32	(34%)	24/134	(18%)
				III				
Severe	10/2	(0%)	3/4	(75%)	5/11	(45%)	8/17	(47%)
Total	9/91	(10%)	11/105	(11%)	19/56	(34%)	39/252	(16%)

The mortality results found by Concato et al. for the conjunction of age and co-morbidity, irrespective of surgery, are shown in Table 20-3. Although the small numbers of patients in the severe co-morbidity group prevent a clear demonstration of a double-gradient phenomenon, the results seemed consistent enough to warrant a suitable consolidation. The nine bivariate categories of Table 20-3 were therefore combined, as shown by the dotted lines, into three composite age–co-morbid stages. The severe co-morbidity row formed Stage III. The remaining oldest age group formed Stage II, and all the other categories formed Stage I.

The respective mortality rates for composite Stages I, II, and III, as shown in the row totals of Table 20-4, were 9%, 31%, and 47%. The total results for two types of surgery showed the anticipated higher 5-year mortality for TURP (17.5%) vs. OPEN (13.5%). Although this difference was not stochastically significant (X^2 = .76; P = .38), it had similar magnitudes and was significant at P < .05 in the larger original study[5] derived from claims data. When examined within the three composite stages, however, the mortality differences essentially vanished. The mortality rates were identical for TURP vs. OPEN in Stage I and slightly lower for TURP in Stage II. (The small numbers in Stage III preclude an effective comparison.)

Table 20-4 5-year mortality rates for two types of surgery in composite stages formed in Table 20-3

Composite Stage	Type of surgery				Total	
	TURP		OPEN			
I	8/89	(9%)	9/101	(9%)	17/190	(9%)
II	7/23	(30%)	7/22	(32%)	14/45	(31%)
III	7/14	(50%)	1/3	(33%)	8/17	(47%)
Total	22/126	(17.5%)	17/126	(13.5%)	39/252	(15%)

Because the Concato group was making a claim of "no difference" in the two treatments, the incremental differences were checked for confidence intervals. The 95% confidence intervals ranged from −.08 to +.08 in Stage I, from −.28 to +.26 in Stage II, and from −.050 to +.129 in the totals. The results within the composite consolidated stages thus offered support for the investigators' contention that 5-year mortality was not affected by the type of prostatic surgery.

20.3 Demonstration in illustrative data set

In Chapter 16, to compare the results with those of Cox regression, a conjunctive consolidation was applied for 5-year survival rates in the illustrative data set of 200 patients with lung cancer. The results were shown in Tables 16-1 and 16-2. To show important operating details that were not described earlier, a complete conjunctive consolidation is presented now for the illustrative data set, with the binary target variable being the **alive/dead** status at 6 months.

20.3.1. Preliminary analyses

After the preliminary analyses discussed (and illustrated) in Chapter 9, *amount of cigarette smoking* was eliminated as a candidate variable because it did not produce a bivariate effect. *Male* was omitted because the female category had only eighteen members. To simplify the illustration here, *cell type* was also omitted, thus confining the analysis to six independent variables.

Before the analysis began, the dimensional variables for *age, percent weight loss, hematocrit,* and *progression interval* had each been ordinally partitioned into three categories as follows:

Age: ≤ 54, 55–69, ≥ 70
Percent weight loss: none, >0 – <10%, ≥ 10%
Hematocrit: < 37, 37– <43, ≥ 44
Progression interval (mos.): ≤ 4.0, 4.1–8.9, ≥ 9.0

The *age* demarcation was chosen to separate the older and younger groups from the middle categories. The splits for percent weight loss and hematocrit were chosen for reasons discussed in Section 8.4.4.5. Progression interval was divided approximately into tertiles, avoiding splits at noninteger cutpoints.

20.3.2. Initial explorations

In the first step of the analysis, the six independent variables were mutually conjoined to form 15 (= 6 × 5/2) two-way tables, with 6-month survival rates appearing in the cells. The largest two-way table had 20 cells formed by conjunction of the five TNM stage and four symptom stage categories. The smallest tables had 9 cells formed by several conjunctions of two variables that each had three categories.

Before the 15 tables were examined, the consolidation of symptom stage and TNM stage seemed particularly appealing clinically as a first step, because both variables reflect anatomic effects of the cancer. (The consolidation of these two variables had also been a desirable first step when similar analyses were done for cancer of the rectum, as shown earlier here, and in studies reported elsewhere when conjunctive consolidation was used to form prognostic staging systems for cancers of the breast,[7] larynx,[8] endometrium,[9] and prostate[10] and for Hodgkin's disease[11] and acute leukemia.)[12]

20.3.3. Special 'three-way' computer program

None of the existing BMDP or SAS PROGRAMS will conveniently produce the type of "internal-cell" 3-way table, introduced in Chapter 3 and shown many times thereafter, that makes 2-way stratifications easy to evaluate. In such a table, the two conjoined independent variables appear in the rows and columns; and rates for the binary target variable are shown in the cells. Consequently, a special computer program, prepared by Carolyn K. Wells, was used for this purpose. The printout for the conjunction of TNM stage and symptom stage is shown in Fig. 20-5.

Because of the small numbers in many cells, not all of the stochastic features were significant in Fig. 20-5. The total chi-square results of 45.20, 18.67, and 30.04, respectively, for the total table, row margins, and column margins had values of P < .001 for the corresponding degrees of freedom, which were 12, 4, and 3. The P values were < .05, however, only for the TNM stage I row, and for the asymptomatic and pulmonic-systemic stage columns. The chi-square test for linear trend, interpreted at 1 degree of freedom, also had P < .05 in these same groupings.

Because of many small cells, below the modicum size of 20 (or even 10), a perfect double gradient is not apparent; but its basic directions are retained whenever the cell sizes are large enough. Because the table was a biologic best choice and seemed to have reasonable statistical credentials, it was selected for the first consolidation.

20.3.4. First consolidation

The two variables in Fig. 20-5 were joined, as outlined in Table 20-5, to form four symptom-TNM composite categories called S-T stages. The five categories in the *distant* symptom stage were first demarcated as the *worst* stage because its low marginal total of 26% was strikingly different from the adjacent 41%. The remaining

```
TARGETTED CONTINGENCY TABLE PROGRAM -- CA LUNG DATA

6 MONTH SURVIVAL
              SXSTAGE
                 ASX        PULSYS      MEDREG       DIST        TOTALS      CHI-SQ
TNMSTAGE

      I        11/  13     23/  27      0/   0      4/  11      38/  51
             (  84.62)   (  85.19)   (*******)   (  36.36)   (  74.51)   (  10.75)

      II       2/   3     11/  21      0/   0      2/   3      15/  27
             (  66.67)   (  52.38)   (*******)   (  66.67)   (  55.56)   (   0.39)

      IIIA     3/   4      4/   8      3/   4      0/   2      10/  18
             (  75.00)   (  50.00)   (  75.00)   (   0.00)   (  55.56)   (   3.82)

      IIIB     2/   6      6/  17      3/   7      2/   6      13/  36
             (  33.33)   (  35.29)   (  42.86)   (  33.33)   (  36.11)   (   0.18)

      IV       0/   2      4/   9      6/  18      8/  39      18/  68
             (   0.00)   (  44.44)   (  33.33)   (  20.51)   (  26.47)   (   3.36)

      TOTALS  18/  28     48/  82     12/  29     16/  61      94/ 200
             (  64.29)   (  58.54)   (  41.38)   (  26.23)   (  47.00)   (  18.67)

      CHI-SQ (   8.65)   (  12.99)   (   2.35)   (   4.65)   (  30.04)   (  45.20)
```

Fig. 20-5 Targeted "3-way" contingency table for 6-month mortality rates in conjunction of TNM stage and symptom stage. Denominators in each cell show number of patients in the bivariate category. Numerators show the corresponding number of 6-month survivors; and parentheses show the survival percentage. The rows and columns marked CHI-SQ show the corresponding values of chi-square (calculated as sum of [observed−expected]^2/expected results in cells) for each row or column. The lower right CHI-SQ value of 45.20 indicates results for all cells in the table. Additional values of chi-square for linear trend, not shown here, were calculated separately. In all instances, the latter values were slightly smaller than those shown for CHI-SQ in each row and column. The TNM stages are labelled with their "official" designations. The symptom stages (SXSTAGE) are abbreviated as ASX = asymptomatic; PULSYS = pulmonic or systemic stage; MEDREG = mediastinal or regional symptoms; and DIST = distant symptoms.

three categories in TNM Stage 1 were demarcated next as the best stage, because they had the highest survival rates (85%). The remaining 12 cells all had small membership that did not individually allow confident decisions. Using statistical principles about spanning the meridian (see Section 8.2.2.2.2), the 12 cells were split into one group with survival rates all above 50%, and another group with rates all below 50%. This split also had biologic coherence because the first group combined TNM Stages II and IIIA; and the second group contained TNM Stages IIIB and IV.

Table 20-5 Pattern of conjunctive consolidation for data in Fig. 20-5

TNM stage	Symptom Stage				Total
	(1) Asymptomatic	(2) Pulmonic-systemic	(3) Mediastinal regional	(4) Distant	
(1) I	11/13 (85%)	23/27 (85%) [a]	—	4/11 (36%)	38/51 (75%)
(2) II	2/3 (67%)	11/21 (52%)	— [b]	2/3 (67%) [d]	15/27 (56%)
(3) IIIA	3/4 (75%)	4/8 (50%)	3/4 (75%)	0/2 (0%)	10/18 (56%)
(4) IIIB	2/6 (33%)	6/17 (35%)	3/7 (43%) [c]	2/6 (33%)	13/36 (36%)
(5) IV	0/2 (0%)	4/9 (44%)	6/18 (33%)	8/39 (21%)	18/68 (26%)
Total	18/28 (64%)	48/82 (59%)	12/29 (41%)	16/61 (26%)	94/200 (47%)

The composite stages in Table 20-5 are marked **a, b, c, d** for identification in subsequent usage. The consolidated results, best shown in the row totals of the subsequent Table 20-6, had a striking intrasystem gradient, from 85% to 26%, with intercategory gradients of 27%, 22%, and 10%; and each of the 4 composite cells is well above modicum size. The results are stochastically significant for the total chi-square score ($X^2 = 38.6$; $P = 5.4 \times 10^{-8}$) and for the chi-square for linear trend ($X_L^2 = 36.8$; $P = 1.06 \times 10^{-6}$). The gradients between adjacent stages show $X^2 = 7.4$ ($P = .007$) for **a** vs. **b**, $X^2 = 4.6$ ($P = .03$) for **b** vs. **c**, and $X^2 = 1.2$ ($P = .27$) for **c** vs. **d**.

20.3.5. Second consolidation

After the first step in the conjunctive consolidation process, the second step can contain a sequential or a parallel addition. In the *sequential* approach, a single variable is added and consolidated, being chosen as the best of the $k - 2$ variables that remain after the first two variables have been joined. The *parallel* approach adds another composite variable, formed from a good conjunction that was noted when the $(k)(k - 1)/2$ pairs of tables were initially examined before the first consolidation. Both approaches are illustrated here.

20.3.5.1. Sequential addition

In Table 20-6, hematocrit is conjoined as a single variable to the composite stages—**a, b, c, d**—that were formed in Table 20-5. The distinctive double gradient in Table 20-6 indicates that hematocrit exercises an independent effect within the composite stages formed by the other two variables. (The ordinal sequence of hematocrit was reversed in Table 20-6 to maintain consistency with other tabular arrangements in which survival decreases as the categories go rightward.)

Table 20-6 Conjunctive consolidation adding hematocrit to TNM-symptom composite stages of Table 20-5

TNM-symptom composite stage	Hematocrit categories			Total
	≥ 44	$\geq 37 - < 44$	< 37	
a	10/11 (91%)	18/21 (86%)	6/8 (75%)	34/40 (85%)
b	9/14 (64%)	13/21 (67%)	1/5 (20%)	23/40 (57%)
c	8/14 (57%)	10/32 (31%)	3/13 (23%)	21/59 (36%)
d	9/22 (41%)	7/29 (24%)	0/10 (0%)	16/61 (26%)
Total	36/61 (57%)	48/103 (47%)	10/36 (28%)	94/200 (47%)

A best conjunctive consolidation is not immediately evident for the 12 cells of Table 20-6. The first two cells in the top row should obviously be joined because of their similar survival rates (91% and 86%). The bottom right-hand cell, with a survival rate of 0% is distinctive but its size is too small—only 10 members—to form a separate stratum. The upper right-hand corner cell, with a survival rate of 75%, was 11% below and 8% above its respective nearest survival "neighbors" (at 86% and 67%). In a judgmental decision, the 75%-survival cell was allowed to remain as part of the top row. To achieve a three-stage stratification, the remaining cells were then divided, according to the meridian principle, into two groups with components having individual survival rates above or below 50%.

The demarcating lines placed on Table 20-6 show the pattern of consolidation that forms the composite S-T-H staging for these three variables. The composite stages and their associated survival rates can be labelled as **best,** 34/40 (85%); **middle,** 30/49 (61%); and **worst,** 30/111 (27%). The last stage is obviously heterogeneous, with component cells having survival rates that range from 0 to 41%. The component rates are homogeneous, however, in that they are all below 50%.

20.3.5.2. Parallel addition

When pairs of the original candidate variables were examined in step 1, an interesting conjunction was produced by *hematocrit* and *percent weight loss,* as shown in Table 20-7.

Table 20-7 Results and pattern of consolidation for 6-month survival rates in conjunction of percent weight loss and hematocrit in illustrative data set

Percent weight loss	Hematocrit			
	≥ 44	37–43	< 37	Total
None	17/25 (68%)	13/24 (54%)	2/6 (33%)	32/55 (58%)
< 10%	12/25 (48%)	23/45 (51%)	3/9 (33%)	38/79 (48%)
≥ 10%	7/11 (64%)	12/34 (35%)	5/21 (24%)	24/66 (36%)
Total	36/61 (59%)	48/103 (47%)	10/36 (28%)	94/200 (47%)

In addition to having a reasonably appropriate double gradient, this pair of variables was attractive clinically because their consolidation could offer a surrogate marker for functional severity of the cancer. Since hematocrit values below 37 indicated anemia, and since weight loss above 10% usually augured an unfavorable prognosis,[2] we initially expected to form a reverse L-shaped consolidation, containing five "bad" categories: two from the anemic group, two from the ≦ 10% weight loss group, and one from both. Surprisingly, however, the "plethoric" patients (with hematocrits ≧ 44) had relatively high survival (64%) in the group with ≧ 10% weight loss. As another surprise, the plethoric patients had a relatively low survival rate (48%) in patients with the "intermediate" weight loss of < 10%. Because these surprises might arise merely from the sampling process with relatively small numbers, we decided to let both of the surprising bivariate categories stay in the "good" group, for which four of the five component survival rates exceeded 50%.

Accordingly, hematocrit and percent weight loss were consolidated in the truncated reverse-L shape shown with dotted lines in Table 20-7. The two categories of the hematocrit weight loss (H-W) composite variable were designated as **bad** and **good** for functional status. The **bad** group contained four consolidated conjoined categories with a total survival rate of 22/70 (31%), and the **good** group contained five, with total survival rate 72/130 (55%).

20.3.6. Third consolidation

In a purely sequential approach, the third step would have occurred after each of the remaining candidate variables was conjoined with the composite S-T-H stages formed in Table 20-6. In those conjunctions, *age* did not show a double gradient. *Percent weight loss* and *progression interval* both showed interesting possibilities, but the apparent improvements in prognostic gradients did not seem striking enough to warrant the complexity of adding another variable to the composite group. With no further variables added in step 3, a composite S-T-H staging system could be formed from consolidations that took two steps.

Table 20-8 Conjunction of composite S-T variable from Table 20-5 and composite H-W variable from Table 20-7

Symptom-TNM composite variable	Hemotocrit–weight loss composite variable		Total	
	Good	Bad		
a	26/29 (90%)	8/11 (73%)	34/40	(85%)
b	17/29 (59%)	6/11 (55%)	23/40	(57%)
c	16/36 (44%)	5/23 (22%)	21/59	(36%)
d	13/36 (36%)	3/25 (12%)	16/61	(26%)
Total	72/130 (55%)	22/70 (31%)	94/200	(47%)

Alternatively, because the composite hematocrit-percent-weight-loss "severity" variable (in Table 20-7) was clinically appealing, a parallel approach was used to conjoin the composite S-T and composite H-W variables in a third consolidation. As shown in Table 20-8, the conjunction of these two composite variables produced a striking complete double gradient. For the consolidation process, the two cells at the extreme corners of Table 20-8 were left intact. They had para-zenith and para-nadir survival rates of 90% and 12%, respectively. To avoid an excessively large number of stages, the remaining cells were partitioned with the meridian principle into those whose survival rates exceeded 50% and those with rates below 50%. The resulting pattern of four composite stages, marked **alpha, beta, gamma,** and **delta,** is shown in Table 20-9. Because four variables were consolidated, the composite stages have tetravariate contents.

Table 20-9 Conjunctive consolidation of Table 20-8 to form a tetravariate S-T-H-W composite staging system

S-T composite	Hematocrit–weight loss composite	
	Good	Bad
a	26/29 **alpha** (90%)	**beta**
b		31/51 (61%)
c	34/95 (36%)	**gamma**
d		**delta** 3/25 (12%)

Each composite stage in Table 20-9 has more than 20 members, and the intrasystem gradient goes from 90% to 12%, with intercategory gradients 29%, 25%, and 24%. All of the diverse stochastic tests that might be applied are significant at $P < .05$.

20.3.7. Subsequent examinations

The composite four-variable stages formed in Table 20-9 were next conjoined with each of the remaining variables, of which none except *progression interval* was effective. The most striking impact of progression interval (not shown here) was in the composite stage **beta**, where the survival rates dropped successively from 16/22 *(73%)* to 9/15 *(60%)* to 6/14 *(43%)* as the progression interval declined, respectively, from ≥ 9.0 months, to 4.1–8.9 months, to ≤ 4.0 months. In the other composite stages, however, the impact of *progression interval* was less distinctive, and the group sizes were relatively small. Because the composite staging already seemed sufficiently complex, *progression interval* was not joined as an additional variable; and the process was ended after the third consolidation.

20.3.8. Statistical summaries

Table 20-10 shows a summary of the statistical accomplishments of the different variables and consolidations for the illustrative data set. The accomplishments

Table 20-10 Summary of statistical accomplishments for bivariate analyses and conjunctive consolidations

	Variable	No. of categories	Total chi-square	Chi-square for linear trend	No. of concordant errors in fit	Intra-system gradient	% 6-month survival and (no. of members) in extreme cells	
							Zenith	Nadir
Bivariate analyses	TNM stage	5	30.0	29.3	64	48%	75% (51)	26% (68)
	Symptom stage	4	18.1	18.0	72	38%	64% (28)	26% (61)
	Age	3	0.6	0.4	94*	8%	49% (43)	41% (32)
	Percent weight loss	3	5.8	5.8	85	22%	58% (55)	36% (66)
	Hematocrit	3	8.9	8.7	86	31%	59% (61)	28% (36)
	Progression interval	3	8.4	6.6	83	22%	54% (81)	32% (65)
Conjunctive consolidations	Symptom-TNM (S-T) composite stages (Table 20-6)	4	38.6	36.8	60	59%	85% (40)	26% (61)
	Hematocrit–weight loss (H-W) composite stages (Table 20-7)	2	10.5	†	80	24%	55% (130)	31% (70)
	S-T-Hematocrit composite stages (Table 20-8)	3	44.6	44.5	55	61%	85% (40)	24% (89)
	S-T-H-W composite stages (Table 20-9)	4	42.2	42.1	60	78%	90% (29)	12% (25)

*No category spanned the meridian of 50% survival. †Same as total chi-square (for two categories)

are shown as total chi-square scores, chi-square scores for linear trend, number of concordant errors in fit, intrasystem gradient, zenith survival rate, and nadir survival rate for the bivariate analyses and for four conjunctive consolidations.

The original number of 94 errors in concordant fit is reduced to 64 errors in the bivariate results for TNM staging alone, yielding a relatively high lambda score of (94 − 64)/94 = .32. Although those 64 errors are thereafter reduced to only 60, the four-variable S-T-H-W composite system raises the intrasystem gradient from 48% to 78%, the total chi-square score from 30.0 to 42.2, and the chi-square linear trend score from 29.3 to 42.1. The three-variable S-T-H system, however, with only three composite categories, had fewer errors (55) for concordant fit than the four-variable S-T-H-W system, and had higher values in both chi-square scores. On the other hand, the S-T-H-W system had a much larger intrasystem gradient.

If a choice must be made between the two latter composite systems, the S-T-H arrangement is probably preferable because it is simpler (having both fewer variables and fewer stages) as well as better scores for fit and stochastic significance. The S-T-H-W system might be used for the relatively high accuracy of predictions for individual patients in its zenith and nadir stages.

20.4. Illustrations in published literature

Because of its relative novelty, conjunctive consolidation has not yet been widely employed for multicategorical stratification. Aside from the examples already cited in the text, the published consolidations have not extended beyond the conjunction of two variables.

20.4.1. Bivariate decision rules

In several publications, an apparent conjunctive consolidation is used to display the decision rule that converts the categories of two variables into a decision about a third.

In one recent example, shown in Fig. 20-6, Wong et al.[13] used a conjunctive pattern to express therapeutic conclusions from a cost-benefit analysis of treatment for chronic stable angina pectoris. Four ordinal categories of coronary vasculature were conjoined with binary splits for mild/severe angina and for normal/depressed ventricular function. Close examination of Fig. 20-6, however, shows that it could be consolidated into 11 zones rather than 14, because the therapeutic recommendations for severe angina are identical for either category of ventricular function.

In another example, shown in Fig. 20-7, a three-level job-strain variable was constructed by consolidating three levels of "psychological demands" with three levels of the individual's "control" over activities in the job. The partition boundaries were formed by demarcating results of a logistic regression for diverse variables affecting cigarette smoking among employees at a chemical plant.[14]

The decision rules for two diagnostic algorithms, shown in Fig. 20-8, were used[15] to diagnose myocardial infarction as **definite, possible,** or **absent** (i.e., **no**)

Multicategorical Stratification 525

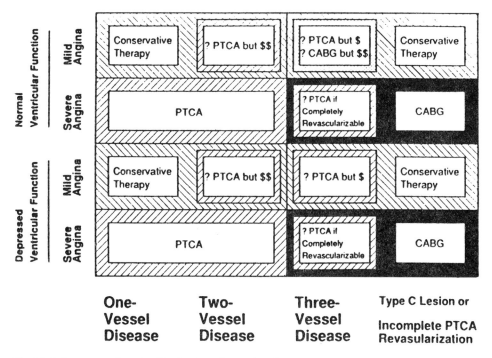

Fig. 20-6 Summary of results. The horizontal axis classifies patients by anatomy, the vertical axis by the clinical severity of angina (as a marker of ischemia) and of ventricular function. The $ symbol denotes a high and the $$ symbol denotes a very high cost-effectiveness ratio. The organization of the figure corresponds to (showing) the mildest disease in the upper left quadrant and the most severe disease in the lower right. PTCA = percutaneous transluminal coronary angioplasty; CABG + coronary artery bypass grafting. (Figure and text of legend are taken from Fig. 3 of the Chapter reference 13.)

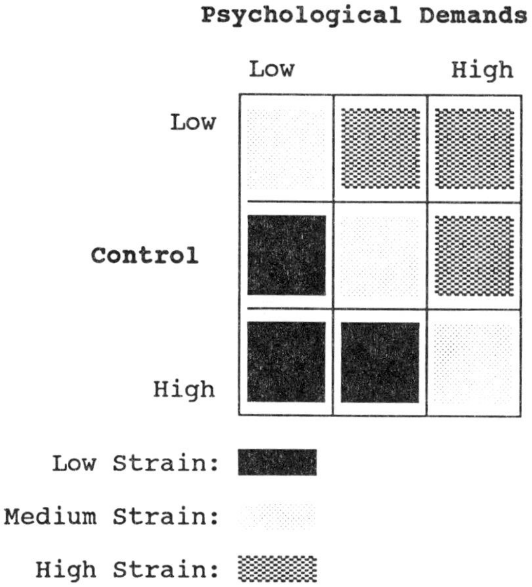

Fig. 20-7 Construction of three-level job-strain composite variable. (Figure is taken from Fig. 1A of Chapter reference 14.)

in patients with and without chest pain. The two variables used for the decisions were four categories of *enzymes* and four categories of classification in the Minnesota code for electrocardiograms. Note that the categorical approach here, unlike the methods of algebraic analysis, allows the use of unknown data for a category called **missing enzymes.**

20.4.2. Display of bivariate stratifications

An alternative purpose of a conjunctive-consolidation format is to display data for a bivariate stratification.

In Fig. 20-9, the incidence of coronary heart disease in the famous Framingham study[16] is shown for participants categorized simultaneously according to three ordinal levels of triglyceride and HDL cholesterol concentrations. The investigators did not actually consolidate the two risk factors, presumably because a double gradient was not present in the results.

In another study, the authors[17] developed a risk score for coronary heart disease by combining features of cigarette smoking, serum cholesterol, and systolic blood pressure. The risk score was then related to five grades of socioeconomic status, which was a composite variable constructed by the conjunctive consolidation of income level and education level, as shown in Fig. 20-10. The investigators do not describe the mechanism used for consolidation, but the risk scores (presumably mean values) were as follows for the five socioeconomic strata: **I,** 5.5; **II,** 7.7; **III,** 9.0; **IV,** 10.8; and **V,** 13.3. The authors concluded that the "risk in Group V (13.3) is more than double that in Group I (5.7)".

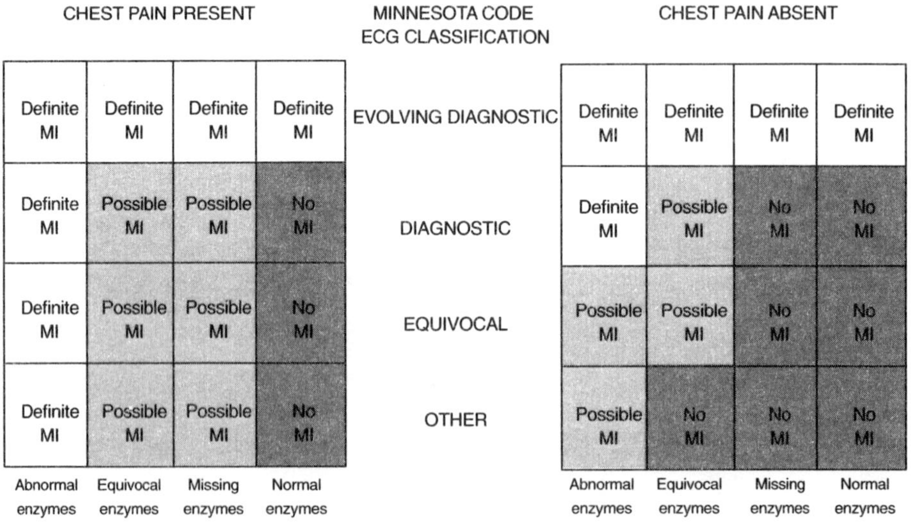

Fig. 20-8 Diagnostic algorithm for myocardial infarction (MI) cases: the Minnesota Heart Survey. If an autopsy is available and shows an acute myocardial infarction within 8 weeks of death, the case is coded as definite MI. ECG, electorcardiogram. (Figures and text of legend taken from Fig. 1 of Chapter reference 15.)

Multicategorical Stratification 527

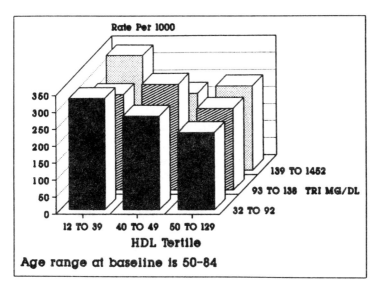

Fig. 20-9 Rate of coronary heart disease development in relation to triglyceride (TRI) and high-density lipoprotein (HDL) levels (age-adjusted 14-year rates in men, Framingham Heart Study). (Figure and text of legend taken from Fig. 5 of Chapter reference 16.)

20.5. Concluding comments

The main goal of this chapter (and the next one) is to give you a reasonably good idea of the available alternatives to the conventional algebraic methods of targeted multivariable analysis. The algebraic methods have become popular because computerized automation has eliminated the horrendous calculations that would have been necessary in the pre-computer era.

Beyond the mere easing of calculations, however, computers offer entirely new methods, which were also not readily available before, to display and examine data. With these displays, an investigator (and reader) can determine exactly

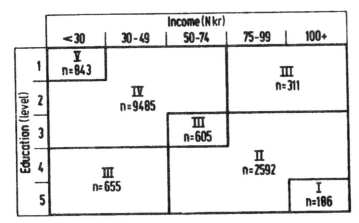

Fig. 20-10 Definition of five groups (I–V) of socioeconomic status by means of certain combinations of income and education in 14,677 symptom-free men aged 40–49. (Figure and text of legend taken from Chapter reference 17.)

what is happening, rather than relying on arbitrary mathematical configurations and on arcane "black box" computations. The gradients that indicate trends in the effects of variables can be shown directly as increments from one category to the next, rather than indirectly as coefficients for an average effect.

The judgmental methods of multicategorical stratification are not perfect and can be improved in various ways. For example, in the conjunctive consolidation process, certain *biologic* guidelines might be automated to produce greater consistency in the judgmental decisions; cells with small sizes (below the modicum) might be easier to interpret for consolidations if their target results were temporarily "stabilized" by imputation from adjacent cells; and consolidation decisions might more often use a cost matrix than the current appraisals of quantitative gradients and stochastic significance. In the existing automated methods of recursive partitioning, discussed in the next chapter, mechanisms might be developed that allow certain biologic judgments and criteria to be incorporated more effectively into the mathematical decisions.

The challenges offer exciting opportunities for collaboration between medical persons, who will articulate the substantive principles of judgment, and statisticians, who will employ or create the appropriate mathematical tools. Many of the disadvantages of the existing algebraic methods occur because they were developed with exclusively mathematical goals. As the biologic and mathematical connoisseurs collaborate to aim at mutually shared goals, the categories or strata can be specified and verified by substantive wisdom; and the best methods to aggregate the categories can be created by statistical sagacity. Both disciplines can gain in sharing the challenge, the intellectual fun, and the achievement.

Chapter References

1. Feinstein, 1975; 2. Feinstein, 1990; 3. American Joint Committee for Cancer Staging and End-Results Reporting, 1978; 4. Winkel, 1973; 5. Roos, 1989; 6. Concato, 1992; 7. Charlson, 1974; 8. Feinstein, 1977; 9. Wells, 1984; 10. Clemens, 1986; 11. Boyd, 1978; 12. Boyd, 1979; 13. Wong, 1990; 14. Green, 1990; 15. McGovern, 1992; 16. Castelli, 1989; 17. Holme, 1976.

21 Recursive Partitioning

Outline ....

21.1 General principles
21.2. Automated interaction detector (AID) program
    21.2.1. Basic operation and first step
    21.2.2. Subsequent steps
    21.2.3. Determining reduction in variance
        21.2.3.1. Dimensional target
        21.2.3.2. Binary target
    21.2.4. 'Terminal unions'
    21.2.5. Illustrations from literature
        21.2.5.1. Diagnosis of headaches
        21.2.5.2. Risk factors for gallstones
        21.2.5.3. Clearance of methotrexate
        21.2.5.4. Energy survey data
21.3. Chi-square automated interaction detection (CHAID)
21.4. Autogroup procedure (AUTOGRP)
21.5. Polarized stratification (STRAT)
21.6. Classification and regression trees (CART)
    21.6.1. Gini diversity index
        21.6.1.1. Illustration for binary targets
        21.6.1.2. Illustration for polytomous targets
    21.6.2. The 'twoing rule'
        21.6.2.1. Illustration for binary targets
        21.6.2.2. Illustration for polytomous targets
    21.6.3. Problems of interpreting polytomous outcomes
    21.6.4. Error penalties and credit scores
    21.6.5. Indexes of evaluation
        21.6.5.1. 'Pruning' of tree
        21.6.5.2. 'Costs of misclassification'
        21.6.5.3. 'Importance of variables'
        21.6.5.4. 'Validation' techniques
    21.6.6. Application of CART to lung cancer data set
    21.6.7. Another portrait of CART printout
    21.6.8. Medically oriented discussions
    21.6.9. American College of Rheumatology series
21.7. Knowledge SEEKER (KS) program
21.8. Additional multicategorical arrangements
    21.8.1. Survival analysis
    21.8.2. Longitudinal analysis
21.9. Concluding comments
....

Recursive partitioning produces multicategorical stratification by forming a tree-like pattern of stepwise branching partitions. Although the process can be carried out judgmentally, its popularity has been increased by statistical algorithms that allow automated operation.

21.1 General principles

As illustrated by the example in Fig. 19-5, each step in recursive partitioning begins with an "ancestral" group, also called a "root node", which can be symbolized as {C}. The members of {C} all have the characteristic described by either a single category or several conjoined categories. $\{C_0\}$ is the original group before partition, and (if the symbol is needed) $\{C_i\}$ can represent any of the different i groups (or subgroups) formed thereafter. At the end of each step in the partitioning process, the initial {C} is either left intact or is divided when conjoined with the parts of a candidate split. When the process is completed, the groups that emerge without further splitting are called terminal nodes.

In the KS (KnowledgeSEEKER) procedure discussed in Section 21.7, the splits can have three or more parts. In all other automated procedures, however, the splits have two parts formed by added candidate variables that have been divided dichotomously. Because the discussion is easier with binary splits, they will be the main operation described in this chapter.

In the explorations that precede each step, the existing {C} is examined for the effects of conjunctive splits with all independent variables available as candidates. For the conjunctions, each candidate variable is dichotomously split into such categories as a_1, a_2 for variable X_1; b_1, b_2 for variable X_2; d_1, d_2 for variable X_3, etc. When the split new variable is conjoined with the ancestral group, {C} is converted to a new "descendant" pair of multivariate categories, each containing attributes of both the ancestral group and the conjoined category. For example, if attributes b_1 and b_2 of variable X_2 are selected for the split, the result is as follows:

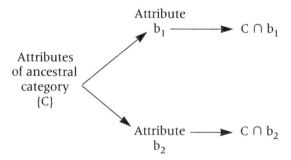

At least four automated recursive-partitioning procedures have been formally described and named with abbreviations or acronyms as AID, CHAID, CART, and STRAT. A fifth procedure, called AUTOGRP, allows judgmental decisions, often applied with the statistical strategies of AID, or CHAID. A sixth procedure, called KS, can be operated in either an automated or interactive (judgmental) mode. As discussed in the next six sections, the diverse recursive-partitioning methods differ mainly in the basic statistical strategy (reducing variance or improving fit) used for the decisions in splitting and evaluating.

In most of the illustrations here, recursive partitioning is aimed at binary target variables, but it can also be applied to dimensional or polytomous targets or to survival curves.

21.2. Automated interaction detector (AID) program

The earlier form of recursive partitioning was developed by social scientists at the University of Michigan. With "interaction" used as a label for conjoined categories, the program was called the *Automated Interaction Detector*,[1] abbreviated as AID. What the program actually produces, however, is conjoined clusters of composite categories from diverse variables, rather than explorations for a complete conjunction of variables.

21.2.1. Basic operation and first step

In the first pass through the data, the AID process examines the effect on the target event of binary splits in all independent variables. The best split is chosen for producing the greatest reduction in variance of the original system. For example, suppose the target rate is 100/200 = 50% in a group with four independent variables, A, B, D, and E, that have two, two, three and four ordinal categories, respectively.

In explorations for the first step, the total group is the root or ancestral node. The outcome rates (or means) in the target variable are checked for the conjunction of $\{C_0\}$ with binary splits of each independent variable. For binary variables A and B, results would be compared for category a_1 vs. category a_2, and for category b_1 vs. b_2. For variables (such as D and E) that have more than two categories, binary splits are created and checked for each successive downward (or upward) level of categories, somewhat like the receiver-operating-curve process used in demarcating a boundary for a diagnostic marker test.

The example here refers to the four independent variables, A, B, D, and E, that were cited in the first paragraph. The explorations would examine seven possible splits (one for variable A, one for B, two for D, and three for E), from which the AID program would choose the one that gives the greatest proportionate reduction in system variance. Suppose this best split shows 65/90 = 72% for category b_1 and 35/110 = 32% for category b_2 in variable B.

21.2.2. Subsequent steps

The subsequent steps are illustrated in Fig. 21-1.

At the second step, each of the first two selected subgroups—$\{b_1\}$ and $\{b_2\}$—would form the substrate or ancestral groups to be examined. For the explorations, each subgroup would be divided into the splits formed by conjunctions with the remaining variables (A, D, and E). With one split available for A, two for D, and three for E, the program would examine twelve splits—six that divide subgroup b_1 and six for subgroup b_2. Suppose the greatest proportionate reduction in variance at this second step was produced by the split of b_1 together with c_3 vs. b_1 together with c_1, c_2. With commas to show the combination of two (or more) cate-

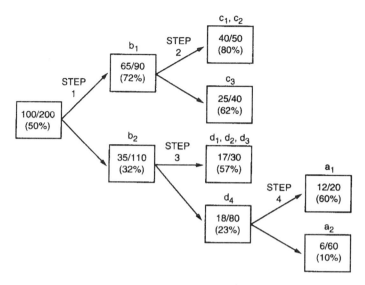

Fig. 21-1 Recursive partitioning with the Automated Interaction Detector (AID) procedure to produce five strata, containing components from variables A, B, C, and D. (For further details, see text.)

gories in a single variable, the split could be written as $b_1 \cap c_3$ vs. $b_1 \cap (c_1, c_2)$. The actual results might have been:

$b_1 \cap (c_1, c_2)$: 40/50 *(80%)*
$b_1 \cap c_3$: 25/40 *(62%)*.

The sums of numerators and denominators for the two groups in a split always add up to the value of the numerator and denominator in the unsplit ancestral group. In this instance, the sums are 65/90, as previously noted for b_1 alone.

At the end of step 2, three strata or nodes will have been formed: the b_2 subgroup, and the two subgroups just produced by the split of b_1. In the next step, these three strata are conjoined with all available categorical splits in other variables to see whether any further significant reductions can be produced in the system variance. The result at the end of step 3 might produce the following four strata:

$b_1 \cap (c_1, c_2)$: 40/50 *(80%)*
$b_1 \cap c_3$: 25/40 *(62%)*
$b_2 \cap (d_1, d_2, d_3)$: 17/30 *(57%)*
$b_2 \cap d_4$: 18/80 *(23%)*

In the next step, these four strata would intersect with whatever else is available for conjunctions that might reduce variance. The result might lead to the following five strata:

$b_1 \cap (c_1, c_2)$: 40/50 *(80%)*
$b_1 \cap c_3$: 25/40 *(62%)*
$b_2 \cap (d_1, d_2, d_3)$: 17/30 *(57%)*
$b_2 \cap d_4 \cap a_1$: 12/20 *(60%)*
$b_2 \cap d_4 \cap a_2$: 6/60 *(10%)*

At this point, these five strata would be conjoined with the available remaining categories, but the process might then be concluded because no other splits can be found to produce stochastically significant reductions in variance.

21.2.3. Determining reduction in variance

The method of measuring reduction in variance depends on whether the target variable contains dimensional or binary data. The different calculations were shown in Chapters 7 and 8, but a simple illustration is presented here to refresh your memory.

21.2.3.1. Dimensional target

For any ancestral group, the data for a dimensional target variable can be symbolized as $\{Y_i\}$, with mean $\bar{Y}$ and group size N. After the split, the data for the target variable in members of the first group will comprise a series of values designated as $\{Y_h\}$, with mean $\bar{Y}_1$ and group size n_1. The corresponding data in the second group will be $\{Y_j\}$, with mean $\bar{Y}_2$ and group size n_2. For the cited symbols, $\{Y_i\} = \{Y_h\} + \{Y_j\}$; $N = n_1 + n_2$; $N\bar{Y} = n_1\bar{Y}_1 + n_2\bar{Y}_2$; and $\Sigma Y_i^2 = \Sigma Y_h^2 + \Sigma Y_j^2$.

For example, the ancestral group might contain the data {1, 2, 3, 4, 6, 7, 8} with mean $\bar{Y} = 4.23$ and N = 7. If the data set is split between items 4 and 6, the first group will contain {1, 2, 3, 4}, with mean $\bar{Y}_1 = 2.5$ and $n_1 = 4$; and the second group will contain {6, 7, 8} with mean $\bar{Y}_2 = 7.0$ and $n_2 = 3$.

The original basic group variance will be $S_{yy_0} = \Sigma(Y_i - \bar{Y})^2$, which, in this example, turns out to be 41.71. The new group variance in the two splits will be

$$S_{yy_1} + S_{yy_2} = \Sigma(Y_h - \bar{Y}_1)^2 + \Sigma(Y_j - \bar{Y}_2)^2.$$

The results here turn out to be $\Sigma(Y_h - \bar{Y}_1)^2 = 5$ and $\Sigma(Y_j - \bar{Y}_2)^2 = 2$, with a sum of 7. The basic group variance will have been reduced from 41.71 to 7.

If the original data set had been split between items 3 and 4, rather than between 4 and 6, the two new groups would have been $\{Y_h\} = 1, 2, 3$ and $\{Y_j\} = 4, 6, 7, 8$. Their group variances would have been $S_{yy_1} = 2$ and $S_{yy_2} = 8.75$. Because 10.75, which is the sum of $S_{yy_1} + S_{yy_2}$, is larger than in the previous split, the first split would be preferred for its lower group variance.

It can be shown algebraically that the reduction in group variance is

$$S_{yy_0} - (S_{yy_1} + S_{yy_2}) = [n_1 n_2 / N][\bar{Y}_1 - \bar{Y}_2]^2 \qquad [21.1]$$

For example, in the first split, this formula produces $[(4)(3)/7][2.5 - 7]^2 = 34.71$, which is the value previously found for $41.71 - 7$. In the second split, the corresponding result is $[(3)(4)/7][2 - 6.25]^2 = 30.96$, which is $41.71 - 10.75$. Because the split that minimizes $S_{yy_1} + S_{yy_2}$ will maximize the value of $[n_1 n_2 / N][\bar{Y}_1 - \bar{Y}_2]^2$, formula [21.1] offers a simple index for assessing variance reduction for splits in groups of dimensional data.

21.2.3.2. Binary target

As noted in Section 8.2.2.1, the basic group variance in the binary target variable will be NPQ for an ancestral group that contains N members with target event rate $P = T/N$ and $Q = 1 - P$. The new group variance will be $n_1 p_1 q_1 + n_2 p_2 q_2$ if the split produces two groups with respective sizes n_1 and n_2, and with respective target rates $p_1 = t_1/n_1$ and $p_2 = t_2/n_2$. If $N = n_1 + n_2$ and $T = t_1 + t_2$, the reduction in variance will be NPQ $- (n_1p_1q_1 + n_2p_2q_2) = \{[T(N - T)]/N\} - \{[t_1(n_1 - t_1)/n_1] + [t_2(n_2 - t_2)/n_2]\}$.

For example, at the first step in Fig. 21-1, NPQ = (100)(100)/200 = 50. After the first split, $n_1p_1q_1 + n_2p_2q_2$, using the shortcut formula, will be [(65)(25)/90] + [(35)(75)/100] = 18.06 + 23.86 = 41.92. The group variance will have been reduced by 50 − 41.92 = 8.08, which is a proportionate reduction of 8.08/50 = .1616.

The chosen split at each step is intended to minimize the value of $n_1p_1q_1 + n_2p_2q_2$, thus maximizing the value of NPQ $- (n_1p_1q_1 + n_2p_2q_2)$. The proportionate reduction in group variance will be

$$\phi^2 = (\text{NPQ} - \Sigma n_h p_h q_h)/\text{NPQ}. \qquad [21.2]$$

For the example in Fig. 21-1, the AID process has produced five terminal strata. The original group variance of the system has been reduced from 50 to 34.9. [The latter value is obtained as $(40 \times 10/50) + (25 \times 15/40) + (17 \times 13/30) + (12 \times 8/20) + (6 \times 54/60)$.] The value of ϕ^2 would be (50 − 34.9)/50 = .302; and so about 30% of the original group variance will have been "explained" by the partition into these five strata.

21.2.4. 'Terminal unions'

As noted in Section 19.5.1.4, some of the terminal strata (or nodes) generated in earlier or later steps of the procedure may have relatively similar results for the target event. For example, in Fig. 21-1, the three terminal strata in the middle of the vertical array have event rates of 62%, 57%, and 60%. If these three strata are combined to form a terminal union, the final set of clusters for the stratification in Fig. 21-1 could be labelled as Group I [containing $b_1 \cap (c_1, c_2)$]. Group III [containing $b_2 \cap d_4 \cap a_2$] and a middle Group II, which contains everyone else. The final stratified outcome rates would be as follows:

Group I: 40/50 *(80%)*
Group II: 54/90 *(60%)*
Group III: 6/60 *(10%)*

The result is a relatively simple and relatively easy-to-remember stratification, containing three groups with a striking prognostic gradient, a proportionately reduced group variance of .302[= (50 − 35)/50], and an error rate reduction of .48[= (100 − 52)/100].

21.2.5. Illustrations from literature

Having been available and automated for many years, the AID procedure has been applied for both medical and nonmedical analyses. Several of the published results are used as illustrations here.

21.2.5.1. Diagnosis of headaches

Diehr et al.[2] used the AID program to separate the characteristics of migraine and tension headaches in 726 patients with headache in an acute care setting.

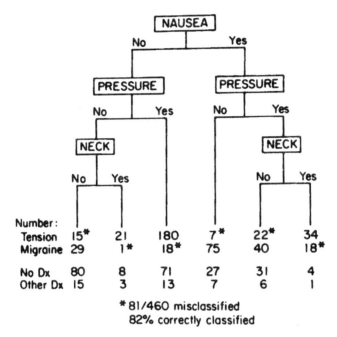

Fig. 21-2 "Final tree" constructed with "best 3 variables" associated with migraine vs. tension diagnoses in 460 patients with acute headaches. (Figure taken from Fig. 3 in Chapter reference 2.)

After a recursive partitioning for five independent variables referring to presence (or absence) or nausea, pressure or tightness in back of neck, pain worst at back of neck, previous use of ergot, and a clinical prodrome, the authors constructed "a final tree", shown here as Fig. 21-2, from only the first three cited variables. If the tree in Fig. 21-2 were suitably arranged with terminal unions and shown in stratified format (for proportions of patients in each category who had migraine), an effective clinical diagnostic system to separate migraine from tension headache might be as follows:

Stage I:	Nausea without pressure (regardless of neck pain)	75/82 *(91%)*
Stage II:	No nausea, pressure, or neck pain; OR nausea and pressure, but no neck pain	69/106 *(65%)*
Stage III:	Nausea and pressure and neck pain	18/52 *(35%)*
Stage IV:	Pressure without nausea, regardless of neck pain; OR neck pain without pressure or nausea	19/220 *(9%)*
	Total	181/460 *(39%)*

The staging system produces an impressive intrasystem gradient of 82%, intercategory gradients of 26%, 30%, and 26%, a reduction in concordant errors from 181 to (7 + 37 + 18 + 19) = 81, and an explained group variance of .46[= (109.78 − 59.62)/109.78]. The diagnostic estimates are particularly accurate, at rates of 91%, in the first and fourth stages.

21.2.5.2. Risk factors for gallstones

Nomura et al.[3] examined three binary variables (presence/absence of *fatty liver, obesity,* and *male sex*) and two dimensional variables *(age and total cholesterol)* as risk factors for the presence or absence of gallstones in a general population in Okinawa.

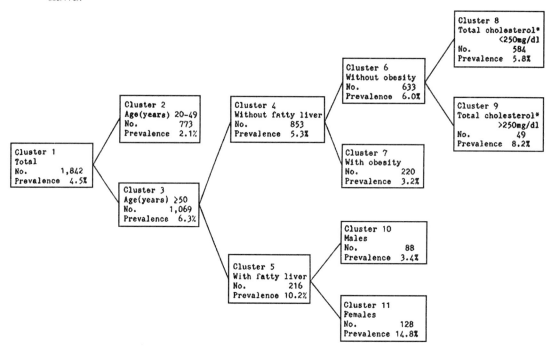

Fig. 21-3 Automated interaction detector analysis of factors associated with gallstone disease in 1,842 persons over 20 years of age, Yaeyama District, Okinawa, Japan, 1984. (*serum total cholesterol levels [mg/dl].) (Figure and legend taken from Fig. 3 of Chapter reference 3.)

Fig. 21-3 shows the main results of the AID analysis of these data. The original gallstone prevalence of 4.5% was highest in partitions for older age, fatty liver, and women, and was lowest (2.1%) in younger people. In older people without fatty liver, prevalence was reduced by obesity and by low total cholesterol values.

Because of large group sizes, the results were stochastically significant for all of the displayed partitions. On the other hand, the results do not show striking gradients. The highest intrasystem gradient in Fig. 21-3 is 14.8 − 2.1% = 12.7%, and it covers four strata that intervene between the lowest and highest values. The AID analysis is thus valuable for demonstrating that the gradient is relatively unimpressive. This distinction would not be clearly shown in algebraic analyses where the coefficients for each variable might indicate its relative impact as an average odds ratio that does not denote the actual occurrence rates or incremental changes in those rates. Thus, in addition to the AID analysis, Nomura et al. reported multiple logistic regression odds ratios for the same five variables shown in Fig. 21-3. Only two of the five odds ratios were stochastically significant; and they had the relatively small values of 2.9 for age $\geq$ 50 and 2.1 for presence of a fatty liver.

The AID analyses also offer a way of checking the algebraic-model odds ratios for several zones of the data. In the first split in Fig. 21-3, the prevalence of gallstones was 6.3% for age $\geq$ 50 and 2.1% for age < 50. The corresponding odds ratio is $(.063/.021)(.979/.937) = 3.13$. In persons at age $\geq$ 50, gallstones were present in 10.2% with fatty liver and in 5.3% without fatty liver. This odds ratio is $(.102/.053)(.947/.898) = 2.03$. In both instances, the logistic regression results (i.e., 2.9 and 2.1, respectively) were reasonably close to the "gold standard" of the direct results.

21.2.5.3. Clearance of methotrexate

The original Sonquist and Morgan technique for the AID procedure[1] was applied by Crom et al.[4] to determine how the clearance of methotrexate in children newly diagnosed with acute lymphocytic leukemia was affected by sex, age (in four ordinal categories), creatinine clearance (in three ordinal categories), and SGPT (serum glutamic pyruvic transaminase level, in four ordinal categories). The overall systemic clearance in 108 patients was a mean of 78.4 ± 15.4 standard deviation units in ml./min./m^2.

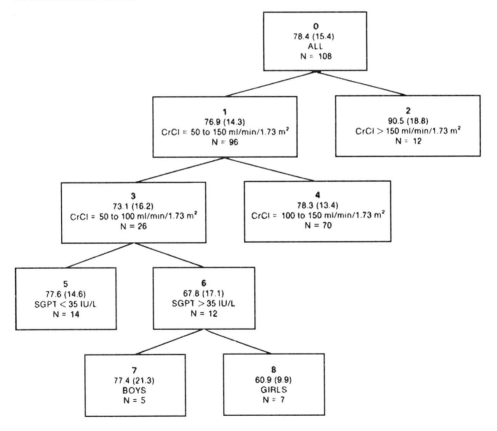

Fig. 21-4 Use of the automated interaction detector method to identify patient characteristics related to methotrexate clearance. The mean (SD) MTX clearance (ml/min/m^2) is shown for each of the "final" groups and their "parent" groups, as determined by the AID analysis. N is the number of patients in each group, and the connecting lines indicate the splits that formed each subgroup. (Figure and almost all of the legend taken from Fig. 2 in Chapter reference 4.)

The AID analysis of the dimensional results, shown in Fig. 21-4, produced five terminal strata, with means ranging from a low of 60.9 to a high of 90.5. Note that in the second step of Fig. 21-4, the first variable, marked "CrCl" (for creatinine clearance), was split a second time before any additional variables were introduced.

21.2.5.4. Energy survey data

In a nonmedical application, the AID procedure was used to analyze "interactions among surveyed variables" in a study of customer utilization of energy.[5] In the first analysis, the investigators found that energy consumption was affected by such "obvious" factors as type of house, electric or nonelectric heating fuel, and number of heated rooms. The authors therefore performed a "second stage" AID analysis on the residual values formed after mean annual kilowatt hours consumption for the appropriate group was subtracted from each case.

The final results showed that energy usage increases when the home is increasingly occupied, by more people, with higher thermostat settings, and more dishwasher usage. You might have suspected these findings without an elaborate survey or multivariable analysis, but if you would like your suspicion documented, there it is.

21.3. Chi-square automated interaction detection (CHAID)

For binary target variables, the AID procedure was modified in a special chi-square arrangement called CHAID,[6] for which each reduction in group variance is algebraically reexpressed as

$$NPQ - (n_1 p_1 q_1 + n_2 p_2 q_2) = \frac{n_1 n_2}{N} (p_1 - p_2)^2. \qquad [21.3]$$

This expression is a direct counterpart of formula [21.1] showing $(n_1 n_2/N)(\bar{Y}_1 - \bar{Y}_2)^2$ for group variance reduction in partition of a mean. When formula [21.3] is divided by NPQ, the proportionate reduction in group variance becomes

$$\phi^2 = [n_1 n_2 (p_1 - p_2)^2]/N^2 PQ$$

Since $X^2 = \phi^2 N$, the formula for the chi-square test is

$$X^2 = \frac{n_1 n_2}{NPQ} (p_1 - p_2)^2. \qquad [21.4]$$

This formula can be used to calculate X^2 for any proposed binary split; and the largest X^2 will suggest the split to be selected.

The CHAID procedure was used to demonstrate the importance of risk factors for different sites of infection in 169,518 patients in the Study of the Efficacy of Nosocomial Infection Control (SENIC).[7] Because of the huge number of patients, the results form an extraordinarily elaborate structure, shown in Fig. 21-5, that resembles a gigantic bush more than a tree.

To help prune the tree during the analysis, the investigators applied several operating constraints. Adjacent categories were sequentially combined if their infection rates were not "significantly different at the 0.001 level" (presumably for

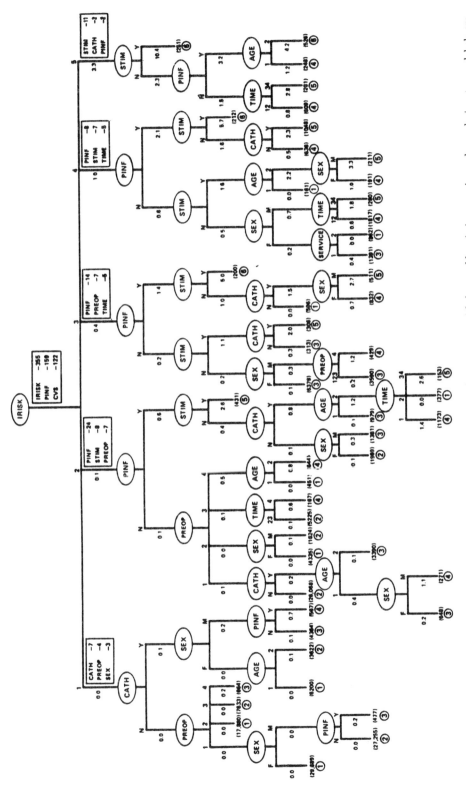

Fig. 21-5 Tree diagram of the interactions of risk factors associated with nosocomial bacteremia. Risk factor abbreviations are in ovals; analysis category labels are on horizontal lines; infection rates are on vertical lines; numbers of patients are in parentheses; and risk strata numbers are in circles. Log p values of the three risk factors having the lowest values at the given stage in the analysis are in boxes. (Figure and text of legend taken from Fig. 5 of Chapter reference 7.)

P values). Adjacent categories of a factor were combined if their infection rates were "not significantly different at the 0.15 level"; and any category "containing fewer than 150 patients was merged with an adjacent category".

21.4. Autogroup procedure (AUTOGRP)

The autogroup procedure is an interactive computer program[8] that uses an AID-like method in "conversational mode" rather than in the conventional "batch processing" approach. (The conversational approach shows the available options, from which the best split at each step is chosen judgmentally by the analyst, rather than by the strict dictates of the variance-reduction algorithm.)

In one medical application, the AUTOGRP method was used to determine a combination of clinical variables that best discriminate between bacterial and aseptic cases of pediatric meningitis.[9] The AUTOGRP procedure is probably most famous (or infamous) in clinical medicine today for its analytic role in constructing the diagnosis-related-group (DRG) taxonomy when patients were partitioned "into medically meaningful and homogeneous groups with differing lengths of hospital stay".[10]

21.5. Polarized stratification (STRAT)

In a generally unknown strategy called *polarized stratification,* the goal is to find polar strata, having target-event rates that are close to either a zenith of 100% or a nadir of 0%. Polar strata are particularly desirable for getting accuracy in concordant fit, because the estimates made from such strata are relatively error-free. For example, in a stratum with 20/20 = 100% survival, we would estimate everyone as having survived; and all estimations would be correct. In a stratum with 19/20 = 95% survival, the estimate that everyone survived would have only one error.

If the analysis is aimed at achieving polarized strata, the selected splits may be different from those chosen when the goal is to minimize variance. For example, consider the target-event rate 50/100 = 50% in an ancestral group that could be divided into the following two partitions:

a_1: 40/50 = 80% b_1: 22/22 = 100%
a_2: 10/50 = 20% b_2: 28/80 = 35%

The group variances will be (40)(10/50) + (10)(40/50) = 16 in the first partition and (22)(0/22) + (28)(52/80) = 18.2 in the second partition. The total errors in the first partition will be 10 + 10 = 20. The total errors in the second partition will be 0 + 28 = 28. Thus, according to reduction in group variance and concordant errors, the first partition would be preferred.

On the other hand, if the goal is to achieve a polar stratum, the second partition would be preferred. Its first stratum has a perfect concordant fit, i.e., no errors; and the second stratum would still be available for further partitioning with other variables.

To use the results pragmatically in predictions for future patients, a clinician might want to work with strata whose outcome rates are as close as possible to either a zenith value of 100% or a nadir value of 0%. If these polar strata contain enough patients to provide stable denominators, the clinician can feel particularly confident in predicting either success or failure if a new patient is in a stratum where the previous outcome rate was either 100% or 0%.

An automated computer program that uses this "polar" strategy was constructed many years ago by Feinstein and Landis, and reported in a brief abstract,[11] but a full description has not been published. The program works in a manner analogous to the AID process, but at each step in the recursive partitioning, the choice of splits is aimed at advancing toward a polarized (zenith or nadir) target rate, rather than reducing variance. As long as the proposed partition is stochastically significant, the preferred split at each step in the process would be in the direction of greater polarization, rather than greater reduction in variance.

Like all other recursive-partitioning procedures, the polarizing program produces arbitrarily formed clusters; and the final set of clusters may need judgmental recombinations into terminal unions, because some of the terminal nodes demarcated in the automated search may have quantitatively similar distinctions.

21.6 Classification and regression trees (CART)

After being inaugurated by social scientists in the 1960s, the AID procedure seldom appeared in clinical research. If known to consultants, the procedure was often disparaged or deplored because it lacked a "modelled" mathematical configuration. In the 1970s, however, tree-like recursive methods of nonalgebraic analysis began to develop mathematical respectability when they were "discovered" and further developed by several statisticians in California universities. According to Breiman, Friedman, Olshen, and Stone, their book on Classification and Regression Trees (CART) had its "conception" in 1980, although published[12] in 1984.

The CART procedure is a highly complex system that is not easy to understand (or explain). It is described in a heavy dose of mathematical symbols and concepts that are seldom expressed in simple prose, and seldom specifically illustrated with worked examples. An accompanying manual, called "An Introduction to CART Methodology,"[13] does not significantly improve communicative lucidity, but offers a few extra illustrations and ostensible explanations.

The CART procedure can be operated in a manner similar to AID, but offers two additional possibilities: (1) the target variable can be polytomous (rather than binary or dimensional); and (2) the main goal can be avoiding misclassifications, rather than reducing variance. If reduction in variance is the splitting criterion, the CART procedure becomes a stratified counterpart of regular regression analysis for dimensional target variables. The CART method, however, is usually applied to target variables that are nondimensional. The latter applications are the focus of discussion here.

To work with binary or polytomous (ordinal or nominal) targets, the CART procedure employs several new statistical tactics for choosing best splits. Two of those tactics—the Gini diversity index and the "twoing rule"—are analogous to methods of variance reduction.

21.6.1. Gini diversity index

Published[14] in Italian at the time of the First World War, Gini's statistical indexes have been essentially unknown in clinical research, although the "Gini coefficient" was recently said by LeClerc et al.[15] to be "widely used in economics". (The coefficient was applied by the LeClerc group as an "inequality measure" for comparing national rates of differential mortality.)

For the CART procedure, the Gini diversity index is formulated as a "measure of node impurity",

$$i(t) = \sum_{i \neq j} p(i|t)p(j|t). \quad [21.5]$$

In this expression, i(t) is the "index of impurity" for a particular group or node, t; p(i|t) is the probability (or rate) of occurrence for category i in node t, and p(j|t) is the probability (or rate) of occurrence for category j. If a particular node contains n_h people collected from a total group of N people, the proportionate amount of "total impurity" in that node is n_h/N.

21.6.1.1. Illustration for binary targets

Suppose a group has a success rate of 30/80 = .375. According to the Gini concept, this group has 30 members in one category and 50 members in the other. The values for formula [21.5] will be p(i|t) = 30/80 and p(j|t) = 50/80, so that Gini index will be (30/80)(50/80) = .234. Thus, for binary outcome data in a single group, the Gini index is simply pq, the variance of the proportion, p.

Now suppose this group is split into two groups, having success rates of 9/49 = .18 and 21/31 = .68. The first split group has 9 successes and 40 failures; the second group has 21 successes and 10 failures. The first group also contains a proportion of 49/80, and the second group contains 31/80, of the previous total. The Gini index for the split will be

$$(49/80)(9/49)(40/49) + (31/80)(21/31)(10/31) = .177,$$

and the previous total impurity will have been reduced from .234 to .177. The formula will be $[(n_1/N)p_1q_1] + [(n_2/N)p_2q_2]$.

For a binary target variable, therefore, the Gini index resembles the variance reduction score. For a binary target, the variance reduction score represents the change from NPQ to $\sum n_h p_h q_h$; the Gini index represents the change from PQ to $(1/N)\sum n_h p_h q_h$.

21.6.1.2. Illustration for polytomous targets

The Gini index has its main mathematical distinction when an ordinary variance reduction score cannot be calculated because the target variable has three or more polytomous categories. If each polytomous category has n_h members, each categorical proportion is n_h/N, where $N = \sum n_h$. A counterpart of variance is then calculated for each proportion as $[n_h/N][(N - n_h)/N]$. The sum of these individual "variances", when adjusted for subgroup sizes, forms the Gini index.

For example, suppose the original group (or node) contains 150 people with 40, 60, and 50, respectively, in the categories of **mild, moderate, severe.** According to formula [21.5] the Gini index will be

$$\left(\frac{40}{150}\right)\left(\frac{60+50}{150}\right)+\left(\frac{60}{150}\right)\left(\frac{40+50}{150}\right)+\left(\frac{50}{150}\right)\left(\frac{40+60}{150}\right)=.658.$$

Now suppose the total group is split so that the two new groups are as follows:

	Mild	Moderate	Severe	Total
Group 1	30	40	20	90
Group 2	10	20	30	60

The new Gini index will be

$$\frac{90}{150}\left[\frac{30(40+20)}{(90)(90)}+\frac{40(30+20)}{(90)(90)}+\frac{20(30+40)}{(90)(90)}\right]$$

$$+\frac{60}{150}\left[\frac{10(20+30)}{(60)(60)}+\frac{20(10+30)}{(60)(60)}+\frac{30(10+20)}{(60)(60)}\right]$$

$$=.385+.244=.629.$$

Suppose an alternative split produced the following results:

	Mild	Moderate	Severe	Total
Group 1	35	23	22	80
Group 2	5	37	28	70

The new Gini index will be

$$\frac{80}{150}\left[\frac{35(23+22)}{(80)(80)}+\frac{23(35+22)}{(80)(80)}+\frac{22(35+23)}{(80)(80)}\right]$$

$$+\frac{70}{150}\left[\frac{5(37+28)}{(70)(70)}+\frac{37(5+28)}{(70)(70)}+\frac{28(5+37)}{(70)(70)}\right]$$

$$=.347+.112=.459.$$

The second split would be preferred because it produces a smaller Gini index and thus less impurity.

21.6.2. The 'twoing rule'

An additional CART strategy uses the "twoing rule" when data are split dichotomously into a left node, t_L, and a right node, t_R. According to this rule, the selected split should maximize the value of

$$\frac{p_L p_R}{4}\Sigma_j[|p(j|t_L)-p(j|t_R)|]^2. \qquad [21.6]$$

In formula [21.6], p_L and p_R are the proportions of the total data in each node. The values of $p(j|t_L)$ and $p(j|t_R)$ are the proportions of each outcome category in each node. The Σ_j symbol means that results are added for each of the outcome categories designated as j in each node.

21.6.2.1. Illustration for binary targets

Consider the previous group (Section 21.6.1.1.) with success rate 30/80 that was split into two nodes having rates of 9/49 and 21/31. The twoing index will be

$$\frac{(49/80)(31/80)[|(9/49)-(21/31)|+|(40/49)-(10/31)|]^2}{4}=.0593[.493+.493]^2=.0577.$$

If some other split produced 10/50 and 20/30 in the success rates for the two new groups, the twoing index would be

$$\frac{(50/80)(30/80)}{4}[|(10/50) - (20/30)| + |(40/50) - (10/30)|]^2 = .0586[.467 + .467]^2 = .0510.$$

Because the goal is to maximize the twoing index, the first split would be preferred over the second.

For a binary target variable, expressed in each group with the symbols n_h, p_h, and q_h, the twoing index is simply

$$\frac{n_1 n_2}{4N^2}[|p_1 - p_2| + |q_1 - q_2|]^2.$$

Since the absolute values of $|p_1 - p_2|$ and $|q_1 - q_2|$ are identical, the twoing index becomes

$$\frac{n_1 n_2}{4N^2}[2(p_1 - p_2)]^2,$$

which is

$$\frac{n_1 n_2}{N^2}(p_1 - p_2)]^2,$$

As noted in Section 21.3, the proportionate reduction in group variance for a two-group split of binary data is

$$\phi^2 = \frac{n_1 n_2 (p_1 - p_2)^2}{N^2 PQ}.$$

Thus, for a binary target, the twoing index is simply $\phi^2 PQ$ and is an immediate counterpart of the variance reduction score. The conventional chi-square value will be (N/PQ) times the twoing index.

21.6.2.2. Illustration for polytomous targets

For the two splits of the polytomous target described in Section 21.6.1.2, the twoing index for the first split will be

$$\left[\left(\frac{90}{150}\right)\left(\frac{60}{150}\right)\left(\frac{1}{4}\right)\right][|(30/90) - (10/60)| + |(40/90) - (20/60)| + |(20/90) - (30/60)|]^2$$
$$= [.06][.167 + .111 + .278]^2 = .0185.$$

The twoing index for the second split will be

$$\left[\left(\frac{80}{150}\right)\left(\frac{70}{150}\right)\left(\frac{1}{4}\right)\right][|(35/80) - (5/70)| + |(23/80) - (37/70)| + |(22/80) - (28/70)|]^2$$
$$= [.0622][.366 + .241 + .125]^2 = .0333.$$

The second split would be preferred because it produces a larger twoing index.

21.6.3. Problems of interpreting polytomous outcomes

If the target event is binary, the Gini and twoing indexes do not produce anything really new. Their results are essentially variance reduction scores. The main apparent advantage of these new indexes is that they allow the counterpart of a variance reduction score to be calculated when the target is a polytomous ordinal or nominal variable. A disadvantage of the calculations, however, is that they obscure the evaluation of gradients, as well as the distinction between ranked ordinal and unranked nominal target variables.

With a dimensional target variable, the gradient from one predictor category to the next is readily perceived by comparing the corresponding means, medians, or other summary expressions in each category. With a binary target valuable, the gradient is readily evident from the binary proportions of the target "rates". The results of polytomous categories, however, cannot be readily summarized with a single value.

The entire mathematical scheme of symbols and ideas becomes displaced for polytomous comparisons, because all of the p_h values in the foregoing formulas refer to proportions of the *target* categories. These proportions cannot be immediately subtracted to form the increments of target rates in the *predictor* categories.

For example, for the first illustration cited in Section 21.6.1.2, the arrangements of the original and the two subsequent split groups were as follows:

Groups		Size of group	Rate of occurrence of target category in group		
			Mild	Moderate	Severe
Original		150	40 (27%)	60 (40%)	50 (33%)
First split	Group 1	90	30 (33%)	40 (44%)	20 (22%)
	Group 2	60	10 (17%)	20 (33%)	30 (50%)
Second split	Group 1	80	35 (44%)	23 (29%)	22 (28%)
	Group 2	70	5 (7%)	37 (53%)	28 (40%)

The only overt method for directly comparing gradients in the two splits is to use an additional separate "split" that would convert the target variable itself into binary categories. If this new split divides the target variable into **mild** and **non-mild** categories, the intergroup gradients would be easy to distinguish. The 2 × 2 table for the first split (of the independent variables) would be as follows:

Group	Mild	Non-mild	Total	Rate of mild outcomes
1	30	60	90	33%
2	10	50	60	17%
Total	40	110	150	27%

The gradient for the target rate in the two groups would be 16% (= 33% − 17%). An analogous 2 × 2 table for the second split of independent variables would yield 44% vs. 7% as the rate of occurrence for mild outcomes in the two groups. The second split would be preferred because it produces the larger gradient.

With a different dichotomous partition that emphasizes the **severe** category in the target variable, however, binary rates of occurrence could be compared for **severe** and **non-severe** outcomes. These rates would be 22% vs. 50% in the first split and 28% vs. 40% in the second split. By producing a larger gradient, the first split would now contradict the previous result.

Furthermore, if expressed in nominal rather than ordinal categories, the target variable could be arranged in diverse binary partitions, leading to many possible gradients that could vary extensively in showing the consequences of different splits for the predictor variables.

For these reasons, although the Gini and twoing indexes offer an attractive method to express statistical results for partition of polytomous outcome variables, the use of such outcomes is fundamentally an unattractive scientific strategy because it confuses an investigator trying to make sense of the results in the target variable. Different splits of independent variables can produce contradictory gradients for different choices of a dichotomous "split" in an ordinal target variable; and results for a nominal target variable will be almost impossible to interpret.

21.6.4. Error penalties and credit scores

Instead of using the Gini and twoing indexes to produce counterparts of reduced variance, the CART procedure can be directed to work with a different goal: reducing misclassifications. Instead of being merely recorded as right or wrong in concordant fit, however, the predictions that emerge from the CART method can be given scores with a cost-matrix strategy—outlined in Section 19.5.4.—that gives credit for correct classifications and penalties for different degrees of error.

For the total group of 150 persons discussed in Section 21.6.1.2, everyone would be predicted as **moderate,** since it is the largest category in the group. With this prediction for everyone, and with the cost matrix listed earlier in Section 19.5.4.1, the credit score would be $40(-1) + (60)(+1) + 50(0) = +20$. For the first split discussed in Section 21.6.1.2, the persons in Group 1 would all be predicted as having a **moderate** outcome, and the persons in Group 2 would all be predicted **as severe.** The credit score would be $30(-1) + 40(+1) + 20(0) = +10$ for Group 1 and $10(-2) + 20(0) + 30(+1) = +10$ for Group 2. The total credit score would be $10 + 10 = 20$. For the second split, everyone would be predicted as **mild** in Group 1 and **moderate** in Group 2. The credit scores would be $35(+1) + (23)(0) + 22(0) = 35$ in Group 1 and $5(-1) + 37(+1) + 28(0) = 33$ in Group 2, so the total credit score would be 68. The second split would therefore be regarded as much better than the first.

21.6.5. Indexes of evaluation

The CART procedure uses several additional statistical indexes to prepare, evaluate, and "validate" its results.

21.6.5.1. 'Pruning' of tree

Unlike other multicategorical programs, which apply the *modicum* criterion for stratum size while the partitions (and consolidations) are being carried out, the CART program allows a "very large tree" to be constructed, and then "prunes" the many terminal nodes. The pruning strategy uses eliminations rather than the consolidations that form terminal unions.

The criteria for pruning are not clearly presented in the descriptions of CART, but seem to involve considerations of entities such as minimum size, "weakest link branches", "complexity threshold of the node", a "1 SE rule", and "cost complexity measures." In the pruning process, the algorithm seems to favor "end cuts", i.e., leading to nadir or zenith groups, rather than "middle cuts".

21.6.5.2. 'Costs of misclassification'

Unless a special cost matrix is constructed and used in the CART activities, the "costs of misclassification" are simply the same as the error rates determined in other programs. With a cost matrix, an appropriately weighted score is calculated for the misclassifications.

21.6.5.3. 'Importance of variables'

At each step in the CART partitions, the split is chosen from the candidate variables that had the highest potential "improvement score". Among the variables that were *not* selected for the split at that step, the highest contender is called the *principal surrogate variable,* and is given an improvement score proportionate to the size of the node. At the end of the process, the sum of potential improvement scores is prepared for each variable that served as a principle surrogate. The variable with the highest score is given the arbitrary value of 100, and all other scores are adjusted proportionately.

21.6.5.4. 'Validation' techniques

The "success" of CART is expressed with two forms of resubstitution. The first approach is simply a conventional calculation of the error rate or costs of misclassification for the particular tree selected as best.

The second approach is called *n-fold cross validation,* with n = 10 being the most common choice. If n = 10, the available data set is divided into ten parts. For each part, the remaining nine parts are used to prepare an optimal CART tree, whose costs of misclassification are calculated. The sum of the 10 costs is then compared with the original cost of misclassification for the unpartitioned group.

21.6.6. Application of CART to lung cancer data set

To show how the illustrative data set for lung cancer would be processed, the CART program was obtained from its providers[16] and applied according to the stated directions. The target variables were the binary *6-month survival* in one "run", and the dimensional *ln survival* in a second. The program was allowed to operate without any imposed constraints, using its own default specifications.

For the 6-month survival target and the three main independent variables in the 200-member illustrative data set, the CART procedure produced only one split. To try to elicit more splits, the procedure was then given the same target and data set, but was allowed to examine the same seven independent variables that had been used in the previous forms of algebraic analysis. This opportunity, however, led only to the same single split as before. The results are demonstrated in the CART printout of Fig. 21.6. The left side of Fig. 21-6 indicates that the program explored 10 patterns of trees beyond the original unsplit terminal node formed by the entire group. The number of terminal nodes in the 10 patterns ranged from 39 to 2. Of these patterns, the one selected as best (marked with * for tree 10) by the CART program had a single split with two terminal nodes. The right side of Fig. 21-6 shows the construction of the tree.

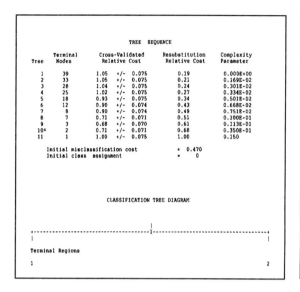
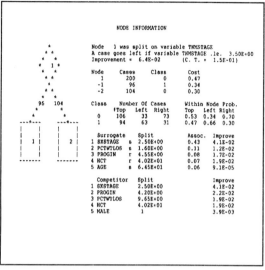

Fig. 21-6 Main printout of CART results for predicting 6-month survival in 200 patients of the illustrative data set. (For further details, see text.)

Because of the somewhat obscure communication of the CART printout, the results are not easy to discern. With suitable effort, one can eventually determine that the right rectangle (marked 2) contains 104 patients, in TNM stages coded 4 or 5, of whom 31 (30%) survived 6 months. The left rectangle (marked 1) contains 96 patients, in TNM stages coded 1, 2, or 3, of whom 63 (66%) survived 6 months. (These results can also be obtained, much more simply and easily, from the top of Table 9-2.)

Although many additional stratification opportunities had been found earlier (throughout Section 20.3) during conjunctive consolidation of the illustrative data set, the CART procedure did not advance beyond this single split of the data. No additional splits were constructed, presumably because no other independent variables seemed to emerge as effective.

When given the dimensional target of *ln survival* and seven independent variables for these data, CART produced no trees. With the three main independent variables, CART produced a single tree, splitting on the *symptom stage* variable. Categories **1** and **2** went to one side (with mean ln survival = 2.2) and categories **3** and **4**, on the other side, had mean ln survival = 0.64. (These results can also be readily discerned in Table 9-2.) The CART printout for the *ln survival* target is not shown here; it resembled what appears in Fig. 21-6.

21.6.7. Another portrait of CART printout

When Hadorn et al.[17] examined "a series of cart models produced by varying the misclassification cost specifications", the analytic goal was to explore fifteen candidate variables in search of the best predictors of 6-month survival in 2853 patients hospitalized with myocardial infarction. The total population was divided randomly so that two thirds served as a "generating" or "learning" set, and one third became the "challenge" set for testing results.

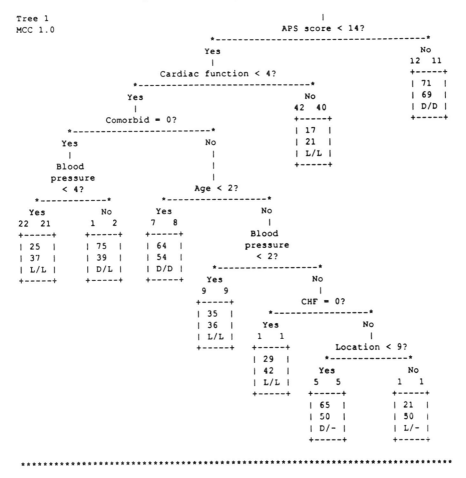

MCC = relative misclassification costs of false positive prediction. Variable abbreviations: APS = APACHE Acute Physiology Score; CHF = congestive heart failure score; Location = location of infarction.

Key to reading CART tree nodes:

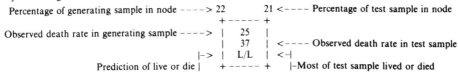

Fig. 21-7 CART results rearranged by Hadorn et al. for predictors of 6-month survival after myocardial infarction. (For further details, see text. Figure taken from Fig. 1 of Chapter reference 17.)

In response to different levels set for the "relative misclassification costs of false positive prediction", the CART system produced five recursive partitioning trees, of which the first is shown here as Fig. 21-7. The format of the CART printout in Fig. 21-7 differs substantially from what is seen in Fig. 21-6. When contacted by telephone, the first author stated that the raw printout (which resembled the arrangement shown in Fig. 21-6) had been "modified to illustrate how the splits were made at each step".

Regardless of the difference in format of printout, the results noted by Hadorn et al. "generally paralleled" what was found with algebraic models and seemed generally sensible. On the other hand, the investigators were unhappy that the CART procedure and several concomitantly tested algebraic models reduced the model-free error rate in prediction by only about 22% in the challenge test. A scientific source of this problem may have been the absence of important clinical predictors, such as arrhthymias and symptoms, and the difficulty of trying to forecast 6-month survival for myocardial infarction without separating short-term acute-phase predictors from those that affect posthospital survival.

Hadorn et al. were also disappointed by "the relatively poor cross-validation performance of the CART models . . . [despite] the careful attention paid to this issue by the developers of the methodology". Nevertheless, since all the analytic procedures gave essentially similar results, Hadorn et al. suggested that "CART (and similar techniques) . . . [may have] a major advantage in the development of . . . clinical practice guidelines [which] are, in essence, identical to recursive partitioning models".

21.6.8. Medically oriented discussions

In an early medical application, Goldman et al.[18] used the CART system, with a cost-matrix strategy, to develop categorical clusters for diagnosing presence or absence of myocardial infarction in emergency room patients with acute chest pain. The investigators developed a decision tree, shown in Fig. 21-8, that was subsequently modified after additional data were collected in validation studies.[19,20] Pleased with recursive partitioning, the investigators later wrote two papers [21,22] discussing its advantages in prognosticating and in "controlling confounding". Results of recursive partitioning were also compared with logistic regression in both papers, and with "propensity scores"[23] in the second.

In an early publication of comparative methods, Gilpin et al.[24] used logistic regression, discriminant analysis, and recursive partitioning to find predictors for 30-day mortality after myocardial infarction. (The investigators also compared results for a fourth method, called *nearest-neighbor procedure,* in which groups are formed from persons having similar multidimensional "distances" from the centroid of the independent variables. Since the target variable is not considered when the groups are formed, the nearest-neighbor procedure is not a targeted method, and will not be further discussed here.) The investigators concluded that all three of the main methods "performed similarly *within* a given population, although each [method] used the information contained in the prognostic variables differently".

In yet another early application of CART and the cost-matrix strategy, Levy, Caronna, Singer, et al.[25] analyzed predictors for the polytomously categorized outcome of 210 patients with hypoxic-ischemic coma. The decision rule for the

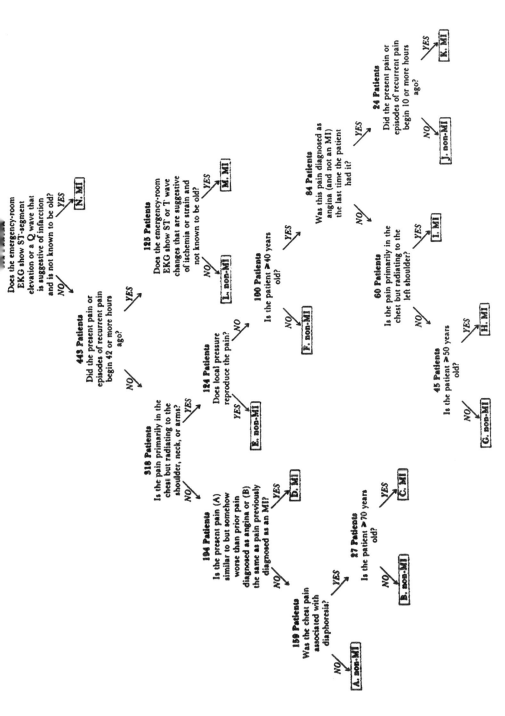

Fig. 21-8 Computer-Derived Decision Tree for the Classification of Patients with Acute Chest Pain. Each of the fourteen letters (A through N) identifies a terminal branch of the tree. For any given patient, start with the first question regarding ST-segment elevation and then trace the patient through the relevant subsequent questions until a terminal branch is reached. In the Yale-New Haven Hospital sample, seven terminal branches (C, D, H, I, K, M, and N) contained all 60 patients with acute myocardial infarction as well as 28 patients with unstable angina and 43 patients with other ultimate diagnoses. ER = emergency room, EKG = electrocardiogram, and MI = myocardial infarction. (Figure and text of legend taken from Fig. 1 of Chapter reference 18.)

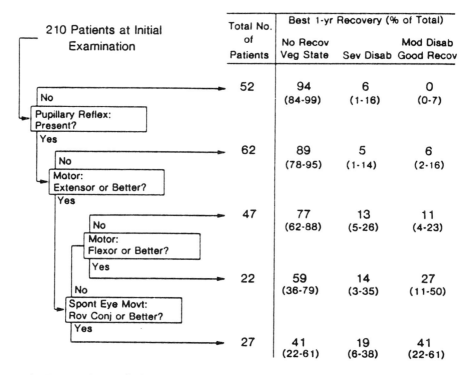

Fig. 21-9 Decision rule derived after cost matrix CART strategy for outcome of 210 patients with hypoxic-ischemic coma. Rules based on neurological examinations that classify patients in hypoxic-ischemic coma in accordance with their best functional state within first year. Figures in parentheses represent the 95% confidence intervals for percentages given immediately above. Abbreviations used in describing outcomes are as follows: recov = recovery; veg = vegetative; disab = disability; sev = severe; and mod = moderate. The following abbreviations are used in describing neurological signs: spont eye movt = spontaneous eye movements; and rov conj = roving conjugate. Initial examination (top left) was obtained within six hours of onset of coma in 55% of patients and within 12 hours in 84%. (Figure and text of legend taken from Fig. 1 of Chapter reference 25.)

patient's state on admission is shown in Fig. 21-9. Separate rules with different component predictors were developed for patients' states at 1 and 3 days and at 1 and 2 weeks after admission.

Stitt et al.[26] used the CART program to find predictors of death in patients with AIDS. The decision tree, shown here in Fig. 21-10, must have been reformatted from the original printout.

Offering a medically oriented discussion to statisticians, Bloch and Segal[27] empirically compared three categorical methods of adjusting for covariates. One method, called *pure-aggregation strata,* forms strata by aggregating cells in a contingency table. The mechanism of aggregation, which is not described in a manner comprehensible to nonmathematicians, apparently depends on minimizing variance between strata. A second method used the direct standardization technique of conventional epidemiologic research. The third method formed classification trees via recursive partitioning. Bloch and Segal also formed strata with the counterpart of a derived scoring system (see Section 19.6.3) by applying logistic regres-

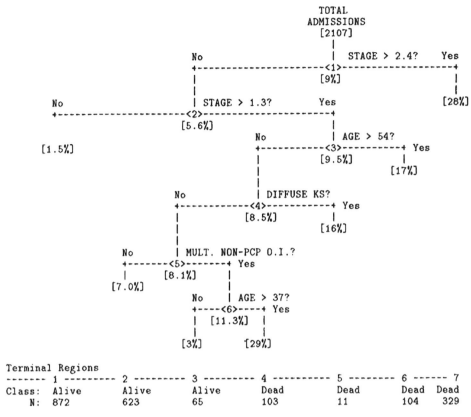

Fig. 21-10 Predicting death rates in patients with AIDS, using Classification and Regression Trees (CART). (Figure taken from Fig. 2 of Chapter reference 26.)

sion to the covariates, and then grouping patients with similar values for the predicted outcome.

21.6.9. American College of Rheumatology series

In an enlightened series of instructive papers appearing in the August 1990 issue of *Arthritis and Rheumatism*, the American College of Rheumatology sponsored the report of an investigation of "classification rules" for diagnosing seven forms of vasculitis.[28] Each diagnostic rule was constructed by at least two methods: (1) recursive partitioning and (2) a "traditional format", which consisted of a simple count (called "number of criteria") of the presence of "discriminating variables". The diagnostic accuracy of the two methods was then compared (and was relatively similar) for each of the seven vasculitic ailments.

The last paper of the series, by Bloch et al.,[29] discusses the four main analytic procedures in the research. In addition to the traditional format and recursive partitioning, the investigators used discriminant function and logistic regression analyses. The recursive-partition tree constructed for the diagnosis of giant-cell (temporal) arteritis is shown in Fig. 21-11.

554 Targeted-Cluster Methods

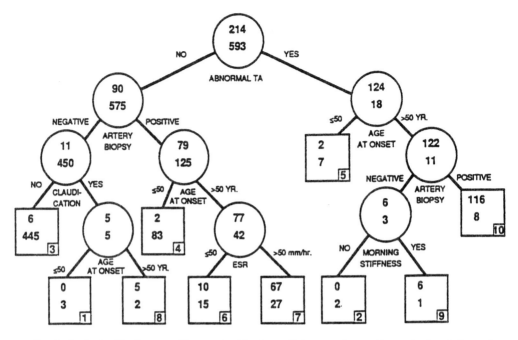

Fig. 21-11 A classification tree, illustrated with giant cell (temporal) arteritis (GCA). The circles and boxes contain the number of patients with GCA (top number) and the number of control patients with other forms of vasculitis (bottom number). The small, square box in the lower right corner of the boxes indicates the rank of the classifying group. TA = temporal artery; ESR = erythrocyte sedimentation rate (Westergren). (Figure and text of legend taken from Fig. 1 of Chapter reference 29.)

The predictive efficacy of the different approaches was compared with receiver-operating-characteristic (ROC) curves, which seemed quite similar for all four procedures. Despite the similarity of analytic results, the investigators said they preferred the recursive-partition method because "describing patient groups ... is informative".

21.7. KnowledgeSEEKER (KS) program

The newest entry in the array of recursive-partitioning methods is called KnowledgeSEEKER. Its manual[30] is a model of descriptive clarity that could well be emulated by other authors of such operational documents. In a good statistical discussion elsewhere, Biggs et al.[31] emphasize the apparent superiorities of KS to the AID and CHAID procedures. The CART procedure is not directly compared, but is acknowledged for its role in giving a "sound statistical footing" to classification tree techniques.

The KS procedure can be operated in a fixed algorithm or with interactive decisions by the user at each step. Although all types of target variables can be managed, the KS method, unlike all the other recursive-partitioning procedures, allows the *independent* variables to be divided into polytomous, not just binary, splits. The statistical decisions at each step are made with either chi-square tests or with F ratios in the analysis of variance. The KS method also has various other options, including a Bonferroni adjustment for the P values examined in multiple comparisons.

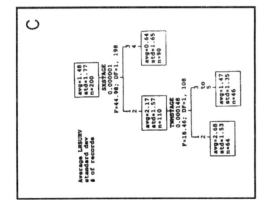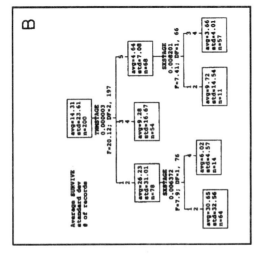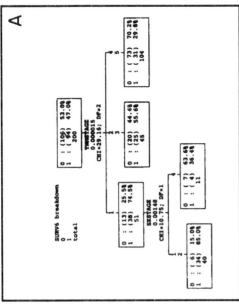

Fig. 21-12 Printout of recursive partitioning with KnowledgeSEEKER (KS) procedure for 200 patients in illustrative data set. The outcome variables are binary survival at 6 months in part A, actual survival time in part B, and the natural logarithm of survival time in part C. In part A, TNM Stage is first split trichotomously, and an extreme split of the two ends of symptom stage is then added to form four terminal nodes. In part B, the first and second splits have the same structure as before, but an additional full binary split of symptom stage is added to TNM Stage 5 to produce five terminal nodes. In part C, the splitting sequence is altered. Symptom stage receives a full binary split at the first step; and a full binary split of TNM stage is added, in the second step, to the symptom-stage 1, 2 categories. For the 6-month survival in part A, the original misclassification error rate is lowered from .47 to .305, a proportionate reduction of 35%. For the dimensional outcomes, the partitioning explained 23.6% of the original variance in part B, and 24.8% in part C.

Because the KS method is easy to use, and the results are easy to understand, it seems preferable to CART, although medical applications have not yet been reported. When aimed at three forms of the outcome variable in the illustrative data set here, KS produced the three slightly different trees shown in Fig. 21-12.

21.8. Additional multicategorical arrangements

As its popularity has increased, multicategorical stratification has been expanded in at least three additional directions. One approach proposes a new strategy, called *decision-tree induction*,[32] for making splits. The second approach brings in a new target variable: survival curves rather than single-state dimensions or alive/dead binary (or other) categorical conditions. The third approach is aimed at longitudinal analysis.

21.8.1. Survival analysis

For application in survival analysis, clustered categories have usually been constructed judgmentally, and the associated survival curves have been compared with suitable statistical indexes. The judgmental method can be conjunctive consolidation, as illustrated throughout Section 20.2, or a modification of recursive partitioning. The process of first evaluating survival curves for categories of individual variables, and then judgmentally forming clusters has been called *recursive-partitioning amalgamation* (abbreviated as RPA).[33] The statistical indexes for evaluating the best splits of survival curves are most commonly the log-rank test or Gehan's generalized Wilcoxon test. The results are usually displayed in two diagrams: one showing the tree structure, and the other showing the corresponding survival curves.

Fig. 21-13 depicts the particularly simple dual-diagram prepared by Kwak et al.[34] for identifying prognostic factors in diffuse large-cell lymphoma. A more elaborate arrangement was developed by Curran et al.[35] to indicate prognostic factors in patients receiving radiation therapy for malignant glioma. The recursive-partitioning-and-judgmental-amalgamation tactic has also been applied to prognosticate for patients with ovarian carcinoma;[33] small-cell lung cancer,[36] and coronary disease.[37] In two[33,37] of these reports, the investigators thoughtfully discussed the recursive-partitioning and algebraic-regression methods, and compared the results of analyzing the same set of data with both types of methods.

In a new proposal for "tree-structured proportional hazards modeling", recursive stratification is done with "a combination of statistical tests and residual analysis".[38] The proposal is illustrated with dual-diagrams for heart-transplant and liver-transplant data.

21.8.2. Longitudinal analysis

Tree-structured analysis has also been proposed[39] for the longitudinal process called *tracking*, which is aimed not at predicting a single outcome, but at describing a persons' state at repeated follow-up serial times. (Pediatricians do this form of tracking when they compare the percentile locations of a child's annual height and

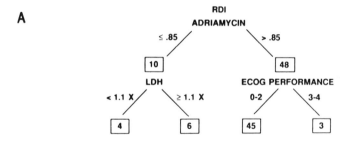
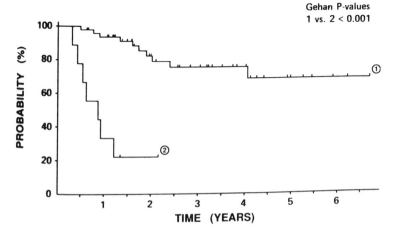

Fig. 21-13 Dual diagram for recursive partitioning and survival curves for predictive factors in patients with diffuse large-cell lymphoma. (Figure and legend taken from Fig. 6 of Chapter reference 34.)

weight in the "growth curves" of a reference population.) The new approach[39] constructs trees for the effects of such variables as age, sex, respiratory symptoms, and exposure to smoking on the longitudinal course of pulmonary function tests.

21.9. Concluding comments

The priority of the AID procedure seems to have been overlooked in current medical literature, where the popularity of recursive partitioning is regularly attributed to CART. The main pragmatic advantages of CART appear to be its ability to deal with polytomous target variables and to make decisions based on a complex cost matrix. Neither of these advantages is needed, however, for most analyses of binary or dimensional targets, or for situations in which the "predictive" errors are counted simply as right or wrong. For these conventional situations, CART offers no real advantages over AID, and produces results that seem excessively difficult to decipher and comprehend. If the investigator really wants to aim at a polytomous target variable (or even at a nonpolytomous target), the KS procedure seems much easier to use than CART, and has the advantage of letting the independent variables be divided (if desired) into polytomous rather than binary splits. The only distinctive advantage of the CART procedure, therefore, is the availability of a complex cost matrix for analysts who want to use such a matrix.

A disadvantage of *all* the automated recursive-partitioning procedures is that a pre-set mathematical algorithm makes the decisions. The composite categories that emerge from the partitioning process will therefore often have heterogeneous constituents with no apparent biologic coherence. To avoid this problem, the results of recursive partitioning with a rigid mathematical algorithm might be best used for screening rather than for making final conclusions. For *scientific* purposes, the conclusions should always be tempered by judgmental substantive evaluations and, if necessary, rearrangements. Procedures such as AUTOGRP, KS, and RPA, which permit judgmental evaluations during the process, may therefore become particularly appealing to data analysts who are concerned with substantive as well as statistical sensibility in the results. Eventually, however, the most attractive approach may be a procedure that combines the substantive emphasis of conjunctive consolidation with the advantages of a well-defined statistical algorithm.

As noted at the end of Chapter 20, the exploration (and solution) of the medicostatistical challenges is a fascinating opportunity in future collaborative research.

Chapter References

1. Morgan, 1963; 2. Diehr, 1981; 3. Nomura, 1988; 4. Crom, 1986; 5. Laschober, 1986; 6. Kass, 1980; 7. Hooton, 1981; 8. Mills, 1976; 9. Meyers, 1978; 10. Thompson, 1975; 11. Feinstein, 1973; 12. Breiman, 1984; 13. An introduction to CART methodology, 1985; 14. Gini, 1914, pp. 1203–48; 15. Leclerc, 1990; 16. CART. Version 1.1., 1988; 17. Hadorn, 1992; 18. Goldman, 1982; 19. Goldman, 1988; 20. Lee, 1991; 21. Cook, 1984; 22. Cook, 1988; 23. Rosenbaum, 1983; 24. Gilpin, 1983; 25. Levy, 1985; 26. Stitt, 1991; 27. Bloch, 1989; 28. Hunder, 1990; 29. Bloch, 1990; 30. KnowledgeSEEKER, 1990; 31. Biggs, 1991; 32. Long, 1993; 33. Ciampi, 1988; 34. Kwak, 1990; 35. Curran, 1993; 36. Albain, 1990; 37. Carmelli, 1991; 38. Ahn, 1994; 39. Segal, 1993.

22 Additional Discussion and Conclusions

Outline . . . .

22.1. Additional methods of analysis
    22.1.1. Analysis of covariance
    22.1.2. Path analysis
    22.1.3. Novel targeted methods
        22.1.3.1. Neural networks
        22.1.3.2. Artificial intelligence
    22.1.4. Nontargeted multivariate methods
    22.1.5. Multitargeted methods
22.2. Comparison of results for different multivariable methods
22.3. Goals and methodologic principles
    22.3.1. Analytic goals
        22.3.1.1. Bivariate confirmation
        22.3.1.2. Forecasting
        22.3.1.3. Predictive groupings
        22.3.1.4. Adjustment for confounders
            22.3.1.4.1. Principles and violations of 'confounding'
            22.3.1.4.2. Principles and problems of standardization
            22.3.1.4.3. Clinical 'staging'
            22.3.1.4.4. Demarcated scores
        22.3.1.5. Quantifying impact of individual variables
    22.3.2. Analytic expressions
        22.3.2.1. Importance of outcome variables
        22.3.2.2. Inadequate choices of predictor variables
            22.3.2.2.1. Neglect of temporal sequence
            22.3.2.2.2. Management decision vs. descriptive variables
        22.3.2.3. Decisions in coding
        22.3.2.4. Expression of accomplishments
            22.3.2.4.1. Individual discrepancies
            22.3.2.4.2. Impact of variables
            22.3.2.4.3. Confusion in appraisals
            22.3.2.4.4. Achievements of total analysis
            22.3.2.4.5. Calibration vs. discrimination
        22.3.2.5. Post-spective vs. pre-spective expressions
22.4. Issues in validation
    22.4.1. Relationship to goals
    22.4.2. External and internal data
    22.4.3. Chronologic splits
    22.4.4. Expression of successful challenge
22.5. Choice of analytic methods
22.6. Requirements for reporting
    22.6.1. Identification of groups and data
    22.6.2. Computer operations

22.6.3. Identification of multiple comparisons
22.6.4. Accomplishments of model
22.6.5. Impact of variables
22.6.6. Fulfillment of requirements

....

This final chapter contains a miscellany of topics that extend and emphasize various points in the previous discussions. Among the new topics are an outline of hitherto undiscussed methods, and a summary of results comparing different multivariable analyses for the same sets of data. The topics to be emphasized (and expanded) are certain issues in goals, principles, expressions, validation, and selection of methods for multivariable analyses. The last section offers a set of guidelines for appraising reports in which multivariable methods were used.

22.1. Additional methods of analysis

Among several multivariable analytic methods that have not yet been discussed, a tactic called *entropy minimax* seems to use stratifications. It might be an alternative[1] to recursive partitioning, but the original description[2] is too inscrutable for further discussion here, and the method does not seem to have been used in medical reports. The rest of this section is concerned with methods that occasionally appear in medical literature. They are the targeted procedures called *analysis of covariance* and *path analysis*, two novel approaches, and several multivariate techniques.

22.1.1. Analysis of covariance

Devised by R. A. Fisher[3] as a "convenient way of arranging the arithmetic," the *analysis of variance* (ANOVA) has become well known and frequently used. In simple (one-way) form, it is a counterpart of regressing a dependent variable against an independent categorical variable, which can be ordinal or nominal.

In two-way or three-way analyses, the ANOVA process contains additional dimensional variables, called covariates, that form adjusted results during the *analysis of covariance* (ANCOVA). The ANOVA and ANCOVA strategies were particularly popular in the days before categorical variables could easily be decomposed and coded for automated methods of multivariable regression analysis. Today, the adjustments that might formerly have been done with ANCOVA are readily achieved with one of the big four multivariable methods discussed in Parts III and IV of the text.

22.1.2. Path analysis

Sometimes used in social science,[4] but rarely in medical literature, *path analysis* resembles what might happen in a recursive rather than sequential exclusion process for linear regression. In the first step of the activities, the investigator finds the particular independent variable, X_1, that most explains the variance in the target variable, Y. The second step is aimed at finding the independent variable, X_2, that most explains X_1. The third step tries to identify the independent variable, X_3, that most explains X_2. The goal is to construct a "causal chain" or "path" in which X_3 explains X_2, which explains X_1, which in turn explains the target, Y. Although conceptually attractive, the mathematical approach does not appeal to scientists who are skeptical of using a simple set of quantitative calculations to unravel complex issues in causality.

Having been devised by Sewall Wright[5] for analyzing human population genetics, path analysis is most likely to appear in genetic epidemiology,[6,7] although it was used in a study of psychophysiological data.[8]

22.1.3. Novel targeted methods

The entire text so far has been devoted to targeted analytic arrangements that use either algebraic models or stratified clusters. With the ingenuity offered by easy computation and automation, two novel additional tactics have been proposed: *neural networks* and *artificial intelligence*.

22.1.3.1. Neural networks

The neural network strategy is not easy to understand, even in discourse[9,10] allegedly aimed at clinical readers. The "model" apparently works in a type of "scavenger hunt," searching for optimal arrangements of variables or components of variables. Unfortunately, the proponents of neural networks seldom report checks of "external" validity to confirm what was found.

Any algebraic model or stratified-cluster analysis could easily achieve spectacularly good "internal" results if allowed an enormous number of variables or strata. In fact, if we used a separate variable for each person's *individual* values of $X_1, X_2, X_3, \ldots$, or a separate stratum for each person's covariate pattern, a perfect fit and perfect "retrospective" predictions could be obtained for the data. The huge arrangement would have no numerical stability, however, and might be useless for any future applications. Because neural networks permit this enormous intricacy, but because the intricacy is not "transparent" or otherwise evident to the reader, the results may seem highly impressive. If the neural network approach really offers a "magic" that cannot be achieved with algebraic or cluster models, the proponents should begin offering demonstrations of external validity, discussed in Section 22.4, to confirm that the reality agrees with the theory.

22.1.3.2. Artificial intelligence

Artificial intelligence (AI) is another form of "magic" that is enthusiastically proposed but almost never suitably validated. The strategy and tactics of artificial intelligence are also hard to understand, and are beyond the scope of this discourse. Clinicians obviously feel uncomfortable with the overt artificiality of AI, but statisticians do not seem to like the methods, either. In commenting on a proposed use of AI in medical diagnosis, the prominent British statistician M. J. R. Healy[11] said,

> It is something of a scandal that this appears to have been done in such an amateurish fashion. . . . Little of this knowledge [about the measurement of uncertainty or a chain of argument] appears to have been utilized by leaders in the AI community. Instead, we have

operationally meaningless "degrees of certainty" propagated by inappropriate formulae to outcomes whose ostensibly quantitative labels are justified only by loaded terminology.

Leaders in the AI community can avoid these criticisms by subjecting the AI methods to the same demands and tests of validity used for other analytic methods. The merits of the AI strategy are currently dubious because it has not yet, despite many years of availability, made substantial realistic contributions to multivariable analysis.

22.1.4. Nontargeted multivariate methods

For reasons noted in Section 4.1.1, the text has emphasized many-to-one multivariable methods that aim at a single target variable. Consequently, little attention has been given to the many truly multivariate procedures in which the analysis has either a many-to-many structure, aimed externally at two or more target variables simultaneously, or a many-internal goal of reducing an array of independent variables, without regard to a selected target. The many-internal multivariate procedures include *factor analysis, principal components analysis, multidimensional scaling, cluster analysis,* and *correspondence analysis.*

Correspondence analysis[12-14] is somewhat analogous to factor analysis, but is regarded as particularly useful for preparing aggregates of categorical (rather than dimensional) data. Like factor analysis, which looks for variables that can be aggregated into new "factors," correspondence analysis seeks attributes that can be grouped to form new "profiles." It is sometimes called the "canonical analysis" of contingency tables.

22.1.5. Multitargeted methods

The many-to-many multivariate procedures that aim at multiple external targets include *canonical correlation analysis* and *multivariate analysis of variance* (MANOVA). *Discriminant-function analysis,* which has been included here as a single-targeted method, is often regarded as multivariate because the single nominal target variable is decomposed into multiple binary categories.

The problems of interpreting the results make the simultaneous analysis of multiple target variables scientifically unappealing (see Section 4.1.1.1). Nevertheless, readers interested in MANOVA can find a useful account of its application in a paper, published thirty years ago by James Grizzle,[15] that reported the analysis of four outcome variables in a randomized trial of five treatments for chronic leukemia. Grizzle's paper also contains the raw data used in the analyses as well as clear descriptions (and illustrations) of the processes used for calculating Mahalanobis D^2 and for the stochastic adjustment of multiple comparisons.

22.2. Comparison of results for different multivariable methods

Many investigators have compared the results obtained when the same set of data was analyzed with different multivariable methods. Most of the comparisons have had the disadvantage of being "data dependent": the results pertain only to the structure and contents of the particular data set used for the analysis. In one study,[16] however, the investigators deliberately rearranged the basic information into different configurations (symmetrical, bimodal, exponential, etc.) to determine how the known biologic attributes of the data would be affected by diverse distributions of the constituents.

Table 22-1 Results of reports in which one multivariable analytic method was compared with another in same set(s) of data

First author	Journal	Year	Compared method**						Comments
			MLR	DFA	LGR	COX	RP	Other	
Halperin[17]	J. Chron. Dis.	1971	x	x					DFA can give "misleading" results regarding "significance"
Reading[18]	Biometrics	1973	x					Four methods of "successive screening"	Nothing obviously superior
Coronary Drug Project[19]	J. Chron. Dis.	1974	x		x				"Similar results . . . with the two models"
Brunk[20]	Chapter in book	1975	x		x	x		Several other algebraic methods	"All nine methods . . . similar with regard to the number of correct classifications"
D'Agostino[21]	Biometrics	1978	x	x					Similar overall accuracy, but substantial differences in false positive and false negative rates of prediction
Eisner[22]	Am. J. Pub. Health	1979	x					Mantel-Haenszel stratification	Similar results for many factors, but "contradictory results for the effect of maternal education"
Dillman*[23]	Cancer Research	1983	x	x			x		Same variables chosen by all three methods. RP slightly better in classifying challenge set
Gilpin*[24]	Cardiology	1983	x	x			x	"Nearest neighbor"	Poor results for "nearest neighbor." Various individual differences but reasonably similar total results for the three other methods
Green[25]	J. Chron. Dis.	1983	x	x					Similar results for short- but not long-term survival†
Schmitz*[26]	Stat. Med.	1983	x	x				DFA compared for "kernel" and quadratic methods	Quadratic poorest; DFA & LGR similar; "kernel" slightly better

First author	Journal	Year	Compared method**						Comments
			MLR	DFA	LGR	COX	RP	Other	
Madsen*[27]	Am. Heart J.	1984		x		x	x		All gave "equally precise prognostic evaluation"
Harrell*[28]	Stat. Med.	1984			x			"Variable clustering" and "incomplete principal components" (IPC)	Variable-clustering method recommended "for clinical predictive problems"
Harrell*[1]	Cancer Treatment Reports	1985			x		x	Clinical "sickness score" and IPC	Relatively similar results, but sickness score and IPC best
Ciampi[29]	Comput. Stat. Data Anal.	1985				x	x	Correspondence analysis	No clear conclusion; no method "universally preferable;" joint use recommended
Brenn[30]	Stat. Med.	1985	x	x	x				No major differences in results
Kannel[31]	Am. Heart J.	1986			x			Cross-stratification	Similar results
Marshall[32]	Stat. Med.	1986					x	Best of all possible partitions	Results similar to previous logistic regression for same data
Peduzzi[33]	J. Chron. Dis.	1987			x	x			Similar results for short-term and uncommon outcomes
Guppy[34]	Med. Dec. Making	1989			x			Sequential Bayes algorithm	Poor performance by algorithm
Segal[35]	Stat. Med.	1989				x	x		Many qualitative similarities but quantitative comparison "not possible"
Walter*[36]	J. Clin. Epidemiol.	1990	x	x	x	x			Similar stepwise choices of first two variables, but many other disparities. See published report.
Sankrithi[37]	Int. J. Epidemiol.	1991	x					"Product-form" (exponential) regression	"Better fits" for product-form method. Different results for variables

First author	Journal	Year	Compared method**					Comments
			MLR	DFA	LGR	COX	RP Other	
Hadorn[38]	Stat. Med.	1992	x		x	x	x 3 "unit-weight" derived models	Logistic model slightly superior, but CART preferred despite poor cross-validation
Long*[39]	Comp. Biomed. Research	1993			x		"Decision-Tree Induction"	Logistic regression superior
Dunn*[40]	Cont. Clin. Trials	1993			x	x		Only minor differences
Liaño*[41]	Nephron	1993	x		x			Linear regression superior

* Results checked in challenge or new data set.
** x = this method was used; MLR = multiple linear regression; DFA = discriminant function analysis; LGR = logistic regression; COX = Cox regression; RP = recursive partitioning.
†Subsequent complaint by Hauck[42] that LGR group was not properly selected.

Table 22-1 chronologically lists twenty-six reports[1,17-41] in which one multivariable analytic method was compared against at least one other, for the same set (or sets) of real-world data. The list excludes comparisons using simulated data, and does not include several potentially eligible reports that could not be obtained when the table was prepared. Of the twenty-six reports, only ten (those marked with asterisks) also tested a challenge or new data set.

The overall conclusions of the comparisons are difficult to summarize. Although certain procedures occasionally seemed superior, and although various disparities were noted in the comparisons, perhaps the most common finding was the relative similarity of results for the different methods. Furthermore, no specific guidelines could be established to indicate the particular empirical circumstances in which one analytic method would be unequivocally superior to another.

Accordingly, the general conclusion of this crude "overview" might be that investigators and data analysts need not worry about "missing" an optimal choice. Almost any of the multivariable methods can be used as long as it does not seem grossly inappropriate. When several methods are "eligible," the investigator can choose the one that seems aimed at the right goal, and that is also easiest to use and to understand.

22.3. Goals and methodologic principles

Probably the most important and most often overlooked issue in multivariable analysis is the investigator's goal. When asked, "Why are you doing the analysis?", some investigators may say, "To see whether anything is significant," or, more commonly, "Because the editorial reviewers asked for it." A more scientifically informed and motivated set of answers, however, would respond to such questions as What is your main *scientific* purpose in doing this analysis? What do you hope to get out of it? What will you do with the results?

Unless these questions are specifically asked and answered, the analysis may be done with a default goal, which is determined by a computer program that aims to fit an algebraic model to the data. To avoid this thought-free decisional mechanism, the scientific goals of the analysis can help determine which program to choose, whether to use an algebraic or some other method, how to evaluate the results, and how to validate them.

22.3.1. Analytic goals

Since scientists seldom want merely to fit an algebraic model to multivariable data, the more specific investigative goals need careful attention. These goals were discussed in Chapter 3, but are reviewed here so that they can be described more specifically, in the next five subsections, as bivariate confirmation, forecasting, predictive grouping, adjustment, or quantifying impact.

22.3.1.1. Bivariate confirmation

If certain variables seem important in the simple, ordinary bivariate analyses, and if the multivariable goal is merely to confirm that these variables maintain their importance in the context of the other variables, an appropriate algebraic model is simple and easy to use. If it provides the desired confirmation, the multivariable details need seldom be analyzed extensively or even reported, since nothing further will be done with them.

22.3.1.2. Forecasting

The term *forecasting* refers here to prognostication for individual persons. Regardless of how the analytic results have been validated (as discussed in Section 22.4), a strategy must be developed for separating the "training set," and the set of persons used for "challenge" testing. Despite the merits (discussed later) of a chronologic partition, the validation is best done with an entirely separate collection of people, preferably at some other institution.

The chronologic or extrainstitutional challenges, however, may lead to suitable groups for testing, but will not solve the problem of how to make the individual forecasts. (An important disadvantage of Cox regression models is that the analytic results cannot be readily used for individual forecasting. Because an individual person's value of $\hat{S}(t)$ can be estimated only for a particular time point [using the convoluted tactics discussed in Section 16.1.1.2.1], a forecast cannot be made unless a specific time is selected.)

As shown in Table 22-1, different analytic methods often give similar average results for the same group of patients, but may have striking differences in the individual predictions. Consequently, few clinical prognosticators would want to make predictions for individual patients by using the single calculated estimate that emerges when individual values of $X_1, X_2, X_3, \ldots$ are entered into an algebraic model. Instead, clinicians would want the greater predictive "security" that is possible when the individual forecasts are made from results in a pertinent "resemblance" clinical subgroup.[43,44] In pragmatic reality, therefore, individual forecasting is usually done not with the additive individual score of an algebraic formula, but with the clustered results of a cogent prognostic subgroup.

22.3.1.3. Predictive groupings

A prominent intellectual defect in late-twentieth-century clinical science has been passive acceptance of the idea that prognosis is mainly a statistical enterprise. Knowledgeable clinicians have failed to protest the gross excesses of the mathematical activities, to articulate the key pathophysiologic and biologic principles that are regularly ignored and omitted, and to identify the clinically cogent groupings and subsets that should be considered. Uninhibited and unenlightened by either the knowledge or constraints of clinical reality, the mathematical models have often been constructed as magnificent aerial castles, founded on quicksand.

The development of cogent prognostic groups is an irreplaceable necessity in modern clinical science. The groups are needed both for individual forecasting (as discussed in the previous section) and for compensatory adjustments (as discussed in the next section). Such groupings should, above all, have the scientific "face validity" of being clinically sensible. Yet clinicians have persistently evaded the job of specifying the composition of those groupings. Boolean-union entities as important and obvious as *severity of congestive heart failure* have been disregarded in favor of either mathematical aggregates formed by principal component analyses, or weighted algebraic combinations, expressed as $\Sigma b_j X_j$ for such individual manifestations as dyspnea, peripheral edema, third heart sounds, etc.

At the most basic level of scientific credibility, the main challenge in prognostic grouping is not to choose a strategy for multivariable analysis, but to identify the cogent variables to be observed, coded, and entered into the analysis. If not identified beforehand, the most important variables will be omitted from the collection of data; and even those variables that have been collected may be improperly analyzed unless they are first joined into clinically appropriate Boolean combinations. A remarkable aberration in twentieth-century science is the current idea that magisterial statistical analyses, after the research is done, will extract trustworthy conclusions from data that were grossly incomplete and inadequately classified for the analysis.

Assuming that satisfactory data have been collected and classified, however, cogent groupings can be formed in at least five ways: (1) by demarcating the mathematically computed scores of an algebraic model; (2) by demarcating a set of clinical scores formed judgmentally (à la Apgar) with or without the aid of a formal mathematical analysis; (3) by constructing clustered stages from the composite variables of a conjunctive consolidation; (4) by constructing clustered stages from simple clinical judgment (as in the TNM stage classification of cancer); or (5) by using arrangements of the "decision-node" clusters that emerge from a recursive-partitioning tree. The respective merits of these methods have been discussed earlier in the text, and different researchers will have different preferences about the best way to form the groups.

Regardless of whatever statistical encomiums accompany the groups, however, they must be clinically sensible. The delineation of criteria for what is meant by *sensible*, however, is another major gap in the intellectual performance of clinicians, who usually passively accept the absence of common sense in many of the mathematical methods. In a recent set of criteria for "the theory and evaluation of sensibility,"[45] twenty specific features were proposed as subdivisions of six main attributes: purpose and framework, comprehensibility, replicability, suitability of scale, face validity, and content validity.

Another requirement of sensibility for individual clinical forecasts, however, is that the method be relatively easy to use. Ease of usage will generally make grouped categories preferable to the arithmetical idiosyncrasy of multi-variable weighted formulas or the arboreal complexity of branching multicategorical trees. If the goal is to make suitable analytic adjustments rather than individual predictions, however, the groupings do not have to be simple; and any of the five techniques cited above can be used as long as it meets *clinical*, not just statistical, requirements. An interesting point to bear in mind is that the two oldest and most widely used forms of multivariate clinical classification—the Apgar score and TNM cancer stages—were both originally developed with pure clinical judgment, unaccompanied by any mathematical analytic formulations.

22.3.1.4. Adjustment for confounders

Another multivariable analytic goal is to do compensatory adjustments that can reduce or eliminate the biased comparisons produced by confounding.

22.3.1.4.1. Principles and violations of confounding

Confounding is constantly discussed and adjusted in statistical work, but is seldom suitably defined in a scientifically precise manner. The term usually refers to a bias that can afflict (and perhaps invalidate) the results ascribed to a particular causal maneuver, which is usually either an etiologic or therapeutic agent. Confounding can arise from biases that occur as inequalities in at least three separate locations along the causal pathway:[46] baseline states, performance of maneuvers, or detection of outcome events for the compared groups. A fourth source of confounding is a biased transfer process when the analytic data are collected or arranged after the outcomes have occurred.

Most efforts to deal with confounding aim only at *susceptibility bias*, which arises before the maneuvers are imposed, if the compared groups have unequal baseline prognostic susceptibilities to the outcome event. The classic example of susceptibility bias (as illustrated in Table 3-9) is a comparison of results in surgical therapy for cancer, given mainly to patients with localized lesions and no major co-morbidity, against the results of radiotherapy or chemotherapy, given mainly to patients with metastases or major co-morbidity or both. One of the main roles of randomization in clinical trials is to prevent the occurrence of this bias when the judgmental assignment of therapy is affected (as it usually is in clinical practice) by the patient's prognostic susceptibility to the outcome event.

A confounding variable must satisfy both of two main criteria: (1) it must have a distinct effect on the outcome event, *and* (2) it must be unequally distributed in the compared groups. (For example, the stage of a cancer can affect survival but is not a confounder if it is equally distributed among the compared treatment groups.) These two criteria, however, are constantly violated when the results are said to be adjusted for confounding in an analysis that merely included a large array of multiple variables. The alleged confounding variables are not confounders unless they were shown to fulfill both requirements: distinctive effects on the target and imbalanced distribution in the compared groups.

A multivariable adjustment that uses nonconfounding "confounders" becomes both a snare and a delusion. It is a delusion because it deceives both the investigators and the readers into thinking that the adjustment process, although inadequate, must have been satisfactory because it included so many variables. The adjustment is also a snare because it distracts attention from the true confounders (which may have been ignored), and because the excess of multiple variables may have created unstable or erroneous results, due to the low events-per-variable problem discussed earlier.

22.3.1.4.2. Principles and problems of standardization

The oldest and best-known form of adjustment for confounding is the epidemiologic strategy called *standardization.* It was developed to allow general mortality rates, when compared in different geographic regions, to be adjusted for disparities in the *age* and *sex* composition of the regions. These two variables are confounders because they have distinct effects on general mortality, and they are seldom equally distributed in the compared geographic regions.

For the adjustment process, the results found in partitioned age-sex strata for each region are mathematically combined into an appropriately weighted *single* value for the region. In the most common forms of standardization, for mortality or other general population data, the mathematical combinations use methods that are called *direct* or *indirect*.[46,47] If the main results are expressed in odds ratios, rather than mortality rates, other mathematical methods, such as the Mantel-Haenszel procedure,[47,48] are used for the aggregation.

The standardization process has been highly successful in demographic epidemiology, where mortality rates are compared mainly in different geographic regions or secular time periods. The adjustments involve only a few variables, such as age, sex, and perhaps race and/or socioeconomic status; the stratified arrangements are easy to construct; the data (coming from regional populations) are usually abundant; and the investigators are content to receive a single adjusted value for each of the compared groups.

The process is less successful in etiologic studies of epidemiologic risk factors, and in clinical epidemiologic studies of therapy. In both sets of studies, the group sizes are usually much smaller than a general population, and the number of pertinent confounding variables may be too extensive for a numerically effective cross-classification.

In etiologic studies, therefore, when the variables are too abundant for successful adjustment with Mantel-Haenszel stratifications, the investigators may turn to algebraic methods that can manage large numbers of variables, while providing a single adjusted result for each selected risk factor. In clinical studies, however, the investigators do not want to examine a single adjusted result; they want to know what the compared treatments have done in the different clinical strata. For this reason, the standardization process is almost never used in clinical work. The process itself is satisfactory, but clinicians usually want to compare results in cogent clinical subgroups, not in a single aggregated average.

Consequently, public health and clinical epidemiologists regularly employ two different approaches to achieve the different goals. The public health investigators will try to get a single adjusted value, using either algebraic methods or combinatorial stratification (à la Mantel Haenszel). The clinical investigators, hoping to apply the results in subsequent practice, will reject the simplicity of a single, adjusted average value. Instead, they will want to see the results in appropriately stratified clinical subgroups.

22.3.1.4.3. Clinical 'staging'

The most common method of getting appropriately stratified clinical subgroups is to use the type of judgmental reasoning that formed TNM stages such as **localized, regionally spread,** and **distantly spread** for the morphologic extensiveness of cancer. This judgmental demarcation is easy for a single concept such as *anatomic spread*, but is more difficult when multiple variables must be included, or when a single variable such as *histologic type* or *pulmonic symptoms* contains many nominal component categories.

These are the types of challenge for which conjunctive consolidation and sequential sequestration are particularly well suited. They organize the variables in a manner that is both clinically and statistically sensible, thus avoiding the biologically heterogeneous mixtures that can arise either from recursive partitioning or from the scoring systems discussed in the next section.

22.3.1.4.4. Demarcated scores

When groups are formed from the demarcation of numerical scores, the original scores can be chosen from clinical judgment alone (as in the Apgar index) or from converting algebraic analytic results by any of the methods described throughout Section 19.6. The boundaries for the scores can be demarcated with quantile procedures (to produce strata of equal size), or with the esthetic or clinical strategies discussed in Section 8.4.4.

Regardless of how the scores are prepared, however, any scoring system has the disadvantage that heterogeneous mixtures can have the same score. For example, for the five variables used in the Apgar index, newborn babies can achieve a score of **6** in such diverse ways as $0 + 0 + 2 + 2 + 2$, or $2 + 2 + 2 + 0 + 0$, or $1 + 1 + 2 + 1 + 1$. Despite similar scores of **6,** however, the clinical implications of a baby's state may differ substantially according to the particular components that received scores of **0, 1,** or **2.** For example, we might be much more alarmed by a score of **0** in *heart rate* than in *reflex responses.*.

Consequently, if a group result is desired for individual forecasting or clinical staging, a system of clinically constructed strata, with well-defined contents, may be preferable to demarcated scores. As noted in Chapter 20, the clinical constructions may have biologically heterogeneous categories if many variables are involved, but the heterogeneity will have been clearly identified, and the components (if formed appropriately) will at least be statistically homogenous in rates of occurrence for the target event.

22.3.1.5. Quantifying impact of individual variables

A fifth goal in multivariable analysis is to quantify the impact of an individual variable. The result is expressed in statements such as, "You live 3 years longer for each 5-point drop in substance X" or "Exposure to substance X raises your risk by a factor of 3.7."

This goal, which can seldom be achieved with multicategorical analysis, has led to the popularity of algebraic methods in epidemiologic studies of etiologic risk factors. When adjusted for the many concomitant covariates in the algebraic model, the associated b_j coefficient indicates the impact of an X_j "risk factor" in the multivariate context. The impact is usually quantified as an adjusted odds ratio or hazard ratio.

Despite its attractive simplicity, however, the quantitative risk is merely an average for the entire range of values, and may differ substantially in different zones of the same variable, or in the interactions within different zones of the covariates. Consequently, to claim a specific quantitative impact for an individual variable, the investigator must demonstrate that the variable has appropriate linear conformity, and that the cited coefficient is unaffected both by concomitant interactions and by instabilities produced by a low events-per-variable ratio. This additional demonstration is either seldom carried out or seldom reported.

The algebraic method of quantifying individual impact has also been frequently applied in recent years not just for appraising etiologic risk factors or pre-therapeutic prognostic factors, but for directly comparing the effects of treatments. This approach has the two major disadvantages discussed earlier: (1) a single quantitative average impact of therapy is grossly unsatisfactory because it fails to indicate effects in pertinent clinical subgroups and (2) a scientific anachronism is committed when the same analysis includes, as temporally similar, both the treatment and the baseline conditions that *precede* treatment.

The improper temporal construction can be avoided if the covariates are analyzed separately, without including treatment. The algebraic results of the covariate analyses can then be used as scores to demarcate strata, within which the effects of treatment can be examined. The latter process, which is well illustrated by Miettinen's confounder score,[49] would also permit "therapeutic interactions" to be analyzed and interpreted in a clear, understandable manner.

22.3.2. Analytic expressions

In any analysis, the most important reason for recognizing the different *scientific* goals is that they can be a prime guide in avoiding problems when the analyst chooses variables, classifies variables, and expresses the analytic results.

22.3.2.1. Importance of outcome variables

A common source of inadequate statistics is the fallacious assumption that standard prognostic predictors can always be established for a particular disease. The "admission" criteria for the disease will delineate the group of patients under analysis, but the prognostic predictors will always depend on the choice of the outcome event. For example, in patients with acute myocardial infarction, the important predictor variables will differ according to whether the outcome event is relief of chest pain, postinfarction survival, or return to work, and whether the outcome event is checked at 1 day, 1 month, 1 year, or 3 years after the acute infarction. Even in etiologic case-control studies, where the outcome event is development of a particular disease, the predictive risk factors will depend on the timing of the disease and its pattern of manifestations (asymptomatic, symptomatic) and of discovery (screening detection, differential diagnosis, etc.).

Investigators who are unaware of these outcome distinctions may often work in the vain hope that the important predictor variables will be revealed not with careful thought, observation, and analysis, but with "time-dependent" or other statistical manipulations.

22.3.2.2. Inadequate choices of predictor variables

Two problems that may lead to inadequate choices of predictor variables are the neglect of temporal sequence in clinical biology, and the recording of management decisions rather than clinical conditions.

22.3.2.2.1. Neglect of temporal sequence

If temporal sequence is neglected, an entity that is a "marker" of disease may be regarded as though it were an antecedent etiologic "determinant." For example, a menopausal woman who develops unexpected irregularities in vaginal bleeding may be pleased when it stops after she is given replacement estrogen therapy. Later on, if she is found to have endometrial cancer, it may be attributed to the preceding therapy, without recognition of the possibility that the bleeding was the cancer's first clinical manifestation. This type of "protopathic bias" has been a relatively common event in etiologic case-control studies.[46,50]

An analogous problem occurs in cohort studies of *combined* therapy when the data analysts do not differentiate between a single clinical decision to give both treatments, and a sequence of two clinical decisions, in which the second treatment is added after the first seems to have failed.

22.3.2.2.2. Management decisions vs. descriptive variables

The variables included in an analysis can be greatly flawed if management decisions are substituted for patients' conditions. For example, *congestive heart fail-*

ure, *chest pain, anemia,* and *metastatic cancer* are all descriptions of patients' conditions, but *hospitalization, discharge,* and *transfer from institutional location A to location B* are all management decisions. These decisions do not describe the status of a patient's condition; they indicate what someone decided to do about it.

Since the management decisions can be affected by many factors other than the patient's medical condition, the subsequent analytic results can become inconsistent or nonreproducible if the main descriptive variables reflect management rather than status. A patient may be hospitalized not because of severity of illness, but for diverse social reasons. Radical surgery may be done not for specific clinical criteria, but because of a rigid therapeutic policy at a particular institution. Intrainstitutional transfers may be evoked not by changes in illness but by availability of beds. Wasson et al.[51] have offered some additional illustrations of the analytic problems created when investigators classify management decisions rather than clinical conditions.

To avoid all these inconsistencies, the best scientific approach, whenever possible, is to classify and analyze clinical conditions. They constitute some of the most important judgmental variables, affecting both prognosis and choice of treatment, that have regularly been omitted from statistical collections of data. The existence and components of these variables can readily be identified and classified for analysis if the investigator regularly seeks the answer to the question, "What was the reason for this management decision?". (For example, many years ago, when I "discovered" co-morbidity,[52] I was trying to determine why surgery had been denied to patients who seemed "obviously" eligible for it.)

22.3.2.3. Decisions in coding

Blettner and Sauerbrei[53] recently offered some enlightening examples of contradictions and inconsistencies that can arise not only from different approaches to the coding decisions (discussed in Chapter 5), but also from decisions in classification of variables, imputation of missing values, and sequential operations in selection of variables.

Another overlooked problem is the occasional (or frequent) desirability of analyzing dimensional variables according to their *rank* values, rather than the actual dimensions. This type of simple "transformation" can regularly eliminate some of the distortions produced when variables that have eccentric non-Gaussian distributions are summarized with means and standard deviations. The ranked, "nonparametric" approach will also be particularly valuable when relative deviations from the median (discussed shortly), rather than squared deviations from the mean, become used to cite discrepancies in algebraic calculations.

22.3.2.4. Expression of accomplishments

The accomplishments of an analysis are usually expressed with three types of statistical indexes that denote (1) individual discrepancies between observed and "predicted" results, (2) the impact of variables, and (3) the total analytic achievement. The choice of these expressions requires thoughtful attention.

22.3.2.4.1. Individual discrepancies

The advent of easy computation has evoked some "revolutionary" new proposals, such as bootstrap and jackknife methods, for analyzing data. Some writers,[54] in fact, have gone so far as to urge the abandonment of any statistical methods, including the entire strategy of parametric analysis, that were developed in the past century before the arrival of ubiquitous, personal digital computers. Most of the new computer-intensive approaches, however, are aimed at analyzing the total collection of data. Much less attention has been given to the expression of individual discrepancies.

For example, examining the absolute deviation from the mean, $|X_i - \bar{X}|$, is more intuitively sensible than examining the squared deviation, $(X_i - \bar{X})^2$, which became well established only because it facilitated the calculations. If absolute deviations[55,56] become acceptable because they can now be easily computed, however, why not examine absolute or percentile deviations from the median, which is a much better general central index than the mean? Almost unique among clinical specialists, pediatricians have effectively worked for several decades with percentile indexes that track the growth of young children. The enlightenment brought by percentiles (and other rank indexes) need not be confined to pediatrics.

Almost twenty years ago, Shapiro[56a] proposed a "coefficient of predictive accuracy" that was essentially an average value of log likelihood; but the coefficient does not seem to have been used subsequently.

22.3.2.4.2. Impact of variables

Suppose variable X_1 has the following mortality rates for three ordinal categories: a_1, 1%; a_2, 2%; a_3, 4%. Suppose variable X_2, in another study, has the corresponding values of b_1, 10%; b_2, 20%; and b_3, 40%.

If the impacts are summarized as gradients, variable X_2 is obviously more impressive. The intrasystem gradient of 30% (= 40% − 10%) is 10 times larger than the 3% (= 4% − 1%) found for variable X_1. If impact is summarized as a risk ratio, however, the two variables have identical values of 2. For each variable, the risk is doubled in the ordinal advance from the first category to the second, and from the second to the third.

If the results were expressed as odds ratios, variable X_2 would be more impressive. Cited as $[(p_2/p_1)(q_1/q_2)]$, the odds ratios for variable X_1 would be $[(.2)/(.1)][(.99)/(.98)] = 2.02$ from the first to the second category and $[(.4)/(.2)][(.98)/(.96)] = 2.04$ from the second category to the third. For variable X_2, the corresponding ratios would be $[(.20)/(.10)][(.90)/(.80)] = 2.25$ between the first category and the second, and $[(.40)/(.20)][(.80)/(.60)] = 2.67$ from the second category to the third. The geometric mean of the two odds ratios would be 2.03 for X_1 and 2.45 for X_2. (The inflation of the true risk ratios is the reason odds ratios can be misleading if the underlying base rates are > 1%.)

Regardless of whether risk ratios or odds ratios are used, however, these results help demonstrate the crucial point that the *gradient* and the *ratio* are two different expressions, and that ratios obscure the magnitude of gradients. Accustomed to evaluating impact by examining gradients, many scientists are confused by unfamiliar or arcane citations of risk ratios and odds ratios. Although developed as a standard mathematical "treatment" for expressing and comparing risks, the ratios commit a scientifically deplorable ablation of the denominator that is the base rate for the risk. The ratios do not indicate the actual magnitude of risks or the increment in compared risks. Instead of showing gradient as an incremental difference of $p_2 - p_1$, the ratios show the relative value of p_2/p_1 for a risk ratio, or $(p_2/p_1)(q_1/q_2)$ for an odds ratio.

22.3.2.4.3. Confusion in appraisals

The custom of using ratios has led to extraordinary confusion in medical appraisals of the impact of etiologic or therapeutic variables. A risk factor that produces a gradient of only 4 per thousand (= .006 − .002) in the occurrence rate of a disease may be publicized as frightful because its risk ratio is 3 (= .006/.002). If the ratio is converted to a proportionate increment, expressed as (.006 − .002/.002, the factor can become particularly menacing, because the increase is cited as 200%. Forrow et al.[57] have recently shown how clinicians can be deluded by the manner in which the compared results are expressed. In a series of clinical vignettes reporting results for the *same* treatment, the treatment was often preferred when the achievement was cited as a proportionate increment, and often rejected when cited as a direct increment.

Laupacis et al.[58] have suggested that the misleading ratios and proportionate increments can be avoided if the results are always cited in a quantity called the *number needed to treat* (NNT), which is simply the reciprocal of the direct-increment gradient. Thus, NNT = $1/|p_2 - p_1|$. For the increment of .004 between .006 and .002, the value of NNT is 250. With the NNT expression, a "scary" factor, having a risk ratio of 3, may become much less frightening when shown to be an alleged incremental cause of disease in only 1 of 250 exposed persons.

The main point of the foregoing discussion is that if the goal of multivariable analysis is to quantify the impact of variables, the use of odds ratios or risk ratios (or their counterpart expressions in proportional increments, "attributable population risks," etc.) has an extraordinary mathematical appeal. An important adverse side effect of this appeal is that the ratios can also produce extraordinary scientific delusions and distortions.

22.3.2.4.4. Achievements of total analysis

Perhaps the worst way to express what has been accomplished at the end of a multivariable analysis is to cite a P value, confidence interval, or other stochastic index (such as chi-square or Z) for "significance" of the total result. If something "significant" has been found, these citations merely indicate that the group of data was large enough for the numerical result to be relatively stable. These indexes actually have their main value when the result is *not* "significant." An appropriate confidence interval, showing the range of possible variation for what was found, is probably the easiest way to provide stochastic confirmation of "insignificance."

The P values and other stochastic indexes are ultimately unsatisfactory because they do not provide a quantitative description of the achievement. (A confidence interval, at best, offers a range of possible variation for the result, but contains no mechanism for appraising its *quantitative* importance.) To choose a best quantitative description requires attention to the main goal of the analysis. Since the scientific goals cited throughout Section 22.3.1 seldom involve a single expression for the fit of an algebraic (or cluster) arrangement, the citation of fit is often relatively useless. As long as the results have not been distorted by the model, the different forms of R^2 and other "explanatory" indexes for reductions in variance (or log likelihood) and for proportions of explained variance (or likelihood) are helpful only during the process of choosing an optimal structure during the exploration of the data.

After the analysis is completed, however, the explanatory indexes are remarkably unhelpful in contributing to the scientific goals. As discussed earlier in the text, these indexes are worthwhile only when they indicate a poor fit for data in which visual and other examinations of conformity suggest that the wrong model was used. In such circumstances, the model should be suitably changed—but the clue that indicates the need for change will come from checking conformity, not from the explanatory index. The results of the model (see Chapter 7) may sometimes be distorted even if the explanatory index is impressively high.

If the goal is bivariate confirmation, the fit of the model is unimportant as long as the desired confirmation was achieved. If not, other problems (such as algebraic distortion or small group sizes) should be addressed. If the goal is forecasting, the fit of the model is important, but should be checked with indexes of concordant fit, not with explanations of variance and/or likelihood. The indexes that offer yes/no results or credit/debit penalty appraisals for correct predictions will contain the prime accounts of what has been accomplished.

If the goal is quantifying impact for an individual variable, and if the impact is determined from the regression coefficient in an algebraic model, the most important issue is distortion, not goodness of fit. Because neither the explanatory nor the concordant indexes are pertinent for demonstrating distortion, it must be checked with the various other procedures (linear conformity, events per variable, constant proportional hazards, etc.) discussed earlier in the text.

Finally, if the main goals are to find and use important variables in predictive groupings and/or adjustment of confounders, the analytic achievements must be assessed specifically for those purposes. Here again, the explanatory or concordant fit of the total arrangement is much less important than the distinctive gradients shown in the outcome event for the chosen groups. These gradients are *not* examined in another new but lamentable statistical strategy, which cites the results of *predictive* groupings by plotting ROC (receiver operating characteristic) curves.

The ROC strategy, which was originally developed for choosing a dichotomous cutpoint in the ordinal or dimensional results of a diagnostic marker test, seems strikingly inappropriate for evaluating predictive groups. First, the ROC curve is formed from binary demarcations and does not show gradients. Second, the ROC curve is based on the isolated proportions that become values for sensitivity and specificity, with no attention given to the size of the participating groups that form those proportions. Offering neither gradients nor estimates of uncertainty, ROC curves are a grossly inadequate way to evaluate predictive groupings. A much more effective approach is to appraise gradients, subgroup sizes, and subgroup patterns with the indexes cited in Chapter 8 and further illustrated in Chapter 20.

22.3.2.4.5. Calibration vs. discrimination

The difference between calibration and discrimination is an important distinction when individual discrepancies are converted to total analytic accomplishments. *Calibration* refers to the overall results obtained for a group rather than for individuals. Thus, if $\hat{Y}_i = .10$ is the estimated probability for death in a group, the estimate is poorly calibrated if 80% of the group die, but has excellent calibration if 10% of the people die, regardless of which persons they may be. *Discrimination* refers to a "batting average" based on individual performances. Thus, if each estimate of $\hat{Y}_i$ is associated with a chosen boundary that leads to a specific prediction of **alive** or **dead** for each person, the discrimination will reflect the number of right and wrong predictions. If a probability of .10 is set as the boundary for predicting **dead** (vs. **alive**), and if death is correctly estimated for everyone with a $\hat{Y}_i$ value of $\geq .10$, the discrimination is perfect, but calibration is poor because $\hat{Y}_i$ values of .12 and .93 would receive the same categorical estimate.

Diamond[59] has pointed out that because calibration really denotes "reliability," a predictive system having poor calibration may be unreliable despite good discrimination. He argues further that the goals of perfect discrimination and perfect reliability cannot be obtained because—like the relationship of sensitivity and specificity in diagnostic marker tests—as one rises, the other must fall, and vice versa. Addressing the same topic, Habbema, Hilden, and Bjerregaard[60] have also concluded that it would be difficult or impossible "to devise a statistic which reflects discriminatory ability and reliability at the same time."

These distinctions help emphasize the need for a suitable analytic focus in evaluating discrepancies. A system used for making individual forecasts should be checked for the discrimination reflected by indexes of concordant fit; a system aimed at separating and adjusting groups should be appraised for the calibration denoted by gradients in the target variable.

If the predicted entity is a category, the most obvious and direct index of discrepancy is the discrimination indicated by concordant accuracy: the estimate is either right or wrong. Many mathematical statisticians dislike this type of index, however, because it requires departing from theoretical distributions of continuous dimensional variables and entering the reality of binary **yes/no** or **0/1** decisions. Such decisions constantly occur in the real medical world when choices are made about diagnoses, prognostic estimates, and treatments. Although the indexes of individual discrepancy have received thoughtful mathematical attention[61] and pragmatic proposals for expressions such as Brier scores,[62] the discussions usually focus on using the indexes to summarize results for a group. For determining what happens directly in individual estimates, however, the best current index is the batting average denoted by concordant accuracy for individual predictions, as shown in classification tables of right and wrong estimates. If the direct batting average is regarded as too coarse, a cost matrix that gives suitable credits and penalties for different degrees of accuracy can be used to produce the counterpart of a "slugging average" or an index of "runs batted in."

As modern computation allows statistical analysis to be liberated from the constraints of now-traditional paradigms in parametric and probabilistic doctrines, newer (and perhaps better) strategies can be developed for the best way to express and interpret discrepancies between a predicted $\hat{Y}_i$ and the observed Y_i.

22.3.2.5. Post-spective vs. pre-spective expressions

A final problem to be avoided is the thoughtless reporting of results with "post-spective" rather than "pre-spective" statistics. In a post-spective citation, the target variable is put in the denominator and the independent variable is put in the numerator to produce a disoriented expression that cannot be used predictively. For example, the mean levels of antecedent bilirubin might be reported in groups of patients who lived and in those who died. In a pre-spective citation, however, the independent variable is put (where it belongs) in the denominator, and the target variable is put in the numerator. This type of citation would show the survival proportions for patients in different zones of antecedent bilirubin levels. Unlike a post-spective citation, the latter pre-spective expression can be used predictively.

Post-spective citations have probably become popular because they avoid the need to demarcate zones, and because a complex predictive comparison can be cited simply with a single pair of numbers: the two post-spective means. Despite the mathematical elegance, however, the post-spective citations are useless for future applications.

22.4. Issues in validation

The requirements of a successful (or credible) validation will also differ according to the purpose of the multivariable analysis.

22.4.1. Relationship to goals

If the analysis is intended merely to confirm that previously identified important variables retain their importance in a multivariate context, the results themselves serve as an act of validation, and further demands may be unnecessary. For any of the other goals cited in Section 22.3.1, however, the multivariable results will be extrapolated beyond the group from which they were derived. Just as a scientific hypothesis is not proved with the same data from which it was generated, the validity of the multivariable results cannot be proved with what was achieved when they were "custom-tailored" to fit the "training set."

22.4.2. External and internal data

The best form of validation would be to test external performance in an entirely different (but pertinent) group of data. A growing literature is accumulating on the "shrinkage,"[63] i.e., a reduced or poor level of performance, that is often found when the external groups are tested. Some of the defects may come from the original mathematical arrangement, but sometimes the problem may arise from an inappropriately chosen external group. The external group is *inappropriate* if it has crucial clinical differences from the original group. For example, a staging system for cancer should not be tested in patients with coronary disease; and an index of disability devised for nursing-home patients is inappropriate for ambulatory people living at home. On the other hand, a broad-spectrum index of disability could be appropriately tested in patients with coronary disease, although originally developed in patients with cancer.

The main problems of shrinkage, however, probably occur when the index (or system) being tested does not account for important clinical distinctions in an *appropriate* challenge group. Despite having the same basic clinical diagnosis as the original generating set, the patients in the new test set may be substantially older, sicker, or have more co-morbidity in critical attributes that were omitted from the statistical system.

Although always desirable, an external test need not be always demanded. After extensive effort and time to assemble a suitable group and data, the investigators would be harshly punished if forced to repeat the study in an entirely new group before any conclusions could be published from the first study. Accordingly, the validation in most studies is done from the available internal data acquired in that study.

Almost everyone agrees that part of the available group can be used as a training or generating set, and the remainder as a challenging test set. There is no consensus, however, on how best to divide the sets. The *predictive jackknife* approach, which sequentially leaves out one person at a time and forms its model from the remaining $N - 1$ persons, has the merit of always predicting for someone who was not included in the generating set; but the jackknife forms N models so that the investigator needs additional new strategies to combine the N models into an optimal single result. With the commonly used *cross-validation* strategy, the data are split into two parts or k parts. With the two-part approach, one set is the generating and the other, the test set. The k-part approach, sometimes called "n-fold cross validation" (see Section 21.6.5.4), is like a coarse jackknife, in which $k - 1$ parts of the data are used each time to generate results for the remaining k^{th} part.

22.4.3. Chronologic splits

Whether the split consists of 2 or k parts, however, a mechanism is needed to choose the constituents of each part. Impressed by the general éclat of random selection, most investigators almost immediately decide on random choices. Although statistically proper, this decision makes the validation process substantially depart from the usual way in which medical information is acquired and used. In most medical decisions, what was learned in the past becomes applied for the present. To recapitulate this process, the patients can be split chronologically rather than randomly, so that the earlier members of the group become the generating set for the challenge offered by the later members. This realistic approach can be used whenever the study group has been accrued in a time sequence, but the chronologic split may not work well if referral patterns or other clinical distinctions have changed the composition of the early and late groups. On the other hand, if the contents of the data or the analytic arrangements are so frail that the results from one set of patients cannot be successfully applied to subsequent patients *at the same institutional setting,* the investigators might be asked both to justify their belief that the analysis is useful, and also to identify future patients who would be an appropriate focus for the results.

22.4.4. Expression of successful challenge

A final validation decision is to choose an index for expressing a successful challenge. As noted in Section 22.3.2.4, the choice will depend on the challenge. For individual forecasting, success should be measured with expressions for individual accuracy. For quantifying the impact of individual variables, the test set's repetition of the previous average value is pleasant, but not really helpful since the impact may still vary in different zones of the data. A preferable way to validate impact of a variable is to show that the impact remains essentially the same in pertinent zones of the test group. If the goal is to form predictive groupings for adjustment of confounders, essentially the same gradients and target values should be found in counterpart groups of the test set.

22.5. Choice of analytic methods

The availability of diverse analytic strategies has evoked many discussions of their comparative merits. In keeping with academic (and perhaps editorial) traditions, the discussions are often theoretical and unaccompanied by real-world data or examples. When empirical performances are directly compared (as in Section 22.2), however, dramatically different methods often produce essentially the same statistical results; and no overt rules have been established for the specific *empirical* circumstances in which one appropriate method is distinctly better (or worse) than another.

When reviewing the actual performances, readers may be startled to discover that a mathematically "inappropriate" method, such as multiple linear regression for a binary target variable, produces exactly the same results as the more "appropriate" discriminant function analysis. Readers may be further startled (or enlightened) to discover that many mathematical warnings—about requirements for Gaussian distributions and homogeneous variances, the impropriety of using ordinal grades as dimensional variables, or the alleged "loss of data" when dimensional variables are cardinally ranked, ordinally graded, or dichotomized—are constantly ignored during the actual analytic operations.

The value of algebraic models is regularly questioned by real-world investigators[64] and regularly defended by mathematical theorists.[65] The conflict was well summarized in the caustic understatement of D. J. Finney:[66] "Amongst the many papers on statistical science published today, some appear to find new outlets for mathematical theory without materially assisting scientific research." An unsystematic review of the literature can reveal intriguing instances of statisticians arguing among themselves[67,68] about the best way to construct and to test models, and of a collaborating microbiologist and statistician urging[69] that the complex analytic formulas of algebraic models for microbial infection be replaced with the decision-rule procedures of "computer simulation algorithms." On the other hand, clinical epidemiologists may receive exhortations to make greater use of algebraic models, such as "Markov process formulation," because the models "have given those of us who are actively engaged in clinical research . . . a tool for thinking critically about our scholarship and clinical practice."[70]

A fertile source of revealing attitudes and understanding (or misunderstanding) is comments made by statisticians when they take opposite sides about the value of different approaches. For example, G. A. Barnard[71] denounced recursive partitioning methods because they lead "to a glut of unsound analyses which already seems to be bringing statistical method . . . into unjustified disrepute . . . by guarantee[ing] to get significance out of any data whatsoever." (This same objection, of course, could be aimed at mathematically sanctified parametric and algebraic-model analyses.) In a dazzlingly heretical comment to fellow-statisticians, Geisser[72] complained that

> stress on the estimation of parameters, made fashionable by mathematical statisticians, has been . . . a comfortable posture . . . [because] exposure by observation of any inadequacy in estimation is rendered virtually impossible . . . Parameters themselves for many statistical applications appear to be artificial constructs foisted upon an unwary experimenter by self-serving statisticians making "precise" statements about non-existent entities.

In a memorable paper in a journal of physiology, Hofacker[73] evoked significant results and spurious statistical conclusions when stepwise linear regression was applied to three sets of randomly generated data. In a somewhat analogous exploration, published in a journal of cardiology, Diamond[74] used bootstrap resampling to test the accomplishments of stepwise logistic regression. He found that "reproducibility . . . decreased from 100% to 0% as a function of sample size and the strictness of the criterion employed." When the criterion for reproducibility required selection of the same three important variables that had been chosen in the reference model, "maximum reproducibility was only 30%." The same problem was noted even in a statistical journal as theoretical as *Biometrika*, when Zhang[75] used a simulated set of data to conclude that "inferences on the regression coefficient are impaired by the variable selection procedure" and that "the size of the nominal confidence sets tend (*sic*) to be inflated if they are derived based on the selected model."

The internal contradictions within authoritative viewpoints were apparent when a prominent statistician,[71] after saying, "It is most important that statisticians should be thought of as being interested in the scientific content of the problems they handle," also revealed a striking intellectual distance from those problems by concomitantly stating that "zero-one classifications are almost never what is required in a practical situation."

A remarkable feature of all the discussions and debates has been the absence of attention to the *scientific* goals of the mathematical analyses. The protagonists will regularly argue about the theoretical principles, constraints, and statistical achievements of different analytic methods, while never considering the question of what the investigator wants or hopes to get. The taxonomy of investigative goals—as discussed in Chapters 2 and 3 of the text, and summarized here in Section 22.3.1—appears to be both unfamiliar and ignored in the mathematical discussions. Because different scientific goals require different analytic procedures, however, many inherently statistical arguments about methods could be avoided by simply checking whether a particular method can indeed provide the desired scientific (rather than statistical) result.

Since all of the methods have advantages and disadvantages, and since all of the reasonably appropriate methods seem to yield reasonably similar results, probably the best way to choose a method is to (1) decide what you want to get, (2) determine whether the method will give it to you, and (3) be sure you understand what it does and produces.

22.6. Requirements in reporting

The next six subsections offer rough guidelines for evaluating published reports in which multivariable methods were used. Two prime requirements of any scientific report refer to reproduction and interpretation. The methods should be presented in enough detail to allow a reader to repeat the study if given the same investigative opportunity; and crucial descriptions should be supplied to allow suitable interpretation of the results.

22.6.1. Identification of groups and data

Before any citation of statistical methods, the reader should receive satisfactory accounts of the mechanism that provided the groups and data under study, and of the criteria used for classifying judgmental categories that are not self-evident. They include diagnostic decisions, as well as subjective ratings such as *good, success, severe,* and *cause of death*.

To be able to interpret results, the reader must know the system of numerical codes for categories, and about any transformations of variables. An important but often overlooked point is the management of missing data. What processes were used for imputation and what proportions of data were imputed? What was the effective group size that remained in the multivariable analysis after exclusion of patients with unknown data? Many reports misleadingly cite an impressively large sample size without mentioning the much smaller effective sample that actually entered the analysis after "unknowns" were excluded by the computer program.

22.6.2. Computer operations

The source, identity, and key operational features of the computer program are needed for both reproducibility and interpretation. Such statements might be, "A full multiple linear regression was done with the SAS PROC REG program Version 6.07", or "Stepwise forward logistic regression was done using the default options of the BMDP2L program, 1990." The reader should also know which variables were submitted to the analytic procedure, and which ones were included in the final model.

22.6.3. Identification of multiple comparisons

The question of how to manage multiple comparisons is an unresolved controversy that has been mentioned several times in the text. Some writers believe that the α level need not be adjusted[76] for individual stochastic decisions, and that the issue is "no problem,"[77] whereas other writers are much more worried[47,78] about fallaciously attributing importance to unimportant variables and to spurious associations.

In the absence of consensus guidelines, you may want to choose whatever policy you think is best (or gives you the results you would like) for managing the multiple-comparison problem. On the other hand, if you resort to the "desperate measure" of considering the *scientific* goals of analysis and communication, at least one crucial principle becomes apparent: the reader must be informed. In particular, if the scientific conclusion of the multiple analyses was not stated as an advance hypothesis in the research, the investigators should always indicate the number of comparisons that were done before "bingo" was achieved for the proposed conclusion. If the number of comparisons is stated, readers become able, if they so desire, to do a Bonferroni (or other) adjustment for the multiplicity. If the number of comparisons is not stated, however, the reader is left uninformed and unempowered; and the investigator becomes guilty of a deception that was usually inadvertent, but that sometimes may be deliberate.

22.6.4. Accomplishments of model

The explanatory fit of the model (for reasons discussed earlier) need seldom be reported but may be sought by statistical reviewers. Indexes of concordant fit are essential if the model is to be used for forecasting.

22.6.5. Impact of variables

If the results are converted to groupings or scores, suitable justification should be cited for the chosen variables and the selected scoring system. Because the effects of the groupings or scores will be demonstrated during the validation process (another requirement!), the mathematical details for each component variable may not be essential. If individual variables, however, are quantified for impact (such as risk ratios or odds ratios), the reporting demands become particularly stringent. They include not only citations (such as confidence intervals) for the uncertainty of each estimated coefficient, but particularly an account of what was done to check for interactions, for linear conformity of the algebraic model, and (in Cox regression) for the proportional hazards assumption. If the events-per-variable ratio was not noted by the author, the constituent entities should be available for a check by the reader.

22.6.6. Fulfillment of requirements

Many of these specifications can easily be supplied (or implied) without requiring additional space, if suitable ingenuity is used in the labelling of tables, graphs, or other visual displays. The additional space needed for other details may require no more than several sentences of text, and can always be put into a small-type Appendix.

These requirements are not excessive or pedantic; and they can be readily fulfilled if editors, reviewers, and readers (as well as authors) will bear in mind that a multivariable analysis is neither magical nor standardized. The results emerge from a complex processing of data in the same way that results for measurement of cholesterol level come from a complex processing of serum. With either type of processing, the key components, operations, and interpretations must be suitably identified and, if necessary, justified, to make the work not merely an act of mathematical artistry, but particularly a useful contribution to science.

Chapter References

1. Harrell, 1985; 2. Christensen, 1981; 3. Fisher, 1934; 4. Duncan, 1966; 5. Wright, 1934; 6. Kramer, 1986; 7. Li, 1991; 8. Ray, 1980; 9. Baxt, 1991; 10. Davis, 1993; 11. Healy, 1984; 12. Hill, 1974; 13. Greenacre, 1984; 14. Higgs, 1990; 15. Grizzle, 1964; 16. Feinstein, 1990; 17. Halperin, 1971; 18. Reading, 1973; 19. Coronary Drug Project Research Group, 1974; 20. Brunk, 1975; 21. D'Agostino, 1978; 22. Eisner, 1979; 23. Dillman, 1983; 24. Gilpin, 1983; 25. Green, 1983; 26. Schmitz, 1983; 27. Madsen, 1984; 28. Harrell, 1984; 29. Ciampi, 1986; 30. Brenn, 1985; 31. Kannel, 1986; 32. Marshall, 1986; 33. Peduzzi, 1987; 34. Guppy, 1989; 35. Segal, 1989; 36. Walter, 1990; 37. Sankrithi, 1991; 38. Hadorn, 1992; 39. Long, 1993; 40. Dunn, 1993; 41. Liaño, 1993; 42. Hauck, 1985; 43. Koss, 1971; 44. Feinstein, 1972; 45. Feinstein, 1987, Chapter 10, pp. 141–66; 46. Feinstein, 1985; 47. Fleiss, 1981; 48. Mantel, 1959; 49. Miettinen, 1976; 50. Horwitz, 1980; 51. Wasson, 1985; 52. Feinstein, 1970; 53. Blettner, 1993; 54. Gifi, 1990, pp. 25–26; 55. Narula, 1993; 56. Morgenthalter, 1992; 56a. Shapiro, 1977; 57. Forrow, 1992; 58. Laupacis, 1988; 59. Diamond, 1992; 60. Habbema, 1981; 61. Van Houwelingen, 1990; 62. Redelmeir, 1991; 63. Copas, 1983; 64. Vandenbroucke, 1987; 65. Robins, 1986; 66. Finney, 1956; 67. Samuels, 1991; 68. Thompson, 1989; 69. Black, 1987; 70. Beck, 1994; 71. Barnard, 1974; 72. Geisser, 1974; 73. Hofacker, 1983; 74. Diamond, 1989; 75. Zhang, 1992; 76. Rothman, 1991; 77. Poole, 1991; 78. Kupper, 1976.

Master List of References

[Numbers in brackets after each reference indicate the chapter(s) where the reference is cited.]

Afifi, A. A. and Elashoff, R. M. Missing observations in multivariate statistics. I. Review of the literature; II. Point estimation in simple linear regression; III. Large sample analysis of simple linear regression; IV. A note on simple linear regression. J. Am. Statist. Assn. 1966;61:595–604; 1967;62:10–29; and 1969;64:337–58, 359–65. [6]

Agresti, A. Categorical data analysis. New York: John Wiley, 1990. [19]

Ahn, H. and Loh, W.-Y. Tree structured proportional hazards regression modeling. Biometrics 1994;50:471–85. [21]

Akaike, H. A new look at the statistical identification model. IEEE Trans. Auto. Control 1974;19:716–23. [13]

Albain, K. S., Crowley, J. J., LeBlanc, M., and Livingston, R. B. Determinants of improved outcome in small-cell lung cancer: An analysis of the 2,580-patient Southwest Oncology Group data base. J. Clin. Oncol. 1990;8:1563–74. [21]

Altman, D. G. Practical statistics for medical research. New York: Chapman and Hall, 1991. [2]

Altman, D. G. and De Stavola, B. L. Practical problems in fitting a proportional hazards model to data with updated measures of the covariates. Stat. in Med. 1994; 13:301–341. [15]

American Joint Committee for Cancer Staging and End-Results Reporting. Manual for staging of cancer 1978. Chicago: American Joint Committee on Cancer, 1978. [20]

American Joint Committee on Cancer Task Force on Lung. Staging of lung cancer 1979. Chicago: American Joint Committee on Cancer, 1979. [5]

American Joint Committee on Cancer. Manual for staging of cancer. 3rd ed. Philadelphia: J. B. Lippincott Co., 1988. [3,4]

Andersen, P. K. Testing goodness of fit of Cox's regression and life model. Biometrics 1982;38:67–77. [16]

———. Survival analysis 1982–1991: The second decade of the proportional hazards regression model. Stat. Med. 1991;10:1931–41. [16]

Apgar, V. A proposal for a new method of evaluation of the newborn infant. Anesth. Analg. 1953;32:260–7. [2,3,4]

Armitage, P. Tests for linear trends in proportions and frequencies. Biometrics 1955;11:375–86. [8]

———. Statistical methods in medical research. New York: John Wiley & Sons, Inc., 1971. [3]

Armstrong, B. G. and Sloan, M. Ordinal regression models for epidemiologic data. Am. J. Epidemiol. 1989;129:191–204. [14]

Ashby, D., Pocock, S. J., and Shaper, A. G. Ordered polytomous regression: An example relating serum biochemistry and haematology to alcohol consumption. Appl. Statist. 1986;35:289–301. [14]

Aube, H., Milan, C., and Blettery, B. Risk factors for septic shock in the early management of bacteremia. Am. J. Med. 1992;93:283–8. [14]

Bailey, W. C., Higgins, D. M., Richards, B. M., and Richards, J. M., Jr. Asthma severity: A factor analytic investigation. Am. J. Med. 1992;93:263–69. [4]

Balakrishan, N. Introduction and historical remarks (page 2), in Handbook of the logistic distribution, N. Balakrishan (ed). New York: Marcel Dekker, Inc., 1992. [13]

Barker, D. J., Meade, T. W., Fall, C. H. D., Lee, A., Osmond, C., Phipps, K., and Stirling, Y. Relation of fetal and infant growth to plasma fibrinogen and factor VII concentrations in adult life. Br. Med. J. 1992;304:148–52. [12]

Barnard, G. A. Discussion of preceding paper by Stone. J. Roy. Statist. Soc. Ser. B Metho. 1974;36:133–35. [22]

Barnard, M. M. The secular variations of skull characters in four series of Egyptian skulls. Ann. Eugenics 1935;6:352–71. [17]

Barnett, V. Comment on "The identification of multiple outliers". J. Am. Statist. Assn. 1993;88:795–96. [12]

Barzilay, J., Warram, J. H., Rand, L. I., Pfeifer, M. A., and Krolewski, A. S. Risk for cardiovascular autonomic neuropathy is associated with the HLA-DR3/4 phenotype in type I diabetes mellitus. Ann. Intern. Med. 1992;116:544–9. [12]

Baxt, W. G. Use of an artificial neutral network for the diagnosis of myocardial infarction. Ann. Intern. Med. 1991;115:843–8. [22]

Beale, E. M. L. Note on procedures for variable selection in multiple regression. Technometrics 1970;12:909–14. [11]

Beck, J. R. and Scardino, P. T. How useful are models of natural history in clinical decision making and clinical research? J. Clin. Epidemiol. 1994;47:1–2. [22]

Begg, C. B. Significance tests of covariate imbalance in clinical trials. Controlled Clin, Trials 1990;11:223–25. [19]

Begg, C. B. and Gray, R. Calculation of polychotomous logistic regression parameters using individualized regressions. Biometrika 1984;71:11–18. [14]

Begg, C. B. and Iglewicz, B. A treatment allocation procedure for sequential clinical trials. Biometrics 1980;36:81–90. [19]

Bellinger, D., Leviton, A., Waternaux, C., and Needleman, H. Response to Letter to Editor. N. Engl. J. Med. 1987;317:896–97. [12]

Bellinger, D., Leviton, A., Waternaux, C., Needleman, H., and Rabinowitz, M. Longitudinal analyses of prenatal and postnatal lead exposure and early cognitive development. N. Engl. J. Med. 1987;316:1037–43. [12]

Belsley, D. A., Kuh, E., and Welsch, R. E. Regression diagnostics: Identifying influential data and sources of collinearity. New York: John Wiley & Sons, 1980. [7,12]

Berk, K. N. Tolerance and condition in regression computations. J. Am. Stat. Assn. 1977;72:863–66. [11]

Berkman, L. F., Leo-Summers, L., and Horwitz, R. I. Emotional support and survival after myocardial infarction. A prospective, population-based study of the elderly. Ann. Intern. Med. 1992;117:1003–9. [14]

Berkson, J. and Gage, R. R. Calculation of survival rates for cancer. Proc. Staff Meetings Mayo Clinic 1950;25:250. [15]

Biggs, D., deVille, B., and Suen, E. A method of choosing multiway partitions for classification and decision trees. J. Appl. Statist. 1991;18:49–62. [21]

Birkett, N. J. Computer-aided personal interviewing: A new technique for data collection in epidemiologic surveys. Am. J. Epidemiol. 1988;127:684–90. [5]

Bishop, Y. M. M., Fienberg, S. E., and Holland, P. W. Discrete multivariate analysis: Theory and practice. Cambridge: MIT Press, 1975. [19]

Black, F. L. and Singer, B. Elaboration versus simplification in refining mathematical models of infectious disease. Annu. Rev. Microbiol. 1987;41:677–701. [22]

Blettner, M. and Sauerbrei, W. Influence of model-building strategies on the results of a case-control study. Stat. Med. 1993;12:1325–38. [22]

Bloch, D. A., Moses, L. E., and Michel, B. A. Statistical approaches to classification. Methods for developing classification and other criteria rules. Arthritis Rheum. 1990;33:1137–44. [21]

Bloch, D. A. and Segal, M. R. Empirical comparison of approaches to forming strata. Using classification trees to adjust for covariates. J. Am. Statist. Soc. 1989;84:897–905. [21]
Bluemenstein, B. A. Verifying keyed medical research data. Stat. Med. 1993;12:1535–42. [5]
Boswick, J. M., Lee, K. L., Califf, R. M., Topol, E. J., et al. Missing response data: To impute or not to impute? Controlled Clin. Trials 1988;9:261. [6]
Boyd, N. F., Clemens, J. D., and Feinstein, A. R. Pretherapeutic morbidity in the prognostic staging of acute leukemia. Arch. Intern. Med. 1979;139:324–28. [20]
Boyd, N. F. and Feinstein, A. R. Symptoms as an index of growth rates and prognosis in Hodgkin's disease. Clin. Invest. Med. 1978;1:25–31. [20]
Bradley, J. V. Distribution-free statistical tests. Englewood Cliffs, N.J.: Prentice-Hall, 1968. [13]
Brazer, S. R., Pancotto, F. S., Long, T. T., III, Harrell, F. E., Jr., Lee, K. L., Tyor, M. P., and Pryor, D. B. Using ordinal logistic regression to estimate the likelihood of colorectal neoplasia. J. Clin. Epidemiol. 1991;44:1263–70. [14]
Breiman, L., Friedman, J. H., Olshen, R. A., and Stone, C. J. Classification and regression trees. Belmont, Cal.: Wadsworth International Group, 1984. [21]
Brenn, T. and Arnesen, E. Selecting risk factors: A comparison of discriminant analysis, logistic regression and Cox's regression model using data from the Tromso Heart Study. Stat. Med. 1985;4:413–23. [22]
Breslow, N. E. and Day, N. E. Statistical methods in cancer research Volume 1—The analysis of case-control studies (1980) and Volume 2—The design and analysis of cohort studies (1987). Lyon (France): International Agency for Research on Cancer. [3,13,14]
Breslow, N. E., Day, N. E., Halvorsen, K. T., Prentice, R. L., and Sabai, C. Estimation of multiple relative risk functions in matched case-control studies. Am. J. Epidemiol. 1978;108:299–307. [14]
Brier, G. W. Verification of forecasts expressed in terms of probability. Monthly Weather Rev. 1950;78:1–3. [13]
Brown, C. C. On a goodness-of-fit test for the logistic model based on score statistics. Communications in Statistics 1982;11:1087–1105. [13]
Brunk, H. D., Thomas, D. R., Elashoff, R. M., and Zippin, C. Computer-aided prognosis. Chapter 3, pages 63–80 in: Elashoff, R. M., ed., Perspectives in Biometrics. Vol. 1. New York: Academic Press, 1975. [22]
Burnand, B., Kernan, W. N., and Feinstein, A. R. Indexes and boundaries for "quantitative significance" in statistical decisions. J. Clin. Epidemiol. 1990;43:1273–84. [7,8]
Campos-Filho, N. and Franco, E. L. Epidemiologic programs for computers and calculators: A microcomputer program for multiple logistic regression by unconditional and conditional maximum likelihood methods. Am. J. Epidemiol. 1989;129:439–44. [14]
Carlberg, B., Asplund, K., and Hagg, E. Factors influencing admission blood pressure levels in patients with acute stroke. Stroke 1991;22:527–30. [12]
Carmelli, D., Halpern, J., Swan, G. E., Dame, A., McElroy, M., Gelb, A. B., and Rosenman, R. H. 27-year mortality in the Western collaborative group study: Construction of risk groups by recursive partitioning. J. Clin. Epidemiol. 1991;44:1341–51. [21]
CART. Version 1.1., 1988. California Statistical Software Inc., 961 Yorkshire Ct., Lafayette, Cal. [21]
CART methodology, An introduction to. Lafayette, Cal.: California Statistical Software Inc., 1985. [21]
Castelli, W. P., Wilson, P. W. F., Levy, D., and Anderson, K. Cardiovascular risk factors in the elderly. Am. J. Cardiol. 1989;63:12H–19H. [20]
Cease, K. B. and Nicklas, J. M. Prediction of left ventricular ejection fraction using simple quantitative clinical information. Am. J. Med. 1986;81:429–36. [12]
Chan, C. K., Feinstein, A. R., Jekel, J. K., and Wells, C. K. The value and hazards of standardization in clinical epidemiologic research. J. Clin. Epidemiol. 1988;41:1125–34. [3]

Charlson, M. E. and Feinstein, A. R. The auxometric dimension. A new method for using rate of growth in prognostic staging of breast cancer. J. Am. Med. Assn. 1974;228:180–85. [20]

Christensen, E. Multivariate survival analysis using Cox's regression model. Hepatology 1987;7:1346–58. [16]

Christensen, R. Entropy minimax sourcebook. Vol. I: General description. Lincoln, Mass.: Entropy Ltd., 1981. [22]

Ciampi, A., Lawless, J. F., McKinney, S. M., and Singhal, K. Regression and recursive partition strategies in the analysis of medical survival data. J. Clin. Epidemiol. 1988;41:737–48. [21]

Ciampi, A., Thiffault, J., Nakache, J.-P., and Asselain, B. Stratification by stepwise regression, correspondence analysis and recursive partition: A comparison of three methods of analysis for survival data with covariates. Comput. Stat. Data Anal. 1986;4:185–204. [22]

Classen, D. C., Evans, R. S., Pestotnik, S. L., Horn, S. D., Menlove, R. L., and Berke, J. P. The timing of prophylactic administration of antibiotics and the risk of surgical-wound infection. N. Engl. J. Med. 1992;326:281–6. [14]

Clemens, J. D., Feinstein, A. R., Holabird, N., and Cartwright, S. C. A new clinical-anatomic staging system for evaluating prognosis and treatment of prostatic cancer. J. Chronic Dis. 1986;39:913–28. [20]

Cochran, W. G. A test of linear function of the deviations between observed and expected numbers. J. Am. Statist. Assn. 1955;50:377–97. [13]

———. Comparison of methods for determining stratum boundaries. Bull. de l'Institut International de Statistique 1961;38:345–58. [8]

———. The effectiveness of adjustment by subclassification in removing bias in observational studies. Biometrics 1968;24:295–313. [8]

Cohen, A. Dummy variables in stepwise regression. Am. Statist. 1991;45:226–28. [11]

Cole, B. F., Gelber, R. D., and Goldhirsch, A. Cox regression models for quality adjusted survival analysis. Stat. Med. 1993;12:975–87. [16]

Concato, J., Feinstein, A. R., and Holford, T. R. The risk of determining risk with multivariable models. Ann. Intern. Med. 1993;118:201–10. [P,5,12]

Concato, J., Horwitz, R. I., Feinstein, A. R., Elmore, J. G., and Schiff, S. F. Problems of comorbidity in mortality after prostatectomy. J. Am. Med. Assn. 1992;267:1077–82. [5,20]

Concato, J., Peduzzi, P., Holford, T. R., and Feinstein, A. R. The importance of "events per independent variable" in proportional hazards and other multivariable analyses. Clin. Res. 1993;41:180A (Abstract). [5,10,12]

Cook, E. F. and Goldman, L. Empiric comparison of multivariate analytic techniques: Advantages and disadvantages of recursive partitioning analysis. J. Chronic Dis. 1984;37:721–31. [21]

———. Asymmetric stratification: An outline for an efficient method for controlling confounding in cohort studies. Am. J. Epidemiol. 1988;127:626–39. [21]

Cook, R. D. and Weisberg, S. Residuals and influence in regression. London: Chapman and Hall, 1982. [7,12]

Copas, J. B. Regression, prediction, and shrinkage. J. Roy. Statist. Soc. Ser. B Metho. 1983;45:311–54. [22]

Coronary Drug Project Research Group. Factors influencing long-term prognosis after recovery from myocardial infarction—Three-year findings of the Coronary Drug Project. J. Chronic Dis. 1974;27:267–85. [22]

Coste, J., Spira, A., Ducimetiere, P., and Paolagg, J.-B. Clinical and psychological diversity of non-specific low-back pain. A new approach towards the classification of clinical subgroups. J. Clin. Epidemiol. 1991;44:1233–45. [4]

Couglin, S. S., Trock, B., Criqui, M. H., Pickle, L. W., Browner, D., and Tefft, M. C. The logistic modeling of sensitivity, specificity, and predictive value of a diagnostic test. J. Clin. Epidemiol. 1992;45:1–7. [14]

Cowie, H., Lloyd, M. H., and Soutar, C. A. Study of lung function data by principal components analysis. Thorax 1985;40:438–43. [4]

Cox, D. R. The analysis of binary data. London: Methuen & Co., 1970. [13]

———. Regression models and life tables (with discussion). J. Roy. Statist. Soc. Ser. B. Metho. 1972;34:187–220. [4,15]

Cox, D. R. and Snell, E. J. A general definition of residuals. J. Roy. Statist. Soc. Ser. B Metho. 1968;30:248–75. [16]

Cramer, J. A. and Spilker, B. (eds.) Patient compliance in medical practice and clinical trials. New York: Raven Press Ltd., 1991. [15]

Crom, W. R., Glynn, A. M., Abromowitch, M., Pui, C., Dodge, R., and Evans, W. E. Use of the automated interaction detector method to identify patient characteristics related to methotrexate clearance. Clin. Pharmacol. Ther. 1986;39:592–7. [21]

Crombie, I. K. and Irving, J. M. An investigation of data entry methods with a personal computer. Comput. Biomed. Res. 1986;19:543–50. [5]

Curran, W. J., Jr., Scott, C. B., Horton, J., et al. Recursive partitioning analysis of prognostic factors in three radiation therapy oncology group malignant glioma trials. J. Nat. Cancer Inst. 1993;85:704–10. [21]

D'Agostino, R. B., Teebagy, N. C., Pozen, M. W., Guglielmeno, J. T., Bielawski, L. I., and Hood, W. B. Comparison of logistic regression and discriminant analysis as emergency room decision models for the diagnosis of acute coronary disease. Biometrics 1978;34:155 (Abstract). [22]

Dalenius, J. The problem of optimum stratification. Skandinavisk Aktuarietidskrift 1950;33:203–13. [8]

Daneshmend, T. K., Hawkey, C. J., Langman, M. J., Logan, R. F., Long, R. G., and Walt, R. P. Omeprazole versus placebo for acute upper gastrointestinal bleeding: Randomized double blind controlled trial. Br. Med. J. 1992;304:143–7. [14]

Davis, G. E., Lowell, W. E., and Davis, G. L. A neural network that predicts psychiatric length of stay. M. D. Comput. 1993,10.87–92. [22]

Diamond, G. A. Future imperfect: The limitations of clinical prediction models and the limits of clinical prediction. J. Am. Coll. Cardiol. 1989;14:12A–22A. [22]

———. Clinical epidemiology of sensitivity and specificity. J. Clin. Epidemiol. 1992;45:9–13. [14]

———. What price perfection? Calibration and discrimination of clinical prediction models. J. Clin. Epidemiol. 1992;45:85–89. [22]

Diehr, P., Wood, R. W., Barr, V., Wolcott, B., Slay, L., and Tompkins, R. K. Acute headaches: Presenting symptoms and diagnostic rules to identify patients with tension and migraine headache. J. Chronic Dis. 1981;34:147–58. [21]

Dillman, R. O. and Koziol, J. A. Statistical approach to immunosuppression classification using lymphocyte surface markers and functional assays. Cancer Res. 1983;43:417–21. [22]

Dixon, W. J. (ed.) BMDP Statistical Software Manual Volumes 1 and 2. Berkeley: University of California Press, 1990. [5,16]

Dolgin, S. M., Schwartz, J. S., Kressel, H. Y., et al. Identification of patients with cholesterol or pigment gallstones by discriminant analysis of radiographic features. N. Engl. J. Med. 1981;304:808–11. [18]

Doll, R. Is cadmium a human carcinogen? Ann. Epidemiol. 1992;2:335–7. [12]

Dougados, M., van der Linden, S., Juhlin, R., et al. The European spondylarthropathy study group preliminary criteria for the classification of spondylarthropathy. Arthritis Rheum. 1991;34:1218–27. [14]

Draper, N. R. and Smith, H. Applied regression analysis. New York: John Wiley & Sons, 1966. [P]

Duncan, O. D. Path analysis: Sociological examples. Am. J. Sociol. 1966;72:1–16. [22]

Dunn, J. A. and Finn, C. B. Cox regression versus discriminant techniques: Can we predict a high risk group in stage I epithelial ovarian cancer? Controlled Clin. Trials 1993;14:463 (Abstract). [22]

Efron, B. and Gong, G. A. A leisurely look at the bootstrap, the jackknife, and cross-validation. Am. Statist. 1983;37:36–48. [12]

Eisner, V., Brazie, J. V., Pratt, M. W., and Hexter, A. C. The risk of low birthweight. Am. J. Public Health 1979;69:887–93. [22]

Eskenazi, B., Fensten, L., and Sidney, S. A multivariate analysis of risk factors for preeclampsia. J. Am Med. Assn. 1991;266:237–41. [14]

Everitt, B. S. Cluster analysis. 3rd ed. London: Edward Arnold, 1993. [4]

Farrington, C. P. Review of PEST 2.1: Planning an evaluation of sequential trials. Appl. Stat. 1993;42:561–63. [5]

Feinstein, A. R. Symptoms as an index of biological behaviour and prognosis in human cancer. Nature 1966;209:241–45. [5,12]

———. Clinical judgment. Baltimore: Williams & Wilkins Co., 1967. [5,19]

———. A new staging system for cancer and a reappraisal of "early" treatment and "cure" by radical surgery. N. Engl. J. Med. 1968;279:747–53. [11,12]

———. The pre-therapeutic classification of co-morbidity in chronic disease. J. Chronic Dis. 1970;23:455–69. [22]

———. Clinical biostatistics: XV. The process of prognostic stratification (Part 1). Clin, Pharmacol. Ther. 1972;13:442–57. [P, 8]

———. On classifying cancers while treating patients. Arch. Intern. Med. 1985;145:1789–91. [3,5,12]

———. Clinical epidemiology: The architecture of clinical research. Philadelphia: W. B. Saunders Co., 1985. [3,5,7,15,22]

———. Clinimetrics. New Haven: Yale University Press, 1987. [2,3,4,5,8,22]

———. Quantitative ambiguities in matched versus unmatched analyses of the 2 × 2 table for a case-control study. Int. J. Epidemiol. 1987;16:128–34. [14]

———. The unit fragility index: An additional appraisal of "statistical significance" for a contrast of two proportions. J. Clin. Epidemiol. 1990;43:201–09. [7]

Feinstein, A. R., Gelfman, N. A., and Yesner, R. The diverse effects of histopathology on manifestations and outcome of lung cancer. Chest 1974;66:225–29. [5]

Feinstein, A. R. and Landis, J. R. A computer program for finding multivariate prognostic clusters. Clin. Res. 1973;21:725 (Abstract). [21]

———. The role of prognostic stratification in preventing the bias permitted by random allocation of treatment. J. Chronic Dis. 1976;29:277–84. [19]

Feinstein, A. R., Pritchett, J. A., and Schimpff C. R. The epidemiology of cancer therapy III. The management of imperfect data. Arch. Intern. Med. 1969;123:448–61. [5]

Feinstein, A. R., Rubinstein, J. F., and Ramshaw, W. A. Estimating prognosis with the aid of a conversational-mode computer program. Ann. Intern. Med. 1972;76:911–21. [22]

Feinstein, A. R., Schimpff, C. R., Andrews, J. F., Jr., and Wells, C. K. Cancer of the larnyx: A new staging system and a re-appraisal of prognosis and treatment. J. Chronic Dis. 1977;30:277–305. [20]

Feinstein, A. R., Schimpff, C. R., and Hull, E. W. (with the technical assistance of H. L. Bidwell). A reappraisal of staging and therapy for patients with cancer of the rectum. I. Development of two systems of staging. Arch. Intern. Med. 1975;135:1441–53. [4,20]

Feinstein, A. R. and Singer, B. H. Forming better unions: an antidote to statistical 'reductionism'. Clin. Res. 1993;41:396A (Abstract). [5]

Feinstein, A. R. and Wells, C. K. A clinical-severity staging system for patients with lung cancer. Medicine 1990;69:1–33. [P,4,5,8,11,12,19,20]

Feinstein, A. R., Wells, C. K., and Walter, S. D. A comparison of multivariable mathematical methods for predicting survival—I. Introduction, rationale, and general strategy. J. Clin. Epidemiol. 1990;43:339–47. [22]

Feskens, E. J., Bowles, C. H., and Krombout, D. A longitudinal study on glucose tolerance and other cardiovascular risk factors: Associations within an elderly population. J. Clin. Epidemiol, 1992;45:293–300. [12]

Fieller, N. Comment on "The identification of multiple outliers". J. Am. Statist. Assn. 1993;88:794–95. [12]

Fienberg, S. E. The analysis of cross-classified categorical data. Cambridge: MIT Press, 1980. [19]

Fieselmann, J. F., Hendryx, M. S., Helms, C. M., and Wakefield, D. S. Respiratory rate predicts cardiopulmonary arrest for internal medicine inpatients. J. Gen. Intern. Med. 1993;8:354–60. [14]

Finney, D. J. Multivariate analysis and agricultural experiments. Biometrics 1956;12:67–71. [22]

———. Letter to Editor. Biometric Bulletin 1992;9:2. [12]

Fisher, R. A. Theory of statistical estimation. Proceedings of the Cambridge Philosophical Society 1925;22:700–25. [13]

———. Discussion on statistics in agricultural research (by J. Wishart). J. Roy. Statist. Soc., Supple. 1934;1:51–53. [22]

———. The use of multiple measurements in taxonomic problems. Ann. Eugenics, 1936;7 (Part II):179–88. [17]

Fleiss, J. L. Statistical methods for rates and proportions. 2nd ed. New York: John Wiley & Sons, 1981. [2,7,8,22]

Forrow, L., Taylor, W. C., and Arnold, R. M. Absolutely relative: How research results are summarized can affect treatment decisions. Am. J. Med. 1992;92:121–24. [22]

Fox, J. Regression diagnostics. Newbury Park Cal.: Sage, 1991. [7]

Freedman, L. S., Parkinson, M. C., Jones, W. G., et al. Histopathology in the prediction of relapse of patients with stage 1 testicular teratoma treated by orchidectomy alone. Lancet 1987;2:294–98. [19]

Furukawa, T. and Sumita, Y. A cluster-analytically derived subtyping of chronic affective disorders. Acta Psychiatr. Scand. 1992;85:177–82. [4]

Gail, M. H., Brinton, L. A., Byar, D. P., Corle, D. K., Green, S. B., Schairer, C., and Mulvihill, J. J. Projecting individualized probabilities of developing breast cancer for white females who are being examined annually. J. Nat. Cancer Inst. 1989;81:1879–86. [16]

Gardner, M. J. and Barker, D. J. P. A case study in techniques of allocation. Biometrics 1975;31:931–42. [19]

Garrison, C. Z., Weinrich, M. W., Hardin, S. B., Weinrich, S., and Wang, L. Post-traumatic stress disorder in adolescents after a hurricane. Am. J. Epidemiol. 1993;138:522–30. [14]

Gehan, E. A. A generalized Wilcoxon test for comparing arbitrarily singly-censored samples; and A generalized two-sample Wilcoxon test for doubly-censored data. Biometrika 1965;52:203–23, 650–52. [16]

Geisser, S. Discussion of preceding paper by Stone. J. Roy Statist. Soc. Ser. B Metho. 1974;36:141–42. [22]

Gifi, A. Nonlinear multivariate analysis. Chichester, U.K.: John Wiley, 1990. [22]

Giles, G. G., Marks, R., and Foley, P. Incidence of non-melanocytic skin cancer treated in Australia. Br. Med. J. 1988;296:13–17. [14]

Gillespie, E. S. An application of multivariate outlier detection in assessing family characteristics for bank advertisements. Statistician 1993;42:231–5. [12]

Gilpin, E., Olshen, R., Henning, H., and Ross, J., Jr. Risk prediction after myocardial infarction: Comparison of three multivariate methodologies. Cardiology 1983;70:73–84. [21,22]

Gini, C. Sulla misura della concentrazione e della variabilità dei caratteri. Atti del Reale Istituto Veneto di Scienze, lettere ed Arti, AA 1913–1914, Tomo LXXIII, parte II, 1914, 1203–48. [21]

Glantz, S. A. and Slinker, B. K. Primer of applied regression and analysis of variance. New York: McGraw-Hill, 1990. [P,3,10]

Glasbey, C. A. Examples of regression with serially correlated errors. Statistician 1988;37:277–91. [12]

Goldbourt, U., Behar, S., Reicher-Reiss, H., Zion, M., Mandelzweig, L., and Kaplinsky, E. Early administration of nifedipine in suspected acute myocardial infarction. Arch. Intern. Med. 1993;153:345–53. [16]

Goldman, L., Cook, E. F., Brand, D. A., et al. A computer protocol to predict myocardial infarction in emergency department patients with chest pain. N. Engl. J. Med. 1988;318:797–803. [21]

Goldman, L., Weinberg, M., Weisberg, M., et al. A computer-derived protocol to aid in the diagnosis of emergency room patients with acute chest pain. N. Engl. J. Med. 1982;307:588–96. [21]

Goodman, L. The analysis of cross-classified data having ordered categories. Cambridge: Harvard University Press, 1984. [19]

Goodman, L. A. and Kruskal, W. H. Measures of association for cross-classifications. New York: Springer-Verlag, 1979. [14]

Gordon, T. J. Hazards in the use of the logistic function with special reference to data from prospective cardiovascular studies. J. Chronic Dis. 1974;27:97–102. [3]

Grambsch, P. M., Dickson, E. R., Wiesner, R. H., and Langworthy, A. Application of the Mayo primary biliary cirrhosis survival model to Mayo liver transplant patients. Mayo Clin. Proc. 1989;64:699–704. [16]

Grambsch, P. and O'Brien, P. C. The effects of transformations and preliminary tests for non-linearity in regression. Stat. Med. 1991;10:697–709. [6,7]

Gray, R. J. Flexible methods for analyzing survival data using splines, with application to breast cancer prognosis. J. Am. Statist. Assn. 1992;87:942–51. [12]

Green, K. L. and Johnson, J. V. The effects of psychosocial work organization on patterns of cigarette smoking among male chemical plant employees. Am. J. Public Health 1990;80:1368–71. [20]

Green, M. S. and Jucha, E. Interrelationships between blood pressure, serum calcium, and other biochemical variables. Int. J. Epidemiol. 1987;16:532–6. [12]

Green, M. S. and Symons, M. J. A comparison of the logistic risk function and the proportional hazards model in prospective epidemiologic studies. J. Chronic Dis. 1983;36:715–24. [22]

Greenacre, M. J. Theory and application of correspondence analysis. New York: Academic Press, 1984. [22]

Greenblatt, D. J., Harmatz, J. S., Stanski, D. R., Shader, R. I., Franke, K., and Koch-Weser, J. Factors influencing blood concentrations of chlordiazepoxide: A use of multiple regression analysis. Psychopharmacology 1977;54:277–82. [12]

Greenland, S., Schlesselman, J. J., and Criqui, M. H. The fallacy of employing standarized regression coefficients and correlations as measures of effect. Am. J. Epidemiol. 1986;123:203–08. [13]

Greenwood, M. Epidemics and crowd diseases. An introduction to the study of epidemiology. London: Williams and Norgate Ltd., 1935. [15]

Grizzle, J. E. Multivariate comparison of results of treatment in chronic lymphocytic and chronic granulocytic leukemia. J. Chronic Dis. 1964;17:127–52. [22]

Grizzle, J. E., Starmer, C. F., and Koch, G. G. Analysis of categorical data by linear models. Biometrics 1969;25:489–504. [19]

Guppy, K. H., Detrano, R., Abbassi, N., Janosi, A., Sandhu, S., and Froelicher, V. The reliability of probability analysis in the prediction of coronary artery disease in two hospitals. Med. Decis. Making 1989;9:181–89. [22]

Guralnick, D. B. (ed) Webster's New World Dictionary, 2nd ed. Toronto: Nelson, Foster & Scott, Ltd., 1972. [7]

Guyatt, G. (Letter to Editor) Transferrin as a predictor of iron deficiency. Am. J. Med. 1992;92:543. [6]

Habbema, D. F., Hilden, J., and Bjerregaard, B. The measurement of performance in probabilistic diagnosis. V. General recommendations. Meth. Inform. Med. 1981;20:97–100. [22]

Hadorn, D. C., Draper, D., Rogers, W. H., Keeler, E. B., and Brook, R. H. Cross-validation performance of mortality prediction models. Stat. Med. 1992;11:475–89. [21,22]

Halperin, M., Blackwelder, W. C., and Verter, J. I. Estimation of the multivariate logistic risk function: A comparison of the discriminant function and maximum likelihood approaches. J. Chronic Dis. 1971;24:125–58. [22]

Hanley, J. A. and McNeil, B. J. The meaning and use of the area under a receiver operating characteristic (ROC) curve. Radiology 1982;143:29–36. [13]

Harrell, F. E., Jr. The LOGIST procedure. In: SUGI Supplemental Library User's Guide, Version 5 Edition, Cary, N.C., SAS Institute, 1986. [13]

———. The PHGLM Procedure. In: SUGI Supplemental Library User's Guide, 5th Edition, Hastings, R. P., ed. Cary, N.C.: SAS Institute, 1986. [15]

Harrell, F. E., Jr., and Lee, K. L. Verifying assumptions of the Cox proportional hazards model. In: Proceedings of the Eleventh Annual SAS User's Group International Conference. Cary, N.C.: SAS Institute, Inc., 1986. [16]

Harrell, F. E., Jr., Lee, K. L., Califf, R. M., Pryor, D. B., and Rosati, R. A. Regression modeling strategies for improved prognostic prediction. Stat. Med. 1984;3:143–52. [13,22]

Harrell, F. E., Jr., Lee, K. L., Matchar, D. B., and Reichert, T. A. Regression models for prognostic prediction: Advantages, problems, and suggested solutions. Cancer Treat. Rep. 1985;69:1071–7. [10,12,22]

Harrell, F. E., Jr., Lee, K. L., and Pollock, B. G. Regression models in clinical studies: Determining relationships between predictors and response. J. Nat. Cancer Inst. 1988;80:198–202. [2,12]

Harris, E. K. and Albert, A. Survivorship analysis for clinical studies. New York: Marcel Dekker, Inc., 1991. [15]

Harvey, E. B., Boice, J. D., Honeyman, M., and Flannery, J. T. Prenatal X-ray exposure and childhood cancer in twins. N. Engl. J. Med. 1985;312:541–45. [14]

Hauck, W. W. (Letter to Editor) A comparison of the logistic risk function and the proportional hazards model in prospective epidemiologic studies. J. Chronic Dis. 1985;38:125–26. [22]

Hauck, W. W. and Miike, R. A proposal for examining and reporting stepwise regressions. Stat. Med. 1991;10:711–15. [12]

Healy, M. J. R. Discussion of preceding paper by Spiegelhalter and Knill-Jones. J. Roy. Statist. Soc. Ser. A Stat. 1984;147:58–59. [22]

Heckerling, P. S., Leiken, J. B., and Maturen, A. Occult carbon monoxide poisoning: Validation of a prediction model. Am. J. Med. 1988;84:251–6. [12]

Heinonen, O. P., Slone, D., Monson, R. R., Hook, E. B., and Shapiro, S. Cardiovascular birth defects and antenatal exposure to female sex hormones. N. Engl. J. Med. 1977;296:67–70. [5]

Helfenstein, U. and Minder, C. The use of measures of influence in epidemiology. Int. J. Epidemiol. 1990;19:197–204. [12]

Henderson, W. G., Fisher, S. G., Cohen, N., et al. Use of principal components analysis to develop a composite score as a primary outcome variable in a clinical trial. Controlled Clin. Trials 1990;11:199–214. [4]

Herbert, J. H. and Kott, P. S. An empirical note on regression with and without a poorly measured variable using gas demand as a case study. Statistician 1988;37:293–98. [12]

Higgins, T. L., Estafanous, F. G., Loop, F. D., Beck, G. J., Blum, J. M., and Paranandi, L. Stratification of morbidity and mortality outcome by preoperative risk factors in coronary artery bypass patients. A clinical severity score. J. Am. Med. Assn. 1992;267:2344–8. [14]

Higgs, N. T. Practical and innovative uses of correspondence analysis. Statistician 1990;40:183–94. [22]

Hill, M. O. Correspondence analysis: A neglected multivariate method. Appl. Statist. 1974;23:340–54. [22]

Hlatky, M. A., Pryor, D. B., Harrell, F. E., Jr., et al. Factors affecting sensitivity and specificity of exercise electrocardiography. Multivariable analysis. Am. J. Med. 1984;77:64–71. [14]

Hoaglin, D. C. and Welsch, R. E. The hat matrix in regression and ANOVA. Am. Statist. 1978;32:17–22. [12]

Hofacker, C. F. Abuse of statistical packages: The case of the general linear model. Am. J. Physiol. 1983;245:R299–R302. [22]

Holford, T. R., White, C., and Kelsey, J. L. Multivariate analysis for matched case-control studies. Am. J. Epidemiol. 1978;107:245–56. [14]

Holme, I., Helgeland, A., Hjermann, I., Lund-Larsen, P. G., and Leren, P. Coronary risk factors and socioeconomic status: The Oslo study. Lancet 1976;2:1396–98. [20]

Holt, V. L., Chu, J., Daling, J. R., Stergachis, A. S., and Weiss, N. S. Tubal sterilization and subsequent ectopic pregnancy: A case-control study. J. Am. Med. Assn. 1991;266:242–6. [14]

Hooton, T. M., Haley, R. W., Culver, D. H., White, J. W., Morgan, W. M., and Carroll, R. J. The joint associations of multiple risk factors with the occurrence of nosocomial infection. Am. J. Med. 1981;70:960–70. [21]

Horwitz, R. I. and Feinstein, A. R. The problem of "protopathic bias" in case-control studies. Am. J. Med. 1980;68:255–58. [22]

Hosmer, D. W. and Lemeshow, S. A goodness-of-fit test for the multiple logistic regression model. Communications in Statistics 1980;A10:1043–69. [13]

———. Applied logistic regression. New York: John Wiley & Sons, 1989. [13,14]

Hubbard, B. L., Gibbons, R. J., Lapeyre, A. C., III, Zinsmeister, A. R., and Clements, I. P. Identification of severe coronary artery disease using simple clinical parameters. Arch. Intern. Med. 1992;152:309–12. [14]

Hunder, G. G., Arend, W. P., Bloch, D. A., et al. The American College of Rheumatology 1990 criteria for the classification of vasculitis. Introduction. Arthritis Rheum. 1990;33:1065–67. [21]

ICD-10. International statistical classification of diseases and related health problems. 10th ed. Geneva: World Health Organization, 1992. [2]

Jackson, J. E. A user's guide to principal components. New York: John Wiley, 1991. [4]

Jacqmin, H., Commenges, D., Letenneur, L., Barberger-Gateau, P., and Dartigues, J.-F. Components of drinking water and risk of cognitive impairment in the elderly. Am. J. Epidemiol. 1994;139:48–57. [12]

Jennrich, R. and Simon, P. Stepwise discriminant analysis. In: Dixon, W. J., ed. BMDP Statistical Software Manual Volume 1. Berkeley: University of California Press, 1990. [18]

Joiner, B. L. Lurking variables: Some examples. Am. Statist. 1981;35:227–33. [12]

Justice, A. C., Feinstein, A. R., and Wells, C. K. A new prognostic staging system for the acquired immunodeficiency syndrome. N. Engl. J. Med. 1989;320:1388–93. [8]

Kahn, H. A. and Sempos, C. T. Statistical methods in epidemiology. New York: Oxford University Press, 1989. [3]

Kalbfleisch, J. D. and Prentice, R. L. The statistical analysis of failure time data. New York: John Wiley & Sons, 1980. [15,16]

Kannel, W. B. Contribution of the Framingham study to preventive cardiology. J. Am. Coll. Cardiol. 1990;15:206–11. [19]

Kannel, W. B., Neaton, J. D., Wentworth, D., Thomas, H. E., Stamler, J., Hulley, S. B., and Kjelsberg, M. O. Overall and coronary heart disease mortality rates in relation to major risk factors in 325,348 men screened for the MRFIT. Am. Heart J. 1986;112:825–36. [22]

Kannel, W. B. and Wolf, P. A. Pulling it all together: Changing the cardiovascular outlook. Am. Heart J. 1992;123:264–67. [12,19]

Kaplan, E. L. and Meier, P. M. Nonparametric estimation from incomplete observations. J. Am. Statist. Soc. 1958;53:457–81. [15]

Kass, G. V. An exploratory technique for investigating large quantities of categorical data. Appl. Statist. 1980;29:119–27. [21]

Kassirer, J. P. and Kopelman, R. I. Diagnosis and the structure of memory. 3. The nature of categories. Hosp. Pract. 1991;26:30–35. [19]

Katz, M. H., Baron, R. B., and Grady, D. Risk stratification of ambulatory patients suspected of *pneumocystis* pneumonia. Arch. Intern. Med. 1991;151:105–10. [19]

Kawazoe, N., Eto T., Abe, I., et al. Pathophysiology in malignant hypertension: With special reference to the renin-angiotension system. Clin. Cardiol. 1987;10:513–8. [12]

Kay, R. Proportional hazard regression models and the analysis of censored survival data. Appl. Statist. 1977;26:277–37. [16]

Kelsey, J. L., Thompson, W. D., and Evans, A. S. Methods in observational epidemiology. New York: Oxford University Press, 1986. [3]

Kendall, M. G. and Buckland, W. R. A dictionary of statistical terms. 3rd ed. New York: Hafner Publishing Co., Inc., 1971. [7]

Kerlinger, F. N. and Pedhauzur, E. J. Multiple regression in behavioral research. New York: Holt, Rinehart and Winston, 1973. [P,5]

Kim, I., Williamson, D. F., Byers, T., and Koplan, J. P. Vitamin and mineral supplement use and mortality in a U.S. cohort. Am. J. Public Health 1993;83:546–50. [16]

Klatsky, A. L., Armstrong, M. A., and Friedman, G. D. Coffee, tea, and mortality. Ann. Epidemiol. 1993;3:375–81. [16]

Kleinbaum, D. G. and Kupper, L. L. Applied regression analysis and other multivariable methods. North Scituate, Mass.: Duxbury Press (Division of Wadsworth Publishing Co., Belmont, Cal.), 1978. [P]

Kleinbaum. D. G., Kupper, L. L., and Muller, K. E. Applied regression analysis and other multivariable methods. Boston: Wadsworth Publishing Co. (PWS-KENT Division), 1988. [P,3,5,10,14,17]

Knowledge Seeker, User's Guide, Decision Making Software, Version 2.0. Ottawa, Canada: FirstMark Technologies Ltd., 1990. [21]

Koehler, A. B. and Murphree, E. S. A comparison of the Akaike and Schwartz criteria for selecting model order. Appl. Statist. 1988;37:187–95. [13]

Koenig, W., Sund, M., Ernst, E., Mraz, W., Hombach, V., and Keil, U. Association between rheology and components of lipoproteins in human blood: Results from the MONICA project. Circulation 1992;85:2197–2204. [12]

Korn, E. L. and Simon, R. Measures of explained variation for survival data. Stat. Med. 1990;9:487–503. [15]

Koss, N. and Feinstein, A. R. Computer-aided prognosis: II. Development of a prognostic algorithm. Arch. Intern. Med. 1971;127:448–59. [P,22]

Kramer, A. A., Green, L. J., Croghan, I. T., Buck, G. M., and Ferer, R. Bivariate path analysis of twin children for stature and biiliac diameter: Estimation of genetic variation and co-variation. Hum. Biol. 1986;58:517–25. [22]

Kramer, M. S. and Feinstein, A. R. Clinical biostatistics: LIV. The biostatistics of concordance. Clin. Pharmacol. Ther. 1981;29:111–23. [2]

Krone, R. J., Gillespie, J. A., Weld, F. M., et al. Low-level exercise testing after myocardial infarction: Usefulness in enhancing clinical risk stratification. Circulation 1985;71:80–89. [19]

Kupper, L. L., Stewart, J. R., and Williams, K. A. A note on controlling significance levels in stepwise regression. Am. J. Epidemiol. 1976;103:13–15. [22]

Kurtzke, J. F. On estimating survival; a tale of two censors. J. Clin. Epidemiol. 1989;42:169–75. [15]

Kushi, L. H., Sellers, T. A., Potter, J. D., and Folsom, A. R. Response to Letter to Editor. J. Nat. Cancer Inst. 1992;84:1667–69. [12]

Kwak, L. W., Halpern, J., Olshen, R. A., and Horning, S. J. Prognostic significance of actual dose intensity in diffuse large-cell lymphoma: Results of a tree-structured survival analysis. J. Clin. Oncol. 1990;8:963–77. [21]

Laberge, F., Fritsche, H. A., Umsawasdi, T., et al. Use of carcinoembryonic antigen in small cell lung cancer. Prognostic value and relation to the clinical course 1. Cancer 1987;59:2047–52. [16]

Lachenbruch, P. A. Discriminant analysis. New York: Hafner Press, 1975. [17]

Lafferty, F. W. Primary hyperparathyroidism: Changing clinical spectrum, prevalence of hypertension, and discriminant analysis of laboratory tests. Arch. Intern. Med. 1981;141:1761–66. [18]

Lamm, S. H., Parkinson, M., Anderson, M., and Taylor, W. Determinants of lung cancer risk among cadmium-exposed workers. Ann. Epidemiol. 1992;2:195–211. [12]

Landesberg, G., Luria, M. H., Cotev, S., et al. Importance of long-duration postoperative ST-segment depression in cardiac morbidity after vascular surgery. Lancet 1993;341:715–19. [14]

Landis, J. R. and Koch, G. G. The measurement of observer agreement for categorical data. Biometrics 1977;33:159–74. [2]

Lane, R. S., Barsky, A. J., and Goodson, J. D. Discomfort and disability in upper respiratory tract infection. J. Gen. Intern. Med. 1988;3:540–6. [12]

Laschober, P. J. and Tamura, H. Application of AID in analysis of energy survey data—a case study. Statistician 1986;35:459–69. [21]

Lashner, B. A., Jonas, R. B., Tang, H.-S., Evans, A. A., Ozeran, S. E., and Baker, A. L. Chronic hepatitis: Disease factors at diagnosis predictive of mortality. Am. J. Med. 1988;85:609–14. [16]

Lauer, K. The risk of multiple sclerosis in the U.S.A. in relation to sociogeographic features: A factor-analytic study. J. Clin. Epidemiol. 1994;47:43–48. [4]

Lauer, M. S., Levy, D., Anderson, K. M., and Plehn, J. F. Is there a relationship between exercise systolic blood pressure response and left ventricular mass? The Framingham Heart Study. Ann. Intern. Med. 1992;116:203–10. [12]

Laupacis, A., Sackett, D. L., and Roberts, R. S. An assessment of clinically useful measures of the consequences of treatment. N. Engl. J. Med. 1988;318:1728–33. [22]

Leclerc, A., Lert, F., and Fabien, C. Differential mortality: Some comparisons between England and Wales, Finland and France, based on inequality measures. Int. J. Epidemiol. 1990;19:1001–10. [21]

Lee, E. T. Statistical methods for survival data analysis. Belmont, Cal.: Lifetime Learning Publications, 1980, [15,16]

Lee, T. H., Juarez, G., Cook, E. F., Weisberg, M. C., Rouan, G. W., Brand, D. A.., and Goldman, L. Ruling out acute myocardial infarction: A prospective multicenter validation of a 12-hour strategy for patients at low risk. N. Engl. J. Med. 1991;324:1239–46. [21]

Levy, D. E., Caronna, J. J., Singer, B. H., Lapinski, R. H., Frydman, H., and Plum, F. Predicting outcome from hypoxic-ischemic coma. J. Am. Med. Assn. 1985;253:1420–26. [P, 21]

Levy, S. M., Lee, J., Bagley, C., and Lippman, M. Survival hazards analysis in first recurrent breast cancer patients: Seven year follow-up. Psychosom. Med. 1988;50:520–27. [16]

Li, C. C. Method of path coefficients: A trademark of Sewall Wright. Hum. Biol. 1991;63:1–17. [22]

Liaño, F., Gallego, A., Pascual, J., et al. Prognosis of acute tubular necrosis: An extended prospectively contrasted study. Nephron 1993;63:21–31. [22]

Lindberg, G., Eklund, G. A., Gullberg, B., and Rastam, L. Serum sialic acid concentration and cardiovascular mortality. Br. Med. J. 1991;302:143–6. [12]

Little, R. J. A. Some statistical analysis issues at the World Fertility Survey. Am. Statist. 1988;42:31–36. [12]

Little, R. J. A. and Rubin, D. B. Statistical analysis with missing data. Chichester: John Wiley, 1987. [6]

Lockwood, C. J., Senyei, A. E., Dische, M. R., et al. Fetal fibronectin in cervical and vaginal secretions as a predictor of preterm delivery. N. Engl. J. Med. 1991;325:669–74. [14]

Long, W. J., Griffith, J. L., Selker, H. P., and D'Agostino, R. B. A comparison of logistic regression to decision-tree induction in a medical domain. Comput. Biomed. Res. 1993;26:74–97. [21,22]

Madsen, E. B., Gilpin, E., and Henning, H. Short-term prognosis in acute myocardial infarction: Evaluation of different prediction methods. Am. Heart J. 1984;107:1241–51. [22]

Maetani, S., Tobe, T., Hirakawa, A., Kashiwara, S., and Kuramoto, S. Parametric survival analysis of gastric cancer patients. Cancer 1980;46:2709–16. [19]

Makuch, R. W. and Rosenberg, P. S. Identifying prognostic factors in binary outcome data: An application using liver function tests and age to predict liver metastases. Stat. Med. 1988;7:843–56. [14]

Mallows, C. L. Some comments on C_p. Technometrics 1973;15:661–75. [11]

Mann, C. R. Consider this: Presentation of an ethics case. Amstat News 1993;201:22. [12]

Mantel, N. Evaluation of survival data and two new rank order statistics arising in its consideration. Cancer Chemother. Rep. 1966;50:163–70. [15,16]

———. Why stepdown procedures in variable selection? Technometrics 1970;12:621–25. [11]

Mantel, N. and Haenszel, W. Statistical aspects of the analysis of data from retrospective studies of disease. J. Natl. Cancer Inst. 1959;22:719–48. [22]

Marantz, P. R., Tobin, J. N., Wassertheil-Smoller, S., Ahn, C., Steingart, R. M., and Wexler, J. P. Prognosis in ischemic heart disease: Can you tell as much at the bedside as in the nuclear laboratory? Arch. Intern. Med. 1992;152:2433–37. [19]

Marler, M. R. and Ernhart, C. B. (Letter to Editor) Lead exposure and cognitive development. N. Engl. J. Med. 1987;317:895–96. [12]

Marshall, R. J. Partitioning methods for classification and decision making in medicine. Stat. Med. 1986;5:517–26. [22]

Mason, J. H., Anderson, J. J., and Meenan, R. F. A model of health status for rheumatoid arthritis: A factor analysis of the Arthritis Impact Measurement Scales. Arthritis Rheum. 1988;31:714–20. [4]

McCullagh, P. Regression models for ordinal data. J. Roy. Statist. Soc. Ser. B Metho. 1980;42:109–42. [14]

McDonald, G. B., Hinds, M. S., Fisher, L. D., et al. Veno-occlusive disease of the liver and multiorgan failure after bone marrow transplantation: A cohort study of 355 patients. Ann. Intern. Med. 1993;4:255–67. [16]

McGovern, P. G., Folsom, A. R., Sprafka, M., et al. Trends in survival of hospitalized myocardial infarction patients between 1970 and 1985. The Minnesota heart survey. Circulation 1992;85:172–79. [20]

McNeil, B. J., Keeler, E., and Adelstein, S. J. Primer on certain elements of medical decision making. N. Engl. J. Med. 1975;293:211–15. [13]

Menotti, A., Mariotti, S., Seccareccia, F., Torsello, S., and Dima, F. Determinants of all causes of death in samples of Italian middle-aged men followed up for 25 years. J. Epidemiol. Community Health 1987;41:243–50. [16]

Merzenich, H., Boeing, H., and Wahrendorf, J. Dietary fat and sports activity as determinants for age at menarche. Am. J. Epidemiol. 1993;138:217–24. [16]

Metz, C. E., Goodenough, D. J., and Rossmann, J. Evaluation of receiver operating characteristic curve data in terms of information theory, with applications in radiography. Radiology 1973;109:297–303. [13]

Meyers, A., Brand, D. A., Dove, H. G., and Dolan, T. F., Jr. A technique for analyzing clinical data to provide patient management guidelines. A study of meningitis in children. Am. J. Dis. Child. 1978;132:25–9. [21]

Miah, K., von Arbin, M., Britton, M., et al. Prognosis in acute stroke with special reference to some cardiac factors. J. Chronic Dis. 1983;36:279–88. [19]

Miettinen, O. S. Stratification by a multivariate confounder score. Am. J. Epidemiol. 1976;104:609–20. [19,22]

Mills, R., Fetter, R. B., Riedel, D. C., and Averill, R. AUTOGRP: An interactive computer system for the analysis of health care data. Med. Care 1976;14:603–15. [21]

Miranda, C. P., Herbert, W. G., Dubach, P., Lehmann, K. G., and Froelicher, V. F. Post-myocardial infarction exercise testing. Non-Q wave versus Q wave correlation with coronary angiography and long-term prognosis. Circulation 1991;84:2357–2365. [16]

Morgan, J. N. and Sonquist, J. A. Problems in the analysis of survey data, and a proposal. J. Am. Stat. Assn. 1963;58:415–34. [19,21]

Morgenthalter, S. Least-absolute-deviations fits for generalized linear models. Biometrika 1992;79:747–54. [22]

Moss, A. J., DeCamilla, J., Davis, H., and Bayer, L. The early posthospital phase of myocardial infarction: Prognostic stratification. Circulation 1976;54:58–64. [19]

Mountain, C. F. The new international staging system for lung cancer. Surg. Clin. North Am. 1987;67:925–35. [5]

Narula, S. C., Sposito, V. A., and Wellington, J. F. Intervals which leave the minimum sum of absolute errors regression unchanged. Appl. Statist. 1993;42:369–78. [22]

Neaton, J. D., Duchene, A. G., Svendsen, K. H., and Wentworth, D. An examination of some quality assurance methods commonly employed in clinical trials. Stat. Med. 1990;9:115–24. [5]

Nobuyoshi, M., Abe, M., Nosaka, H., et al. Statistical analysis of clinical risk factors for coronary artery spasm: Identification of the most important determinant. Am. Heart J. 1992;124:32–8. [12]

Nomura, H., Kashiwagi, S., Hayashi, J., et al. Prevalence of gallstone disease in a general population of Okinawa, Japan. Am. J. Epidemiol. 1988;128:598–605. [21]

Norris, R. M., Brandt, P. W. T., Caughey, D. E., Lee, A. J., and Scott, P. J. A new coronary prognostic index. Lancet 1969;1:274–78. [3]

Nunally, J. C. Psychometric theory. 2nd ed. New York: McGraw-Hill, 1978. [3]

Nyboe, J., Jensen, G., Appleyard, M., and Schnohr, P. Smoking and the risk of first acute myocardial infarction. Am. Heart J. 1991;122:438–47. [16]

O'Connor, G. T., Plume, S. K., Olmstead, E. M., et al. Multivariate prediction of in-hospital mortality associated with coronary artery bypass graft surgery. Northern New England Cardiovascular Disease Study Group. Circulation 1992;85:2110–18. [14]

Ooi, W. L., Budner, N. S., Cohen, H., Madhavan, S., and Alderman, M. H. Impact of race on treatment response and cardiovascular disease among hypertensives. Hypertension 1989;14:227–34. [16]

Overall, J. E., Henry, B. W., Markett, J. R., and Emken, R. L. Decisions about drug therapy. I. Prescriptions for adult psychiatric outpatients. Arch. Gen. Psychiatry 1982;26:140–45. [18]

Pahor, M., Guralnik, J. M., Gambassi, G., Bernabel, R., Carosella, L., and Carbonin, P. The impact of age on risk of adverse drug reactions to digoxin. J. Clin. Epidemiol. 1993;46:1305–14. [14]

Pearl, R. Studies in human biology. Baltimore: Williams and Wilkins, 1924. [13]

Pearl, R. and Reed, L. J. On the rate of growth of the population of the United States since 1790 and its mathematical representation. Proc. Nat. Acad. Sci. 1920;6:275–88. [13]

Peduzzi, P., Hardy, R., and Holford, T. A stepwise variable selection procedure for nonlinear regression models. Biometrics 1980;36:511–16. [15]

Peduzzi, P., Holford, T., Detre, K., and Chan, Y.-K. Comparison of the logistic and Cox regression models when outcome is determined in all patients after a fixed period of time. J. Chronic Dis. 1987;40:761–67. [22]

Persico, M., Luzar, V., Caporaso, N., and Coltorti, M. Riclassificazione istologica delle epatiti virali croniche. Un'analisi *cluster*. (Histologic reclassification of chronic viral hepatitis. A cluster analysis.) Medic 1993;1:23–27. [4]

Peters, M. and Murphy, K. Factor analyses of pooled hand questionnaire data are of questionable value. Cortex 1993;29:305–14. [4]

Peterson, B., and Harrell, F. E., Jr. Partial proportional odds models for ordinal response variables. Appl. Statist. 1990;39:205–17. [14]

Peto, R. and Peto, J. Asymptotically efficient rank invariant test procedures (with discussion). J. Roy. Stat. Soc. Ser. A Stat. 1972;135:185–206. [15,16]

Peto, R., Pike, M. C., Armitage, P., et al. Design and analysis of randomized clinical trials requiring prolonged observations of each patient. Part II. Analysis and examples. Br. J. Cancer 1977;35:1–39. [16]

Pezzella, F., Turley, H., Kuzu, I., et al. *bcl*-2 protein in non-small-cell lung carcinoma. N. Engl. J. Med. 1993;329:690–94. [16]

Pike, M. C., Bernstein, L., and Peters, R. K. (Letter to Editor) Re: Dietary fat and postmenopausal breast cancer. J. Nat. Cancer Inst. 1992;84:1666–67. [12]

Pintor, C., Cella, S. G., Loche, S., Puggioni, R., Corda, R., Locatelli, V., and Muller, E. E. Clonidine treatment for short stature. Lancet 1987;1:1226–30. [18]

Pla, L. Determining stratum boundaries with multivariate real data. Biometrics 1991;47:1409–22. [8]

Pocock, S. J. and Simon, R. Sequential treatment assignment with balancing for prognostic factors in the controlled clinical trial. Biometrics 1975;31:103–15. [19]

Poole, C. Multiple comparisons—No problem! Epidemiol. 1991;2:241–42. [22]

Port, F. K., Wolfe, R. A., Mauger, E. A., Berling, D. P., and Jiang, K. Comparison of survival probabilities for dialysis patients vs. cadaveric renal transplant recipients. J. Am. Med. Assn. 1993;270:1339–43. [16]

Pregibon, D. Logistic regression diagnostics. Ann. Statist. 1981;9:705–24. [14]

Prud'homme, G. J., Canner, P. L., and Cutler, J. A. Quality assurance and monitoring in the Hypertension Prevention Trial. Controlled Clin. Trials 1989;10:84S–94S. [5]

Rao, C. R. Linear statistical inference and its applications. 2nd ed. New York: John Wiley, 1973. [13]

Ray, R. L. Path analysis of psychophysiological data. Psychophysiology 1980;17:401–07. [22]

Reading, J. C. and Klauber, M. R. Extensions of the successive screening method of classification. Biometrics 1973;29:791–800. [22]

Redelmeier, D. A., Bloch, D. A., and Hickman, D. H. Assessing predictive accuracy: How to compare Brier scores. J. Clin. Epidemiol. 1991;44:1141–46. [22]

Reed, L. J. and Berkson, J. The application of the logistic function to experimental data. J. Phys. Chem. 1929;33:760–79. [13]

Reiber, G. E., Pecoraro, R. E., and Koepsell, T. D. Risk factors for amputation in patients with diabetes mellitus: A case-control study. Ann. Intern. Med. 1992;117:97–105. [14]

Reynolds-Haertle, R. A. and McBride, R. Single vs. double data entry in CAST. Controlled Clin. Trials 1992;13:487–94. [5]

Ries, A. L., Kaplan, R. M., and Blumberg, E. Use of factor analysis to consolidate multiple outcome measures in chronic obstructive pulmonary disease. J. Clin. Epidemiol. 1991;44:497–503. [4]

Robins, J. M. and Greenland, S. The role of model selection in causal inference from non-experimental data. Am. J. Epidemiol. 1986;123:392–402. [22]

Roos, N. P., Wennberg, J. E., Malenka, D. J., et al. Mortality and reoperation after open and transurethral resection of the prostate for benign prostatic hyperplasia. N. Engl. J. Med. 1989;320:1120–24. [5,20]

Rosenbaum, P. R. and Rubin, D. B. The central role of the propensity score in observational studies for causal effect. Biometrika 1983;70:41–55. [21]

———. Constructing a control group using multivariate matched sampling methods that incorporate the propensity score. Am. Statist. 1985;39:33–38. [19]

Rothman, K. No adjustments are needed for multiple comparisons. Epidemiol. 1991;1:42–46. [22]

Roweth, B. Statistics for policy: Needs assessment in the rate support grant. Public Admin. 1980;59:173–86. [12]

Rubin, H. R. and Wu, A. W. The risk of adjustment. Med Care 1992;30:973–75. [14]

Rupprecht, C. E., Glickman, L. T., Spencer, P. A., and Wiktor, T. J. Epidemiology or rabies virus variants: Differentiation using monoclonal antibodies and discriminant analysis. Am. J. Epidemiol. 1987;126:298–309. [18]

Ryback, R. S., Eckards, M. J., Felsher, B., and Rawlings, R. R. Biochemical and hematologic correlates of alcoholism and liver disease. J. Am. Med. Assn. 1982;248:2251–65. [18]

Sacree, M. and Kilsby, D. C. The relevance of linear regression for assessing the performance of microbiological counting techniques. Letters in Applied Microbiology 1988;7:1–4. [12]

Sakamoto, Y., Ishiguro, M., and Kitagawa, G. Akaike information criterion statistics. Dordrecht: D. Reidel, 1986. [13]

Samuels, M. L., Casella, G., and McCabe, G. P. Interpreting blocks and random factors. J. Am. Statist. Assn. 1991;86:798–821. [22]

Sankrithi, U., Emanuel, I., and Van Belle, G. Comparison of linear and exponential multivariate models for explaining national infant and child mortality. Int. J. Epidemiol. 1991;20:565–70. [22]

SAS Technical Report P-200, SAS/STAT Software: CALIS and LOGISTIC Procedures, Release 6.04. Chapter 3, The LOGISTIC Procedure. Cary, N.C.: SAS Institute, Inc., 1990. [13,14]

SAS Technical Report P-229, SAS/STAT Software: Changes and enhancements, Release 6.07 Chapter 19. The PHREG Procedure. Cary, N.C.: SAS Institute, Inc., 1992. [16]

SAS/STAT User's Guide, Version 6, Fourth Edition, Volume 1, ANOVA-FREQ and Volume 2, GLM-VARCOMP. Cary, N.C.: SAS Institute Inc., 1990. [5,18]

Savage, I. R. Contributions to the theory of rank order statistics: The two-sample case. Ann. Math. Stat. 1956;27:590–615. [16]

Schachtel, B. P., Fillingim, J. M., Beiter, D. J., Lane, A. C., and Schwartz, L. A. Rating scales for analgesics in sore throat. Clin. Pharmacol. Ther. 1984;36:151–6. [2]

Schemper, M. The explained variation in proportional hazards regression. Biometrika 1990;77:216–8. [15]

Schenker, N., Treiman, D. J., and Weidman, L. Analyses of public use decennial census data with multiply imputed industry and occupation codes. Appl. Statist. 1993;42:545–56. [6]

Schlesselman, J. J. Case-control studies. New York: Oxford University Press, 1982. [3]
Schlundt, D. G., Taylor, D., Hill, J. O., Sbrocco, T., Pope-Cordle, J., Kasser, T., and Arnold, D. A behavioral taxonomy of obese female participants in a weight-loss program. Am. J. Clin. Nutr. 1991;53:1151–8. [4]
Schmitz, P. I. M., Habbema, J. D. F., and Hermans, J. The performance of logistic discrimination on myocardial infarction data, in comparison with some other discriminant analysis methods. Stat. Med. 1983;2:199–205. [22]
Schubert, T. T., Bologna, S. D., Nensey, Y., Schubert, A. B., Mascha, E. J., and Ma, C. K. Ulcer risk factors: Interactions between *helicobacter pylori* infection, nonsteroidal use, and age. Am. J. Med. 1993;94:413–18. [14]
Schwartz, R. Estimating the dimension of a model. Ann. Statist. 1978;6:461–64. [13]
Sclar, D. A., Skaer, T. L., Chin, A., Okamoto, M. P., and Gill, M. A. Utility of a transdermal delivery system for antihypertensive therapy. Part 2. Am. J. Med. 1991;91:57S–60S. [12]
Segal, M. R. and Bloch, D. A. A comparison of estimated proportional hazards models and regression trees. Stat. Med. 1989;8:539–50. [22]
Segal, M. R. and Tager, I. B. Trees and tracking. Stat. Med. 1993;12:2153–68. [21]
Seymour, D. G., Green, M., and Vaz, F. G. Making better decisions: Construction of clinical scoring systems by the Spiegelhalter-Knill-Jones approach. Br. Med. J. 1990;300:223–6. [19]
Shannon, C. E. A mathematical theory of communication. Bell System Tech. J. 1948;27:379–423, 623–56. [8]
Shapiro, A. R., The evaluation of clinical predictions. A method and initial application. New Engl. J. Med. 1977;296:1509–1514. [22]
Siegel, S. and Castellan, N. J., Jr. Non-parametric statistics for the behavioral sciences. 2nd ed. New York: McGraw-Hill Book Co., 1988. [9]
Skene, A. I., Smith, J. M., Dore, C. J., Charlett, A., and Lewis, J. D. Venous leg ulcers: A prognostic index to predict time to healing. Br. Med. J. 1992;305:1119–21. [16]
Smith, D. M., Weinberger, M., Katz, B. P., and Moore, P. S. Postdischarge care and readmissions. Med. Care. 1988;26:699–708. [19]
Sokal, R. R. and Sneath, P. H. A. Principles of numerical taxonomy. San Francisco: W. H. Freeman & Co., 1963. [4]
Solvoll, K., Selmer, R., Loken, E. B., Foss, O. P., and Trygg, K. Coffee, dietary habits, and serum cholesterol among men and women 35–49 years of age. Am. J. Epidemiol. 1989;129:1277–88. [12]
Spiegelhalter, D. J. and Knill-Jones, R. P. Statistical and knowledge-based approaches to clinical decision-support systems, with an application in gastroenterology. J. Roy. Statist. Soc. Ser. A Stat. 1984;147:35–37. [19]
Stablein, D. M., Miller, J. D., Choi, S. C., and Becker, D. P. Statistical methods for determining prognosis in severe head injury. Neurosurgery 1980;6:243–48. [14]
Stayner, L. T., Smith, R., Thun, M., Schnorr, T., and Lemen, R. A dose-response analysis and quantitative assessment of lung cancer risk and occupational cadmium exposure. Ann. Epidemiol. 1992;2:177–94. [12]
Stayner, L., Smith, R., Schnoor, T., Lemen, R., and Thun, M. Letter to Editor. Ann. Epidemiol. 1993;3:114–18. [12]
Stitt, F. W., Frane, M., and Frane, J. W. Mood change in rheumatoid arthritis: Factor analysis as a tool in clinical research. J. Chronic Dis. 1977;30:135–45. [4]
Stitt, F. W., Lu, Y., Dickinson, G. M., and Klimas, N. G. Automated severity classification of AIDS hospitalizations. Med. Decis. Making 1991;11(suppl):S41–S45. [21]
Sturges, H. A. The choice of a class interval. J. Am. Stat. Assn. 1926;21:65. [8]
SUGI Supplemental Library User's Guide, Version 5. Chapter 34. The PROC PHGLM Procedure. Cary, N.C.: SAS Institute, Inc., 1986. [16]

Talley, N. J., McNeil, D., and Piper, D. W. Discriminant value of dyspeptic symptoms: A study of the clinical presentation of 221 patients with dyspepsia of unknown cause, peptic ulceration, and cholelithiasis. Gut 1987;28:40–46. [14]

Taves, D. R. Minimization: A new method of assigning patients to treatment and control groups. Clin. Pharmacol. Ther. 1974;15:443–53. [19]

Therneau, T. M., Grambsch, P. M., and Fleming, T. R. Martingale-based residuals for survival models. Biometrika 1990;77:147–60. [16]

Thompson, J. D., Fetter, R. B., and Mross, C. D. Case mix and resource use. Inquiry 1975;12:300–12. [21]

Thomspon, J. R. Empirical model building. New York: Wiley, 1989. [22]

Tibshirani, R. A plain man's guide to the proportional hazards model. Clin. Invest. Med. 1982;5:63–68. [16]

Tukey, J. W. Exploratory data analysis. Reading, Mass.: Addison-Wesley, 1977. [2,6]

U.S. Bureau of the Census. Statistical Abstract of United States: 1980. 101st ed. Washington, D.C., 1980. [13]

Vach, W. and Blettner, M. Biased estimation of the odds ratio in case-control studies due to the use of ad hoc methods of correcting for missing values for confounding variables. Am. J. Epidemiol. 1991;134:895–907. [6]

Vandenbroucke, J. P. Should we abandon statistical modeling altogether? Am. J. Epidemiol. 1987;126:10–13. [22]

Van Houwelingen, J. C. and Le Cessie, S. Predictive value of statistical models. Stat. Med. 1990;9:1303–25. [22]

Verhulst, P. J. Notice sur la loi que la population suit dans sons accroissement. Corr. Math. et Phys. 1838;10:113–21. [13]

―――. Recherches mathématiques sur la loi d'accroissement de la population. Nouv. Mem. de l'Acad. Roy. des Sci. et Belles-Lett. de Bruxelles 1845;18:1–38. [13]

Verweij, P. J. M. and van Houwelingen, H. C. Cross-validation in survival analysis. Stat. Med. 1993;12:2305–14. [15]

Victora, C. G., Tomasi, E., Olinto, M. T. A., and Barros, F. C. Use of pacifiers and breastfeeding duration. Lancet 1993;341:404–06. [16]

Vlietstra, R. E., Frye, R. L., Kronmal, R. A., et al. Risk factors and angiographic coronary artery disease: A report from the coronary artery surgery study (CASS). Circulation 1980;62:254–61. [18]

Vollertsen, R. S., Nobrega, F. T., Michet, C. J., Jr., Hanson, T. J., and Naessens, J. M. Economic outcome under medicare prospective payment at a tertiary-care institution: The effects of demographic, clinical, and logistic factors on duration of hospital stay and Part A charges for medical back problems (DRG 243). Mayo Clin. Proc. 1988;63:583–91. [12]

Von Korff, M., Wagner, E. H., and Saunders, K. A chronic disease score from automated pharmacy data. J. Clin. Epidemiol. 1992;45:197–203. [14]

Walter, S. D., Feinstein, A. R., and Wells, C. K. Coding ordinal independent variables in multiple regression analysis. Am. J. Epidemiol. 1987;125:319–23. [4,5]

―――. A comparison of multivariable mathematical models in predicting survival. II. Statistical selection of prognostic variables. J. Clin. Epidemiol. 1990;43:349–59. [5,22]

Wasson, J. H., Sox, H. C., Neff, R. K., and Goldman, L. Clinical prediction rules. Applications and methodological standards. N. Engl. J. Med. 1985;313:793–99. [3,22]

Watson, C. J. and Croft, D. J. A multiple discriminant analysis of physician specialty choice. Comput. Biomed. Res. 1978;11:405–21. [18]

Wells, C. K., Stoller, J. K., Feinstein, A. R., and Horwitz, R. I. Co-morbid and clinical determinants of prognosis in endometrial cancer. Arch. Intern. Med. 1984;144:2004–09. [20]

Winkel, P. and Juhl, E. (Letter to Editor) Assumptions in linear discriminant analysis. Lancet 1971;2:435–36. [18]

Winkel, P., Lyngborg, K., Olesen, K. H., Meibom, J., and Hansen, P. F. A method for systematic assessment of the relative prognostic significance of symptoms and signs in patients with a chronic disease. II. Analysis of validity of model and derivation of major prognostic stages. Comput. Biomed. Res. 1973;6:457–64. [20]

Winkel, P., Paldam, M., Tygstrup, N., and The Copenhagen Study Group for Liver Diseases. A numerical taxonomic analysis of symptoms and signs in 400 patients with cirrhosis of the liver. Comput. Biomed. Res. 1970;3:657–65. [4]

Wiseman, R. A. and Dodds-Smith, I. C. Cardiovascular birth defects and antenatal exposure to female sex hormones: A reevaluation of some base data. Teratology 1984;30:359–70. [5]

Wong, J. B., Sonnenberg, F. A., Salem, D. N., and Pauker, S. G. Myocardial revascularization for chronic stable angina: Analysis of the role of percutaneous transluminal coronary angioplasty based on data available in 1989. Ann. Intern. Med. 1990;113:852–71. [20]

Wright, J. G. and Feinstein, A. R. A comparative contrast of clinimetric and psychometric methods for constructing indexes and rating scales. J. Clin. Epidemiol. 1992;45:1201–18. [3]

Wright, S. The method of path coefficients. Ann. Math. Stat. 1934;5:161–215. [22]

Yandell, B. S. Smoothing splines—a tutorial. Statistician 1993;42:317–19. [2]

Yates, A. Multivariate exploratory data analysis: A perspective on exploratory factor analysis. Albany: State University of New York, 1987. [4]

Yates, F. Sampling methods for censuses and surveys. 3rd ed. London: Griffin, 1971. [12]

Yates, F. and Healy, M. J. R. (Letter to Editor) Statistical methods in anthropology. Nature 1951;168:1116–17. [18]

Yeoh, C. and Davies, H. Clinical coding: Completeness and accuracy when doctors take it on. Br. Med. J. 1993;306:972. [5]

Yerushalmy, J. Statistical problems in assessing methods of medical diagnosis, with special reference to X-ray techniques. Public Health Rep. 1947;62:1432–49. [3]

Yesner, R., Gelfman, N. A., and Feinstein, A. R. A reappraisal of histopathology in lung cancer and correlation of cell types with antecedent cigarette smoking. Am. Rev. Respir. Dis. 1973;107:790–97. [5]

Zhang, P. Inference after variable selection in linear regression models. Biometrika 1992;79:741–46. [22]

Index

Absolute deviation from mean, 573–4
Accomplishments, indexes of, 137–144
   of stratification, 523
Accuracy of estimates. *See* Estimates, accuracy of
Adjusted R^2. *See* R^2, adjusted
Adjustment(s), 46, 416
   extra-equation, 289, 416–7
   for susceptibility bias, 53
   intra-equation, 289, 416–7
   mechanisms of, 54–55
AID procedure, 489, 531–538, 541
AIDS, Cart analysis, 552
Akaike index (AIC), 313, 334, 346
Algebraic model. *See* Model, algebraic
Altman, D. G., 395
Amalgamation, Recursive Partitioning, 556
American College of Rheumatology, 553
Analysis, actuarial, 372, 377–9
   canonical, 61, 69
   cluster, 69, 562
   correspondence, 562
   covariance, 560
   discrete multivariate, 488
   factor, 3, 62, 69, 562
   life-table, 372, 376
   longitudinal, 395, 556–7
   path, 560–1
   principal component, 3, 69, 562
   survival, 372–380, 395–6, 499–500, 556
   of variance, 61, 560
   multivariate, 61, 69, 562
Ancestral group, 530
Apgar score, 62, 568
Arthritis, Giant Cell, 553–4
Artificial intelligence, 561–2
ASCC, 440, 464
Association, definition, 20
Autogroup procedure (AUTOGRP), 540
Automated Interaction Detector. *See* AID

B (Dispersion matrix among groups), 439
Barnard, G. A., 581
Barnard, M. M., 434
Baseline condition, 49
Bayes theorem, 326–7, 449
Begg, C. B., 504, 505
Belsley, D. A., 285
Berkson-Gage life-table method, 377–379
Berkson, J., 377

Bias, protopathic, 572
   susceptibility, 53–54, 568
Biggs, D., 554
Biplot, 444
Bivariate, definition, 1
   data, patterns of, 9–10
   displays, examination of, 124–126
Blettner, M., 573
Bloch, D. A., 552
BMDP system, 99–100
Bonferroni adjustment, 554, 583
Boolean unions. *See* Unions, Boolean
Bootstrapping, 186–7
Boundaries, selection of, 182–3
   correlated choices, 183–4
   theoretical choices, 183
Bounds on condition, 248
Box-and-whiskers plot. *See* Box plot
Box plot, 110–113
Breiman, L., 541
Brier index, 305, 577
Brown, C. C., 319

C index of concordance, 315–16, 336
Calibration, 576–8
Cancer, lung, composite staging system, 517–523
Cancer, rectal, staging of, 75–76
Canonical analysis. *See* Analysis, canonical
Canonical correlation analysis, 562
Canonical variables, 444
Caronna, J. J., 550
CART, 541–554
   indexes of evaluation, 546–8
   application to illustrative data set, 547–8
Categorization, 46–47
Category, multivariate, 478
   from 'bar graph,' 501
   from stratification, 501
   from 'tree,' 502
   transcending, 484
CBAR index, 349
Cells, imputation of, 528
Censoring, 373–375
Centroid(s), definition, 439
   role, 445–6, 455
CHAID, 538–40
Chest pain, CART analysis, 550
Chi-square automated interaction detector. *See* CHAID

Chi-square, C. C. Brown, 319, 343
  goodness of fit, 317
  Hosmer-Lemeshow, 319, 343
  improvement, 389–390
  independence, 317
  log-likelihood, 318
  residual, 341
  score calculation, 318, 387, 388
  test for linear trend, 177
  Wald, 318, 334, 341
Classification, 23
  accuracy of, 337
  decisions for illustrative data set, 85–91
  cross-, 481
Classification and regression trees. See CART
Classification function(s), 432, 439, 448–449
  accuracy, 469–470
Clinical boundaries, 182–3
'Close alternatives,' 269
Cluster(s), 18, 30–31
  definition, 478
  formation of, 63–64
  targeted, 165–6
  terminal, 479
Cochran, W. G., 314
Coding, conditional, 95
  of data, principles, 91
  decisions in, 573
  dummy zones, 96
  effect, 94
  intervariable, 96
  marginal, 94
  null-based, 95, 480
  ordinal transformation, 95
  orthogonal, 94
  reference-cell, 94
Coefficient, correlation, 141
Coefficient of predictive accuracy, 574
Coefficients, finding, 136
  fitting, in Cox regression, 385–7
  interpretation of, 98
  standardized, 322–323, 335, 391, 407–8
Coherence, biologic, 73–74, 513, 514
  rules of, 74–75
Collapsibility of strata, criteria for, 179–80
Collapsing categories, criteria, 491–494
Collinear(ity), 39, 241–6
Coma, CART analysis, 550–552
Composite stages, 408–10
Complexity of additional variables, 207–226
Comparison
  of bivariate results, 78–79
  of multivariable results, 80, 562–5
Compression, 23, 37, 65
Concato, J. P., xv
Concordance, 20
  C index. See C index

Conditional methods, in Cox regression, 377
Confidence interval(s), 145–146, 268–9
Configuration, 24–26, 47–48, 65
Confirmation, bivariate, 566
Conformity, linear, 282, 352–354, 366, 394, 422
  of model, 11, 15, 149–151
Confounding, 56–57, 504, 568–9
Confounding variable, criteria, 568–9
Congruence(s), concept, 170–1
Conjunctive consolidation. See Consolidation, conjunctive
Conjunctive effect. See Effect, conjunctive
Consolidation, of categories, 88–89
  conjunctive, x, xiii, 69, 73, 408–10, 488, 490, 512–528
Constancy, of proportional hazards, 394
Constraints
  of independent variables, 68
  of orientation, 66
  of target variable, 66–68
Contrast, 20
Convergence in calculations, 311
Correlation, canonical, average squared (ASCC), 440
Correlation(s), 20
  bivariate, examination of, 156
  coefficients, examination of, 188–195
  coefficients, partial, 237, 251
  multiple interdependent, 48
Correspondence analysis, 562
Cost matrix. See Matrix, cost
Covariance, xii
  analysis of, 560
  calculation of, 211–212
  extra sources, 215–216
Covariates, time-dependent, 395–6
COVRATIO, 285
Cox, D. R., 314, 371
Cox regression. See Regression, Cox
Criteria, collapsing categories, 491–494
  entry-removal, 262–414
Cronbach's alpha, 48
Curran, W. J., 556

D symbols, in multivariable regression, 229–30, 248, 312
D^2. See Distance, Mahalanobis
Data, ambiguous, 88
  bivariate. See Bivariate
  entry, verification of, 102–104
  missing, 88, 114
  missing, hazards of, 114–115
  missing, patterns of, 114
  types of, 6–8
Deceptions, stochastic, 155
Decision, management, 572–3
Decision-tree induction, 556

Decision rule(s), 64, 71–7
    bivariate, 524–5
    nontargeted, 71–73
    targeted, 73–74
Default criteria, problems in, 277–8
Default specifications, 262–263
Degrees of freedom, concepts, 139–40
Density, probability, Gaussian, 325, 450–1
Dependent, definition of, 10
Design, efficiency of, 55
DeStavola, B. L., 395
Determinant, 219
Deviate, standardized normal, 266–7, 320
Deviation, residual, 133–134
DFBETAS, 285, 349
DFFITS, 284
Diagnosis-related group (DRG), constructed via AUTOGRP, 540
Diagnostics, regression, 151–152, 279–286
    Cox models, 393–4, 419–425
    discriminant analysis, 444
    for illustrative data set, 287
    logistic model, 323
    in published literature, 291, 366
    therapy for 'lesions,' 286
Diffuse monovalent patterns, 13
Discriminant function analysis, x, 69, 70, 295, 325, 329, 431–474, 562
    relation to linear regression for two groups, 443, 465–468
Discriminant function, basic concept, 437–8
    conversion from classification functions, 442
Distance, Mahalanobis, 285, 443, 446–7, 455, 562
    Cook's, 285, 349
    absolute, 305
Distinctiveness, persistent, 38
Distortions, in good fits, 153–154
    in poor fits, 154
Distributions, categorical variables, 108–110
    dimensional variables, 110
    eccentric, 118–120
    eccentric, remedies for, 120–122
    Gaussian, 113
    univariate, inspections, 107
Double gradient. *See* Gradient, double
Draper, N. R., xii
Dummy variables. *See* Variables, dummy

Effect(s), additive, 155
    conjunctive, 40
    exaggeration of, 40–41
    multiplicative, 155–56
Eigenvalue, 219
Eigenvector, 219
Energy survey, AID analysis, 538
Entropy minimax, 560

Entry-removal criteria, 262, 414
Error, mean square residual, 220
    penalties, 546
    proportionate reduction in, 168–9
Estimate, 21, 36–37, 65
    accuracy of, 45–46
    with categorical method, 483
    diagnostic, 36–37
    vs. impact, 23
    plurality, 433
    predictive, 22, 37
'Esthetic' boundaries, 182
Ethics, and outliers, 284
Events per variable, 225–6, 269, 354, 423, 429
Explorations, choice of procedures, 263
    in Cox regression, 392
    incremental, 234–241

F, partial, 236, 249
F ratio, 139, 140–141, 222–3, 439, 456
F test, 145
Factor analysis. *See* Analysis, factor
Feinstein, A. R., 541
Finney, D. J., 294, 580
Fisher, R. A., 325, 434, 560
Fit, assessment for target variable, 166–172
    basic, of model, 130–131
    basic, logistic model, 304
    concordant, 166–7
Forecasting, 566
Formats, for algebraic models, 26–29
Forrow, L., 575
Friedman, J. H., 541

Gage, R. R., 377
Gallstones, risk of, AID analysis, 536–7
Gamma index, 317, 336
Geisser, S., 581
General linear model. *See* Model, linear, concepts of
Gilpin, E., 550
Gini, C., 542
Gini index, 541–3
Glantz, S., xiv
Goals, of analysis, 151–2, 289–91, 365–6, 428, 565–571
    scientific, 581–2
Goldman, L., 550
Goodman, L. A., 336
Gradient(s), constancy, 43
    criteria, 174–5
    definition, 42
    double, 43–44, 73, 513
    inspection of, 197
    intercategory, 173–4
    intrasystem, 173
    standard magnitudes, 42

Graunt, J., 376
Grizzle, J. E., 562
Group(s), formation of, 50–51

Hadorn, D. C., 548–550
Hardy, R., 390
Harrell, F. E., xv, 313, 389, 390, 397, 407

'Hat' symbol, definition, 70
   diagonal, 285, 348
   matrix, 285, 348
Hauck, W. W., 269
Hazard, baseline, 403–4
   constant, assumption of, 386, 429
   cumulative, 381–2, 420–2
   function, 372, 380–383
   ratio(s), 391–2
   relative, 98, 415
Headaches, AID analysis of, 534–5
Healey, M. J. R., 474, 561
Hoaglin, D. C., 285
Hofacker, C. F., 581
Holford, T. A., 390
'Homogeneity'
   biologic, 73–74, 491–2
   statistical, 73, 492
Hosmer, D. W., 314, 319
Hotelling-Lawley Trace, 462
Hotelling's T^2, 447, 465

Iglewicz, B. A., 505
Illustrative data set, bivariate evaluations, 188–203
   construction, 85–104
Imbalances, baseline, 504
Impact, 19
   vs. estimate, 23
   modifying, 38
   measured by stratified cluster, 408
   in targeted-cluster analyses, 486
   of variables, 428, 441
Improvement chi-square, 389–90
Imputation. *See* Variables, imputation of
Incremental tests, logistic, 320–1
Independent, definition, 10
Independent variable(s), in illustrative data set, 87–88
Index(es), of deviation, 167
   of diversity, 181–2
   Gini, of diversity. *See* Gini index
   of 'impurity.' *See* Gini index
   of individual discrepancy, 131, 133–5, 304–5, 573
   of misfit, 348–349
   penalty, 238
   of post-deletion change, 349–50
   R values, in Cox regression, 407
   stability of, 225
   of total accomplishment, 575–6
   total discrepancy, 131–2, 305–6
'Influence,' indexes of, 285
Information, alleged loss of, 485
Intelligence, artificial, 561–2
Interaction(s), 40–42, 270–279, 351–2
   additional codings, 271–272
   alternative strategy, 276
   appraisal of, 263
   choosing terms, 272–3
   in Cox regression, 415, 422–3
   examples in published literature, 276–7, 292, 367, 429
   in illustrative data set, 277–9
   problems of interpretation, 273–4
   quantification of, 143–44, 422–44, 571
   recognizing potential, 275
   reverse codings, 278, 352
   'saturated' model, 278
Interaction models, challenges, 354–356
Interdependent, definition, 10
Intervals, in life-table structure, 377–80
Interventional maneuvers. *See* Maneuvers, interventional
Isometry, 492

Jackknife method, 185, 441
Johnson, Robert Wood, Clinical Scholars, xv
Joiner, B. L., 275
Juhl, E., 474

Kaplan, E. L., 378
Kaplan-Meier life-table method, 378–9
Kerlinger, F. N., xiii, 94
Kendall's tau, 195, 317, 337
Kernel method, 461
Kleinbaum, D. G., xiii, 443
Knill-Jones, R. P., 503
Knowledge Seeker system, 530, 554–6
   in illustrative data set, 555
Koehler, A. B., 313
Kruskal, W. H., 336
Kuder-Richardson coefficient, 48
Kuh, E., 285
Kupper, L. L., xii, 443
Kwak, L. W., 556

Lachenbruch, P., xv, 435, 437, 443
Lambda, 168–9
Lambda, 168–9
Lambda, Wilks, 439–40
Landis, J. R., 541
Laupacis, A., 575
Least-squares principle, 136–7
LeClerc, A., 542
Lemeshow, S., 314, 319

'Leverage,' index of, 284–5
Levy, D. E., 550
Likelihood, 171, 295, 306–7
    basic, calculation of, 307
    logistic application, 310
    'explained' proportion, 313
    partial, 372, 386
    for 2 × 2 table, 308
Likelihood ratio, partial maximum, 386
Linear model. *See* Model, linear
Linear regression. *See* Regression, linear (simple or multiple)
Log likelihood, 388, 415–6
Log likelihood ratio, 309
    in Cox regression, 385–388
    as index of accomplishment, 309
    as stochastic test, 309–10
    as test of independence, 318
Log-linear model(s). *See* Model(s), Log-linear
Log minus log survivor function, 420–1
Log rank test, 379, 415–6
Logistic regression. *See* Regression, logistic
Logistic transformation. *See* Transformation, logistic
Logit, 27, 299

Mahalanobis distance. *See* Distance, Mahalanobis
Mantel, N., 416
Mantel-Haenszel procedure, 569
Mallows C_p, 238–9, 248, 256–59
Maldistribution(s), 117–118
Maneuvers, interventional, 49–50
    effects of, 51
Many-internal analysis, 2
Many-to-many analysis, 2
Many-to-one analysis, 2, 61
Matched arrangements, 300–1
Matrix, algebra, 218–9
    cost, 339, 497–9, 546, 550
    variance-covariance, 219
McNeil, B. J., 315
Median survival, 415
Meier, P. M., 378
Meningitis, AUTOGRP analysis, 540
Meridian, spanning, 169–70
Methods, choice of, 580–4
Methotrexate, clearance, AID analysis, 537
Metric, Mahalanobis, 502
Metz, C. E., 315
Miettinen's confounder score, 571
Miike, R., 209
Minimization technique, 505
Model(s)
    algebraic, 11, 16
    centered and non-centered, 386–7, 402–3
    formation of, 132–3
    intra- and extra-adjustments, 416–7

    linear, concepts of, 212–214
    linear, in Cox regression, 371–2, 382–3
    linear, in logistic regression, 299–300
    log-linear, 487–8
Modicum size of categories, 196–7
    of stratum, 178–9, 515, 520
Monotonic patterns, 13
Monovalent patterns, 12
Morgan, J. N., 537
MPLR, 389
MSE (mean-square residual error), 220, 252
Muller, K. E., xiv, 443
Multicategorical stratification. *See* Stratification, multicategorical
Multiple-comparison problem, 225, 582–3
Multiplicative role of coefficients, 321
Multivariable, definition, 1
    distinction vs. Multivariate, 2–3
Multivariate. *See* Multivariable
Murphree, E. S., 313
Myocardial infarction, CART analysis, 550

Nearest neighbor method, 461, 550
Networks, neural, 561
Newton-Raphson procedure, 311
Node(s), 493
    Root, 530
    Terminal, 530
Nominal variables, compound analysis, 160–62
    dummy binary, 259–60
    problems of inclusion, 259–62
Nomogram, predictive, 289
Nondependent, definition, 10
Non-monotonic patterns, 13
Normal probability plot, 113
Nosocomial infection, CHAID analysis, 538–540
Number needed to treat, 575

Odds, application, 327–328
    definition, 299
Odds ratios, 321
Olshen, R. A., 541
Options, in computer programs, 100, 101
Orientation, of arrangement, 60-62
    dependent, 60
    non-targeted, 61–62
    targeted, 60
Outcomes, 49–50
Outliers(s), definition, 283
    scientific concept, 283–4
    statistical measurements, 284

PHH method, 390
PMLR, 386–390
Partial correlation coefficient(s), 237
Partial F values, 236
Partial index R, 390–1

Partial maximum likelihood ratio, 386, 389
Partial regression coefficient(s), 213–4, 266–267
  calculation, 217
  connotation, 216–27
  standardized, 238
Partition, cross-, 482–483
  zonal, 17–18, 30–31
Partitioning, recursive, x, xii, 73, 487, 488–9, 529–558
Path analysis, 560–1
Patterns, covariate, 347, 348
  of data, 11–14, 152–153
  residual, 156–158
  targeted, choice of, 80–81
Pearl, R., 301
Pearson correlation coefficient. *See* r
Pedhazur, E. J., xiii, 92
Peduzzi, P., xv, 390
Penalty indexes, 238–9, 313–4
Peto, R., 416
Phi (ϕ) coefficient, 168, 195
Pillai's trace. *See* Trace, Pillai's
Plurality decisions, 497, 499
Pocock, S. J., 505
Polynomial, 27
Polytomous targets. *See* Target(s), polytomous
Polytomous variables. *See* Variables, polytomous
Polyvalent patterns, 14
Porosity, multivariate, 114
Post-spective expressions, 578
Precision, 56
  of calculations, 262
Predictive groupings, 566–8
Predictor variables, inadequate choices, 572–3
Pre-spective expressions, 578
Principal components analysis. *See* Analysis, principal components
Printout, errors in, 102
  inspection of, 100–101
Probability, classification, calculation, 451–3
  conditional, 450–1
Prognostic groups, 566–8
Prognostic stratification. *See* Stratification, prognostic
Proportional hazards, check of constancy, 417–8, 420
  use of time-interaction, 420
Proportional hazards regression. *See* Regression, Cox
Prostate, hyperplasia, staging system, 516–7

Quadratic discriminant analysis, 474
Quantiles, in choosing boundaries, 183
Quantitative interpretation, coefficients, 321

r, definition of, 141–2
r^2, definition of, 137
  discussion of, 138–139
R index (Cox coefficients), 390–1, 407
R index (logistic coefficients), 323, 325
R^2, definition and usage, 221, 230, 248, 265
  adjusted, 223, 230, 256, 257
  advantages and disadvantages, 221–2
  incremental changes, 237
R^2. *See* R^2, adjusted
R^2, 313–4, 334, 389, 406
Randomization, complete, 504, 505
  stratified, 504, 505
Rao's V^2, 447, 455
Ratio, hazard, 391
  risk, 391
Rectilinear models
  advantages, 32
  disadvantages, 32–33
Recursive partitioning. *See* Partitioning, recursive
Recursive partitioning amalgamation. *See* Amalgamation, recursive partitioning
Reduction, strategies of, 64
  of cells in table, 77
Reed, L. J., 301
Regression, all possible, 255–259, 348
  backward, 240–41
  conditional, 360–2
  Cox, x, xii, 69, 70, 295, 371–430
  Cox, for groups, 414–5
  forced, 254–347
  forced, closed, 254–55
  forced, open 255
  forward, 239–40
  full, 230–234, 333–340, 399–411, 459–463
  logistic, x, xii, 69, 70, 78–80, 295, 297–369, 454
  multiple linear, x, xii, 69, 79, 207–294
  quadratic, 162, 362
  rectilinear, 20
  simple linear, xii, 20, 77–78, 128
  step-down (*see* backward)
  step-up (*see* forward)
  stepwise, 241, 247–253, 341–347, 411–413, 463
Regression coefficient, 135
  standardized, 143, 224–5, 267
Regression diagnostics. *See* Diagnostics, regression
Reporting, requirements of, 582–584
Residual mean square, 252
Residual sums of squares, 229–230
Residual(s), observational, 286
  Studentized, 284
Residuals, in Cox regression, 423
  deviance, 349
  martingale, 423–4
  Pearson, 349
  standardized, 158–9, 349

univariate, 279–280
   for 'YHAT', 280
Reversal of effect, 41
Risk, ratio, 391
   relative, 98
'Robustness', 122
ROC curve, 315, 554, 576
Rosenbaum, P. R., 502
Roweth, B., 293
Roy's Greatest Root, 462
Rubin, D. B., 502
Ryback, R. S., 474

SAS system, 99–100
STDXBETA index, 423
Sauerbrei, W., 573
Savage, H. R., 416
$S_h(t)$, 385, 402
S_M, definition of, 135, 220
$S_o(t)$, 383, 402, 408
S_R, definition of, 135, 220
S_{xx}, definition of, 135
S_{xy}, definition of, 136
S_{yy}, definition of, 131, 135–136, 220
Scales, types of, 6–8
   binary, 7
   dimensional, 7
   nominal, 7
   ordinal, 7
   quasi-dimensional, 8
Scaling, multidimensional, 562
Schwartz index (SC), 313, 334, 346
Score(s), 18
   arbitrary weight points, 426, 501
   chi-square calculation, 318
   complex, 45
   demarcation of, 570
   factor count, 426, 500–1
   formation of, 63
   formation from algebraic models, 425–6
   formulas, 426, 501
   propensity, 502
   sample, 45
   variance reduction, 168
Segal, M. R., 552
'Sensibility', criteria for, 567
Sequestration, sequential, 69, 73, 488, 489–90, 507–12
Seymour, D. G., 503
Shannon's index, 181
Shapiro, A. R., 574
Shape, logistic, 300–1
   of stratified distribution, 180
Significance, boundaries, 268
Significance, quantitative, concept of, 146–7
   criteria for, 172
   interpretation, 148–149, 267–8
   vs. stochastic, 146–149
Significance, 'statistical', 122
Significance, stochastic, concept of, 146–7
   for ordinal variables, 198
Simon, R., 505
Singer, B. H., xiv, xv, 550
Slinker, B. K., xiv
Slope, algebraic model, 175–6
   linear models, 494–5
   quantitative significance, 176–495
   standardized, 176, 495–6
   stochastic tests, 177, 496
Smith, H., xii
Somers D index, 317, 336
Sonquist, J. A., 537
Spiegelhalter, D. J., 503
Spline(s), 27–28, 287
Stability, concepts, 266
   of indexes, 144–146
Stage(s), categorical, 44
   for cancer, 71–72
'Staging', clinical, 570
Symptom stages, sequestration for, 509–11
Standardization, 54, 55, 516, 569–570
Standardized coefficients. *See* Coefficients, standardized
Stem-leaf plot, 110–112
Step function, 27–28
Stitt, F. W., 552
Stochastic significance. *See* Significance, stochastic
Stone, C. J., 541
Strata, demarcation of, 178
   distribution of, 180
   number of, 180
Stratification, as mechanism of adjustment, 54
   cross-, 481–483
   intrinsic value, 505–6
   multicategorical, xii, 507–528
   polarized, xiii, 69, 73, 540–1
   prognostic, xiii
Straight line. *See* Rectilinear
Subgroup effects, 51–53
Substitution, 21, 36
Sum of squares, for model(s), 230
   incremental, 235
   residual, 229–230
   type II, 249
Survival analysis. *See* Analysis, survival
Survival, cumulative, 377–379, 381
   individual estimates, 402–3
Symbols, for basic data, 130
   for bivariate data, 188–190
   for multivariable regression, 229–230

t test, 144, 145, 267, 446
Tabulation, cross-, 481, 483

Target(s)
  in illustrative data set, 86–87
Target(s) (continued)
  multiple variables, 60–61
  nadir, 541
  of orientation, 60
  polytomous, 362–3, 432–4, 473, 497, 541, 542–3, 544–6, 554
  single, 61
  zenith, 541
Target variable, appraisal of, 123
Targeted cluster methods, 475–558
Taves, D. R., 505
Taxonomy, 23–24
  effectiveness of, 97–98
  of multivariable analysis, 68–70
  numerical, 69
Tibshirani, R., 416
Time-dependent covariates. *See* Covariates, time-dependent
Time series, 22
Tolerance, 246–8, 262
Trace, Pillai's, 440, 464
Tracking, 556
Transformation(s), hazards of, 122
  logistic, 27, 299
  in published literature, 291, 428
  to rank values, 573
  for 'sensibility', 155
  of variables, 120–121, 366–7
Trend(s), 35–36, 66
  curvilinear, 31
  in target variable, 172–178
  linear, 175–6
    chi-square test for, 177–8, 496–7, 518
    in dimensional variables, 199
    in ordinal variables, 197–8, 492
Trial-and-error process, 311–312
'Twoing' rule, 541, 543
Tukey, J., 110, 113

U statistic, 440
Union, primary (pulmonic) symptoms, 511
Unions, Boolean, 89–90
  multibinary ordinal, 509
  ordinal scale, 91
  terminal, 492–4, 508, 534, 541
Univariate, definition, 1
  distributions. *See* Distributions, univariate

Validation, challenge set, 579
  challenges in, 578–580
  chronologic splits, 579
  concept of, 184
  cross-, 185
  expression of success, 580
  external, 184
  internal and external, 578
  jackknife, 185
  resubstitution, 185
  "shrinkage", 578
  split group, 185
  V-fold, 186, 547
Validity, "face", 486
Variable(s), binary, 92
  compound nominal group, 260–261
  definition, 1
  dependent, 10
  'design', 356–360
  dimensional, 92
  'dummy', 93, 160–1, 259–60, 356–7, 423
  impact of, 574
  imputation of, 116
  independent, 10
  isolation of, 115–116
  'lurking', 275
  nominal, 93
  omission of important, 57, 287–8
  polytomous, 362–3, 554
  predictor, inadequate choices, 572–3
  reducing, 38
  screening, 38
  segregation of, 116–117
  transcending, 484
  transfer of, 116
Variance, basic group, 135
  distinctive, 235–6
  'explained', 137
  group, 132, 134
  mean group, 230
  model group, 135
  partition of, 135
  proportionate reduction, 137–138, 533–4
  residual group, 135
  unique, 235–6
Vasculitis, classification of, 553–4
Verhulst, P. F., 300

W (variance-covariance matrix), 439
Wald index, 318, 320, 390. *See also* Chi-Square, Wald
Wasson, J. H., 573
Wells, C. K., xvi, 518
Welsch, R. E., 285
Wilks lambda, 439
Winkel, P., 474, 512
Wright, S., 561

XBETA index, 423

'YHAT', Residuals for, 280–81
Yates, F., 293, 474

Z ratio, 266

Z: PH test, 406–7
Zhang, P., 581
Zonal partition. *See* Partition, zonal

Zones, ordinalized, examination of, 159–161
  with dummy variables, 160
  with ordinal variables, 161

BARCODE INSIDE